AF370786

SAR-Based Signal Processing and Target Recognition

SAR-Based Signal Processing and Target Recognition

Editors

Lan Du
Gang Xu
Haipeng Wang

Basel • Beijing • Wuhan • Barcelona • Belgrade • Novi Sad • Cluj • Manchester

Editors

Lan Du
Xidian University
Xi'an, China

Gang Xu
Southeast University
Nanjing, China

Haipeng Wang
Fudan University
Shanghai, China

Editorial Office
MDPI
St. Alban-Anlage 66
4052 Basel, Switzerland

This is a reprint of articles from the Special Issue published online in the open access journal *Remote Sensing* (ISSN 2072-4292) (available at: https://www.mdpi.com/journal/remotesensing/special_issues/sar_signal_processing_target_recognition).

For citation purposes, cite each article independently as indicated on the article page online and as indicated below:

Lastname, A.A.; Lastname, B.B. Article Title. *Journal Name* **Year**, *Volume Number*, Page Range.

ISBN 978-3-0365-8636-6 (Hbk)
ISBN 978-3-0365-8637-3 (PDF)
doi.org/10.3390/books978-3-0365-8637-3

Cover image courtesy of U.S. Sandia National Laboratories

Contents

About the Editors

Lan Du

Lan Du is currently a Professor at the National Laboratory of Radar Signal Processing, Xidian University. She received B.S., M.S. and Ph.D. degrees in Electronic Engineering from Xidian University, Xi'an, China, in 2001, 2004, and 2007, respectively. From 2007 to 2009, she was involved in research as a Postdoctoral Research Associate with the Department of Electrical and Computer Engineering, Duke University, Durham, NC, USA. Her main research interests are in the fields of statistical signal processing and machine learning with applications to radar target recognition. She is a Fellow of CIE and Senior Member of IEEE. She is a Guest Editor of Remote Sensing, Associate Editor of IEEE J-STARS, Committee member for Journal of Radars, and TPC member for the IET International Radar Conference. Her Ph.D. dissertation was granted the National Excellent Doctoral Dissertation of China in 2009. Her research work was supported by the NSFC for Excellent Young Scholars. She is the first or corresponding author of more than 100 papers published in IEEE Trans. SP, JMLR, ESWA, Information Sciences, IEEE Trans. GRS, IEEE Trans. AES, IEEE J-STARS, and NIPS, among others, and received the best paper award at the IET International Radar Conference 2013 and CIE International Conference on Radar 2006, among others.

Gang Xu

Gang Xu is currently an Associate Professor at the State Key Laboratory of Millimeter Waves, School of Information Science and Engineering, Southeast University, Nanjing, China. He received a B.S. degree and Ph.D degree in Electrical Engineering from Xidian University, Xi'an, China, in 2009 and 2015, respectively. From 2015 to 2016, he was a full-time Postdoctoral Research Fellow with the School of Electrical and Electronic Engineering (EEE), Nanyang Technological University (NTU), Singapore. His major research interests include SAR/ISAR imaging, InSAR processing, and sparse radar imaging. He has authored and coauthored more than 60 scientific papers and over 20 papers in conference proceedings. He has guest-edited several Special Issues on remote sensing and frontiers in signal processing.

Haipeng Wang

Haipeng Wang is currently a Professor at the Key Laboratory of Electromagnetic Wave Information Science (MoE), Department of Communication Science and Engineering, School of Information Science and Engineering, Fudan University, Shanghai, China. He received B.S. and M.S. degrees in Mechanical and Electronic Engineering from the Harbin Institute of Technology, Harbin, China, in 2001 and 2003, respectively, and a Ph.D. degree in Environmental Systems Engineering from the Kochi University of Technology, Kochi, Japan, in 2006. He was a Visiting Researcher with the Graduate School of Information, Production and Systems, Waseda University, Fukuoka, Japan, in 2008. His research interests include signal processing, SAR image processing and analysis, speckle statistics, applications to forestry and oceanography, and machine learning and its applications to SAR images. He is a Senior Member of IEEE and CIE. He is a Guest Editor of Remote Sensing, Associate Editor of IEEE GRSL, and TPC member for IGARSS.

Communication

Vehicle Target Detection Network in SAR Images Based on Rectangle-Invariant Rotatable Convolution

Lu Li, Yuang Du and Lan Du *

The National Lab of Radar Signal Processing, Xidian University, Xi'an 710071, China;
luli92@stu.xidian.edu.cn (L.L.); yadu_1@stu.xidian.edu.cn (Y.D.)
* Correspondence: dulan@mail.xidian.edu.cn

Abstract: In recent years, convolutional neural network (CNN)-based methods have been extensively explored for synthetic aperture radar (SAR) target detection. Nevertheless, the convolutional sampling locations of CNNs cannot accurately fit vehicle targets due to the fixed sampling mechanism in the convolutional kernel. In this paper, we focus on the vehicle target detection task in SAR images and propose a novel rectangle-invariant rotatable convolution (RIRConv) to determine more accurately the convolutional sampling locations for vehicle targets. Specifically, this paper considers the shape characteristic of vehicle targets in SAR images, which always retain a rectangular shape despite having varying sizes, aspect ratios, and rotation angles. The proposed RIRConv equips three additional learnable attribute parameters, namely, width, height, and angle attributes, to adaptively adjust the sampling locations in the convolutional kernel according to the targets. In addition, the RIRConv applies a modulation mechanism to focus on the sampling locations that significantly affect the output. Finally, the RIRConv is introduced into the single-shot multibox detector (SSD) to realize SAR vehicle target detection. In this way, the feature representation capability of SSD for vehicle targets can be enhanced, thus leading to higher detection performance. Notably, the proposed RIRConv is "plug-and-play" and can also be used with other existing advanced technologies to achieve higher detection performance. The experiments based on the measured miniSAR data validate the effectiveness of the proposed method.

Keywords: synthetic aperture radar (SAR); vehicle target detection; rectangle-invariant rotatable convolution (RIRConv)

Citation: Li, L.; Du, Y.; Du, L. Vehicle Target Detection Network in SAR Images Based on Rectangle-Invariant Rotatable Convolution. *Remote Sens.* **2022**, *14*, 3086. https://doi.org/10.3390/rs14133086

Academic Editor: Piotr Samczynski

Received: 12 May 2022
Accepted: 24 June 2022
Published: 27 June 2022

Publisher's Note: MDPI stays neutral with regard to jurisdictional claims in published maps and institutional affiliations.

1. Introduction

Synthetic aperture radar (SAR) imaging is not limited by time, illumination, and weather constraints and can obtain large areas of high-resolution radar images with the development of SAR sensor technology. With the acquisition of numerous SAR images in recent years, SAR automatic target detection has become one of the vital issues studied by researchers. An important branch in this field is SAR vehicle target detection, which is of great significance in urban traffic management and military surveillance. During the past few decades, constant false alarm rate (CFAR) methods [1,2], as one of the conventional SAR target detection methods, have been extensively studied. However, the performance of CFAR methods heavily depends on the statistical modeling of clutter. The land scene where the vehicle targets are located is complex, and the interference of buildings, trees, and other clutter is too severe to accurately establish the statistical model, thus restricting the performance of CFAR methods.

With the improvement of computing power, target detection methods based on convolutional neural networks (CNNs) are developing rapidly. These methods can be roughly divided into two-stage and one-stage detectors. The two-stage detectors, such as region-based CNN (R-CNN) [3] and Faster R-CNN [4], must first extract region proposals and then perform further classification and regression. Without the proposal extraction step, one-stage detectors, such as You Only Look Once (YOLO) [5] and the single-shot multibox detector (SSD) [6], are faster and more straightforward. Furthermore, unlike YOLO, SSD combines the predictions in multi-scale feature maps, which is conducive to detecting objects of various sizes. Therefore, SSD has been widely used in the optical target detection task. Apart from optical images, SSD has also been adopted for SAR vehicle target detection and performs better than conventional methods [7,8].

Nevertheless, a square kernel shape (e.g., 3×3) is applied in the regular convolutional layer, in which fixed locations are sampled for the convolution operation. Although a stack of convolutional layers can increase the range of the receptive field, the shape of the receptive field remains square. However, the vehicle targets in the SAR images we are interested in mainly include cars, trucks, vans, and so on, which vary significantly in terms of size, aspect ratio, and rotation angle and do not always maintain a square shape. The convolutional sampling locations of CNNs cannot accurately fit the vehicle targets, resulting in the limited feature representation ability of CNNs for the vehicle targets. With the goal of enhancing the transformation modeling capability of CNNs, Dai et al. [9] proposed the deformable convolution, which adjusts the sampling locations with learned offsets to matching targets. However, it is mainly designed for general geometric transformations without considering the characteristics of vehicle targets in SAR images. As for vehicle target detection in SAR images, there are more rotation and scale variations, but hardly nonrigid deformation. Furthermore, learning these additional offsets without extra constraints makes the network converge more difficult.

Inspired by the idea of matching the sampling locations of convolutions with the corresponding targets, we propose a novel rectangle-invariant rotatable convolution (RIR-Conv) for SAR vehicle target detection. Specifically, we noticed that the vehicle targets in SAR images vary significantly in terms of size, aspect ratio, and rotation angle, but always maintain a rectangular shape. The proposed RIRConv combines this characteristic. It keeps the rectangular sampling shape in the convolution kernel unchanged, while adaptively adjusting the sampling locations by learning three additional attribute parameters (i.e., the width, height, and rotation angle attributes) according to different vehicle targets. The RIRConv makes the sampling locations focus more on the target area of interest. In addition, refer to [10], we apply a modulation mechanism into RIRConv. By multiplying a learnable modulation mask with the pixel values at the sampling locations, the RIRConv can pay more attention to the sampling locations that significantly impact the output. Notably, because the RIRConv always keeps the sampling shape of the rectangle, only three attribute parameters and a modulation mask must be learned to make the sampling process more adaptive. In this way, learning these parameters does not make the network converge more difficult, and the network can still be trained end-to-end. By introducing the RIRConv into SSD, the feature representation capability of SSD for vehicle targets gains a marked improvement, thus improving the detection performance.

The principal contributions of this paper are as follows: (1) the shape-related prior information of vehicle targets in SAR images is fully exploited in the proposed method, and (2) this work proposes a design for a convolution operation (i.e., the RIRConv) for vehicle targets in SAR images. In particular, the RIRConv can extract the features of vehicle targets more accurately by adaptively adjusting the sampling locations, thus improving the detection performance.

2. Proposed Method

2.1. Network Architecture

The architecture of the proposed SAR vehicle target detection network is illustrated in Figure 1. We take the SSD network [6] as our baseline. Based on SSD, the proposed RIRConv is introduced. The framework is briefly introduced as follows.

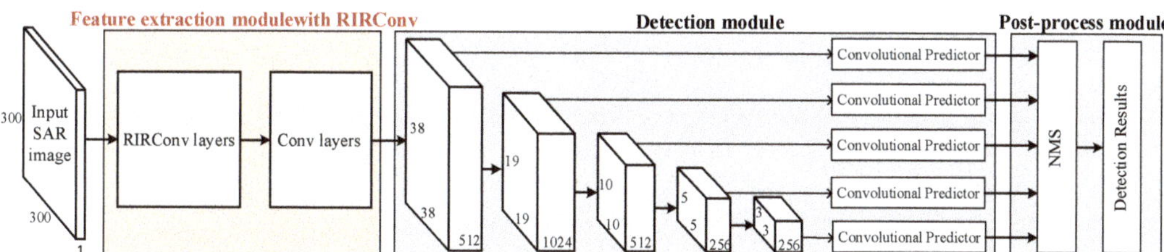

Figure 1. The architecture of the proposed vehicle target detection network for SAR images.

(1) Feature extraction module with RIRConv. This module is used to extract deep features of input SAR images. Its backbone is the truncated VGG16 network [11], which consists of 10 convolutional layers with 3×3 convolutional kernels and three pooling layers. In the feature extraction module with RIRConv, the proposed RIRConv replaces the first two convolutional layers of the truncated VGG16.

(2) Detection module. The detection module is added following the feature extraction module with RIRConv. It first extracts feature maps of five scales, the sizes of which are $38 \times 38, 19 \times 19, 10 \times 10, 5 \times 5$, and 3×3, respectively. Then, these feature maps are fed into the convolutional predictors to output the detected boxes and the confidence score of each box.

(3) Post-process module. After the detection module outputs the detected boxes, we use the non-maximum suppression (NMS) algorithm [12] to reduce the repeated detected boxes positioned to the same vehicle targets. Thus, we are able to obtain the detection results, which are the predicted specific locations of the vehicle targets.

In the following section, the details of the RIRConv are introduced concretely.

2.2. Rectangle-Invariant Rotatable Convolution

The vehicle targets vary significantly in terms of size, aspect ratio, and rotation angle but maintain a rectangular shape. Considering this prior information, we propose a novel RIRConv. Its convolutional kernel is equipped with width, height, and rotation angle attributes to adaptively adjust the sampling locations according to the targets.

First, the regular convolution operation samples the input feature map $\mathbf{X}$ using a fixed square grid G centered on the location $\mathbf{p}_0$. For example, for a 3×3 convolutional kernel, $G = \{(-1,-1), (-1,0), (1,0), \ldots\ldots, (-1,1), (0,1), (1,1)\}$. Then, the sampled values are element-wise multiplied by the learned weight $\mathbf{w}$. Finally, the corresponding output value on the output feature map $\mathbf{Y}$ is obtained through summation. The convolution operation can be formulated as follows:

$$\mathbf{Y}(\mathbf{p}_0) = \sum_{\mathbf{p}_n \in G} w(\mathbf{p}_n) \cdot \mathbf{X}(\mathbf{p}_0 + \mathbf{p}_n) \tag{1}$$

where $\mathbf{p}_n = [p_{nx}, p_{ny}]$ enumerates the locations in the square grid G.

Considering that the vehicle targets maintain a rectangular shape with different sizes and aspect ratios, a scale-transformation matric $\begin{bmatrix} \Delta w & 0 \\ 0 & \Delta h \end{bmatrix}$ is applied to each location $\mathbf{p}_n$ in G; hence, and the sampling is now on the new location:

$$\mathbf{p}_n{}^{st} = \begin{bmatrix} \Delta w & 0 \\ 0 & \Delta h \end{bmatrix} \mathbf{p}_n \tag{2}$$

where Δw and Δh represent the extent of the expansion of the convolution kernel on the width and height, respectively. Based on Δw and Δh, the sampling points can form a rectangle instead of a fixed square.

With the goal of rotating the sampling points to match the angles of targets, a rotation-transformation matric $\begin{bmatrix} \cos\theta & -\sin\theta \\ \sin\theta & \cos\theta \end{bmatrix}$ with the rotation angle θ is adopted. As a result, the sampling is now on location $\mathbf{p}_n{}^{rst}$, which can be formulated as follows:

$$\mathbf{p}_n{}^{rst} = \begin{bmatrix} \cos\theta & -\sin\theta \\ \sin\theta & \cos\theta \end{bmatrix} \mathbf{p}_n{}^{st} = \begin{bmatrix} \cos\theta & -\sin\theta \\ \sin\theta & \cos\theta \end{bmatrix} \begin{bmatrix} \Delta w & 0 \\ 0 & \Delta h \end{bmatrix} \mathbf{p}_n \tag{3}$$

where cos and sin denote the cosine and sine functions, respectively.

Figure 2 presents the sampling locations with different Δw, Δh, and θ values based on a 3×3 convolutional kernel. As expected we can see in Figure 2 that, given the Δw, Δh, and θ values, the sampling locations have a rectangular shape with different widths, heights, and rotation angles, respectively.

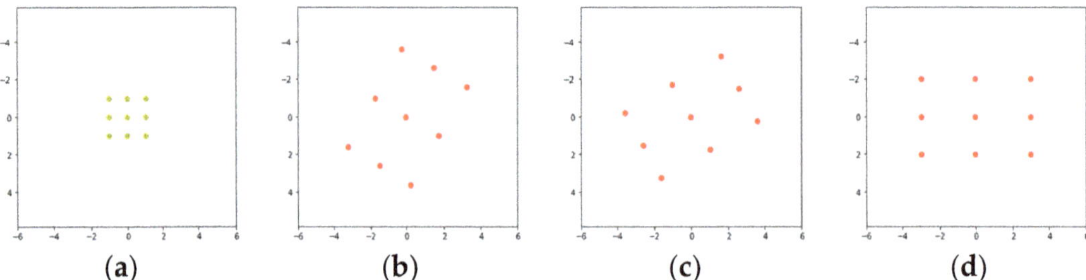

(a)	**(b)**	**(c)**	**(d)**

Figure 2. The sampling locations on a 3×3 kernel: (**a**) original sampling locations; (**b**) the sampling locations when $\Delta w = 2$, $\Delta h = 3$, and $\theta = 30^0$; (**c**) the sampling locations when $\Delta w = 2$, $\Delta h = 3$, and $\theta = 60^0$; (**d**) the sampling locations when $\Delta w = 2$, $\Delta h = 3$, and $\theta = 90^0$.

Through the new sampling locations $\{\mathbf{p}_n{}^{rst} | n = 1, 2, ..., N\}$, where $N = |G|$ represents the number of sampling locations in G, the convolution operation can be formulated as:

$$\mathbf{Y}(\mathbf{p}_0) = \sum_{\mathbf{p}_n \in G} w(\mathbf{p}_n) \cdot \mathbf{X}(\mathbf{p}_0 + \mathbf{p}_n{}^{rst}) = \sum_{\mathbf{p}_n \in G} w(\mathbf{p}_n) \cdot \mathbf{X}(\mathbf{p}_0 + \mathbf{p}_n + \Delta\mathbf{p}_n) \tag{4}$$

where $\Delta\mathbf{p}_n = \mathbf{p}_n{}^{rst} - \mathbf{p}_n = \begin{bmatrix} \cos\theta \cdot \Delta w \cdot p_{nx} - \sin\theta \cdot \Delta h \cdot p_{ny} - p_{nx} \\ \sin\theta \cdot \Delta w \cdot p_{nx} + \cos\theta \cdot \Delta h \cdot p_{ny} - p_{ny} \end{bmatrix}$.

Next, a learnable modulation mask, which can assign modulation weights to sampling locations, is adopted to enable the RIRConv to focus on the sampling locations that significantly impact the output. The modulation weights indicate the importance of different sampling locations and will be multiplied by the pixel values at the corresponding sampling locations. Finally, the RIRConv operation can be formulated as follows:

$$\mathbf{Y}(\mathbf{p}_0) = \sum_{\mathbf{p}_n \in G} w(\mathbf{p}_n) \cdot \mathbf{X}(\mathbf{p}_0 + \mathbf{p}_n + \Delta\mathbf{p}_n) \cdot \Delta m_n \tag{5}$$

As the offset $\Delta\mathbf{p}_n$ may be fractional, the sampling location may not correspond to an integer position. Thus, Equation (5) is implemented via bilinear interpolation as:

$$\mathbf{X}(\mathbf{p}) = \sum_{\mathbf{q}} K(\mathbf{q}, \mathbf{p}) \cdot \mathbf{X}(\mathbf{q}) \tag{6}$$

where $\mathbf{p} = \mathbf{p}_0 + \mathbf{p}_n + \Delta\mathbf{p}_n$, which denotes a fractional location; $\mathbf{q}$ enumerates all integral spatial locations in the feature map $\mathbf{X}$; and $K(\cdot, \cdot)$ represents the bilinear interpolation kernel proposed by [9]:

$$K(\mathbf{q}, \mathbf{p}) = k(q_x, p_x) \times k(q_y, p_y), \tag{7}$$

where $k(a,b) = max(0, 1 - |a - b|)$.

The proposed RIRConv is illustrated in Figure 3. As can be seen, an additional convolutional layer is employed over the input feature map to learn the transformation parameters and mask (i.e., Δw, Δh, θ, and Δm). The offsets can then be computed based on the learned Δw, Δh, and θ values. The offsets and mask are employed to obtain the values in the output feature map, as shown in Equation (6). During training, the transformation parameters, mask, and other network parameters are learned simultaneously and adaptively under supervision. Although an additional convolution layer is added, this only amounts to small additional network parameters and costs.

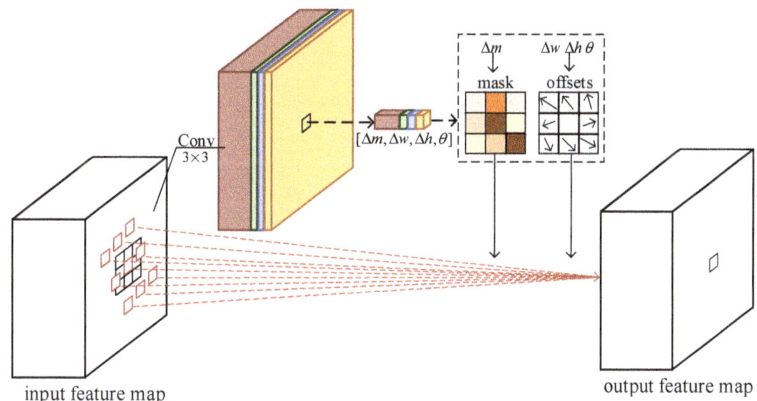

Figure 3. Illustration of the 3×3 RIRConv.

3. Experimental Results and Analysis

3.1. Experimental Data Description

The experimental data comprise the miniSAR data provided by the U.S. Sandia National Laboratories. The miniSAR dataset is an actual measured dataset acquired by the spotlights mode in the Ku band. The resolutions of SAR images in the miniSAR dataset are 0.1 m $\times$ 0.1 m, and the sizes are 1638 $\times$ 2510 pixels [13]. All these SAR images have large-area complex scenes, in which the background clutter includes man-made clutter (e.g., buildings, roads, and streetlights) and natural clutter (e.g., trees and grasslands). The split of training and test SAR images is consistent with [7]. In each SAR image of the miniSAR dataset, the targets, which we are interested in, are all vehicle targets. The labels of these vehicle targets are all marked manually. However, these SAR images are collected by the airborne SAR at different places. Figure 4a shows a SAR image from the training set, which is obtained by imaging a highway area with the airborne SAR. Figure 4b shows a SAR image from the test set, which is obtained by imaging an urban area with the airborne SAR. Therefore, the training and test SAR images have obvious differences in scene types, clutter types, etc., which guarantees the experiments in this paper can verify the generalization of the proposed method.

3.2. Experimental Settings

As described in Section 3.1, the sizes of the original SAR images are too large, not satisfying the size restriction of general network inputs. Therefore, when training, the large original SAR images were cropped into many sub-images with a fixed size of 300 $\times$ 300. Then, we augment the training sub-images by adding noise, filtering, rotating, and flipping. After augmentation, we arrive at a total of 1430 sub-images in the training set. These sub-images and the corresponding labels are used to train the network. Under the weighted loss function (i.e., smooth L1 loss for localization and softmax loss for classification), the proposed network is trained by a stochastic gradient descent (SGD) optimizer in an end-to-end manner. When testing, the original test SAR images are also cropped into many test sub-images by sliding window with overlapping. Then all the test sub-images are fed into the network to process individually. After that, all detection results in test sub-images are stitched together. Finally, we use the NMS algorithm to select the predicted bounding box with the highest confidence and to remove duplicate bounding boxes.

The experiments are implemented with the Pytorch deep learning framework, using a personal computer with Intel Xeon E5-2630 v4 CPU of 2.2 GHz and NVIDIA GTX 1080Ti GPU on an Ubuntu 16.04 Linux system.

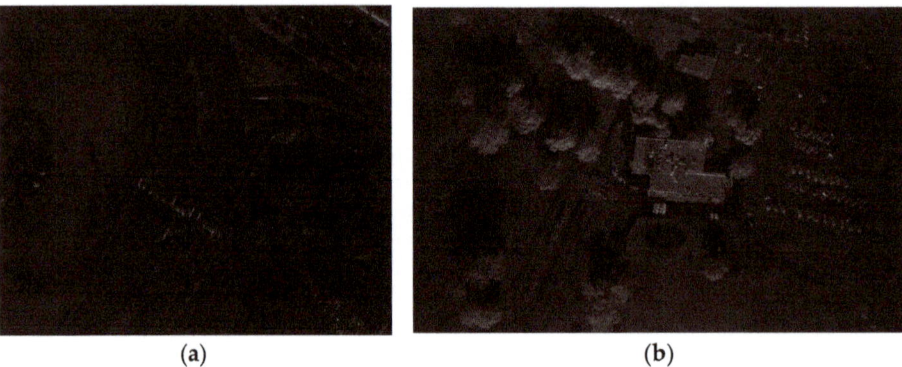

(a)   (b)

Figure 4. Two SAR images in the miniSAR dataset. (**a**) a SAR image from the training set; (**b**) a SAR image from the test set.

3.3. Evaluation Criteria

The quantitative evaluation criteria we used in the experiments include precision, recall, and F1-score, which can be formulated as:

$$\text{precision} = \frac{\text{the number of target chips}}{\text{the total number of chips}}, \tag{8}$$

$$\text{recall} = \frac{\text{the number of detected targets}}{\text{the total number of targets}}, \tag{9}$$

$$\text{F1} - \text{score} = \frac{2 \times \text{precision} \times \text{recall}}{\text{precision} + \text{recall}}, \tag{10}$$

where precision measures the fraction of true positives among all detected results. The recall measures the fraction of positives over the number of ground truth, and the F1-score is the harmonic mean between precision and recall, which is the main reference index to evaluate the detection performance comprehensively. The higher the value of the above three evaluation criteria, the better the performance of the detection method.

3.4. Comparison with Other SAR Target Detection Methods

In this section, we compare the proposed SAR target detection method with conventional SAR detection methods and some deep-learning based target detection methods in SAR images. The conventional SAR target detection methods for comparison are two common CFAR detectors, namely, the two-parameters CFAR detector and the Gamma-CFAR detector. The deep-learning based target detection methods in SAR images include some basic target detection networks (i.e., the Faster R-CNN [4] and the original SSD [6]) and two other target detection networks proposed in the last three years, namely, the DA-TL SSD [8] and the RefineDet [14]. The DA-TL SSD applies subaperture decomposition to acquire three-channel SAR images and then uses the VGG pre-trained model trained on the optical ImageNet dataset to initialize corresponding parameters. Based on SSD architecture, the

RefineDet introduces the idea of two-step cascaded regression used in two-stage target detection methods. At the same time, RefineDet introduces a structure similar to FPN to fuse features. Figure 5 exhibits the target detection results on two test SAR images, while Table I lists the numerical detection results in terms of the evaluation criteria shown in Section 3.3.

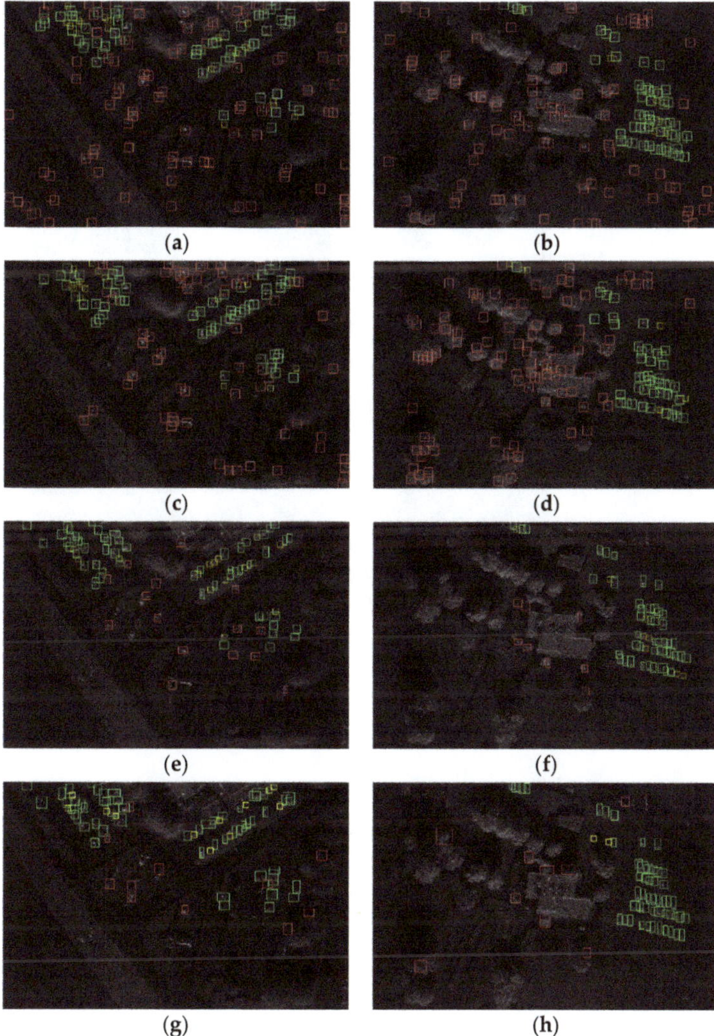

(a)	(b)
(c)	(d)
(e)	(f)
(g)	(h)

Figure 5. *Cont.*

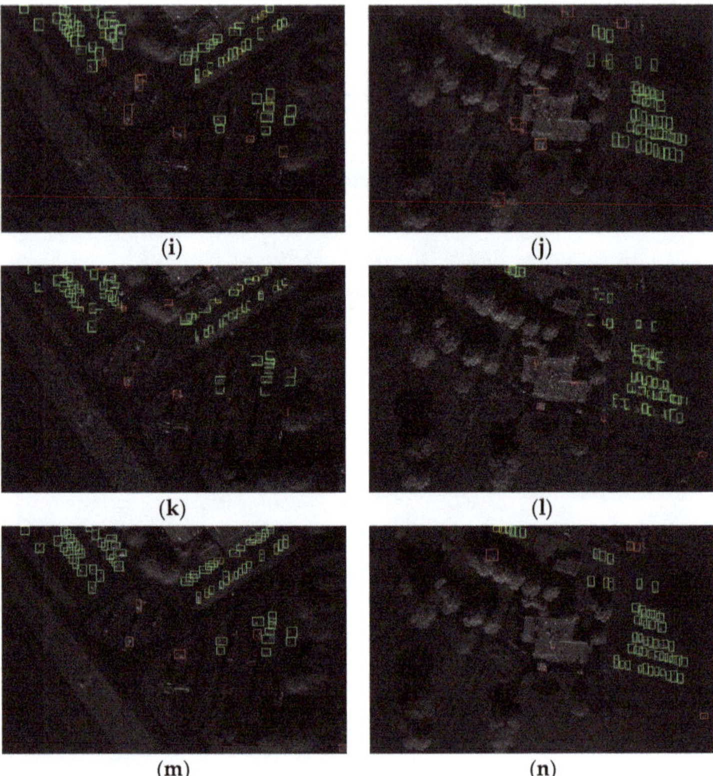

Figure 5. The target detection results compared with other detection methods for Images I and II, where green, red, and yellow rectangles represent the detected target chips, false alarms, and missing alarms, respectively. (**a,b**) the two-parameter CFAR method, (**c,d**) the Gamma CFAR method, (**e,f**) the Faster RCNN, (**g,h**) the original SSD, (**i,j**) the DA-TL SSD, (**k,l**) the RefineDet, (**m,n**) the proposed target detection method.

From Figure 5a–d, we can see that the CFAR methods have numerous false alarms, leading to lower precisions. From rows 2 and 3 in Table 1, we can observe that the precision of the two-parameter CFAR is 0.3789, while that of the Gamma CFAR is a little better, at 0.3931, which significantly reduces the corresponding comprehensive criterion F1-score. The reason is that CFAR methods face difficulties in accurately establishing the statistical model in complex scenes, resulting in poor detection performance. Compared with unsupervised CFAR methods, the deep-learning based target detection methods draw support from the data-driven approach and have better performance. From Figure 5e–n, we can see that the deep-learning based methods have fewer false alarms than the CFAR methods. Moreover, the deep-learning based methods also have fewer missing alarms, leading to higher recalls. In conclusion, the detection performance of the deep-learning based methods is generally better than that of conventional CFAR methods. In the listed deep-learning based target detection networks, the proposed method has the highest precision, recall, and F1-score. Therefore, it can achieve the best detection performance compared to other detection methods. As mentioned in Section 3.1, the training and test SAR images have obvious differences in terms of scene types, clutter types, and so on. Meanwhile, from Figure 5m–n, we can see that the proposed method generates very few false and missing alarms on the test SAR images, which verifies that the proposed method has strong generalization.

Table 1. Overall evaluation of different target detection methods.

	Precision	Recall	F1-Score
Two-parameter CFAR	0.3789	0.7966	0.5135
Gamma CFAR	0.3931	0.8136	0.5301
Faster R-CNN	0.8115	0.8051	0.8083
Original SSD	0.8468	0.8814	0.8638
DA-TL SSD	0.8843	0.8983	0.8912
RefineDet	0.8828	0.9237	0.9027
Proposed method	0.9134	0.9431	0.9280

3.5. Model Analyses

3.5.1. Sampling Locations in the RIRConv

Figure 6 exhibits the SAR chips of vehicle targets and the sampling locations obtained from the RIRConv on three different SAR targets. Considering that the sizes of the SAR images are too large, which cannot show the sampling locations clearly. Thus, we plot the sampling locations on the vehicle target chips of SAR images. Figure 6a–c show the SAR chips of three vehicle targets with different lengths, widths, and rotation angles. Figure 6d–f show the sampling locations obtained from the RIRConv. In Figure 6d–f, the points denote the sampling locations in two levels of 3×3 RIRConv kernels. The green points at the center of the vehicle targets represent the center sampling location of the RIRConv kernel. The modulation weights of the sampling locations are reflected in the colors of the points: the larger the modulation weight, the closer it is to a red color, while the smaller the modulation weight, the closer it is to a blue color. By stacking two layers of 3×3 RIRConv, we can obtain 81 sampling points.

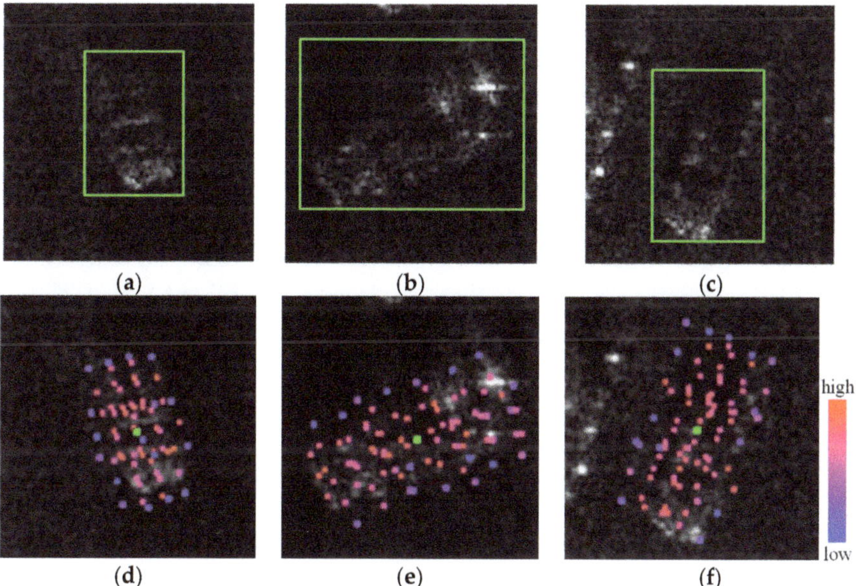

Figure 6. SAR chips of vehicle targets and the sampling locations ($9^2 = 81$ points in each SAR chip) in two levels of 3×3 RIRConv kernels. (**a,b,c**) SAR chips of vehicle targets. (**d,e,f**) Sampling locations in two levels of 3×3 RIRConv kernels.

From Figure 6d–f, we can see that these sampling locations no longer obey a fixed square but can be adaptively adjusted according to the lengths, widths, and rotation angles of the vehicle targets, thus proving that the RIRConv can adjust the sampling locations adaptively. Moreover, by observing the colors of these sampling points, we can see that the sampling points with low modulation weights (i.e., the sampling points close to blue) are mostly distributed outside the vehicle targets. The sampling points with high modulation weights (i.e., the sampling points close to red) are mostly distributed inside the vehicle targets. It proves that the modulation weights can make the RIRConv pay more attention to the sampling locations that significantly impact the output.

3.5.2. Parameters, FLOPs, and Runtime Analysis

We compare the parameters, FLOPs, and runtime of the proposed method and other deep-learning based target detection methods. As shown in Table 2, we can see that the proposed method is significantly less than RefineDet and Faster RCNN in terms of parameters, FLOPs, and runtime, but only slightly higher than SSD. Nevertheless, as described in Section 3.4, the proposed method can obtain the highest detection performance. In conclusion, the proposed method can significantly improve the detection performance by adding a few parameters, FLOPs, and runtime.

Table 2. Complexities of the proposed method and other deep-learning based methods.

	Parameters	FLOPs	Runtime (Seconds/Per Test Sub-Image)
Faster R-CNN	1.36×10^8	18.5×10^{10}	0.102
SSD	2.37×10^7	6.1×10^{10}	0.015
RefineDet	3.39×10^7	7.6×10^{10}	0.054
Proposed method	2.38×10^7	6.7×10^{10}	0.021

4. Conclusions

In this paper, we proposed a novel vehicle target detection network based on RIRConv. For the vehicle target detection task in SAR images, the proposed RIRConv is designed according to the shape prior information of the vehicle targets to determine more accurately the convolutional sampling locations for vehicle targets. The RIRConv is lightweight and can be trained without additional supervision in an end-to-end manner. Finally, we introduced the proposed RIRConv into SSD to realize SAR vehicle target detection. The qualitative and quantitative experimental results based on the measured SAR dataset show the effectiveness of the proposed method. In addition, as the vehicle targets in optical remote sensing images also maintain a rectangular shape, we believe the RIRConv can also be used for optical vehicle target detection. However, the effect of RIRConv is more prominent for SAR vehicle target detection.

Author Contributions: Conceptualization, L.D. and L.L.; methodology, L.L. and Y.D.; software, L.L. and Y.D.; validation, L.L.; data curation, L.L.; writing—original draft preparation, L.L.; writing—review and editing, L.D. and Y.D.; supervision, L.D.; All authors have read and agreed to the published version of the manuscript.

Funding: This research was funded in part by the National Science Foundation of China (Grant Number U21B2039), in part by the 111 Project.

Conflicts of Interest: The authors declare there is no conflict of interest.

References

1. Novak, L.M.; Burl, M.C.; Irving, W.W. Optimal polarimetric processing for enhanced target detection. *IEEE Trans. Aerosp. Electron. Syst.* **1993**, *29*, 234–244. [CrossRef]
2. Alberola-Lopez, C.; Casar-Corredera, J.; de Miguel-Vela, G. Object CFAR detection in gamma-distributed textured-background images. *IEE Proc.-Vis. Image Signal Process.* **1999**, *146*, 130–136. [CrossRef]
3. Girshick, R.; Donahue, J.; Darrell, T.; Malik, J. Rich feature hierarchies for accurate object detection and semantic segmentation. In Proceedings of the IEEE Conference on Computer Vision and Pattern Recognition, Columbus, OH, USA, 23–28 June 2014; pp. 580–587.
4. Ren, S.; He, K.; Girshick, R.; Sun, J. Faster R-CNN: Towards realtime object detection with region proposal networks. In Proceedings of the Neural Information Processing System, Montreal, QC, Canada, 7–12 December 2015; pp. 91–99.
5. Redmon, J.; Divvala, S.; Girshick, R.; Farhadi, A. You only look once: Unified, real-time object detection. In Proceedings of the IEEE Conference on Computer Vision and Pattern Recognition, Las Vegas, NV, USA, 27–30 June 2016; pp. 779–788.
6. Liu, W.; Anguelov, D.; Erhan, D.; Szegedy, C.; Reed, S.; Fu, C.Y.; Berg, A.C. SSD: Single shot multibox detector. In Proceedings of the European Conference on Computer Vision, Amsterdam, The Netherlands, 8–16 October 2016; pp. 21–37.
7. Du, L.; Li, L.; Wei, D.; Mao, J. Saliency-Guided Single Shot Multibox Detector for Target Detection in SAR Images. *IEEE Trans. Geosci. Remote Sens.* **2020**, *58*, 3366–3376. [CrossRef]
8. Wang, Z.; Du, L.; Mao, J.; Liu, B.; Yang, D. SAR target detection based on SSD with data augmentation and transfer learning. *IEEE Geosci. Remote Sens. Lett.* **2018**, *16*, 150–154. [CrossRef]
9. Dai, J.; Qi, H.; Xiong, Y.; Li, Y.; Zhang, G.; Hu, H.; Wei, Y. Deformable Convolutional Networks. In Proceedings of the IEEE International Conference on Computer Vision, Venice, Italy, 22–29 October 2017; pp. 764–773.
10. Zhu, X.; Hu, H.; Lin, S.; Dai, J. Deformable ConvNets V2: More Deformable, Better Results. In Proceedings of the IEEE International Conference on Computer Vision, Long Beach, CA, USA, 15–20 June 2019; pp. 9300–9308.
11. Simonyan, K.; Zisserman, A. Very deep convolutional networks for large-scale image recognition. In Proceedings of the International Conference on Learning Representations, San Diego, CA, USA, 7–9 May 2015.
12. Neubeck, A.; Van Gool, L. Efficient Non-Maximum Suppression. In Proceedings of the 18th International Conference on Pattern Recognition, Hong Kong, 20–24 August 2006; pp. 850–855.
13. SANDIA Mini SAR Complex Imagery. Available online: http://www.sandia.gov/radar/complex-data/index.html (accessed on 10 January 2018).
14. Zhang, S.; Wen, L.; Bian, X.; Lei, Z.; Li, S.Z. Single-Shot Refinement Neural Network for Object Detection. In Proceedings of the 2018 IEEE/CVF Conference on Computer Vision and Pattern Recognition, Salt Lake City, UT, USA, 18–23 June 2018; pp. 4203–4212.

 *remote sensing*

Article

CLISAR-Net: A Deformation-Robust ISAR Image Classification Network Using Contrastive Learning

Peishuang Ni [1], Yanyang Liu [2], Hao Pei [1], Haoze Du [1], Haolin Li [3] and Gang Xu [1,*]

[1] State Key Laboratory of Millimeter Waves, School of Information Science and Engineering, Southeast University, Nanjing 210096, China
[2] Shanghai Institute of Satellite Engineering, Shanghai 200090, China
[3] Institute of Spacecraft Application System Engineering, CAST, Beijing 100094, China
* Correspondence: gangxu@seu.edu.cn

Abstract: The inherent unknown deformations of inverse synthetic aperture radar (ISAR) images, such as translation, scaling, and rotation, pose great challenges to space target classification. To achieve high-precision classification for ISAR images, a deformation-robust ISAR image classification network using contrastive learning (CL), i.e., CLISAR-Net, is proposed for deformation ISAR image classification. Unlike traditional supervised learning methods, CLISAR-Net develops a new unsupervised pretraining phase, which means that the method uses a two-phase training strategy to achieve classification. In the unsupervised pretraining phase, combined with data augmentation, positive and negative sample pairs are constructed using unlabeled ISAR images, and then the encoder is trained to learn discriminative deep representations of deformation ISAR images by means of CL. In the fine-tuning phase, based on the deep representations obtained from pretraining, a classifier is fine-tuned using a small number of labeled ISAR images, and finally, the deformation ISAR image classification is realized. In the experimental analysis, CLISAR-Net achieves higher classification accuracy than supervised learning methods for unknown scaled, rotated, and combined deformations. It implies that CLISAR-Net learned more robust deep features of deformation ISAR images through CL, which ensures the performance of the subsequent classification.

Keywords: inverse synthetic aperture radar (ISAR); image deformation; target classification; unsupervised pretraining; contrastive learning (CL)

Citation: Ni, P.; Liu, Y.; Pei, H.; Du, H.; Li, H.; Xu, G. CLISAR-Net: A Deformation-Robust ISAR Image Classification Network Using Contrastive Learning. *Remote Sens.* **2023**, *15*, 33. https://doi.org/10.3390/rs15010033

Academic Editor: Dusan Gleich

Received: 17 November 2022
Revised: 15 December 2022
Accepted: 17 December 2022
Published: 21 December 2022

1. Introduction

Inverse synthetic aperture radar (ISAR) plays an important role in space target observation, benefiting from its ability to provide high-resolution ISAR images of targets in airspace and aerospace all-day and all-weather [1–5]. The two-dimensional (2D) high-resolution ISAR image contains information about the shape, structure, and electromagnetic (EM) scattering characteristics of the target, so it is usually used for accurate classification of space targets [6,7]. However, the main scattering centers of the target and the ISAR imaging projection plane (IPP) will change rapidly due to the maneuverability of the target during the observation, and the effective rotation vector will also be time-varying. Furthermore, the variation of radar parameters, such as bandwidth, wavelength, imaging accumulation angle, etc., as well as target motion, will bring serious unknown deformations to ISAR images, such as translation, rotation, and scaling [8]. Nowadays, deformation-robust feature extraction and accurate classification for ISAR images are gaining attention [9–11].

With the booming of deep learning, several advanced deep learning methods are being used for synthetic aperture radar (SAR) image detection [12–14] and classification [15–24]. For SAR image classification, some frameworks for few-shot learning [16–18] map SAR images to the embedding space, and then implement classification using distance metric in the embedding space. Bai et al. [19] proposed a sequential SAR image classification network by fusing the temporal and spatial features of multiple SAR images, which improved the classification accuracy. The above data-driven SAR image classification methods cannot overcome the dependence on manually labeled samples. In [20], the authors constructed a hybrid network combining data-driven and model-driven methods. By adding the prior information of SAR images, the hybrid network is able to capture both the distribution and structural features of SAR images simultaneously. The method makes full use of abundant unlabeled SAR images, which not only reduces manual annotation, but outperforms the only data-driven methods. Moreover, the addition of high-level semantic features of optical images further improves the classification accuracy of SAR images [21–24]. However, these SAR image classification techniques are not applicable to ISAR images. This is because the ISAR targets are non-cooperative [25], and the imaging parameters are usually unknown, so it is a little difficult to obtain the accurate ISAR images through parameter estimation. Therefore, it is necessary to research specific classification methods suitable for ISAR images.

For deformation ISAR image classification, the following aspects have been thoroughly investigated in the existing literature: (1) extracting deformation-robust features from the image domain [26–28]; (2) extracting deformation-robust features from the transform domain [8,29,30]; and (3) constructing deformation-robust networks [31–34]. For robust feature extraction from the image domain, Tien et al. [26] assumed the distribution of strong scattering centers is fixed and constructed a template library based on the geometric relationship of scattering centers. However, the above assumption is hard to hold because the main scattering centers will vary with the target motion. For polarimetric ISAR (Pol-ISAR) images, invariant features can be extracted by $\Omega - \Psi - \Phi$ [27] and Cloude–Pottier H/α_{ML} [28] invariant decomposition, and then classification is performed by template matching or convolutional neural network (CNN). However, the 2D shape and size information of the target is not effectively utilized.

For methods that extract deformation-robust features from the transform domain, Lee et al. [8] performed the trace transformation in a small angular region to extract deformation-invariant features. Park et al. [29] extracted translation- and rotation-invariant features from ISAR images by polar mapping of the 2D Fourier transform images. However, both the trace transformation and coordinate system transformation lose the structure and shape information of the target. Lu et al. [30] converted the ISAR images to the log-polar representations by polar transformation. Although the method extracts the robust features for scaling and rotation deformations, the origin of log-polar transformation is difficult to predict accurately.

In the construction of deformation-robust networks, an amount of research has emerged in recent years. The spatial transformer network (STN) [35] can be used for deformation correction of ISAR images. In [31,32], the effect of image deformation is alleviated by affine adjustment of the input ISAR image using a double-layer STN. Although satisfying performance is achieved, inappropriate affine parameters may cause the edges of the STN-adjusted images to exceed the boundaries. In order to better preserve the edge information, the inverse compositional spatial transformer network (IC-STN) [36] is designed to deal with the boundary effect of STN [33]. Although IC-STN can perform better adjustment on deformation ISAR images, the numerous parameters make it difficult to train. For sequential ISAR images, Xue et al. [32] designed a deformed shrink and a deformed affine ConvNet to adjust the image deformation, and then a bidirectional long short-term memory (BiLSTM) network is used to fuse the features of sequential images. Moreover, a hybrid transformer network is proposed for sequential ISAR image classification, which can extract local and global features of an ISAR image sequence [34]. These sequential ISAR image classification networks can obtain more information from deformation ISAR images, but the training process is more time-consuming, and the acquisition of sequential images also requires more stringent observation conditions.

All the above ISAR image deformation-robust networks are deep CNN models based on supervised learning. Due to the complexity of the networks, abundant labeled samples are required to provide supervised information during training to avoid overfitting. In the real world, manual annotation for acquired ISAR images requires extensive engineering experience and theoretical foundation. Currently, the self-supervised learning (SSL) training paradigm without labels is gaining popularity. SSL hopes to learn valuable representations for classification from large amounts of unlabeled data [37]. The encoded representations provided by SSL are more indicative of potential connections between samples than pale human-made annotations. SSL usually contains two phases: unsupervised pretraining and classifier fine-tuning. Generally, the pretraining is implemented by a pretext task, and the deep representations obtained by the pretext task is helpful for downstream classification. Recently, contrastive learning (CL) [38–41] has achieved impressive performance in SSL. Based on the pretext task of instance discrimination [42], CL pretrains an encoder to distinguish samples from different categories by comparing their deep representations in the embedded feature space.

Inspired by SSL, a deformation-roubust ISAR image classification network using CL, i.e., CLISAR-Net, is proposed, which adopts a two-phase training strategy. In the unsupervised pretraining phase, a convolutional encoder is designed using deformable convolution instead of regular convolution. Then, CLISAR-Net will extract discriminative representations of unlabeled ISAR images through the convolutional encoder. The positive sample pairs will be clustered together, while the negative sample pairs will be separated in the feature space using InfoNCE (Noise Contrastive Estimation) loss [43,44]. In the fine-tuning phase, based on the deep representations obtained by the pretrained convolutional encoder, labeled ISAR images are utilized to fine-tune the downstream classifier, so as to realize deformation ISAR image classification. Experimental results indicate that CLISAR-Net achieves better classification accuracy than existing supervised learning methods on the scaled, rotated, and combined deformation ISAR image datasets consisting of four satellites. The main contributions include:

1. Based on CL, the unsupervised ISAR image deep representation learning and classification are explored for the first time. Without manual annotation, we design an unsupervised pretraining encoder to learn transferable deep representations of ISAR images. With the help of deep representations, deformation ISAR image classification can be achieved using labeled training samples.
2. Deformable convolution is applied in the convolutional encoder for contrastive learning. Compared with the regular CNN, the convolutional encoder with the addition of deformable convolution is more adaptable to various deformation modes of ISAR images.
3. In the downstream deformation ISAR image classification task, using only 5% of labeled samples, the classification accuracy of CLISAR-Net is comparable to that of CNN under 100% supervision. This provides strong evidence that the features learned by unsupervised learning are more discriminative than those learned by supervised learning.

The rest of this article is organized as follows. Section 2 elaborates the causes of ISAR image deformation. Section 3 states the structure of the encoder in CLISAR-Net, the loss function of CL, and the optimization process of the encoder. Section 4 presents the experimental details and analyzes the results. Section 5 discusses the superiority of CLISAR-Net. Finally, Section 6 concludes this article and prospects the future work.

2. Causes of ISAR Image Deformation

Based on the EM scattering mechanism, ISAR images usually reflect the structure information of the body and solar panel of the space target [45,46]. However, the change in radar parameters and target motion will lead to the deformation of ISAR images, which makes the space target classification more difficult. In this section, the ISAR target observation geometry is established, and the causes of ISAR image deformation are discussed.

According to the theory of ISAR imaging [47], the target motion can be decomposed into translation and rotation. During translational motion, the Doppler produced by each scattering center is identical, which is not helpful for ISAR imaging. Moreover, the translational motion will lead to the range migration, which needs to be compensated accurately before imaging [48]. When the target rotates around the rotation center, the Doppler of the scattering center will change, which contributes to azimuth imaging. As shown in Figure 1, the target can be described by the turntable model after translation compensation, where O is the rotation center of the target, $\mathbf{r}_S$ is the position vector of the scattering center S, L_{los} is the radar line-of-sight (LOS), and Ω is a three-dimensional (3D) rotation vector, which can be decomposed into Ω_e and Ω_l. The IPP is defined as the plane perpendicular to Ω_e and passing through L_{los}. For complex motion targets, the IPP varies due to the rapid change in Ω, which makes the distribution of scattering centers not fixed. Furthermore, the time-varying effective rotation vector Ω_e also increases the difficulty of azimuth scaling.

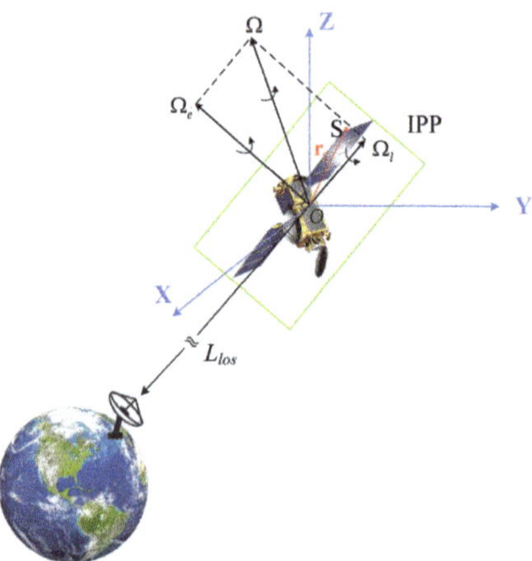

Figure 1. ISAR observation geometry.

For the scattering center S, its 2D ISAR image can be obtained using the range-Doppler (RD) algorithm by fast Fourier transform (FFT):

$$s_{RD}(r, f_d) = A_S \operatorname{sinc}\left(\frac{2B}{c}(r - r_S)\right) \operatorname{sinc}(T_a(f_d - f_{dS})) \tag{1}$$

where

$$\begin{cases} r_S = \mathbf{r}_S \cdot L_{los} \\ f_{dS} = \frac{2}{\lambda}(\Omega_e \times \mathbf{r}_S \cdot L_{los}) \end{cases} \tag{2}$$

where A_S is the amplitude of the scattering center S, c is the speed of light, B indicates the radar bandwidth, r is the range bin, f_d is the Doppler bin, r_S is the projection of $\mathbf{r}_S$ onto L_{los}, f_{dS} is the Doppler of S, T_a is the imaging time interval, $\operatorname{sinc}(x) = \sin(\pi x)/(\pi x)$, λ indicates the wavelength of the carrier frequency, and $\cdot$ and $\times$ denote the inner and cross product, respectively. The ISAR image is a superposition of multiple scattering centers on the target.

For non-cooperative ISAR targets, the parameters B, T_a, L_{los}, Ω_e, etc., always vary with the observation conditions and target motion, resulting in ISAR image deformation. Figure 2 illustrates various deformations for the same target due to different imaging parameters and target motion.

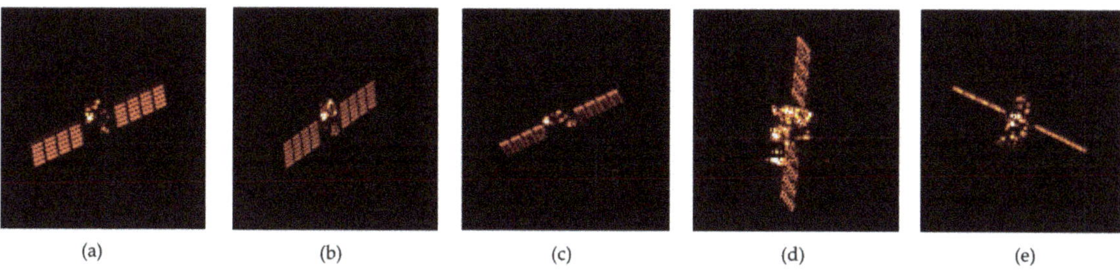

(a)   (b)   (c)   (d)   (e)

Figure 2. Various deformations of ISAR image for the same target. (**a**) Referenced ISAR image, (**b**) ISAR image azimuth scaling caused by accumulation angle, (**c**) ISAR image range scaling caused by radar bandwidth, (**d**) ISAR image rotation caused by target rotational motion, and (**e**) ISAR image deformation caused by the variation of main scattering centers.

Figure 2 shows that different accumulation angles and radar bandwidths will cause scaling deformation of the ISAR image, while the change in target motion direction will produce rotation deformation. During the target rotation, the change in main scattering centers of the target causes obvious fluctuation of the pixel intensity and makes the ISAR image exhibit more significant deformation. Such characteristics of ISAR images pose great difficulties for ISAR image classification. At present, it is an important task to study classification techniques with strong robustness to ISAR image deformation.

3. Proposed Method

In this section, CLISAR-Net is proposed to obtain the deep representations of deformation ISAR images through unsupervised CL, and then realize ISAR image classification based on the deep representations. This method belongs to unsupervised learning and contains two phases: pretraining and fine-tuning. As shown in Figure 3, the encoder in CLISAR-Net is first pretrained with unlabeled data in the pretraining phase to obtain the deep discriminative representations of the deformation ISAR images. In the fine-tuning phase, to accommodate specific downstream classification tasks, the classifier is fine-tuned with labeled training samples on the basis of the obtained deep representations. Actually, the traditional CNN-based deformation ISAR image classification networks are trained directly from the second phase, and they are all supervised learning methods.

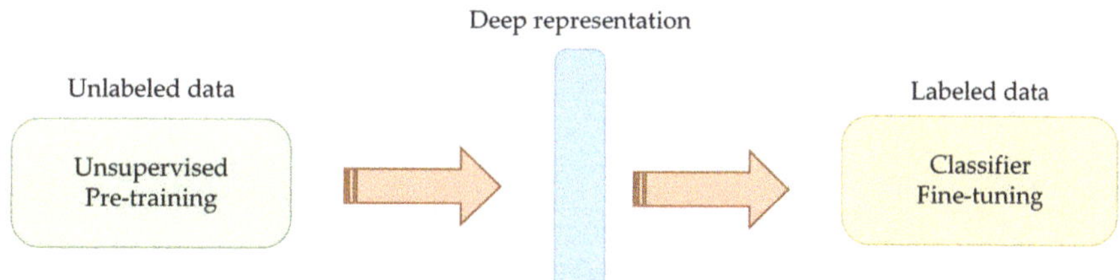

Figure 3. General flow chart of the training of CLISAR-Net.

3.1. Unsupervised Pretraining With Unlabeled Data

Compared with the traditional deformation ISAR image classification networks, CLISAR-Net can be pretrained without any manually labeled samples. This is because CL can motivate the encoder to learn higher-level representations by comparing the similarity between amounts of unlabeled samples and empower it to distinguish samples from different categories. In the pretraining phase of CLISAR-Net, the following issues should be considered: the structure of the encoder, loss function of CL, and the optimization of the encoder.

3.1.1. Structure of the Encoder

The goal of CL is to learn an encoder that can extract the deep representations of samples. In CLISAR-Net, the encoder contains two parts: the convolutional encoder and the projection head, in which the former is what we hope to obtain through unsupervised pretraining. Figure 4 shows the structure of the encoder. The convolutional encoder is stacked with five convolution-pooling blocks, and regular 2D convolutions are used in the former three blocks to learn simple features of ISAR images, such as textures and edges. The latter two blocks are 2D deformable convolutions, the variable convolution kernel can be more adaptive to the structure variation of the deformation ISAR images.

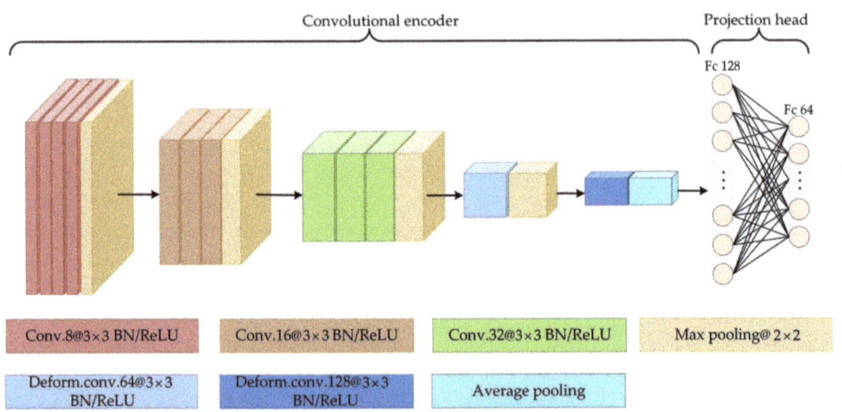

Figure 4. Structure of the encoder for obtaining deep representations of deformation ISAR images in CLISAR-Net.

In the convolutional encoder, *Conv.8@3 × 3BN/ReLU* indicates that there are eight regular convolution kernels sized 3×3 with batch normalization (BN), and ReLU represents rectified linear unit activation. *Max pooling@2 × 2* indicates a max-pool layer with a kernel size of 2×2 and step of 2. For the mth channel of the input feature map $\boldsymbol{I}_m^{(l)}$, the nth channel of the output feature map $\boldsymbol{O}_n^{(l+1)}$ is computed as follows:

$$\boldsymbol{O}_n^{(l+1)} = \sigma\left(BN\left(\sum_m^M \text{conv}\left(\boldsymbol{I}_m^{(l)}, \boldsymbol{K}_n^{(l+1)} \right) + \boldsymbol{b}_n^{(l+1)} \right) \right) \tag{3}$$

where $\boldsymbol{K}_n^{(l+1)}$ and $\boldsymbol{b}_n^{(l+1)}$ are the learnable weights and bias of the $(l+1)$th layer in the nth channel, and M is the number of channels for the input feature map. The pixel value of the output feature map at (x, y) is calculated by:

$$\text{conv}(\boldsymbol{I}, \boldsymbol{K})_{x,y} = \sum_{p=0}^{P-1} \sum_{q=0}^{Q-1} \boldsymbol{I}_{x-p, y-q} \boldsymbol{K}_{p,q} \tag{4}$$

where $P \times Q$ is the kernel size, and $\boldsymbol{I}_{x-p, y-q}$ and $\boldsymbol{K}_{p,q}$ denote the pixel values of the input feature map $\boldsymbol{I}$ and the convolution kernel $\boldsymbol{K}$ at $(x-p, y-q)$ and (p, q), respectively. BN for a mini-batch data $\boldsymbol{X}$ is defined as follows:

$$BN(\boldsymbol{X}) = \alpha \times \frac{\boldsymbol{X} - \mu_B}{\sqrt{\sigma_B^2 + \varepsilon}} + \beta \tag{5}$$

where μ_B and σ_B^2 are mean and variance of the data, ε is a small value that avoids division by zero, and α, β are learnable parameters. Gradient disappearance and explosion can be alleviated by BN. $\sigma(\cdot)$ is a ReLU non-linear activation function, which can be written as:

$$\sigma(\boldsymbol{X}) = \max(0, \boldsymbol{X}) \tag{6}$$

Pooling can usually be seen as a downsampling operation. For the mth channel of $\boldsymbol{I}_m^{(l)}$, after the max-pooling window with a kernel size of 2×2 and step size of 2, the pixel value of the output feature map at (x, y) is:

$$\boldsymbol{P}_n^{(l+1)} = \max\left(\boldsymbol{I}_m^{(l)}(2x + p, 2y + q) \mid p = 0, 1; q = 0, 1 \right) \tag{7}$$

In the convolutional encoder, *Deform conv.64@3 $\times$ 3BN/ReLU* denotes that there are sixty-four deformable convolution kernels sized 3×3, also with BN and a ReLU activation. The deformable convolution adds the learned offset to each sampling position of the receptive field, making the deformable sampling locations able to sample the structural information around the pixel of interest. Additionally, the extracted features are more adaptable to scaling, rotation, and other deformations of ISAR images.

The implementation process of deformable convolution is shown in Figure 5. The offset field is first generated using the offset generator to obtain the offset values, then the input feature maps will be resampled using bilinear interpolation to obtain the deformation feature maps. Finally, the output feature maps are generated by regular 2D convolution on the the deformation feature maps.

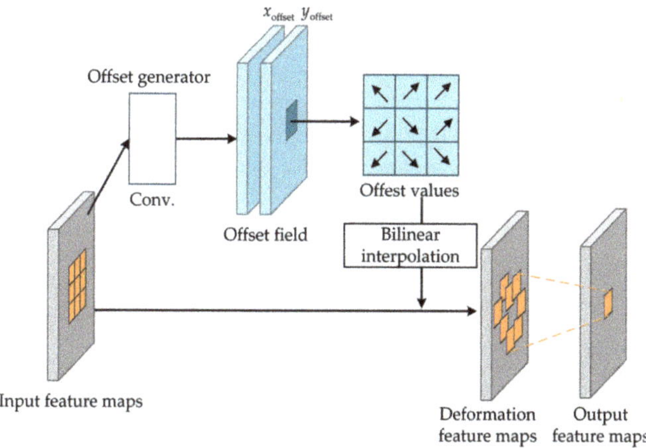

Figure 5. Illustration of 3×3 deformable convolution.

The offset generator is a convolution layer with 3×3 kernels, and the offset field is obtained by convolution on the input feature maps. For any pixel position (x, y) of $\boldsymbol{I} \in \mathbb{R}^{F_x \times F_y}$ ($F_x \times F_y$ is the size of the input feature map), two offsets x_{offset} and y_{offset} are learned in the offset field, i.e., offsets $\Delta F = \{(x_{\text{offset}}, y_{\text{offset}})\} \in \mathbb{R}^{2M \times F_x \times F_y}$, where $2M$ is the number of channels. Since x_{offset} and y_{offset} may be fractional, the new position $(x_{\text{new}}, y_{\text{new}}) = (x + x_{\text{offset}}, y + y_{\text{offset}})$ obtained by resampling usually deviates from integer pixels. Therefore, the bilinear interpolation is required to calculate the pixel value at $(x_{\text{new}}, y_{\text{new}})$ in the deformation feature maps from the pixels around (x, y) in the input feature maps. Assuming that $I_{x,y}$ is the pixel value at (x, y), the output pixel value $\hat{I}_{x_{\text{new}}, y_{\text{new}}}$ is calculated as follows:

$$\hat{I}_{x_{\text{new}},y_{\text{new}}} = \sum_{x}^{F_x} \sum_{y}^{F_y} I_{x,y} \max(0, 1 - |x_{\text{new}} - x|) \max(0, 1 - |y_{\text{new}} - y|) \tag{8}$$

After the above bilinear interpolation to obtain the deformation feature maps, the final output feature maps can be obtained by regular 2D convolution. The whole computation of deformable convolution is summarized as follows:

$$\text{deform.conv}(I, K, \Delta F)_{x,y} = \sum_{p=0}^{P-1} \sum_{q=0}^{Q-1} \hat{I}_{x-p+x_{\text{offet}},y-q+y_{\text{offset}}} K_{p,q} \tag{9}$$

where $\hat{I}$ denotes the deformation feature maps.

After *Deform conv.128@3 $\times$ 3BN/ReLU*, the obtained output feature maps are converted into feature vectors using average pooling. For an input ISAR image sample x_i, the feature vector representation obtained by the above convolutional encoder $f_\psi(\cdot)$ is denoted as $h_i = f_\psi(x_i)$, where ψ denotes all the learnable parameters. Then, the feature vector representation is mapped into the feature space by a projection head $g_\xi(\cdot)$, where ξ is the learnable parameters of $g_\xi(\cdot)$. In this article, the projection head is constructed by two fully connected layers. For the input x_i, the vector representation output by the projection head is:

$$q_i = g_\xi(h_i) = W^{(2)}\sigma\left(W^{(1)}f_\psi(x_i) + b^{(1)}\right) + b^{(2)} \tag{10}$$

where $W, b \in \xi$ indicate the learnable parameters of the projection head. Recent work has shown that calculating the contrastive loss on q_i is more efficient than h_i [38]. The above convolutional encoder $f_\psi(\cdot)$ and projection head $g_\xi(\cdot)$ constitute the base encoder $f_q(\cdot)$ for extracting the deep representations from input ISAR image samples. The loss function of CL is elaborated below.

3.1.2. Loss Function of CL

Traditional supervised learning methods are based on category discrimination, which need to provide the category information manually. CL is a pretext task based on instance discrimination, that is, each sample is regarded as a category, so the samples themselves provide supervised information, and no manual annotation is required. Therefore, the cross-entropy loss function commonly utilized in category discrimination is no longer suitable for instance discrimination, and it is necessary to be modified.

Consider a training set $X = \{x_1, x_2, \cdots, x_Z\} \in \mathbb{R}^{d \times Z}$, where x_i denotes the ith training sample and d is the dimension of the data. Another view of the training set X can be written as $\tilde{X} = \{\tilde{x}_1, \tilde{x}_2, \cdots, \tilde{x}_Z\} \in \mathbb{R}^{d \times Z}$, where x_i and $\tilde{x}_i$ are different views of the same sample, and they treated as a positive sample pair, while x_i and $\tilde{x}_j (j = 1, 2, \cdots, Z, j \neq i)$ form $Z - 1$ negative sample pairs. To facilitate the distinction, we denote $\tilde{x}_i$ as x_i^+ and $\tilde{x}_j$ as x_j^-. For a sample x_i, it is encoded as $q_i = f_q(x_i)$, and similarly, x_i^+ and x_j^- are encoded as $q_i^+ = f_{\tilde{q}}(x_i^+)$ and $q_j^- = f_{\tilde{q}}\left(x_j^-\right)$, respectively, where $f_q(\cdot)$ is the base encoder that needs to be pretrained, and $f_{\tilde{q}}(\cdot)$ is the auxiliary momentum encoder that is necessary for pretraining $f_q(\cdot)$. The contrastive loss needs to realize the following relationship:

$$\text{sim}\left(q_i, q_i^+\right) \gg \text{sim}\left(q_i, q_j^-\right) \tag{11}$$

where $\text{sim}(\cdot, \cdot)$ is a function that measures the similarity of two samples. In CL, the InfoNCE loss function is often used to implement the above relationship. The InfoNCE loss of the deep representation q_i is defined as follows:

$$\mathcal{L}\left(q_i, q_i^+, \left\{q_j^-\right\}\right) = -\log \frac{\exp\left(\text{sim}(q_i, q_i^+)/\tau\right)}{\exp\left(\text{sim}(q_i, q_i^+)/\tau\right) + \sum_{\left\{q_j^-\right\}} \exp\left(\text{sim}\left(q_i, q_j^-\right)/\tau\right)} \tag{12}$$

where τ is a temperature hyperparameter. In the above equation, the number of negative sample pairs is $Z - 1$, which means the upper limit of j is $Z - 1$. As described in the definition, this loss function is a Z-way log loss based on softmax classifier, which tries to classify q_i as q_i^+ [49]. In this article, the vector dot product is utilized to measure the similarity of two samples, i.e.,

$$\text{sim}\left(q_i, q_i^+\right) = q_i^T q_i^+, \text{sim}\left(q_i, q_j^-\right) = q_i^T q_j^- \tag{13}$$

In the InfoNCE loss, each q_i^+ is a positive sample of q_i and also participates in the calculation of InfoNCE loss as a negative sample of all $q_j (j \neq i)$. Actually, a positive sample pair in CL provides supervision to each other.

3.1.3. Optimization of the Encoder

As mentioned above, an auxiliary momentum encoder is also needed for pretraining the base encoder. As shown in Figure 6, the momentum encoder and the base encoder form a parallel dual-stream architecture, and they share the same network structure and hyperparameters [50]. For an ISAR image sample x_i, the data-augmented view of it is x_i^+, they form a positive pair. Similar to Equation (10), x_i^+ is fed into the momentum encoder to obtain its deep representation:

$$q_i^+ = g_{\tilde{\xi}}\left(f_{\tilde{\psi}}(x_i^+)\right) \tag{14}$$

where $f_{\tilde{\psi}}(\cdot)$ and $g_{\tilde{\xi}}(\cdot)$ are the convolutional encoder and projection head of the momentum encoder. Based on the above description, the positive pair, i.e., q_i and q_i^+, can be produced by the two encoders, respectively, and they are both vectors with dimensions of C.

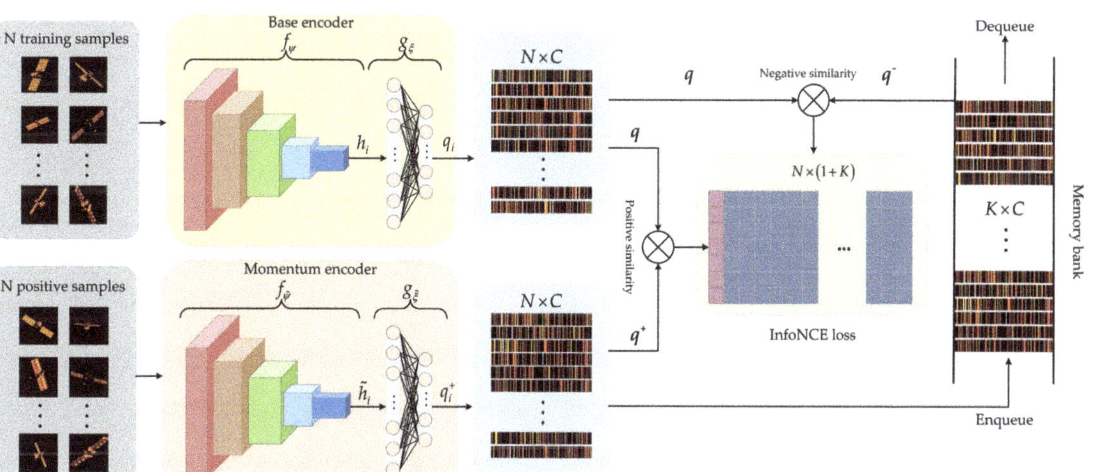

Figure 6. The optimization flow of the encoder in CLISAR-Net. The update of base encoder is realized by backpropagation of the InfoNCE loss, and the momentum encoder is updated in a momentum way.

CL requires an amount of negative pairs when calculating InfoNCE loss; this is because a rich set of negative samples allows the encoder to learn features that are more conducive for discrimination [42]. Consider a mini-batch with N training samples, when CLISAR-Net is optimized by gradient descent on a mini-batch, the number of negative pairs is $N - 1$. To satisfy the InfoNCE loss calculation, a large mini-batch size is needed, but increasing the mini-batch size will bring some adverse effects. For example, as the mini-batch size increases, more memory space is needed, but most standard computing platforms struggle to support such requirements. At the same time, a large mini-batch size will reduce the optimization efficiency. Inspired by previous works [39,42], this article introduces a memory bank to store the negative samples. As shown in Figure 6, the encoded representations of the momentum encoder in the previous mini-batches are stored in the memory bank in turn. In order to utilize more negative pairs, the encoded representations of the momentum encoder in the current mini-batch are not used as negative samples in the InfoNCE loss calculation but are provided by the memory bank. That is, in the current mini-batch, the positive samples are provided by the momentum encoder, while the negative samples come from the memory bank. By calculating the InfoNCE loss, the base encoder can be updated through backpropagation.

During the pretraining of CLISAR-Net, the encoded representations q^+ of the momentum encoder in each mini-batch are stored into the memory bank in turn. Since the samples in each mini-batch have no intersection, for the output representations q of the current mini-batch, the q^+ of all the previous mini-batches can be viewed as negative samples of q. This means that output representations of the momentum encoder for the previous mini-batches can be reused. Supposing that the length of the memory bank is K, then $K = k \times N$. The introduction of the memory bank expands the number of available negative samples, so the number of negative sample pairs is equal to the length of the memory bank and no longer depends on the mini-batch size. Such a design provides more negative sample pairs for the InfoNCE loss calculation and ensures the effectiveness of CL. Moreover, traversing negative samples from the memory bank does not require additional computations, which provides a guarantee for efficient training of CLISAR-Net.

It should be pointed out that the negative samples in the memory bank vary dynamically. In CLISAR-Net, the memory bank is maintained as a queue of negative samples, and the encoded representations output by the momentum encoder will be enqueued into the memory bank after each mini-batch training is completed. When the new representations cannot be enqueued into the memory bank, the representations of the oldest mini-batch will be dequeued from the memory bank, and the new output of the current mini-batch will be enqueued, thereby updating the negative samples dynamically. As the momentum encoder is continuously optimized, and the output encoded representations are gradually updated, the representations of the oldest mini-batch are the most outdated and the most inconsistent with the newest ones. Therefore, in order to maintain the consistency of negative samples, it is necessary to maintain a slowly updated memory bank. In addition, a slowly and dynamically updated memory bank requires less memory space and is more beneficial to maintain the consistency of negative samples than storing the representations of all samples in the dataset [42].

The utilization of the memory bank makes it difficult to update the momentum encoder by loss backpropagation, so the optimization of the momentum encoder needs to be adjusted accordingly. The momentum encoder in CLISAR-Net provides negative samples that support InfoNCE loss calculation, and the negative samples play a partially supervisory role during training, so there should not be too much difference between them. That is, the update of the momentum encoder should be smooth. In this article, we use a momentum update to optimize the momentum encoder, namely:

$$\tilde{\psi} \leftarrow m\tilde{\psi} + (1 - m)\psi, \quad \tilde{\zeta} \leftarrow m\tilde{\zeta} + (1 - m)\zeta \tag{15}$$

where $m \in [0, 1)$ is a momentum coefficient, which is the meaning of the word *momentum*. It shows that only the parameters ψ and ζ of the base encoder are updated by InfoNCE loss backpropagation. The momentum update in Equation (15) makes the parameters $\tilde{\psi}$ and $\tilde{\zeta}$ evolve more smoothly than ψ and ζ. Therefore, although the negative samples in the memory bank are generated by different momentum encoders, the differences between them are small. The experimental results in [39] prove that a larger momentum coefficient will make the momentum encoder update more smoothly, which is also more beneficial to maintain the consistency of negative samples.

3.2. Classifier Fine-Tuning With Labeled Data

The deep representations of deformation ISAR images obtained by pretraining in Section 3.1 cannot be directly used for classification. Therefore, in this section, the pretrained convolutional encoder is transferred to the downstream supervised learning, and a linear classifier is fine-tuned using labeled samples. The diagram of classifier fine-tuning in CLISAR-Net is shown in Figure 7. Specifically, throw away the projection head of the pretrained base encoder, and only the convolutional encoder is retained. In the fine-tuning process, the convolutional encoder is frozen and is directly used to extract the deep representations of labeled training samples. In the fine-tuning phase, only the lightweight classifier will be trained, and the convolutional encoder is not involved. Due to the low complexity of the linear classifier, a small number of labeled samples are sufficient for training.

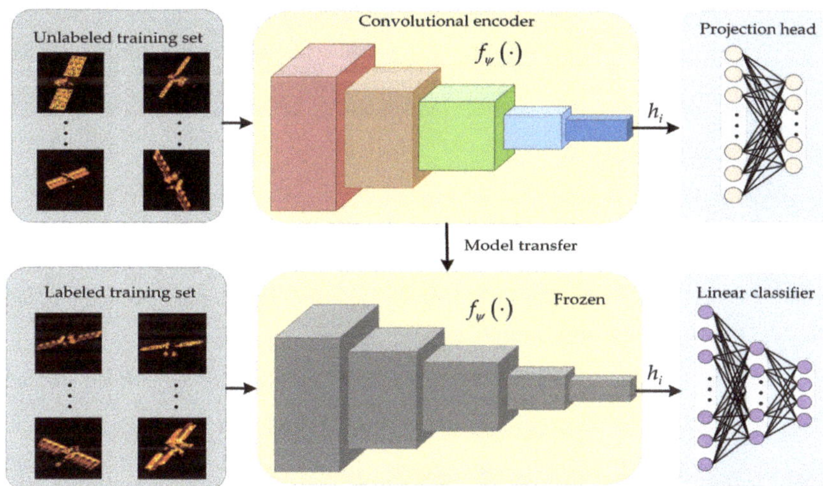

Figure 7. Diagram of classifier fine-tuning in CLISAR-Net.

4. Experiments

4.1. Data Generation

To verify the unsupervised deep representation learning ability of the proposed CLISAR-Net for deformation ISAR images and the classification ability using labeled samples, we conduct experiments on a high-resolution ISAR image dataset including four satellites, i.e., CALIPSO, Cloudsat, Jason-3, and OCO-2 [51]. In experiments, echoes of the target are generated using the shooting and bouncing ray (SBR+) method in the HFSS 2021R2 software developed by Ansoft in the United States.

In the EM calculation, we set radar works in the central frequency of 17 GHz with HV polarization, the imaging bandwidths are 1 GHz (16~17 GHz), 1.5 GHz (16~17.5 GHz), and 2 GHz (16~18 GHz). The azimuth angle γ is 0 °~359 ° with an interval of 0.05°. The accumulating angle $\Delta\theta$ is set to 4°, 5°, and 6°, respectively. Meanwhile, in order to verify the classification performance of CLISAR-Net for satellite targets at different elevation angles, we set two elevation angles for four satellites, respectively, i.e., $\varphi_1 = 55°$ and $\varphi_2 = 60°$ for CALIPSO, $\varphi_1 = 50°$ and $\varphi_2 = 55°$ for Cloudsat and Jason-3, and $\varphi_1 = 65°$ and $\varphi_2 = 70°$ for OCO-2. The RD algorithm is used to perform continuous high-resolution ISAR imaging. For the convenience of classification, each ISAR image is cropped to 128×128 pixels. The optical images, CAD models, and typical ISAR images of CALIPSO, Cloudsat, Jason-3, and OCO-2 are shown in Figure 8.

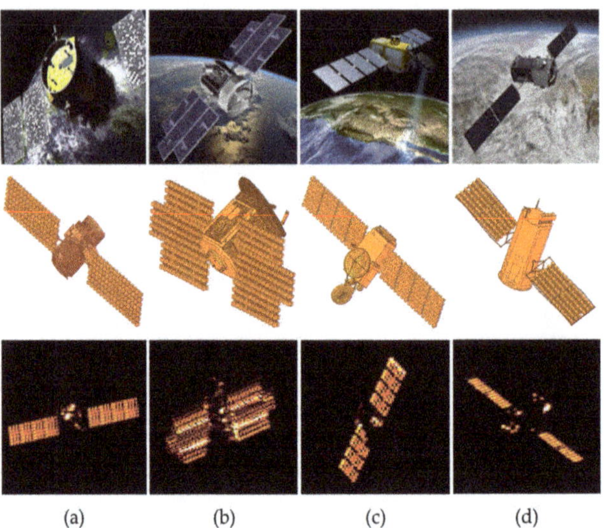

(a)   (b)   (c)   (d)

Figure 8. The optical images, CAD models, and ISAR images of (**a**) CALIPSO, (**b**) Cloudsat, (**c**) Jason-3, and (**d**) OCO-2.

As stated in Section 2, different radar parameters will cause scaling deformation of ISAR images; meanwhile, the target motion will cause rotation deformation of ISAR images. In practice, scaling and rotation deformations usually occur simultaneously during observation, resulting in complex deformation of ISAR images. Therefore, in the experiments, three datasets are generated, namely scaled, rotated, and combined deformation datasets, to evaluate the performance of CLISAR-Net on ISAR images with different deformations. The details of these three datasets are described below.

4.1.1. Scaled Deformation Dataset

To evaluate the classification ability of CLISAR-Net for scaled deformation ISAR images due to range and azimuth scaling, a scaled deformation dataset is generated, as shown in Table 1. The elevation angles of the four satellites in the training and test sets are all φ_1. The other imaging parameters are $\Delta\theta = 5°, B = 1$ GHz and $\Delta\theta = 6°, B = 2$ GHz for the training set, while becoming $\Delta\theta = 6°, B = 1.5$ GHz and $\Delta\theta = 4°, B = 2$ GHz for the test set. The above parameter settings cause the training and test ISAR images to be scaled along both the range and azimuth dimensions. In the scaled deformation dataset, the training sets have 2836 samples, and the test sets have 2840 samples.

Table 1. Detailed parameter settings of training and test sets for scaled deformation dataset.

Class	Training Set		Test Set	
	$\Delta\theta = 5°, B = 1$ GHz $\varphi_1, \gamma = 0°{\sim}359°$	$\Delta\theta = 6°, B = 2$ GHz $\varphi_1, \gamma = 0°{\sim}359°$	$\Delta\theta = 6°, B = 1.5$ GHz $\varphi_1, \gamma = 0°{\sim}359°$	$\Delta\theta = 4°, B = 2$ GHz $\varphi_1, \gamma = 0°{\sim}359°$
CALIPSO	355	354	354	356
Cloudsat	355	354	354	356
Jason-3	355	354	354	356
OCO-2	355	354	354	356
Total number	**2836**		**2840**	

4.1.2. Rotated Deformation Dataset

The rotated deformation ISAR image dataset is shown in Table 2. The same bandwidths and accumulating angles are used in the training and test sets, i.e., $\Delta\theta = 5°, B = 1$ GHz and $\Delta\theta = 6°, B = 2$ GHz. However, the elevation angles of the four satellites in the training set are all φ_1, and in the test set are all φ_2. Furthermore, a quarter of the azimuth angles are lost in the training set, i.e., the azimuth angle γ is $90°{\sim}359°$. Such a setting can verify the ability of CLISAR-Net to extract discriminative representations of the rotated deformation ISAR images when the training samples do not cover all the observation intervals. In the rotated deformation dataset, there are 2116 samples in the training set and 2836 samples in the test set.

Table 2. Detailed parameter settings of training and test sets for the rotated deformation dataset.

Class	Training Set		Test Set	
	$\Delta\theta = 5°, B = 1$ GHz $\varphi_1, \gamma = 90°{\sim}359°$	$\Delta\theta = 6°, B = 2$ GHz $\varphi_1, \gamma = 90°{\sim}359°$	$\Delta\theta = 5°, B = 1$ GHz $\varphi_2, \gamma = 0°{\sim}359°$	$\Delta\theta = 6°, B = 2$ GHz $\varphi_2, \gamma = 0°{\sim}359°$
CALIPSO	265	264	355	354
Cloudsat	265	264	355	354
Jason-3	265	264	355	354
OCO-2	265	264	355	354
Total number	**2116**		**2836**	

4.1.3. Combined Deformation Dataset

In practical observations, ISAR images usually exhibit a combined deformation of scaling and rotation. To analyze the classification robustness of CLISAR-Net in this scenario, a combined deformation ISAR image dataset is constructed, and the parameter settings are shown in Table 3. The elevation angle in the training set is φ_1, while changing to φ_2 in the test set. The other parameters are $\Delta\theta = 5°, B = 1$ GHz and $\Delta\theta = 6°, B = 2$ GHz for the training set, while they are $\Delta\theta = 6°, B = 1.5$ GHz and $\Delta\theta = 4°, B = 2$ GHz for the test set. The combined deformation dataset includes both scaling and rotation deformations, so this group of ISAR images has the most obvious deformations. Same as the rotated dataset, $1/4$ of azimuth angles are lost in the training set for the combined dataset, which increases the difficulty of classification. In the combined deformation dataset, there are 2116 training samples and 2840 test samples.

Table 3. Detailed parameter settings of training and test sets for the combined deformation dataset.

Class	Training Set		Test Set	
	$\Delta\theta = 5°, B = 1$ GHz $\varphi_1, \gamma = 90°{\sim}359°$	$\Delta\theta = 6°, B = 2$ GHz $\varphi_1, \gamma = 90°{\sim}359°$	$\Delta\theta = 6°, B = 1.5$ GHz $\varphi_2, \gamma = 0°{\sim}359°$	$\Delta\theta = 4°, B = 2$ GHz $\varphi_2, \gamma = 0°{\sim}359°$
CALIPSO	265	264	354	356
Cloudsat	265	264	354	356
Jason-3	265	264	354	356
OCO-2	265	264	354	356
Total number	**2116**		**2840**	

4.2. Experimental Setup

4.2.1. Data Augmentations

The data augmentations employed in the training of CLISAR-Net include: (i) random rotation and scaling for the ISAR images with the probability of 0.5, where the rotation angles are uniformly distributed in $\pm 90°$, and the scaling ratio is uniformly distributed in ± 0.2; (ii) random horizontal or vertical flips for the ISAR images with the probability of 0.5; (iii) random center cropping (by $0.8\times \sim 1.2\times$) for the ISAR images and resizing them to 128×128 pixels; and (iv) performing ISAR images amplitude normalization to eliminate the effect of amplitude. The above data augmentations provide more deformation patterns and enable the pretrained base encoder of CLISAR-Net to learn more deformation information. Figure 9 visualizes the data augmentations that we use in this work. During pretraining, two independent data augmentation operators are randomly generated from the above augmentations and applied to each training sample x_i to obtain two different views, and then they are fed into the base encoder and momentum encoder, respectively. Since deep representations of deformation ISAR images were obtained by pretraining, only the normalization operator is used for data normalization in classifier fine-tuning.

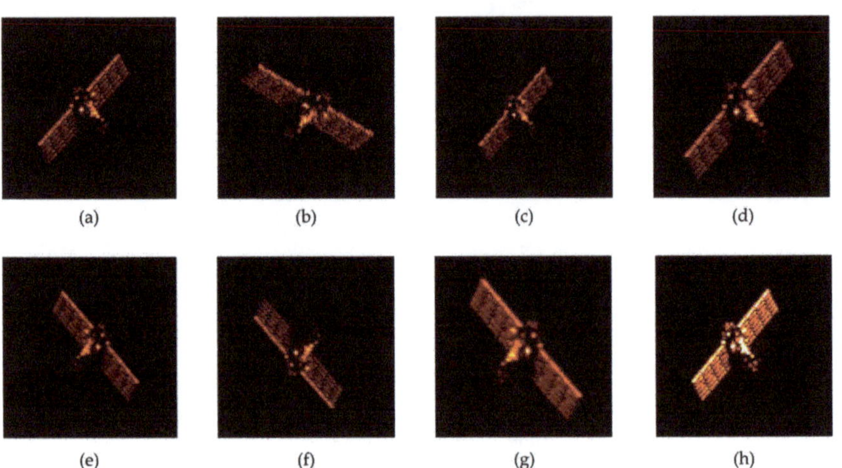

Figure 9. Illustrations of the data augmentation operators. (**a**) Original, (**b**) rotated, (**c**) scaled down, (**d**) scaled up, (**e**) horizontal flip, (**f**) vertical flip, (**g**) center crop, resize, and horizontal flip, and (**h**) normalized ISAR image.

4.2.2. Parameter Settings

In the pretraining phase, the structure of the convolutional encoder is shown in Figure 4. The convolutional kernels are all sized 3×3 with a step size of 1×1. The max pooling layers are all sized 2×2 with a step size of 2. The kernel numbers in five convolution-pooling blocks are 8, 16, 32, 64, and 128, respectively. For the projection head, the nodes of the two fully connected layers are 128 and 64. The momentum encoder has exactly the same structure as the base encoder, which is very important to maintain consistency. For the pretraining , the CosineAnnealing learning rate schedule is used with a maximum learning rate of 0.001 and a half-period of 50 epochs. The mini-batch size is 256, and the Adam optimizer is used to train for 500 epochs. Meanwhile, the momentum coefficient $m = 0.999$, the length of the memory bank is 8192, and the temperature $\tau = 0.1$. The pretraining is implemented by two NVIDIA RTX3090 GPUs using distributed training.

In the fine-tuning phase, the linear classifier connected behind the average pooling layer contains two fully connected layers with 64 and 32 nodes, and the nodes of the softmax layer are 4. The OneCycle learning rate schedule is used to train the linear classifier. Based on the Adam optimization, the classifier is trained for 200 epochs with a maximum learning rate of 0.001 . The mini-batch size is 32. The retraining and the testing process are both performed by one NVIDIA RTX3090 GPU. The entire CLISAR-Net is implemented by Pytorch on Ubuntu20.04 Linux system.

4.2.3. Comparison Methods

In order to verify the ability of CLISAR-Net for deformation ISAR image classification, four supervised and a semi-supervised classifiers are compared in the experiments. Specifically, four CNN-based supervised models are chosen, including CNN, spatial transformer–convolutional neural network (ST-CNN) with a double layer of STN [31], Deform-CNN using the deformable convolutional network (DCN) [32], and CNN connected to BiLSTM (CNN-BiLSTM) [52]. Moreover, the support vector machine (SVM) is chosen as a semi-supervised model. Among them, CNN has the same structure as the convolutional encoder described in Section 3.1.1, except that the deformable convolution in the last two blocks is replaced by regular convolution. The ST-CNN shares the same convolution kernels with CNN, but it adds a double layer of STN before CNN. The structure of Deform-CNN is identical to the convolutional encoder described in Section 3.1.1. It must be pointed out that the Deform-CNN here is trained in a supervised paradigm. A double layer of BiLSTM connected behind CNN constitutes the CNN-BiLSTM, which can be used to process the sequential ISAR images. The input sequential ISAR images of CNN-BiLSTM can be obtained by a continuous sliding window [52]. In the experiments, the length of the sliding window is set to 10, and the step size is 1.

Based on the Adam optimization, the parameter updates of the above networks are all achieved by cross-entropy loss backpropagation. Similarly, these supervised networks for comparison are trained using the CosineAnnealing learning rate schedule with a half-period of 25 epochs. The maximum learning rate is 0.001 for all the networks except ST-CNN, which has a learning rate of 0.0001. This is because STN is sensitive to the learning rate, and a smaller learning rate is helpful for STN to train the ISAR images. The mini-batch size is set to 32, and 200 epochs are trained.

4.3. Classification Results

In the experiment, all the unlabeled training samples are used to pretrain the convolutional encoder in CLISAR-Net to obtain the deep representations of ISAR images. Then, based on the learned representations, the linear classifier in CLISAR-Net is fine-tuned using 5% and 100% of the labeled training samples, respectively, to evaluate the classification ability of CLISAR-Net for deformation ISAR images. Table 4 presents the classification results on the three datasets. We can see that the proposed CLISAR-Net achieves the best classification accuracy under the above conditions. The increase in the number of labeled samples provides more supervised information for training, so better performance is obtained when 100% of labeled samples are used.

Table 4. Classification accuracy (%) of CLISAR-Net and other methods using 5% and 100% of labeled training samples on the three datasets.

Methods	Training with 5% of Labeled Samples			Training with 100% of Labeled Samples		
	Scaled Data.	Rotated Data.	Combined Data.	Scaled Data.	Rotated Data.	Combined Data.
CNN	77.50	73.59	71.87	91.58	87.98	84.96
ST-CNN	82.71	78.77	76.62	93.69	89.77	86.65
Deform-CNN	83.21	78.23	75.96	93.73	88.68	86.58
CNN-BiLSTM	86.34	82.78	80.82	94.33	92.58	91.73
SVM	87.54	84.69	82.64	95.81	93.72	92.39
CLISAR-Net	**91.34**	**86.61**	**84.23**	**98.10**	**96.37**	**95.85**

When only 5% of labeled samples, i.e., 35 samples per category in the scaled dataset and 26 samples per category in the rotated and combined dataset, are used to train the classifiers, CNN has the lowest classification accuracy because there are no mechanisms to deal with ISAR image deformation. The STN network in ST-CNN can adjust the input deformation ISAR image to the view that is easier to recognize. Therefore, the classification accuracy of ST-CNN on the three datasets is improved by 5.21%, 5.18%, and 4.75% over CNN, respectively. Deform-CNN is more adaptable to the deformations of ISAR images through deformable convolution, so it obtains comparable classification ability with ST-CNN. CNN-BiLSTM achieves bidirectional feature extraction and fusion for sequential ISAR images and extracts more deformation-robust features than a single image, thus achieving the best performance among the four supervised models. The classification accuracy of CNN-BiLSTM is 8.84%, 9.19%, and 8.95% higher than CNN on the three datasets. Based on the deep feature representations of deformation ISAR images obtained by CLISAR-Net in the unsupervised pretraining phase, SVM achieves slightly higher classification performance than the other four supervised learning methods. However, the proposed CLISAR-Net achieves the best classification performance. The classification accuracy of CLISAR-Net on the scaled, rotated, and combined deformation datasets reaches 91.34%, 86.61%, and 84.23%, respectively.

When the classifiers are trained using 100% of labeled training samples, the classification accuracy of CLISAR-Net reaches 98.10% for the scaled dataset, 96.37% for the rotated dataset, and 95.85% for the combined deformation dataset. It outperforms the best performing supervised model CNN-BiLSTM by 3.77%, 3.79%, and 4.12%, respectively. Based on the deep representations obtained by unsupervised pretraining, the SVM classifier also performs slightly better than the other four supervised methods. It indicates that in the pretraining phase, based on instance discrimination, CL makes the encoder learn deep discriminative representations of deformation ISAR images, which provides a great help to the classification.

To facilitate the analysis, Figure 10 gives the histograms of the classification results when using 5% and 100% of labeled samples to train the above classifiers. We can summarize that the classification accuracy is the highest for the scaled deformation dataset and the lowest for the combined deformation dataset. Particularly, the classification accuracy for CNN, ST-CNN, and Deform-CNN shows a significant decrease on the rotated and combined deformation datasets. It implies that the traditional end-to-end model for classification based on a single image is difficult to learn deformation-robust features of complex deformation ISAR images when part of the training samples are missing continuously. Comparatively, CNN-BiLSTM maintains a relatively high classification performance on the rotated and combined deformation datasets by fusing the information of multiframe ISAR images. When 100% of the labeled training samples are used to fine-tune the classifier, SVM and CLISAR-Net are less affected by sample misses, especially CLISAR-Net, which shows a more balanced classification performance on the three datasets. Furthermore, from Figure 10a, the classification accuracy of CLISAR-Net exceeds 90% for the scaled deformation dataset, with only 5% of labeled samples, while the CNN is less than 80%. It can be concluded that the smaller the number of labeled training samples, the more obvious the superiority of CLISAR-Net.

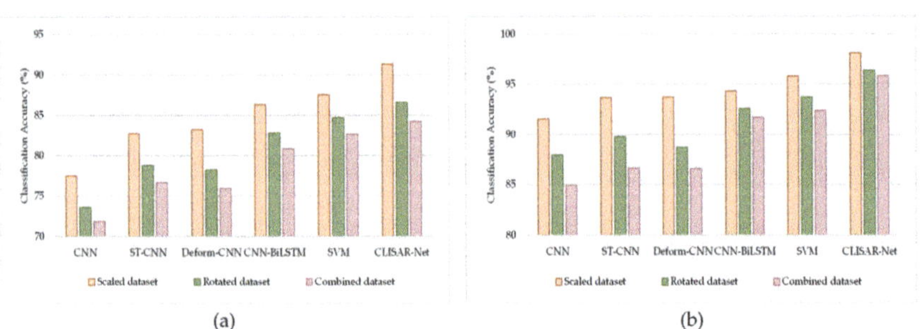

Figure 10. Comparisons of involved methods on three datasets. (**a**) Classification results using 5% of labeled training samples and (**b**) classification results using 100% of labeled training samples.

4.4. Computational Cost

In this subsection, the computational cost of the proposed CLISAR-Net versus other methods is analyzed, including the computational complexity and the inference time, i.e., the average period for acquiring a frame of an image label in the test process. We measure the computational complexity of the model in terms of the number of trainable parameters. As shown in Table 5, the ST-CNN has the largest number of parameters and the longest inference time due to the addition of a double layer of STN. The number of parameters in Deform-CNN is slightly higher than that in CNN because 2D offsets need to be calculated. The double layer of BiLSTM in CNN-BiLSTM similarly brings an additional number of parameters to the model. The SVM itself has no parameters to be retrained, so the inference time is the shortest. For CLISAR-Net, the pretrained encoder adopts the same structure as Deform-CNN, and only the linear classifier needs to be retrained in the downstream classifier, so the number of trainable parameters is medium. Moreover, the average inference time for obtaining a frame of an image label is only 0.1125 ms. The above analysis shows that CLISAR-Net can achieve the optimal classification performance with less computational cost, which also proves the superiority of CLISAR-Net.

Table 5. Comparisons of the number of trainable parameters and inference time between CLISAR-Net and existing methods.

Methods	Number of Parameters (K)	Inference Time (ms)
CNN	124.38	0.0834
ST-CNN	374.41	0.1862
Deform-CNN	146.28	0.1031
CNN-BiLSTM	326.78	0.0932
SVM	/	0.0196
CLISAR-Net	267.73	0.1125

5. Discussion

To demonstrate the superiority of CLISAR-Net for deformation ISAR image classification, Section 5.1 discusses the classification results when different numbers of labeled training samples are used to train the classifiers. Section 5.2 discusses the classification performance of CLISAR-Net when different azimuth angle ranges are included in the training set. Additionally, the features learned by the CLISAR-Net and CNN-BiLSTM are visualized in Section 5.3.

5.1. Effect of Different Training Ratios

To evaluate the effect of a different number of labeled training samples on the deformation ISAR image classification, CNN, ST-CNN, Deform-CNN, CNN-BiLSTM, SVM, and the proposed CLISAR-Net are tested using different ratios of labeled training samples. Figure 11 reports the comparison results on the scaled, rotated, and combined deformation datasets. It can be seen from the results of the three datasets that CLISAR-Net performs the best in all conditions, followed by SVM. Moreover, the classification accuracy increases gradually with the increase in the number of labeled samples, which is consistent with the expectation. When only 5% of labeled samples are used to train CLISAR-Net, the CNN requires 100% of labeled samples to achieve similar classification performance for the three datasets. The performance of ST-CNN and Deform-CNN is always very close, but ST-CNN is more difficult to train. The results of the scaled dataset exemplify the superiority of CLISAR-Net at the ratios of 5%, 10%, and 20%. For the combined dataset, as shown in Figure 11c, when 20% of labeled samples are used, the classification accuracy of CLISAR-Net has a greater improvement than SVM and CNN-BiLSTM. Based on the above analysis, it is reasonable to assume that discriminative deep representations were learned in the unsupervised pretraining phase. As a result, CLISAR-Net can achieve good performance by fine-tuning the classifier with only a small number of labeled samples.

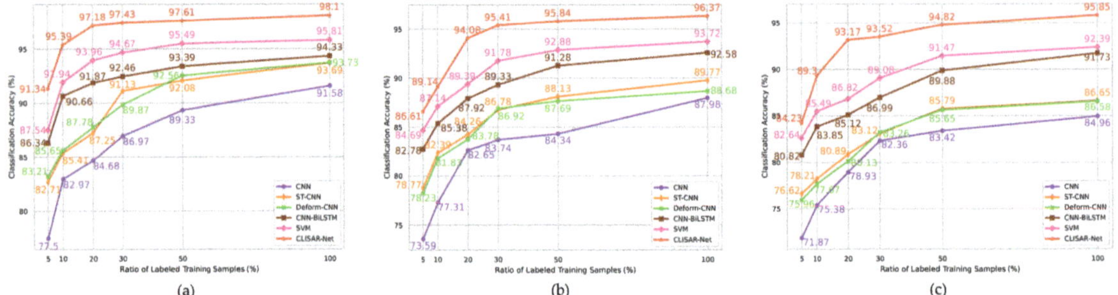

Figure 11. Comparison of the classification accuracy with different ratios of labeled training samples for different methods. (**a**) Results of scaled dataset, (**b**) results of rotated dataset, and (**c**) results of combined dataset.

5.2. Extended to Different Azimuth Angle Ranges

To evaluate the classification robustness of CLISAR-Net for the deformation ISAR images when different azimuth angles are missing in the training set, the azimuth angle γ in the training set of the combined deformation dataset is adjusted to $0°\sim269°$ and $90°\sim269°$, while the test set is kept constant. In the adjusted combined deformation dataset, the numbers of samples in the training sets are 2116 and 1396, respectively, and the number of samples in the test set is still 2840. Figure 12 shows the classification results of CLISAR-Net for the test set when the ranges of the azimuth angle in the training set are $\gamma = 90°\sim359°$, $\gamma = 0°\sim269°$, and $\gamma = 90°\sim269°$. As can be seen in Figure 12, the two polylines corresponding to $\gamma = 90°\sim359°$ and $\gamma = 0°\sim269°$ are intertwined, which indicates that CLISAR-Net has robust classification performance when 1/4 of the azimuth angles are missing in the training set. For the azimuth angle ranges from $90°$ to $269°$, the classification accuracy of CLISAR-Net reaches 93.24% after fine-tuning the downstream classifier using 100% of the labeled training samples, which is only 2.61% and 2.07% lower than that when 1/4 of the azimuth angles are missing, respectively. However, it is not difficult to find that when only 5% and 10% labeled training samples are used to fine-tune the downstream classifier, high classification performance is achieved, even though 1/2 of the azimuth angles are missing in the training set. This reaffirms the conclusion that CLISAR-Net can achieve superior classification accuracy when only a small amount of labeled data are available.

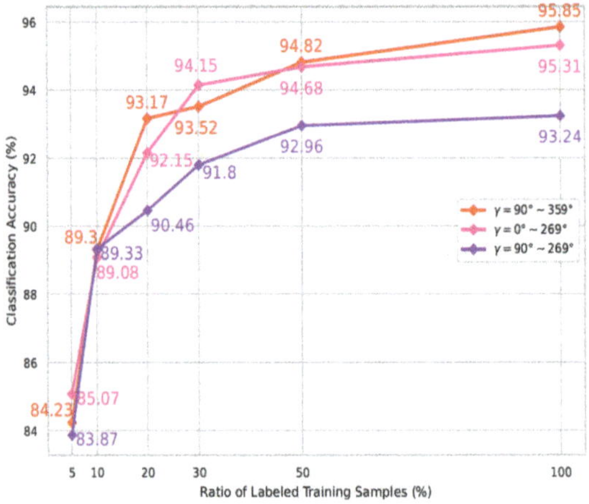

Figure 12. Classification accuracy of CLISAR-Net when different ranges of the azimuth angle are included in the training set of the combined deformation dataset.

5.3. Visualization of Features

The CLISAR-Net performs well on the three datasets, and the essential reason is that the convolutional encoder can extract the deep representations of unlabeled ISAR images during pretraining. To explicitly view the extracted representations, 2D visualizations of the feature representations are performed by t-SNE [53] for the combined deformation dataset. As shown in Figure 13, the distance of points indicates the similarity between samples. Samples of the same satellite are represented by points of the same color, where dark blue points denote CALIPSO, red points indicate Cloudsat, pink points are Jason-3, and cyan points are OCO-2.

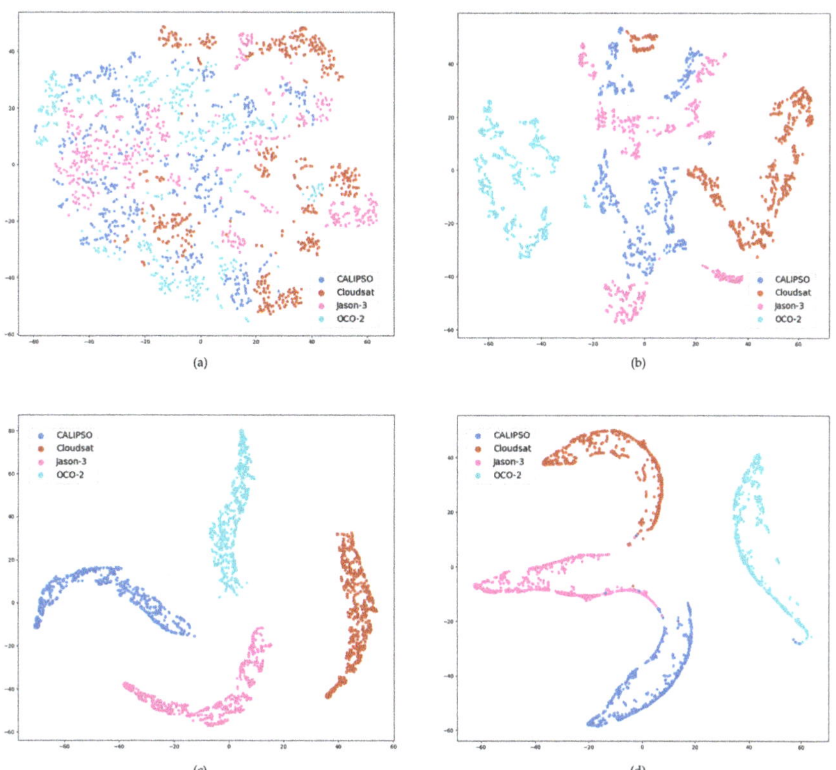

Figure 13. T-SNE visualization of the features for the combined deformation dataset from (**a**) input images, (**b**) pretraining of CLISAR-Net, (**c**) fine-tuning of CLISAR-Net, and (**d**) CNN-BiLSTM.

The features learned by CLISAR-Net in different training phases and the features learned by CNN-BiLSTM, the best performing classification model in the supervised methods, are visualized by t-SNE. In Figure 13, each point represents an ISAR image. From the distribution of the input images, it can be seen that in the combined deformation dataset, the points of different categories are widely distributed in different positions and are completely indistinguishable. Figure 13b shows the distribution of the deep representations obtained by unsupervised pretraining. Compared with Figure 13a, the compactness within each category is increased significantly. OCO-2 is basically separable, and most of the points of the other three categories are separable. The visualization results show that unsupervised pretraining can embed the features of the input images into a more discriminative space. In Figure 13c, the linear classifier in CLISAR-Net is fine-tuned to create more compact category clusters based on Figure 13b and achieves better feature separation for the four satellites. In Figure 13d, although the CNN-BiLSTM improved the feature separability of the four categories, the feature distribution of CALIPSO and Jason-3 still has partial overlap and coverage, so its classification performance is slightly inferior to that of CLISAR-Net, which agrees with the classification results in Table 4.

6. Conclusions

In order to achieve deformation ISAR image classification, an unsupervised ISAR image deep representation learning method is explored based on CL for the first time. The training of CLISAR-Net consists of two phases, i.e., unsupervised pretraining and classifier fine-tuning. In the pretraining phase, the base encoder is optimized by InfoNCE loss back-propagation, and the evolution of the momentum encoder is realized by the momentum update of the parameters. With the help of the discriminative representations obtained by pretraining, high-precision classification of deformation ISAR images can be achieved by retaining the convolutional encoder and fine-tuning the linear classifier. CLISAR-Net demonstrates powerful classification performance in the experiments on scaled, rotated, and combined deformation ISAR image datasets. Compared with traditional CNN-based supervised learning methods, the added unsupervised pretraining phase in CLISAR-Net makes the feature extractor capture more discriminative deep representations of deformation ISAR images, which brings great convenience for downstream classification and enables CLISAR-Net to achieve higher classification performance with a small number of labeled samples.

Although the proposed CLISAR-Net requires two phases to achieve classification, the encoder structure that extracts the feature of deformation ISAR images is simpler than that of the CNN-based methods. Moreover, based on CL, CLISAR-Net opens the door for researches of ISAR image classification based on unsupervised learning. In the future, ISAR image classification will be performed by combining the EM scattering mechanism and semantic information of the target under unsupervised conditions.

Author Contributions: Conceptualization, G.X. and P.N.; methodology, P.N., Y.L. and H.P.; software, P.N., H.P. and H.D.; validation, P.N. and G.X.; investigation, P.N. and Y.L.; resources, Y.L. and H.L.; data curation, P.N.; writing—original draft, P.N. and H.D.; writing—review and editing, P.N. and G.X.; supervision, G.X. All authors have read and agreed to the published version of the manuscript.

Funding: This work was supported by the National Science Foundation of China (NSFC) under Grant 62071113, in part by the Natural Science Foundation of Jiangsu Province under Grant BK20211559.

Conflicts of Interest: The authors declare no conflict of interest.

References

1. Kim, K.T.; Seo, D.K.; Kim, H.T. Efficient Classification of ISAR images. *IEEE Trans. Antennas Propag.* **2005**, *53*, 1611–1621.
2. Liu, L.; Zhou, F.; Bai, X.; Tao, M.; Zhang, Z. Joint Cross-Range Scaling and 3D Geometry Reconstruction of ISAR Targets Based on Factorization Method. *IEEE Trans. Image Process.* **2016**, *25*, 1740–1750. [CrossRef]
3. Wagner, S.; Dommermuth, F.; Ender, J. Detection of Jet Engines via Sparse Decomposition of ISAR Images for Target Classification Purposes. In Proceedings of the 2016 European Radar Conference (EuRAD), London, UK, 5–7 October 2016; pp. 77–80.
4. Huang, Y.; Liao, G.; Xiang, Y.; Zhang, L.; Li, J.; Nehorai, A. Low-rank Approximation via Generalized Reweighted Iterative Nuclear and Frobenius Norms. *IEEE Trans. Image Process.* **2020**, *29*, 2244–2257. [CrossRef]
5. Du, Y.; Jiang, Y.; Wang, Y.; Zhou, W.; Liu, Z. ISAR Imaging for Low-Earth-Orbit Target Based on Coherent Integrated Smoothed Generalized Cubic Phase Function. *IEEE Trans. Geosci. Remote Sens.* **2019**, *58*, 1205–1220. [CrossRef]
6. Xue, B.; Tong, N. Real-World ISAR Object Recognition Using Deep Multimodal Relation Learning. *IEEE Trans. Cybern.* **2019**, *50*, 4256–4267. [CrossRef] [PubMed]
7. Zhang, Y.; Yuan, H.; Li, H.; Chen, J.; Niu, M. Meta-Learner-Based Stacking Network on Space Target Recognition for ISAR Images. *IEEE J. Sel. Top. Appl. Earth Obs. Remote Sens.* **2021**, *14*, 12132–12148. [CrossRef]
8. Lee, S.J.; Park, S.H.; Kim, K.T. Improved Classification Performance Using ISAR Images and Trace Transform. *IEEE Trans. Aerosp. Electron. Syst.* **2017**, *53*, 950–965. [CrossRef]
9. Benedek, C.; Martorella, M. Moving Target Analysis in ISAR Image Sequences With a Multiframe Marked Point Process Model. *IEEE Trans. Geosci. Remote Sens.* **2013**, *52*, 2234–2246. [CrossRef]
10. Islam, M.T.; Siddique, B.N.K.; Rahman, S.; Jabid, T. Image Recognition with Deep Learning. In Proceedings of the 2018 International Cnference on Intelligent Informatics and Biomedical Sciences (ICIIBMS), Bangkok, Thailand, 21–24 October 2018; pp. 106–110.

11. Karine, A.; Toumi, A.; Khenchaf, A.; El Hassouni, M. Radar Target Recognition Using Salient Keypoint Descriptors and Multitask Sparse Representation. *Remote Sens.* **2018**, *10*, 843. [CrossRef]
12. Bai, Q.; Gao, G.; Zhang, X.; Yao, L.; Zhang, C. LSDNet: Light-weight CNN Model Driven by PNF for PolSAR Image Ship Detection. *IEEE J. Miniat. Air Space Syst.* **2022**, *3*, 135–142. [CrossRef]
13. Gao, S.; Liu, H. RetinaNet-based Compact Polarization SAR Ship Detection. *IEEE J. Miniat. Air Space Syst.* **2022**, *3*, 146–152. [CrossRef]
14. Zhang, L.; Gao, G.; Duan, D.; Zhang, X.; Yao, L.; Liu, J. A Novel Detector for Adaptive Detection of Weak and Small Ships in Compact Polarimetric SAR. *IEEE J. Miniat. Air Space Syst.* **2022**, *3*, 153–160. [CrossRef]
15. Sun, Y.; Wang, Y.; Liu, H.; Wang, N.; Wang, J. SAR Target Recognition with Limited Training Data Based on Angular Rotation Generative Network. *IEEE Geosci. Remote Sens. Lett.* **2019**, *17*, 1928–1932. [CrossRef]
16. Wang, L.; Bai, X.; Gong, C.; Zhou, F. Hybrid Inference Network for Few-Shot SAR Automatic Target Recognition. *IEEE Trans. Geosci. Remote Sens.* **2021**, *59*, 9257–9269. [CrossRef]
17. Yang, M.; Bai, X.; Wang, L.; Zhou, F. Mixed Loss Graph Attention Network for Few-Shot SAR Target Classification. *IEEE Trans. Geosci. Remote Sens.* **2021**, *60*, 1–13. [CrossRef]
18. Raj, J.A.; Idicula, S.M.; Paul, B. One-Shot Learning-Based SAR Ship Classification Using New Hybrid Siamese Network. *IEEE Geosci. Remote Sens. Lett.* **2021**, *19*, 1–5. [CrossRef]
19. Xue, R.; Bai, X.; Zhou, F. Spatial–Temporal Ensemble Convolution for Sequence SAR Target Classification. *IEEE Trans. Geosci. Remote Sens.* **2020**, *59*, 1250–1262. [CrossRef]
20. Qian, X.; Liu, F.; Jiao, L.; Zhang, X.; Chen, P.; Li, L.; Cui, Y. A Hybrid Network With Structural Constraints for SAR Image Scene Classification. *IEEE Trans. Geosci. Remote Sens.* **2021**, *60*, 1–17. [CrossRef]
21. Pereira, L.O.; Freitas, C.C.; Sant, S.J.; Reis, M.S. Evaluation of Optical and Radar Images Integration Methods for LULC Classification in Amazon Region. *IEEE J. Sel. Top. Appl. Earth Obs. Remote Sens.* **2018**, *11*, 3062–3074. [CrossRef]
22. Hu, J.; Hong, D.; Zhu, X.X. MIMA: MAPPER-Induced Manifold Alignment for Semi-Supervised Fusion of Optical Image and Polarimetric SAR Data. *IEEE Trans. Geosci. Remote Sens.* **2019**, *57*, 9025–9040. [CrossRef]
23. Huang, Z.; Dumitru, C.O.; Pan, Z.; Lei, B.; Datcu, M. Classification of Large-Scale High-Resolution SAR Images with Deep Transfer Learning. *IEEE Geosci. Remote Sens. Lett.* **2020**, *18*, 107–111. [CrossRef]
24. Zhao, Y.; Jiang, M. Integration of Optical and SAR Imagery for Dual PolSAR Features Optimization and Land Cover Mapping. *IEEE J. Miniat. Air Space Syst.* **2022**, *3*, 67–76. [CrossRef]
25. Xu, G.; Zhang, B.; Chen, J.; Wu, F.; Sheng, J.; Hong, W. Sparse Inverse Synthetic Aperture Radar Imaging Using Structured Low-Rank Method. *IEEE Trans. Geosci. Remote Sens.* **2022**, *60*, 1–12. [CrossRef]
26. Tien, S.C.; Chia, T.L.; Lu, Y. Using Invariants to Recognize Airplanes in Inverse Synthetic Aperture Radar Images. *Opt. Eng.* **2003**, *42*, 200–210.
27. Paladini, R.; Famil, L.F.; Pottier, E.; Martorella, M.; Berizzi, F.; Dalle Mese, E. Point Target Classification via Fast Lossless and Sufficient Ω–Ψ–Φ Invariant Decomposition of High-Resolution and Fully Polarimetric SAR/ISAR Data. *Proc. IEEE* **2013**, *101*, 798–830. [CrossRef]
28. Paladini, R.; Martorella, M.; Berizzi, F. Classification of Man-Made Targets via Invariant Coherency-Mtrix Eigenvector Decomposition of Polarimetric SAR/ISAR Images. *IEEE Trans. Geosci. Remote Sens.* **2021**, *49*, 3022–3034. [CrossRef]
29. Park, S.H.; Jung, J.H.; Kim, S.H.; Kim, K.T. Efficient Classification of ISAR Images Using 2D Fourier Transform and polar Mpping. *IEEE Trans. Aerosp. Electron. Syst.* **2015**, *51*, 1726–1736. [CrossRef]
30. Lu, W.; Zhang, Y.; Yin, C.; Lin, C.; Xu, C.; Zhang, X. A Deformation Robust ISAR Image Satellite Target Rrecognition Method Based on PT-CCNN. *IEEE Access* **2021**, *9*, 23432–23453. [CrossRef]
31. Bai, X.; Zhou, X.; Zhang, F.; Wang, L.; Xue, R.; Zhou, F. Robust Pol-ISAR Target Recognition Based on ST-MC-DCNN. *IEEE Trans. Geosci. Remote Sens.* **2019**, *57*, 9912–9927. [CrossRef]
32. Xue, R.; Bai, X.; Zhou, F. SAISAR-Net: A Robust Sequential Adjustment ISAR Image Classification Network. *IEEE Trans. Geosci. Remote Sens.* **2021**, *60*, 1–15. [CrossRef]
33. Zhou, X.; Bai, X.; Wang, L.; Zhou, F. Robust ISAR Target Recognition Based on ADRISAR-Net. *IEEE Trans. Aerosp. Electron. Syst.* **2022**, *58*, 5494–5505. [CrossRef]
34. Xue, R.; Bai, X.; Cao, X.; Zhou, F. Sequential ISAR Target Classification Based on Hybrid Transformer. *IEEE Trans. Geosci. Remote Sens.* **2022**, *60*, 1–11. [CrossRef]
35. Jaderberg, M.; Simonyan, K.; Zisserman, A. Spatial Transformer Networks. In Proceedings of Advances in Neural Information Processing Systems (NIPS), London, UK, 7–12 December 2015.
36. Lin, C. H.; Lucey, S. Inverse Compositional Spatial Transformer Networks. In Proceedings of the IEEE Conference on Computer Vision and Pattern Recognition (CVPR), Honolulu, HI, USA, 21–26 July 2017; pp. 2568–2576.
37. Misra, I.; Maaten, L. V. D. Self-Supervised Learning of Pretext-Invariant Representations. In Proceedings of the IEEE/CVF Conference on Computer Vision and Pattern Recognition (CVPR), Seattle, WA, USA, 16–20 June 2020; pp. 6707–6717.
38. Chen, T.; Kornblith, S.; Norouzi, M.; Hinton, G. A Simple Framework for Contrastive Learning of Visual Representations. *arXiv* **2020**, arXiv: 2002.05709.

39. He, K.; Fan, H.; Wu, Y.; Xie, S.; Girshick, R. Momentum Contrast for Unsupervised Visual Representation Learning. In Proceedings of the IEEE/CVF Conference on Computer Vision and Pattern Recognition (CVPR), Seattle, WA, USA, 16–20 June 2020; pp. 9729–9738.

40. Grill, J.B.; Strub, F.; Altché, F.; Tallec, C.; Richemond, P.; Buchatskaya, E.; Doersch, C.; Avila Pires, B.; Guo, Z.; Gheshlaghi Azar, M.; et Bootstrap Your Own Latent-A New Approach to Self-Supervised Learning. In Proceedings of the Advances in Neural Information Processing Systems, Vancouver, BC, Canada, 11–14 May 2020; pp. 21271–21284.

41. Li, J.; Zhou, P.; Xiong, C.; Hoi, S.C. Prototypical Contrastive Learning of Unsupervised Representations. *arXiv* **2020**, arXiv:2005.04966.

42. Wu, Z.; Xiong, Y.; Yu, S.X.; Lin, D. Unsupervised Feature Learning via Non-parametric Instance Discrimination. In Proceedings of the IEEE Conference on Computer Vision and Pattern Recognition (CVPR), Salt Lake, UT, USA, 18–22 June 2018; pp. 3733–3742.

43. Oord, A.V.D.; Li, Y.; Vinyals, O. Representation Learning with Contrastive Predictive Coding. *arXiv* **2018**, arXiv:1807.03748.

44. Tian, Y.; Krishnan, D.; Isola, P. Contrastive multiview coding. In *Proceedings of European Conference on Computer Vision (ECCV)*; Springer: Cham, Switzerland, 2020; pp. 776–794.

45. Zhou, Y.; Zhang, L.; Cao, Y. Attitude Estimation for Space Targets by Exploiting the Quadratic Phase Coefficients of Inverse Synthetic Aperture Radar Imagery. *IEEE Trans. Geosci. Remote Sens.* **2019**, *57*, 3858–3872. [CrossRef]

46. Zhou Y.; Zhang L.; Cao Y. Dynamic Estimation of Spin Spacecraft Based on Multiple-Station ISAR Images. *IEEE Trans. Geosci. Remote Sens.* **2020**, *58*, 2977–2989. [CrossRef]

47. Song, D.; Chen Q.; Li, K. An Adaptive Sparse Constraint ISAR High Resolution Imaging Algorithm Based on Mixed Norm. *Radioengineering* **2022**, *31*, 477–485. [CrossRef]

48. Kang, B. S.; Kang, M. S.; Choi, I. O.; Kim, C. H.; Kim, K. T. Efficient Autofocus Chain for ISAR Imaging of Non-Uniformly Rotating Target. *IEEE Sens. J.* **2017**, *17*, 5466–5478. [CrossRef]

49. Sohn, K. Improved Deep Metric Learning with Multi-class N-pair Loss Objective. In Proceedings of the Advances in Neural Information Processing Systems (NIPS), Barcelona, Spain, 5–10 December 2016.

50. Zhang, L.; Zhang, S.; Zou, B.; Dong, H. Unsupervised Deep Representation Learning and Few-Shot Classification of PolSAR Images. *IEEE Trans. Geosci. Remote Sens.* **2020**, *60*, 1–16. [CrossRef]

51. NASA 3D Resource. Available online: https://nasa3d.arc.nasa.gov/models (accessed on 1 January 2020).

52. Bai, X.; Xue, R.; Wang, L.; Zhou, F. Sequence SAR Image Classification Based on Bidirectional Convolution-Recurrent Network. *IEEE Trans. Geosci. Remote Sens.* **2019**, *57*, 9223–9235. [CrossRef]

53. Van der Maaten, L.; Hinton, G. Visualizing Data Using t-SNE. *J. Mach. Learn. Res.* **2008**, *9*, 2579–2605.

Article

Distillation Sparsity Training Algorithm for Accelerating Convolutional Neural Networks in Embedded Systems

Penghao Xiao, Teng Xu, Xiayang Xiao, Weisong Li and Haipeng Wang *

Key Laboratory for Information Science of Electromagnetic Waves (MoE), Fudan University, Shanghai 200433, China; phxiao20@fudan.edu.cn (P.X.); xut22@m.fudan.edu.cn (T.X.); xyxiao20@fudan.edu.cn (X.X.); weisongli20@fudan.edu.cn (W.L.)
* Correspondence: hpwang@fudan.edu.cn

Abstract: The rapid development of neural networks has come at the cost of increased computational complexity. Neural networks are both computationally intensive and memory intensive; as such, the minimal energy and computing power of satellites pose a challenge for automatic target recognition (ATR). Knowledge distillation (KD) can distill knowledge from a cumbersome teacher network to a lightweight student network, transferring the essential information learned by the teacher network. Thus, the concept of KD can be used to improve the accuracy of student networks. Even when learning from a teacher network, there is still redundancy in the student network. Traditional networks fix the structure before training, such that training does not improve the situation. This paper proposes a distillation sparsity training (DST) algorithm based on KD and network pruning to address the above limitations. We first improve the accuracy of the student network through KD, and then through network pruning, allowing the student network to learn which connections are essential. DST allows the teacher network to teach the pruned student network directly. The proposed algorithm was tested on the CIFAR-100, MSTAR, and FUSAR-Ship data sets, with a 50% sparsity setting. First, a new loss function for the teacher-pruned student was proposed, and the pruned student network showed a performance close to that of the teacher network. Second, a new sparsity model (uniformity half-pruning UHP) was designed to solve the problem that unstructured pruning does not facilitate the implementation of general-purpose hardware acceleration and storage. Compared with traditional unstructured pruning, UHP can double the speed of neural networks.

Keywords: neural networks; distillation sparsity training; uniformity half-pruning; general-purpose hardware acceleration

Citation: Xiao, P.; Xu, T.; Xiao, X.; Li, W.; Wang, H. Distillation Sparsity Training Algorithm for Accelerating Convolutional Neural Networks in Embedded Systems. *Remote Sens.* **2023**, *15*, 2609. https://doi.org/ 10.3390/rs15102609

Academic Editor: Andrzej Stateczny

Received: 20 April 2023
Revised: 12 May 2023
Accepted: 14 May 2023
Published: 17 May 2023

1. Introduction

Convolutional neural networks (CNNs) have achieved state-of-the-art results in a range of computer vision tasks, such as image classification [1,2], depth estimation [3,4], and object detection [5,6]. In the field of deep learning, the use of larger neural network models typically leads to higher accuracy in a variety of tasks [7–10]. Although the current CNNs have achieved remarkable results in the field of computer vision, these models depend on many parameters. Modern state-of-the-art models can comprise hundreds of billions of parameters, requiring trillions of computational operations per input sample. This limits their deployment in resource-limited devices and drives the need for model compression techniques. Model compression technology has also developed rapidly with the demand for intelligent terminals. In recent years, model compression has attracted significant research interest and many pertinent approaches have been proposed, such as KD, network pruning, and quantization [11,12].

KD aims to transfer knowledge from an influential teacher network to a smaller, faster student network, in order to expand its performance capability [13]. An obvious way to share the generalization capability of a complex model with a small model is to use the probability of the complex model to generate a target as a "soft label" for training the small model. This soft target has high entropy, providing more information than the "hard label" used during training [14]. Early distillation methods aimed to transfer the last layer of the teacher [3]. However, the lack of oversight by intermediate layers hinders how the information of intermediate layers flows through the student network, reducing the learning potential of the student.

Although KD techniques can be used to transfer features from a complex network of teachers to a smaller network, the initial perception is that a robust teacher network with a high accuracy rate may provide better distillation results. However, when the student network does not have sufficient capacity to learn, further measures can be taken; for example, the student network can be pruned, as a network regularization technique [15,16]. Such approaches can provide more transferable knowledge for student networks with limited capacity. Network pruning [17] is a model compression technique that effectively removes the weights or neurons of a network, while maintaining its accuracy. After the initial training phase of the network, connections with all weights below a threshold are removed. This pruning transforms the dense network layers into sparse layers. The first phase involves learning the network's topology; that is, learning which connections are meaningful and removing the unimportant ones. Then, the sparse network is retrained, such that the remaining links can compensate for the deleted contacts. The pruning and retraining phases can be carried out iteratively, to further reduce the complexity of the network. In effect, this training process learns the network connectivity in addition to the weights, similarly to the case in the mammalian brain [18], where synapses are created in the first few months of a child's development, followed by gradual pruning of little-used connections, falling to typical adult values.

Network pruning is achieved by zeroing specific parameters, to enable the model to achieve sparsity [19,20]. However, it is difficult for the existing pruning algorithms to guarantee both model accuracy and inference performance (speed). Fine-grained sparsity maintains accuracy but is not conducive to memory access and cannot take advantage of general-purpose hardware to accelerate the computation. Therefore, it does not outperform traditional dense models using processor architectures such as GPUs [21]. Coarse-grained sparsity makes better use of processor resources but does not guarantee model accuracy by shearing off convolutional kernels. Overall, fine-grained sparsity ensures accuracy but does not guarantee model speed or efficient storage, while coarse-grained sparsity provides model speed and efficient storage but does not guarantee accuracy. We combine the best of both and design a specific UHP method, to be applied to the proposed algorithm based on general-purpose hardware acceleration conditions.

In this paper, we combine the advantages of KD and model sparsity to design a DST algorithm for accelerating CNNs in embedded systems. KD improves the performance of the student networks, while network sparsity enables efficient model storage and acceleration. The contributions of this paper include the following:

1. We design a unified training framework based on KD and network sparsity for model compression techniques.
2. Combined with general-purpose hardware, a uniform compression format is designed, to implement this pruning matrix for efficient storage and memory indexing.
3. The sparse model can be used for any general-purpose hardware acceleration. We tested the DST algorithm on an embedded system, and the acceleration result was significant.
4. The proposed algorithm was tested on the CIFAR-100, MSTAR, and FUSAR-Ship data sets. Its performance was significant on both SAR and optical data sets.

The remainder of this paper is organized as follows: Section 2 briefly reviews the related literature and summarizes the applications of KD and network pruning for model compression. Section 3 describes the motivation for the research presented in this paper and the limitations of existing algorithms. Section 4 describes the proposed DST algorithm in detail. Section 5 presents the experimental results and performance evaluation, and analytically verifies the DST algorithm. Section 6 concludes the paper and provides an outlook on future spaceborne ATR.

2. Related Work

2.1. KD

KD was initially proposed by Geoffrey Hinton [14], in order to enable a smaller network to learn the correlations between classes from the output of a larger pretrained teacher model. This work was extended in [15], to teach students to use intermediate representations as knowledge. They achieved this by minimizing the L2 distance between the feature maps of the student and teacher. In [15], it was shown that, if the gap between students and teachers is too large, the student's performance decreases. They suggested using intermediate features to distill knowledge between teachers and students. Each model uses a different architecture and has a separate set of weights. Slimmable neural networks [22] can execute on different widths by uniformly compressing the model width through a joint training approach. The smaller models benefited from the shared weights and the implicit KD provided.

In [23], a method to provide additional supervision for students using the feature map of the teacher was proposed. This method transfers only its intermediate layers with the help of encoding feature maps. Similarly, probabilistic knowledge transfer (PKT) [24] transfers the feature map of the penultimate layer (i.e., the layer before the classification layer) by matching its probability distribution. However, sharing the knowledge of an intermediate layer does not capture the critical connections between the layers. Attention transfer (AT) [25] involves an attention mechanism that transfers all intermediate layer representations to address this problem. Similarly, the hierarchical self-supervised augmented knowledge distillation (HSAKD) method [25] employs classifiers at the top of all intermediate layers to supervise the KD procedure. In addition, reference [26] introduced contrast representation distillation (CRD), which uses contrast loss to distill the feature maps from the last convolutional layer. All previous approaches considered the encoding of feature maps to match the width of teachers and students, allowing for a traceable architecture between students and teachers. Therefore, reference [27] introduced a method to efficiently encode the extracted features before the KD; however, encoding is still required.

2.2. Pruning

Neural networks often need to be more balanced, and there may be significant redundancy in deep learning models [28]. Network pruning removes channels and the corresponding weights that contribute insignificantly to the network accuracy. GoogLeNet [29,30] reduces the number of parameters of a neural network by employing an global average pool, instead of a fully connected layer. Network pruning is used to reduce the network's complexity and over-fitting. An early approach to pruning was biased weight decay [31]. Optimal Brain Damage [17] and Optimal Brain Surgeon [32] prune networks to reduce the number of connections based on the Hessian of the loss function, and it has been suggested that such pruning is more accurate than magnitude-based pruning approaches such as weight decay. However, the second-order derivative requires additional computation. There are two main branches of pruning, based on the granularity of the pruning: (1) unstructured pruning, which prunes individual weights; and (2) structured pruning, which prunes neurons (in most cases, channels of convolutional neural networks).

2.2.1. Unstructured Pruning

Unstructured pruning [17] uses a single weight as the basic unit to delete weights and connections in the neural network, while maintaining the number of neurons in the network. General unstructured pruning consists of three steps: (1) training a large network model; (2) removing unnecessary connections (synapses) and weights (neurons), according to custom rules; and (3) finally fine-tuning the entire sparse neural network for updates. The iterative magnitude pruning (IMP) technique [33], which iteratively applies magnitude-based trimming and fine-tuning, results in a significant performance enhancement. Lottery ticket rewinding (LTR) is an iterative magnitude pruning method with entitled repeated rolls [20,34]. IMP with learning rate (LR) rewinding, which recounts the learning rate schedule, has recently been shown to yield better results with more extensive networks [35]. However, the unstructured pruned network structure (i.e., the number of channels per layer) remains the same. As such, it is not easy to accelerate an unstructured pruning network without dedicated hardware [36].

2.2.2. Structured Pruning

Structured pruning [37–40] takes filters (i.e., neurons in channel units) as the basic unit and removes some filters that do not contribute much. Such an approach provides a smaller network with a more efficient network structure. This method can use general-purpose hardware acceleration and does not require specialized hardware and libraries to be designed. Similarly to magnitude-based unstructured pruning, the most straightforward approach is based on weight pruning filters [38,41]. Another method is to add a regularizer that induces sparsity during training [42–44]. Liu et al. [2] and Ye et al. [45] proposed a structured pruning scheme with filter-based batch normalization (BN) scale factors. Zhuang et al. [46] incorporated a polarization regularizer with a BN scale factor into structured pruning. However, the pruned network had more weights (parameters) than the unstructured pruning, due to the limitations of the network structure [34].

2.2.3. KD and Pruning

Previous works [47–50] combined KD and pruning for model compression. However, these approaches coincide with ours, in terms of technology and methodology. In [47], specific layers of the model were pruned, and a separate KD was performed on them. In [48], the pruned model was applied to the distilled student network. In [49], the teacher model was pruned to have the same width as the student network, in order to facilitate the transfer of intermediate layers. In [50], the teacher networks were also pruned, and the authors demonstrated that the pruned teacher networks had a regularization effect.

3. Motivation
3.1. Redundancy in the Student Network

KD [14] transfers knowledge from a strong teacher network to a smaller student network. The student network is trained with soft targets and some intermediate features provided by the teacher network [25,51,52]. Section 2.2.3 investigates and introduces the combined compression method, where knowledge flows from the teacher network to the student network. Even if KD is used to obtain the student network, there is still much redundancy. We propose an algorithm that can improve the network inference speed and provide an efficient storage format, while also improving the accuracy of the network.

3.2. Acceleration and Storage of Unstructured Pruning

As shown in Figure 1, unstructured pruning can make the dense weight tensor sparse. While this reduces the model size and computational effort, it also reduces the computational regularity, making parallelization in hardware more complex. For example, modern hardware architectures have many parallel processors, such as GPUs, with multiple processing unit (PEs). Numerous CUDA cores (also called arithmetic logic units or ALUs) exist in each PE of the general-purpose hardware acceleration unit, which can support multiple computations such as float32, float16, and/or int8 computations. Each PE handles part of the workload in parallel, allowing for faster computation than with serial execution of the workload. A pruning network leads to the potential problem of an unbalanced distribution of nonzero weights when computed in parallel on multiple PEs, resulting in a situation where each PE performs a different number of multiplicative accumulation operations. Therefore, the PE with fewer computational tasks must wait until the PE with the most computational tasks has finished before proceeding to the following computation stage. As such, unbalanced workloads on different PEs may result in a gap between the actual and peak performance of the hardware. Traditional unstructured pruning stores the zero elements (as in Figure 1a) and the masks of the pruning process, resulting in a sizable final model, in terms of storage. This is even more detrimental regarding the deployment and acceleration of embedded systems.

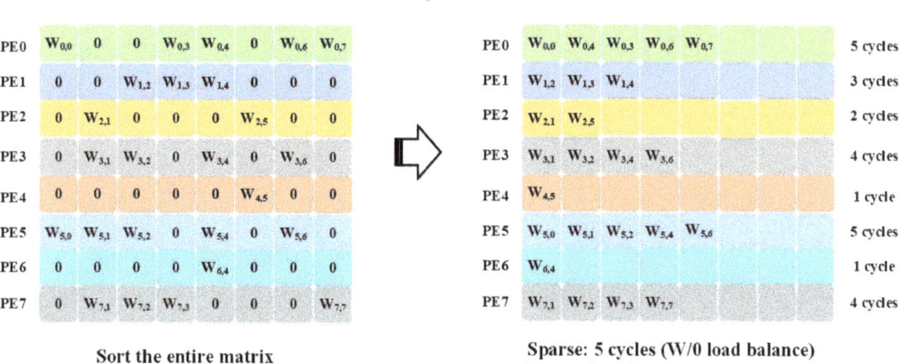

Figure 1. The sparse matrix of the pruning network is depicted in (**a**). Sorting operations performed on the whole matrix, such as that shown in (**b**) PE0–PE7, require at least five clocks for computation.

4. Methodology

4.1. Overview

In this paper, we design a unified end-to-end DST compression framework, as shown in Figure 2. We define a teacher network $f_t(\cdot; w_t)$ and a student network $f_s(\cdot; w_s)$. More formally, $f_t(\cdot; w_t)$ is a cumbersome network that needs to be compressed into a lightweight network $f_s(\cdot; w_s)$. Even after the KD and compression of the student network model $f_s(\cdot; w_s)$, much redundancy remains. This section provides a detailed architectural design based on the pruned student network $f_s(\cdot; w_p)$.

As shown in Figure 2, the design of the pruning algorithm for the student network is added to the KD (with respect to the student and teacher networks). The key idea of the student network pruning algorithm is to reconfigure the unstructured pruning architecture and construct a UHP method. After constructing the student network, each layer is pruned by 50% using the UHP method. Distillation learning is then performed on the pruned student network. Thus, the proposed compression algorithm includes three steps:

1. Train a teacher network and obtain $f_t(\cdot; w_t)$.
2. Apply the UHP algorithm to prune and obtain the pruned student network $f_s(\cdot; w_p)$.
3. Distill the teacher network $f_t(\cdot; w_t)$ to the student network $f_s(\cdot; w_p)$.

4.2. Distillation Formulation

As shown in Figure 2, this paper analyzes DST mathematically. Let $\{(x_i, y_i)\}_{i=1}^{N}$ be the data set, where the value of label y_i comes from $\{1, 2, \ldots, K\}$. For each training instance x, the neural network $f(\cdot; w)$ outputs the probability of each label as $p(k \mid x) = \text{softmax}(z_k) = \dfrac{\exp(z_k)}{\sum\limits_{i=1}^{K} \exp(z_i)}$, where z_i is the logit of the neural network $f(\cdot; w)$. Neural networks are trained by minimizing the cross-entropy loss $H(p_1, p_2) = -\sum\limits_{k=1}^{K} p_1[k] \log p_2[k]$. We are interested in a classification model with a K-dimensional probability distribution of the output. Let $f_{\text{true}}(x_i) \in \mathbb{R}^K$ be a one-hot encoding, where $f_{\text{true}}(x_i)[y_i] = 1$ denotes the real label y_i, and $f_{\text{true}}(x_i)[y'] = 0$ denotes all $y' \neq y_i$. Let $f_t(x; w)$ be the output of the teacher network. When the input is x and the weight is w, we train the teacher $f_t(\cdot; w)$ to achieve w_t such that the cross-entropy loss is minimized.

When the input is $f_s(x; w_s)$ and the weight is w_s, x is the output of the student network. For temperature τ, the knowledge distillation loss is given by:

$$L_{KD}(w_s) = \frac{1}{N} \sum^{N} (1 - \alpha) H(f_{\text{true}}(x_i), f_s(x_i; w_s)) + \alpha H(f_t(x; w_t), f_s(x; w_s)). \tag{1}$$

Further, through analysis of Equation (1):

$$\begin{aligned} L_{KD}(w_s) &= \frac{1}{N} \sum^{N} (1 - \alpha) H(f_{\text{true}}(x_i), f_s(x_i; w_s)) + \alpha H(f_t(x; w_t), f_s(x; w_s)) \\ &= \frac{1}{N} \sum^{N} (1 - \alpha) f_{\text{true}}(x_i) \log f_s(x_i; w_s) + \alpha f_t(x; w_t) \log f_s(x; w_s) \\ &= \frac{1}{N} \sum^{N} f_{\text{true}}(x_i) \log f_s(x_i; w_s) - \alpha f_{\text{true}}(x_i) \log f_s(x_i; w_s) + \alpha f_t(x; w_t) \log f_s(x; w_s) \\ &= \frac{1}{N} \sum^{N} ((1 - \alpha) f_{\text{true}}(x_i) + \alpha f_t(x; w_t)) \log f_s(x_i; w_s). \end{aligned} \tag{2}$$

When $f_\alpha(x; w_t) = (1 - \alpha) f_{\text{true}}(x) + \alpha f_t(x; w_t)$, Equation (2) is equivalent to Equation (3):

$$L_{KD}(w_s) = \frac{1}{N} \sum_{i=1}^{N} H(f_\alpha(x_i; w_t), f_s(x_i; w_s)). \tag{3}$$

As shown by [53] and our derivation of Equations (1) and (3), in addition to teaching students knowledge, KD can be equated to label smoothing regularization (LSR). Thus, KD facilitates network label smoothing regularization training in students.

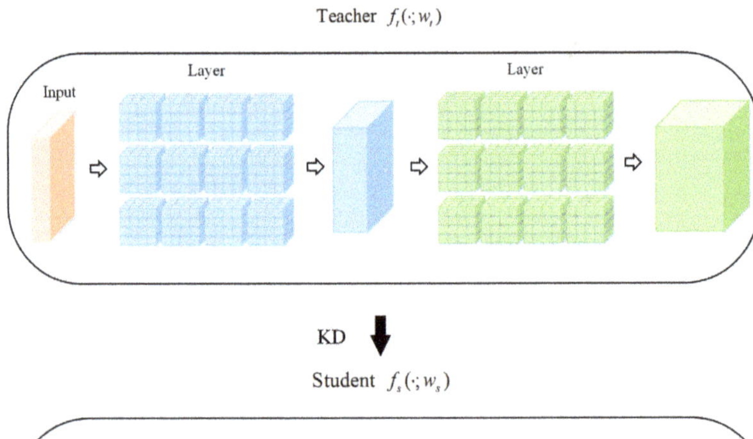

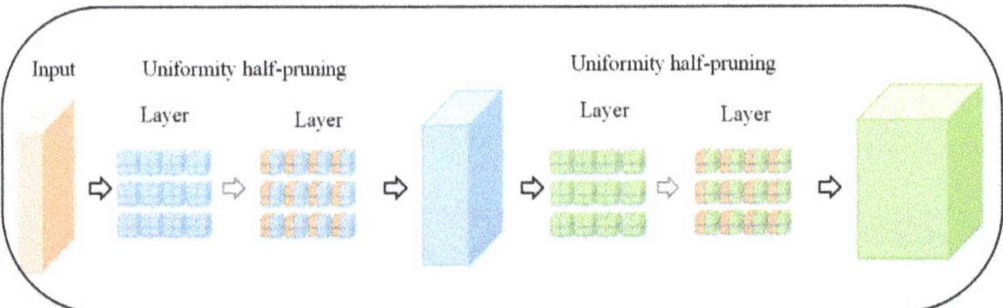

Figure 2. Overview of the DST strategy. The teacher network teaches the pruned student network, which can learn more from the teacher. The pruning algorithm used is the UHP method.

4.3. Pruned Student Distillation

This paper considers a pruned student network $f_s(x; w_p)$ to implement distillation training. The pruned student network $f_s(x; w_p)$ can also learn from the teacher network

$f_t(x; w_t)$. We propose a new network compression framework for unstructured pruning of the student network $f_s(x; w_s)$. The critical challenge is obtaining the pruned student network $f_s(x; w_p)$ to learn knowledge from the teacher network $f_t(x; w_t)$. The pruned student network $f_s(x; w_p)$ is added to the loss function based on the above considerations. Thus, the distillation loss function is redesigned as follows:

$$L_{KD}(w_p) = \frac{1}{N} \sum_{i=1}^{N} H\big(f_\alpha(x_i; w_t), f_s(x_i; w_p)\big), \tag{4}$$

where $f_\alpha(x; w_t) = (1 - \alpha) f_{\text{true}}(x) + \alpha f_t(x; w_t)$.

4.4. Pruning Formulation

The pruned network is obtained using a $m \in \{0, 1\}^{|w|}$ binary mask applied to the student network's weight w_s. Although this mask can be applied to all weights in the network, we restrict our attention to the convolutional layers, as they contribute the most to the overall computational cost. Pruning aims to learn the weight w_p that contributes most to the current objective of achieving a comparable performance to the original model. With the above considerations, we define Equation (5):

$$L(f_s(x, w_s \cdot m)) \approx L(f_s(x, w_p)), \frac{m_0}{|w|} = p, \tag{5}$$

where $p \in [0, 1]$ is a predefined pruning rate that controls the trade-off between the number of weights used, the computational complexity, and the expressiveness of the model.

4.5. Uniformity Half-Pruning

4.5.1. Pruning Structure

The advantage of unstructured pruning is its high pruning ratio (e.g., 90%). Typically, 90% of unstructured sparsity is used for model inference. However, with general-purpose hardware acceleration architectures, such as GPUs and TPUs (tensor cores and systolic arrays, respectively), the gains in inference speed when using unstructured pruning may be more apparent. We start by initializing a random network and training the model with a fixed sparse pattern. The UHP sparse model architecture is obtained, to efficiently utilize hardware during training and inference. Based on the advantages of general-purpose hardware, the GPU kernel is optimized to accelerate this pruning model on the CUDA kernel. This solves the speed problem of unstructured pruning on general-purpose processors at the algorithmic level.

In this paper, we set the sparsity of only two nonzero weights among the four weights, to address the challenge posed by the unstructured pruning sparsity described in Section 3.1. The UHP pattern specifies that, for each group of 4 values, at least 2 must be 0. This results in a 50% sparsity, which makes it more practical to maintain accuracy without hyperparametric exploration than when using a higher sparsity. When accelerating a matrix, the UHP model has the following benefits over other sparsity methods: (1) efficient memory access; (2) low overhead compression format; and (3) $2\times$ higher computational throughput in a general-purpose processor architecture.

4.5.2. Efficient Storage and Accelerating Computation

The UHP matrix (W) storage format is shown in Figure 3. In this mode, only 2 nonzero values from each group of 4 values need to be stored. The metadata in decoded compressed format are stored separately, using 2 bits to encode the location of each nonzero value in the 4 sets of values. For example, the metadata of the first row of the matrix in Figure 1 are given as [[0; 3]; [0; 2]]. When performing matrix multiplication, metadata information is required to obtain the corresponding values from the second matrix.

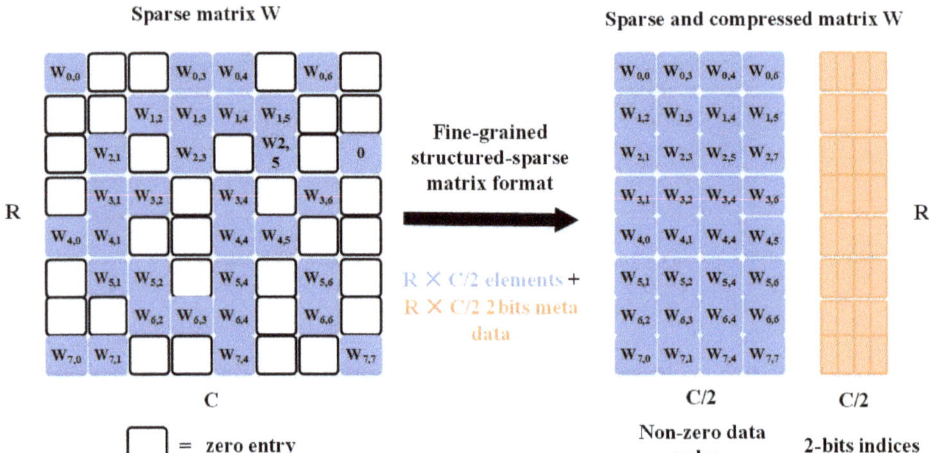

Figure 3. UHP matrix (W) storage format. The size of the uncompressed matrix is $R \times C$, and the size of the compressed matrix is $\frac{R \times C}{2}$.

The UHP matrix (W) storage format allows for efficient memory access. The unstructured sparse pattern results in low utilization of cache lines when accessing memory, therefore promoting low utilization of memory bandwidth. In addition, unstructured schemas usually use CSR/CSC/COO storage formats [54], which leads to data-dependent access and an increased latency of matrix reads. In contrast, each sub-block of the UHP matrix has the same sparsity level, allowing the hardware to take full advantage of large memory reads. Again, as the sparsity is constant throughout the matrix, the location of a nonzero value in the memory can be determined directly from the metadata.

The UHP matrix (W) allows for use of efficient storage formats. As the UHP matrix (W) storage format (shown in Figure 3) requires only 2 bits of metadata per value index; in the case of 32-bit operands, storing the sparse tensor in a compressed format results in a 47% saving in storage space: 4 Bytes require 4×32 bits = 128 bits of storage space, while the sparse UHP matrix (W) results in 2×32 bits + 2×2 bits = 68 bits to store the 2 nonzero weights.

4.5.3. UHP Computation Overhead

In this paper, we designed a convolution operation for the program, as shown in Figure 4. During model inference, the program loads the matrix, indexes the nonzero elements for matrix multiplication operations, and performs accumulation operations. The program unifies the design instructions. These instructions are the basis for the neural network layers involving mathematical operations, mainly matrix convolution operations. A network with 50% sparsity can halve the number of multiplication operations, thus doubling the computational performance of the hardware in the same network.

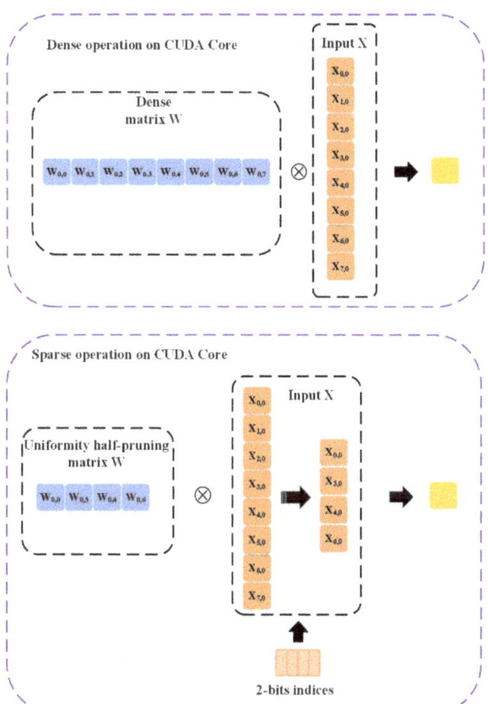

Figure 4. General-purpose hardware acceleration. The algorithm selects elements from the matrix X corresponding to nonzero values of the matrix W for the convolution operation by indexing, skipping unnecessary zero multiplication operations.

4.6. DST Workflow

Figure 5 shows the DST training process. In this method, the network is pruned with a 50% sparse pattern, maintaining the original accuracy and avoiding the need to conduct a hyperparameter search. As the aim is to reduce the size of the neural network and the running time at deployment, we trade higher training costs for smaller models. First, we train a teacher network $f_t(\cdot; w_t)$, which has better performance and generalization ability. Second, the teacher network $f_t(\cdot; w_t)$ and student network $f_s(\cdot; w_s)$ are trained simultaneously (the student network $f_s(\cdot; w_s)$ was not pretrained) and distilled for training. When the student network $f_s(\cdot; w_s)$ reaches a certain accuracy, the student network $f_s(\cdot; w_s)$ is uniformly pruned by half to obtain $f_s(\cdot; w_p)$. Finally, fine-tuning and distillation training are performed on the pruned student network $f_s(\cdot; w_p)$. The algorithm is described in Algorithm 1.

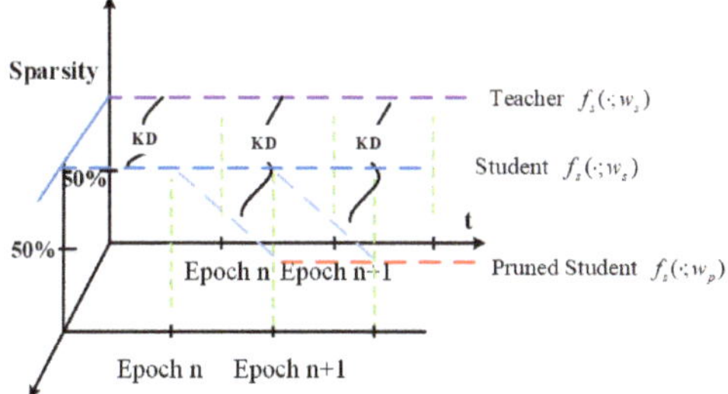

Figure 5. Timing of DST. Purple represents the teacher network $f_t(\cdot; w_t)$. Blue represents the student network $f_s(\cdot; w_s)$. Red represents the student network after pruning $f_s(\cdot; w_p)$.

Algorithm 1: Distillation Sparsity Training (DST)

---

input : teacher network pretrained $f_t(x; w_t)$
student network $f_s(x; w)$

Initialize : $f_t(x; w_t)$ KD $f_s(x; w)$, $L_{KD}(w_s) = \frac{1}{N} \sum\limits_{i=1}^{N} H(f_\alpha(x_i; w_t), f_s(x_i; w_s))$

if $f_s(x; w_s) \leftarrow f_s(x; w)$ **then**
   $f_s(x; w) \leftarrow f_s(x; w_s)$ by UHP
end

Initialize : $f_t(x; w_t)$ KD $f_s(x; w)$, $L_{KD}(w_p) = \frac{1}{N} \sum\limits_{i=1}^{N} H(f_\alpha(x_i; w_t), f_s(x_i; w_p))$

if $f_s(x; w_p) \leftarrow f_s(x; w)$ **then**
   output: $f_s(x; w_p)$
end

---

5. Experiment

5.1. Data Sets

The algorithm proposed in this study was evaluated on both the optical CIFAR-100 data set and the SAR MSTAR and FUSAR-ship data sets, in order to demonstrate the effectiveness of the DST algorithm. Experimental results were obtained to validate the proposed algorithm.

CIFAR-100 [55] consists of labeled subsets of the 80 million tiny image data set, collected by Alex Krizhevsky, Vinod Nair, and Geoffrey Hinton. This data set includes 100 categories, with each category containing 600 images. There are 500 training images and 100 testing images per class. The 100 classes in the CIFAR-100 are grouped into 20 superclasses. Each image comes with a "fine" label (the class to which it belongs) and a "coarse" label (the super-class to which it belongs).

The MSTAR (Moving and Stationary Target Acquisition and Recognition) data set [56] was collected and published by Sandia National Laboratory in the USA. The MSTAR data set can be divided into standard operating conditions (SOCs) and extended operating conditions (EOCs) data sets. There are 10 different ground military targets in the MSTAR data set, including BMP2 (tank), BTR70 (armored vehicle), T72 (tank), BTR60 (armored vehicle), 2S1 (artillery), BRDM (truck), D7 (bulldozer), T62 (tank), ZIL131 (truck), and ZSU234 (artillery). In the SOC, the training set has 2747 samples with a depression angle of $17°$, and the test set has 2425 samples with a depression angle of $15°$.

The FUSAR-Ship data set [57] was constructed and published by the Key Laboratory for Information Science of Electromagnetic Waves (MoE) of Fudan University, Shanghai, China. The FUSAR-Ship data set includes eight different ship targets: Bulk General, General Cargo, Container, Other Cargo, False Alarm, Fishing, Other Ship, and Tanker. We randomly selected 75% as the training set and 25% as the test set, such that the ratio of the training set to the test set was 3:1.

5.2. Experiment Setup and Implementation Details

For this experiment, we used a Jetson AGX Orin embedded device for evaluation of the proposed algorithm. The Jetson AGX Orin is based on an embedded platform with a 12-core ARM® v8.2 64-bit CPU, 32 GB of 256-bit LPDDR5 memory, and a 64 GB eMMC flash device running Ubuntu 20.04. In the evaluation, the inference processes were accelerated using CUDA and Tensor units for computation. We selected different teacher networks for different data sets, to achieve smooth knowledge transfer (e.g., ResNet18, ResNet34, ResNet50). The student networks were MobileNet, MobileNetV2, and MobileNetV3. One teacher network corresponded to three student networks, in order to test the effect of DST separately. For the CIFAR-100 data set, the momentum was 0.9 and the learning rate was 0.1 using the SGD optimizer, where learning rate was decayed by a factor of 5 at the 60th, 120th, and 160th epochs. The network was trained with an input size of 32×32, batch size equal to 128, and weight decay of 0.0005 for 200 epochs. For the MSTAR and FUSAR-ship data sets, the momentum was 0.9 and the learning rate was 0.1 using the SGD optimizer, where the learning rate was decayed by a factor of 5 at the 60th, 120th, and 160th epochs. The network was trained with an input size of 224×224, batch size equal to 128, and weight decay of 0.0005 for 200 epochs. Unstructured pruning (50% pruning ratio) and UHP experiments were performed. To prevent overfitting of the network, we used the Pytorch library to implement dataset augmentation during training; for example, equal scaling of the images, random cropping, and random horizontal and vertical flipping.

5.3. Pruning Analysis

Figures 6–8 show the pruning ratio for the student networks MobileNet, MobileNetV2, and MobileNetV3, respectively, regarding the unstructured pruning (50% pruning ratio) and UHP for each layer. It can be seen that the pruning ratio of each layer under UHP was 50%. From the analysis in Section 4.5, our proposed algorithm is more suitable for general-purpose hardware acceleration. We set the overall pruning ratio for unstructured pruning to 50%; however, the pruning rate differed for each layer. According to the analysis in Section 3.1, the unstructured pruning algorithm does not accelerate well on general-purpose hardware. These results are validated in the experimental evaluation detailed in Section 5.4.

Figures 9–11 show views of the feature maps for the student networks MobileNet, MobileNetV2, and MobileNetV3, respectively, under unstructured pruning (50% pruning ratio) and UHP for each layer. Both sub-figures (a) and (b) show the same channel, where (a) is the channel with an unstructured pruning ratio higher than 50% and (b) is that with UHP. It can be seen that the features extracted under UHP are smoother and more focused than those with unstructured pruning, due to the random pruning per layer in the unstructured case. Figures 10 and 11 show that the relatively high ratio of unstructured pruning led to the direct loss of features in some layers, while the channel features under UHP remained relatively stable. This lack of concentration and the disappearance of features can lead to an inability to learn knowledge well in the process of DST. These results are validated in our analysis of the experimental evaluation in Section 5.5.

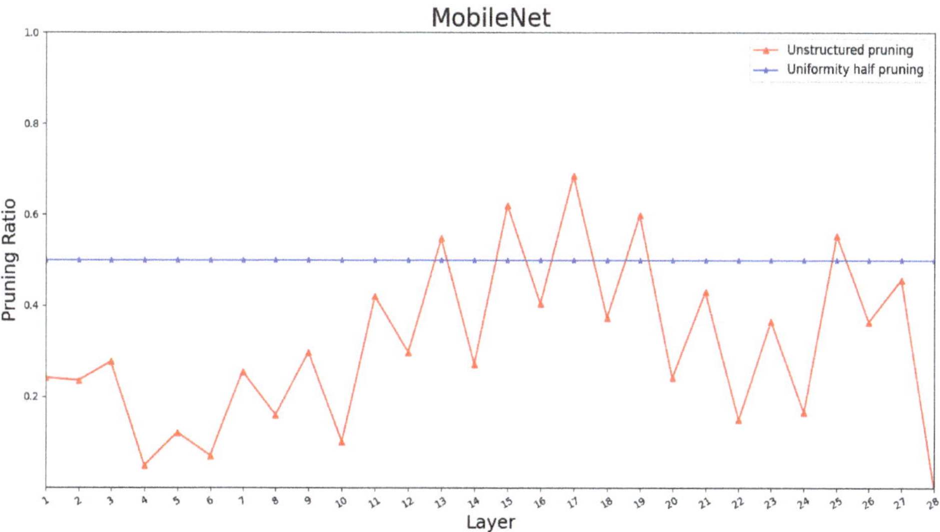

Figure 6. MobileNet network pruning ratio per layer. The blue line denotes UHP, and the red line denotes unstructured pruning.

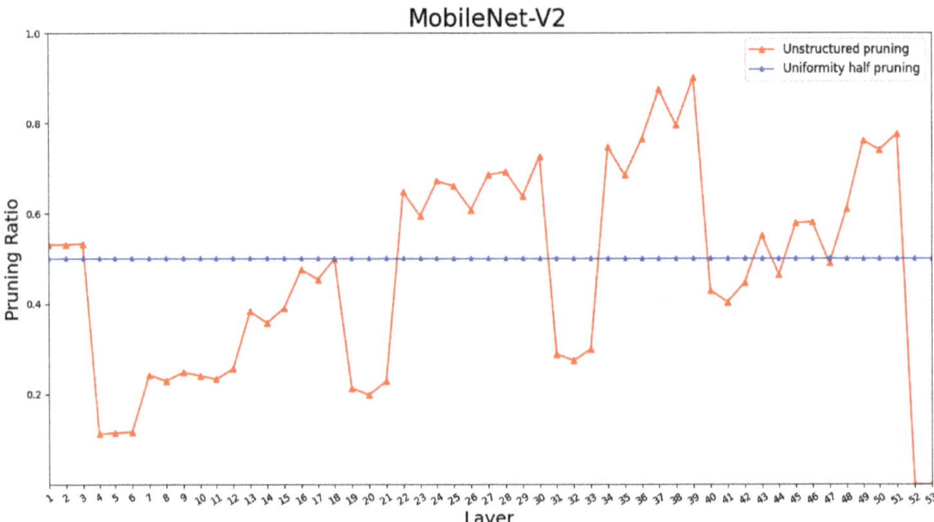

Figure 7. MobileNetV2 network pruning ratio per layer. The blue line denotes UHP, and the red line denotes unstructured pruning.

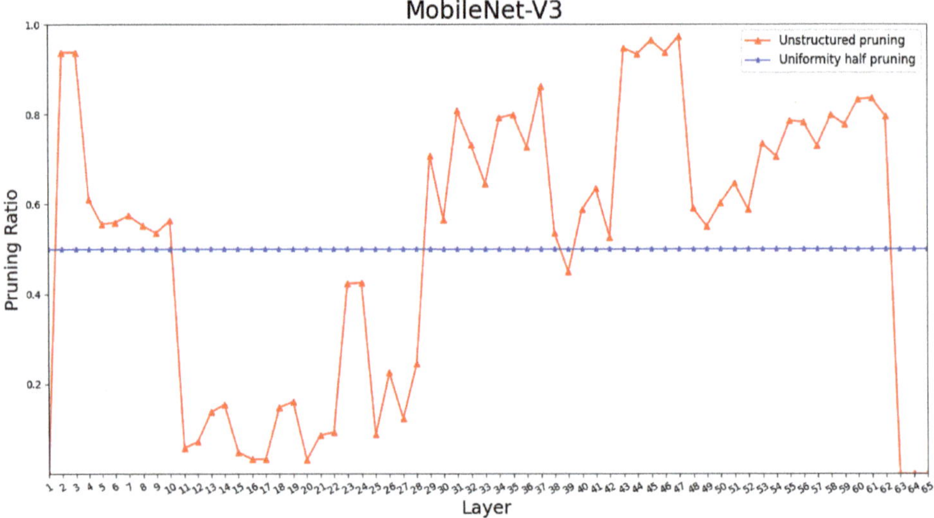

Figure 8. MobileNetV3 network pruning ratio per layer. The blue line denotes UHP, and the red line denotes unstructured pruning.

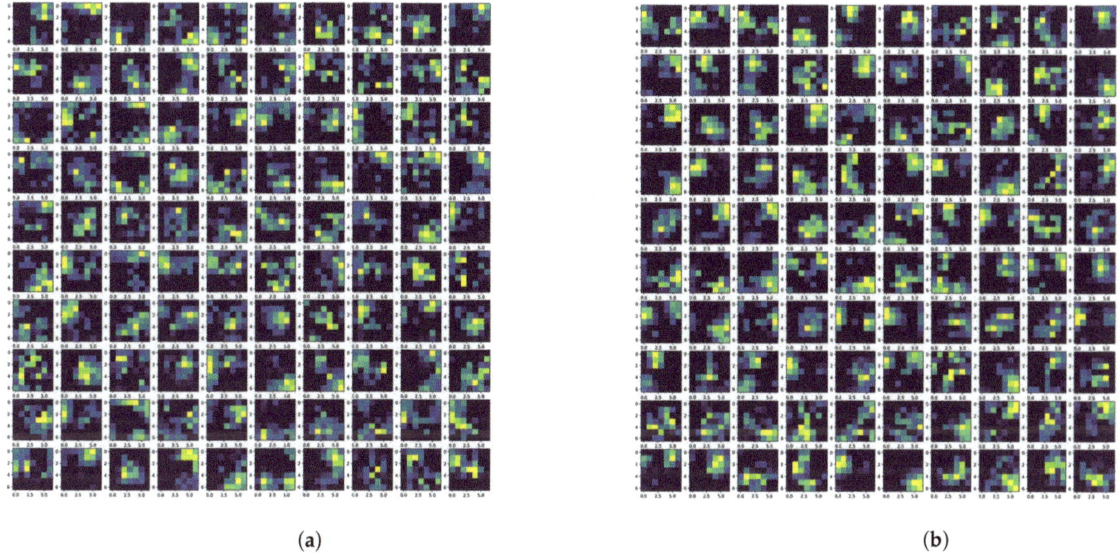

(a)   (b)

Figure 9. The feature maps of MobileNet after pruning: (**a**) Unstructured pruning; and (**b**) UHP.

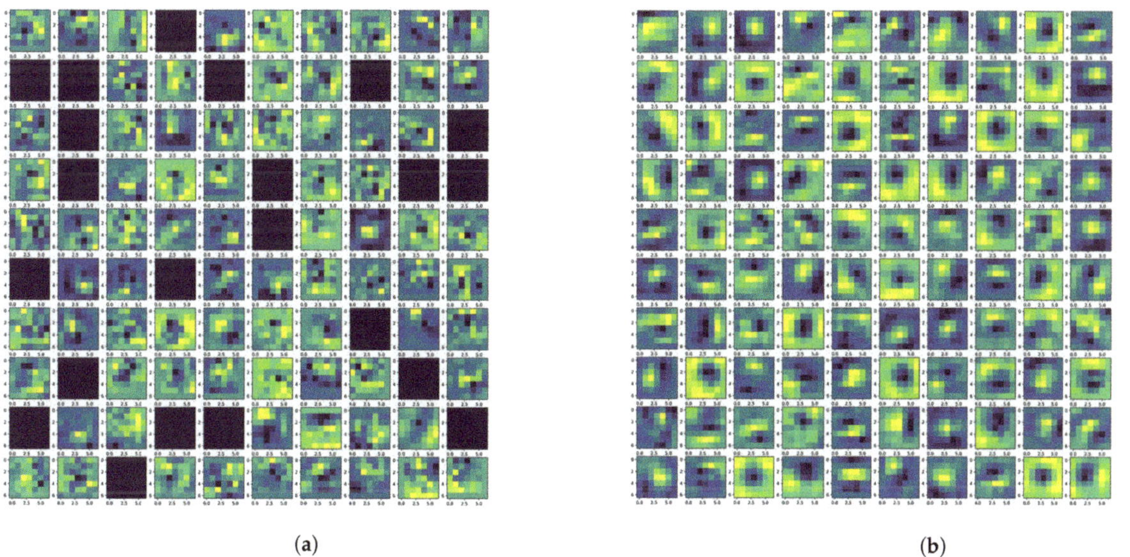

(a)   (b)

Figure 10. The feature maps of MobileNetV2 after pruning: (**a**) Unstructured pruning; and (**b**) UHP.

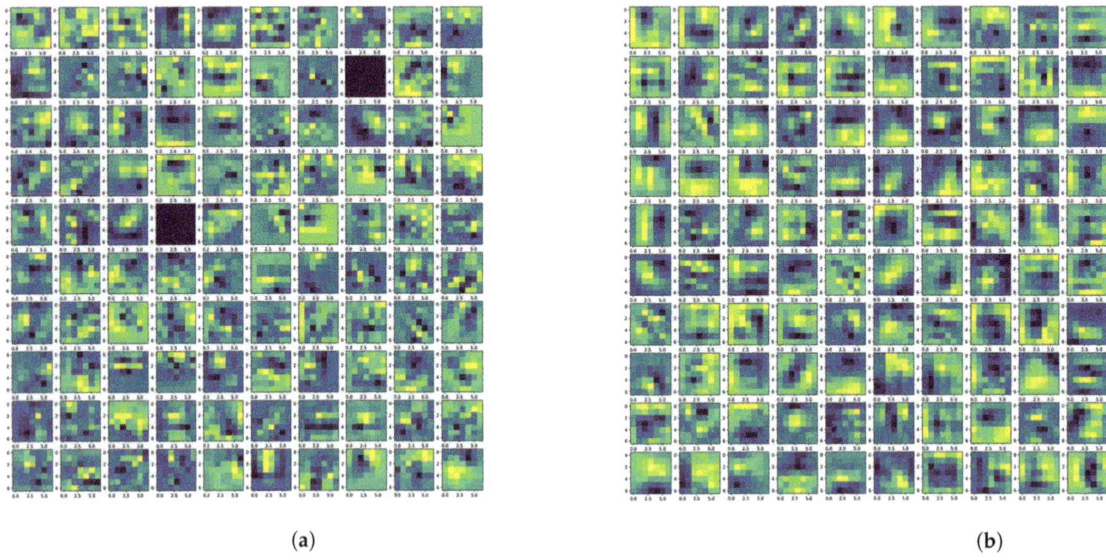

(a)   (b)

Figure 11. The feature maps of MobileNetV3 after pruning: (**a**) Unstructured pruning, and (**b**) UHP.

5.4. Computational Performance Evaluation

The network was deployed in an embedded system (Jetson AGX Orin), in order to validate the inference performance of UHP. In addition to evaluating the accuracy, we further evaluated the computational performance of the network. We validated the efficiency of our algorithm by calculating three metrics on the network. First, we assessed the model size by comparing the size of the pretrained, unstructured pruning, and UHP models. Second, we compared the number of network parameters for the pretrained, unstructured pruning, and UHP models. Third, we tested the frames per second (FPS), which is the computational efficiency of networks in embedded systems (Jetson AGX Orin), achieved by the three models with the embedded system.

Table 1 provides the model size, parameters, and FPS of the student networks. The MobileNet, MobileNetV2, and MobileNetV3 pretrained models were used as benchmarks. For the model size, unstructured pruning saved −1.57%, 5.47%, and 2.99% of the storage, while UHP saved 22.52%, 40.35%, and 37.72% of storage, respectively. For the parameters, both were reduced by 50%. Regarding the FPS of the network, when the input of the network was 32 × 32, unstructured pruning led to an improvement of 6.03%, 2.67%, and 3.22%, while UHP led to an improvement of 130.00%, 110.70%, and 122.83%, respectively. When the network input was 224 × 224, unstructured pruning led to an improvement of 2.81%, 2.13%, and 5.49%, while UHP led to an improvement of 129.90%, 110.00%, and 111.37%, respectively. These experiments further validate the analysis provided in Section 5.3.

5.5. Accuracy Evaluation

5.5.1. Results on CIFAR-100

We conducted nine groups of experiments on the CIFAR-100 data set. The teacher networks ResNet18, ResNet34, and ResNet50 were used to separately teach the student networks MobileNet, MobileNetV2, and MobileNetV3 by implementing DST. The set of experiments included assessment of the accuracy of single teacher and student networks. The accuracy of the networks was compared when implementing unstructured pruning (50% pruning rate) and UHP, when implementing DST, and when implementing unstructured pruning (50% pruning rate) and UHP and performing DST.

Table 1. Evaluation results of the computational performance for the student networks with the embedded system. Distillation sparsity training, DST; unstructured pruning (50% pruning ratio), UP; uniformity half-pruning, UHP. "✓" denotes baseline, and "-" denotes non-baseline.

		MobileNet	MobileNetV2	MobileNetV3	Baseline
Model Size(MB)	Pretrained	12.70	9.69	16.70	✓
	DST-UP [48]	12.90	9.16	16.20	-
	DST-UHP(Ours)	**9.84**	**5.78**	**10.40**	-
	Relative	+1.57%/−**22.52%**	5.47%/−**40.35%**	−2.99%/−**37.72%**	-
Parameters(M)	Pretrained	4.20	3.40	5.40	✓
	DST-UP [48]	2.10	1.70	2.70	-
	DST-UHP(Ours)	**2.10**	**1.70**	**2.70**	-
	Relative	−50.00%/−**50.00%**	−50.00%/−**50.00%**	−50.00%/−**50.00%**	-
FPS	Input 32 × 32 — Pretrained	1160	675	622	✓
	Input 32 × 32 — DST-UP [48]	1230	693	682	-
	Input 32 × 32 — DST-UHP(Ours)	**2668**	**1418**	**1386**	-
	Input 32 × 32 — Relative	+6.03%/+**130.00%**	+2.67%/+**110.70%**	+3.22%/+**122.83%**	-
	Input 224 × 224 — Pretrained	748	422	510	✓
	Input 224 × 224 — DST-UP [48]	769	431	538	-
	Input 224 × 224 — DST-UHP(Ours)	**1720**	**886**	**1078**	-
	Input 224 × 224 — Relative	+2.81%/+**129.90%**	+2.13%/+**110.00%**	+5.49%/+**111.37%**	-

Table 2 shows the accuracy of the student network MobileNet under various conditions. Here, we used MobileNet without KD, UP, or UHP as a benchmark. After DST using the teacher networks ResNet18, ResNet34, and ResNet50, the accuracy increased by 4.00%, 3.21%, and 3.72%, respectively. Table 3 shows the accuracy of the student network MobileNetV2 under various conditions. Here, we took MobileNetV2 without KD, UP, or UHP as a benchmark. After DST using the teacher networks ResNet18, ResNet34, and ResNet50, the accuracy increased by 0.82%, 0.27%, and 0.55%, respectively. Table 4 shows the accuracy of the student network MobileNetV3 under various conditions. Here, we took MobileNetV3 without KD, UP, or UHP as a benchmark. After DST using the teacher networks ResNet18, ResNet34, and ResNet50, the accuracy increased by 1.58%, 1.68%, and 2.70%, respectively. These experiments validated the analysis in Section 5.3.

Table 2. Teacher networks ResNet18, ResNet34, ResNet50, and the student network MobileNet; all implemented the DST results. Unstructured pruning (50% pruning ratio), UP; uniformity half-pruning, UHP. "✓" denotes that the method is used, and "*" denotes that the method is not used.

Teacher	Teacher Acc (Top-1)	Student	KD	UP [58]	UHP (Ours)	Student Acc (Top-1)	Relative
*	*	MobileNet	*	*	*	67.58% [59]	Baseline
			*	✓	*	68.26% [58]	+0.68%
			*	*	✓	67.62%	−0.32%
ResNet18	76.41%		✓	*	*	71.62% [14]	+4.04%
			✓	✓	*	70.75% [48]	+3.17%
			✓	*	✓	**71.48%**	**+4.00%**
ResNet34	78.05%	MobileNet	✓	*	*	70.17% [14]	+2.59%
			✓	✓	*	69.88% [48]	+2.30%
			✓	*	✓	**70.79%**	**+3.21%**
ResNet50	78.87%		✓	*	*	71.25% [14]	+3.67%
			✓	✓	*	70.16% [48]	+2.58%
			✓	*	✓	**71.30%**	**+3.72%**

Table 3. Teacher networks ResNet18, ResNet34, ResNet50, and the student network MobileNetV2; all implemented the DST results. Unstructured pruning (50% pruning ratio), UP; uniformity half-pruning, UHP. "✓" denotes that the method is used, and "*" denotes that the method is not used.

Teacher	Teacher Acc (Top-1)	Student	KD	UP [58]	UHP (Ours)	Student Acc (Top-1)	Relative
*	*	MobileNetV2	*	*	*	68.90% [60]	Baseline
			*	✓	*	69.10% [58]	+0.20%
			*	*	✓	68.33%	−0.57%
ResNet18	76.41%		✓	*	*	70.17% [14]	+1.27%
			✓	✓	*	68.93% [48]	+0.03%
			✓	*	✓	**69.72%**	**+0.82%**
ResNet34	78.05%	MobileNetV2	✓	*	*	69.89% [14]	+0.99%
			✓	✓	*	68.00% [48]	−0.90%
			✓	*	✓	**69.17%**	**+0.27%**
ResNet50	78.87%		✓	*	*	69.75% [14]	+0.85%
			✓	✓	*	69.13% [48]	+0.23%
			✓	*	✓	**69.45%**	**+0.55%**

Table 4. Teacher networks ResNet18, ResNet34, ResNet50, and the student network MobileNetV3; all implemented the DST results. Unstructured pruning (50% pruning ratio), UP; uniformity half-pruning, UHP. "✓" denotes that the method is used, and "*" denotes that the method is not used.

Teacher	Teacher Acc (Top-1)	Student	KD	UP [58]	UHP (Ours)	Student Acc (Top-1)	Relative
*	*	MobileNetV3	*	*	*	71.79% [61]	Baseline
			*	✓	*	72.03% [58]	+0.24%
			*	*	✓	71.75%	−0.04%
ResNet18	76.41%		✓	*	*	73.41% [14]	+1.62%
			✓	✓	*	72.69% [48]	+0.90%
			✓	*	✓	**73.37%**	**+1.58%**
ResNet34	78.05%	MobileNetV3	✓	*	*	73.54% [14]	+1.75%
			✓	✓	*	72.15% [48]	+0.36%
			✓	*	✓	**73.47%**	**+1.68%**
ResNet50	78.87%		✓	*	*	74.52% [14]	+2.73%
			✓	✓	*	73.96% [48]	+2.17%
			✓	*	✓	**74.49%**	**+2.70%**

5.5.2. MSTAR Results

Next, we implemented three groups of experiments using the MSTAR data set. The experimental process was similar to that in Section 5.5.1. As ResNet18 performed well on the MSTAR data set, the other teacher networks were not used. Table 5 shows the accuracy of the student network MobileNet under various conditions, taking MobileNet without KD, UP, or UHP as a benchmark. After DST, by implementing ResNet18 with the teacher network, the accuracy increased by 0.75%. Table 6 shows the accuracy of the student network MobileNetV2 under various conditions, taking MobileNetV2 without KD, UP, or UHP as a benchmark. After DST, by implementing ResNet18 with the teacher network, the accuracy increased by 2.85%. Table 7 shows the accuracy of the student network MobileNetV3 under various conditions, taking MobileNetV3 without KD, UP, or UHP as a benchmark. After DST, by implementing ResNet18 with the teacher network, the accuracy increased by 0.45%. These experiments validated the analysis provided in Section 5.3.

Table 5. Results for the teacher network ResNet18 and the student network MobileNet; both implemented the DST results. Unstructured pruning (50% pruning ratio), UP; uniformity half-pruning, UHP. "✓" denotes that the method is used, and "*" denotes that the method is not used.

Teacher	Teacher Acc (Top-1)	Student	KD	UP [58]	UHP (Ours)	Student Acc (Top-1)	Relative
*	*	MobileNet	*	*	*	98.56% [59]	Baseline
			*	✓	*	98.34% [58]	−0.22%
			*	*	✓	98.21%	−0.35%
ResNet18	99.78%	MobileNet	✓	*	*	99.32% [14]	+0.76%
			✓	✓	*	99.24% [48]	+0.68%
			✓	*	✓	**99.31%**	**+0.75%**

Table 6. Results for the teacher network ResNet18 and the student network MobileNetV2; both implemented the DST results. Unstructured pruning (50% pruning ratio), UP; uniformity half-pruning, UHP. "✓" denotes that the method is used, and "*" denotes that the method is not used.

Teacher	Teacher Acc (Top-1)	Student	KD	UP [58]	UHP (Ours)	Student Acc (Top-1)	Relative
*	*	MobileNetV2	*	*	*	96.54% [60]	Baseline
			*	✓	*	96.55% [58]	+0.01%
			*	*	✓	96.48%	−0.06%
ResNet18	99.78%	MobileNetV2	✓	*	*	99.40% [14]	+2.86%
			✓	✓	*	99.36% [48]	+2.82%
			✓	*	✓	**99.39%**	**+2.85%**

Table 7. Results for the teacher network ResNet18 and the student network MobileNetV3; both implemented the DST results. Unstructured pruning (50% pruning ratio), UP; uniformity half-pruning, UHP. "✓" denotes that the method is used, and "*" denotes that the method is not used.

Teacher	Teacher Acc (Top-1)	Student	KD	UP [58]	UHP (Ours)	Student Acc (Top-1)	Relative
*	*	MobileNetV3	*	*	*	99.13% [61]	Baseline
			*	✓	*	99.06% [58]	+0.07%
			*	*	✓	99.05%	−0.08%
ResNet18	99.78%	MobileNetV3	✓	*	*	99.59% [14]	+0.46%
			✓	✓	*	99.55% [48]	+0.42%
			✓	*	✓	**99.58%**	**+0.45%**

5.5.3. FUSAR-Ship Results

We also conducted nine groups of experiments on the FUSAR-Ship data set. The experimental process was similar to that in Section 5.5.1. Table 8 shows the accuracy of the student network MobileNet under various conditions, taking MobileNet without KD, UP, or UHP as a benchmark. After DST using the teacher networks ResNet18, ResNet34, and ResNet50, the accuracy increased by 3.89%, 4.50%, and 4.20%, respectively. Table 9 shows the accuracy of the student network MobileNetV2 under various conditions, taking MobileNetV2 without KD, UP, or UHP as a benchmark. After DST using the teacher networks ResNet18, ResNet34, and ResNet50, the accuracy increased by 2.88%, 3.31%, and 4.20%, respectively. Table 10 shows the accuracy of the student network MobileNetV3 under various conditions, taking MobileNetV3 without KD, UP, or UHP as a benchmark. After DST using the teacher networks ResNet18, ResNet34, and ResNet50, the accuracy increased by 3.46%, 3.80%, and 4.52%, respectively. These experiments validated the analysis in Section 5.3.

Table 8. Results for the teacher networks ResNet18, ResNet34, ResNet50, and the student network MobileNet; all implemented the DST results. Unstructured pruning (50% pruning ratio), UP; uniformity half-pruning, UHP. "✓" denotes that the method is used, and "*" denotes that the method is not used.

Teacher	Teacher Acc (Top-1)	Student	KD	UP [58]	UHP (Ours)	Student Acc (Top-1)	Relative
*	*	MobileNet	*	*	*	70.42% [59]	Baseline
			*	✓	*	70.35% [58]	−0.07%
			*	*	✓	70.67%	+0.25%
ResNet18	74.38%		✓	*	*	74.82% [14]	+4.40%
			✓	✓	*	73.59% [48]	+3.17%
			✓	*	✓	**74.31%**	**+3.89%**
ResNet34	75.31%	MobileNet	✓	*	*	75.35% [14]	+4.93%
			✓	✓	*	73.82% [48]	+3.40%
			✓	*	✓	**74.92%**	**+4.50%**
ResNet50	75.87%		✓	*	*	75.62% [14]	+5.20%
			✓	✓	*	74.01% [48]	+3.59%
			✓	*	✓	**74.62%**	**+4.20%**

Table 9. Results for the teacher networks ResNet18, ResNet34, ResNet50, and the student network MobileNetV2; all implemented the DST results. Unstructured pruning (50% pruning ratio), UP; uniformity half-pruning, UHP. "✓" denotes that the method is used, and "*" denotes that the method is not used.

Teacher	Teacher Acc (Top-1)	Student	KD	UP [58]	UHP (Ours)	Student Acc (Top-1)	Relative
*	*	MobileNetV2	*	*	*	71.37% [60]	Baseline
			*	✓	*	70.16% [58]	−1.21%
			*	*	✓	69.27%	−2.10%
ResNet18	74.38%		✓	*	*	74.16% [14]	+2.79%
			✓	✓	*	73.29% [48]	+1.92%
			✓	*	✓	**74.25%**	**+2.88%**
ResNet34	75.31%	MobileNetV2	✓	*	*	74.87% [14]	+3.50%
			✓	✓	*	73.74% [48]	+2.37%
			✓	*	✓	**74.68%**	**+3.31%**
ResNet50	75.87%		✓	*	*	74.98% [14]	+3.61%
			✓	✓	*	73.85% [48]	+2.48%
			✓	*	✓	**74.96%**	**+3.59%**

According to the computational performance evaluation in Section 5.4 and the accuracy evaluation in Section 5.5, DST improved the accuracy of the pruned network. Regarding the model size of the student network, UHP saved close to 40% of storage space, compared to unstructured pruning. Considering the FPS of the network, UHP achieved a 2× increase in speed compared to the pretrained model, while the speed increase due to unstructured pruning was insignificant.

Table 10. Results for the teacher networks ResNet18, ResNet34, ResNet50, and the student network MobileNetV3; all implemented the DST results. Unstructured pruning (50% pruning ratio), UP; uniformity half-pruning, UHP. "✓" denotes that the method is used, and "*" denotes that the method is not used.

Teacher	Teacher Acc (Top-1)	Student	KD	UP [58]	UHP (Ours)	Student Acc (Top-1)	Relative
*	*	MobileNetV3	*	*	*	71.12% [61]	Baseline
			*	✓	*	70.19% [58]	−0.93%
			*	*	✓	71.44%	+0.12%
ResNet18	74.38%		✓	*	*	74.57% [14]	+3.45%
			✓	✓	*	73.18% [48]	+2.06%
			✓	*	✓	**74.58%**	**+3.46%**
ResNet34	75.31%	MobileNetV3	✓	*	*	74.98% [14]	+3.86%
			✓	✓	*	74.15% [48]	+3.03%
			✓	*	✓	**74.92%**	**+3.80%**
ResNet50	75.87%		✓	*	*	75.92% [14]	+4.80%
			✓	✓	*	74.13% [48]	+3.01%
			✓	*	✓	**75.64%**	**+4.52%**

6. Conclusions

KD and network pruning are essential technical tools for model compression. In this work, we proposed a model compression algorithm that combines KD and network pruning, which we call DST. In particular, network pruning was applied to the student network to effectively transfer knowledge from the teacher network to the pruned student network and further compress the student network. We theoretically derived and designed a new teacher-pruned student objective function, in order to achieve compression while improving the accuracy of the student network. Considering that unstructured pruning cannot achieve general-purpose hardware acceleration, we reconstructed the unstructured pruning approach, to propose the UHP method. Extensive experiments on the CIFAR-100, MSTAR, and FUSAR-Ship data sets demonstrated that the proposed DST can achieve state-of-the-art compression performance in terms of prediction accuracy, inference acceleration, and storage efficiency. We also obtained the FPS values for the compressed models when using the proposed DST on embedded platforms, in order to demonstrate the real-time application potential of our compressed model on mobile devices.

Deep neural networks include CNNs, transformers, etc. The DST algorithm mainly implements model compression for generic CNNs, but the structure of the transformer differs significantly from CNNs, and the proposed algorithm requires further improvement in its compression structure. Compression techniques for general-purpose deep neural networks (e.g., CNNs, transformers) could be further investigated in the future. Currently, we have only performed tests on a GPU-based Jetson AGX Orin. Test studies have yet to be conducted for other embedded devices (e.g., FPGA, DSP). Spaceborne SAR technology may carry different embedded systems and use various deep neural networks, and we hope the proposed compression algorithm can achieve generality.

Author Contributions: Conceptualization, P.X. and H.W.; methodology, P.X.; software, P.X.; validation, P.X. and T.X.; formal analysis, P.X.; investigation, P.X. and T.X.; resources, P.X.; data curation, X.X. and W.L.; writing—original draft preparation, P.X.; writing—review and editing, P.X.; visualization, P.X. and T.X.; supervision, H.W.; project administration, H.W.; funding acquisition, H.W. All authors have read and agreed to the published version of the manuscript.

Funding: This work was supported in part by the National Natural Science Foundation of China (Grant No. 62271153) and the Natural Science Foundation of Shanghai (Grant No. 22ZR1406700).

Data Availability Statement: Please contact Haipeng Wang (hpwang@fudan.edu.cn) to access the data.

Acknowledgments: The authors would like to thank the anonymous reviewers.

Conflicts of Interest: The authors declare that there are no conflict of interests.

Abbreviations

The following abbreviations are used in this manuscript:

ATR	Automatic Target Recognition
KD	Knowledge Distillation
DST	Distillation Sparsity Training
CNNs	Convolutional Neural Networks
PKT	Probabilistic Knowledge Transfer
AT	Attention Transfer
HSAKD	Hierarchical Self-supervised Augmented Knowledge Distillation
IMP	Iterative Magnitude Pruning
LTR	Lottery Ticket Rewinding
LR	Learning Rate
BN	Batch Normalization
PEs	multiple Processing units
ALUs	Arithmetic Logic Units
FPS	Frames Per Second
UHP	Uniformity Half-Pruning
UP	Unstructured Pruning

References

1. He, K.; Zhang, X.; Ren, S.; Sun, J. Deep Residual Learning for Image Recognition. *arXiv* **2015**, arXiv:1512.03385.
2. Simonyan, K.; Zisserman, A. Very Deep Convolutional Networks for Large-Scale Image Recognition. *arXiv* **2015**, arXiv:1409.1556.
3. Fu, H.; Gong, M.; Wang, C.; Batmanghelich, K.; Tao, D. Deep Ordinal Regression Network for Monocular Depth Estimation. In Proceedings of the 2018 IEEE/CVF Conference on Computer Vision and Pattern Recognition, Salt Lake City, UT, USA, 18–22 June 2018; pp. 2002–2011. [CrossRef]
4. Godard, C.; Aodha, O.M.; Brostow, G.J. Unsupervised Monocular Depth Estimation with Left-Right Consistency. In Proceedings of the 2017 IEEE Conference on Computer Vision and Pattern Recognition (CVPR), Honolulu, HI, USA, 21–26 July 2017; pp. 6602–6611. [CrossRef]
5. Girshick, R.; Donahue, J.; Darrell, T.; Malik, J. Rich Feature Hierarchies for Accurate Object Detection and Semantic Segmentation. In Proceedings of the IEEE Conference on Computer Vision and Pattern Recognition, Columbus, OH, USA, 23–28 June 2014.
6. Lin, T.Y.; Goyal, P.; Girshick, R.; He, K.; Dollar, P. Focal Loss for Dense Object Detection. *arXiv* **2017**, arXiv:1708.02002.
7. Hestness, J.; Narang, S.; Ardalani, N.; Diamos, G.; Jun, H.; Kianinejad, H.; Patwary, M.M.A.; Yang, Y.; Zhou, Y. Deep Learning Scaling Is Predictable, Empirically. *arXiv* **2017**, arXiv:1712.00409.
8. Hestness, J.; Ardalani, N.; Diamos, G. Beyond Human-Level Accuracy: Computational Challenges in Deep Learning. *arXiv* **2019**, arXiv:1909.01736.
9. Patwary, M.; Chabbi, M.; Jun, H.; Huang, J.; Diamos, G.; Church, K. Language Modeling at Scale. *arXiv* **2018**, arXiv:1810.10045.
10. Shazeer, N.; Cheng, Y.; Parmar, N.; Tran, D.; Vaswani, A.; Koanantakool, P.; Hawkins, P.; Lee, H.; Hong, M.; Young, C.; et al. Mesh-TensorFlow: Deep Learning for Supercomputers. *Adv. Neural Inf. Process. Syst.* **2018**, *31*, 1–10.
11. Choudhary, T.; Mishra, V.; Goswami, A.; Sarangapani, J. A Comprehensive Survey on Model Compression and Acceleration. *Artif. Intell. Rev.* **2020**, *53*, 5113–5155. [CrossRef]
12. Cheng, Y.; Wang, D.; Zhou, P.; Zhang, T. Model Compression and Acceleration for Deep Neural Networks: The Principles, Progress, and Challenges. *IEEE Signal Process. Mag.* **2018**, *35*, 126–136. [CrossRef]
13. Gou, J.; Yu, B.; Maybank, S.J.; Tao, D. Knowledge Distillation: A Survey. *Int. J. Comput. Vis.* **2021**, *129*, 1789–1819.
14. Hinton, G.; Vinyals, O.; Dean, J. Distilling the Knowledge in a Neural Network. *arXiv* **2015**, arXiv:1503.02531.
15. Mirzadeh, S.I.; Farajtabar, M.; Li, A.; Levine, N.; Matsukawa, A.; Ghasemzadeh, H. Improved Knowledge Distillation via Teacher Assistant. *AAAI* **2020**, *34*, 5191–5198. [CrossRef]
16. Park, D.Y.; Cha, M.H.; Jeong, C.; Kim, D.S.; Han, B. Learning Student-Friendly Teacher Networks for Knowledge Distillation. *Adv. Neural Inf. Process. Syst.* **2021**, *34*, 13292–13303.
17. LeCun, Y.; Denker, J.S.; Solla, S.A. Optimal Brain Damage. *Adv. Neural Inf. Process. Syst.* **1990**, *2*, 598–605.
18. Walsh, C.A. Peter huttenlocher (1931–2013). *Nature* **2013**, *502*, 172. [CrossRef]
19. Ström, N. Sparse Connection and Pruning in Large Dynamic Artificial Neural Networks. In Proceedings of the 5th European Conference on Speech Communication and Technology (Eurospeech 1997), Rhodes, Greece, 22–25 September 1997; pp. 2807–2810. [CrossRef]
20. Frankle, J.; Carbin, M. The Lottery Ticket Hypothesis: Finding Sparse, Trainable Neural Networks. *arXiv* **2019**, arXiv:1803.03635.

21. NVIDIA. *NVIDIA Turing GPU Architecture Whitepaper*; NVIDIA: Santa Clara, CA, USA, 2018.
22. Russakovsky, O.; Deng, J.; Su, H.; Krause, J.; Satheesh, S.; Ma, S.; Huang, Z.; Karpathy, A.; Khosla, A.; Bernstein, M.; et al. ImageNet Large Scale Visual Recognition Challenge. *Int. J. Comput. Vis.* **2015**, *115*, 211–252. [CrossRef]
23. Gordon, A.; Eban, E.; Nachum, O.; Chen, B.; Wu, H.; Yang, T.J.; Choi, E. MorphNet: Fast & Simple Resource-Constrained Structure Learning of Deep Networks. In Proceedings of the 2018 IEEE/CVF Conference on Computer Vision and Pattern Recognition, Salt Lake City, UT, USA, 18–22 June 2018; pp. 1586–1595. [CrossRef]
24. Passalis, N.; Tefas, A. Learning Deep Representations with Probabilistic Knowledge Transfer. In *Computer Vision–ECCV 2018*; Ferrari, V., Hebert, M., Sminchisescu, C., Weiss, Y., Eds.; Springer International Publishing: Cham, Switzerland, 2018; Volume 11215, pp. 283–299. [CrossRef]
25. Zagoruyko, S.; Komodakis, N. Paying More Attention to Attention: Improving the Performance of Convolutional Neural Networks via Attention Transfer. *arXiv* **2017**, arXiv:1612.03928.
26. Tian, Y.; Krishnan, D.; Isola, P. Contrastive Representation Distillation. *arXiv* **2022**, arXiv:1910.10699.
27. Kim, J.; Park, S.; Kwak, N. Paraphrasing Complex Network: Network Compression via Factor Transfer. *Adv. Neural Inf. Process. Syst.* **2018**, *31*, 1–13.
28. Denil, M.; Shakibi, B.; Dinh, L.; Ranzato, M. Predicting Parameters in Deep Learning. *Adv. Neural Inf. Process. Syst.* **2013**, *26*, 1–9.
29. Lin, M.; Chen, Q.; Yan, S. Network In Network. *arXiv* **2014**, arXiv:1312.4400.
30. Szegedy, C.; Liu, W.; Jia, Y.; Sermanet, P.; Reed, S.; Anguelov, D.; Erhan, D.; Vanhoucke, V.; Rabinovich, A. Going Deeper with Convolutions. In Proceedings of the 2015 IEEE Conference on Computer Vision and Pattern Recognition (CVPR), Boston, MA, USA, 7–12 June 2015; pp. 1–9. [CrossRef]
31. Hanson, S.J.; Pratt, L.Y. Comparing Biases for Minimal Network Construction with Back-Propagation. *Adv. Neural Inf. Process. Syst.* **1988**, *1*, 177–185.
32. Hassibi, B.; Stork, D.G. Second Order Derivatives for Network Pruning: Optimal Brain Surgeon. *Adv. Neural Inf. Process. Syst.* **1992**, *5*, 164–171.
33. Han, S.; Mao, H.; Dally, W.J. Deep Compression: Compressing Deep Neural Networks with Pruning, Trained Quantization and Huffman Coding. *arXiv* **2016**, arXiv:1510.00149.
34. Liu, Z.; Sun, M.; Zhou, T.; Huang, G.; Darrell, T. Rethinking the Value of Network Pruning. *arXiv* **2019**, arXiv:1810.05270.
35. Redmon, J.; Farhadi, A. YOLOv3: An Incremental Improvement. *arXiv* **2018**, arXiv:1804.02767.
36. Han, S.; Liu, X.; Mao, H.; Pu, J.; Pedram, A.; Horowitz, M.A.; Dally, W.J. EIE: Efficient Inference Engine on Compressed Deep Neural Network.*arXiv* **2016**, arXiv:1602.01528.
37. Anwar, S.; Hwang, K.; Sung, W. Structured Pruning of Deep Convolutional Neural Networks. *J. Emerg. Technol. Comput. Syst.* **2017**, *13*, 1–18. [CrossRef]
38. Li, H.; Kadav, A.; Durdanovic, I.; Samet, H.; Graf, H.P. Pruning Filters for Efficient ConvNets. *arXiv* **2017**, arXiv:1608.08710.
39. Liu, Z.; Mu, H.; Zhang, X.; Guo, Z.; Yang, X.; Cheng, K.T.; Sun, J. MetaPruning: Meta Learning for Automatic Neural Network Channel Pruning. In Proceedings of the 2019 IEEE/CVF International Conference on Computer Vision (ICCV), Seoul, Republic of Korea, 27 October–2 November 2019; pp. 3295–3304. [CrossRef]
40. Su, X.; You, S.; Wang, F.; Qian, C.; Zhang, C.; Xu, C. BCNet: Searching for Network Width with Bilaterally Coupled Network. In Proceedings of the 2021 IEEE/CVF Conference on Computer Vision and Pattern Recognition (CVPR), Nashville, TN, USA, 19–25 June 2021; pp. 2175–2184. [CrossRef]
41. He, Y.; Kang, G.; Dong, X.; Fu, Y.; Yang, Y. Soft Filter Pruning for Accelerating Deep Convolutional Neural Networks. *arXiv* **2018**, arXiv:1808.06866.
42. He, Y.; Zhang, X.; Sun, J. Channel Pruning for Accelerating Very Deep Neural Networks. In Proceedings of the 2017 IEEE International Conference on Computer Vision (ICCV), Venice, Italy, 22–29 October 2017; pp. 1398–1406. [CrossRef]
43. Wen, W.; Wu, C.; Wang, Y.; Chen, Y.; Li, H. Learning Structured Sparsity in Deep Neural Networks. *Adv. Neural Inf. Process. Syst.* **2016**, *29*, 1–9.
44. Zhou, H.; Alvarez, J.M.; Porikli, F. Less Is More: Towards Compact CNNs. In *Computer Vision—ECCV 2016*; Leibe, B., Matas, J., Sebe, N., Welling, M., Eds.; Springer International Publishing: Cham, Switzerland, 2016; Volume 9908, pp. 662–677. [CrossRef]
45. Ye, J.; Lu, X.; Lin, Z.; Wang, J.Z. Rethinking the Smaller-Norm-Less-Informative Assumption in Channel Pruning of Convolution Layers. *arXiv* **2018**, arXiv:1802.00124.
46. Zhuang, T.; Zhang, Z.; Huang, Y.; Zeng, X.; Shuang, K.; Li, X. Neuron-Level Structured Pruning Using Polarization Regularizer. *Adv. Neural Inf. Process. Syst.* **2020**, *33*, 9865–9877.
47. Aghli, N.; Ribeiro, E. Combining Weight Pruning and Knowledge Distillation For CNN Compression. In Proceedings of the 2021 IEEE/CVF Conference on Computer Vision and Pattern Recognition Workshops (CVPRW), Nashville, TN, USA, 20–25 June 2021; pp. 3185–3192. [CrossRef]
48. Chen, S.; Zhao, Q. Shallowing Deep Networks: Layer-Wise Pruning Based on Feature Representations. *IEEE Trans. Pattern Anal. Mach. Intell.* **2019**, *41*, 3048–3056. [CrossRef] [PubMed]
49. Sarridis, I.; Koutlis, C.; Kordopatis-Zilos, G.; Kompatsiaris, I.; Papadopoulos, S. InDistill: Information Flow-Preserving Knowledge Distillation for Model Compression. *arXiv* **2022**, arXiv:2205.10003.
50. Park, J.; No, A. Prune Your Model Before Distill It. *arXiv* **2022**, arXiv:2109.14960.

51. Romero, A.; Ballas, N.; Kahou, S.E.; Chassang, A.; Gatta, C.; Bengio, Y. FitNets: Hints for Thin Deep Nets. *arXiv* **2015**, arXiv:1412.6550.
52. Yang, Y.; Qiu, J.; Song, M.; Tao, D.; Wang, X. Distilling Knowledge From Graph Convolutional Networks. In Proceedings of the 2020 IEEE/CVF Conference on Computer Vision and Pattern Recognition (CVPR), Seattle, WA, USA, 13–19 June 2020; pp. 7072–7081. [CrossRef]
53. Yuan, L.; Tay, F.E.; Li, G.; Wang, T.; Feng, J. Revisiting Knowledge Distillation via Label Smoothing Regularization. In Proceedings of the 2020 IEEE/CVF Conference on Computer Vision and Pattern Recognition (CVPR), Seattle, WA, USA, 13–19 June 2020; pp. 3902–3910. [CrossRef]
54. NVIDIA. Sparse Matrix Storage Formats. 2020. Available online: https://www.gnu.org/software/gsl/doc/html/spmatrix.html (accessed on 1 September 2022).
55. Krizhevsky, A. Hinton G. *Learning Multiple Layers of Features from Tiny Images*; Technical report; University of Toronto: Toronto, ON, Canada, 2009.
56. Keydel, E.R.; Lee, S.W.; Moore, J.T. MSTAR Extended Operating Conditions: A Tutorial. In *Aerospace/Defense Sensing and Controls*; Zelnio, E.G., Douglass, R.J., Eds.; Sandia National Laboratory: Albuquerque, NM, USA, 1996; pp. 228–242. [CrossRef]
57. Hou, X.; Ao, W.; Song, Q.; Lai, J.; Wang, H.; Xu, F. FUSAR-Ship: Building a High-Resolution SAR-AIS Matchup Dataset of Gaofen-3 for Ship Detection and Recognition. *Sci. China Inf. Sci.* **2020**, *63*, 140303. [CrossRef]
58. Han, S.; Pool, J.; Tran, J.; Dally, W. Learning Both Weights and Connections for Efficient Neural Network. *Adv. Neural Inf. Process. Syst.* **2018**, *28*, 1–9.
59. Howard, A.G.; Zhu, M.; Chen, B.; Kalenichenko, D.; Wang, W.; Weyand, T.; Andreetto, M.; Adam, H. MobileNets: Efficient Convolutional Neural Networks for Mobile Vision Applications. *arXiv* **2017**, arXiv:1704.04861.
60. Sandler, M.; Howard, A.; Zhu, M.; Zhmoginov, A.; Chen, L.C. MobileNetV2: Inverted Residuals and Linear Bottlenecks. In Proceedings of the 2018 IEEE/CVF Conference on Computer Vision and Pattern Recognition, Salt Lake City, UT, USA, 18–22 June 2018; pp. 4510–4520. [CrossRef]
61. Howard, A.; Sandler, M.; Chen, B.; Wang, W.; Chen, L.C.; Tan, M.; Chu, G.; Vasudevan, V.; Zhu, Y.; Pang, R.; et al. Searching for MobileNetV3. In Proceedings of the 2019 IEEE/CVF International Conference on Computer Vision (ICCV), Seoul, Republic of Korea, 27 October–2 November 2019; pp. 1314–1324. [CrossRef]

Article

Aircraft Detection in SAR Images Based on Peak Feature Fusion and Adaptive Deformable Network

Xiayang Xiao, Hecheng Jia, Penghao Xiao and Haipeng Wang *

Key Laboratory for Information Science of Electromagnetic Waves (MoE), Fudan University, Shanghai 200433, China
* Correspondence: hpwang@fudan.edu.cn

Abstract: Due to the unique imaging mechanism of synthetic aperture radar (SAR), targets in SAR images often shows complex scattering characteristics, including unclear contours, incomplete scattering spots, attitude sensitivity, etc. Automatic aircraft detection is still a great challenge in SAR images. To cope with these problems, a novel approach called adaptive deformable network (ADN) combined with peak feature fusion (PFF) is proposed for aircraft detection. The PFF is designed for taking full advantage of the strong scattering features of aircraft, which consists of peak feature extraction and fusion. To fully exploit the strong scattering features of the aircraft in SAR images, peak features are extracted via the Harris detector and the eight-domain pixel detection of local maxima. Then, the saliency of aircraft under multiple imaging conditions is enhanced by multi-channel blending. All the PFF-preprocessed images are fed into the ADN for training and testing. The core components of ADN contain an adaptive spatial feature fusion (ASFF) module and a deformable convolution module (DCM). ASFF is utilized to reconcile the inconsistency across different feature scales, raising the characterization capabilities of the feature pyramid and improving the detection performance of multi-scale aircraft further. DCM is introduced to determine the 2-D offsets of feature maps adaptively, improving the geometric modeling abilities of aircraft in various shapes. The well-designed ADN is established by combining the two modules to alleviate the problems of the multi-scale targets and attitude sensitivity. Extensive experiments are conducted on the GaoFen-3 (GF3) dataset to demonstrate the effectiveness of the PFF-ADN with an average precision of 89.34%, as well as an F1-score of 91.11%. Compared with other mainstream algorithms, the proposed approach achieves state-of-the-art performance.

Keywords: synthetic aperture radar (SAR); aircraft detection; deep learning; peak feature; adaptive spatial feature fusion; deformable convolution module

Citation: Xiao, X.; Jia, H.; Xiao, P.; Wang, H. Aircraft Detection in SAR Images Based on Peak Feature Fusion and Adaptive Deformable Network. *Remote Sens.* **2022**, *14*, 6077. https://doi.org/10.3390/rs14236077

Academic Editor: Timo Balz

Received: 21 October 2022
Accepted: 27 November 2022
Published: 30 November 2022

Publisher's Note: MDPI stays neutral with regard to jurisdictional claims in published maps and institutional affiliations.

1. Introduction

Because of its capacity to obtain high-resolution SAR images regardless of time and weather, synthetic aperture radar has been widely applied in many fields, such as ocean observations, natural disaster prediction, and battlefield surveillance [1,2]. Automatic target detection in SAR images has been a hot research topic for many years, attracting extensive attention from scholars at home and abroad. The essence of object detection in SAR images is to extract the target from the background via the difference in scattering characteristics, figuring out the location of potential objects. However, it is difficult to interpret SAR images using coherent scattering and imaging mechanisms. Aircraft detection in complicated conditions of SAR images is always a challenging task.

Essentially, traditional SAR target detection algorithms are designed for target extraction based on the differences in the electromagnetic backscattering properties of targets and the clutter background, and the same goes for aircraft detection. The traditional methods applied in SAR images can be grouped into the following four categories: contrast-based, visual attention-based, complex data-based, and multi-feature-based. The methods based on contrast include CFAR [3], various CFAR-derived methods (CA-CFAR [4], SOCA-CFAR [5], GOCA-CFAR [6], OS-CFAR [7], VI-CFAR [8]), GLRT [9] and PR [10]. The advantage of methods based on contrast are easy to implement and can accomplish good performance in simple scenarios. However, it is difficult to select an appropriate statistical clutter model and tackle heterogeneous strong clutter in the practical application. In addition, there are many false alarms and missing targets in object detection under a complex background. The visual attention-based method [11] suppresses the strong hetero enhancement target by introducing prior information, improving the SCR (signal to clutter ratio). Yet, the prior information introduced needs to be concretely analyzed for specific conditions, and the algorithm may be redesigned for different detection tasks. For complex data-based methods [12], the scattering characteristics of the target and clutter, as well as the imaging mechanism, are utilized for detecting the target. It can reflect the difference between man-made targets and natural clutter from the physical mechanism. However, high acquisition costs are required to obtain the raw complex SAR image data, and the capacity to distinguish between artificial clutter interference and artificial targets of interest is yet to be verified. The principal features used in the multi-feature-based approach mainly comprise structural features [13], peak features [4], variance features [14], and extended fractal features [15]. The merit of this solution is that it can detect targets from multi-dimensional features. Nevertheless, due to the poor generalization of this method, it can only achieve good performance for specific data. Among the above detection methods, the CFAR detector is one of the most widely studied and intensively applied. The main components of aircraft are composed of metallic materials with high backscattering coefficients. The airport area has a low backscattering coefficient as it is more capable of absorbing electromagnetic waves. The contrast-based detection algorithm can efficiently detect aircraft in the ROI area of the airport under such conditions. However, in general, traditional methods are ineffective in the face of complex background scattering, owing to the poor robustness and the limited capacity of feature representation.

With the enrichment of SAR images, the traditional algorithm, represented by CFAR, no longer meets the requirements. In recent years, with the development of DL and the improvement of the related hardware device, convolutional neural networks have shown amazing performance in the field of computer vision; and yet these methods are mainly applied to optical images. Detection algorithms based on DL can be divided into the two-stage algorithm and the single-stage algorithm in terms of detection steps. The Faster-RCNN [16] is a typical two-stage algorithm that adopts a trainable region proposal network (RPN) to improve detection efficiency while maintaining high accuracy. However, only high-level features are employed to predict objects, and it is difficult to detect small targets because it loses the low-level features. The single-stage representative algorithm is YOLO [17], which transforms the detection into a regression problem. The algorithm achieves a significant improvement in the detection rate, but the precision becomes poor. An FPN structure [18] is proposed in the literature, which integrates high-resolution information about shallow features with high semantic information on deep features. The detection performance is raised by predicting the different feature layers separately. All the above methods are based on anchor frames. Firstly, the location of potential targets is inferred from the generated anchors in the RPN stage, and the final results are filtered by NMS (Non-Maximum Suppression). After 2019, algorithms based on DL without anchor frames have been developed, such as CornetNet [19], ExtremeNet [20], FCOS [21], and CenerNet [22]. CornerNet is a new one-stage detection method that treats the target as a set of key points. ExtremeNet evolves based on CornerNet, which detects four extremes and one centroid via standard key point estimation. The central heat map of each class is proposed to predict the target center. CenterNet derives the centroid via key point estimation and additional information about the target by regression. Many other methods have been proposed for object detection, e.g., [23,24].

With deep learning (DL) achieving amazing performance in computer vision, DL has also gradually become a standard paradigm in SAR image analysis, including aircraft detection [25], which is detailed in Section 2.1.1. Although many approaches have achieved a certain effect in aircraft detection of SAR images, they still leave much room for improvement.

The imaging mechanisms of SAR and optical satellites display great differences, resulting in an entirely different appearance of the target. The represented features of aircraft in SAR images can be summarized via the following three aspects:

(1) Discrete points. Due to the mechanism of SAR imaging, aircraft appearing in SAR images consist of a few strong scattering points, which form the faint outline of the aircraft. Coupled with the interference of background clutter, it is easy to miss detection;

(2) Multi-scale targets. In this study, the size of most aircraft in the GF3 dataset ranges from 20 to 110 pixels. However, with the deepening of the CNN, the spatial information of small-size aircraft with fewer pixels is easy to lose, which poses a challenge to multi-scale aircraft detection;

(3) Attitude sensitivity. With changes in azimuth angle, the appearance of the same aircraft in different SAR images is not quite identical. It is difficult to acquire the modeling geometric transformation of changeable-appearance aircraft in limited and available training samples.

As is well-known, there is no doubt that abstract features extracted by DL have better representation capacities than traditional manual features. However, should the traditionally established hand-made features be completely abandoned in favor of total reliance on abstract features? This is a meaningful issue. Regretfully, most detection algorithms of SAR based on DL are established by fine-tuning the classical optical network, and do not consider the differences in the imaging mechanism between SAR and optical satellites. Otherwise, compared with optical data, SAR image data are relatively scarce. A large amount of image data is acquired to train the network model, so transfer learning from optical networks alone does not work well. Besides this, many publications are only tested on specific circumstances, e.g., aircraft with clear contours and a single background. Fewer detection networks take the multi-scale and attitude sensitivity of targets in the various and complicated conditions of SAR images into consideration.

Intending to address the above-mentioned problems, an adaptive pyramid deformable network combined with the peak feature fusion (PFF-ADN) is proposed to ameliorate the performance of aircraft detection under complex backgrounds and target dense arrangement conditions. The strong scattering points represent the maximum pixel value of the local scattering region in SAR images, which have superior significance and stability [26]. The distribution of scattering centers can be responded to the peak features, which also can describe the geometric shape of the target in a complex environment. Therefore, combining the peak features is a feasible way to enhance the performance of aircraft detection. (1) In the peak feature extraction stage, because the strong scattering points tend to be distributed on the common components such as nose, fuselage, tail, wings, and engines, the corner points extracted by the Harris detector [27] with rotational invariance, are selected as the basis for modeling the scattering information of aircraft. Then the peak features of images are acquired based on the 8-domain pixel detection of local maxima, which will be sent to the next stage for feature fusion. (2) In the peak feature fusion stage, the image data processed by PFE will be integrated into the image as the G-channel data. This operation can fuse the peak features of aircraft into the image, enhancing the saliency of aircraft under various imaging conditions. Meanwhile, it facilitates the mitigation of variability and provides more effective information to the detection network. (3) In the deep feature extraction stage, an adaptive deformable network for aircraft detection is proposed by incorporating the ASFF and DCM. The ASFF can cope with the inconsistency of multi-scale features, which is beneficial to fully exploit the representational power of the feature pyramid network, improving the ability to detect multi-scale aircraft, especially for small-size aircraft. The DCM is adopted to cope with the attitude sensitivity of aircraft in SAR imaging and various shapes of aircraft, making the detection network accommodate the geometric variations.

Finally, experiments are conducted on the GF3 dataset to validate the effectiveness of each presented module and to demonstrate the superiority of the proposed method.

The main contributions of our work can be summarized as follows:

(1) A novel integrated framework named PFF-ADN is proposed for SAR aircraft detection, which enhances the scattering features of the target by fusing the peak features of images and improving the network structure. This presented method achieves state-of-the-art performance on the GF3 dataset.

(2) A peak feature fusion strategy is designed for enhancing the brightness information of aircraft in the SAR images by extracting and fusing the peak feature information, which has stronger robustness to deal with the variability and obscureness caused by the scattering mechanism of SAR. Compared to the raw images, the aircraft characteristics are highlighted in the enhanced images, providing effective information on aircraft for the subsequent network.

(3) An adaptive deformable network for aircraft detection is designed, which is composed of a Feature Pyramid Network with ASFF structure and deformable convolution module (DCM). The ASSF is introduced to solve the inconsistency of multi-scale features and retain more discrete information about small-size aircraft, which enhances the detectability, especially for small-size aircraft. The DCM is adopted to cope with the attitude sensitivity of aircraft in SAR imaging and various shapes of aircraft, making the detection network accommodate the geometric variations.

The remainder of this paper is organized as follows. Section 2.1 shows a review of the related work, which contains DL methods and traditional methods of aircraft detection in the SAR domain. Section 2.2 describes the proposed PFF-ADN algorithm in detail. In Section 3, the parameter configuration, assessment criteria, experimental results as well as performance evaluation are shown in detail. Finally, we summarize this article and provide the prospect for further research in Section 4.

2. Materials and Methods

2.1. Related Work

2.1.1. Aircraft Detection Based on DL in SAR Images

Recently, with deep learning (DL) achieving amazing performance in computer vision, DL has also gradually become a standard paradigm in SAR image analysis. Guo et al. [25] designed a novel method that combines scattering information enhancement and an attention pyramid network for aircraft detection. The candidate areas containing aircraft are enhanced by scattering information, improving the average precision of aircraft detection. Therefore, inspired by this work, we propose the peak feature can be fused into the image to enhance the information of the target, thus improving the detection performance. Wang et al. [28] proposed a method based on a convolutional neural network (CNN) and data enhancement for aircraft detection. The regions of interest are acquired by the saliency pre-detection approach and then fed into CNN to get the final detection results. Zhang et al. [29] showed a cascaded three-look network to detect aircraft in SAR images. It firstly narrows the inspected image down to the airport area by pixel-based comparison, and it obtains the detection results through Faster-RCNN. Zhao et al. [30] adopted an integration network that contains the dilated attention block and convolution block attention module for aircraft detection, and they confirmed the effectiveness of this method on a mixed SAR aircraft dataset. He et al. [31] exploited the components of aircraft to improve detection performance. To find out the inner link between the aircraft and the components, they established a multi-scale detector including a root filter and part filters. The experimental results indicate this method has higher precision and is able to detect the components of aircraft. Dou et al. [32] came up with an optimized target attitude estimation method wherein prior information about the target shape is acquired by the generative adversarial network. Detection probability is improved with valid prior knowledge of aircraft. Julie et al. [33] suggested a hybrid clustering active learning method to improve the performance of aircraft detection. Han et al. [34] utilized this method, which combines region segmentation and multi-feature decision for aircraft detection in POL-SAR images, and the detection results showed a lower false positive rate. Jia et al. [35] designed a component discrimination network for aircraft detection, which utilizes the dominant component features of aircraft to improve detection and recognition performance. Considering the SAR imaging mechanism, Kang et al. [36] proposed an innovative scattering feature relation network to improve detection performance. It guarantees the completeness of detection results by analyzing and enhancing their correlation. Zhao [37] presented a detector called Attentional Feature Refinement and Alignment Network for detecting aircraft in SAR images, achieving competitive accuracy and speed. Bi et al. [38] present a novel POL-SAR image classification method, achieving promising classification performance and preferable consistency.

2.1.2. Traditional Candidate Feature Methods in SAR Images

Manually designed features based on target structural and scattering features are used for feature representation in the conventional detection approach in the SAR domain. Target structural features include target contour, component shape, and structural distribution. The structural characteristics of the target are mainly manifested by the length and width of the target, the contour, the shape of components, and the structural configuration. Margarit et al. [39], Margarit and Mallorqui [40], and Margarit and Tabasco [41] conducted some work to improve the accuracy of ship classification by modeling the geometric features of the bow, midship and stern sections. Xu [42] et al. proposed a novel structured low-rank and sparse algorithm for high-resolution ISAR imaging, which can effectively deal with various types of sparse sampling and improve the image formation quality greatly. Gao [43] proposed a geometric feature-based aircraft interpretation method. The key geometric parameters of aircraft are obtained for interpretation by extracting the components of the aircraft, such as the engine and nose. In addition, as the SAR image resolution develops, the structural information of objects is presented in more detail, so the detection performance of SAR objects will be facilitated by making full use of radiometric and geometric scattering information. The aircraft in SAR images are usually composed of a series of strong scattered bright spots due to the imaging mechanism of SAR and the material construction of aircraft. Therefore, some scholars have studied how to effectively exploit strong scattering points for aircraft interpretation in SAR. The backscattering characteristics of aircraft are researched to acquire salient point vectors, which are integrated into the template-matching process for aircraft identification in TerraSAR-X images [44]. Zhu et al. [45] put forward a method in which the accuracy of SAR ship classification is improved by integrating the geometric topology of strong spots with amplitude intensity. In conclusion, the scattering features used for aircraft detection in SAR images can be divided into three main categories: peak features, statistical features, and texture features. The distribution of scattering centers can respond to the peak features, which can also describe the geometric shape of the target in a complex environment. Importantly, within a certain range of viewing angles, the position and intensity of strong scattering centers are statistically quasi-invariant [46]. Therefore, the extraction and application of peak features are critical to the orientation of aircraft in SAR images. Here, we focus on the peak features for aircraft detection, which is illustrated in Section 2.2.2.

2.2. The Proposed Method

2.2.1. Overview of the Proposed Method

In this work, the framework of PFF-ADN is illustrated in Figure 1, which mainly comprises three parts: peak feature extraction, peak feature fusion for enhancing the features of aircraft, and the ADN structure for detecting the image processed by PFF. In Section 2.2.2, we first introduce the extraction process of the peak feature, because it is the basis for the next step. Next, the peak feature fusion in Section 2.2.3 is introduced in detail. Finally, The ADN, which incorporates the ASFF and DCM, is described in detail. It is worth mentioning that the DCM and ASFF are introduced to the proposed approach. The two modules can greatly enhance the detector's capacity for modeling geometric transformations and extracting abstract features, respectively. In Section 2.2.4, the principles, expression forms, and realization course of the two modules will be discussed in detail.

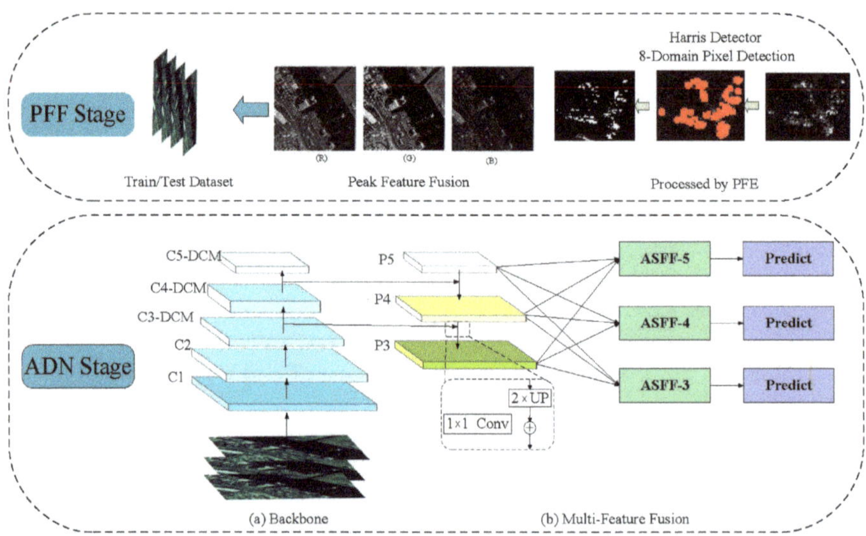

Figure 1. Overview of the PFF-ADN framework. The proposed method mainly comprises the PFF stage and ADN stage. The PFF stage aims to enhance the saliency of the aircraft target in SAR images by fusing the peak features of the aircraft into the image. The ADN is the detection network based on DL, which generates the final detection results.

2.2.2. PFE

Target scattering features are the key information used in understanding SAR images deeply. The peak features of SAR images essentially reflect the scattering center of the target. Meanwhile, peak features can not only depict the geometric features and shape characteristics of the target, but also be very stable in the complex radar imaging system. Therefore, they are one of the most widely used features in SAR image target detection. The structure of an aircraft is relatively complex, the dimensions and details of which can be observed in single-polarization high-resolution SAR images. The strong scattering of aircraft mainly comes from the engine, nose, fuselage, wing, and tail edge. The different parts have different scattering mechanisms, including tip diffraction, cavity scattering, mirror reflection, edge diffraction, multiple reflections and so on [47], as illustrated in Figure 2.

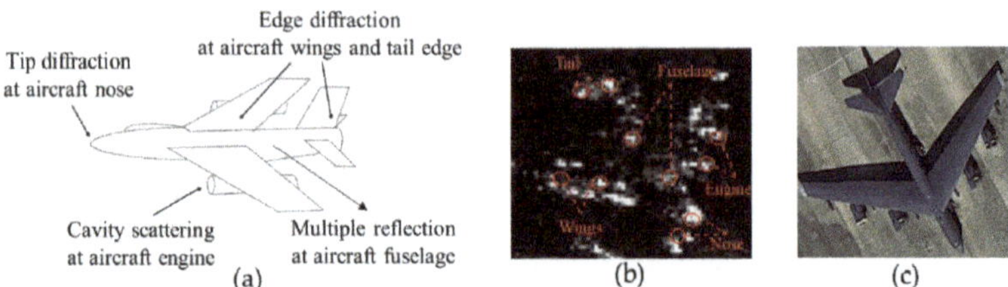

Figure 2. Scattering mechanism of aircraft in SAR images. (**a**) Scattering mechanism analysis of core components of aircraft. (**b**) SAR image of aircraft (**c**) Optical image of aircraft.

Based on the aforementioned analyses, the scattering from aircraft is relatively strong, as exhibited in the highlighted regions in the SAR images. According to experiments, the extraction of peak features is generally stable, regardless of prior knowledge. The specific operation procedure of PFE is shown in Algorithm 1. Therefore, a peak feature enhancement and fusion approach based on the extraction of peak points and multi-channel fusion is proposed to enhance the scattering features of aircraft in SAR images, which facilitates the subsequent detection network.

Algorithm 1: Peak Feature Extraction

Input:	The image $I(x,y)$ the mean μ and variance σ of the background region Hyper-parameter: $k = 0.4$
Output:	Generated peak feature points : $P = (p_1, p_2, \ldots, p_k)$
Main loop:	
1:	Compute the gradient in the direction of x and y:
2:	$I_x = \frac{\partial I}{\partial x} = I \otimes (-1,0,1), I_y = \frac{\partial I}{\partial x} = I \otimes (-1,0,1)^T$
3:	Calculate the product of the gradients of the two directions:
4:	$I_x^2 = I_x \cdot I_x, \ I_y^2 = I_y \cdot I_y, \ I_{xy} = I_x \cdot I_y$
5:	A Gaussian weight is assigned to each gradient acquired:
6:	$A = g(I_x^2) = I_x^2 \otimes \omega, \ C = g(I_y^2) = I_y^2 \otimes \omega,$
7:	$B = g(I_{xy}) = I_{xy} \otimes \omega, \ M = \begin{bmatrix} A\ C \\ C\ B \end{bmatrix}$
8:	for $x \in$ corner-point do
9:	if $R = \left\{ R : \det M - \alpha(traceM)^2 < threshold_H \right\}$
10:	add x into the corner points set $C = C \cup \{x\}$
11:	else
12:	pass
13:	for $\ x \in C$ do
14:	if $\ \ x > \mu + k\sigma$ and $\min(a_{ij} - a_{N(i,j)}) > \sigma$
15:	add x into the peak points set $P = P \cup \{x\}$
16:	else
17:	pass

The peak feature points have strong gradient information, which is stable with changes in the external environment. Thus, it is of great importance to obtain the peak feature points for the subsequent feature enhancement and fusion. Given that the corner points have the advantage of rotation invariance, and insensitivity to noise and gradient features, the Harris detector is utilized to extract corner points first. Then, by comparing the corner points value and threshold, the peak points are automatically searched for. The mathematical expression of Harris is given below (1), where $w(x,y)$ denotes the window function and w is an adjustable parameter, and (x,y) is the pixel position in the image I. $E(\Delta x, \Delta y)$ represents the variation caused by window movement and $(\Delta x, \Delta y)$ is the shift pixel size of window.

$$E(\Delta x, \Delta y) = \sum_{x,y} w(x,y)[I(x + \Delta x, y + \Delta y) - I(x,y)]^2 \tag{1}$$

Referring to Taylor series expansion, Equation (2) is derived by using Equation (1). $g(\sigma)$ stands for the Gaussian convolution kernel with scale σ, which convolves with a matrix, improving the anti-noise ability of the algorithm. I_x is the first derivative of the position x. $\otimes$ represents the convolution operation.

$$M = g(\sigma) \otimes \begin{bmatrix} I_x^2 & I_x I_y \\ I_x I_y & I_y^2 \end{bmatrix} \tag{2}$$

The scattering centers are defined as two kinds of peak feature points, which are comprised of a two-dimensional peak point (vertex) and a one-dimensional peak point (vertex of row and vertex of column). The row (column) vertex is the row (column) local maximum of the pixel within the target region of the SAR image and the vertex is the two-dimensional local maximum of the pixel within the target region of the SAR image. To weaken the influence of noise, when we acquire the peak points, the mean μ and variance σ of the background region need to be estimated first. The peak features of a target (i, j) can be defined by pixels within their domain:

$$p_{ij} = \begin{cases} 0, \ else \\ 1, \ if \ \mathbf{a}_{ij} > \mu + k\sigma \ and \ \min(\mathbf{a}_{ij} - a_{N(i,j)}) > \sigma \end{cases} \tag{3}$$

If $p_{ij} = 1$, it denotes that the current pixel is the peak point. If $p_{ij} = 0$, it represents that the current pixel is not the peak point. $\mathbf{a}_{ij}$ is the pixel value of the current point. $N_{(i,j)}$ is the local area of $\mathbf{a}_{ij}$ The row (column) vertex $N_{(i,j)}$ is the two nearest neighbor pixels in i row (j column) of $\mathbf{a}_{ij}$. k is an empirical parameter determined by conducting multiple experiments. In this paper, we adopt a peak extraction method based on the eight-domain pixel detection of local maxima, which is a classical method with pixel-level accuracy.

For the peak feature extraction of aircraft, three different types of aircraft are selected for a detailed analysis. As shown in Figure 3, Figure 3a is the original image of an aircraft slice; Figure 3b shows the three-dimensional maps of aircraft; Figure 3c shows the extraction results of peak points, which are marked in red to offer more intuitive information. It demonstrates the strong scattering points of aircraft can be more effectively indicated by the peak feature points.

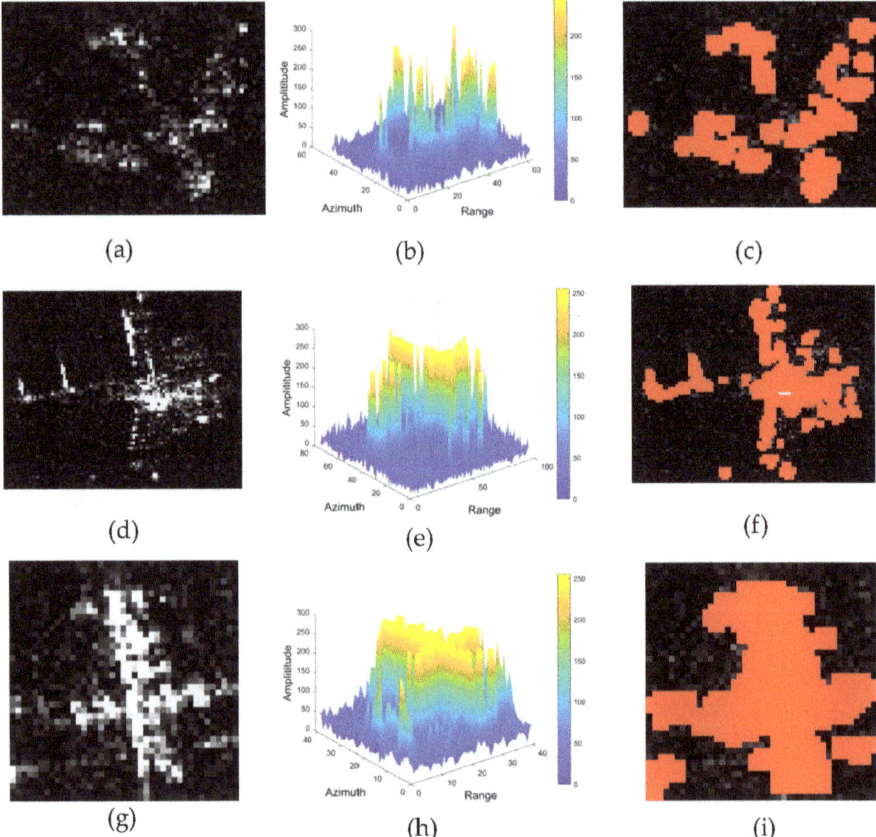

Figure 3. Peak feature extraction of aircraft in SAR images. (**a,d,g**): SAR image of aircraft; (**b,e,h**): 3-dimensional maps of aircraft; (**c,f,i**): Extraction results of peak points.

The peak features of aircraft are scattered all over, mainly in the cavity structure of the engine, the complex electromagnetic structure of the cockpit, the dihedral angle formed by the combined components, the three-sided angle structure, and other strong scattering regions. A fine description of the aircraft structure can be obtained by the extraction of the peak features. The peak points of Figure 3a are mainly concentrated in the engine, tail, and wings and fewer peak points are in the fuselage. In Figure 3b, more peak points are distributed in the junction of fuselage and wing, which is more intuitive to visualize in three-dimensional maps. As shown in Figure 3c, the peak feature distribution is presented as a cross shape, which cannot be used to clearly distinguish the characteristics of each component. It can be inferred that the numbers and locations of peak points vary between types of aircraft. Based on this, peak features are critical in the application of target detection and recognition in SAR images.

It is necessary to handle the SAR images via the PFE method, owing to the aim, which is to detect aircraft in SAR images. By comparing Figure 4a–d, the intensity of the aircraft's position in SAR images is significantly enhanced. The changes are intuitively observed in 2-D and 3-D illustrations of SAR images. As shown in the 3-D maps in Figure 4a–d, the differences in the scattering strength of aircraft in SAR images can be observed more clearly.

2.2.3. PFF

In this study, a very effective and simple method is applied to SAR data preprocessing. k times the average of non-zero data is computed to truncate and normalize the original SAR data, which is abbreviated to KTN, as detailed in Formula (4). For the experimental dataset, the single-channel SAR data is processed by the KTN method.

$$I = unit8\left[\frac{I(I \geq k \cdot mean(I) = k \cdot mean(I)}{\max(I)} \times 255\right] \in [0, 255] \tag{4}$$

In Formula (4), I stands for the data value of each channel. k is a manually tunable parameter. $mean(I)$ and $\max(I)$ denote the average value of all non-zero data for each channel and the max value of data for each channel, respectively. The KTN method is utilized to handle the SAR data and convert them to a uint8 data format. Because of the network pre-trained by the optical samples of COCO [48], the images of COCO present in the RGB format with three-channel. Thus, the SAR data is necessary to be transformed into a three-channel format for detection network training better. In most pre-processing methods, single-channel eight-bit data is often transformed into three-channel data by replication. In this study, one channel of data is replaced with peak feature data I_{PFE} which is obtained in Section 2.2.2, and the other channels adopt different multiples of the mean. By fusing these enhancements in target scattering, the strong scattering characteristics of targets can be learned by the network more effectively. In the process of synthesizing the training image, R and B channels are assigned the data processed by KTN. The data of the G channel is substituted for the data processed by PFE.

$$R = unit8\left[\frac{I(I \geq k \cdot mean(I) = k \cdot mean(I)}{\max(I)} \times 255\right] \in [0, 255] \tag{5}$$

$$G = I_{PFE} \tag{6}$$

$$B = unit8\left[\frac{I(I \geq k \cdot mean(I) = k \cdot mean(I)}{\max(I)} \times 255\right] \in [0, 255] \tag{7}$$

Data visualization based on KNT is shown in Figure 5. Three-times and six-times are set in the KNT, as illustrated in Figure 5a,c. Compared with Figure 5c, Figure 5a shows higher visual lightness. Figure 5b is processed by PFE in Section 2.2.2 and the peak features of aircraft in the image are extracted and enhanced, leading to the information of the target being more distinct.

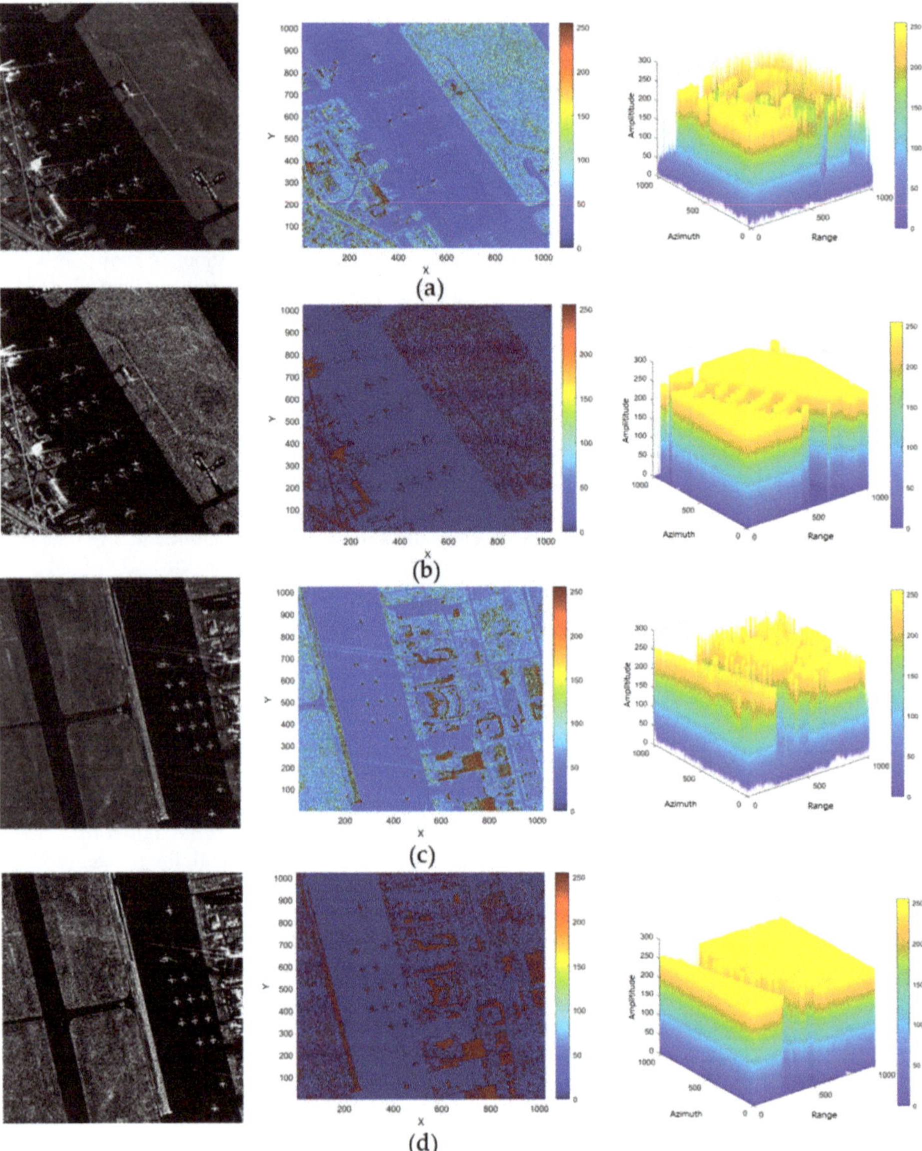

Figure 4. Peak feature extraction and enhancement of SAR images. (**a**,**c**) The three views, from left to right, are the original SAR images, the 2-D illustration of the SAR image, and the 3-D illustration of the SAR image. (**b**,**d**) Corresponding to (**a**,**c**), the three views, the image processed by PFE, 2-D illustration of SAR image, and 3-D illustration of SAR image.

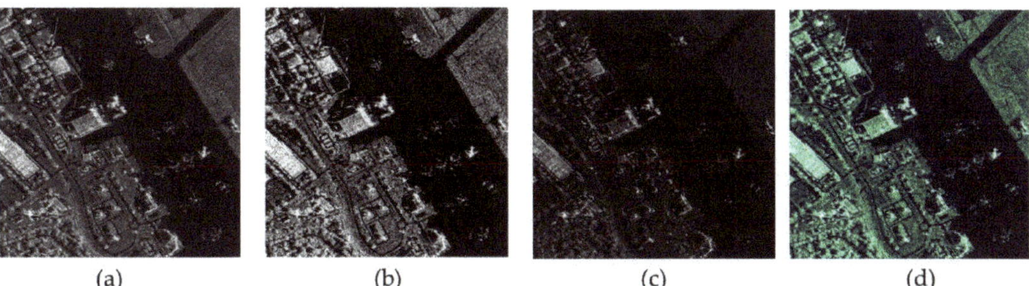

Figure 5. Data visualization based on KNT in a different channel. (**a**) R-channel (**b**) G-channel (**c**) B-channel (**d**) RGB-channel.

2.2.4. ADN

(1)　DCM: The input feature map is sampled at fixed positions in the standard convolution operation, and the sampled pixels are mostly rectangles. This gives rise to obvious issues, such as that the receptive fields are of the same size in the same layer. So, the standard convolution has no ability to handle geometric transformations. Nevertheless, because of the SAR scattering mechanisms, the same target in SAR images appears in various shapes with changes in azimuth angle. In this paper, deformable convolution [49] is adopted to accommodate geometric variations or attitude sensitivity in aircraft viewpoint and the part deformation of aircraft in SAR images.

As shown in Figure 1, ResNet50 [50] is used in this experiment and a deformable convolution kernel is applied in stages 3,4,5 of ResNet50, which is inspired by the work of Zhu [51]. As mentioned below, due to the adaptive receptive field sizes of deformable convolution, the sampling positions of feature maps in stage Conv2D can be adaptively adjusted, which allows the subsequent ASFF module to give more robust representations of geometric transformations. The standard convolution includes two steps. Firstly, the input feature graph X is sampled with a grid R of a specific size. Secondly, the sampling points are dotted with the weight coefficient W and these products are simply summed. As shown in Figure 6a, taking a 3×3 convolution kernel, for example, nine positions p_0 are taken as samples from X to get the output feature map y. For grid R, $(-1, 1)$, and $(1, 1)$ represent the upper left corner and bottom right corner respectively, and define a 3×3 kernel with dilation 1.

$$R = \{(-1, -1), (-1, 0), \dots, (0, 1), (1, 1)\} \tag{8}$$

The convolution output for each point p_0 is y

$$y(p_0) = \sum_{p_n \in R} w(p_n) \times x(p_0 + p_n) \tag{9}$$

In deformable convolution, the offsets $\Delta p_n (n = 1, 2, \dots, N)$ is added to the grid R, So the output y becomes

$$y(p_0) = \sum_{p_n \in R} w(p_n) \times x(p_0 + p_n + \Delta p_n) \tag{10}$$

As illustrated in Figure 6a, the offsets can be obtained by a parallel convolution unit and the sampling points of the input feature map are no longer regular rectangles. The nine cyan boxes indicate the sampling position in 3×3 standard convolution, while the corresponding blue boxes denote the sampling position in 3×3 deformable convolution. Figure 6b shows the receptive field of standard convolution (right) and deformable convolution (left). The sampling positions of standard convolution are fixed and rectangular, while the sampling positions are adaptively adjusted according to the shape and scale of instances. In conclusion, deformable convolution has the ability to learn receptive fields adaptively and enables adaptive part localization for aircraft with different shapes. For target location in a visual field, the deformable convolution proposed solves the problem of geometric transformations and it is conducive to feature extraction and fine localization.

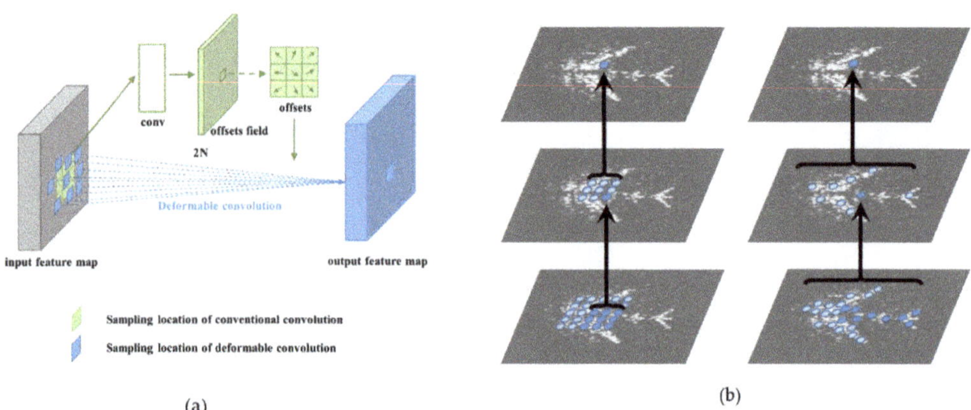

Figure 6. Illustration of DCM (**a**) Schematic diagram of the deformable convolution (**b**) Receptive field of standard convolution and deformable convolution. Top: A activation unit from a 3 × 3 convolution kernel on the top feature map. Middle: The sampling positions from the feature map of the previous layer. Bottom: the sampling positions of two layers of 3 × 3 convolution kernel on the top feature map.

(2) ASFF: To fully exploit the semantic information of deep features and the high-resolution information of shallow features, the structure of the feature pyramid network is often used for feature fusion. However, the representational ability of the feature pyramid is constrained because of the inconsistency between multi-scale features. Inconsistency is reflected in detecting multi-scale aircraft in the same SAR image. The high-resolution information in the shallow layer is beneficial to the detection of small-size aircraft, while large-size aircraft can be detected with semantic information in the deep layer. When large-size aircraft are recognized as true positives, small-size aircraft are easily mistaken as false negatives, resulting in leak detection. Considering that that aircraft in SAR images consist of several discrete points and problems do occur with the detection of multi-scale aircraft, ASFF is introduced to filter conflicting information, suppressing the inconsistency and improving the scale-invariance of features [52]. The essence of ASFF is to adaptively learn the spatial weight of fusion for feature maps at each scale. Firstly, for the features of a given layer, features from other layers are adjusted to the same scale for fusion. Secondly, the best spatial weight for fusion is acquired by subsequent training. Finally, the features of all levels are adaptively aggregated at each level. In other words, some features carrying contradictory information may be filtered out, while other features with cataloged clues are retained.

ASFF-5 is shown in the ADN stage of Figure 1, as an example, the necessary measures of
the feature at the level $l(l = 3, 4, 5)$ is resized to $l(l = 5)$. Secondly, the feature in each layer $(X3, X4, X5)$ is multiplied by the corresponding weight (α, β, γ) and sum the results. The equation is as follows:

$$y_{ij}^l = \alpha_{ij}^l \cdot x_{ij}^{3 \to l} + \beta_{ij}^l \cdot x_{ij}^{4 \to l} + \gamma_{ij}^l \cdot x_{ij}^{5 \to l} \tag{11}$$

where $x_{ij}^{n\to l}$ represents the feature vector of the position (i, j) on the feature maps adjusted from level n to level l. y_{ij}^l denotes the output features, and α_{ij}^l, β_{ij}^l, γ_{ij}^l are respectively the weight parameters of different layers from $n(n = 3, 4, 5)$ to l.

$$\alpha_{ij}^l = \frac{e^{\lambda_{\alpha ij}^l}}{e^{\lambda_{\alpha ij}^l} + e^{\lambda_{\beta ij}^l} + e^{\lambda_{\gamma ij}^l}} \tag{12}$$

Here $\lambda_{\alpha ij}^l$, $\lambda_{\beta ij}^l$, $\lambda_{\gamma ij}^l$ are taken as the control parameters. The 1×1 convolution is utilized to compute the weight scalar maps $(\lambda_{\alpha ij}^l, \lambda_{\beta ij}^l, \lambda_{\gamma ij}^l)$.

3. Experiments and Analyses

In short, experiments on the PFF-ADN we have proposed are conducted for illustration and comparison with GF3 datasets. To confirm the effectiveness of each module, the detection results of ablation experiments are presented and the contributions of different modules in this method are discussed.

3.1. Data Set Description and Parameter Setting

The built datasets contain 69 GF3 SAR images with sufficient desired variations. The GF3 images show C-band HH polarization with 1×1 m resolution, and contain 1903 SAR aircraft. These magnitude images with expert annotation cover a variety of complex scenes in which it is difficult to distinguish targets from the background noise, such as scenes where small aircraft are densely arranged, and scenes where both large and small aircraft exist together. In addition, aircraft adjacent to buildings are common in this dataset. Therefore, GF3 is established to verify the performance of the proposed algorithm in a complex background. We randomly chose 49 images as the training set and the remaining 20 images as the test set. The large-scale images in GF3 are about $20{,}000 \times 20{,}000$ pixels, which need to be cut into slices of 1000×1000 pixels with an overlap of 200 pixels. The reason for the 200-pixel overlap is to avoid cutting off the aircraft. Detailed information on the experimental dataset is given in Table 1. Besides this, the size distribution of aircraft in this dataset also varies considerably, as shown in Figure 7.

Table 1. Detailed information on the experimental dataset.

Dataset	Training	Test	Total
GF-3	495	165	660

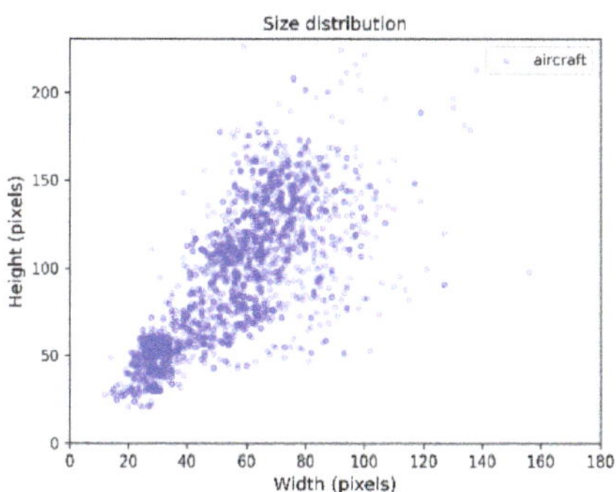

Figure 7. Size distribution of aircraft in the GF3 dataset.

Parameters in the PFF stage: In the PFF step, for the image containing varied brightness information, the mean values of R and B channels are set to three and six empirically when the image data are processed by KTN.

Parameters in the ADN stage: the Resnet50 with DCM is taken as the backbone of the ADN, which is pre-trained on the COCO. The input size for ADN is 1000×1000 pixels and the basic anchor sizes for P3, P4, and P5 are set to 32, 64, and 128, respectively. The stochastic gradient descent (SGD) is applied to train the model with a batch size of eight images. The momentum and weight decay of the optimizer are given fixed values: 0.9, 0.0001, and the training steps are set to 60,000. The initial learning rate is set to 0.001 and decays to 0.0001 after 60,000 steps for the convergence of loss.

In addition, the experiment is performed on an Nvidia Titan 2080 Ti, and the configuration of the operating system environment includes Ubuntu16.04, CUDA10.1, Pytorch1.17, and Tensorflow 2.1.0.

3.2. Evaluation Metrics

To evaluate the performance of the detection task, a variety of evaluation metrics are employed in the experiment, namely, Precision (P), Recall (R), False Alarm Rate (FA), F_1-score, Average Precision (AP), and Running Time. These metrics are computed from four well-established values measured in the experiments: True Positive (TP), False Positive (FP), False Negative (FN), and running time for each scene. TP represents a correctly detected target.

$$P = \frac{TP}{TP + FP} \tag{13}$$

$$R = \frac{TP}{TP + FN} \tag{14}$$

$$FA = \frac{FP}{TP + FP} \tag{15}$$

$$F_1 = \frac{2 \times P \times R}{P + R} \tag{16}$$

$$AP = \int_0^1 P(R)dR \tag{17}$$

TP and FP represent the correctly detected objects and false alarms, respectively. FN denotes a missing target. The intersection over union (IoU) indicates the extent of overlap between the predicted bounding box and the true bounding box in the image. The value of IoU is in the range of 0 to 1. When the two bounding boxes coincide completely, the value of IoU is 1. In this experiment, if the IoU is greater than 0.5, the detected bounding box can be considered as TP.

In addition, the F_1-score is utilized to evaluate the comprehensive performance of the proposed approach, which is defined in Formula (16). Since different precision and recall values arise at different confidence thresholds, the Precision–Recall (PR) curve is introduced to balance the two metrics, where the recall rate is the horizontal coordinate and the precision rate is the vertical coordinate. The value of AP is the area under the PR curve, which can be used for evaluating the overall effectiveness of the algorithm. The method for calculating AP is shown in the Formula (17).

3.3. Ablation Experiments

3.3.1. Effect of PFF

The effective pre-processing algorithm is beneficial to improving the performance of detection in SAR images, especially for the higher recall rate. The effectiveness of the PFF is given in Table 2. As can be seen from the table, the recall rate decreases obviously, but the precision reduces only a little bit. Fortunately, when the PFF is added, the test AP and F1 increase by 5.64% and 3.39%. The two lines in Figure 8 show the implementation results of adding PFF on the number of false alarms and detected targets through a before-and-after comparison. In the first (from left to right) column, two false alarms are suppressed. In the second column, not only are two small false alarm targets suppressed, but a small target in a densely arranged SAR scene is also detected. However, there are still two large-size false alarms that are mistaken for aircraft, even with the added PFF. The reason why the false alarms are not eliminated may be that the features of the background are enhanced along with the target. Another reason may be that the characterization ability of the detection network is insufficient. In the third column, the aircraft at the bottom of the image is also detected. In the fourth column, although the results show three more false alarms of small size, three more aircraft can be assumed to be detected. Since the bounding boxes of the detected aircraft are too large, the IoU between the detected bounding box and the true bounding box is less than 0.5, so these are considered false alarms. It can be inferred that the features from PFF are rich but redundant, leading to serious interference. In a word, the scale-invariance features of aircraft in SAR images are enhanced by adding the PFF, improving the detection performance in SAR images to a certain extent. As shown in Figure 9, the PR curve indicates the superiority of the proposed model.

Table 2. Effectiveness of different components.

Algorithm	*AP* (%)	*P* (%)	*R* (%)	F_1 (%)
Baseline (*)	77.08	88.08	77.81	82.63
* + PFF	82.72	86.34	85.71	86.02

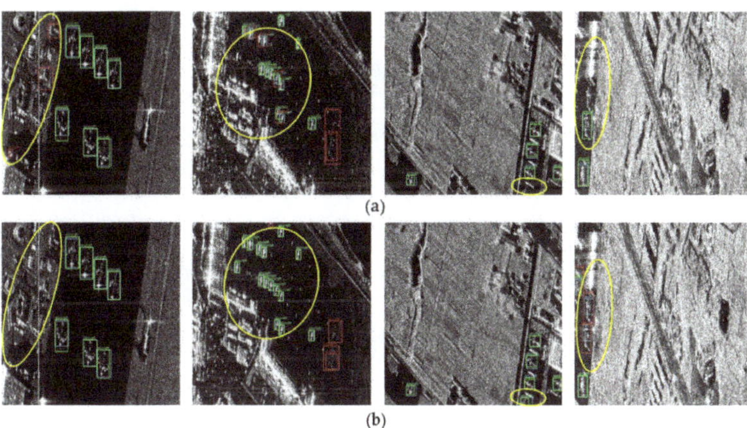

Figure 8. Detection results of two algorithms in four SAR scenes. The cyan rectangles refer to the ground truth. The green and red rectangles denote the detected targets and false alarms, respectively. For a more intuitive comparison of the differences between the two algorithms, a yellow marker is drawn on the detection results. (**a**) Baseline (*) (**b**) * + PFF.

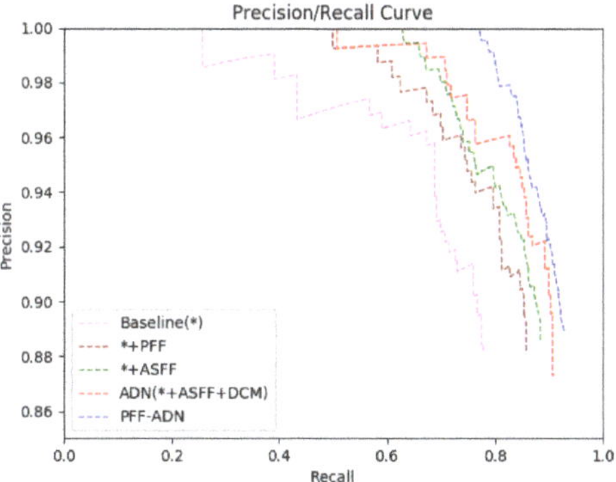

Figure 9. Effectiveness of the proposed PFF-ADN, which includes the PFF, the ASFF, and the ADN (ASFF + PFF). (*) denotes the baseline algorithm.

3.3.2. Effect of ADN

Experiments are designed to verify the effectiveness of ADN, which contains two aspects: the ASFF and the DCM. The quantitative evaluation results of ASFF are displayed in Table 3. The precision rate and recall rate of the baseline algorithm are all increased up to 88.60% and 88.34%, respectively, by adding ASFF. The first two lines in Figure 10 show the results of adding ASFF on the number of false alarms and detected targets through a before-and-after comparison. In the first (from left to right) column, two false alarms are suppressed. In the second column, one more aircraft is detected in dense arrangements of SAR scenes. In the third column, two more aircraft are detected by adding the ASFF. However, one more false alarm appears, because the IoU is less than 0.5. The reason may be that the detection network cannot accurately extract the features of small targets. So, the DCM will be utilized to improve this condition in the next step.

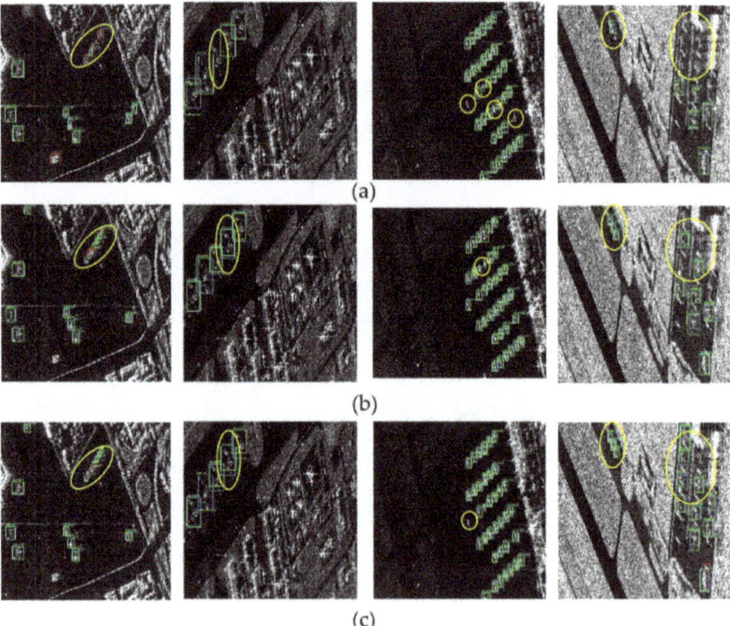

Figure 10. Detection results of three algorithms in four SAR scenes. The cyan rectangles refer to the ground truth. The green and red rectangles denote the detected targets and false alarms, respectively. For a more intuitive comparison of the differences between the two algorithms, a yellow marker is drawn on the detection results. (**a**) Baseline (*), (**b**) * + ASFF, (**c**) ADN: * + ASFF + DCM.

Table 3. Effectiveness of different components.

Algorithm	AP (%)	P (%)	R (%)	F_1 (%)
Baseline (*)	77.08	88.08	77.81	82.63
* + ASFF	84.65	88.60	88.34	88.46
* + ASFF + DCM	86.15	87.36	90.97	89.12

In the fourth column, one more small-size aircraft and two large-size aircraft are detected. It might be inferred that the features of ASFF are rich, leading to more aircraft being detected. In a word, the detection algorithm gives full play to the characterization capabilities of the feature pyramid by adding ASFF, further improving the detection performance of multi-scale aircraft, especially for the small-size aircraft in SAR images.

Furthermore, another experiment is performed to reveal the effectiveness of DCM. As shown in Table 3, the algorithm's performance is greatly enhanced by adding DCM.

Compared to the methods without DCM, namely, * + ASFF, the method (* + ASFF + DCM) acquires a competitive advantage in terms of quantitative indicators. The recall rate is increased up to 88.34% from 90.97%. As displayed in the last two lines of Figure 10, more aircraft against a complicated background are detected in the fourth column of SAR scenes. This confirms that the deformable convolution introduced is beneficial to the fine localization of SAR aircraft. However, in the fourth column, there is still a situation in which noise is recognized as a false alarm. Encouragingly, when the DCM is added, the F_1 value increased to 88.46%, up from 89.12%. The PR curve representing the AP value is shown in Figure 9. It can be concluded from the PR curve that a great improvement in the AP value is achieved by adding the ASFF and the DCM models, respectively, demonstrating the effectiveness of each proposed model.

3.3.3. Performance of PFF-ADN

To validate the superiority of the proposed PFF-ADN further, the results are compared with those of other detectors. Table 4 lists the quantitative assessment results of the algorithms. All the metrics of the proposed approach have been improved greatly. Meanwhile, to better illustrate the performance of each approach, four representative images are selected to visually show the detection results of aircraft in different conditions. This is exemplified in Figure 11 A high false alarm rate and low recall rate appear in the results of lightweight algorithms, such as Yolov4-tiny and Mobile-yolov4. Besides this, in the case of small-size aircraft, the detection results of Yolov4-tiny, RFB, and FastRcnn-resnet50 are not very good. The reason why small-size aircraft are not detected is that these algorithms only utilize the high-semantic information of deeper layers to predict objects. RetinaNet adopts a feature pyramid structure that fuses the semantic information of deep features and the high-resolution information of shallow features, and then makes predictions in different feature layers, significantly improving the recall rate of small-size aircraft. However, as can be seen in the results of other detectors, there are still some aircraft missing. As such, in this paper, the PFF is proposed based on SAR image pre-processing. The PFF is employed to enhance the scattering features of aircraft in SAR images and thus catch the non-obvious features of aircraft. The ADN structure that incorporates the ASFF and the DCM is proposed as the network structure to improve accuracy. The ASFF is introduced to solve inconsistency between multi-scale features, boosting the representational capacity of the feature pyramid structure. The DCM is exploited to cope with the geometric variations or attitude sensitivity of aircraft in SAR images. The last three rows of Figure 11 show that the detection results of PFF-ADN are better than those of RetinaNet in four typical SAR scenes. In the first column, all the small-size aircraft are detected, while small-size aircraft appearing in dense arrangements are not detected in the RetinaNet. In the third column, the small-size aircraft in a dense arrangement are all detected by PFF-ADN, while two aircraft are missed by RetinaNet. In the fourth column, all aircraft are detected by PFF-ADN under complex conditions wherein both large and small aircraft exist.

Table 4. Effectiveness of different components.

Algorithm	AP (%)	P (%)	R (%)	F_1 (%)	FPS (Slice)
RFB	42.24	94.21	42.85	58.91	85
Yolov4-tiny	58.28	81.46	61.27	69.93	341
Mobile-yolov4	68.58	79.47	72.92	76.05	325
FasterRcnn	49.87	65.76	53.75	58.84	13
RetinaNet	78.36	80.04	81.95	80.98	268
ADN	86.15	87.36	90.97	89.12	11
PFF-ADN	89.34	89.44	92.85	91.11	11

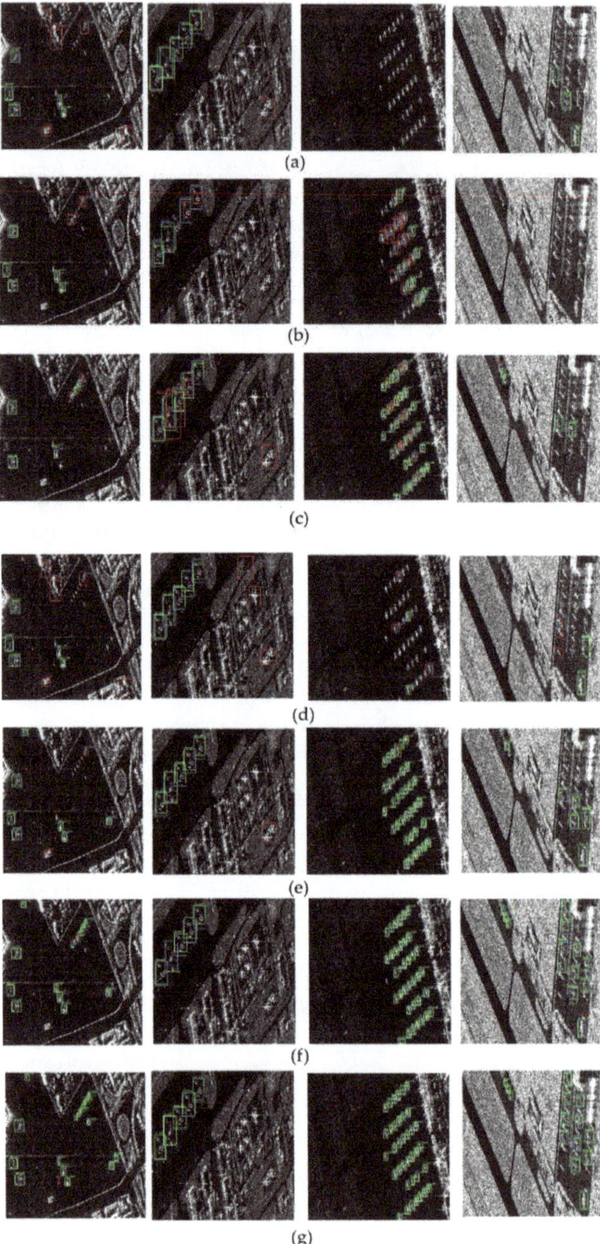

Figure 11. Detection results of each algorithm in four typical SAR scenes. Specifically, the scene in the first (from left to right) column contains mainly medium-sized and small-sized aircraft in a dense arrangement. The scene in the second column represents large-size aircraft against a complex background. The scene in the third column includes lots of small-size aircraft in dense arrangement situations. The scene in the fourth column includes both large and small aircraft and there is a lot of clutter noise in this environment. (**a**) RFB, (**b**) YoloV4-tiny, (**c**) MobileNet-yolov4, (**d**) FasterRcnn, (**e**) RetinaNet, (**f**) ADN, (**g**) PFF-ADN.

To sum up, the validity of the two modules in PFF-ADN has been demonstrated by ablation experiments. All the results are presented in Tables 2–4. The algorithm we propose can not only achieve a higher recall rate, but also suppress false alarms in multiple complex conditions.

The Precision–Recall curve, which measures the detection performance of the network structure, is shown in Figure 12. The larger the area covered by the curve, the better the performance of the network structure. Therefore, in comparison with other detectors, the approach we propose achieves better detection performance in SAR images.

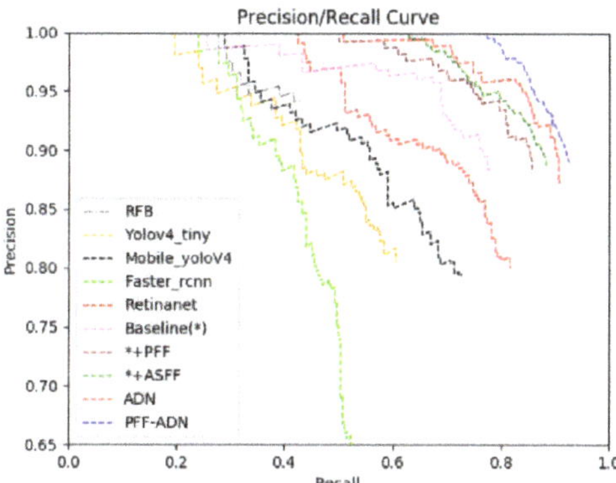

Figure 12. Precision-Recall curves of mentioned methods. (*) denotes the baseline algorithm.

4. Discussion

The experimental results in Section 3.3.1 show the designed PFF is beneficial to enhance the detection performance in SAR images. To better assess the impacts, the quantitative results and visualizations are presented, demonstrating the effectiveness of the PFF. In a word, the pre-processing for SAR images is of great importance. Therefore, image pre-processing methods should be considered in practice. In the future, we can try to introduce some classical SAR image processing methods to explore their effects on SAR object interpretation based on DL.

The experimental results in Section 3.3.2 verified the effectiveness of the DCM and ASFF by ablation experiment. The precision rate and recall rate of the baseline algorithm are all increased up to 88.60% and 88.34%, respectively, by adding ASFF. It confirms that the characterization capability of the detection network is improved greatly by reconciling the inconsistency across different feature scales. The recall rate is increased up to 88.34% from 90.97% by adding the DCM. It demonstrates the DCM is conducive to fine localization by enhancing the geometric modeling abilities of aircraft in various shapes. However, ASFF and DCM requires more computation, which affects the detection efficiency. Therefore, in the future, we can explore how to achieve the balance between precision and speed in specific conditions.

To further verify the superiority of PFF-ADN, the contrast experiment with other classical detectors is conducted. The detailed results are shown in Section 3.3.3. The quantitative results and visualizations all demonstrate that the proposed method can not only achieve a higher recall rate, but also suppress false alarms in multiple complex conditions. In the future, we consider applying the proposed method to remote sensing large-scale images, paving the way for practical application.

5. Conclusions

A novel approach called PFF-ADN for aircraft detection under different arrangements of SAR images is proposed in this paper. This method improves detection performance by combining the classical feature extraction methods with DL approaches. Firstly, an effective image preprocessing approach (PFF) is outlined, in which the features of the target in the image are enhanced by the extraction of peak features and multi-channel data fusion. It is beneficial for handling incomplete scattering and ambiguous characteristics in SAR images. Secondly, in the ADN phase, a new topology of the network composed of ASFF and DCM is proposed to tackle the multiple scales and attitude sensitivity. The ASFF optimizes the characterization capabilities of the network by reconciling the inconsistency across different feature scales, which facilitates the detection of multi-scale aircraft, especially small-size aircraft. The DCM refines the geometric variation capacity of the detection network, which contributes to the handling of the attitude sensitivity and the various shapes of aircraft in SAR images. The satisfactory results of the ablation experiments on the GF3 datasets demonstrate the effectiveness of each proposed module. Meanwhile, detailed comparisons with various methods have also verified the superiority of the proposed algorithm in various circumstances.

In the future, a tight combination between network design and intrinsic properties of aircraft in SAR images need to be considered. In addition, in order to promote the application as soon as possible, it is also very important to carry out SAR object detection under the SAR large-scale scenarios.

Author Contributions: Conceptualization, X.X.; methodology, X.X. and H.J.; software, X.X. and H.J.; validation, X.X.; investigation, X.X.; resources, H.J.; data curation, P.X.; writing—original draft preparation, X.X.; writing—review and editing, H.J. and H.W.; visualization, X.X.; supervision, H.W.; funding acquisition, H.W. All authors have read and agreed to the published version of the manuscript.

Funding: This work was supported in part by the National Natural Science Foundation of China (Grant No. 62271153) and the Natural Science Foundation of Shanghai (Grant No. 22ZR1406700).

Acknowledgments: The authors would like to thanks to the anonymous reviewers.

Conflicts of Interest: The authors declare no conflict of interest.

Nomenclature

SAR	Synthetic Aperture Radar
CFAR	Constant False Alarm Rate
CA	Cell-Averaging
DL	Deep Learning
SCR	Signal to Clutter Ratio
SOCA	Smallest of Cell-Averaging
GOCA	Greatest of Cell-Averaging
OS	Ordered Statistic
VI	Variability Index
PFE	Peak Feature Extraction
PFF	Peak Feature Fusion
ADN	Adaptive Deformable Network
ASFF	Adaptive Spatial Feature Fusion
DCM	Deformable Convolution Module
GLRT	Generalized Likelihood Ratio Test
KTN	K-Times to Truncate and Normalize
FPN	Feature Pyramid Network
GF3	GaoFen-3
IoU	Intersection Over Union
PR	Power Ring

References

1. Zhu, X.X.; Tuia, D.; Mou, L.C.; Xia, G.S.; Zhang, L.P.; Xu, F.; Fraundorfer, F. Deep Learning in Remote Sensing: A Comprehensive Review and List of Resources. *IEEE Geosci. Remote Sens. Mag.* **2017**, *5*, 8–36. [CrossRef]
2. Xu, G.; Zhang, B.; Yu, H.W.; Chen, J.L.; Xing, M.D.; Hong, W. Sparse Synthetic Aperture Radar Imaging from Compressed Sensing and Machine Learning: Theories, Applications and Trends. *IEEE Geosci. Remote Sens. Mag.* **2022**, *12*, 1–26. [CrossRef]
3. Steenson, B.O. Detection performance of a mean-level threshold. *IEEE Trans. Aerosp. Electron. Syst.* **1968**, *AES-4*, 529–534. [CrossRef]

4.   Finn, H.M. Adaptive detection mode with threshold control as a function of spatially sampled clutter-level estimates. *RCA Rev.* **1968**, *29*, 414–465.
5.   Hansen, V.G. Constant false alarm rate processing in search radars. In Proceedings of the IEEE Conference Publication No. 105, "Radar-Present and Future", London, UK, 23–25 October 1973; pp. 325–332.
6.   Trunk, G.V. Range resolution of targets using automatic detectors. *IEEE Trans. Aerosp. Electron. Syst.* **1978**, *AES-4*, 750–755. [CrossRef]
7.   Kuttikkad, S.; Chellappa, R. Non-Gaussian CFAR techniques for target detection in high resolution SAR images. In Proceedings of the 1st International Conference on Image Processing, Austin, TX, USA, 13–16 November 1994; Volume 1, pp. 910–914.
8.   Smith, M.E.; Varshney, P.K. VI-CFAR: A novel CFAR algorithm based on data variability. In Proceedings of the 1997 IEEE National Radar Conference, Syracuse, NY, USA, 13–15 May 1997; pp. 263–268.
9.   Conte, E.; Lops, M.; Ricci, G. Radar detection in K-distributed clutter. *IEE Proc. Radar Sonar Navig.* **1994**, *141*, 116–118. [CrossRef]
10.   Lombardo, P.; Sciotti, M.; Kaplan, L.M. SAR prescreening using both target and shadow information. In Proceedings of the 2001 IEEE Radar Conference (Cat. No. 01CH37200), Atlanta, GA, USA, 3 May 2001; pp. 147–152.
11.   Yu, Y.; Wang, B.; Zhang, L. Hebbian-based neural networks for bottom-up visual attention and its applications to ship detection in SAR images. *Neurocomputing* **2011**, *74*, 2008–2017. [CrossRef]
12.   El-Darymli, K.; Moloney, C.; Gill, E.; McGuire, P.; Power, D.; Deepakumara, J. Nonlinearity and the effect of detection on single-channel synthetic aperture radar imagery. In Proceedings of the OCEANS 2014-TAIPEI, Taipei, Taiwan, 7–10 April 2014; pp. 1–7.
13.   Gu, D.; Xu, X. Multi-feature extraction of ships from SAR images. In Proceedings of the 2013 6th International Congress on Image and Signal Processing (CISP), Hangzhou, China, 16–18 December 2013; Volume 1, pp. 454–458.
14.   Kaplan, L.M.; Murenzi, R.; Namuduri, K.R. Extended fractal feature for first-stage SAR target detection. In Proceedings of the Algorithms for Synthetic Aperture Radar Imagery VI, Orlando, FL, USA, 5–9 April 1999; SPIE: Bellingham, WA, USA, 1999; Volume 3721, pp. 35–46.
15.   Kaplan, L.M. Improved SAR target detection via extended fractal features. *IEEE Trans. Aerosp. Electron. Syst.* **2001**, *37*, 436–451. [CrossRef]
16.   Ren, S.; He, K.; Girshick, R.; Sun, J. Faster R-CNN: Towards Real-Time Object Detection with Region Proposal Networks. *IEEE Trans. Pattern Anal. Mach. Intell.* **2017**, *39*, 1137–1149. [CrossRef] [PubMed]
17.   Redmon, J.; Divvala, S.; Girshick, R.; Farhadi, A. You only look once: Unified, real-time object detection. In Proceedings of the IEEE Conference on Computer Vision and Pattern Recognition, Las Vegas, NV, USA, 27–30 June 2016; pp. 779–788.
18.   Lin, T.Y.; Dollár, P.; Girshick, R.; He, K.; Hariharan, B.; Belongie, S. Feature pyramid networks for object detection. In Proceedings of the IEEE Conference on Computer Vision and Pattern Recognition, Honolulu, HI, USA, 21–26 July 2017; pp. 2117–2125.
19.   Law, H.; Deng, J. Cornernet: Detecting objects as paired keypoints. In Proceedings of the European Conference on Computer Vision (ECCV), Munich, Germany, 8–14 September 2018; pp. 734–750.
20.   Zhou, X.Y.; Zhuo, J.C.; Krhenbühl, P. Bottom-up Object Detection by Grouping Extreme and Center Points. In Proceedings of the IEEE/CVF Conference on Computer Vision and Pattern Recognition, Long Beach, CA, USA, 15–20 June 2019; pp. 850–859.
21.   Tian, Z.; Shen, C.; Chen, H.; He, T. FCOS: Fully Convolutional One-Stage Object Detection. In Proceedings of the 2019 IEEE/CVF International Conference on Computer Vision (ICCV), Seoul, Republic of Korea, 27–28 October 2019; pp. 9627–9636.
22.   Zhou, X.Y.; Wang, D.Q.; Krähenbühl, P. Objects as points. *arXiv* **2019**, arXiv:1904.07850.
23.   He, K.; Gkioxari, G.; Dollár, P.; Girshick, R. Mask R-CNN. In Proceedings of the 2017 IEEE International Conference on Computer Vision (ICCV), Venice, Italy, 22–29 October 2017; pp. 2980–2988. [CrossRef]
24.   Liu, W.; Anguelov, D.; Erhan, D.; Szegedy, C. Ssd: Single shot multibox detector. In Proceedings of the European Conference on Computer Vision, Amsterdam, The Netherlands, 11–14 October 2016; pp. 21–37.
25.   Guo, Q.; Wang, H.; Xu, F. Scattering Enhanced Attention Pyramid Network for Aircraft Detection in SAR Images. *IEEE Trans. Geosci. Remote Sens.* **2020**, *59*, 7570–7587. [CrossRef]
26.   Fu, K.; Fu, J.; Wang, Z.; Sun, X. Scattering-keypoint-guided network for oriented ship detection in high-resolution and large-scale SAR images. *IEEE J. Sel. Top. Appl. Earth Obs. Remote Sens.* **2021**, *14*, 11162–11178. [CrossRef]
27.   Harris, C.; Stephens, M. A combined corner and edge detector. In Proceedings of the Alvey Vision Conference, Manchester, UK, 31 August–2 September 1988; Volume 15, pp. 10–5244.
28.   Wang, S.Y.; Gao, X.; Sun, H.; Zheng, X.W.; Sun, X. An aircraft detection method based on convolutional neural networks in high-resolution SAR images. *J. Radars* **2017**, *6*, 195–203. [CrossRef]
29.   Zhang, L.; Li, C.; Zhao, L.; Xiong, B.; Quan, S.; Kuang, G. A cascaded three-look network for aircraft detection in SAR images. *Remote Sens. Lett.* **2020**, *11*, 57–65. [CrossRef]
30.   Zhao, Y.; Zhao, L.; Li, C.; Kuang, G. Pyramid Attention Dilated Network for Aircraft Detection in SAR Images. *IEEE Geosci. Remote Sens. Lett.* **2020**, *18*, 662–666. [CrossRef]
31.   He, C.; Tu, M.; Xiong, D.; Tu, F.; Liao, M. Adaptive Component Selection-Based Discriminative Model for Object Detection in High-Resolution SAR Imagery. *Int. J. Geo-Inf.* **2018**, *7*, 72. [CrossRef]
32.   Dou, F.; Diao, W.; Sun, X.; Zhang, Y.; Fu, K. Aircraft reconstruction in high-resolution SAR images using deep shape prior. *ISPRS Int. J. Geo-Inf.* **2017**, *6*, 330. [CrossRef]

33. Imbert, J.; Dashyan, G.; Goupilleau, A.; Ceillier, T.; Corbineau, M.C. Improving performance of aircraft detection in satellite imagery while limiting the labelling effort: Hybrid active learning. In Proceedings of the 2021 IEEE International Geoscience and Remote Sensing Symposium IGARSS, Brussels, Belgium, 11–16 July 2021; pp. 220–223.
34. Han, P.; Lu, B.; Zhou, B.; Han, B. Aircraft Target Detection in Polsar Image based on Region Segmentation and Multi-Feature Decision. In Proceedings of the IGARSS 2020—2020 IEEE International Geoscience and Remote Sensing Symposium, Waikoloa, HI, USA, 26 September–2 October 2020; pp. 2201–2204.
35. Jia, H.; Guo, Q.; Chen, J.; Wang, F.; Wang, H.; Xu, F. Adaptive Component Discrimination Network for Airplane Detection in Remote Sensing Images. *IEEE J. Sel. Top. Appl. Earth Obs. Remote Sens.* **2021**, *14*, 7699–7713. [CrossRef]
36. Kang, Y.; Wang, Z.; Fu, J.; Sun, X.; Fu, K. SFR-Net: Scattering Feature Relation Network for Aircraft Detection in Complex SAR Images. *IEEE Trans. Geosci. Remote Sens.* **2022**, *60*, 5218317. [CrossRef]
37. Zhao, Y.; Zhao, L.; Liu, Z.; Hu, D.; Kuang, G.; Liu, L. Attentional Feature Refinement and Alignment Network for Aircraft Detection in SAR Imagery. *arXiv* **2022**, arXiv:2201.07124. [CrossRef]
38. Bi, H.; Yao, J.; Wei, Z.; Hong, D.; Chanussot, J. PolSAR image classification based on robust low-rank feature extraction and Markov random field. *IEEE Geosci. Remote Sens. Lett.* **2020**, *19*, 1–5. [CrossRef]
39. Margt, G.; Mallorqui, J.J.; Rius, J.M.; Sanz-Marcos, J. On the usage of GRECOSAR, an orbital polarimetric SAR simulator of complex targets, to vessel classification studies. *IEEE Trans. Geosci. Remote Sens.* **2006**, *44*, 3517–3526. [CrossRef]
40. Margarit, G.; Mallorqui, J. Assessment of polarimetric SAR interferometry for improving ship classification based on simulated data. *Sensors* **2008**, *8*, 7715–7735. [CrossRef] [PubMed]
41. Margarit, G.; Tabasco, A. Ship classification in single-pol SAR images based on fuzzy logic. *IEEE Trans. Geosci. Remote Sens.* **2011**, *49*, 3129–3138. [CrossRef]
42. Xu, G.; Zhang, B.J.; Chen, J.L.; Hong, W. Structured Low-rank and Sparse Method for ISAR Imaging with 2D Compressive Sampling. *IEEE Trans. Geosci. Remote Sens.* 2022; *early access.*
43. Gao, J.; Gao, X.; Sun, X. Geometrical Features-based Method for Aircraft Target Interpretation in High-resolution SAR Images. *Foreign Electron. Meas. Technol.* **2015**, *34*, 21–28.
44. Chen, J.; Zhang, B.; Wang, C. Backscattering feature analysis and recognition of civilian aircraft in TerraSAR-X images. *IEEE Geosci. Remote Sens. Lett.* **2014**, *12*, 796–800. [CrossRef]
45. Zhu, J.W.; Qiu, X.L.; Pan, Z.X.; Zhang, Y.T.; Lei, B. An improved shape contexts based ship classification in SAR images. *Remote Sens.* **2017**, *9*, 145. [CrossRef]
46. Iii, G.J.; Bhanu, B. Recognizing articulated objects in SAR images. *Pattern Recognit.* **2001**, *34*, 469–485. [CrossRef]
47. Guo, Q.; Wang, H.; Xu, F. Research progress on aircraft detection and recognition in SAR imager. *J. Radars* **2020**, *9*, 497–513.
48. Lin, T.Y.; Maire, M.; Belongie, S.; Hays, J.; Perona, P.; Ramanan, D.; Dollár, P.; Zitnick, C.L. *Microsoft COCO: Common Objects in Context*; Springer International Publishing: Zurich, Switzerland, 2014.
49. Dai, J.; Qi, H.; Xiong, Y.; Li, Y.; Zhang, G.; Hu, H.; Wei, Y. Deformable convolutional networks. In Proceedings of the IEEE International Conference on Computer Vision, Venice, Italy, 22–29 October 2017; pp. 764–773.
50. He, K.; Zhang, X.; Ren, S.; Sun, J. Deep residual learning for image recognition. In Proceedings of the IEEE Conference on Computer Vision and Pattern Recognition, Las Vegas, NV, USA, 27–30 June 2016; pp. 770–778.
51. Zhu, X.; Hu, H.; Lin, S.; Dai, J. Deformable ConvNets V2: More Deformable, Better Results. In Proceedings of the 2019 IEEE/CVF Conference on Computer Vision and Pattern Recognition (CVPR), Long Beach, CA, USA, 16–20 June 2019; pp. 1–13.
52. Liu, S.; Huang, D.; Wang, Y. Learning Spatial Fusion for Single-Shot Object Detection. *arXiv* **2019**, arXiv:1911.09516.

Article

A Novel Topography Retrieval Algorithm Based on Single-Pass Polarimetric SAR Data and Terrain Dependent Error Analysis

Congrui Yang [1,2], Fengjun Zhao [1], Chunle Wang [1], Mengmeng Wang [1,2], Xiuqing Liu [1,*] and Robert Wang [1,2]

[1] Department of Space Microwave Remote Sensing System, Aerospace Information Research Institute, Chinese Academy of Sciences, Beijing 100190, China; yangcongrui18@mails.ucas.ac.cn (C.Y.); fjzhao@mail.ie.ac.cn (F.Z.); clwang@mail.ie.ac.cn (C.W.); wangmengmeng20@mails.ucas.ac.cn (M.W.); yuwang@mail.ie.ac.cn (R.W.)

[2] School of Electronic, Electrical and Communication Engineering, University of Chinese Academy of Sciences, Beijing 100049, China

* Correspondence: lucia@mail.ie.ac.cn

Abstract: Polarimetric synthetic aperture radar (PolSAR) data provide an alternative way for topography retrieval, especially when limited PolSAR data are available. This article proposes a novel topography retrieval algorithm based on the Lambertian backscatter model that further improves the vertical precision of digital elevation model (DEM) generation and requires only one flight. The key idea of the proposed algorithm is to avoid data fluctuations caused by the ratio of the azimuth slope angle to the polarimetric orientation angle (POA). The previous research has confirmed the feasibility of generating a DEM based on single-pass PolSAR data, but its effect on the quality of reference DEM has not been well-explained. To analyze this effect, a large number of experiments on DEM with different resolutions are conducted. In addition, an in-depth analysis of non-linear and terrain-dependent errors is performed. The L-band PolSAR data of NASA/JPL TOPSAR and ALOS-2 PALSAR-2 and interferometric SAR (InSAR) DEM data are used to verify the proposed algorithm. The experimental results show that PolSAR data can be used as an additional reliable information source for DEM fusion under certain conditions to improve the quality of public DEM.

Keywords: Lambertian model; digital elevation model (DEM); polarimetric orientation angle (POA); polarimetric synthetic aperture radar (PolSAR); terrain dependent error analysis

Citation: Yang, C.; Zhao, F.; Wang, C.; Wang, M.; Liu, X.; Wang, R. A Novel Topography Retrieval Algorithm Based on Single-Pass Polarimetric SAR Data and Terrain Dependent Error Analysis. *Remote Sens.* **2022**, *14*, 3176. https://doi.org/10.3390/rs14133176

Academic Editors: Gang Xu, Lan Du and Haipeng Wang

Received: 21 May 2022
Accepted: 28 June 2022
Published: 1 July 2022

Publisher's Note: MDPI stays neutral with regard to jurisdictional claims in published maps and institutional affiliations.

1. Introduction

Polarimetric synthetic aperture radar (PolSAR) data provides another method of terrain reconstruction, especially when radar interferometry cannot be applied. In general, surface topography retrieval can be achieved using orthogonal two-pass or single-pass PolSAR data [1,2]. The polarimetric orientation angle (POA) is a critical polarimetric parameter in topography retrieval, and several classical POA estimation methods have been proposed [3–7]. Jin [8] proposed to use an adaptive threshold method and an image morphology thinning algorithm for linear texture to calculate the azimuth slope when limited single-pass PolSAR data are available. Chen et al. [9] found that the method in [8] was complex and computationally expensive and proposed another slope estimation approach that combines the topography-POA shift model and the refined Lambertian backscattering model, which is used in radarclinometry [10]. Furthermore, Li et al. [11] proposed a new single-pass PolSAR topography retrieval method based on the polarization-dependent intensity ratio, which proved to be more effective than the method in [9].

Instead of the POA, using the polarimetric signatures for the derivation of sea-ice surface topography was discussed in [12], and an accurate DEM of sea-ice surface topography was generated based on co-polarimetric (coPol) coherence. However, this method requires both PolSAR and interferometric SAR (InSAR) data, and only the sea-ice surface topography was verified.

The algorithm proposed in [11] still has two problems, although it can further improve the accuracy of the generated digital elevation model (DEM). First, when the incidence angle is smaller than the ground range slope angle, the azimuth slope direction cannot be correctly determined. Second, the ground range slope angle is easily affected by data fluctuations, especially when the surface of terrain is almost flat. The first problem can be solved by using a large incidence angle for steep relief variation areas, and the second problem can be solved by revising the slopes by the second-order weighted full multigrid (WFMG) algorithm [11]. However, the incidence angle of a SAR image is fixed, which inevitably limits the application range of the algorithm.

This paper proposes a novel topography retrieval algorithm based on single-pass PolSAR data; the main contributions are as follows:

(1) New expressions of azimuth slope and range slope are derived, which can effectively avoid the data fluctuation caused by the ratio of azimuth slope to POA;
(2) The method proposed in [13] is used to correct the irregular points of the slope. The principle is that when the azimuth and ground range pixel size are the same, the integral of the slope along a closed-loop should be zero. After obtaining the azimuth and ground range slope, the method of solving the Poisson equation is the same as the weighted full multigrid algorithm described in [14];
(3) The terrain dependent error was fully analyzed. The experimental results show that the quality of the generated DEM is related to the terrain, and it is especially suitable for terrain with a small elevation variation range and that is free from the interference of vegetation and buildings;
(4) The application potential of single-pass PolSAR data in DEM generation is deeply analyzed. The experimental results show that PolSAR data can be used as an additional reliable information source for DEM fusion under certain conditions to improve the quality of public DEM.

2. Methods

2.1. Background of POA

In the monostatic backscattering case, the 2×2 complex backscattering matrix can be expressed as [15]:

$$[S] = \begin{bmatrix} S_{hh} & S_{hv} \\ S_{vh} & S_{vv} \end{bmatrix} \tag{1}$$

where, for a reciprocal target matrix, $S_{hv} = S_{vh}$.

Lee et al. [5,15,16] have indicated that the POA of terrain surface is geometrically related to the topographical slopes and radar look angle, and has a simple form of

$$\tan \zeta = \frac{\tan \omega}{-\tan \gamma \cos \phi + \sin \phi}, \quad \frac{-\pi}{2} \le \zeta \le \frac{\pi}{2} \tag{2}$$

where ζ is the POA of the ground patch, ω is the azimuth slope angle, γ is the ground range slope angle, and ϕ is the radar look angle.

The SAR imaging geometry of a tilted surface patch is shown in Figure 1a, and the geometric diagrams of the ground azimuth slope $\tan \omega$ and the ground range slope $\tan \gamma$ are shown in Figure 1b.

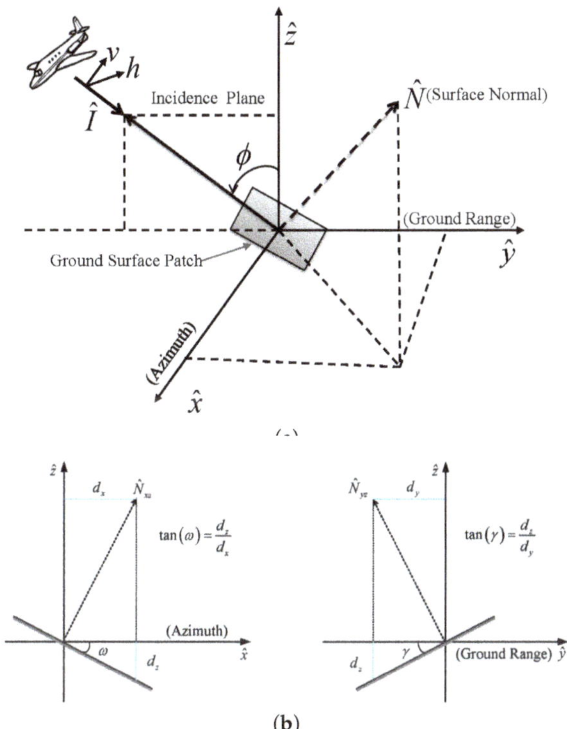

Figure 1. (**a**) SAR imaging geometry of a tilted surface patch. (**b**) The projection of surface normal to the xz and yz planes, and N_{xz} and N_{yz}, which are the corresponding projection components. The blue-marked parts d_x, d_y, and d_z in (**b**) are the changes along the positive direction of the xyz coordinate axis, respectively. The azimuth slope is negative and the ground range slope is positive.

The circular polarization algorithm (CPA) developed by Lee et al. [5,17,18] based on the terrain reflection symmetry hypothesis is one of the most popular algorithms for POA estimation. Based on this hypothesis, for a reflection symmetrical medium, the circular polarization parameter $S_{RR}S_{LL}^{\dagger}$ is real and the circular polarization estimator is expressed as [17]

$$\xi_{cpa} = \begin{cases} \eta, & \eta \le \pi/4 \\ \eta - \pi/2, & \eta > \pi/4 \end{cases} \tag{3}$$

where

$$\eta = \frac{1}{4}\left[\arctan\left(\frac{-4Re\langle (S_{hh} - S_{vv})S_{hv}^{\dagger} \rangle}{-\langle |S_{hh} - S_{vv}|^2 \rangle + 4\langle |S_{hv}|^2 \rangle} \right) + \pi \right] \tag{4}$$

$Re(A)$ denotes the real part of A, and $\arctan(\cdot)$ is computed in the range of $[-\pi, \pi]$. Thus, the POA range is limited to $[-\pi/4, \pi/4]$, but the actual POA range is $[-\pi/2, \pi/2]$, which indicates a phase wrapping problem in steep terrain.

The ambiguous problem of POA estimation on steep terrain can be successfully solved by combining physical scattering characteristics. As for the natural terrain dominated by Bragg scattering, the unambiguous estimation of POA over steep topography can be expressed as [19], which can be called the vertical polarization dominated $E = 0$ deorientation algorithm (VEDA), where E is the Huynen parameter [20].

$$\zeta = \begin{cases} \zeta_{cpa}, & \tilde{C} \leq 0 \\ \zeta_{cpa} + \pi/2, & \tilde{C} \geq 0, -\pi/4 \leq \zeta_{cpa} \leq 0 \\ \zeta_{cpa} - \pi/2, & \tilde{C} \geq 0, 0 \leq \zeta_{cpa} \leq \pi/4 \end{cases} \tag{5}$$

where

$$\tilde{C} = \frac{|\tilde{S}_{hh}|^2 - |\tilde{S}_{vv}|^2}{2} \tag{6}$$

$\tilde{C}$ denotes the deoriented Huynen parameter, $\tilde{S}_{hh}$ and $\tilde{S}_{vv}$ are elements after deorientation of complex backscattering $\mathbf{S}$, and ζ is the unambiguous estimation of POA.

However, it is not clear whether the VEDA is effective for L-band data, because only the real measurement data of the P-band are verified in [19]. To address this issue, an example with NASA/JPL airborne L-band PolSAR data from Camp Roberts is presented, as shown in Figure 2. The pixel size in the azimuth and ground range directions is 10 m. The DEM generated by C-band InSAR of the corresponding area in Figure 2a is shown in Figure 3a, which shows the roughness of the terrain.

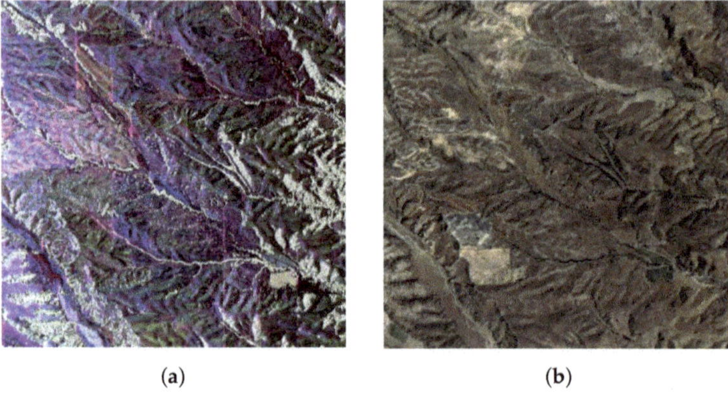

(a)   (b)

Figure 2. (**a**) Pauli image of AIRSAR L-band data from Camp Roberts, 768 rows (azimuth) $\times$ 768 columns (ground range). (**b**) Corresponding area optical image from Google Earth. The direction of flight is from top to bottom.

The POA estimated by the DEM is used as a reference, shown in Figure 3b. The POAs estimated from the L-band PolSAR data using the classic CPA algorithm and the VEDA algorithm are shown in Figure 3c,d, respectively. The L-band PolSAR data have been filtered using a 5×5 boxcar filter to reduce the speckle noise.

The comparison of Figures 2a and 3c show that the POA noise mainly appears in the lower-left corner and right edge of the image, where there are valleys and trees. This is because the volume scattering does not fit the tilted surface model [16]. The image in Figure 3d is smoother than that in Figure 3c, but there is still noise due to the volume scattering. An additional 3×3 boxcar filter has been applied in Figure 3d [19].

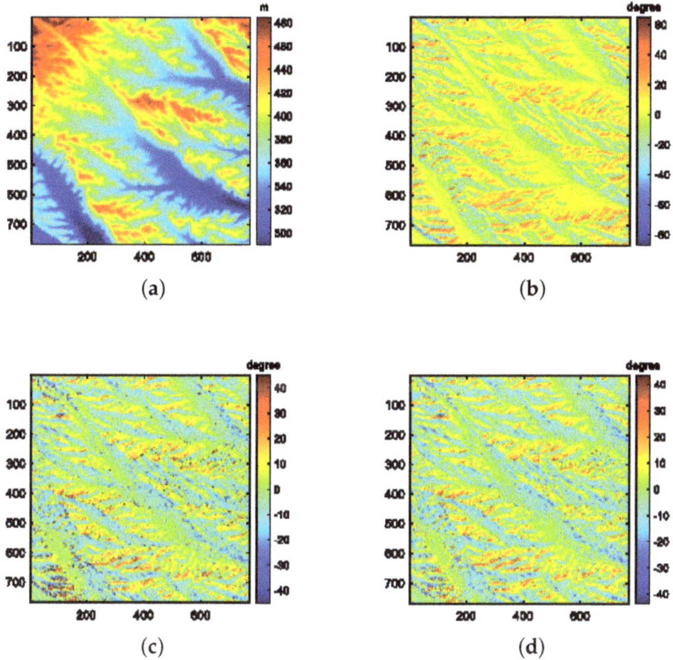

Figure 3. (**a**) C-band AIRSAR InSAR-generated DEM with approximately 10 m resolution. (**b**) POA from the DEM of (**a**). (**c**) POA from L-Band PolSAR data estimated by CPA. (**d**) POA from L-Band PolSAR data estimated by VEDA.

Next, the root mean square difference (RMSD) between the POA from InSAR and POLSAR data is calculated for the whole study area, with Figure 3b as the ground truth value. The results are shown in Table 1, which proves that the VEDA is effective for L-band PolSAR data.

Table 1. RMSD of the POA estimates and ground truth results.

	CPA	VEDA
RMSD (°)	11.24	10.10

2.2. Two-Dimensional Slope Estimation and Correction

To further solve the two problems in [11], that is, the fact that the azimuth slope direction cannot be uniquely determined when the range slope angle is greater than the incident angle and the ground range slope angle can be easily affected by data fluctuations, this paper proposes a novel two-dimensional slope estimation method.

For a terrain that fits the Lambertian backscatter model, the intensity of a SAR image and the slope of a local terrain can be expressed as [10,21,22],

$$I(\omega, \gamma) = K\sigma R_g R_a \frac{\cos\phi \sin^2(\gamma + \phi)}{\cos(\gamma + \phi)} \cos\omega \tag{7}$$

where K is a calibration constant of a SAR system, σ is a backscatter coefficient of the ground surface, and R_g and R_a indicate the ground range and azimuth resolution of a SAR image, respectively.

The estimate of the ground range slope angle in [11,23] is

$$\gamma = \arctan\left(\frac{\sin\phi - \tan\omega/\tan\xi}{\cos\phi}\right). \tag{8}$$

Obviously, the ground range slope angle is easily affected by data fluctuations, especially when the surface is almost flat. The new estimate of ground range slope angle, which is derived in detail in Appendix A, can be expressed as

$$\gamma_c = \arcsin\left(\sqrt{\frac{\sqrt{P^4 + 4P^2} - P^2}{2}}\right) - \phi \tag{9}$$

where P is given by

$$P = \frac{I(\omega,\gamma)\sin^2\phi}{I(0,0)\cos\phi\cos\omega}. \tag{10}$$

and $I(0,0)$ can be estimated by the flat region in the image.

The term $\langle|S_{hh} - S_{vv}|^2\rangle$ in the coherency matrix shows a consistently increasing trend after compensation, so $\cos\omega$ can be obtained by [23]

$$\cos\omega = \frac{\langle|S_{hh} - S_{vv}|^2\rangle}{\langle|\tilde{S}_{hh} - \tilde{S}_{vv}|^2\rangle}. \tag{11}$$

where $\langle|\tilde{S}_{hh} - \tilde{S}_{vv}|^2\rangle$ denotes the results after compensation.

However, ω cannot be uniquely determined only according to (11). Therefore, combining (2) with the estimated γ_c in (9), the corrected azimuth slope angle is expressed as

$$\omega_c = \arccos\left(\frac{\langle|S_{hh} - S_{vv}|^2\rangle}{\langle|\tilde{S}_{hh} - \tilde{S}_{vv}|^2\rangle}\right) \cdot sign(\omega_s) \tag{12}$$

where the ω_s is

$$\omega_s = \arctan(\tan\xi(\sin\phi - \tan\gamma_c\cos\phi)). \tag{13}$$

Finally, ω_c and γ_c denote the azimuth slope angle and the ground range slope angle estimated by the proposed algorithm, respectively. The detailed processing flow is shown in Figure 4. Moreover, the improved algorithm also avoids the problem that the ground range slope in [9] cannot be uniquely determined due to the arccosine equation.

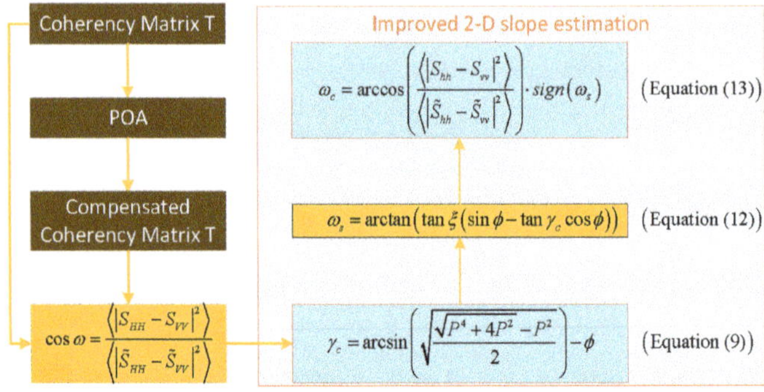

Figure 4. The slope estimation flow chart proposed in this paper.

However, the intensity $I(0,0)$ still needs to be estimated. This is because the ill-posed slopes equation cannot be solved directly in the case of single-pass PolSAR data. Nevertheless, the following experimental data analysis shows that the cost is worthwhile.

In addition, since in areas with speckle noise and shadows, radar information cannot truly reflect the undulating characteristics of the actual terrain, this paper uses the method proposed in [13] to eliminate the irregular points of a slope. The principle is that if the grid spacing of the azimuth direction and the ground range direction is the same, the integral value of the slope along the closed loop should be zero.

2.3. Elevation Estimation and POA Correction

After estimating the azimuth and ground range slope, the surface elevation can also be estimated by solving the Poisson equation [1,2].

The Poisson equation can be expressed as,

$$\nabla^2 \phi(x,y) = \delta(x,y) \tag{14}$$

where ∇^2 is the Laplacian operator $\partial^2/\partial x^2 + \partial^2/\partial y^2$, $\phi(x,y)$ is the height value at (x,y), and the source function $\delta(x,y)$ on the right side is the surface curvature value at (x,y).

Therefore, the discretized Poisson equation is as follows

$$\begin{aligned} \delta(x,y) = &\ (\phi(x+1,y) - 2\phi(x,y) + \phi(x-1,y)) \\ &+ (\phi(x,y+1) - 2\phi(x,y) + \phi(x,y-1)) \end{aligned} \tag{15}$$

where $\delta(x,y)$ is given by

$$\begin{aligned} \delta(x,y) = &\ \Delta_g(x+1,y) - \Delta_g(x,y) \\ &+ \Delta_a(x,y+1) - \Delta_a(x,y) \end{aligned} \tag{16}$$

and

$$\begin{aligned} \Delta_g(x,y) &= R_g \tan(\omega(x,y)), \\ \Delta_a(x,y) &= R_a \tan(\gamma(x,y)). \end{aligned} \tag{17}$$

The WFMG algorithm [14] is used to solve the Poisson equation, and the weight is obtained by the method proposed in [13]. The key idea of the WFMG algorithm is to convert low-frequency components of error into high-frequency components and then to remove them by the Gauss–Seidel relaxation method [24].

To obtain absolute rather than relative elevation values, at least one elevation tie-point must be known for each integrated slope profile [3]. The tie-points can be obtained by down-sampling the public DEM, such as in the shuttle radar topography mission (SRTM) DEM [11]. Then, the DEM can be obtained according to the terrain retrieval processing map, as shown in Figure 5.

Moreover, according to (2), the DEM retrieved from full PolSAR data can also be used to calculate the POA, which provides an effective method to improve the estimation precision of POA without using high-resolution DEM. This will be explained in Section 3.4.

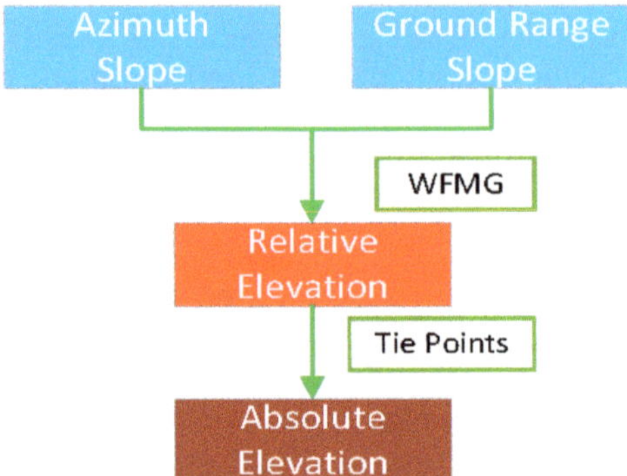

Figure 5. Single-pass PolSAR topography retrieval processing diagram after the azimuth slope and the ground range slope are estimated.

3. Results

3.1. Algorithm Performance Analysis

First, the method proposed in [11] was used to generate the tie-point matrix and the C-band InSAR DEM presented in Figure 3a was downsampled to the resolution of 640 m, which was the same as the resolution in [11], as shown in Figure 6. Meanwhile, for better performance analysis, the DEM in Figure 3a is repeated in Figure 6a and a 5×5 boxcar filter is used to filter the L-band PolSAR data.

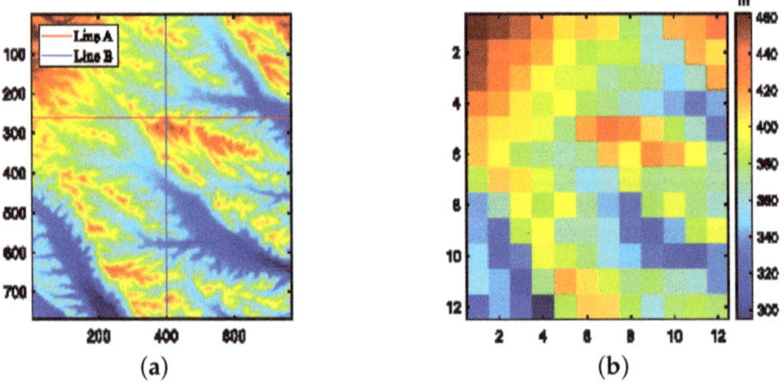

Figure 6. Point matrix image. (**a**) C-band InSAR DEM. (**b**) Downsampled C-band InSAR DEM with approximately 640 m resolution as initial height.

Then, the DEM inversion was performed using the proposed method; the DEM map derived from L-band PolSAR data is shown in Figure 7. Compared with Figure 6a, it can be found that the main features of the topography can be reconstructed through single-pass PolSAR data.

Further, the differences between the PolSAR-derived DEM and the C-band AIRSAR InSAR-derived DEM were deeply analyzed. The scatter plot of the DEM maps obtained from the overall InSAR, as the ground truth value, and PolSAR data is shown in Figure 8a. The cumulative distribution function of the elevation difference is shown in Figure 8b and we can see that 90.35% of the points have an elevation difference of less than 18.9 m. Two profiles were extracted from C-band InSAR DEM and L-band PolSAR DEM along lines A and B in Figure 6a; the comparison results are presented in Figure 8c,d. The overall trend of elevation change is the same, but it can be found that the error is greater in steep areas, which can be obtained by combining the profile at the intersection of lines A and B.

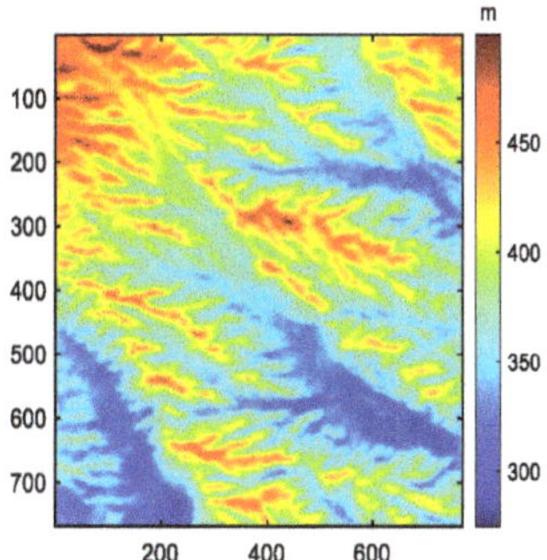

Figure 7. DEM derived from L-band PolSAR data using the proposed method.

Moreover, for quantitative comparison with previous research results, the data involved are resampled to 36 m × 36 m resolution. The RMSD of slope and height between C-band AIRSAR InSAR DEM, as the ground truth value, and L-band PolSAR DEM are shown in Table 2 and Table 3, respectively. The comparison with the results of Chen's method [9] and Li's method [11] shows that the performance of the proposed method is effective. However, the terrain-dependent errors shown by Figure 8c,d require further analysis.

Table 2. RMSD of slope estimates and ground truth results (degrees).

	Azimuth Slope	**Ground Range Slope**
Chen's method	9.87	9.77
Li's method	5.03	6.26
Proposed method	3.50	6.04

Table 3. RMSD of the height estimates and ground truth results.

	Chen's Method	**Li's Method**	**Proposed Method**
RMSD (m)	28.66	14.75	10.87

3.2. Non-Linear and Terrain Dependent Error Analysis

To better assess the impacts of terrain on the retrieval of azimuthal and ground slopes, the NASA/JPL airborne L-band PolSAR data from Camp Roberts was analyzed in detail, including catchments in multiple directions, as shown in Figure 9. In particular, the data of Figures 2 and 9 are referred to as PD1 and PD2, respectively.

Likewise, a 5 × 5 boxcar filter is used to filter the L-band PolSAR data and the data involved are resampled to 36 m × 36 m resolution. However, the case where the resolution of the tie-points is 90m is additionally analyzed, as shown in Figure 10. The inversion results corresponding to the two resolutions are shown in Figure 11 and the height and slope RMSD between the PolSAR-derived DEM and C-band InSAR DEM are given in Tables 4 and 5, respectively. The effectiveness of the proposed method is further verified by comparing Tables 4 and 5 with Tables 2 and 3.

However, the accuracy of PD2 is lower than that of PD1, although the processing methods and data sources are the same. Comparing Figure 2 and Figure 9, the terrain of PD2 includes buildings, highways, vegetation, etc., which is more complex than PD1. Therefore, it can be considered that the elevation accuracy of topography retrieval is related to the terrain. In addition, comparing Figure 11a and Figure 11b, it is also related to the resolution of the tie-points. This will be analyzed in detail in the next section.

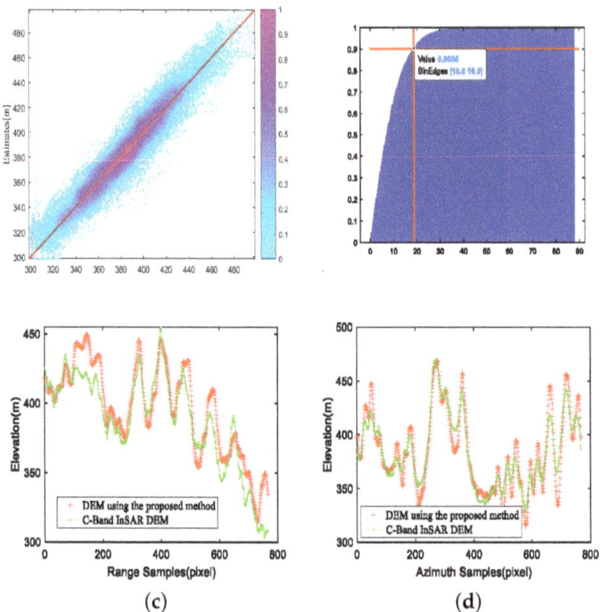

Figure 8. (**a**) Scatter plot between C-band AIRSAR InSAR-derived DEM (truths) and POLSAR derived DEM (estimates). (**b**) The cumulative distribution function of elevation difference between Figures 6a and 7. (**c**) and (**d**) are the comparison of DEM values measured along line A and line B in Figure 6a.

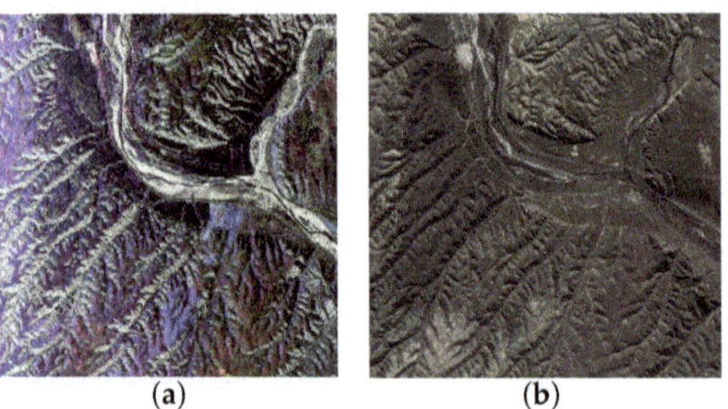

Figure 9. (**a**) RGB Pauli color-coded image with AIRSAR L-band data from Camp Roberts, 1024 rows (azimuth) × 1024 columns (ground range). (**b**) Corresponding area optical image from Google Earth. The direction of flight is from top to bottom.

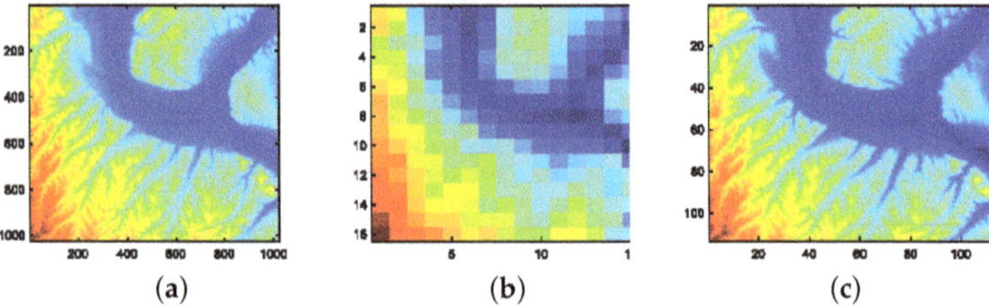

Figure 10. (**a**) C-band AIRSAR InSAR generated DEM. Downsampled 640 m (**b**) and 90 m (**c**) resolution images.

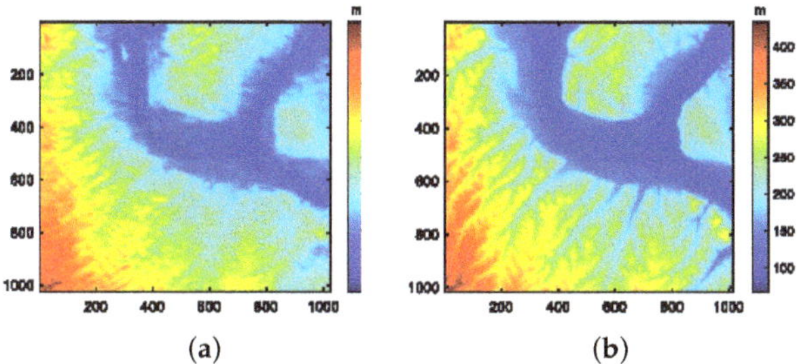

Figure 11. DEM generated from L-band POLSAR data based on tie-points with resolutions of 640 m (**a**) and 90 m (**b**), respectively.

Table 4. RMSD of the height estimates and ground truth results.

Tie Points Res.	640 (m)	90 (m)
RMSD (m)	13.00	2.30

Table 5. RMSD of slope estimates and ground truth results (degrees).

Tie Points Res.	Azimuth Slope	Ground Range Slope
640 (m)	4.82	5.75
90 (m)	1.63	2.00

For further analysis, three regions of interest (ROI) are selected, as shown in Figure 12a, and the tie-points are shown in Figure 12b–d with a resolution of 90 m. The red (ROI1)- and green (ROI2)-marked areas correspond to relatively flat terrain in different directions and the blue (ROI3)-marked areas correspond to relatively steep terrain.

The topography retrieval results of the red-, green-, and blue-marked areas in Figure 12a are shown in Figure 13. Comparing the difference images of the three ROIs in the fourth column, it can be found that the error of steep terrain is greater. A careful analysis of the first two rows shows that the high error areas in the difference image are mainly distributed at the boundary positions of different terrains, which is due to the non-homogeneous scattering of the ground cover at the boundary. This indicates that the proposed method is more suitable for terrain dominated by a single scatterer.

Moreover, the elevation profiles marked by line A–line D in the three ROI regions were extracted, as shown in Figure 14. Combining line A and line D in Figure 13a and line A, line B, and line D in Figure 13e, it can be seen that the poor estimation results in a, j, b, e, and k in Figure 14 are due to the fact that these profiles all span multiple terrains.

The height and slope RMSD between the PolSAR-derived DEM and C-band AIRSAR InSAR DEM are given in Tables 6 and 7, respectively. It can be seen that ROI3, the blue-marked area, has the largest estimation error. Combining Figures 13 and 14, the estimation accuracy of DEM generated by PolSAR data is related not only to the terrain, but also to the homogeny of ground covering scattering. Therefore, these factors should be considered in practice.

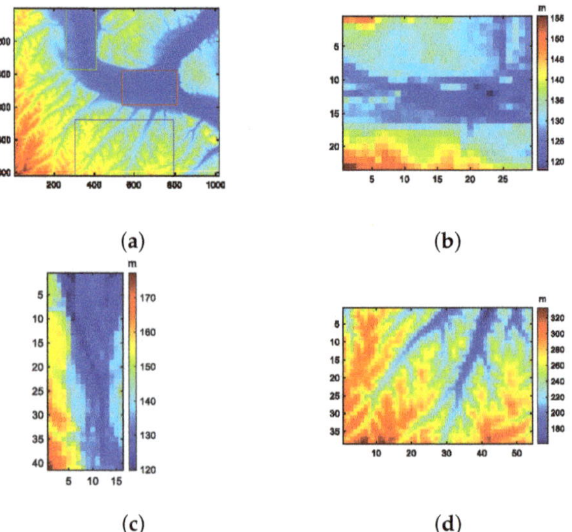

(a)

(b)

(c)

(d)

Figure 12. Three regions of interest (ROI) are selected, as shown in (**a**); the tie-points are shown in (**b–d**) with a resolution of 90 m. (**b–d**) correspond to the red-, green-, and blue-marked areas in (**a**), respectively.

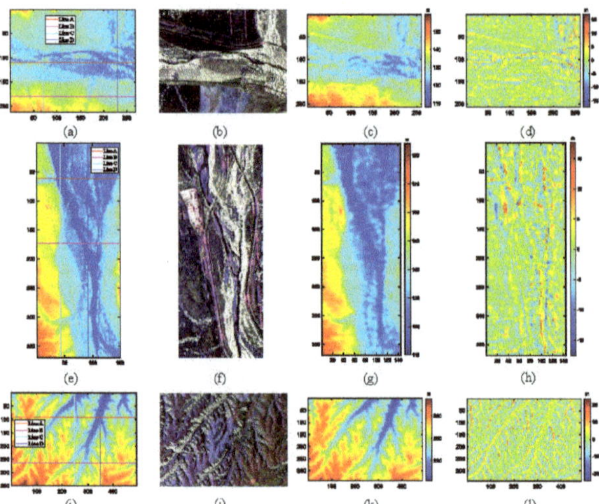

Figure 13. The first and second columns are the corresponding C-band InSAR DEM and Pauli images, respectively, the third column is the DEM generated from PolSAR data and the fourth column is the difference image between the first and third columns. The first to third rows correspond to the red-, green-, and blue-marked areas in Figure 12a, respectively. Further, combining the results of the first to third rows, it can be seen that the error of flat terrain is smaller, and the error is larger at the boundaries of different terrains.

Table 6. RMSD of the height estimates and ground truth results.

Data	ROI1	ROI2	ROI3
RMSD (m)	1.04	1.64	2.40

Table 7. RMSD of slope estimates and ground truth results (degrees).

Data	Azimuth Slope	Ground Range Slope
ROI1	1.09	0.89
ROI2	0.97	1.60
ROI3	1.56	2.10

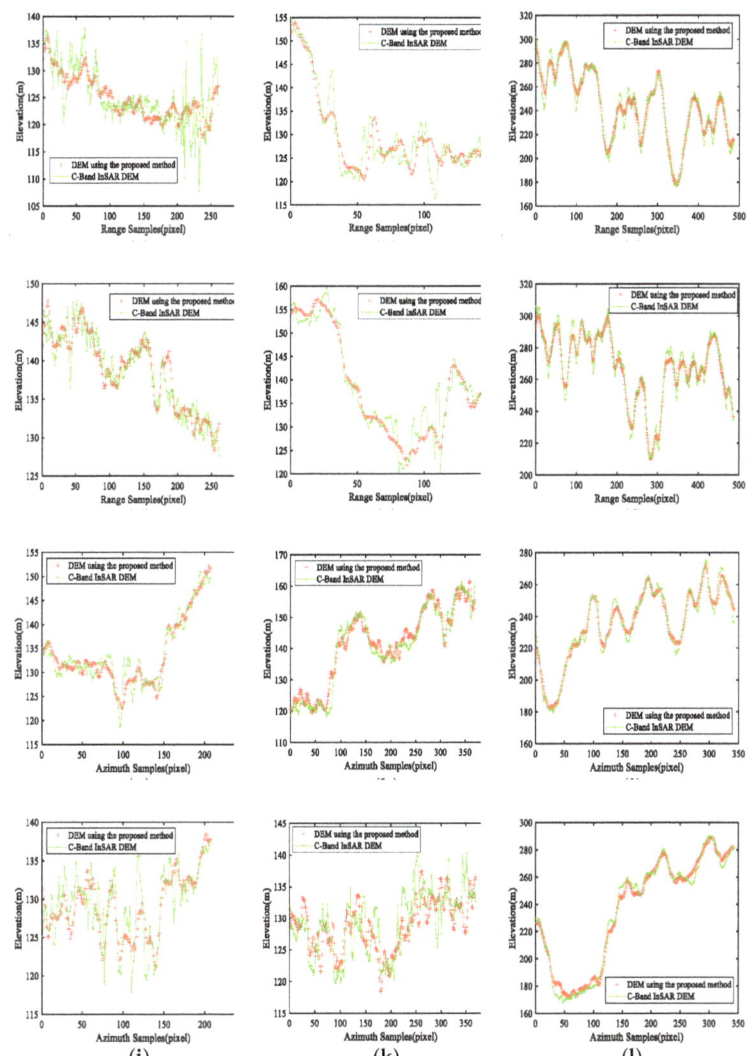

Figure 14. The first to third columns correspond to Figure 13a, e, and i, respectively, and the first to fourth rows correspond to the elevation profiles marked by lines A and D, respectively. It can be seen that the error depends on the terrain, and the error is larger where the elevation changes sharply. The more complex the terrain, the larger the error dynamic range.

3.3. Analysis of Low-Resolution DEM Matching High-Resolution PolSAR Data

To evaluate the effect of low-resolution DEM matching the high-resolution polarization SAR data, the proposed method was used to analyze the situation when tie-points had the resolutions of 90 m and 30 m, as shown in Figure 15. The inversion results corresponding to the two resolutions are shown in Figure 16, where it can be seen that the higher the resolution of the tie-point matrix, the higher the similarity between the PolSAR-derived DEM and InSAR-derived DEM.

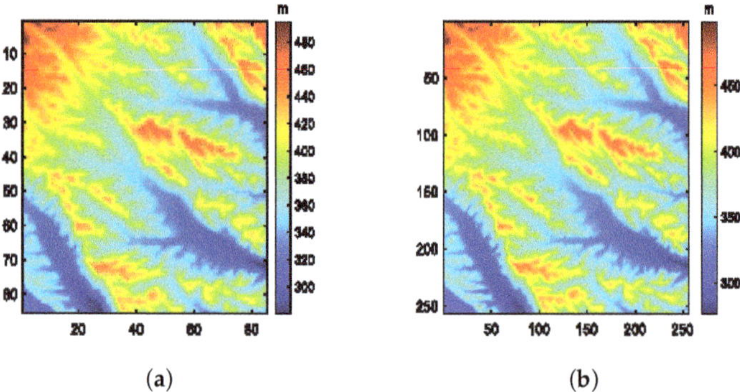

(a)

(b)

Figure 15. Point matrix image. (**a**) Downsampled C-band InSAR DEM to 90 m. (**b**) 30 m resolution as the initial height.

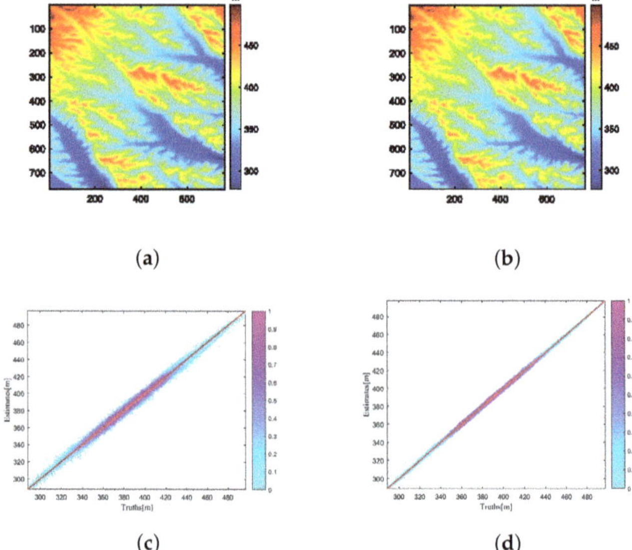

(a)

(b)

(c)

(d)

Figure 16. The first row is the DEM image derived from L-band PolSAR data and the second row is the scatter diagram between C-band InSAR DEM (truths) and PolSAR-derived DEM (estimates). Furthermore, the first and second columns correspond to the cases where the tie-points have 90 m and 30 m resolution, respectively. It can be seen that the higher the resolution of the tie-points, the closer the elevation estimate is to the ground truth.

A quantitative analysis was also conducted, and the height and slope RMSD between the PolSAR-derived DEM and C-band InSAR DEM are given in Tables 8 and 9, respectively. It should be noted that in order to analyze the high-resolution situation, the data involved are not resampled here, but a 5×5 boxcar filter is still used.

Table 8. RMSD of the height estimates and ground truth results.

Tie Points Res.	90 (m)	30 (m)
RMSD (m)	3.20	1.14

Table 9. RMSD of slope estimates and ground truth results (degrees).

Tie Points Res.	Azimuth Slope	Ground Range Slope
90 (m)	3.10	2.33
30 (m)	0.83	0.75

Experimental results show that PolSAR data can integrate terrain textures into low-resolution DEMs and improve the resolution of DEMs without interpolation.

3.4. POA Correction Effect Analysis

According to (2), the DEM retrieved from PolSAR data was used to calculate the POA, and the results are shown in Figure 17a. Comparison of the results in Figure 17a and those in Figure 3c,d shows that after correction, the POA in Figure 17a is smoother and more similar to the POA obtained from the C-band InSAR DEM.

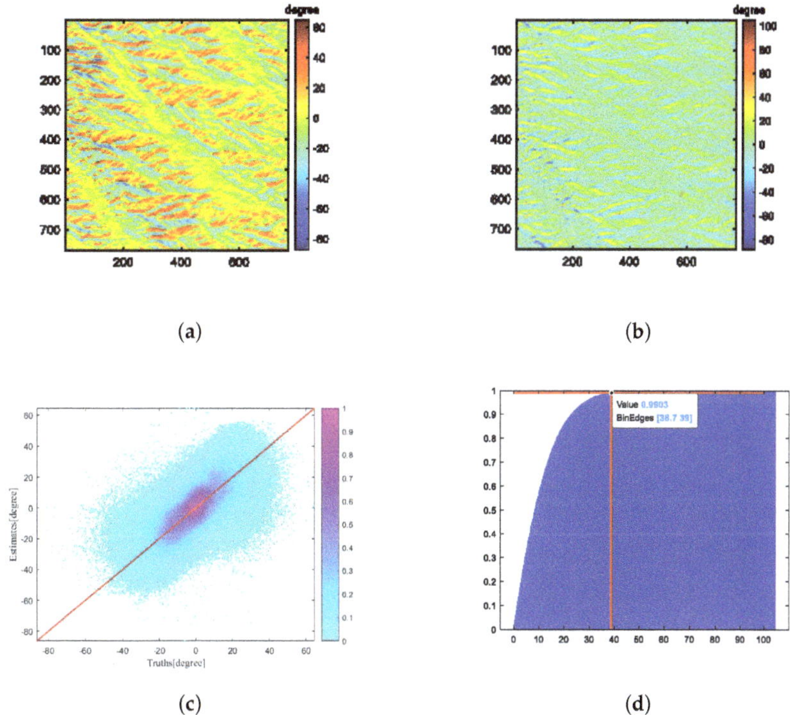

Figure 17. (**a**) POA derived from PolSAR-derived DEM. (**b,c**) are the difference image and scatter plot between (**a**) and Figure 3a, respectively, with the POA derived from INSAR-derived DEM as the ground truth. (**d**) is a cumulative distribution function plot of the absolute value of (**b**).

Combining Figure 17b,d, 99.03% of the absolute values in the difference image are less than 39°, and most are within 20°. Furthermore, it can be clearly seen from Figure 17c that the estimated value and the real value are symmetrically distributed on both sides of the diagonal, and the point density in the middle pink area is the largest, indicating that the corrected POA is highly consistent with the POA derived from DEM.

The experimental results show that the DEM derived from L-band PolSAR data has high application potential for POA correction.

3.5. Application Potential of PolSAR Data in DEM Generation

To further verify the reliability of PolSAR data for DEM generation, a field scientific expedition was conducted in Salt Lake, Qinghai Province, China. The L-band fully PolSAR data obtained by ALOS-2 PALSAR-2 in this area in August 2020 was used for analysis. The RGB Pauli color-coded image is shown in Figure 18a and the aerial view of the same area obtained from Google Earth is shown in Figure 18b. The flight direction was from left to right, with a pixel size of 2.79 m in the azimuth and 2.86 m in the slant range. The DEM data acquired by the Shuttle Radar Topography Mission (SRTM) at a resolution of 1 arc-second (30 m) are displayed in Figure 18c. The reference DEM was transformed into SAR slant range geometry.

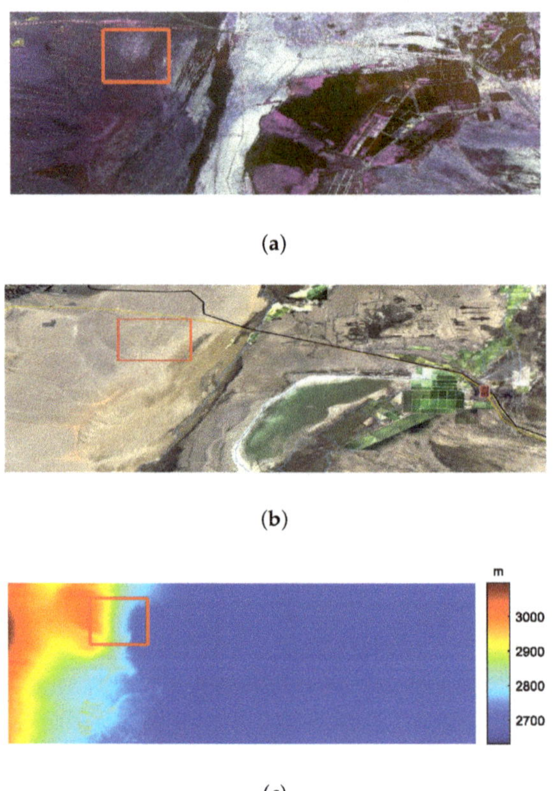

(a)

(b)

(c)

Figure 18. (**a**) RGB Pauli color-coded image with ALOS-2 PALSAR2 L-band data from Qinghai Salt Lake, 25,960 columns (azimuth) × 8624 rows (slant range). (**b**) Corresponding area optical image from Google Earth. (**c**) DEM data acquired by the Shuttle Radar Topography Mission (SRTM) at a resolution of 1 arc-second (30 m) in the corresponding area. The area marked by the red box is the region of interest.

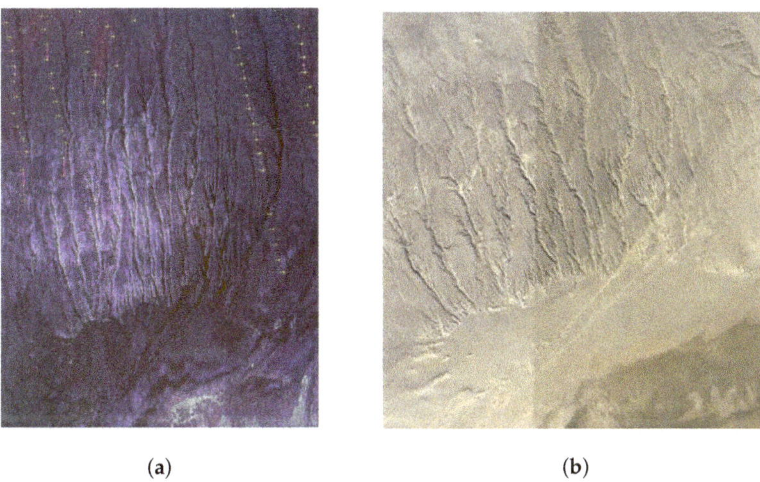

(a)   (b)

Figure 19. The region of interest selected in the experiment. (**a**) corresponds to the boxed area in Figure 18a, 3200 rows (azimuth) × 2486 columns (slant range). (**b**) Corresponding area optical image from Google Earth.

The ROI marked by the red box in Figure 18 is shown in Figure 19. Similarly, the ROI includes a variety of terrain.

The proposed method was used to analyze the situation when tie-points had resolutions of 90 m, and 30 m, as shown in Figure 20. The inversion results corresponding to the two resolutions are shown in Figure 21, and the height and slope RMSDs between the PolSAR-derived DEM and SRTM DEM are given in Table 10 and Table 11, respectively.

Table 10. RMSD of the height estimates and ground truth results.

Tie Points Res.	90 (m)	30 (m)
RMSD (m)	2.23	0.92

Table 11. RMSD of slope estimates and ground truth results (degrees).

Tie Points Res.	Azimuth Slope	Ground Range Slope
90 (m)	3.32	5.03
30 (m)	1.54	3.08

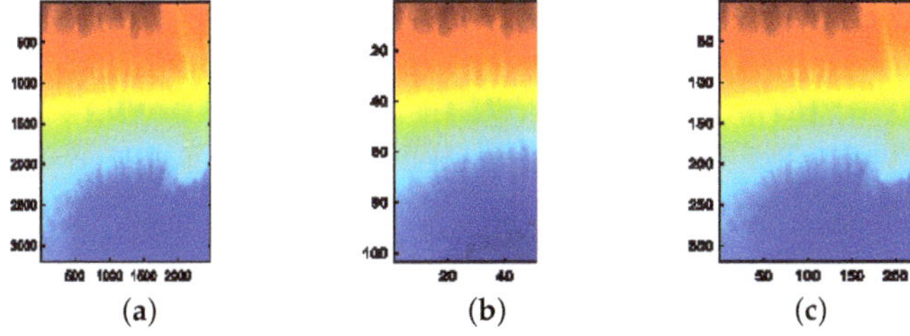

(a)   (b)   (c)

Figure 20. (**a**) DEM from SRTM. Downsampled 90 m (**b**) and 30 m (**c**) resolution images.

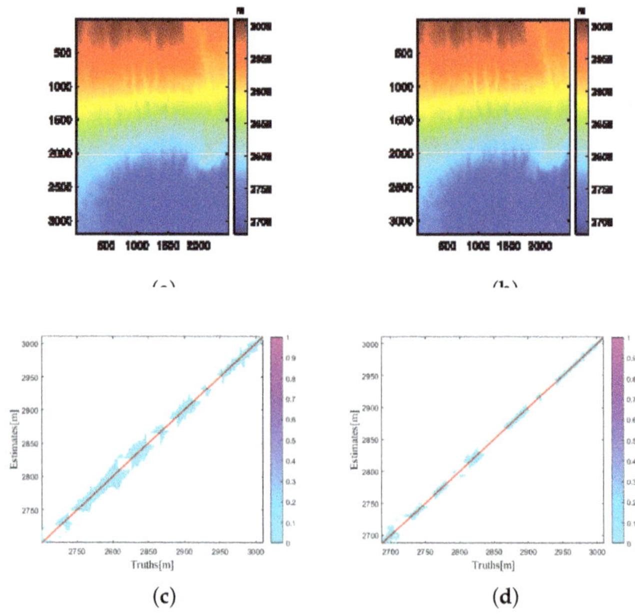

Figure 21. The first row is the DEM image derived from L-band PolSAR data and the second row is the scatter diagram between SRTM DEM (truths) and PolSAR-derived DEM (estimates). The first and second columns correspond to the cases where the tie-points have 90 m and 30 m resolution, respectively. It can be seen that the higher the resolution of the tie-points, the closer the elevation estimate is to the ground truth.

The experimental results of ALOS2-PALSAR2 and AIRSAR data show that the elevation accuracy of DEM generated from POLSAR data depends on terrain, and the elevation accuracy is higher for relatively flat terrain without vegetation and artificial target interference.

4. Discussion

The experimental results in Section 3.1 show that the error of topography retrieval is terrain-dependent. To better assess the impacts of terrain on the retrieval of azimuthal and ground slopes, the PD2 data from Camp Roberts was analyzed in detail. The experimental results in Section 3.2 show that the error of steep terrain is greater. Moreover, the estimation accuracy of DEM generated by PolSAR data is related not only to the terrain, but also to the homogeny of ground covering scattering. Therefore, these factors should be considered in practice.

To evaluate the effect of low-resolution DEM matching the high-resolution polarization SAR data, tie-point matrices with resolutions of 90 m and 30 m, respectively, are analyzed in depth in Section 3.3. It was found that the higher the resolution of the tie-point matrix, the higher the similarity between the PolSAR-derived DEM and InSAR-derived DEM. Further, it can be concluded that PolSAR data can integrate terrain textures into low-resolution DEMs and improve the resolution of DEMs without interpolation.

Moreover, the experimental results in Section 3.4 show that the DEM derived from L-band PolSAR data has high application potential for POA correction. This will contribute to the application of base POA.

To further verify the reliability of PolSAR data for DEM generation, a field scientific expedition was conducted in Salt Lake, Qinghai Province, China. Combining the experimental results of ALOS2-PALSAR2 in Section 3.5 and AIRSAR data, it can be further concluded that the elevation accuracy of DEM generated from POLSAR data depends on terrain and that the elevation accuracy is higher for relatively flat terrain without vegetation and artificial target interference.

In addition, we know that the surface of the moon is bare and dry, and there is no vegetation interference. Therefore, the technology for generating DEM based on PolSAR data may be applied to the topographic mapping of the lunar surface, especially lunar permanently shadowed regions (PSR). Then, similar to the idea of combining the Lunar Orbiter Laser Altimeter (LOLA) and SELENE Terrain Camera (TC) with data adopted in [25], it may be possible to combine LOLA with PolSAR data for topographic mapping of lunar PSR.

5. Conclusions

In this paper, a novel topography retrieval algorithm based on single-pass PolSAR data is proposed. First, a new ground range slope estimation equation is derived and the azimuth slope estimation is also improved. Therefore, data fluctuations caused by the ratio of the azimuth slope angle to the POA are avoided.

Furthermore, the L-band PolSAR data from AIRSAR and ALOS2-PALSAR2 containing a variety of terrain are deeply analyzed. We find that the quality of DEM generated is related to terrain, and the error will be greater at the boundary of different terrains. Therefore, in practice, more attention should be paid to ROIs where surface scattering is homogeneous and the elevation dynamic range is small.

In addition, a detailed comparative analysis of the enhancement effect of PolSAR data on the DEM is conducted. The analysis results demonstrate that PolSAR data can not only be used to generate DEM, but also can further improve DEM resolution under certain elevation accuracy. The application potential of DEM generated from PolSAR data for POA estimation is also analyzed.

In the future, different scattering models need to be optimized according to different terrain characteristics to further improve the precision of topography retrieval.

Author Contributions: Conceptualization, C.Y.; funding acquisition, F.Z., X.L., and R.W.; investigation, C.Y. and M.W.; methodology, C.Y.; project administration, F.Z. and R.W.; resources, F.Z.; supervision, F.Z., C.W., X.L., and R.W.; validation, C.Y. and X.L.; writing—original draft, C.Y.; writing—review and editing, F.Z., C.W., M.W., X.L., and R.W. All authors have read and agreed to the published version of the manuscript.

Funding: This work was funded by the National Key Research and Development Program of China under Grant 2017YFB0502700, the Beijing Municipal Natural Science Foundation (No. 4192065), and the National Natural Science Foundation of China under Grant 61901445.

Data Availability Statement: The AIRSAR data from NASA/JPL are avaiable on https://search.asf. alaska.edu/#/?dataset=AIRSAR&resultsLoaded=true&granule=ts554-CTIF&zoom=7.272¢er=-1 18.993,33.636&end=1998-05-31T15:59:59Z&start=1998-04-30T16:00:00Z.

Acknowledgments: The authors would like to thank the National Aeronautics and Space Administration's Jet Propulsion Laboratory (NASA/JPL) and the Japan Aerospace Exploration Agency (JAXA) for providing the outstanding AIRSAR and ALOS2-PALSAR2 polarimetric SAR data. The authors would also like to thank all colleagues who participated in this experiment.

Conflicts of Interest: The authors declare no conflict of interest.

Appendix A

According to (7), we have

$$\frac{I(\omega,\gamma)\sin^2\phi}{I(0,0)\cos(\gamma)\cos(\phi)} = \frac{\sin^2(\gamma+\phi)}{\cos(\gamma+\phi)}. \tag{A1}$$

Next, we set $P = \frac{I(\omega,\gamma)\sin^2\phi}{I(0,0)\cos(\gamma)\cos(\phi)}$, $\theta = \gamma + \phi$, and obtain

$$\sin^2\theta - P\sqrt{1 - \sin^2\theta} = 0. \tag{A2}$$

Therefore, according to the root formula of the univariate quadratic equation, we can obtain

$$\sin^2\theta = \frac{-P^2 + \sqrt{P^4 + 4P^2}}{2}. \tag{A3}$$

Finally, the new estimate of the ground range slope angle can be expressed as

$$\gamma_c = \arcsin\left(\sqrt{\frac{-P^2 + \sqrt{P^4 + 4P^2}}{2}}\right) - \phi. \tag{A4}$$

References

1. Schuler, D.L.; Ainsworth, T.; Lee, J.; De Grandi, G. Topographic mapping using polarimetric SAR data. *Int. J. Remote Sens.* **1998**, *19*, 141–160. [CrossRef]
2. Schuler, D.L.; Lee, J.S.; Ainsworth, T.L.; Grunes, M.R. Terrain topography measurement using multipass polarimetric synthetic aperture radar data. *Radio Sci.* **2000**, *35*, 813–832. [CrossRef]
3. Schuler, D.L.; Lee, J.S.; De Grandi, G. Measurement of topography using polarimetric SAR images. *IEEE Trans. Geosci. Remote Sens.* **1996**, *34*, 1266–1277. [CrossRef]
4. Lee, J.; Krogager, E.; Schuler, D.; Ainsworth, T.; Boerner, W. On the estimation of polarization orientation angles induced from azimuth slopes using polarimetric SAR data. In Proceedings of the IEEE 2000 International Geoscience and Remote Sensing Symposium. Taking the Pulse of the Planet: The Role of Remote Sensing in Managing the Environment, Honolulu, HI, USA, 24–28 July 2000; Volume 3, pp. 1310–1312.
5. Lee, J.S.; Schuler, D.L.; Ainsworth, T.L. Polarimetric SAR data compensation for terrain azimuth slope variation. *IEEE Trans. Geosci. Remote Sens.* **2000**, *38*, 2153–2163.
6. Lee, J.S.; Schuler, D.L.; Ainsworth, T.L.; Boerner, W.M. Polarization orientation estimation and applications: A review. In Proceedings of the 2003 IEEE International Geoscience and Remote Sensing Symposium, Toulouse, France, 21–25 July 2003; Volume 1, pp. 428–430.
7. Pottier, E.; Schuler, D.; Lee, J.; Ainsworth, T. Estimation of the terrain surface azimuthal/range slopes using polarimetric decomposition of POLSAR data. In Proceedings of the IEEE 1999 International Geoscience and Remote Sensing Symposium, Hamburg, Germany, 28 June–2 July 1999; Volume 4, pp. 2212–2214.
8. Jin, Y.; Luo, L. Terrain topographic inversion using single-pass polarimetric SAR image data. *Sci. China Ser. F Inf. Sci.* **2004**, *47*, 490. [CrossRef]
9. Chen, X.; Wang, C.; Zhang, H. DEM generation combining SAR polarimetry and shape-from-shading techniques. *IEEE Geosci. Remote Sens. Lett.* **2008**, *6*, 28–32. [CrossRef]
10. Paquerault, S.; Maitre, H. New method for backscatter model estimation and elevation map computation using radarclinometry. In Proceedings of the SAR Image Analysis, Modeling, and Techniques. International Society for Optics and Photonics, Barcelona, Spain, 20 November 1998; Volume 3497, pp. 230–241.
11. Li, Y.; Hong, W.; Pottier, E. Topography retrieval from single-pass POLSAR data based on the polarization-dependent intensity ratio. *IEEE Trans. Geosci. Remote Sens.* **2014**, *53*, 3160–3177. [CrossRef]
12. Huang, L.; Hajnsek, I. PolSAR based Scattering Characterization for Sea-Ice Topographic Retrieval using TanDEM-X Data. EUSAR 2021; In Proceedings of the 13th European Conference on Synthetic Aperture Radar, Online Event, 29 March–1 April 2021; pp. 1–5.
13. Wang, M. Terrain mapping using full polarimetric SAR images of two cross track. *Signal Process.* **2009**, *25*, 48–51.
14. Pritt, M.D. Phase unwrapping by means of multigrid techniques for interferometric SAR. *IEEE Trans. Geosci. Remote Sens.* **1996**, *34*, 728–738. [CrossRef]
15. Lee, J.S.; Pottier, E. *Polarimetric Radar Imaging: From Basics to Applications*; CRC Press: Boca Raton, FL, USA, 2017.
16. Lee, J.S.; Ainsworth, T.L.; Wang, Y. Polarization orientation angle and polarimetric SAR scattering characteristics of steep terrain. *IEEE Trans. Geosci. Remote Sens.* **2018**, *56*, 7272–7281. [CrossRef]
17. Lee, J.S.; Schuler, D.L.; Ainsworth, T.L.; Krogager, E.; Kasilingam, D.; Boerner, W.M. On the estimation of radar polarization orientation shifts induced by terrain slopes. *IEEE Trans. Geosci. Remote Sens.* **2002**, *40*, 30–41.
18. Lee, J.S.; Ainsworth, T.L. The effect of orientation angle compensation on coherency matrix and polarimetric target decompositions. *IEEE Trans. Geosci. Remote Sens.* **2010**, *49*, 53–64. [CrossRef]

19. Liang, L.; Zhang, Y.; Li, D. A novel method for polarization orientation angle estimation over steep terrain and comparison of deorientation algorithms. *IEEE Trans. Geosci. Remote Sens.* **2020**, *59*, 4790–4801. [CrossRef]
20. Huynen, J.R. Phenomenological Theory of Radar Targets. Ph. D. Thesis, Technical University Delft, Delft, The Netherlands, 1970 .
21. Frankot, R.T.; Chellappa, R. Estimation of surface topography from SAR imagery using shape from shading techniques. *Artif. Intell.* **1990**, *43*, 271–310. [CrossRef]
22. Paquerault, S.; Maitre, H.; Nicolas, J.M. Radarclinometry for ERS-1 data mapping. In Proceedings of the 1996 International Geoscience and Remote Sensing Symposium, Lincoln, NE, USA, 27–31 May 1996; Volume 1, pp. 503–505.
23. Li, Y.; Hong, W.; Cao, F.; Wang, Y.p.; Wu, Y.r. Estimation of terrain slope using a Compensation-Lambertian method from single-pass POLSAR data. In Proceedings of the IGARSS 2008–2008 IEEE International Geoscience and Remote Sensing Symposium, Boston, MA, USA, 8–11 July 2008; volume 2, pp. II-1310–II-1313 .
24. Zhang, J. Acceleration of five-point red-black Gauss-Seidel in multigrid for Poisson equation. *Appl. Math. Comput.* **1996**, *80*, 73–93.
25. Barker, M.; Mazarico, E.; Neumann, G.; Zuber, M.; Haruyama, J.; Smith, D. A new lunar digital elevation model from the Lunar Orbiter Laser Altimeter and SELENE Terrain Camera. *Icarus* **2016**, *273*, 346–355. [CrossRef]

Article

An Improved RFI Mitigation Approach for SAR Based on Low-Rank Sparse Decomposition: From the Perspective of Useful Signal Protection

Hengrui Zhang [1,2,3], Lin Min [1,2,3], Jing Lu [4], Jike Chang [1,5,*], Zhengwei Guo [1,2,3] and Ning Li [1,2,3]

[1] College of Computer and Information Engineering, Henan University, Kaifeng 475004, China; zhr_henu@henu.edu.cn (H.Z.); mlin@henu.edu.cn (L.M.); gzw@henu.edu.cn (Z.G.); hedalining@henu.edu.cn (N.L.)
[2] Henan Engineering Research Center of Intelligent Technology and Application, Henan University, Kaifeng 475004, China
[3] Henan Key Laboratory of Big Data Analysis and Processing, Henan University, Kaifeng 475004, China
[4] Land Satellite Remote Sensing Application Center, Ministry of Natural Resources, Beijing 100048, China; luj@lasac.cn
[5] School of Software, Henan University, Kaifeng 475004, China
* Correspondence: changjike@henu.edu.cn

Abstract: As an open system, synthetic aperture radar (SAR) inevitably receives radio frequency interference (RFI) generated by electromagnetic equipment in the same band. The existence of RFI seriously affects SAR signal processing and image interpretation. In recent years, many algorithms and models related to RFI mitigation have been proposed. However, most of that focus on effectively mitigating the RFI is insufficient to protect the useful signals. This article proposes a mitigation method of RFI with a signal-protected capability. (1) The kurtosis coefficient is used to detect RFI pulse-by-pulse, and the echoes containing RFI are stored in matrix form. (2) The preliminary extraction of RFI is complete by low-rank sparse decomposition of the echo matrix containing RFI. (3) For the secondary separation of RFI, the accurate position of RFI in the preliminary extraction results is located by the fuzzy C-means clustering; then, we separate the RFI and the remaining useful signals again and reconstruct the useful signals to complete the mitigation work. The proposed method can further protect useful signals while effectively removing interference through the secondary separation of RFI. Experimental results based on simulated and measured data verify the performance and potential of the proposed method.

Keywords: synthetic aperture radar; radio frequency interference; interference mitigation; low-rank sparse decomposition

Citation: Zhang, H.; Min, L.; Lu, J.; Chang, J.; Guo, Z.; Li, N. An Improved RFI Mitigation Approach for SAR Based on Low-Rank Sparse Decomposition: From the Perspective of Useful Signal Protection. *Remote Sens.* **2022**, *14*, 3278. https://doi.org/10.3390/rs14143278

Academic Editors: Haipeng Wang, Gang Xu and Lan Du

Received: 30 May 2022
Accepted: 5 July 2022
Published: 7 July 2022

Publisher's Note: MDPI stays neutral with regard to jurisdictional claims in published maps and institutional affiliations.

1. Introduction

1.1. Background

Synthetic aperture radar (SAR), as an active microwave sensor, can obtain a high-resolution and a continuous coverage of ground targets all-day/night and all-weather through the combination of wideband signals in range and a synthetic aperture in azimuth. These characteristics make SAR widely used in military reconnaissance, resource exploration, terrain mapping, environmental monitoring, disaster warning and assessment, and other related fields [1–4].

However, SAR is susceptible to various and complex electromagnetic interferences in the operating frequency band as an open broadband system. In general, the signals emitted by other radiation sources in the same frequency band are called radio frequency interference (RFI) to SAR [5–8], which can be divided into narrowband interference (NBI) and wideband interference (WBI) according to the bandwidth of RFI. The existence of RFI will seriously restrict the effect of SAR high-resolution imaging and further affect the subsequent application of SAR data, such as crop monitoring and natural disaster assessment [9]. It has become a significant research trend to uncover how SAR can survive in this complex electromagnetic environment while maintaining its excellent performance as much as possible [10,11].

With the development of global radio communications, the electromagnetic environment is becoming more and more complex, and RFI cases in SAR systems are becoming more and more common. Domestic and foreign scholars have proposed various RFI mitigation algorithms and models in recent years [12–14]. Among them, the low-rank-sparse model separates interference and useful signals based on RFI's low-rank-sparse characteristics in the transform domain and has a high RFI mitigation accuracy. However, due to the variety of RFI types, the model will inevitably cause the loss of some useful signals while removing RFI, resulting in the loss of details in the SAR image after RFI removal. Therefore, researching mitigation methods of RFI that can consider both the mitigation accuracy of RFI and the adequate protection of useful signals has significant practical application requirements.

1.2. Previous Work

Effectively mitigating RFI in SAR data has always been an important research topic in the SAR field. Since the 1990s, various RFI mitigation methods have been proposed [15], which can be summed up as parametric, non-parametric, and semi-parametric methods [16].

1.2.1. Parametric Methods

The parametric methods complete the RFI mitigation work by establishing the mathematical model of RFI and adjusting the model's parameters. In [17], aiming at the NBI existing in airborne SAR data, the maximum likelihood estimation method was used to mitigate NBI in SAR data by establishing a sine wave model. In [18,19], the NBI model was established based on the narrow-band property of the RFI signal, and a series of RELAX and its improved methods were used to estimate the model parameters, which achieved a satisfactory RFI mitigation effect. In [20], an RFI mitigation method based on the iterative adaptive approach (IAA) and orthogonal subspace projection (OSP) was proposed, which can estimate the RFI power spectrum adaptively and iteratively without a parameter search and model order estimation.

The form of WBI is more complex and diverse than NBI, so it is difficult to establish an accurate parametric model. In [21], IAA was successfully applied to WBI mitigation by improving instantaneous frequency (IF) resolution in short-time Fourier transform (STFT) and filtering WBI based on the OSP method. In [22], a WBI mitigation method combining IF estimation and regularized time–frequency filtering was proposed, which completes the extraction and mitigation of sinusoidally frequency-modulated WBI.

Since it is challenging to establish parametric models of RFI in practical applications, and the mitigation performance of such methods relies on the estimation of model parameters, the application of parametric methods is not sufficiently extensive.

1.2.2. Non-Parametric Methods

The non-parametric method is based on RFI features in the time domain or transform domain for mitigation without establishing a parametric model. In [23], the frequency domain notch filtering (FNF) was simple to implement and had a strong robustness. However, the FNF will lose the useful signal while eliminating the RFI. In order to solve the above problems, a two-step notch method based on a linear prediction model to compensate for the missing spectrum was proposed [24]. In addition, a time domain notch method by constructing a notch filter in the time domain and using the IAA to recover missing signals was proposed [25], which improved the accuracy of interference mitigation and reduced the loss of useful signals.

The adaptive filtering method separates useful signals from interference by constructing an adaptive filter. In [26], an adaptive spectrum line enhancer was proposed, which can improve the mitigation of time varying NBI. An adaptive Wiener filter was proposed in [27], which achieved better performance than the LMS adaptive filter.

The matrix decomposition-based methods, such as eigen-subspace projection (ESP) [28], independent component analysis [29], complex empirical mode decomposition [30], and independent subspace analysis [31], decompose the echo data into useful signal components and RFI components, thereby realizing the mitigation of RFI. The matrix decomposition-based methods can effectively mitigate the NBI and WBI, but these methods are ineffective in the case of weak RFI.

Different from the above methods that are mainly applied to SAR level-0 products (raw data), the sub-band spectrum cancellation (SSC) method [32] and the block subspace filtering (BSF) method [33] are non-parametric methods for level-1 products (single look complex data). The SSC method is based on the assumption that the distance spectrum is strictly symmetric. Otherwise, the RFI mitigation performance would drop significantly. BSF is based on the assumption that the RFI-free SAR image conforms to the Gaussian distribution, and the intensity of RFI is higher than the useful signal to ensure mitigation accuracy. In [34], a mitigation method of RFI for SAR images with joint change detection and sub-band spectrum cancellation was proposed, effectively preserving target information while mitigating interference.

1.2.3. Semi-Parametric Methods

With the development of the low-rank sparse decomposition (LRSD) algorithm, robust principal component analysis (RPCA) has been used in SAR signals for various applications, such as clutter suppression and moving target detection by separating moving and stationary targets in SAR images [35–39]. In recent years, the theory of LRSD has been successfully applied to mitigate RFI in SAR data. This method converts complex signal separation problems into hyperparameter optimization problems and forms a semi-parametric mitigation method of RFI. In [40], the concept of semi-parametric interference mitigation was proposed, which completed the mitigation of NBI by solving the sparse reconstruction optimization problem. In [41], the sparse reconstruction algorithm was extended to WBI, providing a new idea for the extended application of these methods. In [42–44], a series of low-rank-sparse decomposition models were proposed, further improving the related theories. In [45] a dictionary-based SAR RFI-suppression method under the framework of RPCA was proposed. In [46], a two-dimensional RFI mitigation method was proposed and applied to simulated and measured data successfully. In [47], a graph Laplacian clustering algorithm was proposed to mitigate RFI. In [48], a mitigation algorithm of RFI that combines low-rank and double-sparse features was proposed based on RFI's low-rank and sparse features.

In general, the semi-parametric approaches utilize optimized models to constrain and separate the RFI. However, due to the various RFI types, these models will inevitably cause the loss of some useful signals while removing RFI, resulting in the loss of details in the SAR image.

Moreover, with the development of deep learning methods, intelligent learning methods have become an emerging trend in signal processing. Methods based on neural networks, especially combined with RPCA and applied to RFI mitigation in SAR data, have shown superior performance and potential. In [49], a method to mitigate RFI using a special type of convolutional neural network, the U-Net, was proposed. In [50], a mitigation algorithm of NBI and WBI based on the deep residual network (ResNet) was proposed. In [51], a hybrid model-constrained deep learning approach for RFI extraction and mitigation by fusing the classical model-based and advanced data-driven method was proposed. However, these methods are inherently black box, lacking a certain degree of interpretability, and since their performance relies on access to large amounts of data and computational resources, this limits their applicability to scenarios with limited samples.

1.3. Solution and Contributions of This Article

Given the above problems, this study makes full use of the low-rank characteristics of RFI in the range frequency domain and proposes a mitigation method of RFI with a signal-protected capability. First, the kurtosis coefficient method is used to detect RFI pulse-by-pulse in the range frequency domain, and the echoes containing RFI are stored in the form of a matrix. Second, the RFI is preliminary extracted, which is completed by the LRSD of the echo matrix containing RFI; after decomposition, the low-rank matrix representing RFIs can be preliminary separated. Lastly, for the secondary separation of RFI, we use the fuzzy C-means (FCM) clustering algorithm to pinpoint the RFI in the preliminary extraction results, then, we separate the RFI from the remaining useful signal in the low-rank matrix again and reconstruct the useful signal to complete the mitigation. The proposal can further protect useful signals while removing interference through the secondary separation of RFI effectively. Experimental results based on simulated and measured data verify the performance and potential of the proposed method.

The main contributions of this article are summarized as follows:

1. An idea of implementing secondary separation for RFI is proposed, which can solve the insufficient protection of useful signals in traditional mitigation methods of RFI. By separating the extraction results of RFI again, the loss of useful signals can be effectively reduced. In particular, this idea of secondary separation can be extended to other mitigation methods of RFI, which provides a new perspective of RFI mitigation in SAR data;

2. This study proposes a mitigation approach for RFI with a signal protection capability based on the low-rank property of RFI in the range frequency domain. Specifically, the method includes three steps: the detection, extraction, and secondary separation of RFI. Compared with traditional methods, this proposal pays more attention to the preservation of SAR image details while effectively detecting and mitigating RFI, avoiding the situation that the interpretation of SAR images is more difficult due to excessive loss of useful signals after RFI mitigation.

Theoretical discussions are validated through extensive experiments. Specifically, in experiments with simulated data, the mitigation effect and useful signal protection capability of the proposed method under different bandwidths of RFI and different SINRs are discussed. Experiments with the measured data verify the effectiveness and superiority of the proposed method.

1.4. Organization of This Article

The remainder of this article is organized as follows. Section 2 introduces the geometric and signal models and analyzes the sparse and statistical characteristics of RFI in the range frequency domain. Section 3 shows the detailed workflow of the RFI mitigation method proposed in this article. Section 4 gives the experimental results and performance analysis of the method. Lastly, Section 5 concludes this article and discusses the further application of the method.

2. Problem Formulation

2.1. Geometric Model of RFI

Various electromagnetic devices operating in the same frequency band as SAR systems are major sources of RFI. In recent years, many RFI cases have been observed in spaceborne and airborne SAR data, among which the main sources are not only space-based sources such as airborne radars and co-frequency satellites but also ground-based sources such as ground base stations and ground-based radars. In general, the spatial relationship between the RFI sources and the irradiated area of the spaceborne SAR beam is shown in Figure 1. For ground-based sources, RFI is usually received in the form of direct waves by the main or side lobes of the SAR. For space-based sources, RFI is usually received by the SAR in the form of direct or scattered waves.

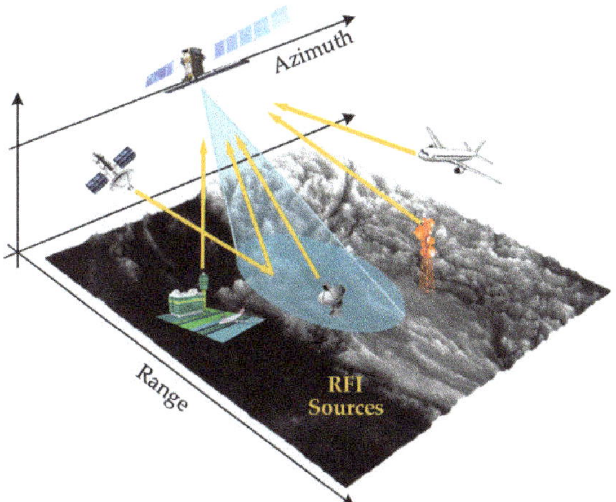

Figure 1. Common sources and geometric model of RFI.

2.2. Signal Model of RFI

When the SAR system is working, the raw signal is usually superimposed into the azimuth (i.e., slow time) and the range (i.e., fast time) domain. Since the RFI signal is independent of the SAR echo signal, in each azimuth echo [25], the SAR echo containing RFI can be expressed as

$$S(\tau,\eta) = X(\tau,\eta) + I(\tau,\eta) + N(\tau,\eta) \tag{1}$$

where $X(\tau,\eta)$, $I(\tau,\eta)$, and $N(\tau,\eta)$ represent the useful signal, RFI, and system noise, respectively. τ and η represent the range fast time and the azimuth slow time, respectively.

Although there are many different forms of RFI, in general, RFI is considered to be a linear combination of multiple single frequencies caused by radiating sources. According to its bandwidth, it can be divided into two categories: NBI and WBI. As mentioned in [15], NBI can be regarded as the superposition of a series of sinusoidal signals, and its model can be expressed as

$$I_{NBI}(\tau,\eta) = \sum_{n=1}^{N} A_n(\eta)exp(2j\pi f_n\tau + \varphi_n) \tag{2}$$

where N represents the number of interference signals. $A_n(\eta)$, f_n, and φ_n represent the amplitude, frequency, and phase of the nth interference signal, respectively. NBI generally does not have a complex frequency modulation, and often appears in the form of bright lines in the image.

In contrast, WBI has a larger bandwidth and a more complex frequency modulation, occupying more frequency cells in the range frequency domain. Typically, WBI can be modeled as linear frequency modulation (LFM) and sinusoidal frequency modulation (SFM) [25]. Its signal model can be expressed as

$$I_{WBI}(\tau,\eta) = \sum_{n=1}^{N} A_n(\eta)exp(j\phi) \tag{3}$$

where the phase term ϕ is divided into two modulation terms, ϕ_{LFM} and ϕ_{SFM}, which represent LFM and SFM, respectively. The specific form is

$$\phi_{LFM} = 2\pi f_n\tau + \pi K_n\tau^2 \tag{4}$$

$$\phi_{SFM} = \beta_n sin(2\pi f_n\tau + \varphi_n) \tag{5}$$

where K_n represents the chirp rate of the nth WBI signal, and β_n represents the modulation factor.

2.3. Low-Rank Sparse Model

The linear reversibility of the fast Fourier transform (FFT) ensures that this transform does not affect the linear superposition characteristics of the SAR echo. Therefore, according to (5), the SAR echo data can be expressed as

$$S(f_\tau, \eta) = X(f_\tau, \eta) + I(f_\tau, \eta) + N(f_\tau, \eta) \tag{6}$$

where f_τ represents the range frequency cells after FFT.

The RFI in the range frequency domain has a relatively stable frequency in the slow time direction, and its amplitude appears as some parallel straight lines, as shown in Figure 2a. It is clear that RFI has low-rank properties in the slow time direction. For further verification, the eigenvalue decomposition of Figure 2a is performed, and the corresponding results are shown in Figure 2b. The eigenvalues reflect the energy of different components in the SAR echo and the structural redundancy of the matrix. As can be observed, only a few large eigenvalues are related to RFI, which further illustrates the low-rank feature of RFI in the range frequency domain.

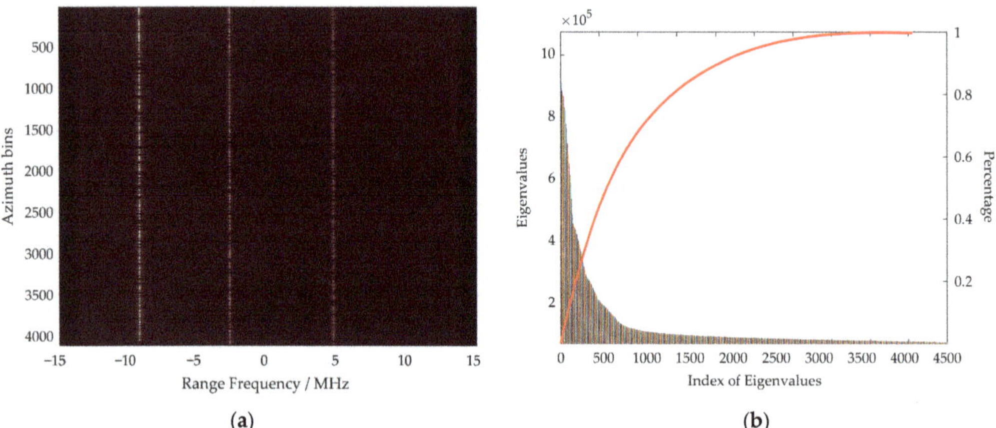

Figure 2. Structural analysis of RFI in range frequency domain: (**a**) spectrogram of SAR echoes contaminated by RFI; (**b**) Eigenvalue sequences and analysis corresponding to (**a**).

According to the low-rank characteristic of RFI in the frequency domain, RFI can be initially extracted by solving the following LRSD problem.

$$\begin{aligned} \min_{\mathbf{I},\mathbf{X}} \quad & rank(\mathbf{I}) + \lambda\|\mathbf{X}\|_0 \\ s.t. \quad & \mathbf{S} = \mathbf{I} + \mathbf{X} \end{aligned} \tag{7}$$

where $rank(\cdot)$ represents the rank of the matrix, $\|\cdot\|_0$ represents the ℓ_0 norm of the matrix, that is, the number of non-zero elements in the matrix, and $\lambda > 0$ is a compromise factor.

The rank and ℓ_0 norm of matrices can be convexly relaxed, providing a way to solve the above issues. Since the kernel norm of the matrix is the convex envelope of the rank, and the ℓ_1 norm of the matrix is the optimal convex approximation of ℓ_0, (7) can be relaxed as the following convex optimization problem.

$$\begin{aligned} \min_{\mathbf{I},\mathbf{X}} \quad & \|\mathbf{I}\|_* + \lambda\|\mathbf{X}\|_1 \\ s.t. \quad & \mathbf{S} = \mathbf{I} + \mathbf{X} \end{aligned} \tag{8}$$

where $\| \cdot \|_*$ is the kernel norm that can represent the sum of the singular values of the matrix, $\| \cdot \|_1$ represents the ℓ_1 norm of the matrix, that is, the sum of the absolute values of each element in the matrix, as mentioned in [12], the compromise factor λ is set as:

$$\lambda = \frac{1}{\sqrt{\max(m, n)}} \tag{9}$$

where m and n represent the number of rows and columns of matrix S, respectively.

2.4. Statistical Model

Generally speaking, under the assumption of complex Gaussian distribution of a SAR echo without RFI, the real and imaginary parts of its range frequency domain spectrogram obey Gaussian distribution. When there is RFI in the echo, the histogram will deviate from the Gaussian distribution and be more concentrated in the low-amplitude region, resulting in sharper peaks on the left and tails on the right.

Figure 3a,b sequentially shows the frequency amplitude images of RFI-free and RFI-containing echoes, and Figure 3c shows the frequency amplitude images of the RFI extracted by LRSD. It can be clearly observed that some useful signals remain in Figure 3c. The amplitude histograms of Figure 3a–c are separately counted for further analysis, as shown in Figure 3d–f. It can be seen that Figure 3d–f all obey the Rayleigh distribution. Figure 3e has a sharper left peak than Figure 3d, while Figure 3f has the sharpest left peak, and there are remaining useful signals in the right tail.

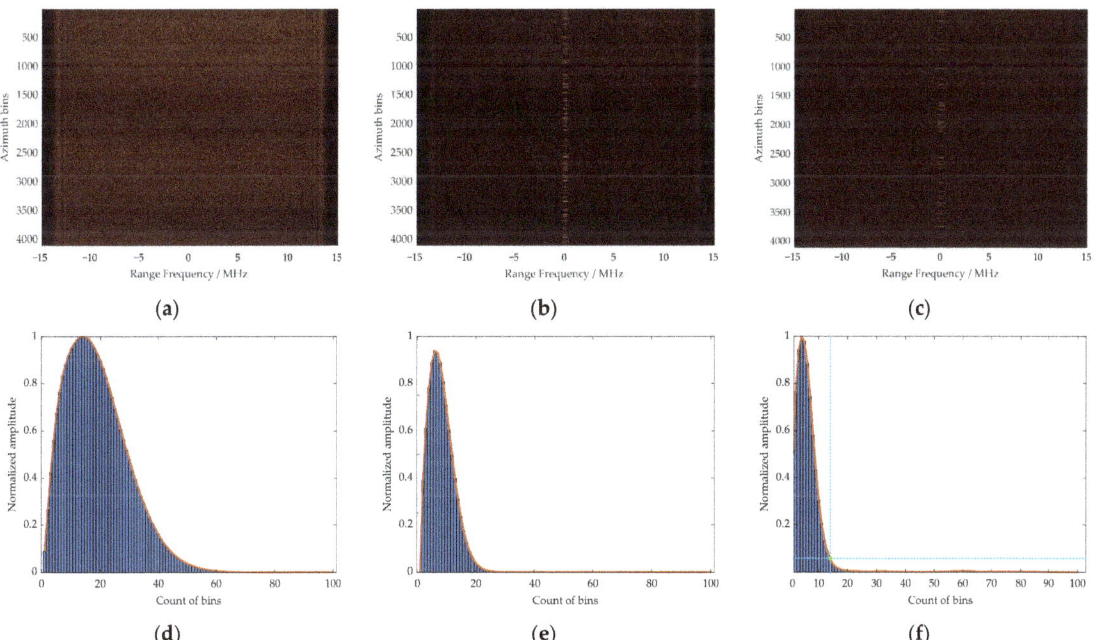

Figure 3. Statistical analysis of SAR range frequency spectrogram: (**a**) RFI-free signal; (**b**) RFI-containing signal; (**c**) Extraction of RFI; and (**d–f**) The amplitude histograms correspond to (**a–c**), respectively.

Through the above analysis, this study performs LRSD based on the low-rank characteristic of RFI to realize the extraction of RFI. On the basis of RFI extraction, a binary masking matrix is generated for the low-rank matrix, so as to complete the secondary separation of RFI.

3. Methodology

In order to solve the problem of RFI detection and mitigation in SAR data, a mitigation method of RFI with signal protection capability is proposed. Specifically, the method includes three steps: RFI detection based on kurtosis, RFI extraction based on LRSD, and RFI secondary separation based on binary masking. The proposal can detect and mitigate RFI robustly while protecting useful SAR signals and effectively reducing the loss of details in SAR images. The specific flow chart is shown in Figure 4.

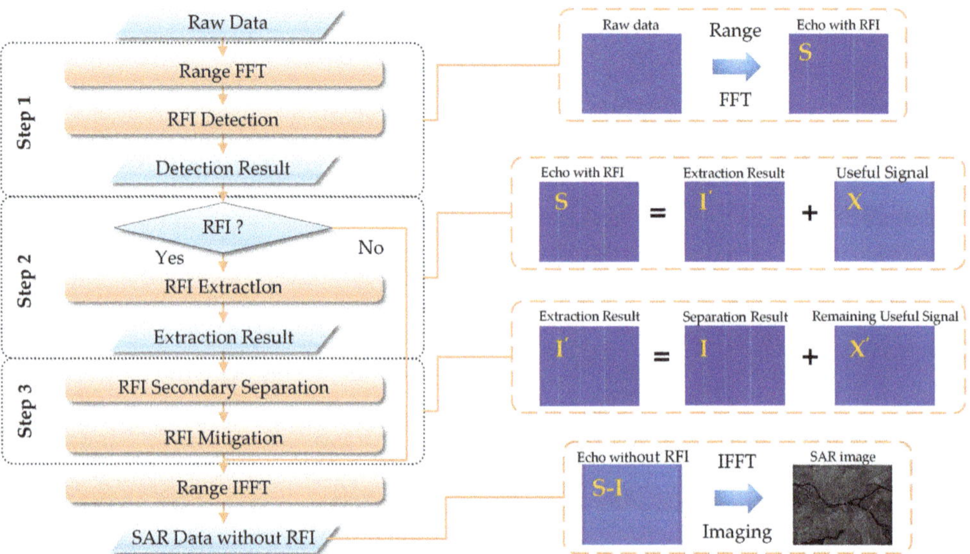

Figure 4. Flow chart of the proposed method.

3.1. RFI Detection Based on Kurtosis

In order to accurately and efficiently mitigate the RFI and protect the useful signal as much as possible, it is necessary to perform detection of RFI pulse-by-pulse. Unlike the traditional mitigation method, which needs to detect the position of RFI in the frequency domain accurately, the method proposed in this study only needs to distinguish between the RFI-containing echo and the useful signal. Considering the difference in statistical characteristics between the SAR signal and RFI, the presence of RFI can be quickly detected by calculating the kurtosis of the frequency domain echoes [52].

Assuming that S is a random variable, the mean is u, and σ is the standard deviation, the kurtosis can be defined as:

$$K(S) = \frac{E[S - u]^4}{\sigma^4} = \frac{\frac{1}{n}\sum_{i=1}^{n}(s_i - \bar{s})^4}{\left(\frac{1}{n}\sum_{i=1}^{n}(s_i - \bar{s})^2\right)^2} \tag{10}$$

where $E[\cdot]$ denotes the expectation operator. Kurtosis characterizes the steepness of distribution, which is usually a statistic relative to a normal distribution. If the kurtosis is greater than 3, the sample has a steep distribution, and conversely, it has a flat distribution.

It is supposed that N_r and N_a denote the number of samples for range and azimuth, respectively. By calculating the kurtosis of each echo in a pulse-by-pulse manner, a sequence of kurtosis values can be obtained, which can be expressed as

$$\Phi(i) = [K_1, K_2, \ldots, K_{N_r}] \tag{11}$$

The robust K-means algorithm [53] is used to classify $\Phi(i)$ into two categories, one of which indicates the presence of RFI and the other indicates the absence. The operation of threshold segmentation can be expressed as

$$\overline{\Phi}(i) = \begin{cases} 0, & \Phi(i) < \alpha, \quad \Rightarrow \textit{Without RFI} \\ 1, & \Phi(i) \geq \alpha, \quad \Rightarrow \textit{With RFI} \end{cases} \tag{12}$$

where $\overline{\Phi}(i)$ is the detection result, α is the threshold, and constants 1 and 0 indicate the presence and absence of RFI, respectively.

3.2. Extraction of RFI Based on LRSD

After the detection work, it is necessary to perform LRSD on the RFI-containing echo in the frequency domain to complete RFI extraction, which can be transformed to obtain the optimal solution to the convex problem (8) based on the analysis in Section 2. The alternating direction multiplier method (ADMM) adopts the idea of divide and conquer, which can be applied to the solution of large-scale optimization problems and obtain the optimal solution of the problem. The augmented Lagrangian function is constructed as follows.

$$L(\mathbf{I}, \mathbf{X}, \mathbf{Y}, \mu) = \|\mathbf{I}\|_* + \lambda \|\mathbf{X}\|_1 - \langle \mathbf{Y}, \mathbf{I} + \mathbf{X} - \mathbf{S} \rangle + \frac{\mu}{2} \|\mathbf{I} + \mathbf{X} - \mathbf{S}\|_F \tag{13}$$

where μ is the penalty factor, $\mathbf{Y}$ is the Lagrange multiplier, $\langle \cdot \rangle$ represents the matrix inner product, and $\|\cdot\|_F$ represents the Frobenius norm. When $\mathbf{Y} = \mathbf{Y}_k$, $\mu = \mu_k$, use the ADMM method to solve the block optimization problem.

$$\min_{\mathbf{I}, \mathbf{X}} \quad L(\mathbf{I}, \mathbf{X}, \mathbf{Y}_k, \mu_k) \tag{14}$$

First, fix the variables $\mathbf{X}$ and $\mathbf{Y}$, and the operation of updating the variable $\mathbf{I}$ is as follows.

$$\begin{aligned}
\mathbf{I}_{k+1} &= \arg\min_{\mathbf{I}} L(\mathbf{I}, \mathbf{X}_{k+1}, \mathbf{Y}_k, \mu_k) \\
&= \arg\min_{\mathbf{I}} \|\mathbf{I}\|_* - \langle \mathbf{Y}_k, \mathbf{I} + \mathbf{X}_{k+1} - \mathbf{S} \rangle + \frac{\mu_k}{2} \|\mathbf{I} + \mathbf{X}_{k+1} - \mathbf{S}\|_F \\
&= \arg\min_{\mathbf{I}} \|\mathbf{I}\|_* + \frac{\mu_k}{2} \|\mathbf{I} - \left(\mathbf{S} - \mathbf{X}_{k+1} + \frac{\mathbf{Y}_k}{\mu_k}\right)\|_F \\
&= G_{\frac{1}{\mu_k}}\left(\mathbf{S} - \mathbf{X}_{k+1} + \frac{\mathbf{Y}_k}{\mu_k}\right)
\end{aligned} \tag{15}$$

where $G_{\frac{1}{\mu_k}}(\cdot)$ is the shrinkage operator of singular threshold, and the specific operation can be described as:

$$\mathbf{W}' = G_{\frac{1}{\mu_k}}(\mathbf{W}) = \begin{cases} [\mathbf{U}, \mathbf{\Sigma}, \mathbf{V}] = svd(\mathbf{W}) \\ \mathbf{\Sigma} = sgn(\mathbf{\Sigma}) \cdot *max(abs(\mathbf{\Sigma}) - \frac{1}{\mu_k}, 0) \\ \mathbf{W}' = \mathbf{U} * \mathbf{\Sigma} * \mathbf{V}^H \end{cases} \tag{16}$$

where $[\mathbf{U}, \mathbf{\Sigma}, \mathbf{V}] = svd(\mathbf{W})$ is the singular value decomposition of $\mathbf{W}$.

Then, fix the variables $\mathbf{I}$ and $\mathbf{Y}$, and the operation of updating the variable $\mathbf{X}$ is as follows.

$$\begin{aligned}
\mathbf{X}_{k+1} &= \arg\min_{\mathbf{X}} L(\mathbf{I}_{k+1}, \mathbf{X}, \mathbf{Y}_k, \mu_k) \\
&= \arg\min_{\mathbf{X}} \lambda \|\mathbf{X}\|_1 - \langle \mathbf{Y}_k, \mathbf{I}_{k+1} + \mathbf{X} - \mathbf{S} \rangle + \frac{\mu_k}{2} \|\mathbf{I}_{k+1} + \mathbf{X} - \mathbf{S}\|_F \\
&= \arg\min_{\mathbf{X}} \lambda \|\mathbf{X}\|_1 + \frac{\mu_k}{2} \|\mathbf{X} - \left(\mathbf{S} - \mathbf{I}_{k+1} + \frac{\mathbf{Y}_k}{\mu_k}\right)\|_F \\
&= F_{\frac{\lambda}{\mu_k}}\left(\mathbf{S} - \mathbf{I}_{k+1} + \frac{\mathbf{Y}_k}{\mu_k}\right)
\end{aligned} \tag{17}$$

where $F_{\frac{\lambda}{\mu_k}}(\cdot)$ is the shrinkage operator of soft threshold, and the specific operation can be described as:

$$F_{\frac{\lambda}{\mu_k}}(\mathbf{W}) = sign(\mathbf{W}) \cdot * max(abs(\mathbf{W}) - \frac{1}{\mu_k}, 0) \tag{18}$$

when $\mathbf{I} = \mathbf{I}_{k+1}$, $\mathbf{S} = \mathbf{S}_{k+1}$, the update formula of matrix $\mathbf{Y}$ is

$$\mathbf{Y}_{k+1} = \mathbf{Y}_k - \mu_k(\mathbf{I}_{k+1} + \mathbf{S}_{k+1} - \mathbf{R}) \tag{19}$$

the penalty factor μ_k can be updated as follows.

$$\mu_{k+1} = \begin{cases} \rho\mu_k & \frac{\mu_k\|\mathbf{X}_{k+1}-\mathbf{X}_k\|_F}{\|\mathbf{S}\|_F} < \varepsilon \\ \mu_k & otherwise \end{cases} \tag{20}$$

therefore, the optimization problem of (8) can be solved iteratively by (15), (17), and (19). It can provide extraction results of RFI until convergence, that is

$$I'(f_\tau, \eta) = \mathbf{I}_k \tag{21}$$

3.3. Secondary Separation of RFI

Since there are various forms of RFI in the actual scene, and the result of LRSD is an approximate solution to the convex optimization problem, some useful signals will inevitably remain in the extracted RFI matrix. It can be expressed as

$$I'(f_\tau, \eta) = I(f_\tau, \eta) + X'(f_\tau, \eta) \tag{22}$$

where $I(f_\tau, \eta)$ is the RFI actually present in the SAR echo and $X'(f_\tau, \eta)$ represents the remaining useful signal.

In order to further reduce the loss of useful signals after RFI mitigation, this study proposes a secondary separation of RFI, the process is shown in Figure 5.

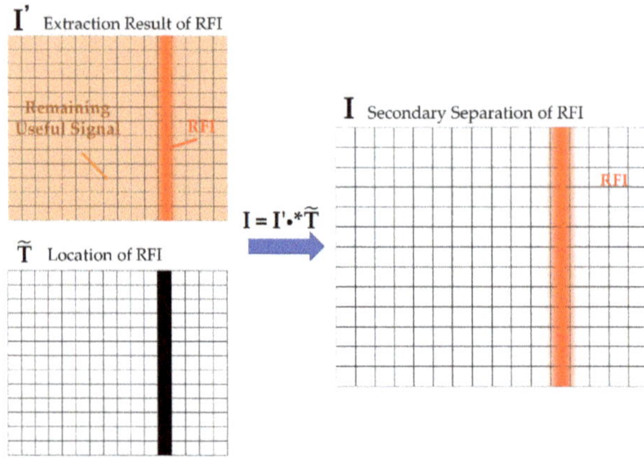

Figure 5. The secondary separation process of RFI.

The secondary separation of RFI is done by using the FCM algorithm to generate a binary masking matrix $\tilde{\mathbf{T}}(f_\tau, \eta)$ from the extraction result of RFI. Then, the secondary separation can be expressed as

$$I(f_\tau, \eta) = I'(f_\tau, \eta) * \tilde{\mathbf{T}} \tag{23}$$

Finally, the useful signal can be reconstructed by removing the RFI, which can be expressed as

$$X_{total}(f_\tau, \eta) = S(f_\tau, \eta) - I(f_\tau, \eta) \tag{24}$$

Through the above detection, extraction, and secondary separation of RFI, the SAR echo without RFI can be obtained. The specific steps of the proposed method are summarized in Algorithm 1, and the detailed performance will be presented in the following subsection.

Algorithm 1 Extraction and Secondary Separation of RFI

Input: $\mathbf{S} = S(f_\tau, \eta), \lambda > 0, \mu_0 > 0, \rho > 1, \delta > 0$
Initialization: $\mathbf{I}_0 = 0, \mathbf{X}_0 = 0, \mathbf{Y}_0 = \dfrac{\mathbf{S}}{max(\|\mathbf{S}\|_2, \sqrt{mn}\|\mathbf{S}\|_\infty)}, k = 0$
While $\|\mathbf{S} - \mathbf{I}_{k+1} + \mathbf{X}_{k+1}\|_F^2 / \|\mathbf{I}\|_F^2 > \delta$ **do**

    Low rank matrix: $\mathbf{I}_{k+1} = G_{\frac{1}{\mu_k}}\left(\mathbf{S} - \mathbf{X}_{k+1} + \frac{\mathbf{Y}_k}{\mu_k}\right)$

    Sparse matrix: $\mathbf{X}_{k+1} = F_{\frac{\lambda}{\mu_k}}\left(\mathbf{S} - \mathbf{I}_{k+1} + \frac{\mathbf{Y}_k}{\mu_k}\right)$

    Update variable Y: $\mathbf{Y}_{k+1} = \mathbf{Y}_k - \mu_k(\mathbf{I}_{k+1} + \mathbf{X}_{k+1} - \mathbf{S})$

    Update penalty factor μ: $\mu_{k+1} = \begin{cases} \rho\mu_k & \frac{\mu_k\|\mathbf{X}_{k+1}-\mathbf{X}_k\|_F}{\|\mathbf{S}\|_F} < \varepsilon \\ \mu_k & otherwise \end{cases}$

    $k = k + 1$
End while
Extraction of RFI: $I'(f_\tau, \eta) = \mathbf{I}_k$
Generate binary masking matrix: $\tilde{\mathbf{T}}(f_\tau, \eta)$
RFI secondary separation: $I(f_\tau, \eta) = I'(f_\tau, \eta) \cdot * \tilde{\mathbf{T}}$
Remove RFI: $X_{total}(f_\tau, \eta) = S(f_\tau, \eta) - I(f_\tau, \eta)$
Output: $I(f_\tau, \eta), X_{total}(f_\tau, \eta)$

---

4. Experimental Results and Discussion

Extensive experiments are performed in this section using simulated SAR data and Sentinel-1 level-0 raw data. Specifically, based on real SAR data and simulated RFI, the mitigation performance of the proposed method is quantitatively analyzed by comparing the root mean square error (RMSE) under different signal-to-interference and noise ratios (SINR) and different RFI bandwidths. Experiments based on measured SAR data verify the effectiveness of the method, and the mitigation performance is evaluated by calculating the gray level entropy and average gradient of the SAR image after RFI mitigation.

4.1. Experimental Results Based on Simulated SAR Data

4.1.1. On the Performance of RFI Mitigation for Different SINR Cases

Simulated experiments based on different SINRs are used to verify the RFI mitigation performance of the proposed method. Furthermore, by comparing the proposal with ESP [28], RPCA [8], and complex tensor RPCA (CT-RPCA) [44] methods, the potential of the proposed method for useful signal protection is further verified. Table 1 summarizes the main system parameters of the simulation.

The experimental results of the proposed method compared with ESP, RPCA, and CT-RPCA are shown in Figure 6. It can be seen from the first row that the RFI under different SINRs has different degrees of contamination to the SAR image. At the SINR of –30, the artifacts caused by RFI significantly suppress scene information. The following four rows are the RFI mitigation results after using ESP, RPCA, CT-RPCA, and the proposed method. As can be seen from Figure 6, all four methods can effectively mitigate RFI under different SINR conditions but have different performances in the protection of useful signals. The magnified region marked with a red box in Figure 6 is the region of interest (ROI), as shown in Figure 7. It can be seen from Figure 7a–d that there are abnormal sidelobe effects on the point targets of the image after RFI suppression using the ESP method, which is due to the loss of useful signals caused by the method over-penalizing the eigenvalues when reconstructing the RFI. Figure 7e–l show the corresponding ROIs after using the RPCA and

CT-RPCA methods. Since these two methods do not further separate the remaining useful signals after the low-rank and sparse decomposition of the SAR signal, the high sidelobe phenomenon still exists to varying degrees. Figure 7m–p show the corresponding ROIs after the proposed method mitigates RFI, and it can be seen that the high sidelobe effect of the point target is no longer present. This indicates that this method can further protect the useful signal while effectively removing RFI.

Table 1. Main parameters of measured SAR data and simulated RFI.

Parameters	Values
Carrier frequency	5.300 GHz
Sampling frequency	32.317 MHz
Efficient velocity	7000 m/s
Slant range	988,647 m
PRF	1256.98 Hz
Pulse width	41.74 μm
Pulse bandwidth	30 MHz
Carrier frequency of RFI	5.305 GHz
Bandwidth of RFI	1 MHz

To quantitatively evaluate the performance of the interference mitigation method, the RMSE was selected as the evaluation index to evaluate the mitigation results in the experiment. RMSE is defined as

$$RMSE(\mathbf{X}, \mathbf{S}) = \frac{\|\mathbf{S} - \mathbf{X}\|_F}{\|\mathbf{S}\|_F} \tag{25}$$

RMSE describes the normalized difference between the original and recovered SAR data. The smaller the RMSE, the better the mitigation performance. Table 2 summarizes the RMSE comparison results of ESP, RPCA, CT-RPCA, and the proposed method under different SINR conditions. The results show that the RMSE of the proposed method is lower under different SINR conditions, which means that the proposed method has a more prominent signal protection performance while mitigating RFI.

Table 2. Evaluation metrics for the four methods in the different SINR cases.

Metric	Method	ESP	RPCA	CT-RPCA	Proposed Method
RMSE	SINR = 0 dB	0.1851	0.1926	**0.1648**	0.1695
	SINR = −10 dB	0.2295	0.2198	0.2221	**0.2126**
	SINR = −20 dB	0.2691	0.2797	0.2536	**0.2450**
	SINR = −30 dB	0.2926	0.3050	0.2878	**0.2816**

The best results in each metric are highlighted in bold.

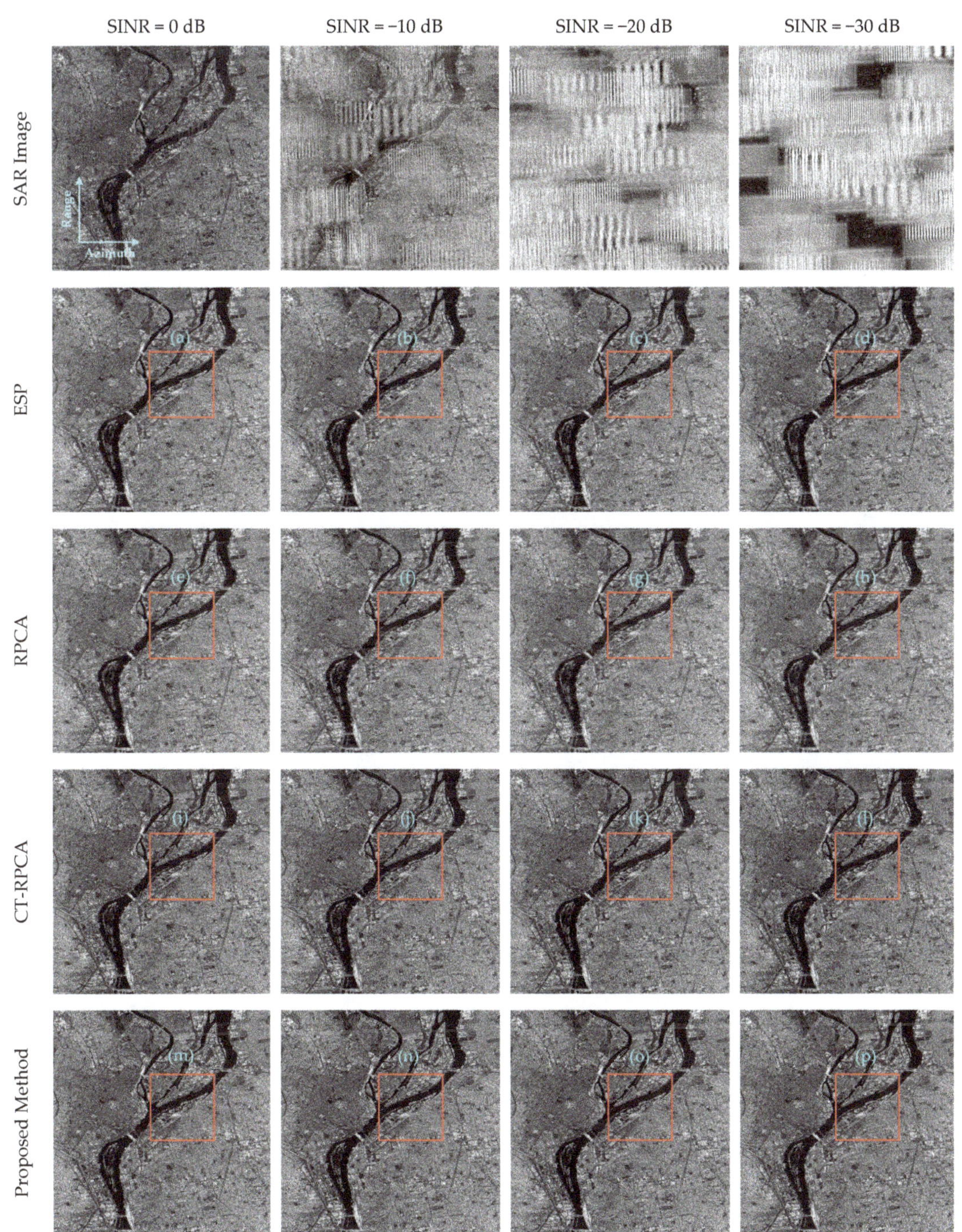

Figure 6. RFI mitigation performance for the ESP, RPCA, CT-RPCA, and proposed approach under different SINRs. (**a**–**p**) ROIs for RFI mitigation results of different methods.

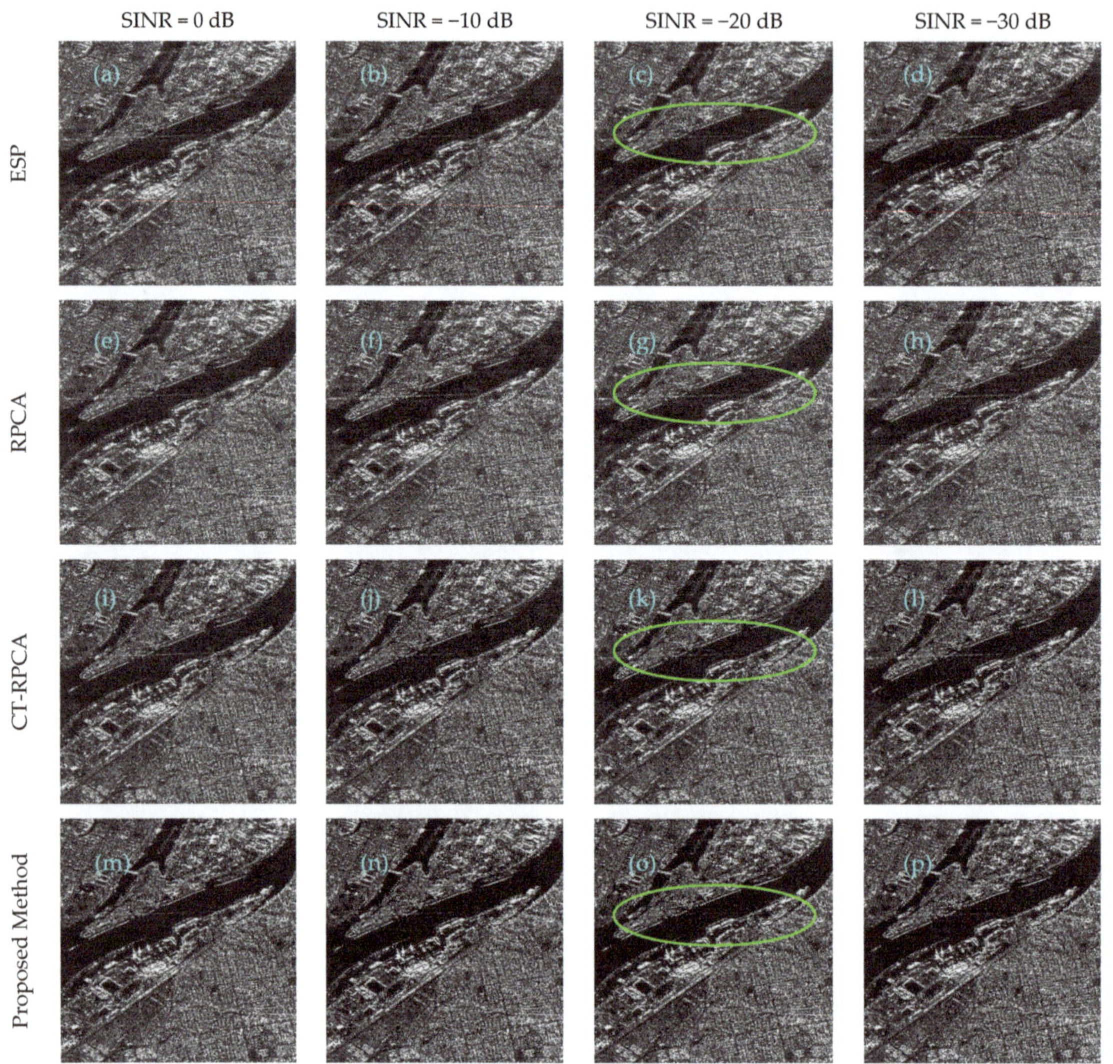

SINR = 0 dB   SINR = −10 dB   SINR = −20 dB   SINR = −30 dB

Figure 7. ROIs in Figure 6. (**a–d**) Mitigation results for the ESP method. (**e–h**) Mitigation results for the RPCA method. (**i–l**) Mitigation results for the CT-RPCA method. (**m–p**) Mitigation results for the proposed approach.

4.1.2. On the Performance of Mitigation for Different Bandwidths of RFI

Simulation experiments were used to explore the proposed method's mitigation performance under different RFI bandwidths and compared with ESP, RPCA, and CT-RPCA methods. The experimental results are shown in Figure 8. From the first row, it can be seen that RFI with different bandwidths has different degrees of influence on the SAR image. When the bandwidth of RFI is 2 MHz, the four methods can effectively mitigate RFI. While the details in the SAR images using the proposed method are clearer, indicating that it is more effective in protecting useful signals, which is verified by subsequent quantitative analysis. When the bandwidth is increased to 4 MHz, the high sidelobe effect caused by the ESP and RPCA methods is more obvious in the SAR images, and the proposed method still shows excellent performance. When the bandwidth is increased to 6 MHz, the RFI mitigation performance of the proposed method degrades.

The magnified view of the ROI marked by the red box in Figure 8 is shown in Figure 9. It is evident from Figure 9a–c that the loss of useful signal caused by ESP methods becomes more severe as the RFI bandwidth increases. Figure 9d–i show the corresponding ROIs for the RPCA and CT-RPCA method, and it can be found that when the RFI bandwidth is increased from 2 M to 6 M, the high sidelobe effect caused by the loss of useful signal still exists since these two methods only rely on LRSD to extract the RFI. Figure 9j–l shows the corresponding ROI of the proposed method. It can be seen that when the bandwidth of the RFI exceeds 4 M, although the interference mitigation effect of the method begins to decline, the useful signal can still be effectively protected.

Table 3 summarizes the RMSE comparison results of ESP, RPCA, CT-RPCA, and the proposed method under different RFI bandwidths.

Table 3. Evaluation metrics for the four methods in the different bandwidths of RFI.

Metric	Method	ESP	RPCA	CT-RPCA	Proposed Method
	Bandwidth = 2 MHz	0.1868	0.2202	0.1853	**0.1819**
RMSE	Bandwidth = 4 MHz	0.2419	0.2377	0.2260	**0.2138**
	Bandwidth = 6 MHz	0.3477	0.3534	**0.3305**	0.3340

The best results in each metric are highlighted in bold.

4.2. Experimental Results Based on Measured Data

Experiments are implemented using Sentinel-1 (TOPS mode) level-0 raw data to verify the method's effectiveness. The performance of different methods for RFI mitigation was evaluated by calculating the gray level entropy and average gradient of SAR images after RFI mitigation.

4.2.1. Experimental Data Description

Figure 10 shows a pseudo color image of the measured Sentinel-1 level-0 raw data. The data were acquired on 12 February 2020. As shown in Figure 10a, the SAR image is heavily contaminated by RFI due to multiple potential sources of interference in the illuminated area. The RFI-corrupted burst is shown in Figure 10b. It can be observed that the artifacts generated by RFI are all over the image.

4.2.2. Experimental Results Based on Spaceborne SAR Data

Figure 11 shows the RFI mitigation results for Sentinel-1 Level-0 raw data. Figure 11a–d show the RFI mitigation results for ESP, RPCA, CT-RPCA, and the proposed method. Figure 11e–h show the ROIs in Figure 11a–d. It is obvious that after using the ESP method, the artifacts generated by RFI in the image have been effectively removed, while the image quality is significantly degraded due to the raised sidelobes of the strongly scattering targets. When using the RPCA and CT-RPCA methods, the image still has certain sidelobe effects due to the loss of part of the useful signal during RFI mitigation. While the proposed method pays more attention to signal protection, it can avoid the loss of image details after RFI mitigation.

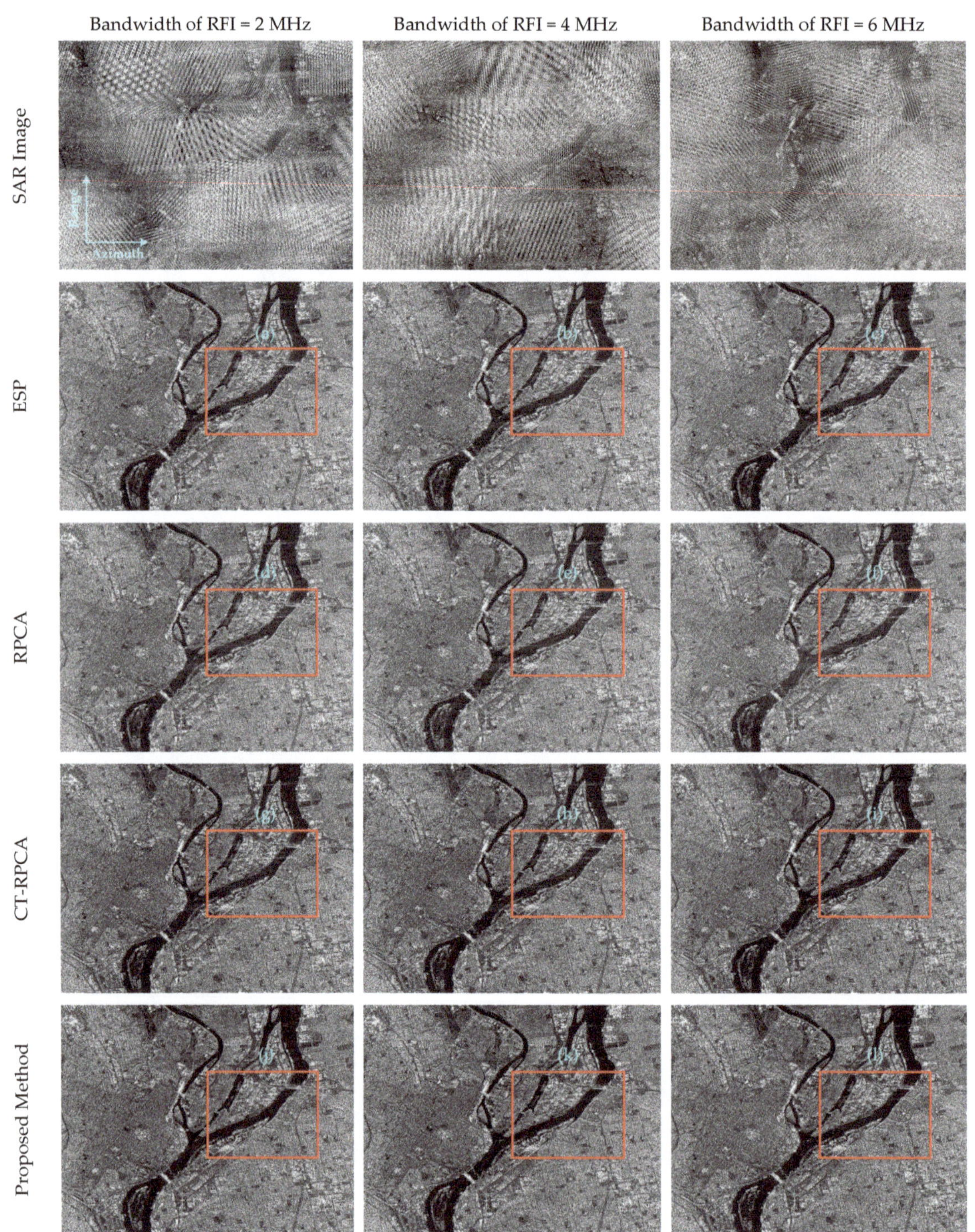

Figure 8. Mitigation performance for the ESP, RPCA, CT-RPCA, and proposed approach under different bandwidths of RFI. (**a–l**) ROIs for RFI mitigation results of different methods.

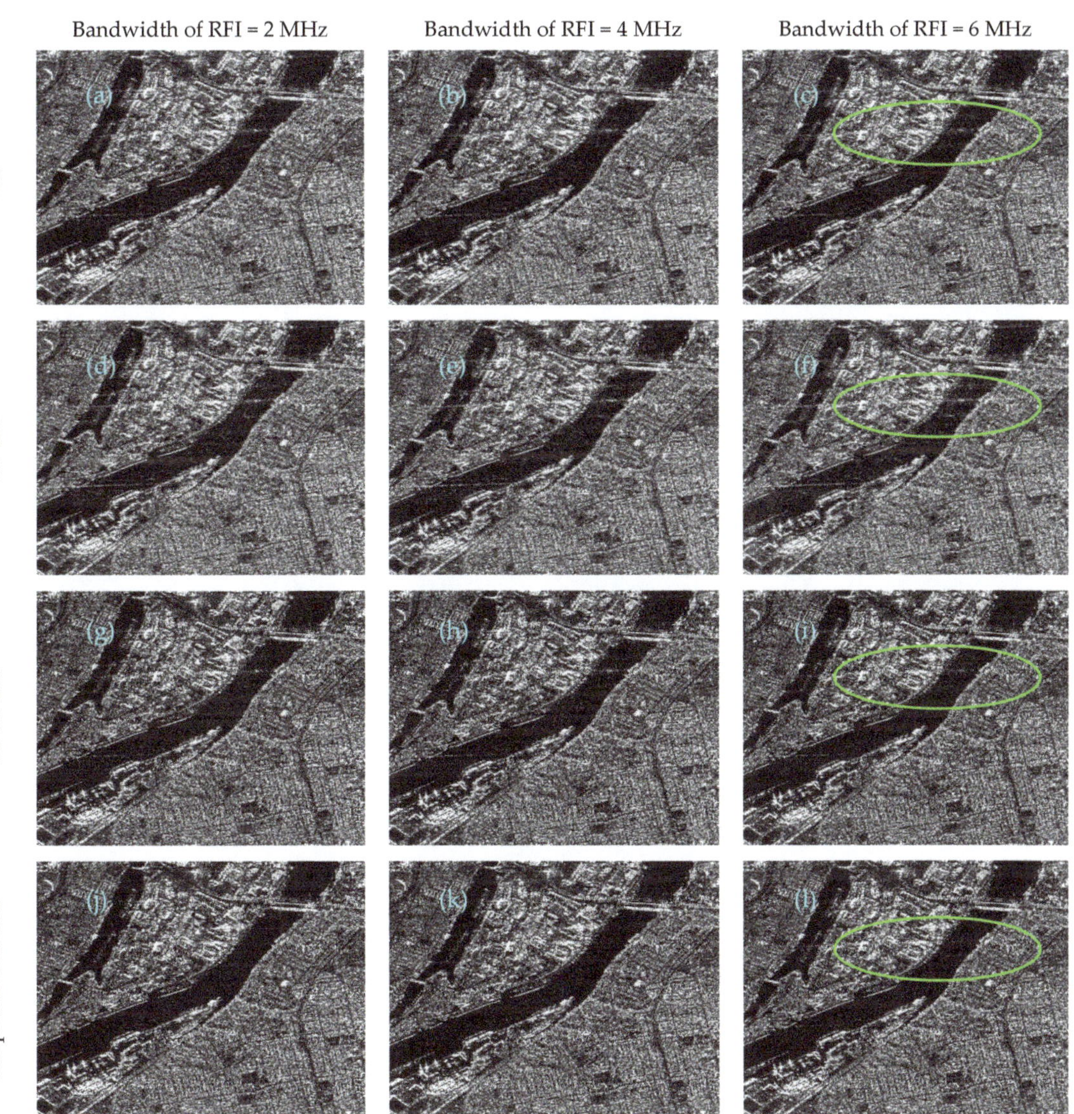

Figure 9. ROIs in Figure 8. (**a–c**) Mitigation results for the ESP method. (**d–f**) Mitigation results for the RPCA method. (**g–i**) Mitigation results for the CT-RPCA method. (**j–l**) Mitigation results for the proposed approach.

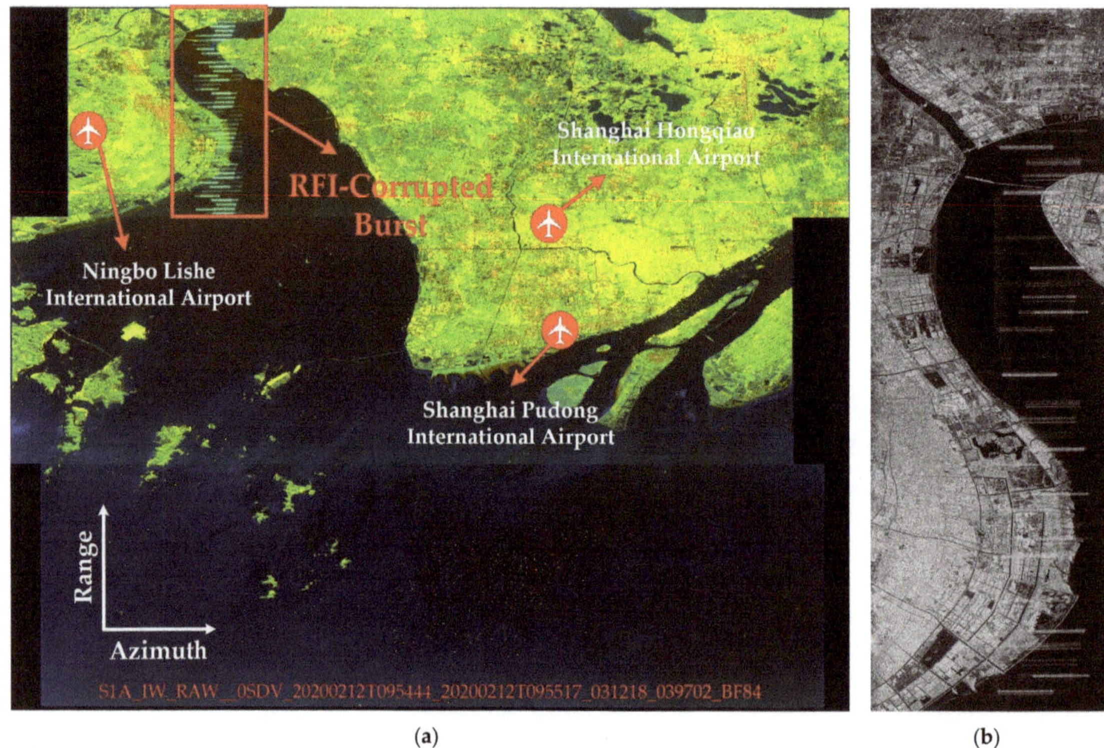

(a)   (b)

Figure 10. RFI-contaminated image of the measured Sentinel-1 level-0 raw data: (**a**) The pseudo color image of the measured Sentinel-1 level-0 raw data; (**b**) The RFI-corrupted burst.

Further, the mitigation performance of the proposed method was analyzed using gray level entropy and average gradients. The entropy of the image can represent the distribution characteristics of the gray level in the image, indicating the texture complexity, which is defined as

$$E = -\sum_{k=1}^{L} P_k log_2(P_k) \tag{26}$$

where E represents the image entropy, L is the total gray level of the image, and P_k represents the probability of the occurrence of a pixel with a gray value of k. The average gradient is defined as

$$AG = \frac{\sum_{\tau=1}^{N_r} \sum_{\eta=1}^{N_a} \frac{1}{4}\sqrt{\left(\frac{\partial S(\tau,\eta)}{\partial \tau}\right)^2 + \left(\frac{\partial S(\tau,\eta)}{\partial \eta}\right)^2}}{(N_r - 1)(N_a - 1)} \tag{27}$$

where $S(\tau,\eta)$ represents the position of the pixel in the SAR image, and $\partial S(\tau,\eta)/\partial \tau$ and $\partial S(\tau,\eta)/\partial \eta$ represent the grayscale gradient of the image in the vertical and horizontal directions, respectively.

The above experimental results show that all four methods can mitigate RFI with different performances. The ESP and RPCA methods are prone to abnormal side lobe effects when effectively removing RFI. The CT-RPCA method adds constraints when performing LRSD and can extract and mitigate RFI more effectively. However, it only decomposes the RFI in SAR signals once and has a limited ability to protect useful signals. Compared with the above three methods, the method proposed in this study has a better signal protection ability while effectively removing RFI, consistent with the previous results in this section. Table 4 shows the evaluation metrics for the four methods. It can be seen that the gray level entropy and the average gradient of the proposed method are higher than ESP, RPCA, and CT-RPCA, which indicates that the proposed method can more effectively protect the useful signal while mitigating RFI.

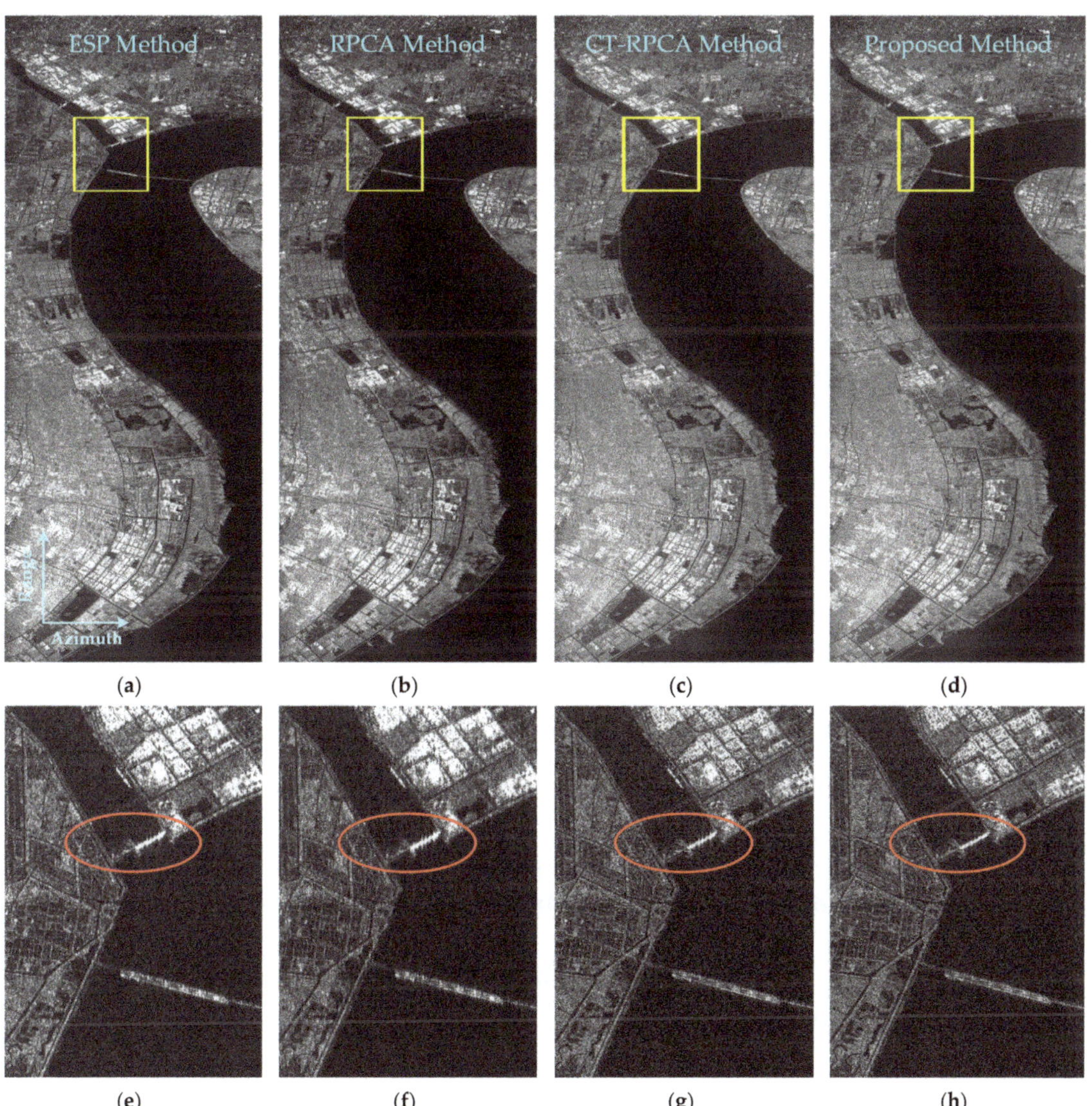

Figure 11. Experimental results for Sentinel-1 level-0 raw data: (**a**) Mitigation result by the ESP method; (**b**) Mitigation result by the RPCA method; (**d**) Mitigation result by the CT-RPCA method; (**c**) Mitigation result by the proposed approach; and (**e**–**h**) ROIs with the yellow box in (**a**–**d**).

Table 4. Gray level entropies and average gradients after ESP, RPCA, CT-RPCA, and proposed approach.

Metric \ Method	ESP	RPCA	CT-RPCA	Proposed Method
Gray Level Entropy	2.7351	2.2473	3.0627	**3.1969**
Average Gradient	2517.2639	2384.9916	2521.8028	**2550.0796**

The best results in each metric are highlighted in bold.

5. Conclusions

The existence of RFI seriously hinders SAR signal processing and image interpretation. However, most existing methods focused on the effective mitigation of RFI, but the protection of the useful signal was insufficient. To solve this problem, this article analyzed the characteristics of RFI in the frequency domain in detail. A mitigation method of RFI with a signal protection capability was proposed, including three steps, i.e., RFI detection, RFI extraction, and RFI secondary separation.

In the proposed method, the kurtosis coefficient was used to detect the RFI pulse-by-pulse in the range frequency domain. The preliminary RFI extraction was realized by performing LRSD on the echo matrix containing RFI. Then, a binary masking matrix was generated based on the extraction result of RFI, and the RFI and remaining useful signal in the low-rank matrix were separated again. The useful signal was reconstructed to complete the interference mitigation work. Compared with the traditional RFI mitigation method, the proposal considered both the RFI mitigation effect and the protection of useful signals simultaneously. The experimental results based on simulated and measured SAR data showed that this method can protect the useful signal more effectively than the ESP, RPCA, and CT-RPCA methods while mitigating RFI.

It is worth noting that protecting useful signals during the mitigation process is as necessary as the RFI mitigation effect. Otherwise, the loss of useful signals will reduce the resolution of SAR images. The proposed method can effectively reduce the loss of useful signals by secondary separation of the RFI extraction results and has great potential in RFI mitigation and useful signal protection. The authors hope that the proposed method can become a valuable tool to solve the RFI problem.

Author Contributions: Formal analysis, H.Z.; methodology, H.Z., J.C. and N.L.; validation, H.Z., L.M., J.L. and J.C.; resources, H.Z., Z.G. and N.L.; writing-original draft, H.Z. and L.M.; writing-review and editing, L.M., J.L., J.C., Z.G. and N.L.; funding acquisition, N.L.; All authors have read and agreed to the published version of the manuscript.

Funding: This work was supported in part by the National Natural Science Foundation of China under Grant 61871175; in part by the Foundation of Key Laboratory of Radar Imaging and Microwave Photonics, Ministry of Education, under Grant RIMP2020003; and in part by the Graduate Education Innovation and Quality Improvement Program of Henan University under Grant SYL20060144.

Data Availability Statement: Not applicable.

Acknowledgments: Thanks to all anonymous reviewers and editors for their comments and suggestions, making this article's content more rigorous and meaningful. In the meantime, the authors would like to express their gratitude to ESA for their open-source data of Sentinel-1 spaceborne SAR, which underpinned the validation experiments in this study.

Conflicts of Interest: The authors declare no conflict of interest.

References

1. Reigber, A.; Scheiber, R.; Jager, M.; Prats-Iraola, P.; Hajnsek, I.; Jagdhuber, T.; Papathanassiou, K.P.; Nannini, M.; Aguilera, E.; Baumgartner, S. Very-High-Resolution Airborne Synthetic Aperture Radar Imaging: Signal Processing and Applications. *Proc. IEEE* **2013**, *101*, 759–783. [CrossRef]
2. Moreira, A.; Prats-Iraola, P.; Younis, M.; Krieger, G.; Hajnsek, I.; Papathanassiou, K.P. A tutorial on Synthetic Aperture Radar. *IEEE Geosci. Remote Sens. Mag.* **2013**, *1*, 6–43. [CrossRef]
3. Deng, Y.; Yu, W.; Zhang, H.; Wang, W.; Liu, D.; Wang, R. Forthcoming Spaceborne SAR Development. *J. Radars* **2020**, *9*, 1–33.

4. Zhou, F.; Tao, M. Research on Methods for Narrow-Band Interference Suppression in Synthetic Aperture Radar Data. *IEEE J. Sel. Top. Appl. Earth Obs. Remote Sens.* **2015**, *8*, 3476–3485. [CrossRef]
5. Li, N.; Lv, Z.; Guo, Z. Observation and Mitigation of Mutual RFI Between SAR Satellites: A Case Study Between Chinese GaoFen-3 and European Sentinel-1A. *IEEE Trans. Geosci. Remote Sens.* **2022**, *60*, 5112819. [CrossRef]
6. Li, N.; Lv, Z.; Guo, Z. Pulse RFI Mitigation in Synthetic Aperture Radar Data via a Three-Step Approach: Location, Notch, and Recovery. *IEEE Trans. Geosci. Remote Sens.* **2022**, *60*, 5225617. [CrossRef]
7. Lv, Z.; Zhang, H.; Li, N.; Guo, Z. A Two-Step Approach for Pulse RFI Detection in SAR Data. In Proceedings of the 2021 IEEE Sensors, Sydney, Australia, 31 October–3 November 2021; pp. 1–4.
8. Su, J.; Tao, H.; Tao, M.; Wang, L.; Xie, J. Narrow-band Interference Suppression via RPCA-Based Signal Separation in Time–Frequency Domain. *IEEE J. Sel. Top. Appl. Earth Obs. Remote Sens.* **2017**, *10*, 5016–5025. [CrossRef]
9. Deng, Y.; Zhao, F.; Wang, Y. Brief Analysis on The Development and Application of Spaceborne SAR. *J. Radars* **2012**, *1*, 1–10. [CrossRef]
10. Lu, X.; Su, W.; Yang, J.; Gu, H.; Zhang, H.; Yu, W.; Yeo, T.S. Radio Frequency Interference Suppression for SAR via Block Sparse Bayesian Learning. *IEEE J. Sel. Top. Appl. Earth Obs. Remote Sens.* **2018**, *11*, 4835–4847. [CrossRef]
11. Yang, H.; He, Y.; Du, Y.; Zhang, T.; Yin, J.; Yang, J. Two-Dimensional Spectral Analysis Filter for Removal of LFM Radar Interference in Spaceborne SAR Imagery. *IEEE Trans. Geosci. Remote Sens.* **2022**, *60*, 5219016. [CrossRef]
12. Huang, Y.; Liao, G.; Zhang, L.; Xiang, Y.; Li, J.; Nehorai, A. Efficient Narrowband RFI Mitigation Algorithms for SAR Systems with Reweighted Tensor Structures. *IEEE Trans. Geosci. Remote Sens.* **2019**, *57*, 9396–9409. [CrossRef]
13. Huang, Y.; Wen, C.; Chen, Z.; Chen, J.; Liu, Y.; Li, J.; Hong, W. HRWS SAR Narrowband Interference Mitigation Using Low-Rank Recovery and Image-Domain Sparse Regularization. *IEEE Trans. Geosci. Remote Sens.* **2022**, *60*, 5217914. [CrossRef]
14. Huang, Y.; Liao, G.; Li, J.; Xu, J. Narrowband RFI Suppression for SAR System via Fast Implementation of Joint Sparsity and Low-Rank Property. *IEEE Trans. Geosci. Remote Sens.* **2018**, *56*, 2748–2761. [CrossRef]
15. Tao, M.; Su, J.; Huang, Y.; Wang, L. Mitigation of Radio Frequency Interference in Synthetic Aperture Radar Data: Current Status and Future Trends. *Remote Sens.* **2019**, *11*, 2438. [CrossRef]
16. Huang, Y.; Zhao, B.; Tao, M.; Chen, Z.; Hong, W. Review of Synthetic Aperture Radar Interference Suppression. *J. Radars* **2020**, *9*, 86–106.
17. Braunstein, M.; Ralston, J.; Sparrow, D. Signal Processing Approaches to Radio Frequency Interference (RFI) Suppression. In Proceedings of the Algorithms for Synthetic Aperture Radar Imagery, Proceedings of the SPIE 2230, Orlando, FL, USA, 9 June 1994; pp. 190–208.
18. Huang, X.; Liang, D. RFI Suppression in UWB-SAR based on RELAX. *Natl. Univ. Def. Technol. J.* **2000**, *22*, 55–59.
19. Huang, X.; Liang, D. Parametric Methods of RFI Suppression in UWB-SAR. *Syst. Eng. Electron.* **2000**, *22*, 94–97.
20. Liu, Z.; Liao, G.; Yang, Z. Time Variant RFI Suppression for SAR Using Iterative Adaptive Approach. *IEEE Geosci. Remote Sens. Lett.* **2013**, *10*, 1424–1428. [CrossRef]
21. Yang, Z.; Du, W.; Liu, Z.; Liao, G. WBI Suppression for SAR Using Iterative Adaptive Method. *IEEE J. Sel. Top. Appl. Earth Obs. Remote Sens.* **2016**, *9*, 1008–1014. [CrossRef]
22. Han, W.; Bai, X.; Fan, W.; Wang, L.; Zhou, F. Wideband Interference Suppression for SAR via Instantaneous Frequency Estimation and Regularized Time-Frequency Filtering. *IEEE Trans. Geosci. Remote Sens.* **2022**, *60*, 5208612. [CrossRef]
23. Cazzaniga, G.; Guarnieri, A.M. Removing RF Interferences from P-band Airplane SAR Data. In Proceedings of the IGARSS' 96, Lincoln, NE, USA, 31 May 1996; pp. 1845–1847.
24. Xu, W.; Xing, W.; Fang, C.; Huang, P.; Tan, W. RFI Suppression Based on Linear Prediction in Synthetic Aperture Radar Data. *IEEE Geosci. Remote Sens. Lett.* **2021**, *18*, 2127–2131. [CrossRef]
25. Li, N.; Lv, Z.; Guo, Z.; Zhao, J. Time-Domain Notch Filtering Method for Pulse RFI Mitigation in Synthetic Aperture Radar. *IEEE Geosci. Remote Sens. Lett.* **2021**, *19*, 4013805. [CrossRef]
26. Vu, V.T.; Sjögren, T.K.; Pettersson, M.I.; Håkansson, L.; Gustavsson, A.; Ulander, L.M.H. RFI Suppression in Ultrawideband SAR Using an Adaptive Line Enhancer. *IEEE Geosci. Remote Sens. Lett.* **2010**, *7*, 694–698. [CrossRef]
27. Lamont-Smith, T.; Hill, R.; Hayward, S.; Yates, G.; Blake, A. Filtering Approaches for Interference Suppression in Low-Frequency SAR. *IEE Proc.-Radar Sonar Navig.* **2006**, *153*, 338–344. [CrossRef]
28. Zhou, F.; Wu, R.; Xing, M.; Bao, Z. Eigensubspace-Based Filtering with Application in Narrow-Band Interference Suppression for SAR. *IEEE Geosci. Remote Sens. Lett.* **2007**, *4*, 75–79. [CrossRef]
29. Zhou, F.; Tao, M.; Bai, X.; Liu, J. Narrow-Band Interference Suppression for SAR Based on Independent Component Analysis. *IEEE Trans. Geosci. Remote Sens.* **2013**, *51*, 4952–4960. [CrossRef]
30. Zhou, F.; Xing, M.; Bai, X.; Sun, G.; Bao, Z. Narrow-band Interference Suppression for SAR based on Complex Empirical Mode Decomposition. *IEEE Geosci. Remote Sens. Lett.* **2009**, *6*, 423–427. [CrossRef]
31. Tao, M.; Zhou, F.; Liu, J.; Liu, Y.; Zhang, Z.; Bao, Z. Narrow-Band Interference Mitigation for SAR Using Independent Subspace Analysis. *IEEE Trans. Geosci. Remote Sens.* **2014**, *52*, 5289–5301.
32. Feng, J.; Zheng, H.; Deng, Y.; Gao, D. Application of Subband Spectral Cancellation for SAR Narrow-Band Interference Suppression. *IEEE Geosci. Remote Sens. Lett.* **2012**, *9*, 190–193. [CrossRef]
33. Yang, H.; Li, K.; Li, J.; Du, Y.; Yang, J. BSF: Block Subspace Filter for Removing Narrowband and Wideband Radio Interference Artifacts in Single-Look Complex SAR Images. *IEEE Trans. Geosci. Remote Sens.* **2022**, *60*, 5211916. [CrossRef]

34. Li, N.; Lv, Z.; Guo, Z. SAR Image Interference Suppression Method by Integrating Change Detection and Subband Spectral Cancellation Technology. *Syst. Eng. Electron.* **2021**, *43*, 2484–2492.
35. Guo, Y.; Liao, G.; Li, J.; Chen, X. A Novel Moving Target Detection Method Based on RPCA for SAR Systems. *IEEE Trans. Geosci. Remote Sens.* **2020**, *58*, 6677–6690. [CrossRef]
36. Yang, D.; Yang, X.; Liao, G.; Zhu, S. Strong Clutter Suppression via RPCA in Multichannel SAR/GMTI System. *IEEE Geosci. Remote Sens. Lett.* **2015**, *12*, 2237–2241. [CrossRef]
37. Guo, Y.; Liao, G.; Li, J.; Gu, T. A Clutter Suppression Method Based on NSS-RPCA in Heterogeneous Environments for SAR-GMTI. *IEEE Trans. Geosci. Remote Sens.* **2020**, *58*, 5880–5891. [CrossRef]
38. Oveis, A.H.; Sebt, M.A. Dictionary-Based Principal Component Analysis for Ground Moving Target Indication by Synthetic Aperture Radar. *IEEE Geosci. Remote Sens. Lett.* **2017**, *14*, 1594–1598. [CrossRef]
39. Leibovich, M.; Papanicolaou, G.; Tsogka, C. Low Rank Plus Sparse Decomposition of Synthetic Aperture Radar Data for Target Imaging. *IEEE Trans. Comput. Imaging* **2020**, *6*, 491–502. [CrossRef]
40. Nguyen, L.H.; Tran, T.; Do, T. Sparse Models and Sparse Recovery for Ultra-Wideband SAR Applications. *IEEE Trans. Aerosp. Electron. Syst.* **2014**, *50*, 940–958. [CrossRef]
41. Liu, H.; Li, D.; Zhou, Y.; Truong, T.K. Joint Wideband Interference Suppression and SAR Signal Recovery Based on Sparse Representations. *IEEE Geosci. Remote Sens. Lett.* **2017**, *14*, 1542–1546. [CrossRef]
42. Huang, Y.; Liao, G.; Xiang, Y.; Zhang, Z.; Li, J.; Nehorai, A. Reweighted Nuclear Norm and Reweighted Frobenius Norm Minimizations for Narrowband RFI Suppression on SAR System. *IEEE Trans. Geosci. Remote Sens.* **2019**, *57*, 5949–5962. [CrossRef]
43. Huang, Y.; Liao, G.; Zhang, Z.; Xiang, Y.; Li, J.; Nehorai, A. Fast Narrowband RFI Suppression Algorithms for SAR Systems via Matrix-Factorization Techniques. *IEEE Trans. Geosci. Remote Sens.* **2019**, *57*, 250–262. [CrossRef]
44. Huang, Y.; Zhang, L.; Li, J.; Hong, W.; Nehorai, A. A Novel Tensor Technique for Simultaneous Narrowband and Wideband Interference Suppression on Single-Channel SAR System. *IEEE Trans. Geosci. Remote Sens.* **2019**, *57*, 9575–9588. [CrossRef]
45. Yang, H.; Chen, C.; Chen, S.; Xi, F.; Liu, Z. A Dictionary-Based SAR RFI Suppression Method via Robust PCA and Chirp Scaling Algorithm. *IEEE Geosci. Remote Sens. Lett.* **2021**, *18*, 1229–1233. [CrossRef]
46. Joy, S.; Nguyen, L.H.; Tran, T.D. Joint Down-Range and Cross-Range RFI Suppression in Ultra-Wideband SAR. *IEEE Trans. Geosci. Remote Sens.* **2021**, *59*, 3136–3149. [CrossRef]
47. Zhang, H.; Huang, Y.; Li, J.; Chen, Z.; Cai, L.; Hong, W. Time-Varying RFI Mitigation for SAR Systems via Graph Laplacian Clustering Techniques. *IEEE Geosci. Remote Sens. Lett.* **2022**, *19*, 4010805. [CrossRef]
48. Ding, Y.; Fan, W.; Zhang, Z.; Zhou, F.; Lu, B. Radio Frequency Interference Mitigation for Synthetic Aperture Radar Based on the Time-Frequency Constraint Joint Low-Rank and Sparsity Properties. *Remote Sens.* **2022**, *14*, 775. [CrossRef]
49. Akeret, J.; Chang, C.; Lucchi, A.; Refregier, A. Radio Frequency Interference Mitigation Using Deep Convolutional Neural Networks. *Astron. Comput.* **2017**, *18*, 35–39. [CrossRef]
50. Fan, W.; Zhou, F.; Tao, M.; Bai, X.; Rong, P.; Yang, S.; Tian, T. Interference Mitigation for Synthetic Aperture Radar Based on Deep Residual Network. *Remote Sens.* **2019**, *11*, 1654. [CrossRef]
51. Tao, M.; Li, J.; Su, J.; Wang, L. Characterization and Removal of RFI Artifacts in Radar Data via Model-Constrained Deep Learning Approach. *Remote Sens.* **2022**, *14*, 1578. [CrossRef]
52. Zhou, C.; Li, F.; Li, N.; Zheng, H.; Wang, R. Improved Eigensubspace-based Approach for Radio Frequency Interference Filtering of Synthetic Aperture Radar Images. *J. Appl. Remote Sens.* **2017**, *11*, 025004. [CrossRef]
53. Shang, R.; Lin, J.; Jiao, L.; Li, Y. SAR Image Segmentation Using Region Smoothing and Label Correction. *Remote Sens.* **2020**, *12*, 803. [CrossRef]

Article

Crop Classification Based on GDSSM-CNN Using Multi-Temporal RADARSAT-2 SAR with Limited Labeled Data

Heping Li [1,2,3], Jing Lu [4], Guixiang Tian [1,2,3], Huijin Yang [1,2,3,*], Jianhui Zhao [1,2,3] and Ning Li [1,2,3]

1   School of Computer and Information Engineering, Henan University, Kaifeng 475004, China
2   Henan Engineering Research Center of Intelligent Technology and Application, Henan University, Kaifeng 475004, China
3   Henan Key Laboratory of Big Data Analysis and Processing, Henan University, Kaifeng 475004, China
4   Land Satellite Remote Sensing Application Center, Ministry of Natural Resources, Beijing 100048, China
*   Correspondence: huijiny@henu.edu.cn

Abstract: Crop classification is an important part of crop management and yield estimation. In recent years, neural networks have made great progress in synthetic aperture radar (SAR) crop classification. However, the insufficient number of labeled samples limits the classification performance of neural networks. In order to solve this problem, a new crop classification method combining geodesic distance spectral similarity measurement and a one-dimensional convolutional neural network (GDSSM-CNN) is proposed in this study. The method consisted of: (1) the geodesic distance spectral similarity method (GDSSM) for obtaining similarity and (2) the one-dimensional convolutional neural network model for crop classification. Thereinto, a large number of training data are extracted by GDSSM and the generalized volume scattering model which is based on radar vegetation index (GRVI), and then classified by 1D-CNN. In order to prove the effectiveness of the GDSSM-CNN method, the GDSSM method and 1D-CNN method are compared in the case of a limited sample. In terms of evaluation and verification of methods, the GDSSM-CNN method has the highest accuracy, with an accuracy rate of 91.2%, which is 19.94% and 23.91% higher than the GDSSM method and the 1D-CNN method, respectively. In general, the GDSSM-CNN method uses a small number of ground measurement samples, and it uses the rich polarity information in multi-temporal fully polarized SAR data to obtain a large number of training samples, which can quickly improve the accuracy of classification in a short time, which has more new inspiration for crop classification.

Keywords: crop classification; synthetic aperture radar (SAR); deep learning; fully polarimetric; multi-temporal; sample limited

Citation: Li, H.; Lu, J.; Tian, G.; Yang, H.; Zhao, J.; Li, N. Crop Classification Based on GDSSM-CNN Using Multi-Temporal RADARSAT-2 SAR with Limited Labeled Data. *Remote Sens.* **2022**, *14*, 3889. https://doi.org/10.3390/rs14163889

Academic Editors: Haipeng Wang, Gang Xu and Lan Du

Received: 7 July 2022
Accepted: 10 August 2022
Published: 11 August 2022

Publisher's Note: MDPI stays neutral with regard to jurisdictional claims in published maps and institutional affiliations.

1. Introduction

Crops are the basis for maintaining human civilization and are of great significance to human diet and social stability [1,2]. Obtaining crop planting distribution information is very important for growth monitoring, yield estimation and food security [3–6]. The ground survey is a traditional way to obtain information on crop planting distribution, which consumes a lot of time and money. Agricultural remote sensing technology, which offers the advantages of a large monitoring range and low cost, has gradually replaced the traditional methods [7,8].

Optical sensors have been widely used in crop classification for a long time [9–11]. However, optical sensors are limited by weather conditions. Images cannot be created in sufficient quality in the case of bad weather (such as cloudy, rainy or foggy) or during the nighttime [12,13]. Crops will experience significant changes to their shape at important phenological periods. However, many farmlands are obscured by clouds, making it more likely that the optical sensor will fail to capture images during the crucial phenological period for crops [14,15]. Synthetic aperture radar (SAR) is an active microwave sensor, which cannot be affected by cloudy and foggy weather [16]. Moreover, SAR is sensitive to the dielectric properties and structure of plants and is very suitable for monitoring and classifying crops [17,18]. In addition, polarimetric SAR (PolSAR), which can provide rich information, can extract a large number of polarimetric features. These polarimetric features are very valuable in crop classification [19,20]. Therefore, the multi-temporal RadarSat-2 image is used as the main data source of crop classification in this study.

There is a long history of extracting polarimetric features from PolSAR data for crop classification [21]. Polarization decomposition technology is the main method of extracting polarization characteristics [22–24] and has been used by researchers in agricultural remote sensing applications [25,26]. Guo et al. [26] used various features of H/α decomposition and support vector machine (SVM) to classify crops. The results show that the features obtained by polarization decomposition provide accurate polarization information, and the classification accuracy of crops exceeds 80%. The method of polarization decomposition is helpful for crop classification. However, the polarization decomposition method is sensitive to the target orientation and uses a specific volume model, which is rarely applicable to all phenological stages of different crops [27,28]. Therefore, there is great potential to classify crops by using polarization characteristics that correlate the relevant physical scattering mechanisms of vegetation canopy with phenology. It has been proved that the generalized volume scattering model which is based on the radar vegetation index (GRVI) can well reflect the relevant physical scattering mechanism of the crop canopy and follow the growth and development of crops [29,30].

Multi-temporal PolSAR has very rich characteristics [31]. In order to improve the accuracy of crop classification, people have proposed various methods to use the rich information in PolSAR, such as support vector machine [32] and random forest (RF) [33]. SVM and RF have attracted much attention because of their good performance and ease of use in limited training samples, but the feature extraction of these methods is still limited [34]. Because of the exceptional results in the area of computer vision, deep learning has become the mainstream method in the field of SAR image processing. In recent years, many scholars have proposed various deep learning methods for crop classification [35,36]. Zhang et al. [35] proposed a novel crop discrimination network with multi-scale features (MSCDN), which proved that the neural network can efficiently extract features in multi-temporal PolSAR, with an overall accuracy rate of crop classification of 99.33%. Chang et al. [36] proposed a convolutional long short-term memory rice field classifier (ConvLSTM-RFC) and demonstrated the effectiveness of the convolutional block attention module (CBAM) in handling multi-temporal features, with the highest accuracy of 98.08%. However, because sample sampling requires great time and money costs, the sample is often limited in the process of crop classification. These deep learning methods easily overfit and degrade the classification performance when the samples are limited [37–39]. Therefore, it is still a challenge to develop a multi-temporal PolSAR crop classification method in the case of limited samples.

The motivation of this study is to propose a multi-temporal and fully polarized crop classification method that can accurately classify crops with limited samples. Firstly, the geodesic distance spectral similarity measure (GDSSM) method is used to calculate the similarity between unlabeled pixels and the samples. Secondly, the open mask product Global Food Security-Support Analysis Data at 30 m (GFSAD) is used to generate a farmland mask in combination with GDSSM to extract farmland areas and non-farmland areas in the study area. Thirdly, multi-temporal backscattering coefficient and GRVI are extracted as features. Fourthly, the data with GDSSM greater than the threshold are selected as training data. Finally, the one-dimensional convolutional neural network (1D-CNN) is used to classify crops. The main contributions of this paper are summarized as follows:

1. For multi-temporal and fully polarimetric SAR images, a crop classification method combining the geodesic distance spectral similarity measure and a one-dimensional convolution neural network (GDSSM-CNN) is proposed. This method can still have excellent classification ability when the sample is limited.

2. The backscatter coefficient and GRVI are extracted from multi-temporal and fully polarized images and combined.
3. For multi-temporal and fully polarimetric SAR images, a farmland region extraction method combining GFSAD data and GDSSM is proposed. Farmland and non-farmland areas can be extracted by this method delicately.

The rest of this paper is arranged as follows: In Section 2, the research area, data and methodology are introduced. The experimental results of crop classification are given in Section 3. The discussion of the proposed method and its performance is introduced in Section 4. Section 5 concludes this paper.

2. Materials and Methods

In this section, the materials and methods used for the experiments are described. Section 2.1 briefly introduces the study area, Section 2.2 describes the data used and the preprocessing of the data and Section 2.3 details the GDSSM-CNN method.

2.1. Study Area

The study area (114°27′40″E–114°48′57″E, 34°35′34″N–35°54′7″N) is located in the east of Kaifeng City, Henan Province, on the south bank of the Yellow River, as shown in Figure 1. Kaifeng has a temperate monsoon climate. It is cold and dry in winter and hot and humid in summer. The annual rainfall is 635 mm, and the annual average temperature (1981–2010) is 14.52 °C. Kaifeng is located on a plain with many rivers and moist soil, which is suitable for planting crops. In addition, Kaifeng City, with a farmland area of 3933 square kilometers and a total grain output of 3.06 million tons in 2021, is an important agricultural city in China. Crops are mainly divided into summer harvest crops and autumn harvest crops. In Kaifeng, summer harvest crops refer to the crops grown from October to June of the next year, mainly including winter wheat and a small amount of garlic; autumn harvest crops refer to crops grown from June to October, mainly including corn, peanuts and a small amount of rice and vegetables. This study focuses on summer crops; the planting area of winter wheat in 2021 was about 3098 square kilometers, and the garlic planting area was about 706 square kilometers.

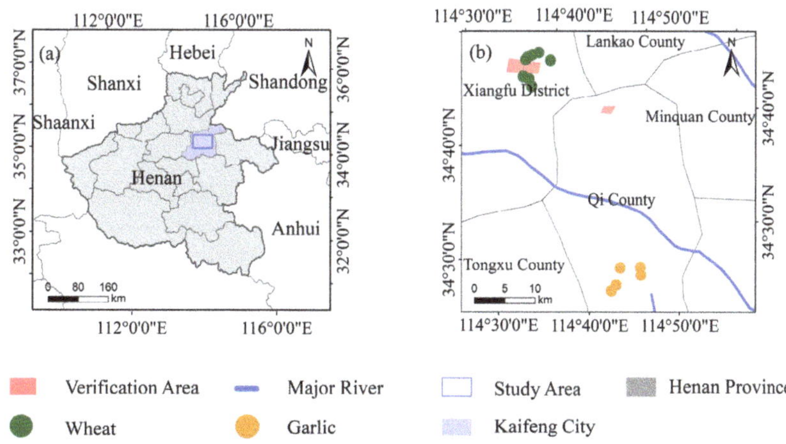

Figure 1. (**a**) The location of the field survey sites in Kaifeng, Henan Province, China. (**b**) The location of the study area within Henan Province.

2.2. Datasets and Preprocessing

RadarSat-2 is a ground imaging sensor equipped with C-band full polarimetric SAR [40]. It supports applications in various fields, including crop monitoring and pollution detection, target recognition and geological mapping. RadarSat-2 can perform imaging tasks all day, and the revisit cycle is 24 days. The specific parameters of the data are shown in Table 1. In this study, the RadarSat-2 data of SLC (Single Look Complex) of the fine quad polarization mode are used. The five-scene data of winter wheat growth in 2020 were obtained, which were 20 February, 15 March, 8 April, 2 May and 26 May, respectively. The time basically covered the complete life cycle of winter wheat from hibernation to maturity in the area.

Table 1. Parameters for RadarSat-2.

Parameters	RadarSat-2	Parameters	RadarSat-2
Product type	SLC	Pass direction	Ascending
Imaging mode	Fine quad polarization	Center frequency	5.4 GHz
Polarization	VV&VH&HH&HV	Look direction	Right
Resolution	8×8 m	Incidence angle	$29°$–$31°$
Band	C	Dates	20 February 2020, 15 March 2020, 8 April 2020, 2 May 2020, 26 May 2020

RadarSat-2 data are preprocessed using the Sentinel Application Platform (SNAP) software provided by the European Space Agency (ESA) and The Environment for Visualizing Images (ENVI), a product of American Exelis Visual Information Solutions company. Pretreatment mainly includes the following: (1) orbit file application; (2) calibration, in which the backscattering coefficient of each pixel can be obtained; (3) polarimetric generation, in which the polarization coherency T matrix of each pixel can be obtained; (4) multilooking; (5) refined Lee filtering; (6) range Doppler terrain correction; (7) conversion of the backscatter coefficient from linear to dB scale; (8) study area extraction; (9) registration. Each pixel in the multi-temporal SAR image corresponds to the same resolution unit. After preprocessing, ArcGIS software extracts the backscattering coefficient and polarization coherence matrix through the longitude and latitude coordinates of the sampling points.

The field survey data are essential information for crop classification and accuracy verification. In order to ensure the authenticity and accuracy of the ground field survey data, field surveys were conducted at the time of each satellite transit from July 2019 to June 2020 to collect and record the sample data of crops in the study area. During the field survey, Jisibao UG905 GPS (positioning accuracy is 1 to 3 m) was used outdoors to record the longitude and latitude information of the sampling point, and photos were taken to record the crop category information of the sampling point. Finally, 10 winter wheat sampling sites and 5 garlic sampling sites were randomly selected as samples. Two areas in the study area were selected randomly as the verification area through Google Earth software, and the types of crops in the verification area were determined through ground field investigation. The validation area contains 32,361 pixels in total, including 18,473 pixels for winter wheat, 6859 pixels for garlic and 7029 pixels for non-farmland areas. The distribution of the verification area and sampling points are shown in Figure 1a.

2.3. Methodology

The flow chart of this method is shown in Figure 2. The method combines the similarity calculation method GDSSM and the deep learning method 1D-CNN (GDSSM-CNN) to increase the number of training data and classify crops when the samples are limited. After collecting data and preprocessing, the method of crop type recognition is mainly divided into five steps. The first step is to calculate the GDSSM of pixels and samples, which can comprehensively compare the similarity of multi-temporal and fully polarimetric SAR data. In step 2, more refined farmland areas are extracted through the open farmland data GFSAD and GDSSM to prevent the interference of non-farmland factors. In step 3, the threshold method is used to extract the pixels that are very similar to the sample category as training data, which can fully train the neural network when the samples are very limited. In step 4, the GRVI of the pixel is calculated and combined with the backscatter coefficient as the feature. GRVI reflects the physical scattering characteristics and phenological characteristics of crops and can improve the performance and efficiency of classification. In step 5, the 1D-CNN neural network model is used for training, which can effectively extract the features of multi-temporal data. Finally, the classification results of the model are analyzed and verified.

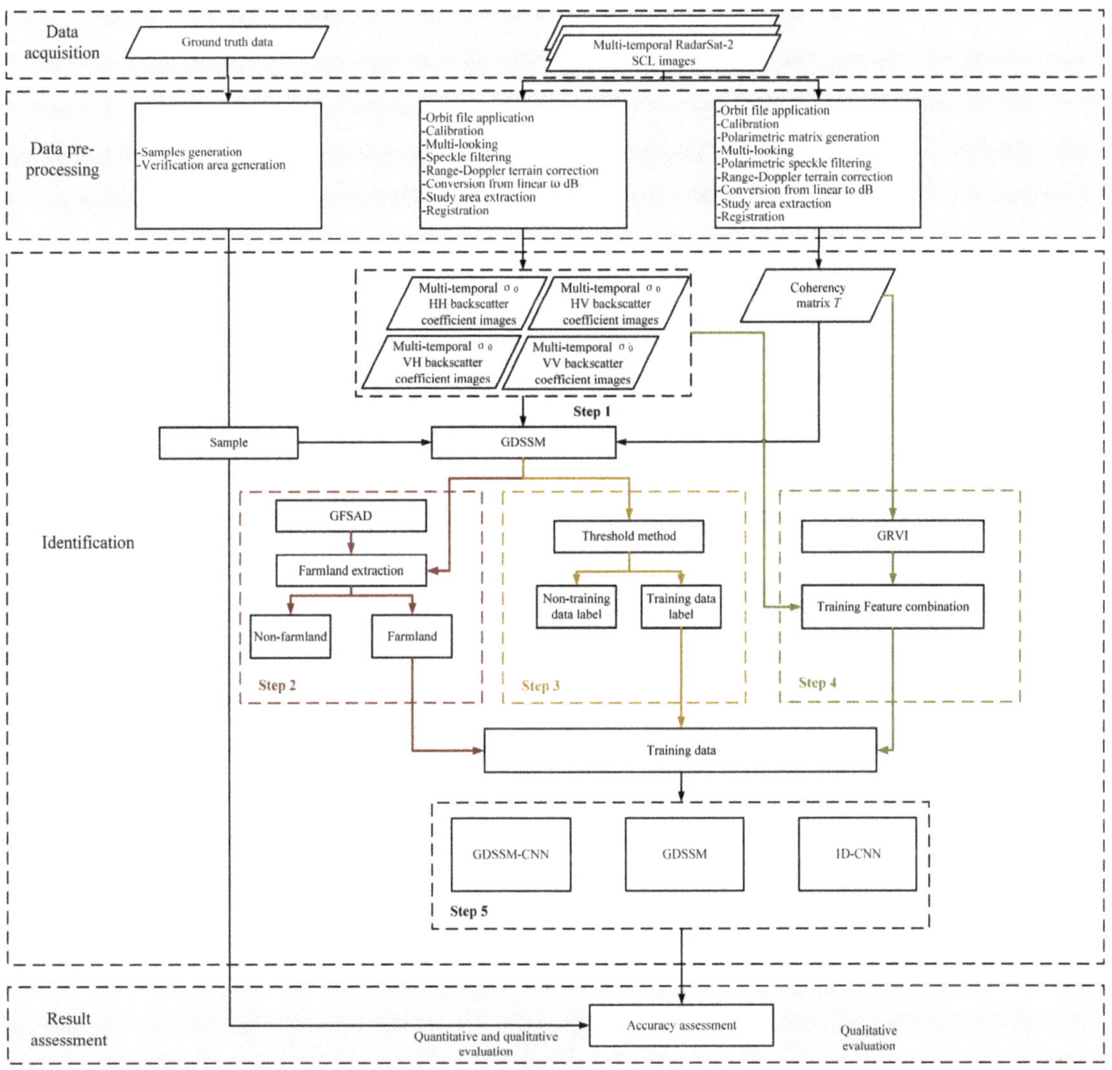

Figure 2. Flow chart of the proposed method.

2.3.1. Geodesic Distance Spectral Similarity Measure

GDSSM (shown in Equation (8)) defines the similarity between the reference target and unknown pixels; it is composed of geodesic distance similarity (GDS) (shown in Equations (1)–(3)), spectral correlation similarity (SCS) (shown in Equations (4) and (5)) and Euclidean distance similarity (EDS) (shown in Equations (6) and (7)). The specific descriptions of SCS and EDS appear in [41,42], and the specific description of GDS appears in [29]. GDS compares the scattering similarity of the polarization coherence matrix T of a reference target and an unknown pixel. In the polarization scattering theory, the Kennaugh matrix of the scattering of incoherent targets reflects the dependence of the radar received power on the transceiver antenna, which is shown by the coherence matrix T:

$$K = \begin{bmatrix} \frac{T_{11}+T_{22}+T_{33}}{2} & \Re(T_{12}) & \Re(T_{13}) & \Im(T_{23}) \\ \Re(T_{12}) & \frac{T_{11}+T_{22}-T_{33}}{2} & \Re(T_{23}) & \Im(T_{13}) \\ \Re(T_{13}) & \Re(T_{23}) & \frac{T_{11}-T_{22}+T_{33}}{2} & -\Im(T_{12}) \\ \Im(T_{23}) & \Im(T_{13}) & -\Im(T_{12}) & \frac{-T_{11}+T_{22}+T_{33}}{2} \end{bmatrix} \tag{1}$$

where $\Re$ and $\Im$ represent the real and imaginary parts of the complex number. The geodesic distance similarity value (GDSV) describes the similarity of geodesic distances between two Kennaugh matrices on the unit sphere. The equation used to calculate GDSV is as follows:

$$\text{GDSV} = \text{GD}(K_p, K_q) = \frac{2}{\pi} \cos^{-1} \frac{Tr\left(K_q^T K_p\right)}{\sqrt{Tr\left(K_q^T K_q\right)} \sqrt{Tr\left(K_p^T K_p\right)}} \tag{2}$$

where K_p and K_q denote the unknown pixel Kennaugh matrix and reference target Kennaugh matrix, respectively. Tr represents the matrix trace, K^T represents the transposition of the of pixel Kennaugh matrix, and the range of GDSV is [0, 1]. GDS is the average of GDSV for five dates. The equation used to calculate GDS is as follows:

$$\text{GDS} = \frac{1}{n} \sum_{i=1}^{n} \text{GDSV}_i = \frac{1}{n} \sum_{i=1}^{n} \text{GD}(K_{pi}, K_{qi}) \tag{3}$$

where n is the number of Radarsat-2 data images, K_{pi} and K_{qi} are the Kennaugh matrices of unknown pixel and reference target in the image, respectively.

SCS and EDS compare the similarity of backscattering coefficient vectors of a reference target and an unknown pixel. The objects of SCS and EDS are the spectral information in optical remote sensing. However, in this study, the multi-phase characteristics of the backscattering coefficient are used to replace the spectral information; that is, the fully polarized backscattering coefficient at a certain time represents a spectral value. The multi-temporal fully polarized backscattering coefficient of each crop in the sample set is used as the vector of the reference target. The multi-temporal backscattering coefficient of a pixel in the dataset to be classified is taken as the vector of the unknown target. The equation used to calculate SCS is as follows:

$$\text{SCS}_{\text{orig}} = \frac{1}{n-1} \left(\frac{\sum_{i=1}^{n}(P_i - \mu_P)(Q_i - \mu_Q)}{\sigma_P \sigma_Q} \right) \tag{4}$$

and

$$\text{SCS} = \frac{\text{SCS}_{\text{orig}} - m_s}{M_s - m_s} \tag{5}$$

where μ and σ are the mean and standard deviation of the vector; P and Q are the backscattering coefficient vectors of the unknown pixel and the reference pixel, respectively; and P_i and Q_i are the backscattering coefficients of the unknown pixel and the reference pixel at a certain time, respectively. m_s and M_s are the minimum and maximum values of SCS_{orig}. The range of SCS is [0, 1]. The larger the value, the more similar the reference pixel is to the target pixel.

The Euclidean distance measure is used to measure the degree of separation or proximity between two vectors. The equation used to calculate EDS is as follows:

$$\text{Ed}_{\text{orig}} = \sqrt{\sum_{i=1}^{n}(P_i - Q_i)^2} \tag{6}$$

and

$$\text{EDS} = \frac{\text{Ed}_{\text{orig}} - m_E}{M_E - m_E} \tag{7}$$

where m_E and M_E are the minimum and maximum values of Ed_{orig}, respectively. The range of EDS is [0, 1]. The smaller the value, the more similar the reference pixel is to the target pixel. GDSSV is composed of GDS, SCS and EDS, and its equation is as follows:

$$\text{GDSSV} = \sqrt{\text{SCS}^2 + (1 - \text{Ed})^2} \times \text{GDS} \tag{8}$$

The range of GDSSV is [0, 1]. The larger the GDSSV, the more similar the unknown pixel is to the reference pixel. When the GDSSV of a pixel and the reference pixel is greater than 0.80, it is considered that the unknown pixel and the reference pixel belong to the same category of crops.

2.3.2. Generalized Volume Scattering Model-Based Radar Vegetation Index

GRVI describes the similarity between Kennaugh matrix K and generalized volume scattering model K_v of pixels. It has been proved to have the potential to monitor the growth of rice at different phenological stages [29,30]. K_v is related to the generalized volume scattering model [43], which is shown as follows:

$$K_v = \frac{1}{\frac{3(1+\gamma)}{4} - \frac{\sqrt{\gamma}}{6}} \begin{bmatrix} \frac{3}{2}(1+\gamma) - \frac{\sqrt{\gamma}}{3} & \gamma - 1 & 0 & 0 \\ \gamma - 1 & \frac{1}{2}(1+\gamma) + \frac{\sqrt{\gamma}}{3} & 0 & 0 \\ 0 & 0 & \frac{1}{2}(1+\gamma) + \frac{\sqrt{\gamma}}{3} & 0 \\ 0 & 0 & 0 & \frac{1}{2}(1+\gamma) - \sqrt{\gamma} \end{bmatrix} \tag{9}$$

where γ indicates the co-polarized ratio. Geodesic distance f_v between K and K_v is as follows:

$$f_v = 1 - \text{GD}(K, K_v) \tag{10}$$

where GD describes the geodesic distance between two Kennaugh matrices. GRVI (shown in Equation (2)) is composed of geodesic distance f_v and modulation parameters β. The equation used to calculate GRVI and β is as follows:

$$\beta = \left(\frac{p}{q}\right)^{2GD(K,K_v)} \tag{11}$$

$$p = \min \begin{bmatrix} \text{GD}(K, K_t) \\ \text{GD}(K, K_c) \\ \text{GD}(K, K_d) \\ \text{GD}(K, K_{nd}) \end{bmatrix}, q = \max \begin{bmatrix} \text{GD}(K, K_t) \\ \text{GD}(K, K_c) \\ \text{GD}(K, K_d) \\ \text{GD}(K, K_{nd}) \end{bmatrix} \tag{12}$$

$$\text{GRVI} = \beta f_v \tag{13}$$

where p and q are the minimum and maximum geodesic distances between K and elementary targets, respectively. K_t is the Kennaugh matrix of a trihedral; K_c is the Kennaugh matrix of a cylinder; K_d is the Kennaugh matrix of a dihedral; K_{nd} is the Kennaugh matrix of a narrow dihedral. This Kennaugh matrix of elementary targets is shown in the work of Ratha et al. [44].

2.3.3. Farmland Extraction

Since the scattering characteristics of the non-farmland area are so complicated, it is possible that parts of the structure and scattering characteristics of crops at different growth stages are similar to those of non-farmland areas, which could lead to incorrect results in the classification findings. In this study, the open dataset GFSAD is used to provide a farmland mask with a resolution of 30 m. It has been widely used in various agricultural applications [45]: farmland type identification, spatial and temporal changes in farmland extent, cropping intensities, etc. However, farmland and villages in Kaifeng are distributed in a scattered way, so there are only dense non-farmland areas such as villages, roads and waters in GFSAD, and many small roads and scattered buildings have not been identified. Unknown pixels with GDSSM similarity less than 0.3 for all crop categories are marked as non-farmland pixels. These non-farmland pixel sets in GDSSM and non-farmland pixel sets in GFSAD form the non-farmland areas in the study area, which are shown as follows:

$$NF = NF_{\text{GDSSM}} \cap NF_{\text{GFSAD}} \tag{14}$$

where NF_{GDSSM} and NF_{GFSAD} are the sets of non-farmland pixels in GDSSM and GFSAD, respectively. NF is the set of non-farmland pixels in the study area.

The farmland mask can be obtained by removing all non-farmland areas in the study area, which is shown as follows:

$$FM = SM - NF \tag{15}$$

where SM is the set of all pixels in the study area. FM is the set of all farmland pixels in the study area. This method, which combines the geodesic distance spectral similarity measure and GFSAD data, can accurately extract the farmland and non-farmland areas in the study area and finally remove the non-farmland areas in the study area.

2.3.4. Training Data Extraction

Deep learning models usually require sufficient data training. Because the dataset of SAR image is less than that of natural scene and SAR image is very sensitive to the incident angle and other information, current optical images and data enhancement operations used in natural scenes are not reasonable for SAR images [46]. The crops planted in different areas may have large morphological differences. For example, the wheat in Henan Province is generally 80 cm high after ripening, while the wheat in Zhejiang Province is generally 65 cm high after ripening. Due to the morphological differences between crops, transfer learning, meta-learning and other sample migration methods are not suitable for crop applications. In this study, GDSSM and the threshold method are used to obtain a large number of samples. GDSSM describes the similarity between the reference target and unknown pixels. The larger the value, the more similar the two pixels are. When the GDSSM of the unknown pixel and the reference pixel is greater than the threshold, the unknown pixel is regarded as the same training data as the reference target category. Through the threshold method, a large number of training data can be selected, which can be used for the training of the neural network. In order to ensure that the training data and samples have a very high degree of similarity, the threshold value needs to be greater than or equal to 0.8. Finally, four thresholds are selected, namely 0.95, 0.9, 0.85 and 0.8, and the effects of different thresholds on classification results are compared.

2.3.5. One-Dimensional Convolutional Neural Network

In order to make effective use of the information in multi-temporal and fully polarimetric SAR images, the backscatter coefficients and GRVI of the four polarization modes are combined into features in the order of HH, HV, VH, VV and GRVI. These combined features are arranged according to the time of five SAR images. In order to use feature combinations effectively, a convolution neural network (CNN) model is used to classify crops. The convolutional neural network is a widely used method in deep learning and has been applied in various applications of agricultural remote sensing. It has been proved to have excellent feature extraction ability. Among CNNs, the one-dimensional convolutional neural network model has more advantages in analyzing sequence characteristic data [47]. Three convolutional filters are set in the 1D-CNN model. Three-layer convolution can comprehensively extract the features of sequence data integration. The maximum pool layers and batch normalization layers are set after the convolution layer, which can reduce the size of the model and improve the calculation speed. Finally, the dimension of the data is transformed by using the flatten layer, and the data are sent to the Softmax layer for classification. The 1D-CNN model is created through the Keras framework.

A RadarSat-2 image has four polarizations. Five features can be obtained from each SAR image. They are four backscattering coefficients ($\sigma0$ HH, $\sigma0$ HV, $\sigma0$VH and $\sigma0$ VV) and a generalized volume scattering model based on radar vegetation index. Since five RadarSat-2 images at different times are used, the characteristic shape of each sample is (1, 25). To facilitate the input of samples into the 1D-CNN model, the GRVI values of five dates are copied, filled into the features and reshaped. The shape of each sample changes to (15, 2), as shown in Figure 3.

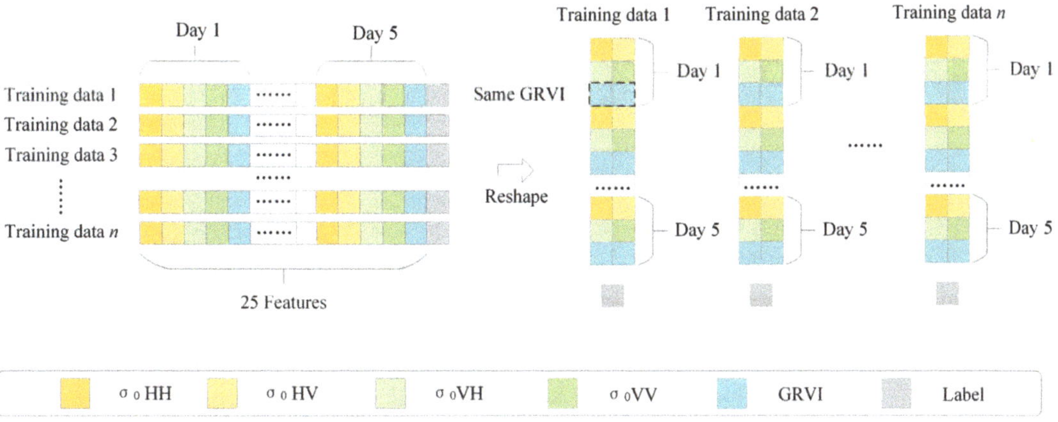

Figure 3. Backscattering coefficient and GRVI reshaping.

The 1D-CNN model is used to identify crop types. The detailed parameters of 1D-CNN are shown in Figure 4. By using the training data obtained by GDSSM, the trained 1D-CNN model is used to classify the crops in the study area.

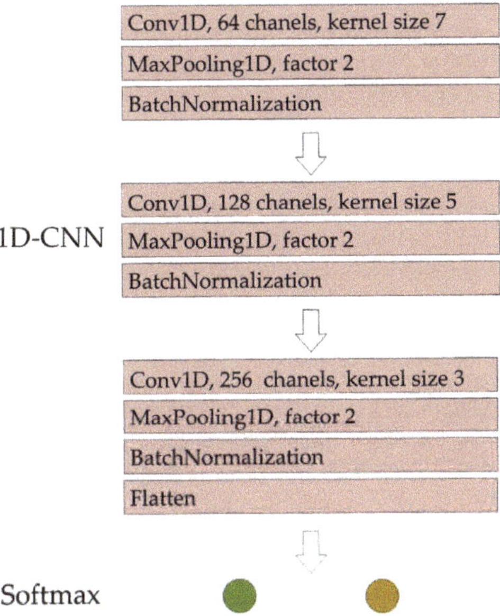

Figure 4. 1D-CNN network structure and parameters.

3. Results

To evaluate the proposed method, this section will present the results of the experiments. Firstly, Section 3.1 presents the backscattering coefficients and GRVI for multitemporal phases. Section 3.2 presents GDSSM for different crops. After that, Section 3.3 shows the farmland masks obtained by the GDSSM method. Finally, Section 3.4 presents the classification results of different methods.

3.1. Temporal Profiles of the Backscatter Coefficient and GRVI

The backscattering coefficient represents the reflectivity of the target unit cross-sectional area in the incident direction of the radar electromagnetic wave and can reflect the geometric and physical characteristics of the illuminated object, such as surface roughness and water content, which can help identify the types of crops. Figure 5 shows the time profile of four polarization modes, where the value of each point represents the average backscatter coefficient of each crop type sample.

The radar vegetation index based on the generalized lifting scattering model is a vegetation index that reflects the scattering mechanism of crops. The vegetation index changes with the growth and development of crops and has been proved to have potential in rice monitoring. This parameter is helpful in classifying crops. Figure 5 shows the time profile of GRVI, where the value of each point represents the average GRVI for each crop type sample.

3.2. GDSSM for Different Crops

Figure 6 shows the GDSSM between unknown pixels and winter wheat samples and between unknown pixels and garlic samples in the study area. As can be seen from Figure 6, GDSSM values of non-farmland pixels such as towns and paths are very low (within a range of 0–0.4). In Figure 6a, the farmland mainly planted with winter wheat has a very high GDSSM with the winter wheat sample (within a range of 0.8–1), while the farmland mainly planted with garlic has a very low GDSSM with the winter wheat sample (within a range of 0–0.2). The opposite is true in Figure 6b. Areas where winter wheat and garlic are grown were identified during the field visit. This proves that GDSSM can accurately show the similarity between unknown pixels and reference samples and is very helpful for distinguishing crop types and farmland and non-farmland areas. Moreover, it is proved that when GDSSM is greater than or equal to 0.8, the unknown pixel has a very high similarity with the reference sample, which is conducive to selecting the pixel with high similarity as the training data.

3.3. Farmland Mask

The extraction results of the farmland mask in some study areas and the mask of GFSAD are shown in Figure 7. It can be seen from the figure that GFSAD only contains a few dense buildings and large villages. After combining GDSSM, many paths and scattered buildings are identified. The reason may be that the paths are very narrow and the image resolution is very low, resulting in a path and a small part of the farmland beside the road being in the same pixel. Through the GDSSM method, it can be judged that the similarity between these mixed pixels and crops is not high, allowing the identification of paths and discrete buildings. This fully shows that the farmland extraction method of GDSSM can identify smaller non-farmland areas and finer farmland areas.

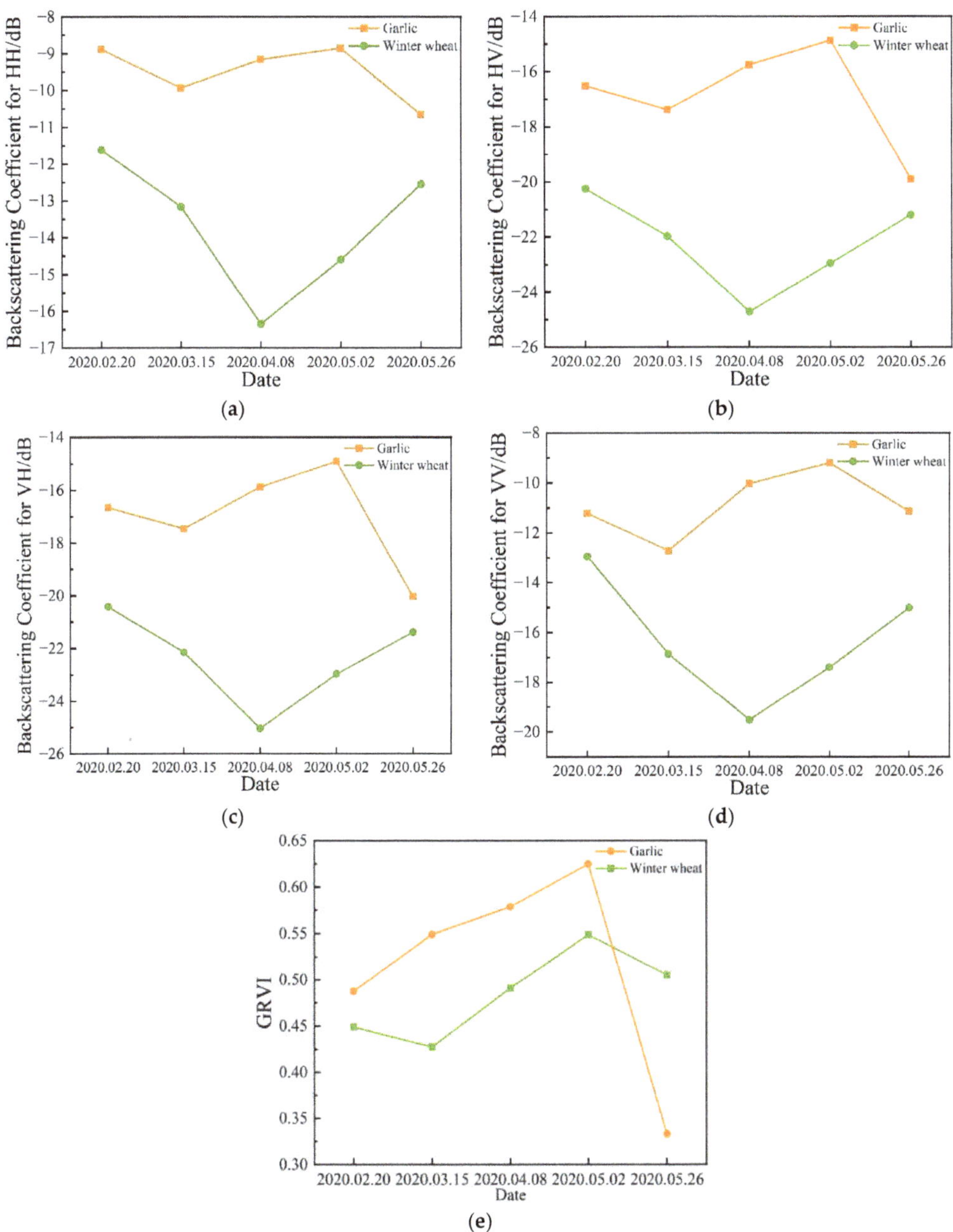

Figure 5. Temporal profiles of the two different crop types with respect to the backscatter coefficient: (**a**) HH (**b**) HV (**c**) VH and (**d**) VV; (**e**) temporal profiles of the two different crop types with respect to GRVI.

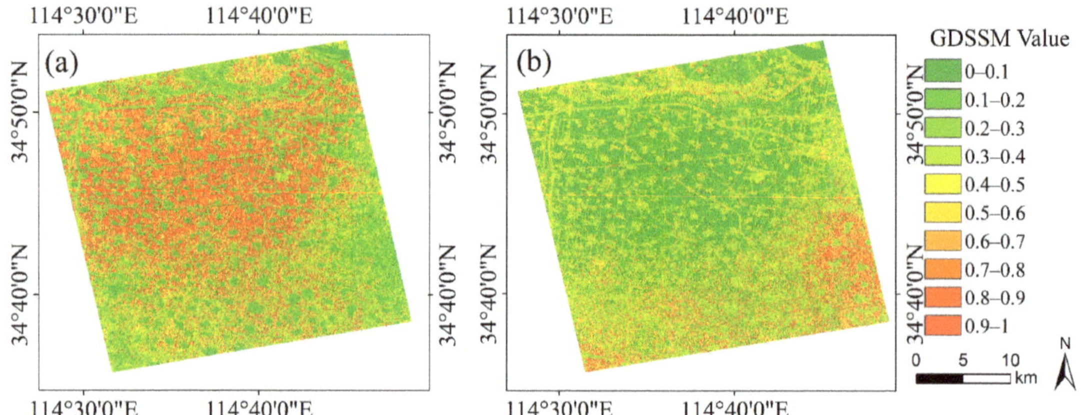

Figure 6. GDSSM value images of study area for (**a**) winter wheat sample and (**b**) garlic sample.

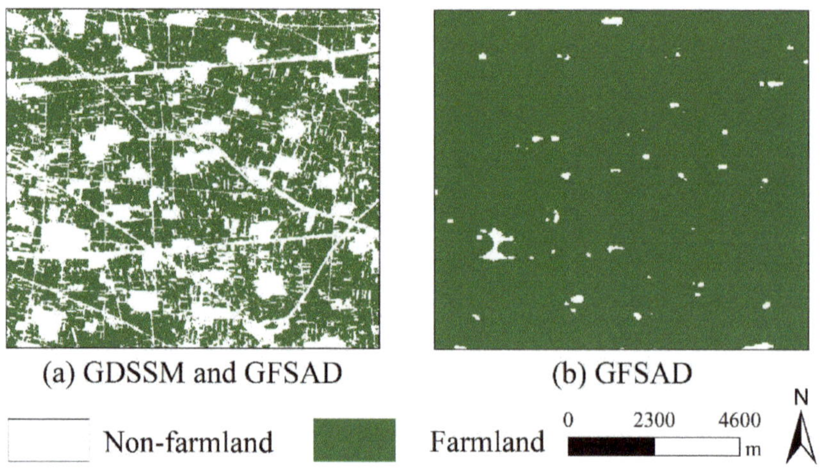

Figure 7. Analysis of the farmland mask for (**a**) GDSSM and GFSAD and (**b**) GFSAD.

3.4. Crop Classification

In order to prove the effectiveness of the model combining geodesic distance spectral similarity measure and one-dimensional convolutional neural network (GDSSM-CNN), the crop recognition method using GDSSM and the crop recognition method using the 1D-CNN model are compared. The use of similarity and threshold methods to classify crops has been proved to have a good effect [48,49]. In addition, in order to study the influence of different thresholds (in the training data generation part) on the classification results, GDSSM-CNN methods with thresholds of 0.95, 0.9, 0.85 and 0.8 are used for comparison. Pixels whose threshold value is greater than or equal to 0.8 are considered to have very high similarities with reference samples (shown in Section 4.2). Therefore, only thresholds greater than or equal to 0.8 are used.

In order to compare the performance of various methods more precisely, the output results of each model and the ground data labels are generated into a confusion matrix for analysis, as shown in Figure 8. From these confusion matrices, it can be seen that GDSSM-CNN models show high accuracy, while the accuracy of using 1D-CNN and GDSSM methods alone is very low. The results are presented in Table 2. The GDSSM-CNN method with a threshold of 0.8 for training data has the highest accuracy, with an accuracy rate of 91.20% and a Kappa coefficient of 0.8509. Experiments show that the combined performance of the GDSSM classifier and 1D-CNN classifier is better than that of either of them alone. The accuracy increased by 19.94% and 23.91%, respectively, and the Kappa coefficient increased by 0.3018 and 0.5497, respectively.

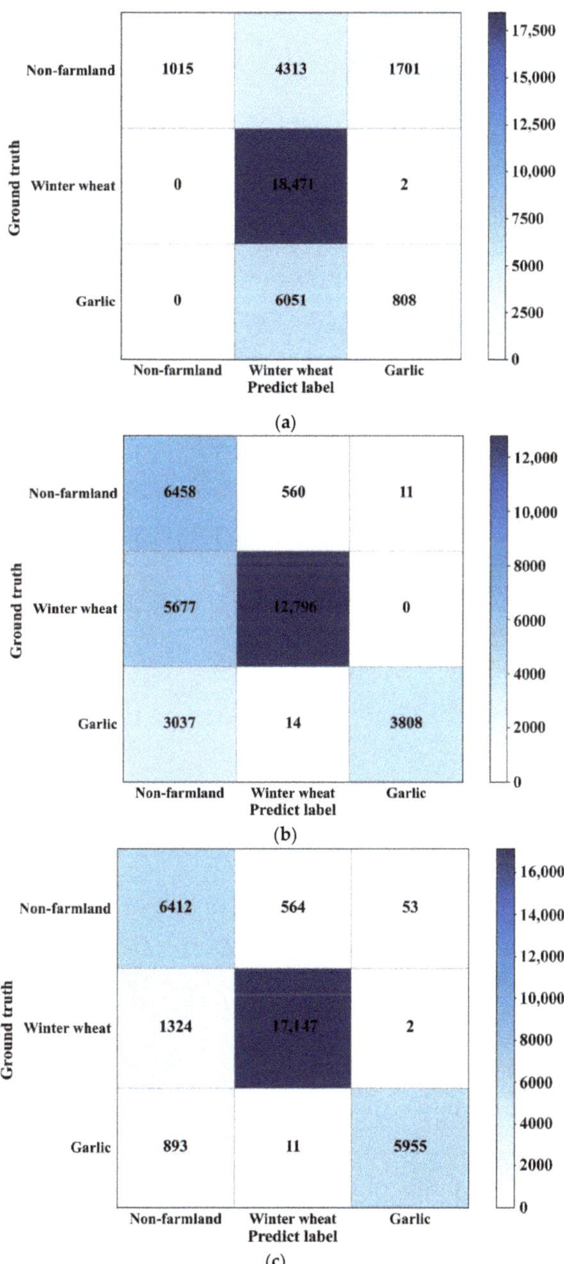

Figure 8. Confusion matrices for models generated by different classifiers. (**a**) 1D-CNN; (**b**) GDSSM; (**c**) GDSSM-CNN.

Table 2. Accuracy evaluation for different classifiers.

	1D-CNN		GDSSM		GDSSM-CNN (Threshold of 0.8)	
	Producer's Accuracy (%)	User's Accuracy (%)	Producer's Accuracy (%)	User's Accuracy (%)	Producer's Accuracy (%)	User's Accuracy (%)
Non-farmland	14.44	1	91.88	42.57	91.22	74.31
Winter wheat	99.99	64.06	69.27	96.36	92.82	96.76
Garlic	11.78	32.18	55.52	99.70	86.82	99.08
Accuracy (%)	67.29		71.26		91.20	
Kappa	0.3012		0.5491		0.8509	

In order to intuitively prove the differences between methods, some classification results are used for comparison, and the comparison results shown in Figures 9 and 10 show the final classification results of the study area obtained by the GDSSM-CNN model.

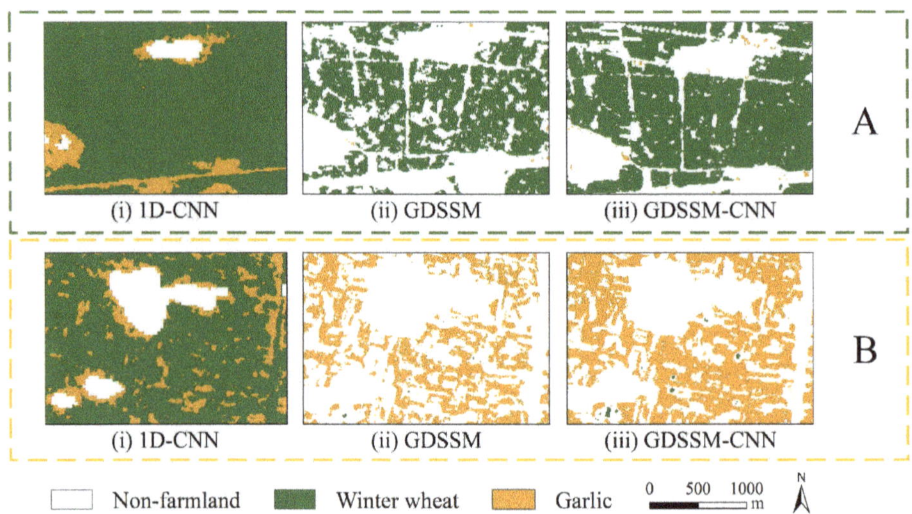

Figure 9. Comparison between crop discrimination results using (i) 1D-CNN, (ii) GDSSM and (iii) GDSSM-CNN in areas (**A**,**B**).

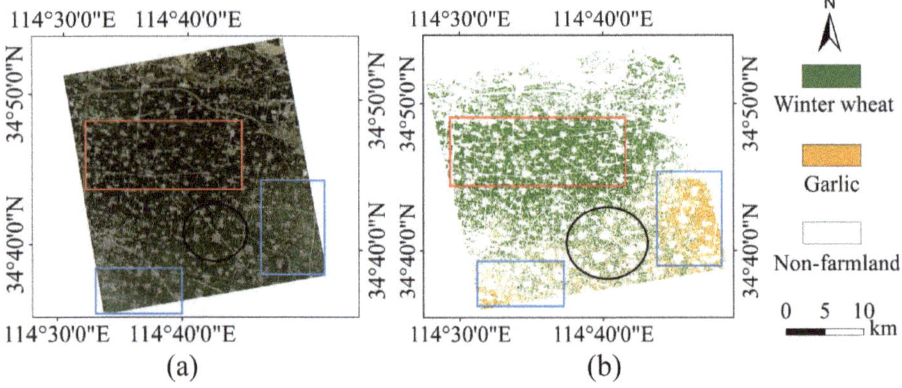

Figure 10. (**a**) Sentinel-2 image of the study area on 17 April 2020. (**b**) Spatial distribution map of crops in Kaifeng in 2020.

4. Discussion

To demonstrate the superiority of the GDSSM-CNN method, this section will discuss the results. Firstly, Section 4.1 discusses the need to combine the backscatter coefficient and GRVI. After that, Section 4.2 explores the effect of different amounts of training data on the classification results. Finally, the performance of the three methods is discussed in Section 4.3.

4.1. Combination of Backscatter Coefficient and GRVI

By analyzing the backscattering coefficients of garlic and winter wheat, it can be seen that the backscattering coefficient of winter wheat is always smaller than that of garlic. From 15 March to 2 May, the backscattering coefficient of winter wheat is significantly different from that of garlic, because during this period, the length of winter wheat increases rapidly and the morphology changes significantly. However, the backscattering coefficients of winter wheat and garlic are very close at some times, such as those on 20 February and 26 May, which means that it is difficult to distinguish crops by only using backscattering coefficients.

Through the analysis of GRVI of garlic and winter wheat, it can be seen that GRVI of winter wheat and garlic both show an upward trend with the growth of crops, but there is little difference between GRVI values of winter wheat and garlic from 20 February to 2 May, which means that it is difficult to distinguish crops by only using GRVI. Different from the backscattering coefficients, the GRVI values of winter wheat and garlic differ greatly on 26 May, which is caused by the harvesting of garlic. The backscattering coefficient and GRVI provide two different kinds of information, which is very necessary. This not only shows that combining the backscatter coefficient and GRVI can be beneficial in identifying crops but also shows the importance of using multi-temporal features. In order to better identify crops, the backscatter coefficient and GRVI are combined in the study.

4.2. Extraction of Training Data

In order to explore the impact of the number of training data on the performance of the classifier, the number of training data generated by different thresholds and the classification accuracy are compared, as shown in Table 3. By analyzing Table 3, it can be found that:

1. From the number of training data, it can be seen that the number of training data increases rapidly with the decrease in the threshold, and the model training time increases rapidly with the increase in the amount of training.
2. From the number of training data and the accuracy of the classifier, it can be seen that the accuracy of the classifier increases with the increase in training data. When the threshold is 0.9, the increase in training data has little impact on the accuracy. This means that the 1D-CNN model has been fully trained by the training data at this time.
3. As the number of training data increases, the number of noise data (training data with wrong labels) increases. However, the accuracy of the classifier becomes higher and higher. This means that when the threshold is greater than or equal to 0.8, the number of noisy data is far less than the number of training data with correct labels, and the increase in low-noise data does not affect the performance of the classifier. When the threshold value is less than 0.8, the unknown pixel does not have a high similarity with the reference target. Therefore, when the threshold value is in the range of 0.8–0.9, the classifier can achieve very high accuracy.

Table 3. Training data and accuracy of different thresholds.

Threshold	Winter Wheat Training Data	Garlic Training Data	Accuracy (%)	Kappa	Training Time
Use samples only	10	5	67.29	0.3012	10 s
0.95	994	259	89.13	0.8136	14 s
0.90	51,179	8183	91.18	0.8506	2 min 17 s
0.85	381,002	52,001	91.19	0.8507	14 min 1 s
0.80	1,040,634	149,326	91.20	0.8509	38 min 11 s

4.3. Comparison of Three Methods

By analyzing the confusion matrix, it can be found that the user's accuracy for non-farmland area is very low in the results using the GDSSM method. The reason is that when only the threshold method is used for classification, the classification performance is affected by the selection of threshold, the proportion of mixed pixels, and image noise, which is also reflected in other classification results of the GDSSM model. In the results using the 1D-CNN method, the user's accuracy and producer's accuracy for garlic are very low. The reason for this result is that the sample size is too small, which makes the model unable to fully learn the types of crops in the training stage. The GDSSM-CNN model has very high user accuracy and producer accuracy, which proves that this model also has very good performance when there is a limited number of samples.

From the difference diagram of the classification results of the three methods, it can be clearly seen that the effect of the GDSSM-CNN model is better than that of other models. In area A, the GDSSM model mistakenly identifies farmland as non-farmland. The 1D-CNN model mistakenly identifies non-farmland as garlic and does not identify some paths. The reason is that the model training is inadequate due to insufficient samples. Many farmlands in the GDSSM-CNN model are correctly identified, which proves that the GDSSM-CNN method has better identification ability and is rarely affected by the number of samples. In area B, the GDSSM model can distinguish the areas where wheat and garlic are mixed, but it still recognizes much farmland as non-farmland. A large amount of garlic was incorrectly identified as winter wheat by the 1D-CNN model. The GDSSM-CNN model still has very good classification results in the area where winter wheat and garlic are mixed, which means that the GDSSM model can select the correct pixels as training data, thus improving the performance of the 1D-CNN. It can be seen from Figure 10 that the planting area of winter wheat gradually shrinks from north to south (shown in the red rectangular boxes in Figure 10), and garlic is mainly planted in the southeast and southwest of the study area (shown in the blue rectangular boxes in Figure 10). The southern part of the study area is mainly planted with winter wheat and garlic (shown in the black circle in Figure 10). Through comparison with optical images, these planting distributions are confirmed.

Based on the analysis and verification of classification results, the following conclusions can be drawn: The GDSSM model can identify winter wheat and garlic, and it is less affected by the limited samples. However, the classification performance of this model is subpar. The 1D-CNN model has a strong feature extraction ability, but the number of samples easily limits its performance. The GDSSM-CNN model, which combines the two models, not only is free from the limitation of the number of samples but also has very high accuracy and robustness.

5. Conclusions

In this study, a crop classification method based on GDSSM-CNN is proposed. This method uses multi-temporal RadarSat-2 images, and the performance of the classifier is not limited by the number of samples. The details are as follows: Firstly, the geodesic distance spectral similarity between unknown pixels and crop samples is calculated. Secondly, GDSSM is used in combination with the disclosed GFSAD farmland mask to remove non-farmland areas and is compared with the GFSAD mask only. Thirdly, backscatter coefficients and GRVI features are extracted from multi-temporal images and combined to generate crop classification models. Fourthly, GDSSM and the threshold method are used to extract unlabeled data with very high similarity with samples as training data, and the effects of different thresholds are compared. The optimal threshold selection range is determined to be 0.8–0.9. Finally, the GDSSM classifier combining GDSSM and 1D-CNN is used to train the model using the feature combination of backscatter coefficient and GRVI.

GDSSM, 1D-CNN and GDSSM-CNN models with different thresholds were used for comparison. Only 10 winter wheat samples and 5 garlic samples were used. The GDSSM-CNN method shows strong crop classification performance. The accuracy of the GDSSM-CNN method with different thresholds is above 89%, which is higher than that of the GDSSM method (the accuracy is 71.26%) and 1D-CNN method (the accuracy is 67.29%), which proves the accuracy of the GDSM-CNN method. Some classification results were selected for comparison, proving that the GDSSM-CNN method has a satisfactory classification effect. In a word, analysis and verification of three methods have proved that the GDSSM-CNN model has incomparable advantages in crop classification.

In addition, the combination of backscatter coefficient and GRVI has proved to be very useful for crop classification results. In addition, compared with the non-farmland mask only using GFSAD, the non-farmland mask added with GDSSM can more finely identify small-scale buildings and roads. This study shows that the use of rich polarization information and multi-temporal SAR images brings new possibilities to crop classification and also provides a new idea for the sample size limitation of neural networks in SAR image applications.

Author Contributions: Conceptualization, H.L., G.T. and N.L.; data curation, H.Y., J.L. and N.L.; investigation, H.L., G.T., J.Z. and N.L.; methodology, H.L., G.T. and H.Y.; supervision, H.L., J.Z. and N.L.; writing—original draft, H.L., J.L. and N.L.; writing—review and editing, H.L., G.T. and N.L. All authors have read and agreed to the published version of the manuscript.

Funding: This research was funded by the National Natural Science Foundation of China (42101386), the College Key Research Project of Henan Province (22A520021, 21A520004), the Plan of Science and Technology of Henan Province (212102210093, 222102110439) and the Plan of Science and Technology of Kaifeng City (2102005).

Data Availability Statement: All data and model presented in this study are available.

Acknowledgments: The authors would like to thank NASA for providing Global Food Security-Support Analysis Data at 30 m.

Conflicts of Interest: The authors declare no conflict of interest.

References

1. Kwak, G.-H.; Park, N.-W. Impact of texture information on crop classification with machine learning and UAV images. *Appl. Sci.* **2019**, *9*, 643. [CrossRef]
2. Zhong, L.; Hu, L.; Zhou, H. Deep learning based multi-temporal crop classification. *Remote Sens. Environ.* **2019**, *221*, 430–443. [CrossRef]
3. Yang, H.; Li, H.; Wang, W.; Li, N.; Zhao, J.; Pan, B. Spatio-temporal estimation of rice height using time series Sentinel-1 images. *Remote Sens.* **2022**, *14*, 546. [CrossRef]
4. Xie, Y.; Huang, J. Integration of a crop growth model and deep learning methods to improve satellite-based yield estimation of winter wheat in Henan Province, China. *Remote Sens.* **2021**, *13*, 4372. [CrossRef]
5. Pôças, I.; Calera, A.; Campos, I.; Cunha, M. Remote sensing for estimating and mapping single and basal crop coefficientes: A review on spectral vegetation indices approaches. *Agric. Water Manag.* **2020**, *233*, 106081. [CrossRef]
6. Sun, C.; Bian, Y.; Zhou, T.; Pan, J. Using of multi-source and multi-temporal remote sensing data improves crop-type mapping in the subtropical agriculture region. *Sensors* **2019**, *19*, 2401. [CrossRef]
7. Shi, S.; Ye, Y.; Xiao, R. Evaluation of food security based on remote sensing data—Taking Egypt as an example. *Remote Sens.* **2022**, *14*, 2876. [CrossRef]
8. Leroux, L.; Castets, M.; Baron, C.; Escorihuela, M.-J.; Bégué, A.; Lo Seen, D. Maize yield estimation in West Africa from crop process-induced combinations of multi-domain remote sensing indices. *Eur. J. Agron.* **2019**, *108*, 11–26. [CrossRef]
9. Sun, Y.; Luo, J.; Wu, T.; Zhou, Y.; Liu, H.; Gao, L.; Dong, W.; Liu, W.; Yang, Y.; Hu, X.; et al. Synchronous response analysis of features for remote sensing crop classification based on optical and SAR time-series data. *Sensors* **2019**, *19*, 4227. [CrossRef]

10. Li, C.; Li, H.; Li, J.; Lei, Y.; Li, C.; Manevski, K.; Shen, Y. Using NDVI percentiles to monitor real-time crop growth. *Comput. Electron. Agric.* **2019**, *162*, 357–363. [CrossRef]
11. French, A.N.; Hunsaker, D.J.; Sanchez, C.A.; Saber, M.; Gonzalez, J.R.; Anderson, R. Satellite-Based NDVI crop coefficients and evapotranspiration with eddy covariance validation for multiple durum wheat fields in the US Southwest. *Agric. Water Manag.* **2020**, *239*, 106266. [CrossRef]
12. Wei, S.; Zhang, H.; Wang, C.; Wang, Y.; Xu, L. Multi-temporal SAR data large-scale crop mapping based on U-Net model. *Remote Sens.* **2019**, *11*, 68. [CrossRef]
13. Adrian, J.; Sagan, V.; Maimaitijiang, M. Sentinel SAR-Optical fusion for crop type mapping using deep learning and Google Earth Engine. *ISPRS J. Photogramm. Remote Sens.* **2021**, *175*, 215–235. [CrossRef]
14. Bouchat, J.; Tronquo, E.; Orban, A.; Verhoest, N.E.C.; Defourny, P. Assessing the potential of fully polarimetric Mono- and bistatic SAR acquisitions in L-Band for crop and soil monitoring. *IEEE J. Sel. Top. Appl. Earth Obs. Remote Sens.* **2022**, *15*, 3168–3178. [CrossRef]
15. Nasirzadehdizaji, R.; Balik Sanli, F.; Abdikan, S.; Cakir, Z.; Sekertekin, A.; Ustuner, M. Sensitivity analysis of multi-temporal Sentinel-1 SAR parameters to crop height and canopy coverage. *Appl. Sci.* **2019**, *9*, 655. [CrossRef]
16. Wang, H.; Magagi, R.; Goïta, K.; Trudel, M.; McNairn, H.; Powers, J. Crop phenology retrieval via polarimetric SAR decomposition and random forest algorithm. *Remote Sens. Environ.* **2019**, *231*, 111234. [CrossRef]
17. Xu, L.; Zhang, H.; Wang, C.; Zhang, B.; Liu, M. Crop classification based on temporal information using Sentinel-1 SAR time-series data. *Remote Sens.* **2019**, *11*, 53. [CrossRef]
18. Gao, H.; Wang, C.; Wang, G.; Fu, H.; Zhu, J. A novel crop classification method based on PpfSVM Classifier with time-series alignment kernel from dual-polarization SAR datasets. *Remote Sens. Environ.* **2021**, *264*, 112628. [CrossRef]
19. Liao, C.; Wang, J.; Xie, Q.; Baz, A.A.; Huang, X.; Shang, J.; He, Y. Synergistic use of multi-temporal RADARSAT-2 and VENµS data for crop classification based on 1D convolutional neural network. *Remote Sens.* **2020**, *12*, 832. [CrossRef]
20. Xie, Q.; Lai, K.; Wang, J.; Lopez-Sanchez, J.M.; Shang, J.; Liao, C.; Zhu, J.; Fu, H.; Peng, X. Crop monitoring and classification using polarimetric RADARSAT-2 time-series data across growing season: A case study in Southwestern Ontario, Canada. *Remote Sens.* **2021**, *13*, 1394. [CrossRef]
21. Li, M.; Bijker, W. Vegetable classification in indonesia using dynamic time warping of Sentinel-1A dual polarization SAR time series. *Int. J. Appl. Earth Obs. Geoinf.* **2019**, *78*, 268–280. [CrossRef]
22. Yamaguchi, Y.; Moriyama, T.; Ishido, M.; Yamada, H. Four-Component scattering model for polarimetric SAR image decomposition. *IEEE Trans. Geosci. Remote Sens.* **2005**, *43*, 1699–1706. [CrossRef]
23. Singh, G.; Yamaguchi, Y. Model-based six-component scattering matrix power decomposition. *IEEE Trans. Geosci. Remote Sens.* **2018**, *56*, 5687–5704. [CrossRef]
24. Singh, G.; Malik, R.; Mohanty, S.; Rathore, V.S.; Yamada, K.; Umemura, M.; Yamaguchi, Y. Seven-component scattering power decomposition of POLSAR coherency matrix. *IEEE Trans-Actions Geosci. Remote Sens.* **2019**, *57*, 8371–8382. [CrossRef]
25. Guo, J.; Wei, P.-L.; Liu, J.; Jin, B.; Su, B.-F.; Zhou, Z.-S. Crop classification based on differential characteristics of H/α scattering parameters for multitemporal Quad- and Dual-Polarization SAR images. *IEEE Trans. Geosci. Remote Sens.* **2018**, *56*, 6111–6123. [CrossRef]
26. Chen, Q.; Cao, W.; Shang, J.; Liu, J.; Liu, X. Superpixel-based cropland classification of SAR image with statistical texture and polarization features. *IEEE Geosci. Remote Sens. Lett.* **2022**, *19*, 1–5. [CrossRef]
27. Ratha, D.; Bhattacharya, A.; Frery, A.C. Unsupervised classification of PolSAR data using a scattering similarity measure derived from a geodesic distance. *IEEE Geosci. Remote Sens. Lett.* **2018**, *15*, 151–155. [CrossRef]
28. Phartiyal, G.S.; Kumar, K.; Singh, D. An improved land cover classification using polarization signatures for PALSAR 2 data. *Adv. Space Res.* **2020**, *65*, 2622–2635. [CrossRef]
29. Mandal, D.; Kumar, V.; Ratha, D.; Lopez-Sanchez, J.M.; Bhattacharya, A.; McNairn, H.; Rao, Y.S.; Ramana, K.V. Assessment of rice growth conditions in a semi-arid region of india using the generalized radar vegetation index derived from RADARSAT-2 polarimetric SAR data. *Remote Sens. Environ.* **2020**, *237*, 111561. [CrossRef]
30. Ratha, D.; De, S.; Celik, T.; Bhattacharya, A. Change detection in polarimetric SAR images using a geodesic distance between scattering mechanisms. *IEEE Geosci. Remote Sens. Lett.* **2017**, *14*, 7. [CrossRef]
31. Xie, Q.; Dou, Q.; Peng, X.; Wang, J.; Lopez-Sanchez, J.M.; Shang, J.; Fu, H.; Zhu, J. Crop classification based on the physically constrained general model-based decomposition using multi-temporal RADARSAT-2 Data. *Remote Sens.* **2022**, *14*, 2668. [CrossRef]
32. Small, D. Flattening gamma: Radiometric terrain correction for SAR Imagery. *IEEE Trans. Geosci. Remote Sens.* **2011**, *49*, 3081–3093. [CrossRef]
33. Orynbaikyzy, A.; Gessner, U.; Conrad, C. Spatial transferability of random forest models for crop type classification using Sentinel-1 and Sentinel-2. *Remote Sens.* **2022**, *14*, 1493. [CrossRef]
34. Liu, Y.; Zhao, W.; Chen, S.; Ye, T. Mapping crop rotation by using deeply synergistic optical and SAR time series. *Remote Sens.* **2021**, *13*, 4160. [CrossRef]
35. Zhang, W.-T.; Wang, M.; Guo, J.; Lou, S.-T. Crop classification using MSCDN classifier and sparse auto-encoders with non-negativity constraints for multi-temporal, Quad-Pol SAR data. *Remote Sens.* **2021**, *13*, 2749. [CrossRef]
36. Chang, Y.-L.; Tan, T.-H.; Chen, T.-H.; Chuah, J.H.; Chang, L.; Wu, M.-C.; Tatini, N.B.; Ma, S.-C.; Alkhaleefah, M. Spatial-temporal neural network for rice field classification from SAR images. *Remote Sens.* **2022**, *14*, 1929. [CrossRef]

37. Shang, R.; Wang, J.; Jiao, L.; Stolkin, R.; Hou, B.; Li, Y. SAR targets classification based on deep memory convolution neural networks and transfer parameters. *IEEE J. Sel. Top. Appl. Earth Obs. Remote Sens.* **2018**, *11*, 2834–2846. [CrossRef]
38. Huang, Z.; Pan, Z.; Lei, B. Transfer learning with deep convolutional neural network for SAR target classification with limited labeled data. *Remote Sens.* **2017**, *9*, 907. [CrossRef]
39. Rostami, M.; Kolouri, S.; Eaton, E.; Kim, K. Deep transfer learning for few-shot SAR image classification. *Remote Sens.* **2019**, *11*, 1374. [CrossRef]
40. Xie, Q.; Wang, J.; Liao, C.; Shang, J.; Lopez-Sanchez, J.M.; Fu, H.; Liu, X. On the use of neumann decomposition for crop classification using Multi-Temporal RADARSAT-2 polarimetric SAR data. *Remote Sens.* **2019**, *11*, 776. [CrossRef]
41. Granahan, J.C.; Sweet, J.N. An evaluation of atmospheric correction techniques using the spectral similarity scale. In Proceedings of the IGARSS 2001. Scanning the Present and Resolving the Future. In Proceedings of the IEEE 2001 International Geoscience and Remote Sensing Symposium (Cat. No.01CH37217), Sydney, NSW, Australia, 9–13 July 2001; Volume 5, pp. 2022–2024.
42. Yang, H.; Pan, B.; Wu, W.; Tai, J. Field-based rice classification in Wuhua county through integration of multi-temporal Sentinel-1A and Landsat-8 OLI data. *Int. J. Appl. Earth Obs. Geoinf.* **2018**, *69*, 226–236. [CrossRef]
43. Antropov, O.; Rauste, Y.; Hame, T. Volume scattering modeling in PolSAR decompositions: Study of ALOS PALSAR data over boreal forest. *IEEE Trans. Geosci. Remote Sens.* **2011**, *49*, 3838–3848. [CrossRef]
44. Ratha, D.; Gamba, P.; Bhattacharya, A.; Frery, A.C. Novel techniques for built-up area extraction from polarimetric SAR images. *IEEE Geosci. Remote Sens. Lett.* **2020**, *17*, 177–181. [CrossRef]
45. Yadav, K.; Congalton, R.G. Accuracy Assessment of Global Food Security-Support Analysis Data (GFSAD) cropland extent maps produced at three different spatial resolutions. *Remote Sens.* **2018**, *10*, 1800. [CrossRef]
46. Sun, X.; Lv, Y.; Wang, Z.; Fu, K. SCAN: Scattering characteristics analysis network for few-shot aircraft classification in high-resolution SAR images. *IEEE Trans. Geosci. Remote Sens.* **2022**, *60*, 1–17.
47. Guo, Z.; Qi, W.; Huang, Y.; Zhao, J.; Yang, H.; Koo, V.-C.; Li, N. Identification of crop type based on C-AENN using time series Sentinel-1A SAR data. *Remote Sens.* **2022**, *14*, 1379. [CrossRef]
48. Yang, H.; Pan, B.; Li, N.; Wang, W.; Zhang, J.; Zhang, X. A systematic method for spatio-temporal phenology estimation of paddy rice using time series Sentinel-1 images. *Remote Sens. Environ.* **2021**, *259*, 112394. [CrossRef]
49. Li, N.; Li, H.; Zhao, J.; Guo, Z.; Yang, H. Mapping winter wheat in Kaifeng, China using Sentinel-1A time-series images. *Remote Sens. Lett.* **2022**, *13*, 503–510. [CrossRef]

remote sensing

Article

A Repeater-Type SAR Deceptive Jamming Method Based on Joint Encoding of Amplitude and Phase in the Intra-Pulse and Inter-Pulse

Dongyang Cheng [1,2,3], Zhenchang Liu [1,2,3], Zhengwei Guo [1,2,3], Gaofeng Shu [1,2,3,*] and Ning Li [1,2,3]

[1] Henan Engineering Research Center of Intelligent Technology and Application, Henan University, Kaifeng 475004, China
[2] Henan Key Laboratory of Big Data Analysis and Processing, Henan University, Kaifeng 475004, China
[3] College of Computer and Information Engineering, Henan University, Kaifeng 475004, China
* Correspondence: gaofeng.shu@henu.edu.cn

Abstract: Due to advantages such as low power consumption and high concealment, deceptive jamming against synthetic aperture radar (SAR) has received extensive attention in electronic countermeasures. However, the false targets generated by most of the deceptive jamming methods still have limitations, such as poor controllability and strong regularity. Inspired by the idea of waveform coding, this paper proposed a repeater-type SAR deceptive jamming method through the joint encoding of amplitude and phase in intra-pulse and inter-pulse, which can generate a two-dimensional controllable deceptive jamming effect. Specifically, the proposed method mainly includes two parts, i.e., grouping and encoding. The number of groups determines the number of false targets, and the presence of the phase encoding produces false targets. The amplitude encoding affects the amplitude of the false targets. For the intra-pulse cases, the proposed method first samples the intercepted SAR signal. Meanwhile, the sampling points are grouped in turn. For the inter-pulse cases, the grouped objects are the pulses. Subsequently, the joint encoding of amplitude and phase is performed on each group, which generates jamming signals with deceptive effects. In this paper, the imaging effect of the generated jamming signals is analyzed in detail, and the characteristics of false targets, including numbers, position, and amplitude, are derived. The simulation and experimental results verify the correctness of the theoretical analysis. In addition, the superiority of the proposed method is verified by comparing it with other methods.

Keywords: deceptive jamming; synthetic aperture radar (SAR); electronic countermeasures (ECM); repeater-type jamming

Citation: Cheng, D.; Liu, Z.; Guo, Z.; Shu, G.; Li, N. A Repeater-Type SAR Deceptive Jamming Method Based on Joint Encoding of Amplitude and Phase in the Intra-Pulse and Inter-Pulse. *Remote Sens.* **2022**, *14*, 4597. https://doi.org/10.3390/rs14184597

Academic Editor: Timo Balz

Received: 28 July 2022
Accepted: 12 September 2022
Published: 14 September 2022

Publisher's Note: MDPI stays neutral with regard to jurisdictional claims in published maps and institutional affiliations.

1. Introduction

1.1. Background

Synthetic aperture radar (SAR) is an active microwave remote sensing imaging system [1–3]. Compared with optical remote sensing, SAR has unique advantages that can work at all times of day and under complex weather conditions and is widely used in various fields, such as environmental remote sensing [4–7] and military reconnaissance [8,9]. Especially in the military field, SAR has progressed into becoming the core sensor of strategic reconnaissance and battlefield surveillance in modern warfare. Meanwhile, to protect the information of important targets and areas, various jamming technologies against SAR have been paid more and more attention in the field of electronic countermeasures (ECM) [10–12].

SAR jamming technology actually prevents the enemy from obtaining effective information from SAR imaging [13]. In general, from the perspective of the jamming effect, SAR jamming can be classified into barrage and deceptive jamming. Barrage jammers increase the noise level of the echo to bury the target of interest by transmitting random noise over a frequency band wider than the SAR bandwidth [14]. Although the implementation of barrage jamming is simple, the power requirements of the jammer are relatively high, which means that it will face an expensive cost budget in actual ECM. Deceptive jamming implants an elaborately designed false scene or target into the SAR image by sending a false target signal with similar characteristics to the real target signal, which affects the interpretation process of the SAR image and achieves the jamming effect of "hard to distinguish real from imitation" [15]. Deceptive jamming, as opposed to barrage jamming, does not require high transmit power and has the advantages of strong concealment and low consumption. In addition, even if the deceptive target is discovered, it is difficult to eliminate by common anti-jamming measures. As a result, the study of deceptive jamming obtains more appealing and promising results for future ECM [16,17].

1.2. Previous Work

In the past few decades, various deceptive jamming methods have been proposed. According to the workflow of jamming implementation, these methods can be classified into direct-type and repeater-type deceptive jamming [18].

1.2.1. Direct-Type Deceptive Jamming

According to the parameter information of the reconnaissance SAR signal, the direct-type deceptive jamming directly generates jamming signals and transmits them to the enemy, which creates false targets in the SAR image. The key to implementation depends on the accuracy of electronic reconnaissance. Reference [19] analyzed the influence of direct-type jamming with different parameter errors on the jamming effect through simulation experiments. Qu et al. developed a reconnaissance equation for SAR satellite reconnaissance, which realized the requirements for grounding the reconnaissance equipment in the reconnaissance of satellite signals [20].

Although direct-type jamming has the advantage of being one-way, it relies heavily on the accuracy of electronic reconnaissance. Meanwhile, the direct-type jamming needs to accurately calculate the position of the SAR flight to transmit the false target jamming signal to the corresponding pulse moment, which is usually difficult to achieve in the environment of high-speed, changing warfare.

1.2.2. Repeater-Type Deceptive Jamming

For repeater-type deceptive jamming, the jammer intercepts the SAR signal in the protected area and applies a specific modulating method to generate the jamming signal. Subsequently, the generated jamming signal is retransmitted back to the SAR device. Compared to direct-type deceptive jamming, repeater-type deceptive jamming is simpler to implement, has lower parameter accuracy requirements, and does not require calculating the radar flight position. Given these benefits, repeater-type deceptive jamming has received a great deal of attention. At present, the main methods include time-delay phase modulation, convolution modulation, frequency diverse array (FDA), and interrupted sampling repeater jamming (ISRJ).

(1) *Time-delay phase modulation:* The jamming method of time-delay phase modulation is a traditional SAR deceptive jamming method that treats the jammer as a linear time-invariant (LTI) system within a single pulse repetition interval (PRI). After passing through the LTI system, the intercepted SAR is modulated with a time-delay phase to generate false targets in the SAR image. However, it faces two problems. One is to accurately calculate the propagation delay between jammers and false scatterers at each PRI. The other is that it requires a large amount of computation. To overcome these deficiencies, Lin. et al. proposed a fast generation method for SAR deceptive jamming signals based on the inverse Range Doppler (RD) algorithm [21]. Reference [22] proposed a method for segmented modulation. From the frequency-domain, Yang et al. proposed a method of frequency-domain pre-modulation [23]. Reference [24] proposed a frequency-domain three-stage algorithm method. Liu. et.al. proposed an inverse Omega-K algorithm method [25].

Subsequently, Yang et al. proposed a large-scene deceptive jamming method based on the idea of module division [26]. On this basis, Yang et al. proposed a large scene deceptive jamming algorithm based on time-delay and frequency-shift template segmentation [27]. Although the above methods alleviate the insufficiency to a certain extent, a large amount of computational burden and the accurate calculation of the propagation delay are still unresolved problems.

(2) *Convolution modulation:* Convolution modulation jamming obtains the jamming signal by convolving the pre-designed jamming function and the intercepted SAR signal. Wang et al. first proposed the principle of convolution modulation jamming implementation [28]. Zhou et al. proposed a large-scene deceptive algorithm based on the convolution modulation theorem [29]. However, it was only suitable for spaceborne SAR in broadside mode. Zhao et al. proposed a jamming method that used an electromagnetic model of the target, which could reflect more details such as scatterer changes and shielding effects [30]. However, a lot of computational resources were still required. Tai et al. proposed a parallel convolutional deceptive jamming method [31].

Although SAR deceptive jamming based on convolution modulation does not need to accurately calculate the propagation delay between jammers and false scatterers, constrained to the convolution theorem, the false targets often lag behind the real targets. Meanwhile, a high level of computational complexity is also required, which is difficult to maintain in real-time processing.

(3) *FDA:* The FDA-based jamming method has small frequency increments between adjacent antenna elements, which produces a transmitted beam pattern related to angle, range, and even time. Zhu et al. first proposed a spaceborne SAR deceptive jamming technology based on the characteristics of FDA [32], which could generate evenly distributed false targets in the SAR image. Bang et al. proposed FDA-based space–time–frequency deceptive jamming against SAR [33]. By controlling the array number and frequency increment of FDA, the false targets could be generated, but the imaging effect of edge false targets was low. Huang et al. proposed a spaceborne scattered-wave, FDA-based SAR deceptive jamming method [34] and analyzed the imaging effect in detail. However, the false targets generated only exist in the range direction. On this basis, Huang et al. applied the FDA-based SAR jamming method to move the deceptive target.

FDA-based SAR deceptive jamming has low computational complexity and can control the number of false targets [35]. However, restricted by equipment, FDA-based SAR deceptive jamming still has the following deficiencies. First, if an excessively large frequency increment is used, additional range ambiguity and Doppler ambiguity will be generated, which results in poor imaging quality in the false targets. Second, the number of false targets depends on the number of FDA array antennas, which will increase the hardware cost and system complexity if a large number of false targets are to be generated. Third, the false target only exists in the range direction and cannot achieve two-dimensional (2D) deceptive jamming.

(4) *ISRJ:* Based on the working method of sending and receiving time-sharing data, ISRJ has received extensive attention in recent years. Wang et al. first proposed the idea of ISRJ and expounded the mathematical principles [36]. Wu et al. applied ISRJ to SAR for the first time and proposed an implementation scheme for ISRJ in the range and azimuth directions [37]. Subsequently, Zhong et al. proposed an improved ISRJ algorithm, which was based on the digital radio frequency memory (DRFM) technique and could obtain multiple false targets before the real targets [38]. Feng et al. summarized the ISRJ jamming technical framework and analyzed, in detail, the key jamming factors that determine the characteristics of false targets [39]. Sun et al. proposed a jamming method combining phase modulation and ISRJ, but it only simulated one-dimensional (1D) signals [40].

Although ISRJ can produce 2D deceptive jamming effects, the amplitude of false targets is distributed in the shape of a sin c function. In addition, the number of false targets is jointly determined by a variety of parameters, and it is difficult to precisely control the number of false targets [39].

In brief, although the above methods can produce deceptive jamming effects in SAR images, they still have their limitations. Meanwhile, with the continuous development of ECM, some composite methods have been proposed [41,42]. However, most of these jamming methods are a combination of existing jamming methods. Therefore, for enriching the theory of ECM and realizing the protection of sensitive targets, it is of great significance to develop diversified and controllable deceptive jamming methods.

1.3. Main Contributions of This Paper

To overcome the limitations of existing jamming methods, inspired by the idea of waveform coding, this paper proposes a 2D, controllable, diversified SAR deceptive jamming method. The core of the proposed method is to encode the intercepted SAR signal. By applying different encoding forms, diverse deceptive jamming effects can be achieved in the SAR image. Specifically, the intra-pulse coding controls the range direction jamming effect, and the inter-pulse coding controls the azimuth direction jamming effect. In intra-pulse coding cases, the method first samples the intercepted SAR signal and the special grouping of sampling points. Then, amplitude and phase joint encoding is performed for each group separately. The number of false targets is determined by the number of groups, and the joint encoding of phase and amplitude affects the amplitude ratio of the false target to the real target. In the inter-pulse coding case, the principle is similar to that of intra-pulse coding. The difference is that the encoding objects of the inter-pulse are pulse trains. More specifically, the major contributions of this paper can be summarized as.

- A repeater-type deceptive jamming model is established, and a deceptive jamming method through the joint coding of amplitude and phase in intra-pulse and inter-pulse (CAPII) is proposed. The proposed method can generate false targets with different amplitudes and numbers in 2D.
- The imaging results of the proposed algorithm are analyzed and deduced in detail. Specifically, false target characteristics, including the number, position, and amplitude of the false target, are discussed separately.
- The feasibility of the proposed method is verified by simulation experiments. Using ISRJ and FDA as a comparative experiment, the superiority of the proposed method is verified.

1.4. Organization of This Paper

The remainder of this article is organized as follows. The model of repeater-type deceptive jamming and methods, including the FDA-based jamming method, ISRG, and the proposed method in this paper, are introduced in Section 2. Section 3 provides a detailed analysis of the imaging effects of the proposed method. The simulation experiments are provided in Section 4. Section 5 discusses the proposed method. Section 6 provides the conclusion.

2. Model and Methods

2.1. SAR Jamming Model

Taking Stripmap SAR as an example, when a repeater-type jammer is against a SAR system, the geometric relationship between the SAR, the jammer, the swath, and internal protected targets is shown in Figure 1. Without a loss of generality, point O represents the origin of the coordinates, and H represents the SAR flight height. Assume that the SAR platform is flying at a constant speed, v, along the Y-direction. P represents SAR platforms. R_0 represents the slant range of the closest approach between the protected scene center and the SAR platform. The jammer is placed at point $M(x_j, 0, 0)$. The red dashed line represents the swath width. R_r represents the instantaneous slant range between the protected scene center and the SAR platform, i.e.,

$$R_r(\eta) = \sqrt{x_j^2 + (v\eta)^2 + H^2},\tag{1}$$

where η stands for azimuth time.

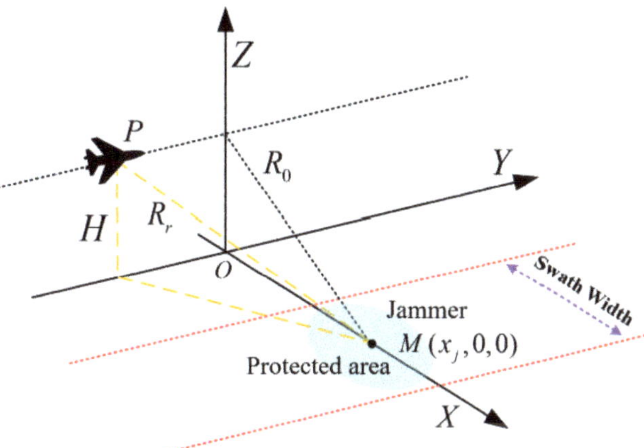

Figure 1. Geometry model of repeater-type jamming.

The working mechanism of repeater-type deceptive jamming is based on "intercept–modulate–retransmit", as shown in Figure 2. The jammer performs a series of operations on the intercepted SAR signal, including amplification, down converter, and analog-to-digital (A/D) converter, to obtain a digital baseband signal. The jammer performs intra-pulse and inter-pulse coding on the digital baseband signals to generate jamming signals. By applying different encoding forms, different jamming effects will be produced in the SAR image. Finally, after digital-to-analog (D/A) converter, upconverter, and gaining control, the jamming signal is retransmitted to the SAR device. The false targets can be generated in the SAR image after the jamming signal is 2D-filtered by SAR.

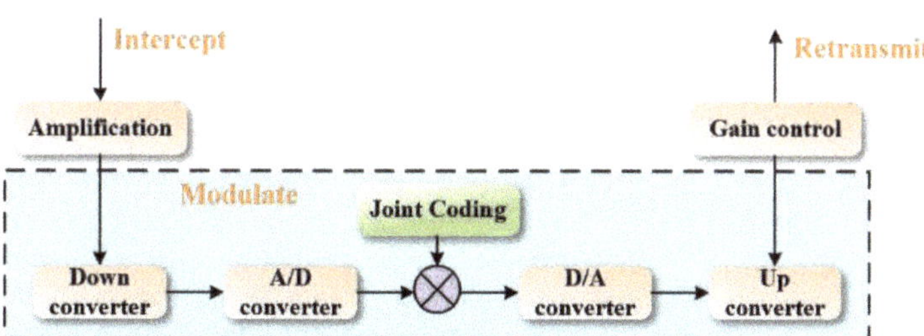

Figure 2. Working flow chart of a jammer.

2.2. FDA-Based Deceptive Jamming Method

FDA-based SAR deceptive jamming is implemented by retransmitting the intercepted SAR signal through the FDA. The carrier frequencies of the array elements in an FDA radar are slightly distinguished, which can produce a range-angle-dependent beampattern. Therefore, the jamming signal transmitted through the FDA produces false targets in the range direction of the SAR image [38]. Figure 3 shows the signal model of an FDA, where d is the spacing among the array elements. θ is the azimuth angle of the jammer to the boresight of the FDA. N stands for the number of FDA arrays. Δf represents the carrier frequency increment among the FDA antennas. f_0 represents the carrier frequency of the intercepted SAR signal.

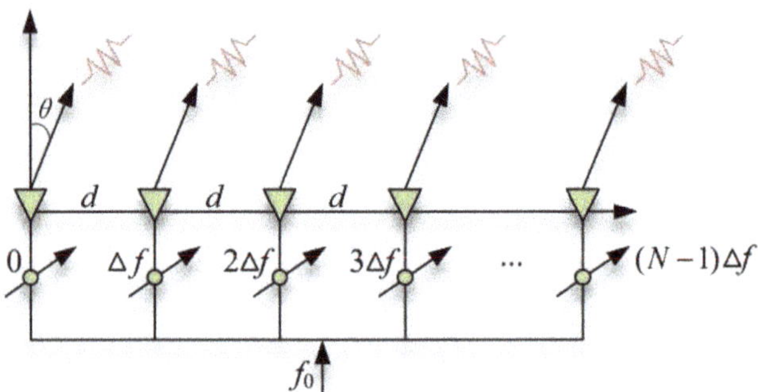

Figure 3. Schematic diagram of the FDA transmitter.

The roundtrip of the SAR signals transmitting through the n-th arrays of the FDA to the SAR receiver can be expressed as

$$R_{Jn} = R_r + \sqrt{x_j^2 + [v\eta + (n-1)d]^2 + H^2},\tag{2}$$

where R_r represents the instantaneous slant range between the FDA receiving antenna and the SAR. The jamming signal received by the SAR platform can be expressed as

$$S_{FDA}(t_r, \eta) = \sum_{n=1}^{N-1} A_0 w_r(t - R_{Jn}/c) w_a(\eta - \eta_c) \exp\{-j2\pi[f_0 + (n-1)\Delta f]R_{Jn}\} \\ \cdot \exp(j2\pi n\Delta f) \exp\left[j\pi K_r(t - R_{Jn}/c)^2\right]\tag{3}$$

where A_0 represents the amplitude modulation coefficient of the jammer. w_r is the SAR pulse complex envelope. w_a represents the azimuth envelope. η_c represents the Doppler center frequency. c is the speed of electromagnetic waves in free space. The η stands for azimuth time, and K_r is the chirp rate of the intercepted SAR signal.

SAR deceptive jamming based on FDA, as shown in Figure 3, is equivalent to the superposition of multiple echoes of different carrier frequency components. It is because of carrier frequency difference that false targets appear in the range direction of the SAR image [38].

2.3. ISRG Method

The ISRJ samples a small part of the signal with high fidelity and then retransmits, with sampling and retransmitting working alternately. As shown in Figure 4, the sampling signals in the azimuth and range directions are assumed to be rectangular envelope pulse trains.

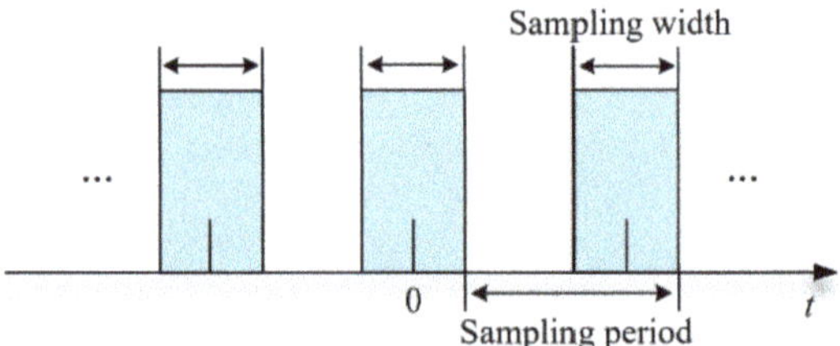

Figure 4. Schematic diagram of ISRJ.

The range direction interrupting the sampling pulse signal, $Q_1(t)$, can be expressed as

$$Q_1(t) = \text{rect}\left(\frac{t}{T_{w1}}\right) \otimes \sum_{n_r=-\infty}^{+\infty} \delta(t - n_r T_{s1}),\tag{4}$$

where $\otimes$ represents the convolution operation. The rect represents the rectangular window function. δ is the impulse function. T_{w1} represents the sampling width, and T_{s1} represents the sampling period in the range direction. The T_{w1}/T_{s1} is defined as the range sampling duty cycle.

The azimuth interrupting the sampling pulse signal $Q_2(t)$ can be expressed as

$$Q_2(t) = \mathrm{rect}\left(\frac{t}{T_{w2}}\right) \otimes \sum_{n_a=-\infty}^{+\infty} \delta(t - n_a T_{s2}), \tag{5}$$

where T_{w2} represents the pulse width, usually $T_{w1} < T_{w2}$, and T_{s2} represents the azimuth sampling period. T_{w2}/T_{s2} is defined as the azimuth sampling duty cycle. After 2D ISRJ, the jamming signal received by SAR can be expressed as

$$\begin{aligned} S_{\mathrm{ISRG}}(t_r, \eta) &= w_r[t_r - 2R_r/c]w_a(\eta - \eta_c)\exp\left[j\pi K_r(t_r - 2R_r/c)^2\right] \\ &\quad \cdot \exp(-j4\pi f_0 R_r/c)Q_1(t_r)Q_2(\eta). \end{aligned} \tag{6}$$

For $S_{\mathrm{ISRG}}(t_r, \eta)$, after pulse compression (PC), a 2D deceptive jamming effect can be generated. Meanwhile, as in (6), the jamming effect of ISGR is affected by various factors, such as sampling duty cycle, sampling period, sampling width [39], etc.

2.4. The Proposed Method

Waveform coding technology can change the time-domain SAR waveform at different times with a high degree of freedom [43]. Based on waveform coding technology, this paper proposes a repeater-type SAR deceptive jamming method through the joint coding of amplitude and phase in intra-pulse and inter-pulse, i.e., CAPII, which can achieve a 2D controllable deceptive jamming effect. The core of the CAPII method is to carry out the special encoding of the intercepted SAR signals between the intra-pulse and inter-pulse, which mainly includes the following two aspects. One is to carry out the special grouping of the intercepted SAR signals between the intra-pulse and inter-pulse. The second is to encode different groups. The encoding objects within the pulse are different sampling points, and the encoding objects between pulses are different pulses. In addition, the encodings within and among the pulses are independent of each other and have no sequence.

Figure 5 is a schematic diagram of the principle (taking 3 groups as an example). The solid lines in Figure 5 represent the transmitted pulses by the jammer, where different colors represent different encoding forms. Solid lines of the same color represent that these pulses are encoded in the same form, i.e., the same group in inter-pulse. The markers on the solid line represent different encodings of sample points within the pulse. Similarly, the same marker represents the sample points having the same encoding value, i.e., the same group in intra-pulse. The arrangement of different groups within the inter-pulse and intra-pulse is shown in Figure 5, in which they are alternately arranged.

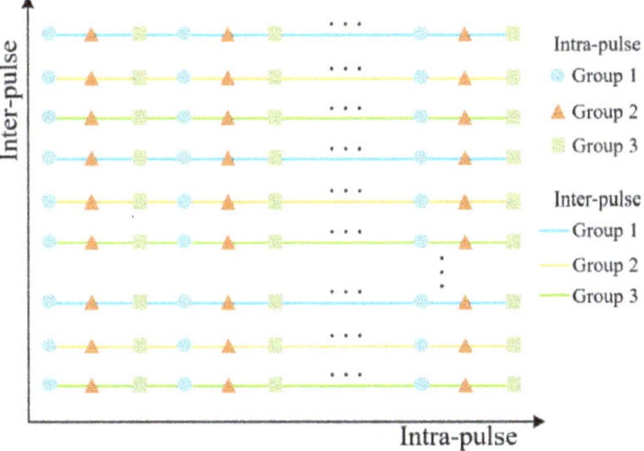

Figure 5. Schematic diagram of the proposed method (taking 3 groups as an example).

The jammer is deployed in the protected area, which intercepts the radiated SAR signal, and performs high-speed full-pulse sampling through the D/A converter. Moreover, the sampling rate follows the Nyquist sampling theorem. The intercepted baseband SAR signal can be expressed as

$$
\begin{aligned}
S(t) &= \sum_{n_1=-\infty}^{+\infty} \mathrm{rect}\left(t - n_1 T_r - \frac{R_r(t)}{c}\right) \exp\left(-j2\pi f_c \frac{R_r(t)}{c}\right) \\
&\quad \cdot \exp\left\{-j\pi K_r \left[t - n_1 T_r - \frac{R_r(t)}{c}\right]^2\right\},
\end{aligned}
\tag{7}
$$

where f_c is the carrier frequency of the intercepted SAR signal, and T_r is the pulse duration time.

Then, intra-pulse and inter-pulse coding are performed on the intercepted baseband SAR signal. Encoding is the core of the CAPII method, and different jamming effects can be obtained by changing encoding forms. For intra-pulse encoding, the encoding object is sampling points, and all pulses are encoded in the same form within the pulse. Firstly, a special grouping of sample points is required. Assuming that the impulse function is adopted for grouping, it can be expressed as

$$
\delta(t) = \begin{cases} 1, & t = 0 \\ 0, & t \neq 0 \end{cases}.
\tag{8}
$$

Then, amplitude and phase joint encoding are performed separately for each group. It is noteworthy that encoding values are the same within the same group, but different among the different groups, i.e.,

$$
P(t) = \sum_{m_r=0}^{M_r-1} \sum_{l_r=-\infty}^{\infty} \delta\left(t - l_r M_r - m_r - \frac{R_r(t)}{c}\right) \exp(j\varphi_{m_r}) \gamma_{m_r},
\tag{9}
$$

where M_r represents the number of intra-pulse groups, m_r represents the index of different groups in intra-pulse, and l_r represents the index of the sample point position. $\varphi_{m_r} \epsilon [-\pi, \pi]$ and γ_{m_r} represent the phase and amplitude encoding values of the m_r-th group, respectively. The phase encoding values are different for all groups, and the amplitude encoding can be the same or different. By performing grouping and coding for each intercepted pulse, the intra-pulse coding is realized, which produces a deceptive jamming effect in the fast time-domain.

For inter-pulse coding, the coding principle is similar to that of intra-pulse coding, but the object of inter-pulse encoding is the pulse. First, different pulses need to be grouped according to the intercepted SAR signal. Then, amplitude and phase encoding are performed on each group. Regarding inter-pulse grouping and coding principles, the same regulations and limitations apply as in intra-pulse amplitude and phase joint coding. Specifically, the inter-pulse phase and amplitude joint coding principle can be expressed as

$$
Q(t) = \sum_{m_a=0}^{M_a-1} \sum_{l_a=-\infty}^{\infty} \mathrm{rect}\left[\frac{t - (l_a M_a - m_a) T_r}{T_r}\right] \exp(j\varphi_{m_a}) \gamma_{m_a},
\tag{10}
$$

where M_a represents the number of groupings in the inter-pulse, m_a represents the index of different groups within inter-pulse, and l_a represents the index of the different pulses.

The jamming signal, after the joint coding of amplitude and phase in the intra-pulse and inter-pulse, can be expressed as

$$
S_{\mathrm{Jamming}} = S(t) P(t) Q(t).
\tag{11}
$$

According to (11), the jamming signal is equivalent to modulating the intercepted SAR signal with two jamming terms. Therefore, the major computational complexity of the proposed method is $\mathcal{W}_a \cdot \mathcal{W}_r$, where $\mathcal{W}_a$ and $\mathcal{W}_r$ represent the number of pulses in the intercepted SAR signal and the number of sampling points of a single pulse, respectively. Then, the jamming signal is retransmitted through the jammer. After the jamming signal, S_{Jamming}, is sent to the SAR platform, it is sampled in 2D. Here, the working delay of the jammer is ignored. Decomposing the received jamming signal into 2D, i.e., slow time-domain, η, and fast time-domain, t_r, it can be expressed as

$$
\begin{aligned}
S_U(t_r, \eta) &= \text{rect}\left(\tfrac{t_r - 2\tau}{T_r}\right) \text{rect}\left(\tfrac{\eta}{T_s}\right) \exp(-j 4\pi f_c \tau) \\
&\quad \cdot \exp\left[j\pi K_r (t_r - 2\tau)^2\right] P(t_r - \tau) Q(\eta),
\end{aligned}
\tag{12}
$$

where T_s represents synthetic aperture time. The τ is the time delay in the fast time-domain, which can be expressed as

$$
\tau = \frac{R_r(\eta)}{c}.
\tag{13}
$$

$R_r(\eta)$ is the instantaneous slant range from the jammer to the SAR platform. After expanding using the Taylor formula, $R_r(\eta)$ can be expressed as

$$
R_r(\eta) = \sqrt{x_j^2 + (v\eta)^2 + H^2} \approx R_0 + \frac{v^2}{R_0}\eta^2.
\tag{14}
$$

The $P(t_r)$ and $Q(\eta)$ in (12) represent the jamming modulation terms performed by the jammer on the intercepted SAR signal, i.e., the joint coding of the intra-pulse and inter-pulse amplitude and phase. From (9) and (10), the main parameters that affect the jamming modulation terms include the number of groups, phase encoding values, and amplitude encoding values. By controlling these parameters, a variety of deceptive jamming effects can be generated in SAR images.

3. Signal Imaging Process and Analysis

To analyze the performance of the proposed method in this paper, ignoring the coupling between the azimuth and the range directions [41], the echo of the jamming signal can be divided into the range component, $S_R(t_r)$, and the azimuth component, $S_A(\eta)$, i.e.,

$$
\left\{
\begin{aligned}
S_R(t_r) &= \text{rect}\left(\tfrac{t_r - 2\tau}{T_r}\right) \exp\left[j\pi K_r(t_r - 2\tau)^2\right] \sum_{m_r=0}^{M_r-1} \sum_{l_r=-\infty}^{+\infty} \delta(t_r - l_r M_r - m_r - 2\tau) \exp(j\varphi_{m_r})\gamma_{m_r} \\
S_A(\eta) &= \text{rect}\left(\tfrac{\eta}{T_s}\right) \exp\left(-j\pi K_a \eta^2\right) \sum_{m_a=0}^{M_a-1} \sum_{l_a=-\infty}^{+\infty} \delta(\eta - l_a M_a - m_a) \exp(j\varphi_{m_a})\gamma_{m_a},
\end{aligned}
\right.
\tag{15}
$$

where K_a represents the azimuth chirp rate, i.e.,

$$
K_a = \frac{2v^2}{\lambda R_0},
\tag{16}
$$

where $\lambda = c / f_c$ stands for the carrier wavelength.

At present, there are many algorithms for processing the SAR signal, such as the RD algorithm, the chirp scaling algorithm, and the Omega-k algorithm, which were proposed based on different application scenarios. In this paper, the RD algorithm is chosen as the jamming imaging algorithm since the RD algorithm decomposes the echo into the range and the azimuth echo, which is convenient for analysis. Meanwhile, the RD algorithm implements 2D PC in the frequency domain and reaches a high level of integration. Generally, the RD algorithm requires two steps, i.e., range compression and azimuth compression. This section will discuss them separately.

3.1. Range Compression

From the above, the echo components in the range direction are divided into M_r groups. In this subsection, for simplicity, each group is analyzed separately, and the values of other groups are set to 0. As shown in Figure 6, the sum of M_r groups, i.e., the range echo components, can be expressed as

$$S_R(t_r) = S_R^{(1)}(t_r) + S_R^{(2)}(t_r) + S_R^{(3)}(t_r) + \cdots + S_R^{(M_r-1)}(t_r), \tag{17}$$

where the m_r-th group $S_R^{(m_r)}$ can be expressed as

$$
\begin{aligned}
S_R^{(m_r)}(t_r) &= \text{rect}\left(\tfrac{t_r-2\tau}{T_r}\right) \exp\left[j\pi K_r(t_r - 2\tau)^2\right] \\
&\cdot \sum_{l_r=-\infty}^{+\infty} \delta(t_r - l_r M_r - m_r - 2\tau) \exp(j\varphi_{m_r})\gamma_{m_r}.
\end{aligned}
\tag{18}
$$

when processing digital signals, the spectrum of the signal is obtained by discrete Fourier transform (FT). For convenience, the digital frequency is represented by the analog frequency, so the spectrum of $S_R^{(m_r)}(t_r)$ can be expressed as

$$
\begin{aligned}
S_R^{(m_r)}(f_r) &= \text{rect}\left(\tfrac{f_r}{B_r}\right)\rho_{m_r} \exp\left(-j\pi\tfrac{f_r^2}{K_r}\right)\ \exp(-j4\pi f_r\tau) \\
&+ \rho_{m_r} \sum_{G=0}^{1} \sum_{k=1}^{M_r-1} \text{rect}\left\{ \frac{\left\{[f_r-\tfrac{1}{2}\beta_r\ (G,k)]-\tfrac{1}{4}(-1)^G(f_{sr}-B_r)\right\}}{(-1)^{G+1}\beta_r\ (G,k)+B_r/2+f_{sr}/2} \right\} \\
&+ \exp(-j2\pi km_r/M_r) \exp\left[-j\pi\frac{[f_r-\beta_r\ (G,k)]^2}{K_r}\right] \exp(-j4\pi f_r\tau),
\end{aligned}
\tag{19}
$$

where B_r is the bandwidth. f_{sr} represents range sampling frequency, and the value of G is 0 or 1; this is because the frequency spectrum range that can be observed is $[-f_{sr}/2,\ f_{sr}/2]$. When G takes other values, the spectrum component does not fall within $[-f_{sr}/2,\ f_{sr}/2]$ after shifting to the right or to the left of kf_{sr}/M_a, as shown in Figure 7.

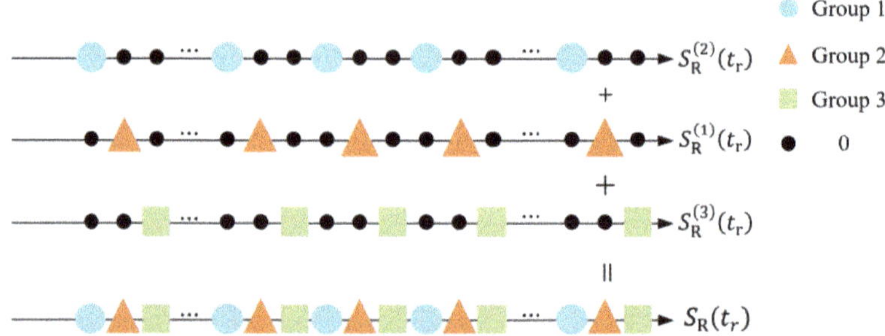

Figure 6. Range direction echo component split diagram (taking 3 groups as an example).

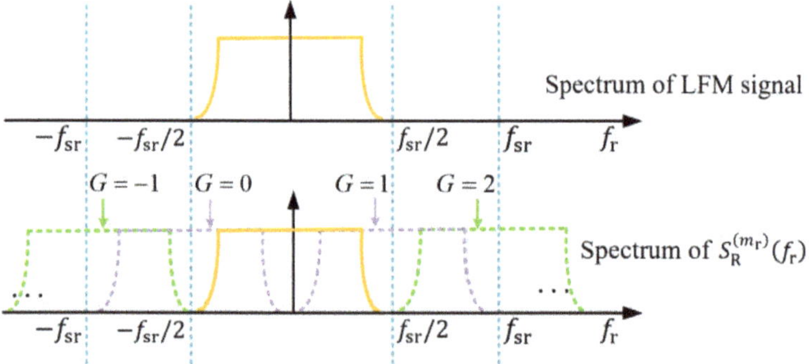

Figure 7. Schematic diagram of range echo component spectrum.

$\rho_{m_r}(f_r)$ represents the amplitude and phase encoding terms, i.e.,

$$\rho_{m_r} = \exp(j\varphi_{m_r})\gamma_{m_r}. \tag{20}$$

$\beta_r(G,k)$ is a function introduced for shorthand and can be represented as

$$\beta_r(G,k) = kf_{sr}/M_r - Gf_{sr}. \tag{21}$$

Thus, the frequency-domain expression of the range echo component, $S_R(t_r)$, can be expressed as

$$S_R(f_r) = S_R^{(1)}(f_r) + S_R^{(2)}(f_r) + S_R^{(3)}(f_r) + \cdots + S_R^{(M_r)}(f_r). \tag{22}$$

To obtain the PC results in the range direction, we use the range frequency domain matched filter, which can be expressed as

$$H_r(f_r) = \operatorname{rect}\left(\frac{f_r}{B_r}\right)\exp\left(j\pi\frac{f_r^2}{K_r}\right). \tag{23}$$

According to (22) and (23), the fast time-domain PC result can be deduced as

$$
\begin{aligned}
S_{\text{out\_R}}(t_r) \;=\;& \sum_{m_r=0}^{M_r-1} \rho_{m_r} B_r \operatorname{sinc}[B_r(t_r - 2\tau)] \\
&+ \sum_{m_r=0}^{M_r-1} \rho_{m_r} \sum_{G=0}^{1}\sum_{k=1}^{M_r-1} \exp\left(\frac{-j2\pi m_r k}{M_r}\right)\exp\left[\frac{-j\pi\beta_r(G,k)^2}{K_r}\right] \\
&\cdot\left[(-1)^{G+1}\beta_r(G,k) + \frac{B_r+f_{sr}}{2}\right] \\
&\cdot\operatorname{sinc}\left\{\left[(-1)^{G+1}\beta_r(G,k) + \frac{B_r+f_{sr}}{2}\right]\left(t_r + \frac{\beta_r(G,k)}{K_r} - 2\tau\right)\right\} \\
&\cdot\exp\left[j2\pi\left(\tfrac{1}{2}\beta_r(G,k) + \tfrac{1}{4}(-1)^{G}(f_{sr} - B_r)\right)\left(t_r + \frac{\beta_r(G,k)}{K_r} - 2\tau\right)\right].
\end{aligned}
\tag{24}
$$

From (24), after intra-pulse coding, the PC result of the jamming signal in the range direction consists of multiple sinc functions with different amplitude coefficients. This forms the basis for deceptive jamming in the range direction. By adjusting the number, M_r, of $\rho_{m_r}(f_r)$, diversified deceptive jamming effects can be generated in the SAR image.

3.2. Azimuth Compression

The azimuth PC is similar to the range direction, which can be regarded as a linear frequency-modulated (LFM) signal with a chirp rate of K_a. Similar to Figure 7, the azimuth echo can be split into M_a groups.

As is expressed in (19), the azimuth echo component can also be expressed as

$$S_A(\eta) = S_A^{(1)}(\eta) + S_A^{(2)}(\eta) + S_A^{(3)}(\eta) + \cdots + S_A^{(M_a-1)}(\eta), \tag{25}$$

where the m_a-th group can be expressed as

$$S_A^{(m_a)}(\eta) = \operatorname{rect}\left(\frac{\eta}{T_s}\right)\exp\left(-j\pi K_a\eta^2\right)\sum_{m_a=-\infty}^{+\infty}\delta(\eta - l_a M_a - m_a)\exp(j\varphi_{m_a})\gamma_{m_a}. \tag{26}$$

Through the analysis of (26), the azimuth echo component is constituted by three items, i.e., the quadratic phase term, $\exp\left(-j\pi K_a\eta^2\right)$; the impulse function term, $\delta(\eta - l_a M_a - m_a)$; and the encoding item, $\exp(j\varphi_{m_a})\gamma_{m_a}$. Then, the frequency-matched filter is used for PC, which can be expressed as

$$H_a(f_a) = \operatorname{rect}\left(\frac{f_a}{B_a}\right)\exp\left(-j\pi\frac{f_a^2}{K_a}\right), \tag{27}$$

where B_a represents the Doppler bandwidth. Similar to range direction, the PC result in the azimuth echo, $S_A(\eta)$, can be deduced as

$$
\begin{aligned}
S_{\text{out\_A}}(\eta) \;=\; & \sum_{m_a=0}^{M_a-1} \rho_{m_a} B_a \sin c(B_a \eta) \\
& + \sum_{m_a=0}^{M_a-1} \rho_{m_a} \sum_{L=0}^{1} \sum_{J=0}^{M_a-1} \exp\left(\frac{-j2\pi m_a J}{M_a}\right) \exp\left[\frac{-j\pi \beta_a \, (L,J)^2}{K_a}\right] \\
& \cdot \left[(-1)^{L+1}\beta_a \, (L,J) + \frac{B_a + f_{sa}}{2}\right] \\
& \cdot \sin c\left\{\left[(-1)^{L+1}\beta_a \, (L,J) + \frac{B_a + f_{sa}}{2}\right]\left(\eta + \frac{\beta_a \, (L,J)}{K_a}\right)\right\} \\
& \cdot \exp\left[j2\pi\left(\tfrac{1}{2}\beta_a \, (L,J) + \tfrac{1}{4}(-1)^L (f_{sa} - B_a)\right)\left(\eta + \frac{\beta_a \, (L,J)}{K_a}\right)\right],
\end{aligned}
\tag{28}
$$

where f_{sa} represents the azimuth sampling rate. $\rho_{m_a}(\eta)$ stands for the phase encoding terms of the m_a-th group, i.e.,

$$
\rho_{m_a} = \exp(j\varphi_{m_a})\gamma_{m_a}.
\tag{29}
$$

The $\beta_a \, (L,J)$ is also a function introduced for shorthand, which can be represented as

$$
\beta_a \, (L,J) = J f_{sa} / M_a - L f_{sa},
\tag{30}
$$

where f_{sa} is the azimuth sampling rate. The value of L is 0 or 1. Similar to the one shown in Figure 6, it is because the azimuth of the frequencies that can be observed is $[-f_{sa}/2, f_{sa}/2]$.

From (28), the PC results in the azimuth direction are similar to those in the range direction, which are also composed of multiple $\sin c$ functions with different amplitude coefficients. This forms the basis for azimuth deceptive jamming.

3.3. False Target Feature Analysis

3.3.1. The Number of False Targets

Through the analysis of (24) and (28), the following relationship exists between the number of false targets and groups, i.e.,

$$
\begin{cases}
N_r = 2(M_r - 1), & \text{azimuth} \\
N_a = 2(M_a - 1), & \text{range,}
\end{cases}
\tag{31}
$$

where N_r and N_a represent the number of false targets in the range and azimuth, respectively. The total number of false targets can be represented as

$$
N = (2M_r - 1)(2M_a - 1) - 1.
\tag{32}
$$

From (31) and (32), the number of false targets can be precisely controlled in the range and azimuth directions by the number of groups.

3.3.2. The Position of False Targets

Through the analysis of (24), if $G = 0$, the value of $\beta_r \, (G,k)$ is greater than 0. At this time, if the value of $\beta_r \, (G,k)$ is brought into Equation (24). The $\sin c$ function will shift left. Relatively, if the value of $\beta_r \, (1,k)$ is less than 0, the $\sin c$ function will shift to the right, i.e., the false target moves to the right of the real target. In other words, the value of G determines the position of the false target in the SAR image. $G = 0$ means the false target is shifted to the left of the real target. $G = 1$ means the false target is shifted to the right of the real target. In addition, the value of k determines the distance between the false target and the real target, i.e., the offset. If $G = 0$, the offset increases with the increase in k. If $G = 1$, the offset reduces with the increase in k. In summary, as shown in Figure 8, in the range, the position of false targets can be represented by the values of G and k. From Figure 8, the false targets are evenly distributed on both sides of the real target. Then, the false target at the (G, k) position can be deduced as

$$\text{POSR}_{l,k}(t_r) = t_r - 2\tau - (G\alpha_s - k\alpha_s/M_r)T_r, \tag{33}$$

where α_s represents the oversampling rate. To facilitate the verification, the time position represented by (33) is converted into a space position in the SAR image, which can be expressed as

$$\Omega_R(G, k) = X + (G\alpha_s - k\alpha_s/M_r)T_r c/2, \tag{34}$$

where X represents the range direction position of the jammer in the SAR image.

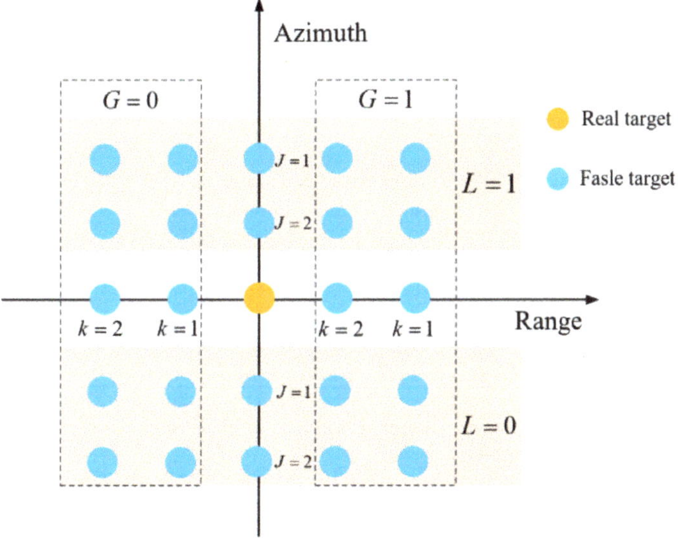

Figure 8. Derivation schematic diagram of false target position distribution (taking $M_a = M_r = 3$ as an example).

Analogously, by analyzing (28), $L = 0$ means that the false target is behind the real target, and $L = 1$ means that the false target is in front of the real target. The value of J determines the offset between the false target and the real target in azimuth. If $L = 0$, the offset increases with the increase in J. If $L = 1$, the offset reduces with the increase in J. In summary, in the azimuth, the different false targets can be represented by the values of L and J. Then, in the azimuth, the false target at the (L, J) position, $\text{POSA}_{L, J}(\eta)$, can also be deduced as

$$\text{POSA}_{L,J}(\eta) = \eta - \left(L\alpha_s - \frac{J\alpha_s}{M_a} \right) T_s. \tag{35}$$

The time position represented by (35) is converted into a space position in the SAR image, which can be expressed as

$$\Omega_A(L, J) = Y + \left(L\alpha_s - \frac{J\alpha_s}{M_r} \right) T_s v, \tag{36}$$

where Y represents the azimuth direction position of the jammer in the SAR image.

Figure 8 shows the derivation results of false targets' positions in the SAR image. Subsequently, according to the values of G, k, L, and J, the position of the false target in the SAR image can be determined by Formulas (34) and (36). This provides greater flexibility for the implementation of deceptive jamming.

3.3.3. The Amplitude of the False Target

The amplitude of the false target is an important factor in implementing deceptive jamming. According to (24), the main factor affecting the amplitude of false targets is ρ_{m_r}. After analyzing (24) and (28), the amplitude of the false target at the (G, k, L, J) position in logarithmic form can be calculated as

$$A_{G,k,L,J} = A_{RG,k} \cdot \alpha_{L,J}, \tag{37}$$

where $\alpha_{L,J}$ denotes the ratio of the amplitude of the false targets at the (L, J) position to the amplitude of the false target with the same azimuth position as the jammer. $A_{RG,k}$ represents the amplitude of the false targets at a range position of (G, k) and can be expressed as

$$A_{RG,k} = \begin{cases} 20 \log_{10} \left\{ \sum\limits_{m_r=0}^{M_r-1} \rho_{m_r} B_r \mathrm{sinc}[B_r(t_r - 2\tau)] \right\}, & \text{Same range position as the jammer} \\ \Psi(G,k), & \text{others} \end{cases}, \tag{38}$$

where $\Psi(G, k)$ can be represented as

$$\begin{aligned} \Psi(G,k) \quad &= 20 \log_{10} \Bigg\{ \bigg[(-1)^{G+1} \beta_r\,(G,k) + \tfrac{B_r+f_{sr}}{2} \bigg] \\ &\quad \cdot \mathrm{sinc} \bigg\{ \bigg[(-1)^{G+1} \beta_r\,(G,k) + \tfrac{B_r+f_{sr}}{2} \bigg] \bigg(t_r + \tfrac{\beta_r\,(G,k)}{K_r} - 2\tau \bigg) \bigg\} \\ &\quad \cdot \sum\limits_{m_r=0}^{M_r-1} \gamma_{m_r} \exp(j\varphi_{m_r}) \exp(-j2\pi m_r k / M_r) \Bigg\} \Bigg\}. \end{aligned} \tag{39}$$

$\alpha_{L,J}$ is classified into two cases; if the false target and the jammer are located in the same azimuth, then its value is 1. Otherwise, the value of $\alpha_{L,J}$ can be expressed as

$$\begin{aligned} \alpha_{L,J} \quad &= 20 \log_{10} \left\{ \frac{\left| \sum_{m_a=0}^{M_a-1} \gamma_{m_a} \exp(j\varphi_{m_a}) \exp(-j2\pi m_a J / M_a) \right|}{\left| \sum_{m_a=0}^{M_a-1} \gamma_{m_a} \exp(j\varphi_{m_a}) \right|} \right\} \\ &\quad + 20 \log_{10} \left(\frac{(-1)^{L+1} \beta_a\,(L,J) + (B_a + f_{sa})/2}{B_a} \right). \end{aligned} \tag{40}$$

From (37), the amplitude of the false target at position (G, k, l, J) can be calculated. In addition, the main factors that affect the amplitude of false targets are $\gamma_{m_a} \exp(j\varphi_{m_a})$. The presence of phase encoding creates false targets. The amplitude encoding value affects the amplitude of false targets. Thus, if the phase encoding value is determined, the false targets of different amplitudes will be generated by adjusting the amplitude encoding value.

Based on the above analysis, the number of false targets in the range and azimuth can be controlled by the number of groups. By controlling the amplitude encoding value, the amplitude of the false target can be adjusted. Meanwhile, the positions of the false target in the SAR image were also derived, which provides greater controllability and flexibility for deceptive jamming implementation.

4. Simulation and Results

In this section, point and area target simulations are performed to prove the effectiveness of the theoretical analysis. Meanwhile, to demonstrate the superiority of the proposed method, the FDA-based jamming method [38] and the ISRJ method [43] are selected as comparative experiments. Table 1 shows the main parameters of the simulation.

Table 1. The main parameters of the simulation.

Parameter	Value
Carrier frequency	5.4 GHz
Pulse width	25 us
Bandwidth	60 MHz
Pulse repetition frequency	1410 Hz
Platform velocity	7568.94 m/s
Closest slant range	815.9 km
Antenna length	15 m
Oversampling ration (α_s)	1.2

4.1. Point Target Simulation

The size of the simulation scene is set up as 16×10 km in size. The jammer is placed in the center of the scene, i.e., point $(5,\ 8)$ km. The range and azimuth resolutions can be calculated as 2.4983 m and 7.5 m from Table 1.

From the analysis in Section 3.3.1, the number of false targets is controlled by the number of groups. Now, we set the phase encoding value as $\exp(j\vartheta\pi)$, $\vartheta = 1 - (m-1)/6$, where $m = 1,\ 2,\ \ldots, M_a = M_r$ represents the phase encoding value in the m-th group in the range or azimuth. Taking $M_a = M_r = 2$ as an example, the phase encoding values of different groups in the azimuth and range directions are $\exp(j\pi)$, $\exp[j(5/6)\pi]$, and $\exp[j(2/3)\pi]$. Figure 9 shows the imaging results for different numbers of groups without amplitude encoding.

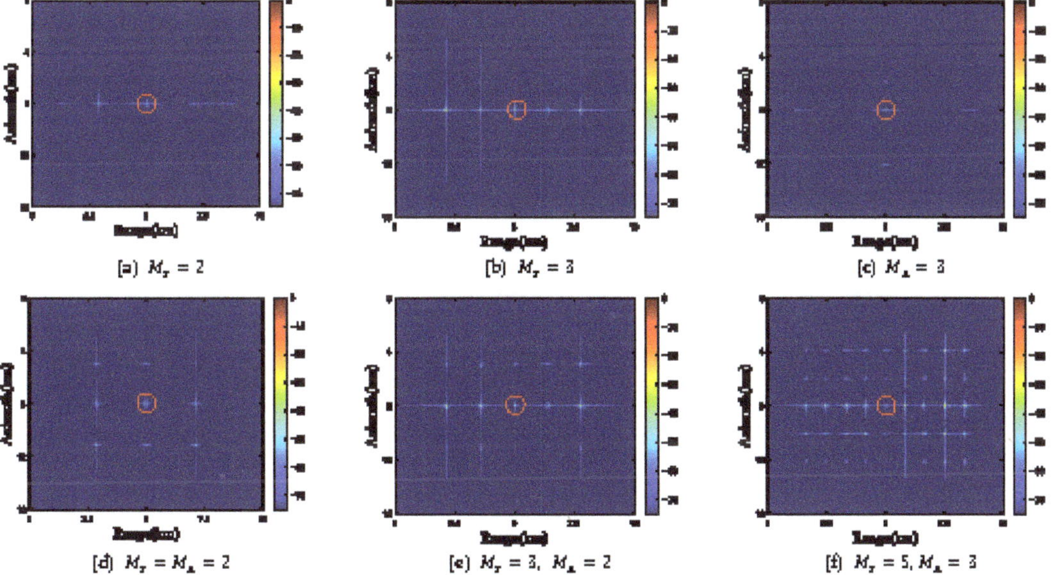

Figure 9. Simulation results of point targets with different grouping numbers (without amplitude encoding), where the real target is inside the red circle. Results with only intra-pulse encoding (a) $M_r = 2$, (b) $M_r = 3$, and (c) results with only inter-pulse encoding ($M_r = 3$). Results with intra-pulse and inter-pulse encoding (d) $M_r = M_a = 2$, (e) $M_r = 3, M_a = 2$, and (f) $M_r = 5, M_a = 3$.

If the number of intra-pulse groups is two and three; the number of false targets in the range direction can be calculated according to Formula (31), i.e., two and four, which can be seen in Figure 9a,b. This is consistent with the analysis results. Figure 9c shows the simulation results for only inter-pulse encoding, in which the number of false targets also satisfies Formula (31). Figure 9d–f shows the imaging results with intra-pulse and inter-pulse 2D phase encoding. The number of false targets in Figure 9d–f is 8, 14, and 44, respectively, which agrees with Equation (32). Therefore, the accurate control of the number of false targets can be achieved by the number of groups within the intra-pulse and inter-pulse. In addition, the distribution of false targets is consistent with the results derived in Figure 8.

Figure 10a shows the simulation results with only phase encoding ($M_r = M_a = 3$). The real target is inside the red circle, i.e., the position of the jammer. To evaluate the false target spatial position and amplitude, four points within the yellow rectangle in Figure 10a were selected for validation. Taking point P_3 as an example, the spatial position of the false target can be calculated with Equations (34) and (36). According to the distribution of false targets in Figure 8, the value of G and L are both equal to 0 at this time for the false target, P_3. In addition, the values of k and j can both be derived as 1. Subsequently, the above values are, respectively, input into (34) and (36), which can obtain the theoretical derivation position of the false target. The theoretical derivation and the simulated value of the different false target space locations are shown in Table 2. It can be concluded that the simulated values are basically consistent with the theoretical values.

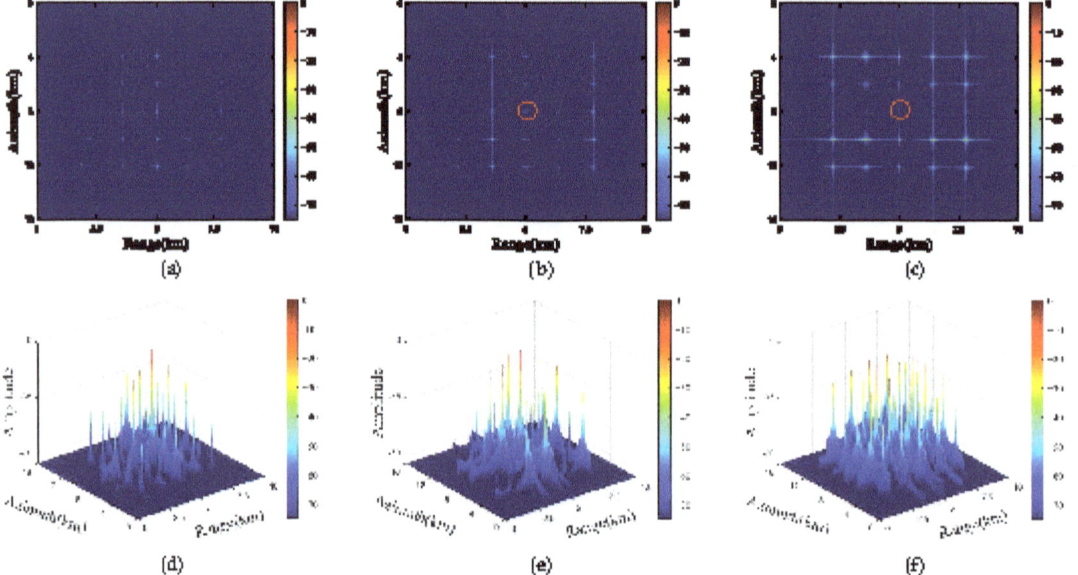

Figure 10. Simulation results of point targets with $M_r = M_a = 3$ (The real target is inside the red circle). (**a**) The results with intra-pulse and inter-pulse encoding without amplitude encoding. (**b**) The result with amplitude coding values of 1, 0.1, and 1. (**c**) The result with amplitude coding values of 0.1, 5, and 0.22. (**d–f**) The 3D imaging results of (**a**), (**b**), and (**c**), respectively.

Table 2. Peak value and position of false targets.

Point	Range Position (km)		Azimuth Position (km)		Amplitude Peak Value (dB)	
	Theoretical	Simulated	Theoretical	Simulated	Theoretical	Simulated
P_1	5	5	6.7921	6.7243	−17.19	−18.56
P_2	6.499	6.672	8	8	−14.34	−13.22
P_3	3.501	3.4280	9.2079	9.1023	−36.17	−38.34

For ease of analysis, the maximum amplitude value (i.e., peak value) of the false target is selected as the evaluation index to verify the correctness of the amplitude analysis.

Since the simulation results of Figure 10a do not perform amplitude encoding, the value of amplitude encoding can be regarded as 1. For point P_1, its range position is the same as the jammer, and the phase encoding values of intra-pulse and inter-pulse are $\exp(j\pi)$, $\exp[j(5/6)\pi]$, and $\exp[j(2/3)\pi]$. Therefore, the value of $A_{RG,k}$ is calculated with (38). For point P_2, its azimuth position is the same as the jammer. Thus, the value of $\alpha_{L,J}$ is 1. The amplitude peak value of different false targets in Figure 10a is also shown in Table 2. It can be seen from Table 2 that the theoretical value and the actual value are basically consistent. Therefore, the analysis of false target amplitude is validated.

The characteristics of the false target, including the number, space position, and amplitude ratio, are analyzed. However, the energy of false targets is always lower. Figure 10b,c show the simulation results for amplitude encoding. Figure 10b is a simulation result of amplitude encoding only in the second group, i.e., the encoding amplitude values are 1, 1.1, and 1 in the range and azimuth directions. Obviously, the amplitude ratio of some false targets is changed. Figure 10c is the simulation result of amplitude encoding for all groups, and the encoding amplitude values are 2.1, 5, and 1.22 in the range and azimuth. At this time, the amplitude ratio of almost all false targets changes. Therefore, the amplitude of the false target is affected by the amplitude encoding value.

To further study the effects of different amplitude encoding, we take point P_3 in Figure 10a as an example, i.e., $G = L = 0$ and $k = J = 1$, and set the amplitude encoding value of the second group within the intra-pulse and inter-pulse to $\varepsilon \in (1, 9]$. The amplitude code values for other groups are set to 1. Figure 11 shows the amplitude corresponding to different amplitude encoding values. The blue line represents the theoretical value calculated according to (37), and the red triangle represents the simulated value. Obviously, the simulated value basically agrees with the theoretical value. Therefore, the amplitude of the specific false target can be adjusted by the amplitude coding value. Meanwhile, the amplitude of every false target can also be calculated.

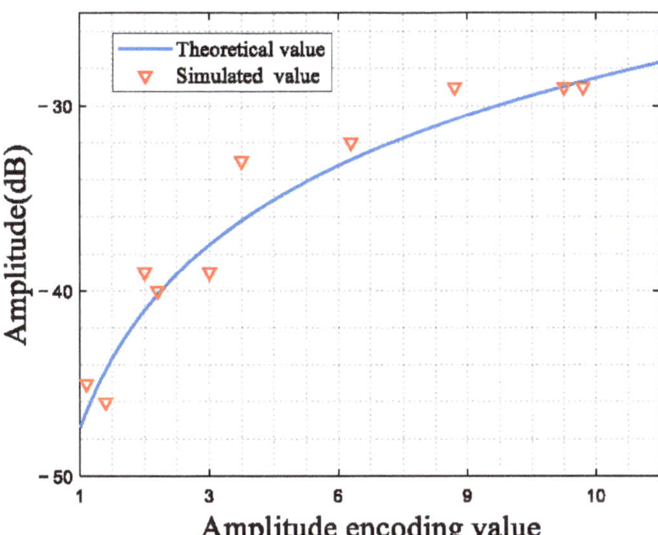

Figure 11. The amplitude corresponds to different amplitude encoding values for point P_3.

For comparison, the FDA-based jamming and ISRJ are selected. Figure 12a,c show the point target simulation results of the FDA-based jamming [38], where the number of FDA elements is five. Obviously, the FDA-based jamming generates false targets only in the range direction. The point target of the simulation results of ISRJ is shown in Figure 12b,d [43]. From the results, although ISRJ can produce 2D false targets, the energy of the false target gradually decreases with the increase in the distance from the real target.

In addition, the false targets generated by the ISRJ method are jointly determined by multiple parameters and are difficult to control. Figure 13a,f show the simulation results of the CAPII method in this paper, where the number of groups is $M_r = M_a = 3$, and the amplitude and phase encoding values are random. Obviously, the CAPII method can not only produce 2D false targets but the amplitude distribution of false targets is also random when amplitude encoding values are random. In particular, Figure 13b,e show the simulation results, where only the second group's range direction value is set to 0, and the other groups are set to 1. Compared to Figure 12a, two columns of false targets corresponding to $k = 2$ are hidden. Figure 13c,f show the simulation results, where the second group's range and azimuth direction value is set to 0, and the other groups are set to 1. Obviously, the two columns of false targets corresponding to $k = 2$ and the two rows of false targets corresponding to $J = 2$ are hidden. That is to say that the false targets of the specific group will disappear if the amplitude encoding value of the corresponding group is set to 0. This reduces the regularity of false target distribution to some extent.

In summary, CAPII generates a 2D false target, and the amplitude of the false target can be adjusted by the amplitude encoding value. Particularly, if the amplitude encoding value of a certain group is set to 0, then its corresponding false targets will be hidden. Meanwhile, the number of false targets can be precisely controlled, which provides greater controllability for deceptive jamming implementations. Therefore, compared with the FDA-based method and ISRJ method, the CAPII method has great advantages.

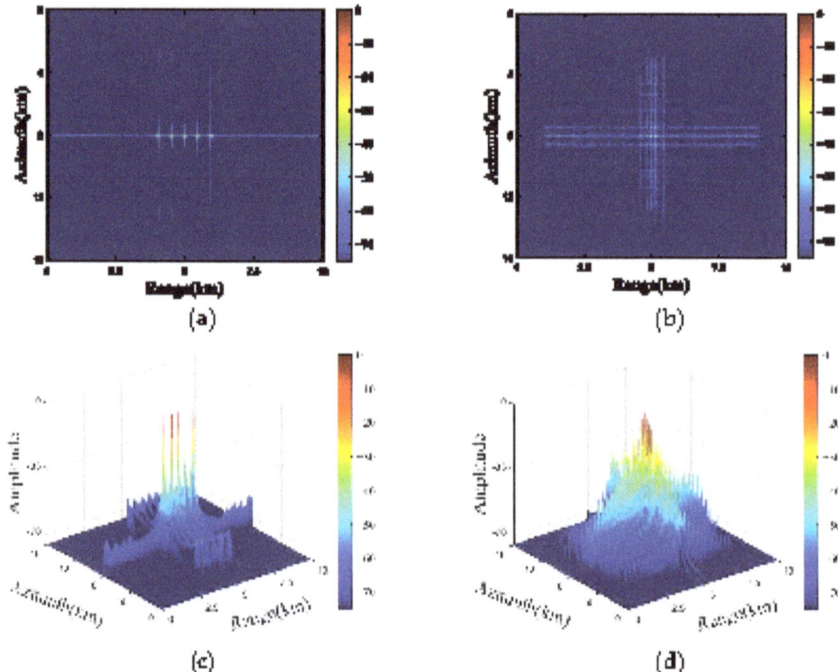

Figure 12. Simulation results of point targets with (**a**) the FDA-based jamming and (**b**) the ISRJ. (**c**) and (**d**) are the 3D imaging results of (**a**) and (**b**), respectively.

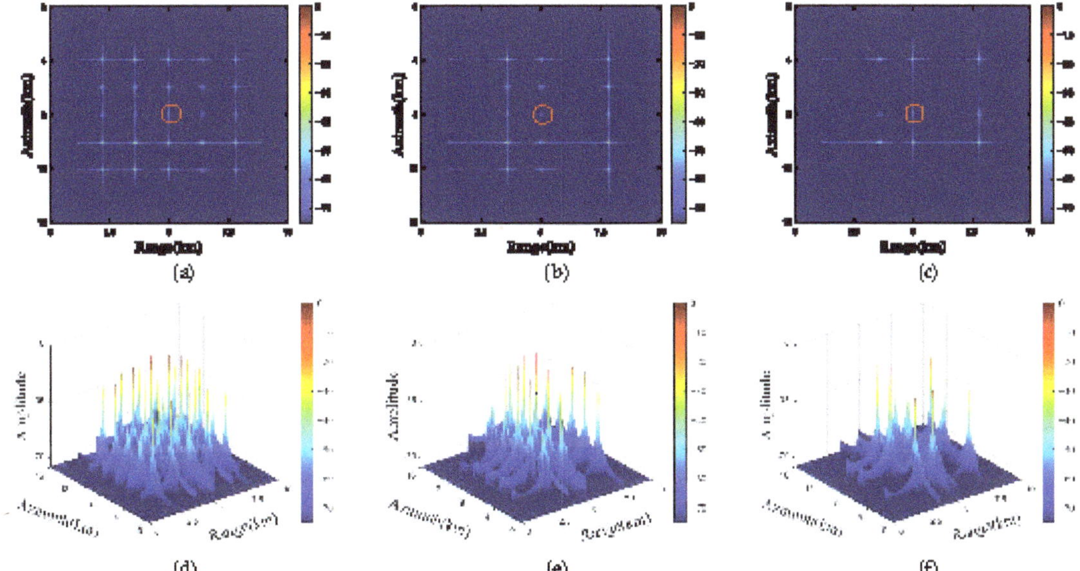

Figure 13. Point target simulation results of CAPII (the real target is inside the red circle). (**a**) Results with intra-pulse and inter-pulse encoding without amplitude encoding. (**b**) The result with amplitude coding values 1, 0, and 1 in the intra-pulse (the amplitude encoding values of inter-pulse are all 1). (**c**) Result with amplitude coding values 1, 0, and 1 in the intra-pulse and inter-pulse. (**d–f**) The 3D imaging results of (**a–c**).

4.2. Area Target Simulation

To further analyze the proposed method, Figure 14 shows the simulation results of the area target. Figure 14c–f are the simulation results of the CAPII method, where all phase encoding is random. Figure 14c is the simulation result of only inter-pulse encoding with $M_a = 3$, in which the false targets are generated in the azimuth. Meanwhile, the number of false targets satisfies Equation (31). Figure 14d is the simulation result of amplitude encoding where the amplitude encoding value is 0 in the second group. Obviously, the corresponding false target is hidden. Figure 14c shows the simulation results of the joint intra-pulse and inter-pulse coding ($M_a = M_r = 3$), which generates a 2D deceptive jamming effect. Simultaneously, the corresponding false targets are hidden. Compared with Figure 14d, the number of false targets in Figure 14f increases significantly. In addition, the false targets of the third group of intra- and inter-pulses are hidden. This reduces the regularity of the distribution of false targets to a certain extent.

Figure 14a shows the area target simulation results of the FDA-based jamming method. The false targets are only generated in the range. Figure 14b shows the area target simulation of the ISRJ method. Although ISRJ can produce 2D false targets, the energy of false energy gradually decreases. Meanwhile, the number of false targets is difficult to control. However, the proposed method, i.e., CAPII, can generate a flexible and controllable 2D deceptive jamming effect. Specifically, the number of false targets can be precisely controlled by the number of groups. The amplitude of the false targets can be modulated by the amplitude encoding value. If the amplitude and phase encoding values are random, the energy of the false targets is randomly distributed. In particular, if the amplitude encoding value of a particular group is 0, its corresponding false targets will be hidden. Therefore, the CAPII method has stronger controllability. The area target simulation experiments fully verify the superiority of the proposed method.

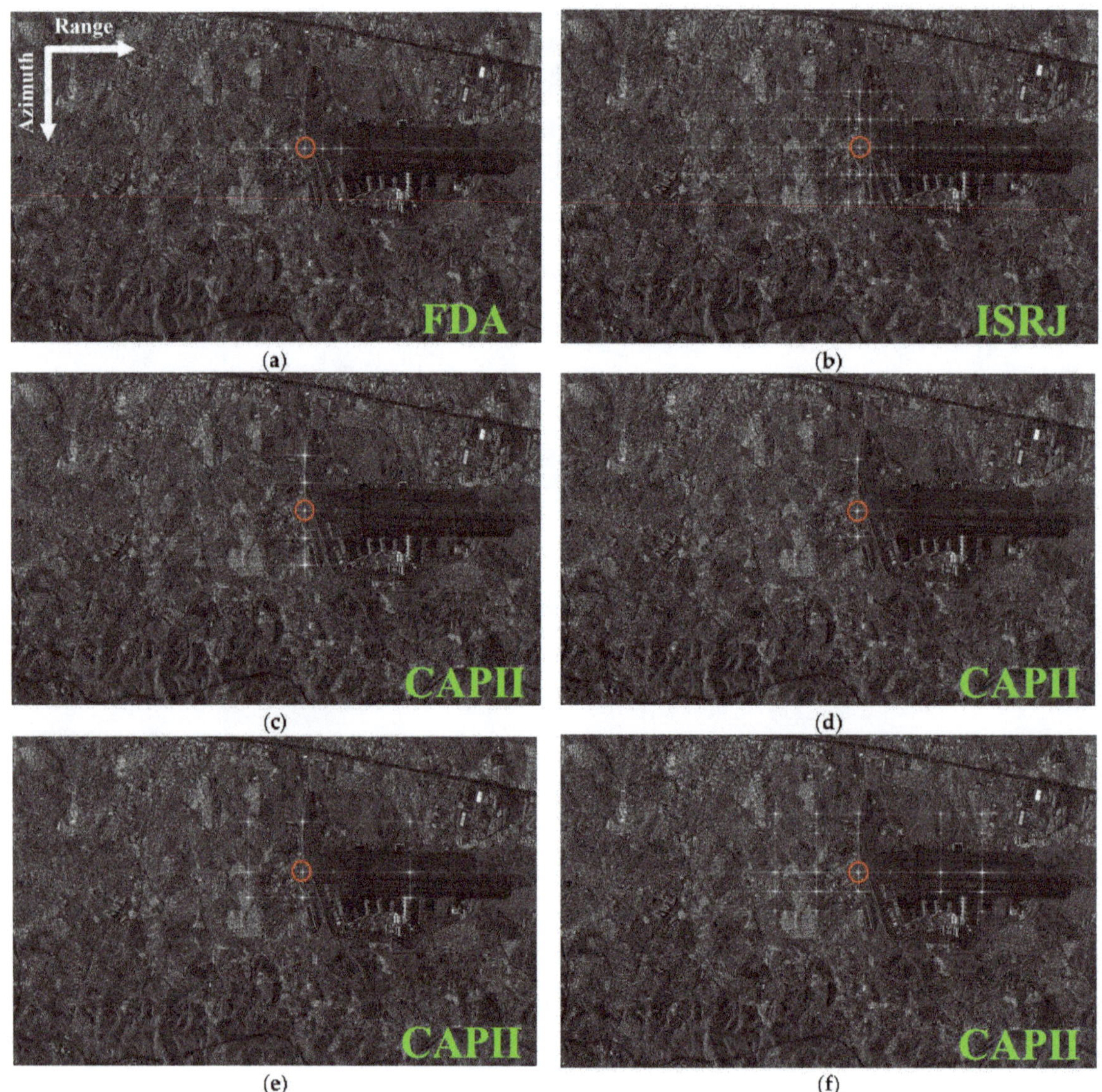

Figure 14. Area target simulation results (the real target is inside the red circle). (**a**) The results of the FDA-based jamming method. (**b**) The results of the ISRJ method. (**c**) The results of CAPII with $M_a = 3$ (encoding value randomly). (**d**) The results of CAPII with $M_a = 3$ (the amplitude encoding value of the second set is 0). (**e**) The results of CAPII with $M_a = M_r = 3$ (the second set of amplitude encoding values within inter-pulse is set to 0). (**f**) The results of CAPII with $M_a = M_r = 4$ (the third set of amplitude encoding values within the inter-pulse is set to 0).

5. Discussion

From the experimental results, it can be seen that the proposed deception jamming method has more flexible and controllable advantages over other methods. However, there are still some problems with the proposed method. For example, the positions of false targets are relatively fixed and are always evenly distributed on both sides of the real targets. In addition, the imaging quality of the false target at the far end is poor. From target identification, uniformly distributed false targets increase the risk of being identified. Therefore, these are problems that we need to further study.

6. Conclusions

With the continuous upgrading of ECM, various jamming technologies for SAR have become a research hotspot. Due to low power consumption and higher concealment, deceptive jamming has attracted more and more attention in the field of ECM. However, the existing deceptive jamming implementation methods are relatively singular. Based on waveform coding theory, this article proposed a repeater-type SAR deceptive jamming method through the joint coding of amplitude and phase in intra-pulse and inter-pulse, which can generate a 2D controllable deceptive jamming effect. The core of the proposed CAPII method is to encode the intercepted SAR signals, and various jamming effects can be achieved by controlling different coding parameters. For intra-pulse coding, the CAPII method first samples the intercepted SAR signal and groups the samples. Second, the joint coding of amplitude and phase is performed for each group respectively. The number of groups determines the number of false targets. The amplitude encoding value affects the energy of false targets. In addition, false target characteristics, including number, position, and amplitude, are analyzed. The simulation results verify the correctness of the theoretical analysis. In addition, FDA-based deceptive jamming and ISRJ were used as comparative methods, which fully demonstrates the superiority of the CAPII method. Therefore, the CAPII method can realize the protection of sensitive targets or areas.

Furthermore, in this paper, we only provided deceptive jamming implementations in stationary target and Stripmap mode. However, the CAPII method is also appropriate for moving targets and other working modes, such as Spotlight SAR, ScanSAR, Tops-SAR, etc.

Our future works are as follows:

- Realizing the control of the positions of the false targets.
- Applying the CAPII method to moving targets.
- Applying the CAPII method to a variety of modes.

Author Contributions: Conceptualization, D.C. and G.S.; Data curation, Z.G.; Formal analysis, G.S. and D.C.; Funding acquisition, G.S. and N.L.; Investigation, G.S., D.C. and Z.L.; Methodology, G.S., D.C. and Z.G.; Project administration, G.S. and N.L.; Resources, N.L.; Software, D.C., Z.L. and Z.G.; Supervision, N.L.; Validation, G.S. and D.C.; Visualization, D.C.; Writing–original draft, G.S. and D.C.; Writing–review and editing, D.C., G.S. and N.L. All authors have read and agreed to the published version of the manuscript.

Funding: This work was supported in part by the National Natural Science Foundation of China under Grants 61871175, in part by Natural Science Foundation of Henan under grant number 222300420115, and in part by the Foundation of Key Laboratory of Radar Imaging and Microwave Photonics, Ministry of Education, under Grant RIMP2020003.

Acknowledgments: The authors would like to thank the anonymous reviewers for their valuable and detailed comments that were crucial to improving the quality of this paper.

Conflicts of Interest: The authors declare no conflict of interest.

References

1. Reigber, A.; Scheiber, R.; Jager, M.; Prats-Iraola, P.; Hajnsek, I.; Jagdhuber, T.; Papathanassiou, K.P.; Nannini, M.; Aguilera, E.; Baumgartner, S. Very-High-Resolution Airborne Synthetic Aperture Radar Imaging: Signal Processing and Applications. *Proc. IEEE* **2013**, *101*, 759–783. [CrossRef]
2. Moreira, A.; Prats-Iraola, P.; Younis, M.; Krieger, G.; Hajnsek, I.; Papathanassiou, K.P. A tutorial on Synthetic Aperture Radar. *IEEE Trans. Geosci. Remote Sens. Mag.* **2013**, *1*, 6–43. [CrossRef]
3. Deng, Y.; Yu, W.; Zhang, H.; Wang, W.; Liu, D.; Wang, R. Forthcoming Spaceborne SAR Development. *J. Radars* **2020**, *9*, 1–33.
4. Tao, M.; Su, J.; Huang, Y.; Wang, L. Mitigation of Radio Frequency Interference in Synthetic Aperture Radar Data: Current Status and Future Trends. *Remote Sens.* **2019**, *11*, 2438. [CrossRef]
5. Huber, S.; Almeida, F.; Villano, M.; Younis, M.; Krieger, G.; Moreira, A. Tandem-L: A Technical Perspective on Future Spaceborne SAR Sensors for Earth Observation. *IEEE Trans. Geosci. Remote Sens.* **2018**, *56*, 4792–4807. [CrossRef]

6. Stofan, E.; Evans, D.; Schmullius, C.; Holt, B.; Plaut, J.; Zyl, J.; Wall, S.; Way, J. Overview of results of Spaceborne Imaging Radar-C, X-Band Synthetic Aperture Radar (SIR-C/X-SAR). *IEEE Trans. Geosci. Remote Sens.* **1995**, *33*, 817–828. [CrossRef]

7. Li, Z.; Wang, H.; Su, T.; Bao, Z. Generation of wide-swath and high-resolution SAR images from multichannel small spaceborne SAR systems. *IEEE Geosci. Remote Sens. Lett.* **2005**, *2*, 82–86. [CrossRef]

8. Ji, P.; Xing, S.; Dai, D.; Pang, B. Deceptive Targets Generation Simulation Against Multichannel SAR. *Electronics* **2020**, *9*, 597. [CrossRef]

9. Ji, P.; Dai, D.; Xing, S.; Pang, B. A New Three-Dimension Deceptive Scene Generation against Single-Pass Multibaseline InSAR Based on Multiple Transponders. *Sensors* **2020**, *20*, 1053. [CrossRef]

10. Huang, Y.; Zhao, B.; Tao, M.; Chen, Z.; Hong, H. A Review of Synthetic Aperture Radar Anti-Jamming Technique. *J. Radars* **2020**, *9*, 89–106.

11. Zhou, F.; Sun, G.; Bai, X.; Bao, Z. A Novel Method for Adaptive SAR Barrage Jamming Suppression. *IEEE Geosci. Remote Sens. Lett.* **2011**, *9*, 292–296. [CrossRef]

12. Wang, H.; Jiang, J.; Pu, J.; Wu, Y.; Ran, D. Low Sidelobe Multi-Phase Segmented Modulation Jamming Method Based on Cosine Amplitude Weighting. *J. Syst. Eng. Electron.* **2021**, *43*, 3185–3193.

13. Li, Y.; Huang, D.; Xing, S.; Wang, X. A Review of Synthetic Aperture Radar Jamming Technique. *J. Radars* **2020**, *9*, 753–764.

14. Huang, L.; Dong, C.; Shen, Z.; Zhao, G. The Influence of Rebound Jamming on SAR GMTI. *IEEE Geosci. Remote Sens. Lett.* **2015**, *12*, 399–403. [CrossRef]

15. Chang, X.; Li, Y.; Zhao, Y. An Improved Scattered Wave Deceptive Jamming Method Based on a Moving Jammer Beam Footprint Against a Three-Channel Short-Time SAR GMTI. *IEEE Sens. Lett.* **2020**, *21*, 4488–4499. [CrossRef]

16. Zhao, B.; Huang, L.; Li, J.; Liu, M.; Wang, J. Deceptive SAR Jamming Based on 1-bit Sampling and Time-Varying Thresholds. *IEEE J. Sel. Top. Appl. Earth Obs. Remote Sens.* **2018**, *11*, 939–950. [CrossRef]

17. Wu, Z.; Xu, H.; Li, J.; Liu, W. Research of 3-D Deceptive Interfering Method for Single-Pass Spaceborne InSAR. *IEEE Trans. Aerosp. Electron. Syst.* **2015**, *51*, 2834–2846. [CrossRef]

18. Deng, Y.; Zheng, Y.; Hu, Y. Analysis of Synthetic Aperture Radar Repeater Jamming. *J. Electron. Inf. Technol.* **2010**, *32*, 69–74. [CrossRef]

19. Zhang, N.; Kuang, L.; Shen, F.; Wan, Q.; Yang, W. A Jamming Technique against Airborne SAR. In Proceedings of the 2006 CIE International Conference on Radar (RADAR), Guangzhou, China, 16–19 October 2006; pp. 1–3.

20. Qu, W.; Jia, X.; Wu, Y. Modeling and Simulation Analyses of SAR Signal Reconnaissance. *J. Stat. Comput. Simul.* **2007**, *10*, 2341–2345.

21. Lin, X.; Liu, P.; Xue, G. Fast Generation of SAR Deceptive Jamming Signal Based on Inverse Range Doppler Algorithm. In Proceedings of the IET International Radar Conference, Xi'an, China, 14–16 April 2013; pp. 1–4.

22. Sun, Q.; Shu, T.; Zhou, S.; Tang, B.; Yu, W. A Novel Jamming Signal Generation Method for Deceptive SAR Jammer. In Proceedings of the IEEE Radar Conference, Cincinnati, OH, USA, 19–23 May 2014; pp. 1174–1178.

23. Sun, Q.; Shu, T.; Yu, K.; Yu, W. Efficient Deceptive Jamming Method of Static and Moving Targets Against SAR. *IEEE Sens. J.* **2018**, *18*, 3601–3618. [CrossRef]

24. Liu, Y.; Wang, W.; Pan, X.; Dai, D.; Feng, D. A Frequency-Domain Three-stage Algorithm for Active Deceptive Jamming Against Synthetic Aperture Radar. *IET Radar Sonar Navig.* **2014**, *8*, 639–646. [CrossRef]

25. Liu, Y.; Wang, W.; Pan, X.; Fu, Q.; Wang, G. Inverse omega-K algorithm for the electromagnetic deceptive of synthetic aperture radar. *IEEE J. Sel. Top. Appl. Earth Obs. Remote Sens.* **2016**, *9*, 3037–3049. [CrossRef]

26. Yang, K.; Ye, W.; Wu, X.; Ma, F.; Li, F. Fast Generation of Deceptive Jamming Signal Against Space-Borne SAR. *IEEE J. Sel. Top. Appl. Earth Obs. Remote Sens.* **2020**, *13*, 5580–5596. [CrossRef]

27. Yang, K.; Ye, W.; Ma, F.; Li, G.; Tong, Q. A Large-Scene Deceptive Jamming Method for Space-Borne SAR Based on Time-Delay and Frequency-Shift with Template Segmentation. *Remote Sens.* **2020**, *12*, 53. [CrossRef]

28. Wang, S.; Yu, L.; Ni, J.; Zhang, G. A Study on the Active Deceptive Jamming to SAR. *Acta Electron. Sin.* **2003**, *12*, 1900–1902.

29. Zhou, F.; Zhao, B.; Tao, M.; Bai, X.; Chen, B.; Sun, G. A Large Scene Deceptive Jamming Method for Space-Borne SAR. *IEEE Trans. Geosci. Remote Sens.* **2013**, *51*, 4486–4495. [CrossRef]

30. Zhang, B.; Zhou, F.; Shi, X.; Wu, Q.; Zheng, B. Multiple Targets Deceptive Jamming Against ISAR Using Electromagnetic Properties. *IEEE Sens. J.* **2015**, *15*, 2031–2038.

31. Tai, N.; Wang, Y.; Han, H.; Xu, X.; Wang, C.; Zeng, Y.; Wang, L. Deceptive Jamming Against ISAR Based on Convolution and Sub-Nyquist Sampling. *IEEE Sens. J.* **2020**, *20*, 1807–1820. [CrossRef]

32. Zhu, Y.; Wang, H.; Zhang, S.; Zheng, Z.; Wang, W. Deceptive Jamming on Space-Borne Sar Using Frequency Diverse Array. In Proceedings of the IGARSS 2018-2018 IEEE International Geoscience and Remote Sensing Symposium, Valencia, Spain, 22–27 July 2018; pp. 605–608.

33. Bang, H.; Wang, W.; Zhang, S.; Liao, Y. FDA-Based Space–Time–Frequency Deceptive Jamming Against SAR Imaging. *IEEE Trans. Aerosp. Electron. Syst.* **2022**, *58*, 2127–2140. [CrossRef]

34. Huang, B.; Wang, W.; Zhang, S.; Wang, H.; Gui, R.; Lu, Z. A Novel Approach for Spaceborne SAR Scattered-Wave Deceptive Jamming Using Frequency Diverse Array. *IEEE Geosci. Remote Sens. Lett.* **2020**, *17*, 1568–1572. [CrossRef]

35. Huang, L.; Zong, Z.; Zhang, S.; Wang, W. 2-D Moving Target Deceptive Against Multichannel SAR-GMTI Using Frequency Diverse Array. *IEEE Geosci. Remote Sens. Lett.* **2022**, *19*, 1–5.

36. Wang, X.; Liu, J.; Zhang, W.; Fu, Q.; Liu, Z.; Xie, X. Mathematical Principle of Intermittent Sampling Repeater Jamming. *Sci. Sin. (Technol.)* **2006**, *8*, 891–901.
37. Wu, X.; Bo, Z.; Dai, D.; Wang, X. Azimuth intermittent Sampling Repeater to SAR. *J. Signal Process.* **2010**, *26*, 1–6.
38. Li, C.; Su, W.; Gu, H.; Ma, C.; Chen, J. Improved Interrupted Sampling Repeater Jamming based on DRFM. In Proceedings of the 2014 IEEE International Conference on Signal Processing, Communications and Computing (ICSPCC), Guilin, China, 5–8 August 2014; pp. 254–257.
39. Feng, D.; Xu, L.; Pan, X.; Wang, X. Jamming Wideband Radar Using Interrupted-Sampling Repeater. *IEEE Trans. Aerosp. Electron. Syst.* **2017**, *53*, 1341–1354. [CrossRef]
40. Sun, J.; Wang, C.; Shi, Q.; Ren, W.; Yao, Z.; Yuan, N. Interrupted Sampling Repeater Signal Based on Phase Modulation. In Proceedings of the International Conference on Microwave and Millimeter Wave Technology (ICMMT), Nanjing, China, 23–26 May 2021; pp. 1–3.
41. Hang, D.; Xing, S.; Li, Y.; Liu, Y.; Xiao, S. Smart Jamming Method Against SAR Based on Multiplication Modulation. *Syst. Eng. Electron.* **2021**, *43*, 3160–3168.
42. Sun, G.; Xing, S.; Huang, D.; Li, Y.; Wang, X. Jamming method of intermittent sampling against SAR-GMTI Based on Noise Multiplication Modulation. *Syst. Eng. Electron.* **2022**, *43*, 1–14.
43. Jin, G.; Deng, Y.; Wang, W.; Zhang, H.; Long, Y.; Zhang, Y.; Wang, R. On the SAR Imaging Performance Analysis of Alternate Transmitting Mode Based on Waveform Diversity: Theory and Simulation. *IEEE Geosci. Remote Sens. Lett.* **2020**, *17*, 1553–1557. [CrossRef]

Article

Proportional Similarity-Based Openmax Classifier for Open Set Recognition in SAR Images

Elisa Giusti [1], **Selenia Ghio** [1], **Amir Hosein Oveis** [1,2,*] and **Marco Martorella** [1,2]

[1]   National Laboratory of Radar and Surveillance Systems (RaSS), National Inter-University Consortium for Telecommunication (CNIT), 56124 Pisa, Italy
[2]   Department of Information Engineering, University of Pisa, 56126 Pisa, Italy
*   Correspondence: aoveis@cnit.it

Abstract: Most of the existing Non-Cooperative Target Recognition (NCTR) systems follow the "closed world" assumption, i.e., they only work with what was previously observed. Nevertheless, the real world is relatively "open" in the sense that the knowledge of the environment is incomplete. Therefore, unknown targets can feed the recognition system at any time while it is operational. Addressing this issue, the Openmax classifier has been recently proposed in the optical domain to make convolutional neural networks (CNN) able to reject unknown targets. There are some fundamental limitations in the Openmax classifier that can end up with two potential errors: (1) rejecting a known target and (2) classifying an unknown target. In this paper, we propose a new classifier to increase the robustness and accuracy. The proposed classifier, which is inspired by the limitations of the Openmax classifier, is based on proportional similarity between the test image and different training classes. We evaluate our method by radar images of man-made targets from the Moving and Stationary Target Acquisition and Recognition (MSTAR) dataset. Moreover, a more in-depth discussion on the Openmax hyper-parameters and a detailed description of the Openmax functioning are given.

Keywords: open set recognition; radar imaging; Synthetic Aperture Radar (SAR); machine learning; deep learning; automatic target recognition

Citation: Giusti, E.; Ghio, S.; Oveis, A.H.; Martorella, M. Proportional Similarity-Based Openmax Classifier for Open Set Recognition in SAR Images. *Remote Sens.* **2022**, *14*, 4665. https://doi.org/10.3390/rs14184665

Academic Editors: Lan Du, Gang Xu and Haipeng Wang

Received: 16 August 2022
Accepted: 15 September 2022
Published: 19 September 2022

Publisher's Note: MDPI stays neutral with regard to jurisdictional claims in published maps and institutional affiliations.

1. Introduction

Radar imaging has been largely investigated in the literature as a means of equipping a radar system with automatic target recognition (ATR) capability. Many papers have demonstrated that radar systems can not only provide kinematic information (position, speed, and course) of land, sea, and air targets during day and night in all weather conditions but can also provide electromagnetic images of the targets, which can be used for recognition purposes [1–6]. Many algorithms have been proposed in the past decades showing that radar images of moving targets can be used for NCTR purposes. Few-shot target classification algorithms in Synthetic Aperture Radar (SAR) have also been intensively studied in recent years [7,8]. The most promising algorithms are based on a training step and, therefore, require a set of data from known targets. Furthermore, many recent papers have shown that deep networks (DN) can provide high-performance recognition [4–6]. However, they require a priori knowledge about the targets. While it is possible to train such a system with terabytes of data, it is impossible to anticipate and train with all possible inputs that a classifier may encounter in a real-world scenario. The real data are inherently dynamic and hence difficult to predict. So far, most state-of-the-art NCTR systems have followed a "closed world" assumption, meaning that the system model is complete and the system can reason using what was observed previously [9–14]. However, this assumption is not realistic and leads to fragile systems that can fail at inference time [15]. The real world contains an open set of targets, and any system knowledge (as also our knowledge) is incomplete. This problem has been more thoroughly investigated in the computer vision field rather than the radar field. Some studies tackled the above-mentioned problem, which is known as open set recognition (OSR), by applying a threshold to the Softmax function [16,17]. Softmax, which is a typical classifier in convolutional neural networks (CNN), maps the activation vector into a probability domain. The activation vector refers to the output of the last fully connected (FC) layer [18]. Note that imposing a threshold on Softmax's outputs is not a practical solution, since CNNs may generate incorrect large scores in the case of open set inputs.

To overcome this issue, the Openmax algorithm [18] has been recently proposed in the optical domain that drops the restriction for the output scores to sum to one. Therefore, it allows the model to recognize the input as an unknown image without necessarily requiring any threshold. Using a conditional generative adversarial network (GAN) to synthesize mixtures of unknowns, Ge et al. [19] proposed the Generative Openmax (G-Openmax) algorithm, which enables a classifier to locate the decision margin according to the knowledge of known classes and the generated unknown samples. However, such unknowns are limited to the subspace of the known classes [20]. Zheng et al. [21] proposed a model based on an autoencoder and an auxiliary classifier to generate pseudo samples. They then used the generated samples to improve out-of-distribution detection performance in natural language understanding by optimizing the entropy regularization term in the training stage. It should be noted that adversarial images cannot fully represent the open set environment. Lee et al. [22] proposed a method for detecting either out-of-distribution or adversarial samples by measuring the Mahalanobis distance between the test sample and the closest class-conditional Gaussian distribution. Inkawhich et al. [23] showed that a large, unlabeled, and unrelated SAR dataset can be used to improve the out-of-distribution detection in SAR-ATR applications. Note that ATR using SAR images is more challenging than optical images since SAR images have lower signal-to-noise ratios (SNR) and lower spatial resolution than optical images. In this regard, in our previous studies [24,25], we have first investigated the applicability of the Openmax classifier for SAR images and then analyzed the possibility of having class-wise hyper-parameter (tail size) and distribution type to optimize the tail-fitting procedure in the Openmax approach. It is worth noting that the standard Openmax approach takes only one tail size value, which should be set heuristically in the calibration phase, and assumes only one distribution type (Weibull), i.e., the same tail size and the same distribution type for all the classes. However, we noticed that each class has a distinctive distance distribution, and if we change the algorithm and carefully set the tail size and the distribution type in each class separately, the overall accuracy will improve significantly. This implies that there exists an imbalance between the classes. However, such an optimization problem (on the hyper-parameters) requires a priori knowledge about the classes that is hard to be achieved in a real-world SAR scenario where a target observed by a slightly different aspect angle may change significantly.

In this paper, with the aim of improving the overall accuracy and robustness of OSR in SAR images, we base our proposed method on the limitations of the Openmax algorithm. The reason why we choose the Openmax classifier and not any of the adversarial-based solutions for this application is that Openmax exploits the statistical information of the training dataset to recognize possible unknown inputs, and it is not based on training with some counterfactual image that cannot technically represent all possible open set inputs. There are two types of errors that the Openmax classifier may encounter: some closed set images may mistakenly be recognized as unknown and some open set images may mistakenly be classified as one of the closed set classes. We have studied different aspects of the Openmax classifier when applied to the target recognition in SAR images and identified the following items as the main sources of two aforementioned errors: feature extraction, tail-fitting, and activation vector modification. We then propose a substitute for the tail-fitting procedure that relies on the interrelations between the test image and different training classes. It is worth noting that the standard Openmax modifies each element of the activation vector based on the similarity between the test image and only the corresponding class of the training set and not other training classes. However, our new method considers also the relationship that exists among the training classes.

In other words, the new method makes use of the similarity between the test image and the different training classes in proportion to the similarity between the training classes to modify the activation vector. Therefore, the proposed approach is hereafter called proportional similarity-based Openmax or simply PS-Openmax. In the end, we used the Moving and Stationary Target Acquisition and Recognition (MSTAR) dataset [26] for the experimental verification. The contributions of the paper are threefold:

- A thorough examination of the Openmax classifier and a detailed discussion on the tail-fitting procedure in different OSR scenarios.
- An analysis of the Openmax limitations and source of errors to effectively avoid the situations where either a known or an unknown target is always misclassified.
- Proposing the proportional similarity-based approach, which makes use of the similarity between the test image and different training classes in proportion to the similarity between the training classes, to increase the robustness and the accuracy.

This paper is organized as follows. Section 2 presents the materials and methods. Section 3 shows the experimental results. Section 4 provides discussion and analysis of the methods. Section 5 concludes this paper.

2. Materials and Methods

In this section, the materials and methods required for the real data experiments are described. First, the Openmax approach is briefly introduced in Section 2.1. Subsequently, the proposed approach is comprehensively explained in Section 2.2. Finally, the experimental setup and materials are introduced in Section 2.3.

2.1. The Openmax Approach

In classification tasks, the Softmax layer is typically used at the end of the network to map the output of the last FC layer, namely "activation vector (AV)", into scores that sum to one. Defining x as the input image and N as the number of closed-set classes, the Softmax score of the class c can be computed by its corresponding activation score $AV_c(x)$, which is divided by a summation over all activation scores, as follows:

$$s_{soft_c} = \frac{e^{AV_c(x)}}{\sum_{i=1}^{N} e^{AV_i(x)}} \tag{1}$$

Given the softmax scores for all the classes as $s_{soft} = [s_{soft_1}, \cdots, s_{soft_c}, \cdots, s_{soft_N}]$, the easiest way to address the open set problem is to impose a certain threshold to $max(s_{soft})$. In other words, if none of s_{soft} scores of the test image exceeds a certain threshold s_{th}, the test image will be recognized as an unknown.

$$\begin{cases} \text{if } max(s_{soft}) > s_{th} & class = argmax(s_{soft}) \\ \text{else} & class = unknown \end{cases} \tag{2}$$

Openmax [18], as an alternative to the Softmax threshold approach, modifies the definition of the Softmax function to include an unknown class. The Openmax procedure is composed of two phases: the model calibration and the calculation of the Openmax scores. The model calibration phase is described in Algorithm 1, and it takes two input types: (1) the outputs of the last FC layer of the network and (2) the scalar η, as a hyper-parameter for the 'tail size' in the calibration process. Note that only the correctly classified training images will be used again in the calibration phase. In fact, AV_1^{train} and AV_N^{train} in the input line denote the activation vectors of the images in the training classes 1 and N, respectively.

The Openmax classifier is employed for a pre-trained CNN where the last layer of the CNN is an FC layer with N neurons. It modifies the final output, i.e., AV, and generates a modified AV to have N + 1 elements where the last element represents the unknown class, and it then maps the modified AV to the probability domain. By a pre-trained CNN, we mean that the model should be first trained using the training dataset and then the statistical features of the training data, i.e., known targets, are extracted to design the classifier that is able to recognize unknown test data. Considering one of the training classes as an example and computing the mean activation vector (MAV) of this class, see line 2 of Algorithm 1, the Openmax fits a Weibull distribution to the tail of Euclidean distances between AVs and MAV of this class; see lines 3 and 4 of Algorithm 1. LibMR, which is a publicly available library (https://github.com/Vastlab/libMR (accessed on 15 August 2022)), provides the FitHigh function for the maximum likelihood estimation using the Weibull distribution.

More specifically, in line 3 of Algorithm 1, the Euclidean distance values among MAV and all AVs of each class are computed and sorted. Next at line 4, a Weibull distribution is fitted to the η largest distances. The outputs of Algorithm 1 are the Weibull model and MAV measured for each training class. The Weibull distribution is commonly used since it has been demonstrated to be the most suitable distribution for statistical meta-recognition [27,28]. Nonetheless, a deep analysis considering different types of distributions is also included in this work; see Section 3.4.

Algorithm 1 Model Calibration

Input: $AV_1^{train}, \cdots, AV_N^{train}, \eta$
Output: $(Weibull_1^{train}, MAV_1^{train}), \cdots, (Weibull_N^{train}, MAV_N^{train})$

1: **for** j = 1, 2, ..., N **do**
2: $\quad MAV_j^{train} = \text{mean}(AV_j^{train})$
3: $\quad ED_j^{train} = \text{sort}\left(\text{EuclideanDistance}(AV_j^{train}, MAV_j^{train})\right)$
4: $\quad Weibull_j^{train} = \text{FitHigh}(ED_j^{train}, \eta)$

Considering a test image, the Openmax second phase is summarized in Algorithm 2.

Algorithm 2 Openmax scores calculation

Input: $(Weibull_1^{train}, MAV_1^{train}), \cdots, (Weibull_N^{train}, MAV_N^{train})$, AV of the test image, N_α
Output: Openmax scores

1: $ord = \text{argsort}(AV, \text{"descending"})$
2: **for** i = 1, $\cdots$, N_α **do**
3: $\quad j = ord(i)$
4: $\quad CD = \text{EuclideanDistance}(AV - MAV_j^{train})$
5: $\quad \tau, \kappa, \lambda = Weibull_j^{train}$
6: $\quad w = 1 - e^{-\left(\frac{\|CD-\tau\|}{\lambda}\right)^\kappa}$
7: $\quad \alpha = (N_\alpha - i + 1)/N_\alpha$
8: $\quad modAV(j) = AV(j)(1 - w \times \alpha)$
9: $unk = \sum_{j=1}^{N}(AV(j) - modAV(j))$
10: $modAV(N + 1) = unk$
11: **for** j = 1, 2, ..., N + 1 **do**
12: $\quad Sopen_j = \frac{e^{modAV(j)}}{\sum_{k=1}^{N+1} e^{modAV(k)}}$
13: $Sopen = [Sopen_1, Sopen_2, ..., Sopen_{N+1}]$

In short, the Openmax subtracts a portion from each element of AV based on the similarity of the test image and the respective training class, sums the subtracted values, and forms a modified AV with one more element appended to its end to represent the unknown class. The algorithm takes three input types: (1) MAV and the Weibull model pair for each training class from Algorithm 1 together with (2) the activation vector of the test image, i.e., AV, and (3) the scalar N_α as another hyper-parameter. By computing the Openmax scores, i.e., $Sopen = [Sopen_1, \cdots, Sopen_c, \cdots, Sopen_{N+1}]$, the corresponding index to the maximum value determines the class assigned to the test image. Note that only the top N_α values of AV will be modified and the rest of the $N - N_\alpha$ elements of AV will be untouched. To select the changeable elements, AV is sorted at line 1 of Algorithm 2, and the corresponding indexes are used at line 3 to modify the jth element of AV. More in detail, Openmax calculates two factors, i.e., 'α' and 'w', to modify each element of AV. In order to calculate 'w' for the modification of the jth element of AV, Openmax first calculates the channel distance (CD) scalar value, see line 4 of Algorithm 2, based on the distance between AV and MAV of the class j. It then evaluates the value of Weibull cumulative distribution function (CDF) of the class j, from Algorithm 1, on the channel distance point; see lines 5 and 6 of Algorithm 2. The other factor for the modification of the jth element of AV is α; see the rule at line 7 of Algorithm 2. For instance, if we assume $N_\alpha = N = 8$, i.e., the scenario where we have eight known classes and we want to modify all eight elements of AV, then $\alpha = 1, 0.875, 0.75, 0.625, 0.5, 0.375, 0.25, 0.125$ will be generated iteratively. Afterward, as we have mentioned before, w together with α are used to modify the j*th* element of the activation vector AV; see line 8 of Algorithm 2. Note that the new element of AV to represent the unknown class is made up of the subtracted values. In other words, the difference between the original activation vector AV and the modified activation vector *modAV* is summed up, see line 9 of Algorithm 2, and it is then appended to the modified activation vector *modAV*, at line 10, as the activation score of the unknown class. In the end, the *modAV*, i.e., the one with N + 1 elements, is mapped to the probability domain to generate the Openmax scores; see lines 12 and 13 of Algorithm 2.

2.2. The Proposed Approach

In this section, first, we highlight some of the inherent limitations of the Openmax classifier, and then, we propose our proportional similarity-based classifier, which obviates these limitations. As illustrated in Figure 1, there are two types of errors that Openmax may encounter: (1) recognizing a known input image as an unknown and (2) recognizing an unknown input image as a known class. The main source of these two errors should be searched in features extraction, tail-fitting, and AV modification:

1. Feature extraction:
   In the Openmax classifier, the raw outputs of the last FC layer are directly used for the scores calculations. However, in the new method, a supplementary activation function is used to map the AV into another domain that is more suitable for the OSR problem. It should be noted that the supplementary activation function will be only used during the inference and not in the training. In fact, only the Softmax activation function is applied to AV in the training phase.
2. Tail-fitting procedure:
   The distance values and their distributions can contain useful information for the OSR solution. The most critical hyper-parameter of Openmax is η by which it analyzes only the tail of distance values. However, a more accurate OSR solution can be designed by exploiting full information of distance values.
3. Activation vector modification:

   (a) The choice of N_α:
       It is worth mentioning that N_α is another hyper-parameter in the original Openmax, and similar to η, it has to be carefully chosen beforehand. By modifying only the top N_α values of AV, i.e., subtracting different portions from those elements, Openmax generates an extra class dedicated to the unknown inputs. In fact, Openmax takes $N_\alpha < N$ to reduce the number of changeable neurons in AV, especially in the case of CNNs that generate some negative scores in their AV. Therefore, the aim of $N_\alpha < N$ is to discard some of the negative values of AV, in other words, to exclude $N - N_\alpha$ smallest values of AV, in order to avoid the new element from having a possibly large negative value. This large negative value forces the classifier not to reject the unknown input image and ends up with the second error shown in Figure 1. Note that by discarding some of the negative values of AV using $N_\alpha < N$, the original Openmax lets

the new element have the largest value in the case of an unknown input image. However, choosing $N_\alpha < N$ in the original Openmax implies a priori knowledge. We will introduce our PS-based classifier that obviates this limitation and has an improved performance toward unknown images.

(b)   Class-independent subtraction:

According to the CNN model and the input test image, it is also quite probable that the new element of AV ends up being a very large positive value and the first error shown in Figure 1, i.e., rejecting a known image, happens. This problem is likely to happen in CNNs that do not generate negative scores in their AV. Therefore, even by reducing the number of changeable neurons in AV, i.e., $N_\alpha < N$, it is still probable that the new element becomes the greatest one, and this forces the classifier to reject the known image. Note that in the original Openmax classifier, the subtraction in each element of AV is performed independently from the others, and the relationship between different classes is not studied. By exploiting this aspect, the PS-based classifier provides an improved accuracy toward the input images of the known classes.

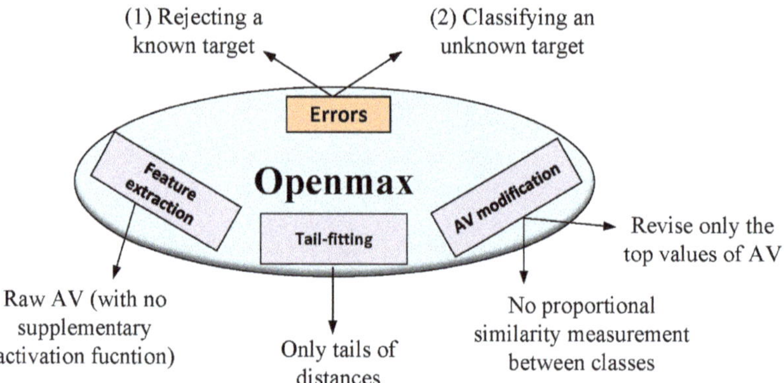

Figure 1. Openmax limitations, source of errors.

Considering the aforementioned limitations, we propose our PS-based classifier, as an extension to Openmax, to improve its robustness and accuracy toward both known and unknown classes. Similar to the original Openmax classifier, the PS-based classifier will also be employed for the pre-trained CNN, and it modifies the AV from N elements to N + 1 elements. The distinctive feature of the proposed method is to consider the relationship between the test image and all the training classes when modifying each element of the AV. The overall framework of the proposed method is illustrated in Figure 2, and its pseudo-code is formulated in Algorithm 3.

The proposed method makes use of a supplementary activation function, takes advantage of all distance information rather than the tails, and provides a different perspective for modifying AV. The inputs of the proposed method, as shown in Algorithm 3, are the MAVs of known classes and the AV of the test image. MAVs, which are illustrated in Figure 2a, are calculated in the same way as in the original Openmax method.

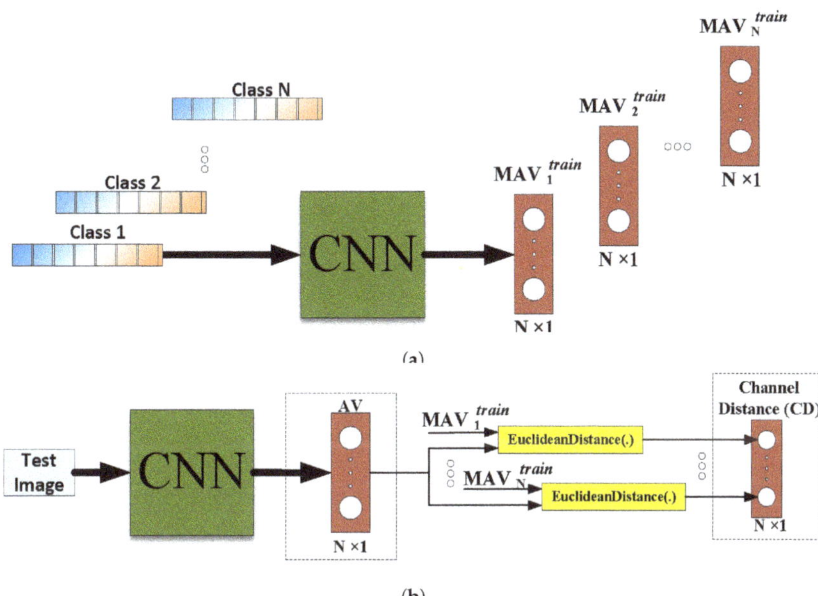

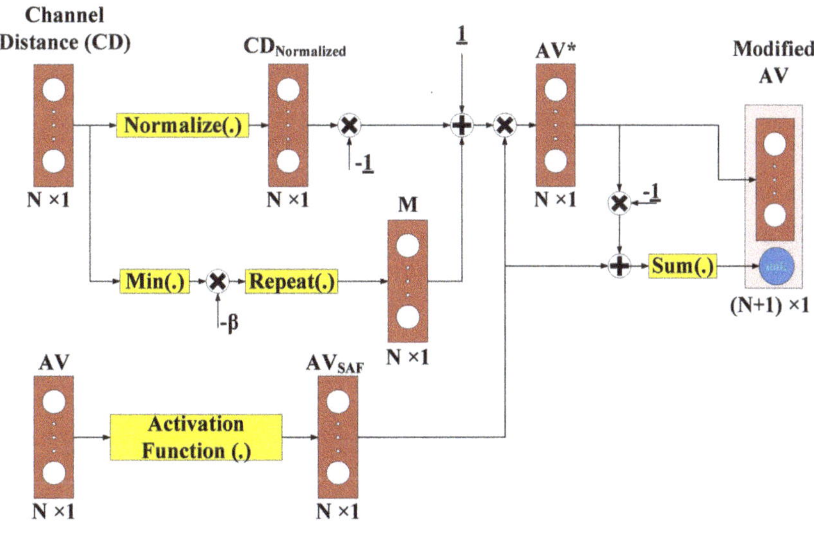

Figure 2. The overall framework of the PS-based Openmax classifier: (**a**) Calculation of MAVs. (**b**) Activation vector of the test image and Channel Distance vector. (**c**) Modification of the AV.

Algorithm 3 PS-Openmax scores calculation

Input: $(\text{MAV}_1^{train}, \cdots, \text{MAV}_N^{train})$, AV of the test image
Output: PS-Openmax scores

1: **for** i $= 1, \cdots, N$ **do**
2: $\quad CD_i = \text{EuclideanDistance}(AV - MAV_i^{train})$
3: $CD_{Normalized} = \dfrac{CD}{\sum_{i=1}^{N} CD_i}$
4: $M = -\beta[min(CD), \cdots, min(CD)]^T$
5: $AV_{SAF} = AV\text{-}min(AV)$
6: $AV^* = [1 + M\text{-}CD_{Normalized}] \circ AV_{SAF}$
7: $unk = \sum_{j=1}^{N}(AV_{SAF}(j) - AV^*(j))$
8: $Modified\,AV = [AV^*, unk]$
9: **for** j $= 1, 2, ..., N + 1$ **do**
10: $\quad s_{ps_j} = \dfrac{e^{Modified\,AV(j)}}{\sum_{k=1}^{N+1} e^{Modified\,AV(k)}}$
11: $s_{ps} = [s_{ps_1}, s_{ps_2}, ..., s_{ps_{N+1}}]$

Considering the test image, we form the Channel Distance vector from all the classes at lines 1 and 2 of Algorithm 3. This part has been illustrated in Figure 2b where the Channel Distance vector and the AV of the test image are surrounded by dashed boxes to be used in Figure 2c. Note that in the original Openmax (see line 4 of Algorithm 2), channel distance is scalar and is only computed for the top N_α elements of AV.

In line 3 of Algorithm 3, we calculate the normalized Channel Distance vector, i.e., $CD_{Normalized}$, to evaluate the proportional relationship among its elements and to effectively avoid the situation that the absolute value of the last element of the AV is always too large. This aspect, i.e., the interrelation between different channel distance values, has not been studied in the original Openmax, and channel distance values were separately (independently) used to modify their corresponding elements of AV. Next, the minimum value of the Channel Distance vector is also taken into account as a parallel measure at line 4 of Algorithm 3. Note that $min(CD)$ is repeated to form a $N \times 1$ vector and $\beta \in (0, 1)$ as a hyper-parameter is used to balance between the two factors: minimum channel distance and the normalized Channel Distance vector.

Typical activation functions, such as Sigmoid, Tanh, Softmax, ReLU, and its variants are employed in an element-wise way. We instead propose the supplementary activation function AV-min(AV) as a vector form activation function to make sure that none of the elements in AV is negative. This supplementary activation function is applied at line 5 of Algorithm 3 to generate AV_{SAF}. This supplementary activation function has been chosen through an extensive series of experiments performed on different activation functions.

In the proposed method, we modify the AV_{SAF} in an element-wise manner by using two factors: M, i.e., the minimum channel distance, and $CD_{Normalized}$, i.e., the normalized Channel Distance vector, using the Hadamard product ($\circ$); see line 6 of Algorithm 3. By calculating the sum of the differences between AV_{SAF} and AV^*, a new element (shown in blue in Figure 2c) is formed; see line 7 of Algorithm 3. Similar to the original Openmax, the new element is appended to the end of Modified AV, see line 8 of Algorithm 3, and the Modified AV is mapped to the probability domain to generate PS-Openmax scores; see lines 9–11 of Algorithm 3.

In summary, the tail-fitting procedure is substituted with a measure of proportional similarity in the PS-Openmax approach, and using the relationship among the Channel Distance values prevents the new element from having always a too large negative value or a too large positive value. Note that there is neither a need to limit the number of changeable neurons in AV, i.e., choosing N_α, nor to calibrate tail size.

2.3. Experimental Setup and Materials

In this part, the experimental setup is defined. First, the CNN structure is described in Section 2.3.1. Afterward, the dataset is introduced in Section 2.3.2, and finally, the performance indexes are defined in Section 2.3.3.

2.3.1. CNN Structure

The CNN structure used to test the proposed approach is shown in Table 1.

It consists of three convolutional layers (Conv1, Conv2, and Conv3), two pooling layers (MaxPooling1 and MaxPooling2), and two FC layers (FC1 and FC2). All the convolution layers are followed by ReLU functions. After a convolution layer, a pooling layer is introduced to reduce the dimensions of convolution layer outputs. The last FC layer is followed by an 8-class Softmax classifier. The Softmax classification module maps the output of the last FC layer to the probability domain as mentioned in (1). The kernel size of the first two convolutional layers is 3×3, whereas that of the last convolutional layer is 5×5 and all pooling layers have 2×2 kernels. Moreover, the stride is set to zero, and zero-padding on borders is applied to avoid the shrinking of feature maps after the convolutional layers. The CNN is implemented by Keras, where the cross-entropy loss function is minimized via the Adam optimization algorithm with a learning rate equal to 0.0001 and the batch size of 8.

Table 1. Structure of the CNN.

Layer	Name	Output Size	Act. Func.	Param.
0	Input	$64 \times 64 \times 1$	-	0
1	Conv1	$64 \times 64 \times 16$	ReLU	160
2	MaxPooling1	$32 \times 32 \times 16$	-	0
3	Conv2	$32 \times 32 \times 16$	ReLU	2320
4	MaxPooling2	$16 \times 16 \times 16$	-	0
5	Conv3	$16 \times 16 \times 64$	ReLU	25,664
6	Flattening	16,384	-	0
7	FC1	50	-	819,250
8	FC2	8	-	408
9	Classifier	8	Softmax	0

2.3.2. Dataset Description

We perform our real data experiments using the well-known MSTAR dataset [26]. MSTAR has been widely used by so many scholars for the training and evaluation of SAR-ATR applications, specifically for deep learning methods [5,28–32]. To be more specific, a part of the MSTAR dataset, which has also been chosen by [29], is used for training and test. The dataset consists of the real SAR images of ten different ground targets of air defense unit (ZSU-23-4), armored personnel carrier (BMP-2, BTR-70), tank (T-72, M-60, M-1, M-2), rocket launcher (2S1), military cargo carrier (M548) and light utility truck (M35). They have been collected by an X-band SAR sensor (9.6 GHz) with 0.3 m by 0.3 m resolution in spotlight mode and full 360° aspect angles coverage.

Each chip has an approximate size of 128 × 128 pixels, although we reshape all the images to the size of 64 by 64 pixels. Under the standard operating conditions (SOC), two depression angles of 17° and 15°, as a matter of routine, are used separately for the train and test sets, respectively [33]. The electro-optical images corresponding to the MSTAR dataset used in this paper are shown in Figure 3.

Let us assume that there are eight known classes (0–7) and two unknown classes (8, 9). In fact, we want to train a CNN with only eight classes (0–7) and test it with all ten classes (0–9) and see if the model is not only able to correctly classify images of the known classes (0–7) but is also able to reject images that belong to open set or unknown classes (8–9). Based on the concept of "Openness", which has been introduced and formulated by Scheirer et al. [34], the complexity of an open set recognition problem can be expressed by the number of target classes to be identified, the number of classes used in training, and the number of classes used in testing as

$$Openness = 1 - \sqrt{\frac{2 \times |training\ classes|}{|testing\ classes| + |target\ classes|}} \tag{3}$$

where $|.|$ represents the number of classes in each respective set. Note that the problem is completely closed when the *Openness* equals to zero whereas larger openness, i.e., close to one, corresponds to more open problems [20]. In our scenario, $|training\ classes|$ = 8 and $|testing\ classes|$ = $|target\ classes|$ = 10; therefore, *Openness* is equal to 0.1. Note that in an extreme case such as training with only one class and testing with ten classes of our dataset, *Openness* reaches 0.68. It is also possible to add more unknown classes from the optical domain, such as the one we will analyze later in Section 3.2; however, these scenarios do not constitute a challenge for the classifier. Moreover, there is a large discrepancy between SAR and optical images, and because of the high cost of SAR image acquisition, a large-scale annotated dataset for the network training is rare [35].

Figure 3. Electro-optical images corresponding to the MSTAR dataset used in this paper; (**a**) BTR70, (**b**) M1, (**c**) M2, (**d**) M35, (**e**) M60, (**f**) M548, (**g**) T72, (**h**) ZSU23-4, (**i**) 2S1, (**j**) BMP2.

2.3.3. Performance Indexes

The classifier performance will be assessed in terms of both the confusion matrices and a number of performance indexes that are commonly used. The well-known confusion matrix is typically a square matrix $CM \in \mathbb{R}^{N \times N}$ that measures the classifier capability in processing the test dataset. Differently, in this case, the unknown class is included in the confusion matrix to measure the ability of the classifier to correctly recognize both known targets and unknown targets. In particular, there are two input unknown classes and one output unknown class. Therefore, in our scenario, $CM \in \mathbb{R}^{N+2 \times N+1}$, and the confusion matrices reported in Section 3 are rectangular and not square. Four different performance indexes based on our non-square CM can be defined: namely, the Precision, Pr, the Recall, Re, the F1-score, $F1$, and the total Accuracy, Ac, as follows:

$$Pr[k] = \frac{CM[k,k]}{\sum_{n=1}^{N_r} CM[n,k]} \tag{4}$$

$$Re[k] = \frac{CM[k,k]}{\sum_{n=1}^{N_c} CM[k,n]} \tag{5}$$

$$F1[k] = \frac{2Pr[k]\,Re[k]}{Pr[k] + Re[k]} \tag{6}$$

$$Ac = \frac{CM[N+2, N+1] + \sum_{n=1}^{N+1} CM[n, n]}{\sum_{n=1}^{N_r} \sum_{i=1}^{N_c} CM[n, i]} \tag{7}$$

where $N_c = N + 1$ and $N_r = N + 2$ are the number of columns and rows of the confusion matrix, respectively. In addition, $N = 8$ is the number of known classes, and $k = 1, \cdots, N + 1$ is the class index. Pr is the ratio of correctly predicted positive observations to the total predicted ones. High values of Pr mean a low false positive rate. Re is the ratio of correctly predicted positive observations to the all observations in the true class. It is also known as the sensitivity. It can be interpreted as a measure of the missed detection. $F1$ is a useful metric, which takes both Pr and Re into account and can be defined as the harmonic mean of the Pr and Re. The total Ac is defined as the mean rate of correctly classified samples [36].

3. Results

This section shows the main results achieved by Softmax, Openmax, and PS-based Openmax classifiers applied to the dataset described in Section 2.3.2. More specifically, the Openmax calibration procedure is described in Section 3.1. Openmax provides calculations for three different input image: one known image from the class 0, one unknown image from the MSTAR dataset and one optical unknown image are summarized in Section 3.2. The classification results of Openmax and Softmax are deeply analyzed in Section 3.3. The discussions on the choice of the tail size and the CDF are reported in Section 3.4. In the end, the results of the proposed PS-based approach are analyzed in Section 3.5. The reason why the proposed PS-based approach has been analyzed separately is that it does not contain tail-fitting procedure, so it is out of the tail size analysis and the CDF type discussion.

3.1. Openmax Pre-Processing

The results shown in this section have been obtained by training the network shown in Table 1 using the image dataset shown in Table 2.

The trained network is evaluated by the training dataset to determine the corrected-classified image set, T_{train}. In this scenario, all the training images are classified correctly, as expected. For each image of T_{train}, the outputs of layers eight and nine, i.e., activation vector (AV), and the Softmax output, are taken into account. The matrix $\mathbf{M} \subset \mathbb{R}^{8 \times 8}$ is then computed as follows:

$$\mathbf{M} = \begin{bmatrix} MAV_1 \\ \vdots \\ MAV_i \\ \vdots \\ MAV_8 \end{bmatrix} \tag{8}$$

where $i = 1, 2, \cdots, N$, $N = 8$ is the number of known classes, and $MAV_i \subset \mathbb{R}^{1 \times 8}$ represents the mean vector of the AV vectors of the class i. The matrix $\mathbf{M}$ is computed by the training set, and it is illustrated in Table 3. Then, the Euclidean distance between the AV of each image and the MAV of its class (one row of $\mathbf{M}$) is computed. As a result, each image in the training set has an associated Euclidean distance value. As explained previously, the η largest Euclidean distances in each class is then used to fit a Weibull distribution. In this part, the MAV and Weibull CDF of each class are considered as the outputs.

Table 2. MSTAR dataset.

	Label	Name	Serial Number	N Train	N Test
known	0	BTR 70	C71	41	51
	1	M1	0AP00N	78	51
	2	M2	MV02GX	75	53
	3	M35	T839	75	54
	4	M60	3336	122	54
	5	M548	C245HAB	69	59
	6	T72	812	55	53
	7	ZSU23-4	D08	115	59
unknown	8	2S1	B01	0	52
	9	BMP2	9563	0	52
total	-	-	-	630	538

Table 3. Matrix **M** shows the mean values of AVs in each class.

Class								
0	25.8774	0.592523	14.0206	9.63976	-19.1118	0.505812	1.15104	-13.4153
1	-9.37342	16.4302	2.61673	-1.71236	5.15303	-2.03516	4.97492	-0.208769
2	3.53298	4.8171	16.9942	-0.0574321	-5.61154	-1.5206	6.13859	-8.09596
3	-3.97235	0.510664	-5.95034	24.1043	-6.40276	11.6505	4.77716	-2.15785
4	-13.0375	9.81838	-1.27891	0.350657	18.859	-6.83463	6.97438	0.575225
5	-5.16936	0.803306	-4.19961	18.1028	-12.6156	29.3176	4.1984	1.5873
6	-4.98337	5.71393	4.89013	1.9776	3.75332	-3.92143	15.4734	-7.05234
7	-9.56859	5.69773	-2.10884	1.12619	-0.957903	2.00531	-4.04335	19.9382

3.2. Openmax Preliminary Test

After performing all the above pre-processing steps, the test images will be processed. In order to measure the channel distance (CD) values, the Euclidean distance between the AV of the test image and each row of **M** is calculated. Based on the Weibull CDFs generated in the previous part and the CD values, w is subsequently calculated, as also specified in line 6 of Algorithm 2, to modify the AV of the test image. Considering that $0 \leq w \leq 1$, it is obvious that $mod\,AV$ can contain values equal to or smaller than those in AV and the differences are reserved for the unknown class. Note that the length of the $mod\,AV$ vector is $N + 1$, such that $\sum_{i=1}^{(N+1)} mod\,AV = \sum_{i=1}^{N} AV$, and the new element represents the activation score of the unknown class.

To make it more clear, the overall classification process of three different images is described in detail hereinafter. Let us consider a test image from the class 0. The classification results and their intermediate steps are reported in Table 4. The tail size has been set to 10 in this case. As can be seen at line 2 of Table 4, the CD value is minimum for the class 0. This shows that by having $CD = 5.53$, the AV of the test image is closer to the MAV of class 0 compared with other MAVs. As a consequence, w or the corresponding Weibull CDF evaluated on the CD is also minimum for class 0. Note that the Softmax classifier classifies this image correctly. Based on AV, α values are calculated at line 5 of Table 4 as described mathematically at line 7 of Algorithm 2. Then, the AV vector is decomposed into the $mod\,AV$ and $(AV - mod\,AV)$. This means that for each class, a certain amount of AV is subtracted and devoted to the unknown class. In fact, the sum of the elements of the $(AV - mod\,AV)$ vector constitutes the new element at the end of AV. The Openmax scores are then calculated, and the target is classified correctly at the last line of Table 4.

Let us now consider the case of having an unknown image as the input. This image is a radar image from the MSTAR dataset, but its class has not been used in the training procedure. The results are shown in Table 5. As it can be noted, all the CD values are much greater than those in Table 4. Consequently, the values from the Weibull CDFs are all large and close to 1. Based on this, larger portions are subtracted from the elements of AV and put aside for the unknown class. The Openmax classifier correctly classifies this input as an unknown image, whereas the Softmax classifier classifies it as class 2 with a very high score, 0.991.

As the final case study, the optical image shown in Figure 4 is the input to the classifier. This test is meant to evaluate Softmax and Openmax classifiers when being fed with a completely different image from those used for training and test. The CD values reflect the differences that are visually evident. The computed channel distances are $CD = [81.1, 128.1, 106.6, 113.7, 137.4, 114.3, 119.3, 135.2]$. These values are extremely larger than those in Table 5. The Openmax classifier, therefore, identifies this image as an unknown. On the other hand, Softmax misclassifies this image as class 0. Both Openmax and Softmax decisions are associated with high levels of certainty. This means that not only does the Softmax misclassify the unknown images as known, but it also assigns a high certainty to the decision. Conversely, the Openmax classifier is able to recognize it correctly as an unknown. For the sake of brevity, the intermediate parameters of the Openmax, given this optical image, are not shown.

Table 4. Intermediate parameters of the Openmax algorithm given an input image from the class 0.

Channel	0	1	2	3	4	5	6	7	
Channel Distance (CD)	5.53	45.82	24.63	41.77	55.79	48.64	39.15	51.35	
w = Weibull CDF (on CD)	0	1	0.99	1	1	1	1	1	
Softmax	0.99	5.68×10^{-10}	2×10^{-4}	5.79×10^{-7}	6.01×10^{-18}	2.39×10^{-11}	1.41×10^{-9}	2.52×10^{-15}	
α	1	0.5	0.875	0.75	0.125	0.375	0.625	0.25	
AV	22.38	1.09	13.92	8.01	-17.27	-2.07	1.99	-11.23	
modAV (@ line 8 Algorithm 2)	22.38	0.55	1.74	2	-15.11	-1.3	0.75	-8.43	
AV-modAV	0	0.54	12.18	6.01	-2.15	-0.77	1.24	-2.8	
modAV (@ line 10 Algorithm 2)	22.38	0.55	1.74	2	-15.11	-1.3	0.75	-8.43	14.24
Openmax	0.99	3.28×10^{-10}	1.08×10^{-9}	1.41×10^{-9}	5.2×10^{-17}	5.19×10^{-11}	4.03×10^{-10}	4.17×10^{-14}	2.94×10^{-4}

Table 5. Intermediate parameters of the Openmax algorithm given an unknown input radar image.

Channel	0	1	2	3	4	5	6	7	
Channel Distance (CD)	19.15	44.92	20.26	45.42	53.99	51.04	34.5	55.67	
w = Weibull CDF (on CD)	0.99	1	0.99	1	1	1	1	1	
Softmax	8.35×10^{-3}	3.09×10^{-10}	0.991	1.2×10^{-8}	1.87×10^{-15}	1.43×10^{-9}	1.01×10^{-4}	8.8×10^{-18}	
α	0.875	0.375	1	0.625	0.25	0.5	0.75	0.125	
AV	16.17	-0.93	20.95	2.72	-12.94	0.59	11.76	-18.31	
modAV (@ line 8 Algorithm 2)	2.12	-0.58	0.01	1.0	-9.71	0.3	2.94	-16.02	
AV-modAV	14.05	-0.35	20.94	1.7	-3.23	0.29	8.82	-2.2	
modAV (@ line 10 Algorithm 2)	2.12	-0.58	0.01	1.0	-9.71	0.3	2.94	-16.02	39.94364
Openmax	3.75×10^{-17}	2.5×10^{-18}	4.52×10^{-18}	1.24×10^{-17}	2.72×10^{-22}	6.05×10^{-18}	8.51×10^{-17}	4.94×10^{-25}	0.99

3.3. Classification Results: Openmax vs. Softmax

In this section, considering the performance indexes introduced in Section 2.3.3, the classification results of Openmax and Softmax are deeply analyzed. The confusion matrix and the corresponding classification reports obtained by using the Openmax approach when $\eta = 5$ are shown in Figure 5a, whereas those of the Softmax classifier are shown in Figure 5b. It can be seen that Softmax correctly classifies all the test images from classes 0 to 7 but it is not able to detect the unknowns. In fact, the images from classes 8 and 9 are classified as class 2, since they have a higher degree of similarity with this class, as can also be seen in Table 5. Note that the last column in the confusion matrices, i.e., class support size, represents the number of test images in the corresponding class. The term "support" is used by scikit-learn library [37], which is a widely used library in the machine learning community for different applications such as classification, regression [38] and so on. In addition, in order to show the discrimination between known and unknown classes clearly, a red box over unknown classes has been drawn.

Figure 4. Testing the Openmax approach with a very different image in comparison with the other open set images in MSTAR radar dataset.

To further improve the capability of classifiers to reject the unknown images, a threshold has been applied to their scores. The total accuracy of both Softmax and Openmax under different thresholds is shown in Figure 6a. Moreover, from Figure 6a, it can be noted that a threshold $\lambda \leq 0.38$ for Softmax and $\lambda \leq 0.5$ for Openmax does not produce any change in the total accuracy. When $\lambda \geq 0.5$, Openmax starts rejecting known images, while most of the unknown images were already rejected correctly. On the other hand, by increasing the threshold on Softmax, unknown images start to be rapidly rejected. Note that the Softmax scores are very close to 1, even in the case of unknown images. To achieve a comparable rejection performance with Openmax, the threshold on Softmax must be chosen carefully in the range of $\lambda \in (0.9, 1)$. Setting such a threshold is not easy, since the performance changes very quickly in this range and degrades when the threshold is too close to one. An incorrect value of the threshold may result in poor total accuracy. Conversely, the Openmax approach does not necessarily need a threshold, as it has the total accuracy equal to 90% even when $\lambda = 0$.

Additionally, Figure 6b shows the *Re* performance index for the two unknown classes and further confirms the Softmax and Openmax behaviors against the unknown targets. Softmax starts rejecting images of classes eight and nine by $\lambda \geq 0.38$. On the other hand, 87.5% of the unknown images are rejected by Openmax even by $\lambda = 0$, and Openmax rejects the remaining unknown images by $\lambda \geq 0.5$. As the threshold approaches 1, the *Re* of the unknown set also tends to 100%, since all the images in classes 8 and 9 will be marked as unknown by both classifiers. The *Re* of all the classes, including known and unknown ones, using the Openmax approach are shown in Figure 6c. It is possible to note that the *Re* values of the known classes are all in the range of $(79\%, 95\%)$. Moreover, the higher the threshold, the lower the *Re* of the known classes and the higher the *Re* of the unknown class.

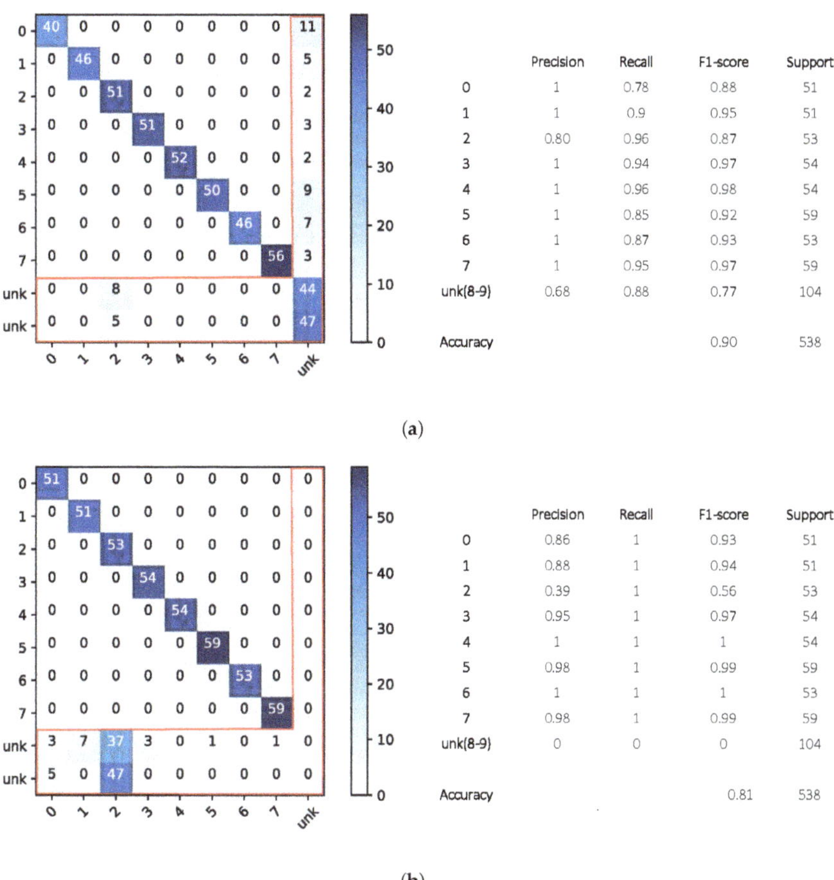

Figure 5. Confusion matrix and the corresponding classification reports using (**a**) Openmax, (**b**) Softmax.

Turning our attention to the precision index, Figure 6d shows the Pr of the unknown class. In particular, we can see from the Openmax diagram (the red curve) that Pr remains constant at 68% as long as the threshold does not exceed 0.5. Subsequently, it begins to decrease as a threshold larger than 0.5 is used and more images from known classes are rejected. Differently, we can see from the Softmax diagram that using thresholds between 0.38 and 0.48, the Pr is equal to 100%. Next, when $\lambda \geq 0.48$, the Pr is unstable and changes in the range $(75\%, 92\%)$. Finally, when λ approaches 1, both classifiers tend to the same value, $Pr = \frac{104}{538} \approx 19.3\%$. Note that the blue curve is not drawn for thresholds lower than 0.38, as we would obtain $\frac{0}{0}$ in Equation (4), which is due to the inability of the Softmax classifier to reject unknown images by such thresholds. Let us now evaluate the Pr of Openmax in known and unknown classes as shown in Figure 6e. We can see that apart from class 2 and Unk, the Pr of other classes is 100% and the threshold does not have any influence on them. The dissimilar behavior of class 2 compared to the other known classes is due to the fact that when unknown images are misclassified, they are assigned to the 2 class. However, by increasing the threshold, the Pr of this class increases and tends gradually to 100%.

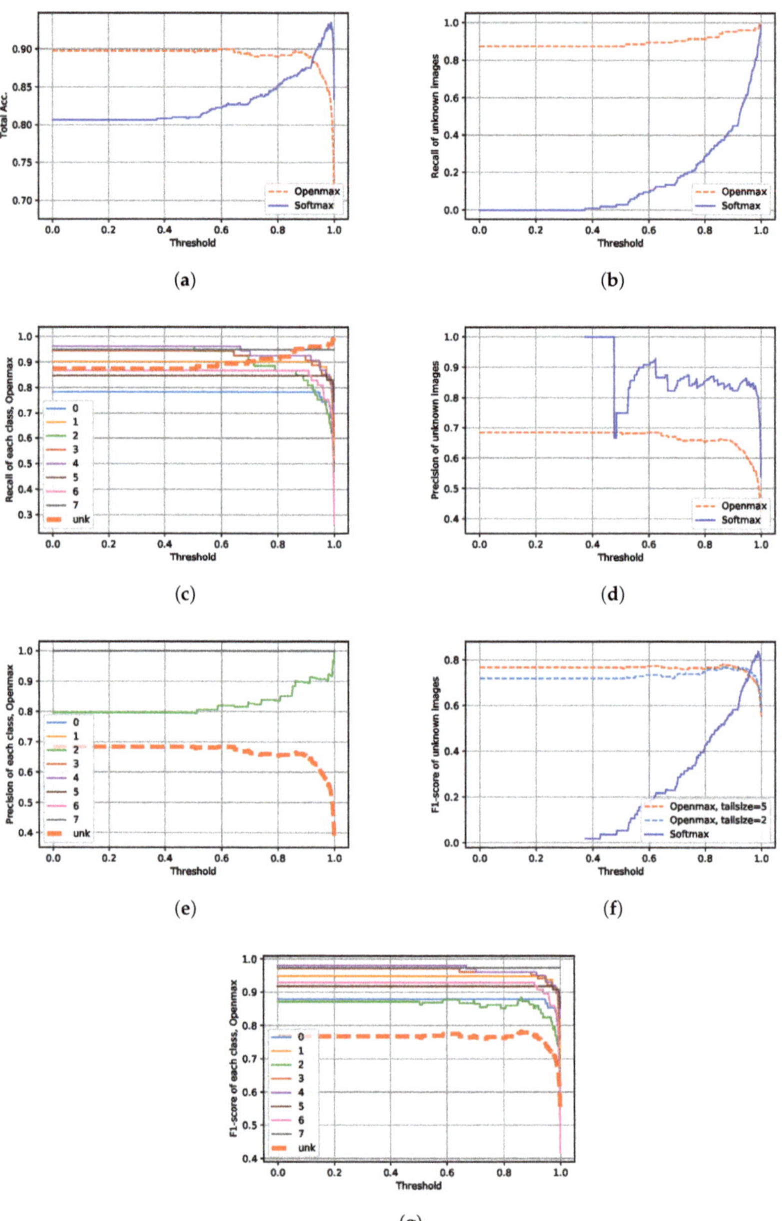

Figure 6. Applying a threshold on the output scores of the classifiers. (**a**) Total accuracy of Softmax and Openmax ($\eta = 5$); (**b**) Recall of the unknown images using Softmax and Openmax ($\eta = 5$); (**c**) Recall of the known classes together with the unknown one using Openmax ($\eta = 5$); (**d**) Precision of unknown images using Softmax and Openmax ($\eta = 5$); (**e**) Precision of the known classes together with the unknown one using Openmax approach; ($\eta = 5$) (**f**) F1-Score of unknown images using Softmax and Openmax ($\eta = 2, \eta = 5$); (**g**) F1-Score of the known classes together with the unknown one using Openmax ($\eta = 5$).

Taking both *Pr* and *Re* of the unknown classes into account, it is possible to calculate the F1-score of unknown images, which is another informative diagram shown in Figure 6f. This diagram shows that using a threshold on Softmax increases the *F1* of the unknown class. Conversely, applying a threshold on Openmax does not severely affect *F1* as long as the threshold does not exceed 0.9. Similar to *Re* and *Pr*, the behavior of the *F1* also depends on the tail size. For showing this, the *F1* obtained using a tail size of $\eta = 2$ has been added to the same plot. In Figure 6g, the *F1* diagram is shown also for the remaining classes using Openmax with $\eta = 5$.

To better analyze the Openmax performance, other experiments have been conducted by changing the unknown classes. However, to limit the number of all possible combinations of known and unknown classes, only class 8 has been changed in each experiment. Therefore, beside the case of $Unk = (8,9)$, eight different experiments have been conducted in which $Unk = (k,9)$ where $k = 0,1,2,\cdots,7$. Obviously, every time the unknown class changes, the target labels need to be changed. For instance, when $k = 3$, the classes are labeled as shown in Figure 7, and the same network is trained again with the new training set.

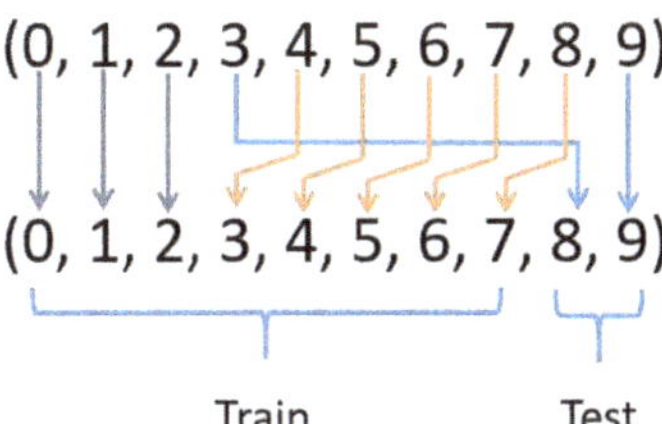

Figure 7. How to rearrange class labels in order to have a different Unk class and generalize the experiment.

The metrics *training accuracy, training loss, validation accuracy* and *validation loss* of each of the nine scenarios are shown in Figure 8a,d to demonstrate that the network converges and it is not overfitted. The confusion matrices and classification reports of these nine scenarios are illustrated in Figure 9. It can be seen that the highest and the lowest accuracies are achieved, respectively, in the case of $Unk = (6,9)$ and $Unk = (1,9)$. The average *Ac* of all scenarios here is 85.9%. The highest and the lowest *Re* of the unknown class are $Re^{Unk(6,9)} = 88\%$ and $Re^{Unk(5,9)} = 52\%$, respectively. The average *Re* of the unknown class is 72.8%. In order to make the comparison fair, all the results shown here were obtained by setting the tail size equal to 5 for each experiment. It is highly likely that the results would be further improved by adapting the tail size in each experiment.

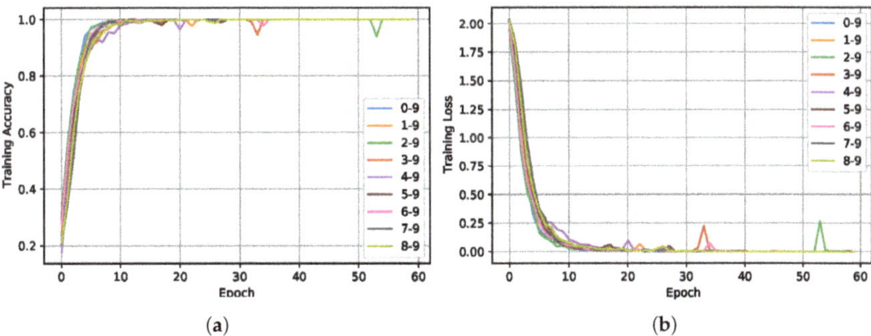

Figure 8. *Cont.*

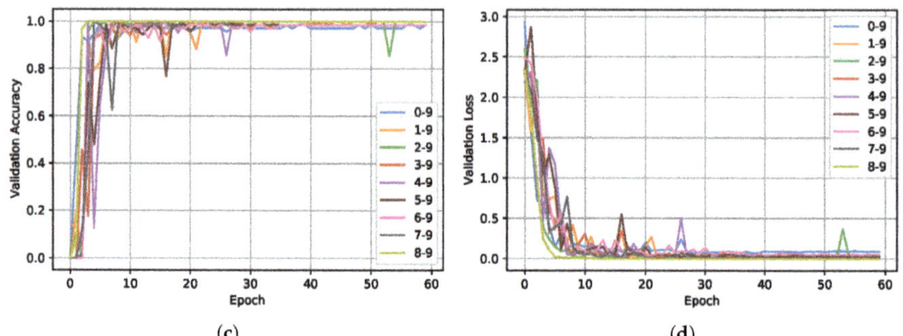

(c)   (d)

Figure 8. Metrics used to demonstrate the goodness of fitting: (**a**) training accuracy, (**b**) training loss, (**c**) validation accuracy, (**d**) validation loss.

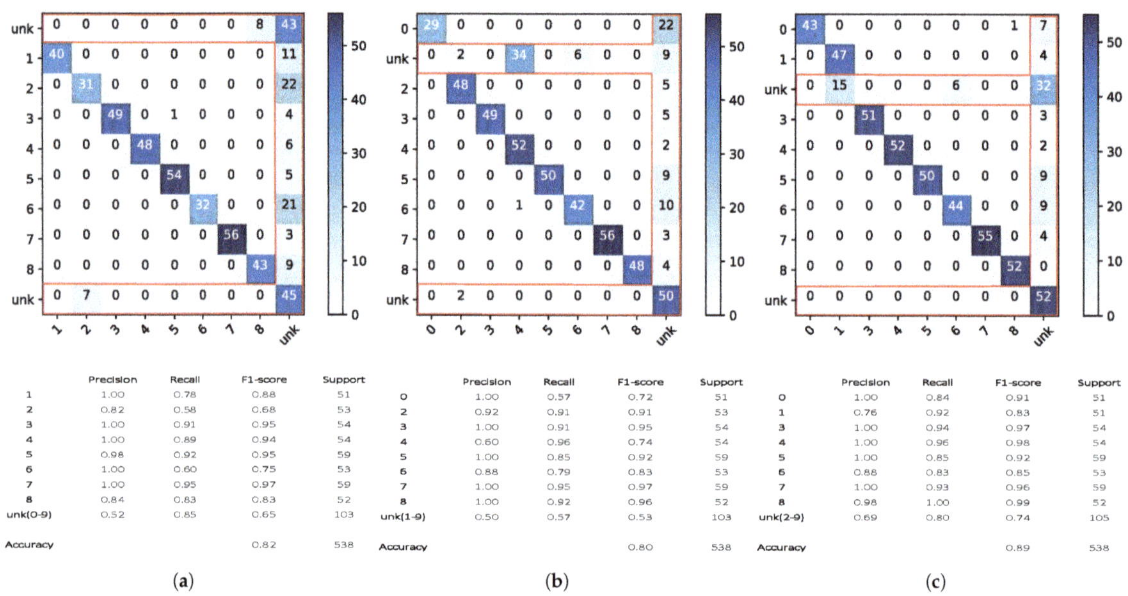

	Precision	Recall	F1-score	Support
1	1.00	0.78	0.88	51
2	0.82	0.58	0.68	53
3	1.00	0.91	0.95	54
4	1.00	0.89	0.94	54
5	0.98	0.92	0.95	59
6	1.00	0.60	0.75	53
7	1.00	0.95	0.97	59
8	0.84	0.83	0.83	52
unk(0-9)	0.52	0.85	0.65	103
Accuracy			0.82	538

(a)

	Precision	Recall	F1-score	Support
0	1.00	0.57	0.72	51
2	0.92	0.91	0.91	53
3	1.00	0.91	0.95	54
4	0.60	0.96	0.74	54
5	1.00	0.85	0.92	59
6	0.88	0.79	0.83	53
7	1.00	0.95	0.97	59
8	1.00	0.92	0.96	52
unk(1-9)	0.50	0.57	0.53	103
Accuracy			0.80	538

(b)

	Precision	Recall	F1-score	Support
0	1.00	0.84	0.91	51
1	0.76	0.92	0.83	51
3	1.00	0.94	0.97	54
4	1.00	0.96	0.98	54
5	1.00	0.85	0.92	59
6	0.88	0.83	0.85	53
7	1.00	0.93	0.96	59
8	0.98	1.00	0.99	52
unk(2-9)	0.69	0.80	0.74	105
Accuracy			0.89	538

(c)

Figure 9. *Cont.*

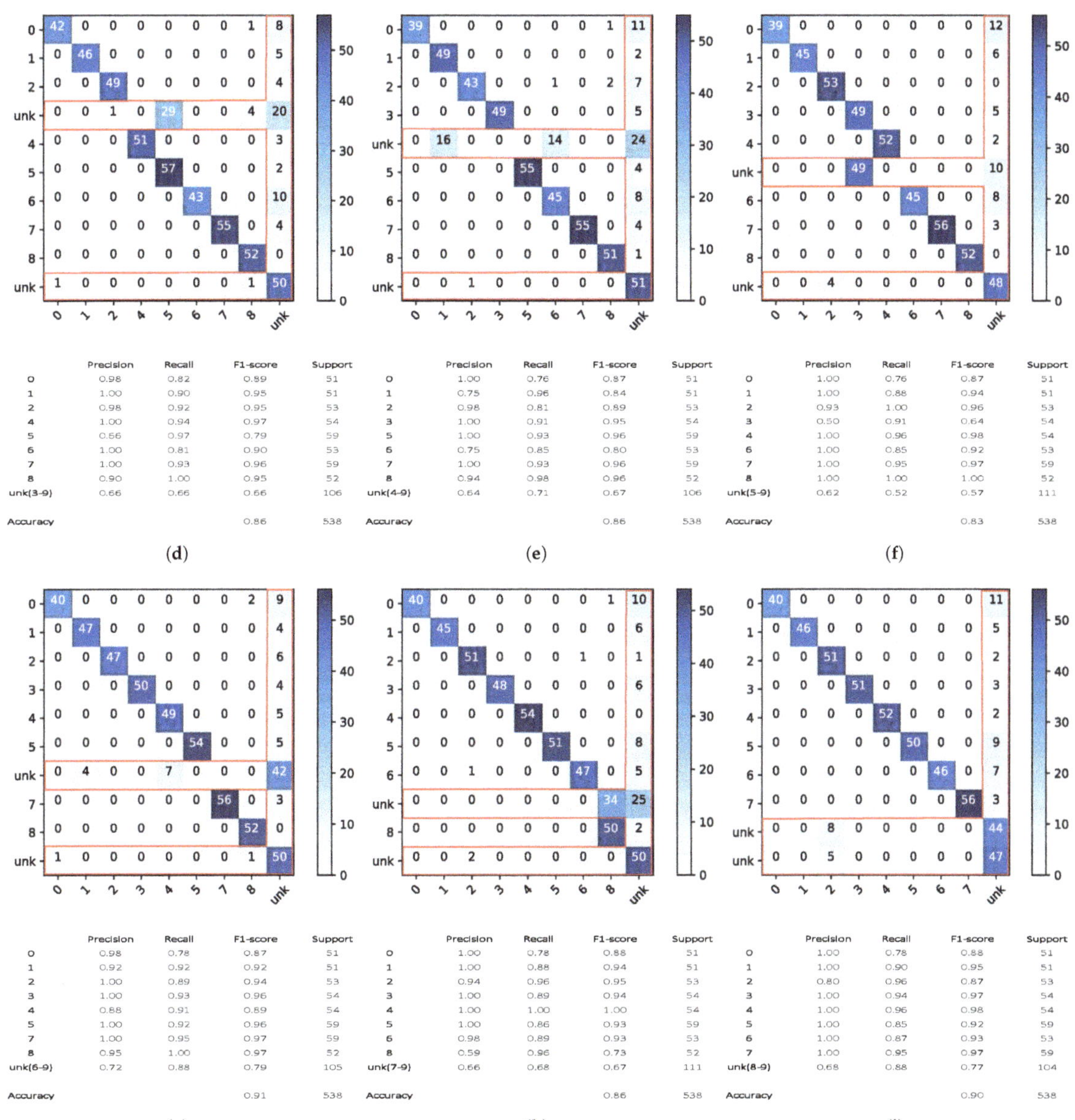

Figure 9. Confusion matrices and classification reports of Openmax when the first unknown class is changed while the second one is fixed: from (**a–i**) correspond to scenarios of having two unknown classes of $(k,9)$ where $k = 0, ..., 8$.

3.4. Statistical Analysis: The Effects of the Tail Size and of the CDF Type on Openmax

In this section, the effects of the tail size and the CDF type on the Openmax classification performance are deeply analyzed. Considering all the training images in class 0, the histogram of Euclidean distances between all AVs$\subset \mathbb{R}^{1 \times 8}$ and the MAV$\subset \mathbb{R}^{1 \times 8}$ of this class is shown in Figure 10. As previously mentioned, the η largest Euclidean distance values are used to fit a Weibull distribution. The choice of tail size is analyzed first. In particular, three different tail size values have been considered, and the relative Weibull CDFs are shown in Figure 11. More specifically, the green, orange and blue lines have been obtained using tail sizes equal to 40, 10, and 2, respectively. As it can be observed, the effect of decreasing the tail size reflects mostly a shift of the CDF toward the right direction, i.e., higher values of CD. To better understand the effect of such behavior, let us reconsider the aforementioned test image from class 0. The channel distance between the AV of this image and MAV of class 0 (the first row of Table 3) is $CD = 5.53$ as shown in Table 4. By looking at Figure 11, it is easy to observe that the Weibull function on this point is equal to 0 when the tail size is either 10 or 2. As a consequence, the corresponding w will be 0. Instead, when using the tail size of 40, Weibull CDF on $CD = 5.53$ gives a value greater than 0. This sets the corresponding w to 0.213, and since the corresponding element in AV is very large, 22.38, a considerable amount of that AV score is subtracted and reserved for the unknown class. As a consequence, by using a tail size equal to 40, Openmax classifies this image as an unknown image with 80.4% certainty. As a general rule, the larger the CD, the larger the w, and it means a higher probability of being recognized as an unknown. It can be concluded that the tail size plays an important role in the Openmax performance and it should be properly chosen, taking into consideration the available training set and the desired classifier performance to detect the unknowns. In fact, small tail size values mean a reduced capability in the detection of the unknowns. Conversely, large tail size values imply an improved capability in the detection of the unknowns but in some cases to the detriment of the known classes and resulting in lower total accuracy.

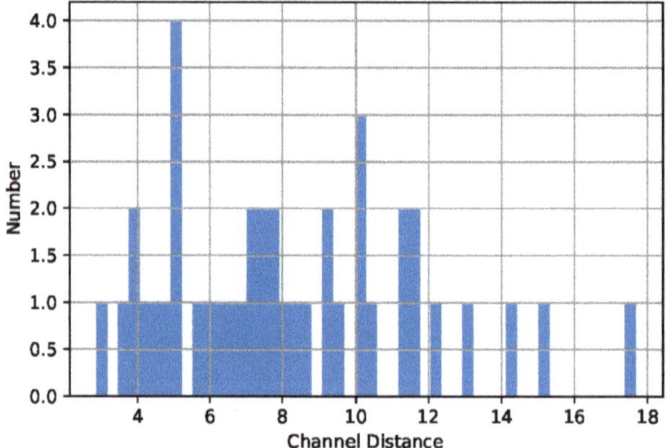

Figure 10. The histogram of Euclidean distances between AVs and MAV of class 0 of training images.

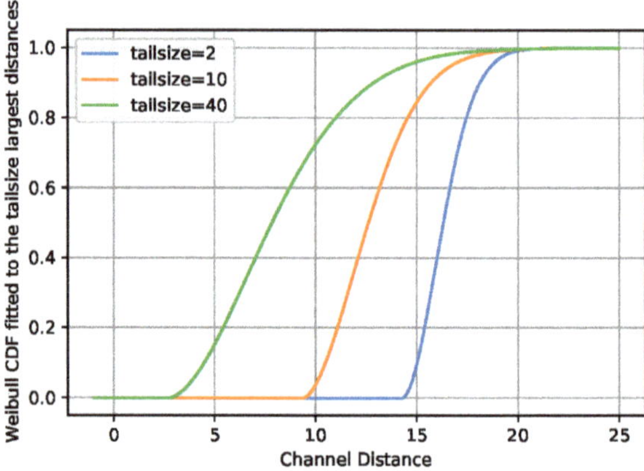

Figure 11. CDF of Weibull function fitted to η = 2, 10, 40 largest distance values between AVs and MAV of the class zero in the training images.

Another point to take into consideration is the choice of the probability function. The original Openmax classifier [18] uses the Weibull CDF. We analyze two other CDFs in this part, namely Uniform and Empirical distributions, to understand how much the CDF shape affects the Openmax performance. The mathematical expressions of probability and cumulative of the Uniform distribution are shown in (9) and (10), respectively.

$$\hat{f}(x) = \begin{cases} \frac{1}{b-a} & a \leq x \leq b \\ 0 & e.w \end{cases} \tag{9}$$

$$\hat{F}(x) = \begin{cases} \frac{x-a}{b-a} & a \leq x \leq b \\ 0 & x \leq a \\ 1 & x \geq b \end{cases} \tag{10}$$

In this regard, $a \leq x \leq b$ in each class, which are used for tail-fitting, are the Euclidean distances between all AVs of the class and their own MAV, whereas $b - a + 1$ is the tail size. The Empirical cumulative distribution function, from M-ordered independent observations of $X_1 < X_2 < X_3 < \cdots < X_M$, can be defined as [39]:

$$\hat{F}(x) = \begin{cases} \frac{k}{M} & X_k \leq x < X_{k+1} \\ 0 & x \leq X_1 \\ 1 & x \geq X_M \end{cases} \tag{11}$$

$\hat{F}(x)$ is in fact the consistent estimator of the true $F(x)$, since when the number of observations tends to infinity, $\hat{F}(x)$ converges to $F(x)$ for all x [40]. In our experiment, M corresponds to the tail size and values of $X_1 \leq x \leq X_M$ in each class, which denotes the Euclidean distance between all AVs of that class and their own MAV. The CDFs of the three models computed with tail size $\eta = [2, 5, 10, 30]$ are shown in Figure 12.

Note that in this scenario, $\eta = 5$ is optimal, and $\eta = 10$ is reasonable, while $\eta = 2$ and $\eta = 30$ are inserted to highlight the differences and similarities between the CDFs using extreme values of η.

It can be seen that when η is small, the difference between the CDFs is not prominent. The Uniform model can be considered as a linear approximation, while the Empirical model acts similar to the zero-order hold approximation of the standard Weibull CDF. However, by increasing the tail size (η), we can see that the Empirical CDF approaches the Weibull model asymptotically and the Uniform distribution deviates from the other two CDFs and gives smaller CDF values. Therefore, smaller values of w are expected from Uniform CDF, and this means a reduced sensitivity toward unknown targets.

The behavior of the Openmax performance against tail size and CDF models has been numerically assessed by means of diagrams shown in Figure 13. The terms Kno and Unk in Figure 13, which are used for the sake of brevity, refer to the known and unknown classes, respectively. In fact, Kno denotes the weighted average of parameters, i.e., *Re*, *Pr*, and *F1*, over all known classes. The value of the tail size is also shown in parenthesis. It is possible to note that by increasing the tail size, the *Re* of the unknown class increases, while its *Pr* decreases. Therefore, the classifier identifies a larger number of inputs as unknown independently of the CDF model. Interestingly, when $\eta = 30$, the results of the Empirical and Weibull models are very similar, both having $Ac = 65\%$, confirming that for a large tail size, the Empirical CDF approaches the Weibull CDF, whereas the Uniform outperforms the others having $Ac = 72\%$. In this scenario, according to the total accuracy, the Uniform model is recommended independently of the tail size. However, the CDF model type has less impact on the classifier performance compared with the tail size choice.

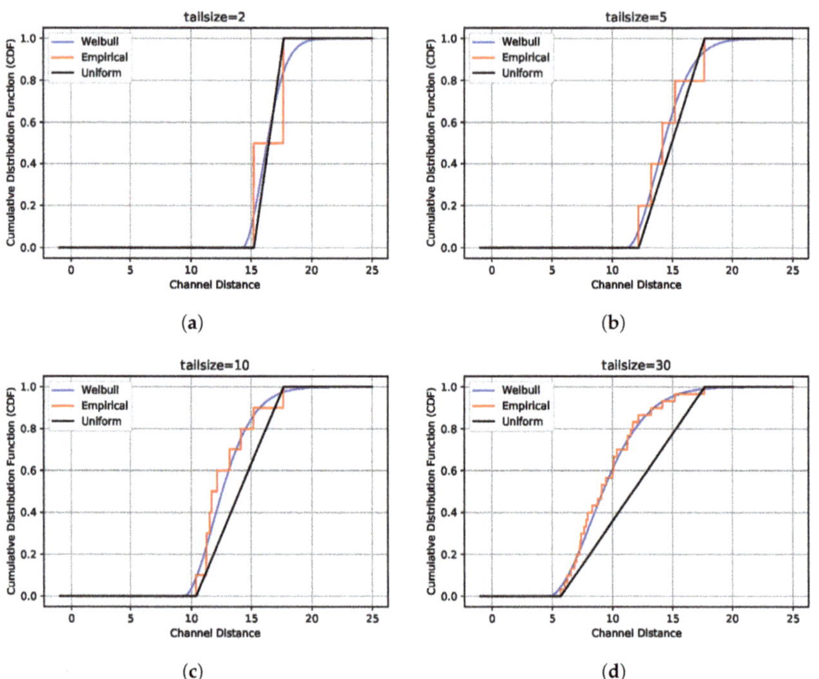

Figure 12. Three different CDFs, i.e., Uniform, Empirical and Weibull distributions, fitted to the tails of distance values of images in class zero (**a**) $\eta = 2$, (**b**) $\eta = 5$, (**c**) $\eta = 10$, (**d**) $\eta = 30$.

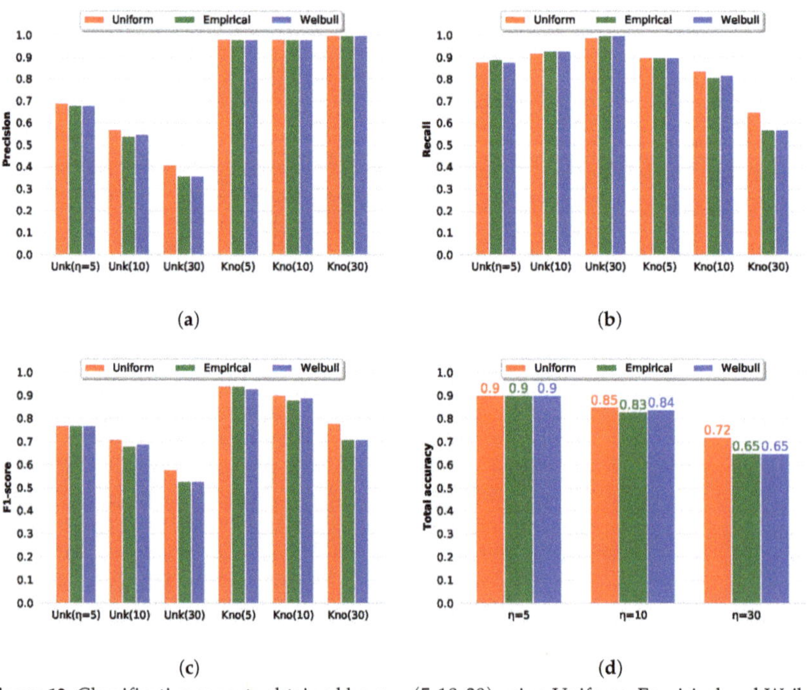

Figure 13. Classification reports obtained by $\eta = (5, 10, 30)$ using Uniform, Empirical and Weibull CDFs: (**a**) Precision, (**b**) Recall, (**c**) F1-score, and (**d**) Total accuracy.

3.5. The Proposed Approach

In this part, we present the results achieved by our proposed PS-based approach that has been described in Section 2.2. The confusion matrix along with the performance indexes of the proposed PS-Openmax for the same scenario of Figure 5, i.e., when classes eight and nine are considered unknown, are shown in Figure 14.

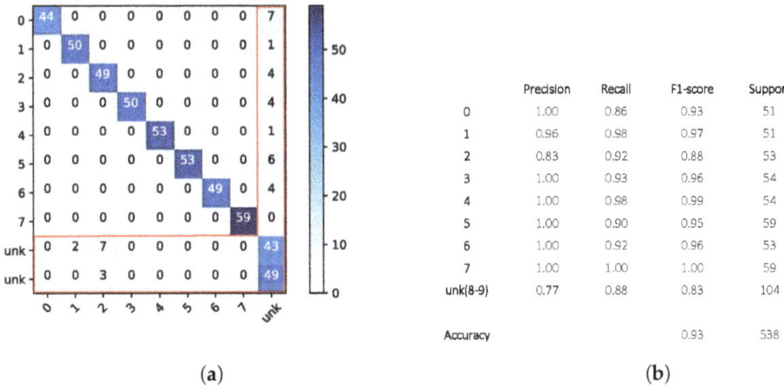

	Precision	Recall	F1-score	Support
0	1.00	0.86	0.93	51
1	0.96	0.98	0.97	51
2	0.83	0.92	0.88	53
3	1.00	0.93	0.96	54
4	1.00	0.98	0.99	54
5	1.00	0.90	0.95	59
6	1.00	0.92	0.96	53
7	1.00	1.00	1.00	59
unk(8-9)	0.77	0.88	0.83	104
Accuracy			0.93	538

(a) (b)

Figure 14. The results of the proposed PS-Openmax approach: (**a**) the confusion matrix and (**b**) the corresponding classification metrics.

We can notice that the total accuracy of the proposed approach is 3% higher than that of the original Openmax in Figure 5a.

4. Discussion

The salient features of the proposed approach are the use of a vector form supplementary activation function and interrelations between channel distances. Note that since the tail-fitting procedure is omitted in the PS-Openmax, there is no need to select and calibrate tail size and statistical distribution type. Moreover, N_α is equal to the number of known classes, so we do not need any a priori information to limit the number of changeable neurons of AV. Therefore, all of AV's elements can easily be modified while the second error shown in Figure 1, i.e., not rejecting an unknown image, is less likely to happen. This means that the proposed approach provides more robust performance.

Considering the performance indexes in Figure 14b, the weighted average of *Pr*, *Re*, and *F1* for the known and unknown classes can be computed. For the sake of an easier comparison, we illustrate these weighted average parameters of the PS-Openmax in Figure 15, together with some selected ones from Figure 13, which were achieved by the original Openmax using $\eta = 5$ and $\eta = 30$ with the optimal statistical models. From Figure 15b, we can see that no matter what tail size and what distribution type are chosen in the original Openmax, the PS-based approach gives a higher *Re* in the known classes. This also corresponds to the higher *Pr* in the unknown classes by the PS-based approach shown in Figure 15a. Taking both of these two parameters into account, we can see from the *F1* bar chart shown in Figure 15c that the PS-based approach outperforms the openmax classifier with $\eta = 30$ (and with the optimal distribution model that is Uniform) in both known and unknown classes. In fact, the scenario with $\eta = 30$ has been only conducted to have a higher recall in unknown classes, i.e., to reject all the unknown images. However, it rejects portions of the known images, as we can see the decreased *Re* of the known classes in Figure 15b or correspondingly the decreased *Pr* of unknown classes in Figure 15a. In the end, the results shown in Figure 15d indicate that the proposed PS-Openmax outperforms the original Openmax classifier in terms of the total accuracy.

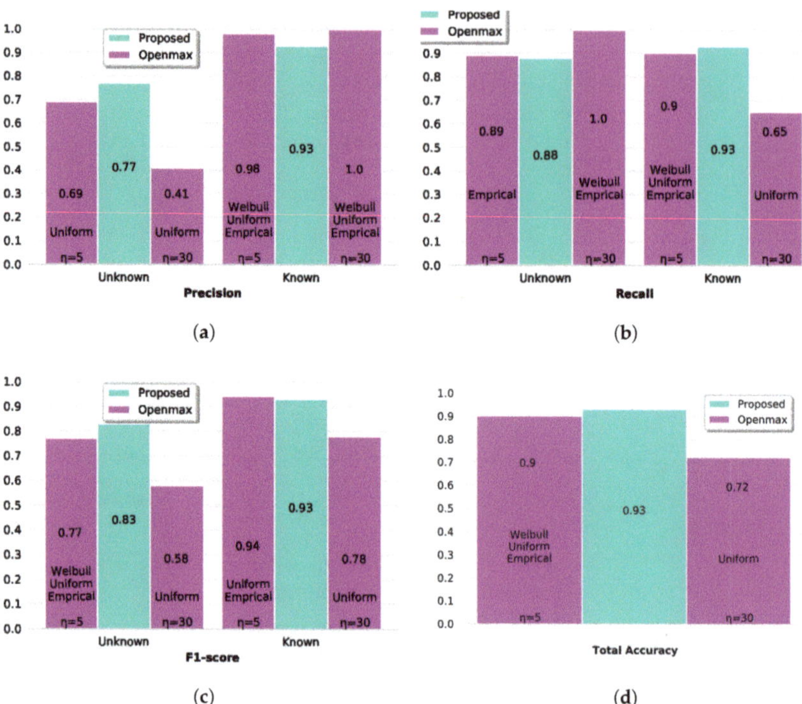

Figure 15. Comparison between the proposed PS-Openmax approach and the original Openmax classifier: (**a**) Precision, (**b**) Recall, (**c**) F1-Score, and (**d**) Total accuracy.

5. Conclusions

In this paper, we have proposed an approach to improve the performance of the Openmax classifier for the problem of open set recognition with radar images. The ability to detect unknown targets contributes to reducing the error rate and enhancing the precision when compared to the classical Softmax approach. This feature may be a desirable behavior, especially in military applications when false alarms need to be kept under control. In the proposed PS-based approach, for modifying the final activation vector of a well-trained CNN, we have exploited a vector form supplementary activation function, interrelations between channel distances, and the minimum channel distance. Although the proposed approach stems from the Openmax, it is technically different in some basic aspects such as substituting the most important part of Openmax, i.e., the tail-fitting procedure, for the aforementioned processing steps. Therefore, there is no necessity to select and calibrate the tail size and the statistical distribution type in the proposed approach. Furthermore, no restriction or in other words no a priori information is required about the number of changeable neurons in the final activation vector. Even by using different tail sizes and different statistical distribution models, which had not been addressed in the Openmax paper, the proposed approach outperforms the original Openmax classifier in terms of performance robustness and total accuracy.

This paper also provides a detailed description of the Openmax functioning that may be useful to an interested reader to try to implement it with any other DN for recognizing out-of-distribution inputs without necessarily imposing a threshold. In particular, the roles of both the tail size and the statistical model have been deeply investigated, showing that the tail size has to be carefully chosen during the model calibration phase. Conversely, the choice of the statistical model has less impact on the performance of the classifier compared with that of the tail size, meaning that the Weibull CDF is applicable to radar images as well. As a future research direction, combinations of the proposed approach and other DN is likely to produce high-performance classifiers.

Author Contributions: Conceptualization, E.G., S.G. and A.H.O.; methodology, A.H.O.; software, A.H.O. and S.G.; validation, E.G.; data curation, E.G.; writing, A.H.O., S.G. and E.G.; draft preparation A.H.O. and S.G.; review and editing, E.G., M.M., S.G. and A.H.O.; project administration, E.G. and M.M. All authors have read and agreed to the published version of the manuscript.

Funding: This research was funded by the National Inter-University Consortium for Telecommunications (CNIT) in Italy.

Acknowledgments: The authors would like to thank the anonymous reviewers for their time and helpful comments that improved the quality of this article.

Conflicts of Interest: The authors declare no conflict of interest.

References

1. Huizing, A.; Heiligers, M.; Dekker, B.; de Wit, J.; Cifola, L.; Harmanny, R. Deep Learning for Classification of Mini-UAVs Using Micro-Doppler Spectrograms in Cognitive Radar. *IEEE Aerosp. Electron. Syst. Mag.* **2019**, *34*, 46–56. [CrossRef]
2. Martorella, M.; Giusti, E.; Capria, A.; Berizzi, F.; Bates, B. Automatic Target Recognition by Means of Polarimetric ISAR Images and Neural Networks. *IEEE Trans. Geosci. Remote Sens.* **2009**, *47*, 3786–3794. [CrossRef]
3. Martorella, M.; Giusti, E.; Demi, L.; Zhou, Z.; Cacciamano, A.; Berizzi, F.; Bates, B. Target Recognition by Means of Polarimetric ISAR Images. *IEEE Trans. Aerosp. Electron. Syst.* **2011**, *47*, 225–239. [CrossRef]
4. Wagner, S.A. SAR ATR by a combination of convolutional neural network and support vector machines. *IEEE Trans. Aerosp. Electron. Syst.* **2016**, *52*, 2861–2872. [CrossRef]
5. Ding, J.; Chen, B.; Liu, H.; Huang, M. Convolutional Neural Network With Data Augmentation for SAR Target Recognition. *IEEE Geosci. Remote Sens. Lett.* **2016**, *13*, 364–368. [CrossRef]
6. Feng, S.; Ji, K.; Zhang, L.; Ma, X.; Kuang, G. SAR target classification based on integration of ASC parts model and deep learning algorithm. *IEEE J. Sel. Top. Appl. Earth Observ. Remote Sens.* **2021**, *14*, 10213–10225. [CrossRef]
7. Yang, M.; Bai, X.; Wang, L.; Zhou, F. Mixed loss graph attention network for few-shot SAR target classification. *IEEE Trans. Geosci. Remote Sens.* **2022**, *60*, 1–13. [CrossRef]
8. Wang, S.; Wang, Y.; Liu, H.; Sun, Y. Attribute-guided multi-scale prototypical network for few-shot SAR target classification. *IEEE J. Sel. Top. Appl. Earth Observ. Remote Sens.* **2021**, *14*, 12224–12245. [CrossRef]
9. Li, Y.; Du, L.; Wei, D. Multiscale CNN based on component analysis for SAR ATR. *IEEE Trans. Geosci. Remote Sens.* **2022**, *60*, 1–12. [CrossRef]
10. Lang, P.; Fu, X.; Feng, C.; Dong, J.; Qin, R.; Martorella, M. LW-CMDANet: A Novel Attention Network for SAR Automatic Target Recognition. *IEEE J. Sel. Top. Appl. Earth Observ. Remote Sens.* **2022**, *15*, 6615–6630. [CrossRef]
11. Li, C.; Du, L.; Li, Y.; Song, J. A novel SAR target recognition method combining electromagnetic scattering information and GCN. *IEEE Geosci. Remote Sens. Lett.* **2022**, *19*, 1–5. [CrossRef]
12. Zeng, Z.; Sun, J.; Han, Z.; Hong, W. SAR Automatic Target Recognition Method based on Multi-Stream Complex-Valued Networks. *IEEE Trans. Geosci. Remote Sens.* **2022**, *60*, 1–18. [CrossRef]
13. Zhang, M.; An, J.; Yu, D.H.; Yang, L.D.; Wu, L.; Lu, X.Q. onvolutional neural network with attention mechanism for SAR automatic target recognition. *IEEE Geosci. Remote Sens. Lett.* **2020**, *19*, 1–5.
14. Choi, J.H.; Lee, M.J.; Jeong, N.H.; Lee, G.; Kim, K.T. Fusion of Target and Shadow Regions for Improved SAR ATR. *IEEE Trans. Geosci. Remote Sens.* **2022**, *60*, 1–17. [CrossRef]
15. Boult, T.E.; Cruz, S.; Dhamija, A.R.; Gunther, M.; Henrydoss, J.; Scheirer, W.J. Learning and the Unknown: Surveying Steps Toward Open World Recognition. *Proc. AAAI* **2019**, *33*, 9801–9807. [CrossRef]
16. Hendrycks, D.; Gimpel, K. A baseline for detecting misclassified and out-of-distribution examples in neural networks. In Proceedings of the International Conference on Learning Representations, Toulon, France, 24–26 April 2017; pp. 1–12. [CrossRef]
17. Guisti, E.; Ghio, S.; Oveis, A.H.; Martorell, M. Transfer Learning-Based Fully-Polarimetric Radar Image Classification with a Rejection Option. In Proceedings of the 18th European Radar Conference (EuRAD), London, UK, 5–7 April 2022; pp. 357–360.
18. Bendale, A.; Boult, T.E. Towards open set deep networks. In Proceedings of the 2016 IEEE Conference on Computer Vision and Pattern Recognition (CVPR), Las Vegas, NV, USA, 27–30 June 2016; pp. 1563–1572. [CrossRef]
19. Ge, Z.Y.; Demyanov, S.; Chen, Z.; Garnavi, R. Generative openmax for multi-class open set classification. *arXiv* **2017**, arXiv:1707.07418.
20. Geng, C.; Huang, S.-J.; Chen, S. Recent advances in open set recognition: A survey. *IEEE Trans. Pattern Anal. Mach. Intell.* **2020**, *43*, 3614–3631. [CrossRef]
21. Zheng, Y.; Chen, G.; Huang, M. Out-of-domain detection for natural language understanding in dialog systems. *IEEE ACM Trans. Audio Speech Lang. Process.* **2020**, *28*, 119–1209. [CrossRef]

22. Lee, K.; Lee, K.; Lee, H.; Shin, J. A simple unified framework for detecting out-of-distribution samples and adversarial attacks. In Proceedings of the 32nd International Conference on Neural Information Processing Systems, Montréal, QC, Canada, 3–8 December 2018; Volume 31, pp. 7167–7177.
23. Inkawhich, N.A.; Davis, E.K.; Inkawhich, M.J.; Majumder, U.K.; Chen, Y. Training sar-atr models for reliable operation in open-world environments. *IEEE J. Select. Top. Appl. Earth Observ. Remote Sens.* **2021**, *14*, 3954–3966. [CrossRef]
24. Guisti, E.; Ghio, S.; Oveis, A.H.; Martorella, M. Open Set Recognition in Synthetic Aperture Radar Using the Openmax Classifier. In Proceedings of the 2022 IEEE Radar Conference (RadarConf22), New York, NY, USA, 21–25 March 2022.
25. Oveis, A.H.; Guisti, E.; Ghio, S.; Martorella, M. Extended Openmax Approach for the Classification of Radar Images with a Rejection Option. *IEEE Trans. Aerosp. Electron. Syst.* **2022**. [CrossRef]
26. The Air Force Moving and Stationary Target Recognition Database. Available online: www.sdms.afrl.af.mil/index.php?collection=mstar (accessed on 5 April 2014).
27. Scheirer, W.J.; Rocha, A.; Micheals, R.J.; Boult, T.E. Meta-recognition: The theory and practice of recognition score analysis. *IEEE Trans. Pattern Anal.* **2011**, *33*, 1689–1695. [CrossRef] [PubMed]
28. Chen, W.; Wang, Y.; Song, J.; Li, Y. Open set HRRP recognition based on convolutional neural network. *J. Eng.* **2019**, *21*, 7701–7704. [CrossRef]
29. Lewis, B.; Scarnati, T.; Sudkamp, E.; Nehrbass, J.; Rosencrantz, S.; Zelnio, E. A SAR dataset for ATR development: The Synthetic and Measured Paired Labeled Experiment (SAMPLE). In Proceedings of the Algorithms for Synthetic Aperture Radar Imagery, Baltimore, MD, USA, 18 April 2019; pp. 39–54.
30. Chen, S.; Wang, H. SAR target recognition based on deep learning. In Proceedings of the 2014 International Conference on Data Science and Advanced Analytics (DSAA), Shanghai, China, 30 October–1 November 2014; pp. 541–547.
31. Du, K.; Deng, Y.; Wang, R.; Zhao, T.; Li, N. SAR ATR based on displacement- and rotation-insensitive CNN. *Remote Sens. Lett.* **2016**, *7*, 895–904. [CrossRef]
32. Wang, L.; Bai, X.; Zhou, F. SAR ATR of ground vehicles based on ESENet. *Remote Sens.* **2019**, *11*, 1316. [CrossRef]
33. Mossing, J.C.; Ross, T.D. Evaluation of sar atr algorithm performance sensitivity to mstar extended operating conditions. *Proc. SPIE Int. Soc. Opt. Eng.* **1998**, *13*, 554–565.
34. Scheirer, W.J.; Rocha, A.d.; Sapkota, A.; Boult, T.E. Toward open set recognition. *IEEE Trans. Pattern Anal. Mach. Intell.* **2013**, *35*, 1757–1772. [CrossRef]
35. Oveis, A.H.; Guisti, E.; Ghio, S.; Martorella, M. A Survey on the Applications of Convolutional Neural Networks for Synthetic Aperture Radar: Recent Advances. *IEEE Aerosp. Electron. Syst. Mag.* **2021**, *37*, 18–42. [CrossRef]
36. Giannakopoulos, T.; Pikrakis, A. *Introduction to Audio Analysis: A MATLAB® Approach*; Academic Press: Cambridge, MA, USA, 2014.
37. Pedregosa, F.; Varoquaux, G.; Gramfort, A.; Michel, V.; Thirion, B.; Grisel, O.; Blondel, M.; Prettenhofer, P.; Weiss, R.; Dubourg, V.; et al. Scikit-learn: Machine learning in Python. *J. Mach. Learn. Res.* **2011**, *12*, 2825–2830.
38. Oveis, A.H.; Guisti, E.; Ghio, S.; Martorella, M. CNN for Radial Velocity and Range Components Estimation of Ground Moving Targets in SAR. In Proceedings of the 2021 IEEE Radar Conference (RadarConf21), Atlanta, GA, USA, 7–14 May 2021.
39. Coles, S. *An Introduction to Statistical Modeling of Extreme Values*; Springer: New York, NY, USA, 2001.
40. Van der Vaart, A.W. *Asymptotic Statistics*; Cambridge University Press: Cambridge, UK, 1998.

remote sensing

Article

Retrieval of Farmland Surface Soil Moisture Based on Feature Optimization and Machine Learning

Jianhui Zhao [1,2,3], Chenyang Zhang [1,2,3], Lin Min [1,2,3], Zhengwei Guo [1,2,3,*] and Ning Li [1,2,3]

[1]   College of Computer and Information Engineering, Henan University, Kaifeng 475004, China
[2]   Henan Engineering Research Center of Intelligent Technology and Application, Henan University, Kaifeng 475004, China
[3]   Henan Key Laboratory of Big Data Analysis and Processing, Henan University, Kaifeng 475004, China
*   Correspondence: gzw@henu.edu.cn

Abstract: Soil moisture is an important parameter affecting environmental processes such as hydrology, ecology, and climate. Synthetic aperture radar (SAR) microwave remote sensing is an important means of farmland surface soil moisture (SSM) measurement. The inversion of farmland SSM by microwave remote sensing is greatly affected by vegetation cover. To address this problem, a multisource remote sensing inversion method of farmland SSM based on feature optimization and machine learning is proposed in this paper. Six typical machine learning algorithms suitable for small sample training, including random forest, radial basis function neural network, generalized regression neural network, support vector regression, genetic algorithm–back propagation neural network, and extreme learning machine, were selected in this paper. The features extracted from Sentinel-1/2 and Radarsat-2 remote sensing data were analyzed by Pearson correlation, and those with high correlation coefficients were selected to form the optimal feature subset as the input for the subsequent machine learning models. Then, the SSM collaborative inversion models under different machine learning algorithms were constructed, and comparative experiments were set up to select the optimal prediction model. The models' accuracy under different feature parameters were studied, and the difference in the performance between the dual-polarization SAR data and the quad-polarization SAR data in SSM inversion was explored. The experimental results showed that among the six models, the random forest model had a higher inversion accuracy, with a coefficient of determination of 0.6395 and a root mean square error of 0.0264 cm^3/cm^3. Meanwhile, the inversion accuracy could be greatly improved after feature optimization, and the inversion accuracy of the quad-polarization SAR data combined with optical remote sensing data, was better than that of the dual-polarization SAR data combined with optical remote sensing data.

Keywords: surface soil moisture; multisource remote sensing; feature optimization; machine learning

Citation: Zhao, J.; Zhang, C.; Min, L.; Guo, Z.; Li, N. Retrieval of Farmland Surface Soil Moisture Based on Feature Optimization and Machine Learning. *Remote Sens.* **2022**, *14*, 5102. https://doi.org/10.3390/rs14205102

Academic Editors: Haipeng Wang, Gang Xu and Lan Du

Received: 8 August 2022
Accepted: 9 October 2022
Published: 12 October 2022

Publisher's Note: MDPI stays neutral with regard to jurisdictional claims in published maps and institutional affiliations.

1. Introduction

Although surface soil moisture (SSM) cannot be directly extracted and utilized, it is closely related to human life and can impact plant growth, the meteorological environment, and even the ecosystem cycle [1]. On a global scale, SSM is closely related to the global climate. It influences the entire terrestrial water cycle, and it is a key parameter in scientific research in many fields such as meteorology, hydrology, and agriculture [2,3]. On a regional scale, SSM affects the growth of crops, and knowledge of the spatial and temporal distributions and dynamics of SSM can guide agricultural production. Therefore, achieving large-scale, higher spatial resolution and more accurate SSM inversion can be of great help to crop production, hydrological research, and drought monitoring [4,5].

Traditional methods of monitoring SSM, for example, using SSM meters or ground observation stations, are mainly based on point measurements. The number of sampling points is limited, and it is difficult to accurately and efficiently monitor the SSM information for a large area. The advantage of using an SSM meter is that the result is accurate, but it requires a large amount of labor and time, which is time consuming and laborious [6,7]. The establishment of ground observation stations can monitor SSM in real time without manpower, but the cost of large-scale monitoring is high [8,9]. Remote sensing technology is becoming an important tool for monitoring spatial and temporal SSM information, with the advantage of large-scale, efficient, and dynamic monitoring in real-time [10]. In recent years, as many synthetic aperture radar (SAR) sensor satellites have been updated and put into use, an increasing number of experts and scholars have started to use SAR data for SSM inversion in practical applications [11,12]. The Sentinel-1/2 satellites, the data from which are free to download and have extremely wide coverage areas, provide data support for SSM inversion studies [13–16]. The Radarsat-2 satellite is one of the most mature commercial satellites available, with quad-polarization and multimode imaging capabilities to meet more individual data requirements [17]. The surface information reflected by microwave signals varies with different polarization modes, frequencies, and angles. Thus, multi-polarization SAR data can reflect surface information more comprehensively and be more fully applied in SSM inversion. The backscattering coefficient of SAR is not only related to its polarization mode, incidence angle and SSM, but is also directly affected by vegetation and surface roughness. SSM retrieval needs to effectively suppress the impact of vegetation coverage and surface roughness [18,19]. In areas with high vegetation cover, eliminating the effects caused by vegetation is a top priority. A large amount of vegetation information can be extracted from remote sensing data. Liujun Zhu et al. [18] used remote sensing data and surface observation station data to retrieve SSM based on the improved change detection method, partially eliminating the impact of vegetation and roughness changes in the experiment. However, these methods require a large number of sample data from ground observation stations. For those areas without ground observation stations, it is difficult to obtain a large number of samples. From the perspective of a small sample data drive, it is still difficult to conduct high-precision SSM inversion studies in vegetation-covered areas and needs to be explored in depth. Therefore, the SSM retrieval method suitable for small samples from the perspective of the machine learning model and feature optimization was studied in this paper.

Due to the nonlinear relationship between surface parameters, the vegetation index, and the radar backscattering coefficient, the need to improve inversion accuracy often leads to numerous parameters and complex structures in SSM inversion models. The machine learning method has a strong nonlinear fitting ability and can learn independently. It can help solve nonlinear problems in the process of SSM inversion, and it has been used widely to retrieve SSM. Rains et al. [20] used the support vector regression (SVR) method and the water cloud model to retrieve SSM and obtained a higher accuracy. Abbes et al. [21] used SMAP and MODIS data and an LSTM neural network to retrieve SSM, and the inversion results were consistent with the actual situation. Tsagkatakis et al. [22] used SMAP radiometer data and Sentinel-1 SAR data to construct a convolutional neural network (CNN) for SSM inversion and achieved a high degree of accuracy. Greifeneder et al. [23] used a gradient boosted regression trees (GBRT) algorithm, remote sensing data, and measured data to inverse SSM in the study area, and the R2 of the experimental results could reach 0.81. El Hajj et al. [24] proposed an artificial neural network (ANN) method based on remote sensing data and measured data to retrieve SSM, and achieved a high degree of accuracy. Although these machine learning methods perform well in a practical application, most of them require a large number of sample data to ensure sufficient training. For the research based on a small data size, there are still many problems to address. For instance, there are many parameters in the process of soil moisture inversion, and the relationship between surface parameters, vegetation index and radar backscatter coefficient is complex. It is of great importance to find out the optimal machine learning model and further optimize the model parameters in the case of small samples.

In order to address the problem that the inversion of farmland SSM is greatly affected by vegetation cover, based on feature optimization and machine learning methods, a multi-source remote sensing inversion method of farmland SSM is proposed in this paper. The feature parameters, including vegetation index, surface roughness, backscatter coefficient and its combination mode, polarization characteristic parameters, etc., were extracted from Sentinel-1, Radarsat-2, and Sentinel-2 remote sensing data, providing a more comprehensive reference for soil moisture inversion research. The feature parameters were then optimized using the Pearson correlation analysis method. Six typical machine learning models suitable for small sample training, including random forest (RF), radial basis function neural network (RBF), generalized regression neural network (GRNN), support vector regression (SVR), genetic algorithm-back propagation neural network (GA-BP), and extreme learning machine (ELM), were constructed and compared to select the optimal prediction model for the study area in this paper. The difference between the performance of the dual-polarization SAR data and the quad-polarization SAR data in SSM inversion was explored simultaneously.

2. Materials and Methods

2.1. Materials

2.1.1. Study Area and In Situ SSM

The study area was located in Xiangfu District, Kaifeng city, Henan Province, China, with an area of approximately 500 km2, as shown in Figure 1a. It was located in the North China Plain, which is cold and dry in the winter and hot and rainy in the summer, with a sufficient annual precipitation and a long frost-free period, suitable for crop growth. The crops mainly include winter wheat, corn, and peanut, with winter wheat being the main crop during the experiment. The experiment was conducted at the standing, jointing, and filling stages of winter wheat. During these three phenological periods, the winter wheat plants were more abundant; the vegetation cover was higher; the ground vegetation cover did not change much; field activities, such as plowing and sowing, which affect surface roughness, did not occur. Therefore, the modeling and analysis in this paper were uniformly carried out for these three similar phenological periods.

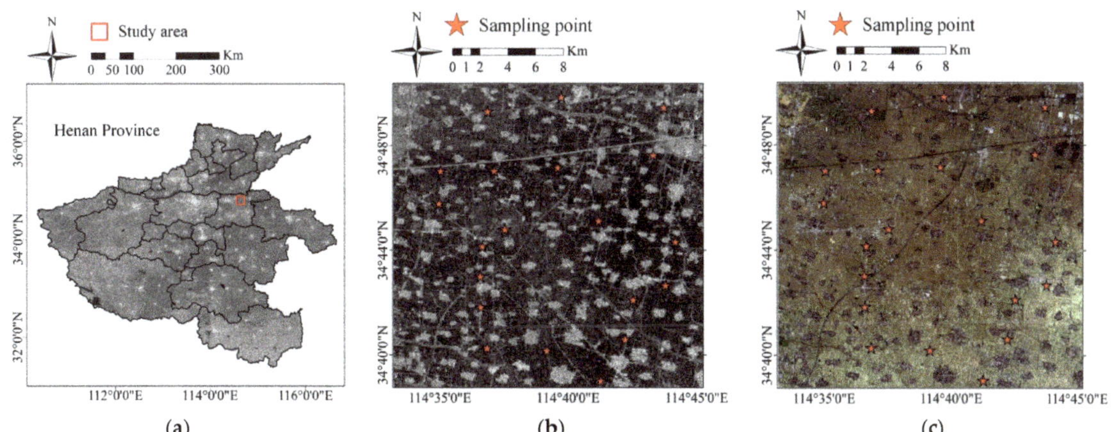

Figure 1. Location and Sentinel-1/2 images of the study area and sampling points: (**a**) location of the study area; (**b**) Sentinel-1 image of the study area and sampling points; (**c**) Sentinel-2 image of the study area and sampling points.

There are two main sources of in situ SSM data used in SSM remote sensing inversion studies. One is from ground-based observation stations or automatic observation networks, which are easy to obtain data from, and these data are usually collected more frequently and in larger quantities. The other is from the traditional manual measurement method, which relies on manual ground sampling and measurement on the date of satellite transit; it is often difficult to obtain data with this method, and these data are usually collected a limited number of times and in small quantities. Since there were no ground-based observation sites or automatic observation networks in the study area, manual measurements were used to obtain ground-based data and to carry out SSM inversion studies based on small sample sizes of measured data.

A total of 20 sampling points, as shown in Figure 1b,c, were set-up in the study area. The SSM values and latitude and longitude coordinates of all of the sampling points were collected in the field at the time of the Sentinel-1 and Radarsat-2 satellite transits, and a total of 60 sets of valid measured data for each satellite were collected for subsequent experiments. A TDR350 soil moisture meter with a probe length of 3.8 cm was used to measure the volumetric soil moisture content of the surface layer of the farmland. The SSM value of five points were measured using the cross-measurement method at each sampling point, and the measurement points were distributed in a "+" shape, as shown in Figure 2. Their average value was recorded as the in situ SSM value of each sampling point. The sampling points were located using an outdoor handheld UG905 locator, with a positioning accuracy of 1 to 3 m, and the WGS84 coordinate system was used to record the coordinates of the sampling points.

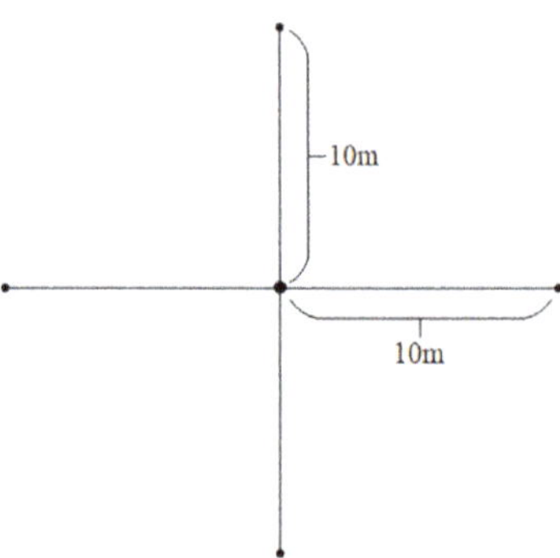

Figure 2. Measurement distribution of a single sampling point.

2.1.2. Remote Sensing Data and Preprocessing

The details and acquisition dates of the Sentinel-1 and Radarsat-2 SAR image data and Sentinel-2 optical image data used in this paper are shown in Table 1. The data used were all in the same phenological periods with the same vegetation growth conditions, allowing for uniform modeling and inversion.

Table 1. Remote sensing data information.

Data Source	Acquisition Date	Phenological Period	Product Type	Polarization Mode
Sentinel-1 (SAR Image Data)	22 March 2020 4 April 2020 21 May 2020	Standing Stage Jointing Stage Filling Stage	IW SLC	Dual-Polarization
Radarsat-2 (SAR Image Data)	15 March 2020 8 April 2020 26 May 2020	Standing Stage Jointing Stage Filling Stage	Standard Quad-Polarization	Quad-Polarization
Sentinel-2 (Optical Image Data)	23 March 2020 12 April 2020 22 May 2020	Standing Stage Jointing Stage Filling Stage	L2A	

The Sentinel-2 optical data selected for the experiments were 3 quasi-synchronous L2A level products of similar dates with Sentinel-1 and Radarsat-2 transits, as the L2A level data were products that were preprocessed with atmospheric corrections, etc. [25].

The acquired SAR images were preprocessed using the Sentinel Application Platform (SNAP) software to perform preprocessing operations such as radiometric calibration, multi-viewing, Refined Lee filtering, and terrain correction. The multi-viewing operation is a very important step in SAR image preprocessing. Multi view images improve the radiation resolution and reduce the spatial resolution. Refined Lee filtering operation can suppress speckle noise of the SAR image. In order to better compare the geometric and radiometric characteristics of the SAR images, it is necessary to use terrain correction to convert SAR data from an oblique range or ground distance projection to a geographic coordinate projection. After preprocessing, ArcGIS software was used to extract the backscatter coefficients using the latitude and longitude coordinates of the sampling points [26].

2.2. Methods

To eliminate the influence of vegetation cover on the accuracy of the SSM inversion results and to select the optimal model for inversion, six typical machine learning models, including RF, RBF, GRNN, SVR, GA-BP, and ELM, were selected for comparison experiments. The difference in the SSM inversion accuracy between the dual-polarization and quad-polarization SAR data was also explored. The technology roadmap for the experiments in this paper is shown in Figure 3, with the main steps as follows.

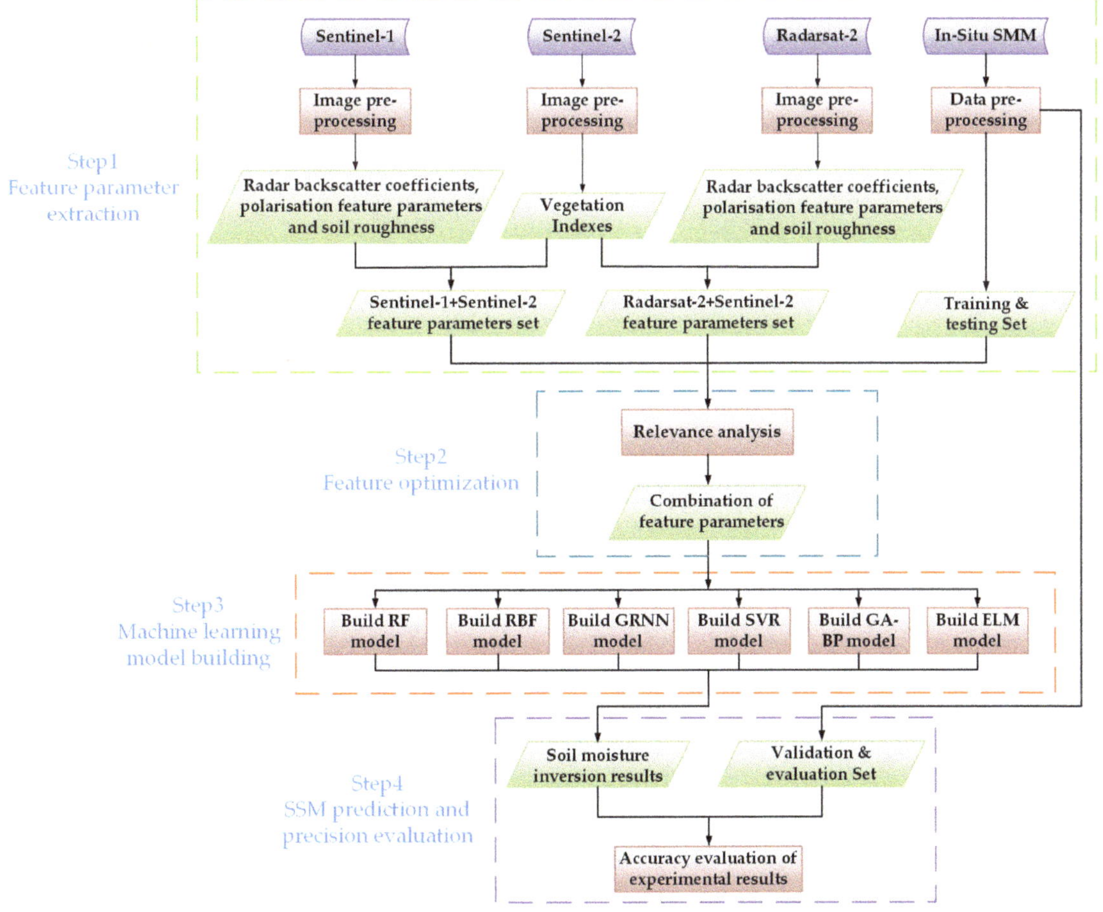

Figure 3. Technology roadmap.

- Step 1. Feature parameter extraction

The Sentinel-1 data were subjected to H/A/α polarization decomposition and combined with Sentinel-2 optical data to calculate the relevant vegetation indices. A total of 22 parameters were extracted from this combination. The Radarsat-2 data were subjected to H/A/α polarization decomposition and Freeman-Durden three-component decomposition and combined with the Sentinel-2 optical data. A total of 37 parameters were extracted from this combination, and the details are shown in Section 2.2.1.

- Step 2. Feature optimization

The Pearson correlation analysis method was carried out using the extracted feature parameters. According to the correlation between the feature parameters and the SSM values measured, the optimal features subset was selected. The details are shown in Section 2.2.2.

- Step 3. Construction of the machine learning model

The RF, RBF, GRNN, SVR, GA-BP, and ELM models were constructed and used, and each model was tuned to ensure the training and inversion accuracy of the model. The details are shown in Section 2.2.3.

- Step 4. SSM prediction and precision evaluation

For each machine learning model, two sets of multisource remote sensing data, Sentinel-1 dual-polarization data combined with Sentinel-2 optical data and Radarsat-2 quad-polarization data combined with Sentinel-2 optical data, were used for comparison. The in situ measured SSM values from 60 sampling points were randomly divided into two groups: one with 50 samples used for training, and the other with 10 samples used for validation and accuracy evaluation. The details are shown in Section 2.2.4.

2.2.1. Polarization Feature Parameter Extraction

SAR detects feature characteristics by transmitting microwave beams to and receiving echo signals from objects. Radar system parameters, such as wavelength, angle of incidence and polarization mode, and feature parameters, such as the dielectric constant and physical structure of the target object, have a direct impact on radar information. Similarly, many vegetation indices can be extracted from optical remote sensing data to describe the surface vegetation information [27].

- Polarization feature parameter extraction from the Sentinel-1 dual-polarization data

The Sentinel-1 single look complex (SLC) dual-polarization data used in this paper have the advantage of wide coverage, high spatial and temporal resolutions, and are publicly available free of charge. The polarization characteristics are mainly the backscatter coefficients and the polarization decomposition characteristics.

Active microwave remote sensing for SSM inversion is mainly based on the information reflected by the backscatter coefficients. Based on the latitude and longitude of each sampling point, the incident angle (θ), VV polarization backscatter coefficient (σ^0_{VV}), and VH polarization backscatter coefficient (σ^0_{VH}) were extracted from the preprocessed Sentinel-1 SAR data as the feature parameters for subsequent experiments. Since $\cos(\theta)$ and $\sin(\theta)$ are also related to SSM [10], and the $\sigma^0_{VH}/\sigma^0_{VV}$ backscattering coefficient is only related to the surface roughness for a certain radar incidence angle [28], $\cos(\theta)$, $\sin(\theta)$, and $\sigma^0_{VH}/\sigma^0_{VV}$ were also used as feature parameters. Meanwhile, the combination of different polarization backscattering coefficients in the forms of polarization sum ($\sigma^0_{VV} + \sigma^0_{VH}$), polarization difference ($\sigma^0_{VV} - \sigma^0_{VH}$) and polarization multiplication ($\sigma^0_{VH} \times \sigma^0_{VV}$) were also added. A total of 9 feature parameters related to the radar backscatter coefficients were extracted from the Sentinel-1 SAR data.

Polarization decomposition allows the more complex scattering process of an object to be broken down into several simple scattering mechanisms. Using polarization decomposition, more feature parameters can be extracted from SAR remote sensing data.

The eigenvalue decomposition of the coherence matrix or covariance matrix of the target feature was performed using H/A/α decomposition for Sentinel-1 dual-polarization data, from which the scattering entropy (H), the complementary parameter to the scattering entropy–inverse entropy (A), the mean scattering angle (α), and the eigenvalues (λ_1 and λ_2) could be extracted. A total of 5 polarization parameters were extracted from the Sentinel-1 SAR data [29].

- Polarization feature parameter extraction from the Radarsat-2 quad-polarization data

The Radarsat-2 quad-polarization data contained more scattering information than the dual-polarization data. The incident angle (θ) at the corresponding position was extracted from the preprocessed Radarsat-2 SAR data based on the latitude and longitude of each sampling point, and the $\cos(\theta)$ and $\sin(\theta)$ were extracted in the same way as the dual-polarization feature parameters. The backscatter coefficients of each polarization and their polarization combinations were then extracted including the VV polarization backscatter coefficients (σ^0_{VV}), VH polarization backscatter coefficients (σ^0_{VH}), HH polarization backscatter coefficients (σ^0_{HH}), HV polarization backscatter coefficients (σ^0_{HV}), cross-polarization sums ($\sigma^0_{HV} + \sigma^0_{HH}$ and $\sigma^0_{VH} + \sigma^0_{VV}$), cross-polarization differences ($\sigma^0_{HV} - \sigma^0_{HH}$ and $\sigma^0_{VH} - \sigma^0_{VV}$), cross-polarization ratios ($\sigma^0_{HV}/\sigma^0_{HH}$ and $\sigma^0_{VH}/\sigma^0_{VV}$), co-polarization ratio ($\sigma^0_{HH}/\sigma^0_{VV}$), and polarization multiplication ($\sigma^0_{HV} \times \sigma^0_{HH}$, and $\sigma^0_{VH} \times \sigma^0_{VV}$). A total of 16 feature parameters related to the radar backscatter coefficients were extracted from the Radarsat-2 SAR data.

The H/A/α decomposition and Freeman–Durden three-component decomposition were used to decompose the quad-polarization data, from which the scattering entropy (H), inverse entropy (A), mean scattering angle (α), and eigenvalues (λ_1, λ_2 and λ_3) could be extracted. The Freeman–Durden three-component decomposition allowed for the extraction of the body scattering (*Freeman_Vol*), secondary scattering (*Freeman_Dbl*), and surface scattering (*Freeman_Odd*), as well as the calculation of the total power (*Span*). A total of 10 polarization feature parameters were extracted from the Radarsat-2 SAR data [30–32].

Unlike the dual-polarization data, the quad-polarization data allowed for the extraction of the radar vegetation index (RVI), which describes vegetation information. Three radar vegetation indices (i.e., *Van_RVI*, *Freeman_RVI*, and *Kim_RVI*) were available and were calculated as shown in Equations (1)–(3) [33–35],

$$Van\_RVI = \frac{4 \times \lambda_3}{\lambda_1 + \lambda_2 + \lambda_3} \tag{1}$$

$$Freeman\_RVI = \frac{f_v}{f_s + f_d + f_v} \tag{2}$$

$$Kim\_RVI = \frac{8 \times \sigma_{HV}}{2 \times \sigma_{HV} + \sigma_{HH} + \sigma_{VV}} \tag{3}$$

- Vegetation index and surface roughness

The backscattering coefficient of SAR is not only related to its own polarization mode, incident angle, and SSM, but the vegetation cover and surface roughness also have a direct influence on the surface scattering information. The effects of the vegetation cover and surface roughness need to be suppressed in SSM inversion.

In areas covered by crops, the coverage by the crops makes most of the microwaves unable to penetrate the vegetation to reach the surface. This greatly reduces the closeness between SSM and microwave signals and increases the difficulty of retrieving SSM covered by crops. The main feature parameter that can be extracted from optical remote sensing data is the vegetation index. The vegetation index is a combination of the operation of the ground reflectance in two or more wavelength ranges to enhance a certain characteristic or detail of the vegetation. At present, there are more than 100 vegetation indices proposed in the field of remote sensing [36], but only a few of them have been tested in practice. Limited by the type of sensor and the combination of bands used, different vegetation indices have different band application ranges and application fields.

Based on the multi band data provided by the Multispectral Imager(MSI) carried by Sentinel-2 and the actual vegetation coverage situation in the study area, 7 vegetation indexes [37–40] commonly used in SSM inversion research were finally selected for this experiment including the normalized difference vegetation index (NDVI), normalized difference water index (NDWI), ratio vegetation index (RVI), moisture stress index (MSI), water band index (WBI), fusion vegetation index (FVI), and enhanced vegetation index (EVI). Their calculation formulas are shown in Equations (4)–(10),

$$\text{NDVI} = \frac{\rho_{842} - \rho_{665}}{\rho_{842} + \rho_{665}} \tag{4}$$

$$\text{NDWI} = \frac{\rho_{842} - \rho_{1610}}{\rho_{842} + \rho_{1610}} \tag{5}$$

$$\text{RVI} = \frac{\rho_{842}}{\rho_{665}} \tag{6}$$

$$\text{MSI} = \frac{\rho_{1610}}{\rho_{842}} \tag{7}$$

$$\text{WBI} = \frac{\rho_{865}}{\rho_{945}} \tag{8}$$

$$\text{FVI} = \frac{2\rho_{842} - \rho_{665} - \rho_{1610}}{2\rho_{842} + \rho_{665} + \rho_{1610}} \tag{9}$$

$$\text{EVI} = 2.5 \times \frac{\rho_{842} - \rho_{665}}{\rho_{842} + 6 \times \rho_{665} - 7.5 \times \rho_{490} + 1} \tag{10}$$

where ρ_{842}, ρ_{665}, ρ_{1610}, ρ_{865}, ρ_{945}, ρ_{665}, and ρ_{490} represent the band values corresponding to 842, 665, 1610, 865, 945, 665, and 490 nm in the Sentinel-2 data, respectively.

Soil roughness is an important factor affecting the microwave backscatter coefficient. Removing the effect of surface roughness can improve the accuracy of the SSM inversion results. It is meaningful to extract the feature parameters that can characterize the surface roughness.

The extracted surface roughness information varies with the frequency of the waveband, the incident angle and the polarization method, which makes it difficult to simulate the surface roughness. Based on the existing research theory that there is a relationship between the surface roughness and the difference in the cross-polarization backscattering coefficient, a combined roughness model was established from the SAR data [41], as shown in Equations (11)–(13),

$$Z_s = \exp\left(\frac{\sigma^0_{HV} - \sigma^0_{VV} - B_v(\theta)}{A_v(\theta)}\right) \tag{11}$$

$$A_v = \begin{array}{l} -2.640\,8\sin^3(\theta) + 5.293\sin^2(\theta) \\ -3.838\sin(\theta) + 2.2042 \end{array} \tag{12}$$

$$B_v = \begin{array}{l} 4.152\,2\sin^3(\theta) - 13.1\sin^2(\theta) \\ +16.947\,2\sin(\theta) - 16.422\,8 \end{array} \tag{13}$$

where Z_s is the combined roughness: A_v and B_v are the coefficients relating only to the incident angle. The coefficients of the A_v and B_v were obtained using nonlinear least squares and linear regression fitting, and they were only applicable to the combined roughness model under C-band SAR data.

2.2.2. Feature Optimization

The training accuracy of machine learning models is closely related to the size and quality of the training data. If the size of the training data is too large, the model will converge too slowly. In severe cases, a data disaster will occur, affecting the model's autonomous learning, causing misjudgments of the prediction results, and reducing the accuracy of the model. Analyzing the feature parameter set and selecting the feature parameters with the high correlations as the input data for the machine learning model, can improve the prediction accuracy of the model and reduce the consumption.

After preprocessing, the 22 feature parameters extracted from the Sentinel-1 and Sentinel-2 data were labeled as feature parameter set A, as shown in Table 2. The 37 feature parameters extracted from the Radarsat-2 and Sentinel-2 data were labeled as feature parameter set B, as shown in Table 3. The Pearson correlation analysis method was carried out between the extracted feature parameters and the in situ measured SSM values to obtain their correlation coefficients, which were between -1 and 1. The higher the absolute value, the stronger the correlation. The correlation coefficients were ranked from largest to smallest, as shown in Tables 2 and 3. The top 10 feature parameters in each set with the highest correlation were selected as the optimal feature subset to participate in the subsequent inversion experiments. The optimal feature subset selected from feature parameter set A included the NDWI, α, σ^0_{VV}, θ, A, FVI, σ^0_{VH}, NDVI, MSI, and $\cos(\theta)$. The optimal subset of features selected from the set of feature parameters B included α, NDWI, Van_RVI, σ^0_{VH}, $Freeman\_Dbl$, λ_1, A, λ_3, θ, and λ_2.

Table 2. Feature parameters extracted from the Sentinel-1 and Sentinel-2 data and their correlation coefficients with the in situ measured SSM values.

No.	Parameter	CC	No.	Parameter	CC
1	NDWI	0.5834 **	12	$\sigma^0_{VH}/\sigma^0_{VV}$	-0.236
2	α (Scattering Angle)	0.4143 *	13	λ_1	-0.219
3	σ^0_{VV}	0.3992 *	14	H (Scattering Entropy)	0.1424
4	θ	-0.3971 *	15	Z_s	-0.1401
5	A (Anisotropy)	-0.3723 *	16	EVI	0.1331
6	FVI	-0.3321 *	17	$\sigma^0_{VH} \times \sigma^0_{VV}$	-0.0751
7	σ^0_{VH}	-0.3281 *	18	λ_2 (Eigenvalue)	-0.0685
8	NDVI	-0.3134 *	19	RVI	0.0642
9	MSI	-0.2987	20	$\sigma^0_{VH}-\sigma^0_{VV}$	-0.0604
10	$\cos(\theta)$	0.2700	21	$\sigma^0_{VH}+\sigma^0_{VV}$	0.0511
11	$\sin(\theta)$	-0.2699	22	WBI	-0.0501

* Indicates a significant correlation at the 0.05 level; ** indicates a significant correlation at the 0.01 level.

Table 3. Feature parameters extracted from the Radarsat-2 and Sentinel-2 data and their correlation coefficients with the in situ measured SSM values.

No.	Parameter	CC	No.	Parameter	CC
1	α (Scattering Angle)	0.4961 **	20	Zs	-0.1231
2	NDWI	0.4102 *	21	$\sin(\theta)$	-0.1197
3	Van_RVI	-0.3843 *	22	$\sigma^0_{VH}/\sigma^0_{VV}$	0.1076
4	σ^0_{VH}	0.3821 *	23	$\cos(\theta)$	-0.1031
5	$Freeman\_Dbl$	-0.3694 *	24	$\sigma^0_{HH} \times \sigma^0_{HV}$	-0.0767
6	λ_1 (Eigenvalue)	-0.3691 *	25	$\sigma^0_{HV} + \sigma^0_{HH}$	0.0720
7	A (Anisotropy)	0.3639 *	26	Kim_RVI	-0.0643
8	λ_3 (Eigenvalue)	-0.3513 *	27	σ^0_{HH}	0.0638
9	θ	-0.3387 *	28	RVI	0.0576
10	λ_2 (Eigenvalue)	-0.3141 *	29	$\sigma^0_{VH} + \sigma^0_{VV}$	0.0553
11	MSI	-0.3140 *	30	$\sigma^0_{VH} \times \sigma^0_{VV}$	-0.0537
12	$\sigma^0_{HH}/\sigma^0_{VV}$	0.2986	31	WBI	-0.0486
13	FVI	0.2548	32	$Freeman\_RVI$	-0.0324
14	NDVI	0.1736	33	$\sigma^0_{HV}/\sigma^0_{HH}$	-0.0230
15	σ^0_{HV}	-0.2565	34	$\sigma^0_{HV} - \sigma^0_{HH}$	0.0112
16	H (Scattering Entropy)	-0.1534	35	σ^0_{VV}	0.0070
17	$\sigma^0_{VH} - \sigma^0_{VV}$	0.1483	36	$Freeman\_Vol$	0.0012
18	$Freeman\_Odd$	0.1442	37	*span*	0.0011
19	EVI	0.1253			

* Indicates a significant correlation at the 0.05 level; ** indicates a significant correlation at the 0.01 level.

2.2.3. Construction of the Machine Learning Model

Machine learning has a strong nonlinear fitting ability. It is helpful for solving the problem of too many parameters and too complex a structure of SSM inversion models, caused by the nonlinear relationship between the surface parameters, vegetation index, and radar backscatter coefficient. Due to the small number of samples collected in the study area, in order to avoid overfitting, six typical machine learning models suitable for small sample training (including RF, RBF, GRNN, SVR, GA-BP, and ELM models) were selected for the experiment.

- RF model

A random forest model is basically a bunch of regression equations. Each equation is built for each decision tree branch. DT branches are distinguished from each other by the most significant differences in features. RF is an integration algorithm based on the decision tree. Each decision tree is a classifier, and randomness is introduced into the training process of the decision tree to randomly select samples and features. For each input sample, n trees will have n classification results. RF integrates all classification voting results and specifies the one with the most votes as the final output. This process embodies randomness and integration. The RF model has the advantages of improving prediction accuracy, reducing overfitting, and being insensitive to missing data and multicollinearity [42].

In the experiment, the number of leaves was adjusted to 5, 10, 20, 50, and 100, and the optimum number of leaves was obtained by comparing the mean square error (MSE) of different leaves. The results showed that the optimum number of leaves was 10.

- RBF model

The radial basis function neural network model can approach any nonlinear function and deal with laws that are difficult to analyze in the system. It has a good ability for generalization and a fast learning convergence speed. The RBF model was composed of an input layer, a hidden layer, and an output layer. The role of the input layer was to input the training data into the network, and its nodes were composed of the input samples. The hidden layer used the activation function to perform nonlinear transformation on the input data. The common activation function of the RBF hidden layer was the Gaussian radial basis function, and the output layer used the linear optimization strategy, which was a linear combination of the first two [43].

- GRNN model

The generalized regression neural network is a type of RBF. It has a strong nonlinear mapping ability and learning speed, and its advantages are greater than the RBF. Research shows that a GRNN has certain advantages for small sample prediction, and it can also manage unstable data. The GRNN is different from the RBF in structure, and is composed of an input layer, a mode layer, a summation layer, and an output layer [44].

- SVR model

Support vector regression is the application of a support vector machine (SVM) in regression analysis. SVM determines a hyperplane by maximizing the interval, so that most of the sample points are located outside the two decision boundaries. Different from SVM, SVR also considers maximizing the interval, but considers the points within the decision boundary to make as many sample points as possible within the interval [45]. Its advantages are that it supports multidimensional space, different kernel functions are used for different decision functions, and small sample datasets can also be trained.

- GA-BP model

Neural network approaches are total black boxes of rules based on multilayer combinations of forecasting factors. Back propagation (BP) neural networks have been widely used in many fields, but easily fall into local minima and depend on the design structure. Sometimes it cannot find the global optimal value. Although a genetic algorithm (GA) does not have a self-learning ability, it has the ability of global optimization. Therefore, using a GA to optimize a BP neural network can improve the shortcomings of the neural network. It not only gives play to the nonlinear mapping ability of the neural network and the global optimization ability of a GA, but also accelerates the learning speed of the neural network and comprehensively improves the accuracy and fitting ability of the whole prediction model [46]. During the construction of the GA-BP neural network, the Kolmogorov theorem [41] can be used to determine the number of hidden layer nodes, as shown in Equation (14),

$$s = \sqrt{0.43mn + 0.12n^2 + 2.54m + 0.77n + 0.35} + 0.51 \tag{14}$$

where s, m, and n are the numbers of hidden layers, input layers, and output layers, respectively.

- ELM model

An extreme learning machine is a kind of machine learning model based on a feedforward neural network. The characteristic of this algorithm is that it can randomly generate weights and thresholds. Unlike a BP neural network, it does not need to continuously reverse adjust, and it only needs to set the number of hidden layer nodes to obtain the optimal solution, which greatly improves the training speed. The commonly used activation functions of its hidden layer are the radial basis function, gaussian function, trigonometric function, and sigmoid function. It has the advantages of a fast learning rate and a good generalization performance [47].

2.2.4. SSM Prediction and Precision Evaluation

To better eliminate the influence of the vegetation cover on the accuracy of the SSM inversion results, six typical machine learning models were selected in this paper for comparative experiments, from which the optimal prediction model suitable for the study area was chosen. In this experiment, two sets of comparative experiments were set up based on feature parameter extraction. All feature parameters and the optimal feature subset were used as the input data for each model, to compare and analyze the impact of the feature parameter selection on the accuracy of the SSM reversion results.

3. Results

In this paper, four precision evaluation indexes were used to evaluate the experimental results, and the spatial distribution of SSM was obtained.

3.1. Accuracy of the Experimental Results

To verify the effectiveness of the proposed method, a verification experiment of SSM inversion was carried out on winter wheat farmland in the study area. Using the in situ measured SSM data, a comparative experiment was implemented to explore the application performance of different data sources, different machine learning models, and different input parameters in SSM inversion. Their influences on the experimental results were qualitatively and quantitatively analyzed, and several meaningful conclusions were obtained.

In this experiment, four precision evaluation indexes, which were bias, root mean square error (RMSE), unbiased root mean square error (ubRMSE), and coefficient of determination (R^2), were used to evaluate the inversion accuracy. To reduce the randomness of the experimental results, the average values obtained after multiple repeated experiments were recorded as the experimental results, as shown in Table 4.

Table 4. Comparison of the accuracy of the inversion results.

No.	Method	Parameter	Bias	RMSE	ubRMSE	R^2
Sentinel-1 + Sentinel-2	RF	22	0.0138	0.0371	0.0365	0.5912
		10	0.0086	0.0311	0.0306	0.6282
	RBF	22	0.0211	0.0463	0.0451	0.5007
		10	0.0171	0.0358	0.0346	0.5697
	GRNN	22	0.0183	0.0422	0.0418	0.5525
		10	0.0134	0.0350	0.0338	0.6087
	SVR	22	0.0165	0.0416	0.0408	0.5414
		10	0.0146	0.0367	0.0353	0.5931
	GA-BP	22	0.0118	0.0391	0.0376	0.5893
		10	0.0086	0.0337	0.0329	0.6167
	ELM	22	0.0203	0.0387	0.0372	0.5516
		10	0.0173	0.0327	0.0304	0.6012
Radarsat-2 + Sentinel-2	RF	37	0.0132	0.0332	0.0294	0.5954
		10	0.0091	0.0271	0.0264	0.6395
	RBF	37	0.0199	0.0403	0.0394	0.4976
		10	0.0167	0.0546	0.0490	0.6113
	GRNN	37	0.0139	0.0371	0.0369	0.5675
		10	0.0113	0.0399	0.0373	0.6536
	SVR	37	0.0155	0.0433	0.0424	0.5674
		10	0.0126	0.0376	0.0361	0.6076
	GA-BP	37	0.0147	0.0341	0.0326	0.6039
		10	0.0114	0.0324	0.0289	0.6343
	ELM	37	0.0197	0.0389	0.0366	0.5709
		10	0.0148	0.0317	0.0308	0.6186

3.2. Spatial Distribution of SSM

According to the above conclusions, based on the Radarsat-2 SAR data and the Sentinel-2 optical data, the distribution and frequency distribution maps of SSM in the study area were obtained by using the optimal feature subset and the selected optimal prediction model—the RF model. The results are shown in Figures 4–6. To highlight the farmland areas in the SSM distribution map, non-farmland areas such as buildings, roads, and rivers, were prescreened and filled with white pixels. The average SSM inversion values in the three phases of the study area were 0.0772, 0.0537 and 0.0213 cm^3/cm^3, respectively, and the average in situ measured SSM values at the sampling points were 0.0908, 0.0639, and 0.0165 cm^3/cm^3, respectively. The inversion results of SSM were consistent with the in situ measured values.

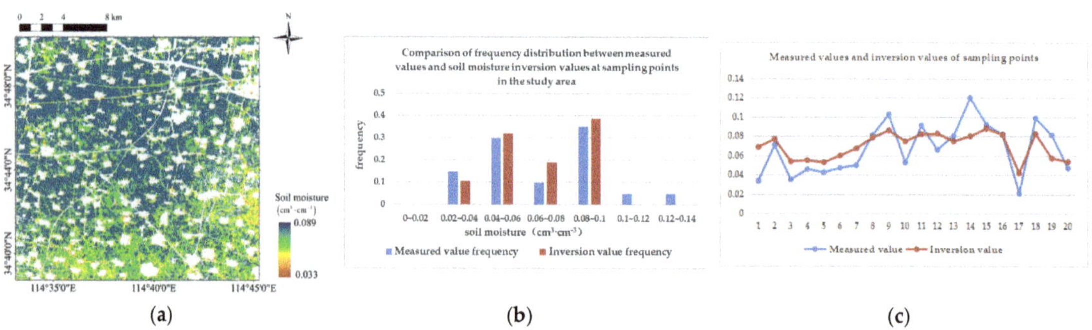

Figure 4. Retrieval results of SSM in the study area on 15 March 2020: (**a**) distribution map of retrieved SSM; (**b**) frequency distribution of retrieved and in situ measured SSM; (**c**) comparison of the differences between the measured values and the inversion values at the sampling points.

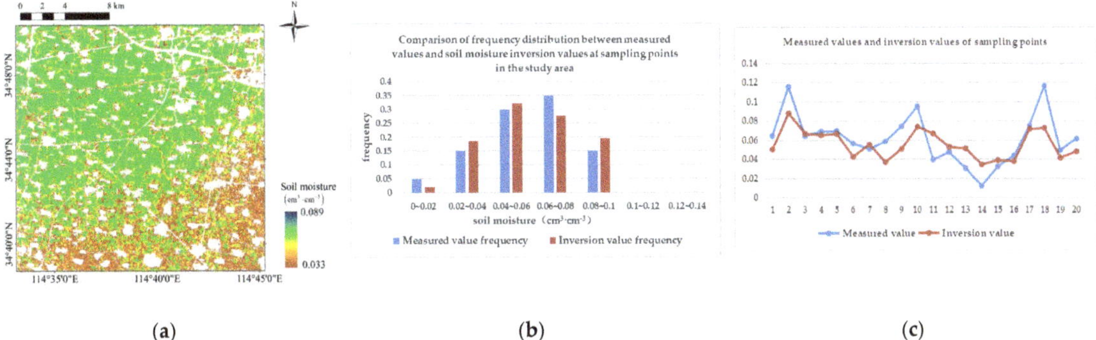

Figure 5. Retrieval results of SSM in the study area on 8 April 2020: (**a**) distribution map of retrieved SSM; (**b**) frequency distribution of retrieved and in situ measured SSM; (**c**) comparison of the differences between the measured values and the inversion values at sampling points.

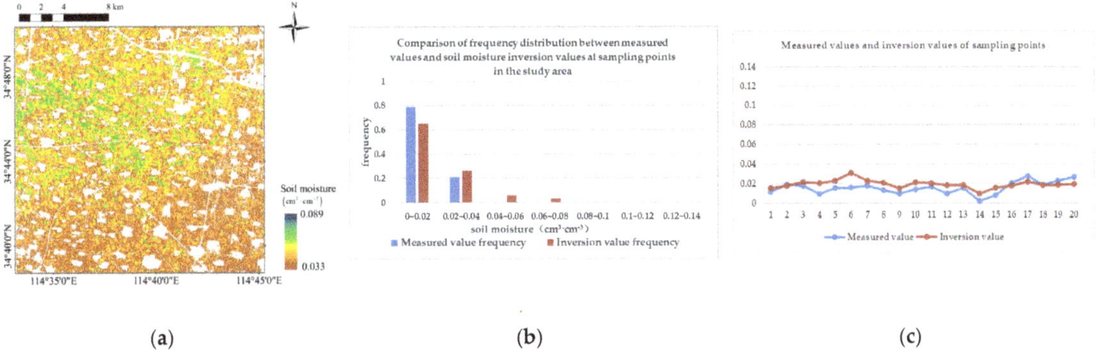

Figure 6. Retrieval results of SSM in the study area on 26 May 2020: (**a**) distribution map of retrieved SSM; (**b**) frequency distribution of retrieved and in situ measured SSM; (**c**) comparison of the differences between the measured values and the inversion values at sampling points.

4. Discussion

4.1. Accuracy Evaluation of the Experimental Results

From the perspective of the data source, the experimental results of the Radarsat-2 quad-polarization SAR data combined with the Sentinel-2 optical remote sensing data, were more accurate than the Sentinel-1 dual-polarization SAR data combined with the Sentinel-2 optical remote sensing data, and they had a better performance in the six models and four indicators. However, the Sentinel-1 dual-polarization data could also achieve a certain accuracy. In the absence of quad-polarization data, dual-polarization data combined with optical remote sensing data could also be used for SSM inversion to obtain an acceptable accuracy.

From the perspective of the model, it was found that the comprehensive performance of the RF model was the best. Among the four indexes, only its R^2 of the Radarsat-2 combined with the Sentinel-2 data was slightly lower than that of the GRNN model. Among the six models, the RBF model had the worst accuracy among the experimental results. While the GA-BP, SVR, and ELM models also had better inversion accuracy, they were not as good as the RF and GRNN models. Combining the four evaluation indexes, the RF model was selected as the optimal prediction model for the subsequent experiments.

From the perspective of the input data, the experimental results were more accurate when the optimal feature subset was used as the input data. It can be concluded that after removing the redundant feature parameters, the inversion accuracy could be improved and the SSM inversion value was closer to the in situ measured value, which illustrates the effectiveness and superiority of the proposed method.

4.2. Spatial Distribution Analysis of SSM

The frequency distribution of the measured and retrieved SSM values in the study area are shown in Figures 4b, 5b and 6b, respectively. The frequency distribution of the retrieved values on 15 March 2020 was mainly in the ranges of 0.04~0.06 and 0.048~0.1 cm^3/cm^3. The frequency distribution of the retrieved values on 8 April 2020 was mainly in the range of 0.04~0.08 cm^3/cm^3. The frequency distribution of the retrieved values on 26 May 2020 was mainly in the range of 0~0.04 cm^3/cm^3. They were relatively consistent with the frequency distribution of in situ measured values at the sampling points on the same days, which shows that the proposed method had strong applicability in the study area. The difference comparison between the measured values and the inversion values at the sampling points is shown in Figures 4c, 5c and 6c. At 20 sampling points, the inversion values and the measured values had a relatively consistent change trend. The results show that the inversion values can display whether the sampling points are dry, as well as the reliability of the soil moisture distribution.

5. Conclusions

A collaborative SSM inversion method based on machine learning and feature optimization was proposed by combining Sentinel-1 and Radarsat-2 microwave remote sensing data with Sentinel-2 optical remote sensing data. Six typical machine learning models were compared, and the differences in the performance between the dual-polarization SAR data and the quad-polarization SAR data in SSM inversion were explored. The experimental results showed that the quad-polarization SAR data performed better in SSM inversion, and the optimization of the feature parameters could greatly improve the accuracy of SSM inversion. Among the six typical machine learning models, RF, RBF, GRNN, SVR, GA-BP, and ELM, which are suitable for small sample training, the RF model had a higher accuracy with an R^2 of 0.6395 and an RMSE of 0.0264 cm^3/cm^3. The retrieved SSM values from the study area using the proposed method were consistent with the in situ measured SSM values, demonstrating the application potential of the proposed SSM inversion method. The proposed method provides a reference for SSM inversion in the surface layer of agricultural fields from multisource remote sensing data, and will be further discussed in terms of its applicability to other farmland surface types in the future.

Author Contributions: Methodology, J.Z., C.Z., L.M., Z.G. and N.L.; formal analysis, L.M., Z.G. and N.L.; investigation, J.Z. and C.Z.; resources, L.M., Z.G. and N.L.; data curation, C.Z. and N.L.; validation, J.Z., C.Z. and N.L.; writing—original draft, J.Z. and C.Z.; writing—review and editing, J.Z., C.Z., L.M., Z.G. and N.L.; visualization, C.Z.; supervision, L.M. and Z.G.; funding acquisition, N.L. All authors have read and agreed to the published version of the manuscript.

Funding: This work was supported in part by the National Natural Science Foundation of China under grants 61871175 and 42101386, in part by the College Key Research Project of Henan Province under grants 22A520021 and 21A520004, in part by the Plan of Science and Technology of Henan Province under grants 212102210093 and 222102110439, in part by the Plan of Science and Technology of Kaifeng City under grant 2102005, and in part by the Foundation of Key Laboratory of Land Satellite Remote Sensing Application, Ministry of Natural Resources of the People's Republic of China under grant KLSMNR-202204.

Institutional Review Board Statement: Not applicable.

Informed Consent Statement: Not applicable.

Data Availability Statement: Not applicable.

Acknowledgments: The authors would like to thank the anonymous reviewers for their valuable and detailed comments, that are crucial in improving the quality of this paper. The authors would also like to thank all the teachers and students of the "SAR information processing" team for their help in this paper.

Conflicts of Interest: The authors declare no conflict of interest.

References

1.   Chen, S.L.; Liu, Y.B.; Wen, Z.M. Review on soil moisture retrieval by satellite remote sensing. *Prog. Geosci.* **2012**, *27*, 1192–1203.
2.   Wang, H.Q.; Magagi, R.; Goita, K. Potential of a two-component polarimetric decomposition at C-band for soil moisture retrieval over agricultural fields. *Remote Sens. Environ.* **2018**, *217*, 38–51. [CrossRef]
3.   Anagnostopoulos, V.; Petropoulos, G.P.; Ireland, G.; Carlson, T.N. A modernized version of a 1D soil vegetation atmosphere transfer model for improving its future use in land surface interactions studies. *Environ. Model. Softw.* **2017**, *90*, 147–156. [CrossRef]
4.   Zhang, X.; Yuan, X.; Liu, H.; Gao, H.; Wang, X. Soil Moisture Estimation for Winter-Wheat Waterlogging Monitoring by Assimilating Remote Sensing Inversion Data into the Distributed Hydrology Soil Vegetation Model. *Remote Sens.* **2022**, *14*, 792. [CrossRef]
5.   Chen, S.; Yan, Q.; Jin, S.; Huang, W.; Chen, T.; Jia, Y.; Liu, S.; Cao, Q. Soil Moisture Retrieval from the CyGNSS Data Based on a Bilinear Regression. *Remote Sens.* **2022**, *14*, 1961. [CrossRef]
6.   Wang, S.N.; Li, R.P.; Wu, Y.J.; Zhao, S.X.; Wang, X.Q. Soil moisture inversion based on environmental variables and machine learning. *J. Agric. Mach.* **2022**, *53*, 332–341.
7.   Ma, H.Z.; Liu, S.M.; Peng, A.H.; Sun, L.; Sun, G.Y. Active and passive cooperative algorithm at L-Band for bare soil moisture inversion. *Trans. Chin. Soc. Agric. Eng.* **2016**, *32*, 133–138.
8.   Wang, S.G.; Ma, C.F.; Zhao, Z.B.; Wei, L. Estimation of Soil Moisture of Agriculture Field in the Middle Reaches of the Heihe River Basin based on Sentinel-1 and Landsat 8 Imagery. *Remote Sens. Technol. Appl.* **2020**, *35*, 13–22.
9.   Wang, Y.T. Remote Sensing Retrieval of Soil Moisture in Ordos Blown-Sand Region Based on SVR. Master's Thesis, Chang'an University, Xi'an, China, 2019.
10.   Lin, L.B. Soil Moisture Retrieval under Vegetation Cover Using Multi-Source Remote Sensing Data. Master's Thesis, Nanjing University of Information Science and Technology, Nanjing, China, 2018.
11.   Zhang, W.F.; Chen, E.X.; Li, Z.Y.; Yang, H.; Zhao, L. Review of applications of radar remote sensing in agriculture. *J. Radars* **2020**, *9*, 444–461.
12.   Xu, J.X.; Li, X.; Zhu, Y.C.; Fang, S.B.; Wu, D.; Wu, Y.J. Progress of the Methods of Remote Sensing Monitoring the Soil Moisture. *Adv. Meteorol. Sci. Technol.* **2019**, *9*, 17–23.
13.   Gong, R. Overview of "Sentinel" satellite family. *Space Int.* **2014**, *7*, 23–28.
14.   Attema, E.; Davidson, M.; Floury, N.; Levrini, G.; Snoeij, P. Sentinel-1 ESA's New European Radar Observatory. In Proceedings of the 7th European Conference on Synthetic Aperture Radar, Friedrichshafen, Germany, 2–5 June 2008; pp. 1–4.
15.   Yang, K.; Yang, J.B.; Jiang, B.R. Sentinel-1 Satellite Overview. *Urban Geotech. Investig. Surv.* **2015**, *2*, 24–27.
16.   Yang, B.; Li, D.; Gao, G.S.; Chen, C.; Wang, L. Processing analysis of Sentinel-2A data and application to arid valleys extraction. *Remote Sens. Land Resour.* **2018**, *30*, 128–135.
17.   Huang, S.; Ding, J.L.; Zhang, J.Y.; Chen, W.Q. Backscattering Coefficient Research Based on Microwave Remote Sensing of Radarsat-2 Satellite. *Acta Opt. Sin.* **2017**, *37*, 317–327.
18.   Zhu, L.; Si, R.; Shen, X.; Walker, J.P. An advanced change detection method for time-series soil moisture retrieval from Sentinel-1. *Remote Sens. Environ.* **2022**, *279*, 113137. [CrossRef]
19.   Sun, J.X.; Zhang, D.Y.; Hou, Y.C. Multi-source Remote Sensing Data Cooperates to Retrieve Forest Surface Soil Moisture. *Remote Sens. Technol. Appl.* **2021**, *36*, 564–570.
20.   Rains, D.; Lievens, H.; Lannoy, G.J.M.D.; Mccabe, M.F.; Miralles, D.G. Sentinel-1 Backscatter Assimilation Using Support Vector Regression or the Water Cloud Model at European Soil Moisture Sites. *IEEE Geosci. Remote Sens. Lett.* **2022**, *19*, 1–5. [CrossRef]
21.   Abbes, A.B.; Magagi, R.; Goita, K. Soil Moisture Estimation from Smap Observations Using Long Short- Term Memory (LSTM). In Proceedings of the 2019 IEEE International Geoscience and Remote Sensing Symposium, Yokohama, Japan, 28 July–2 August 2019; pp. 1590–1593.
22.   Tsagkatakis, G.; Moghaddam, M.; Tsakalides, P. Multi-Temporal convolutional neural networks for satellite-derived soil moisture observation enhancement. In Proceedings of the 2020 IEEE International Geoscience and Remote Sensing Symposium, Waikoloa, HI, USA, 26 September–2 October 2020; pp. 4602–4605.
23.   Greifeneder, F.; Notarnicola, C.; Wagner, W. A Machine Learning-Based Approach for Surface Soil Moisture Estimations with Google Earth Engine. *Remote Sens.* **2021**, *13*, 2099. [CrossRef]
24.   El Hajj, M.; Baghdadi, N.; Zribi, M.; Bazzi, H. Synergic Use of Sentinel-1 and Sentinel-2 Images for Operational Soil Moisture Mapping at High Spatial Resolution over Agricultural Areas. *Remote Sens.* **2017**, *9*, 1292. [CrossRef]
25.   Pan, Y.Y.; Li, C.C.; Ma, X.X.; Wang, B.S.; Fang, X. Atmospheric Correction Method of Sentinel-2A Satellite and Result Analysis. *Remote Sens. Inf.* **2018**, *33*, 41–48.
26.   Christiansen, M.P.; Teimouri, N.; Laursen, M.S.; Mikkelsen, B.F.; Jorgensen, R.N. Preprocessed Sentinel-1 data via a web service focused on agricultural field monitoring. *IEEE Access* **2019**, *7*, 65139–65149. [CrossRef]

27. Wang, C.; Zhang, H.; Chen, X. *Quad Polarization Synthetic Aperture Radar Image Processing*; Science Press: Beijing, China, 2008.
28. Guan, Y.T.; Li, J.P. Soil moisture inversion based on genetic optimization neural network and multi-source remote sensing data. *J. Water Resour. Water Eng.* **2019**, *30*, 255–259.
29. Cloude, S.R.; Pottier, E. A review of target decomposition theorems in radar polarimetry. *IEEE Trans. Geosci. Remote Sens.* **1996**, *34*, 498–518. [CrossRef]
30. Li, Z.; Liao, J.J. *Inversion Model and Method of Surface Parameters of Synthetic Aperture Radar*; Science Press: Beijing, China, 2011.
31. Freeman, A.; Durden, S.L. A three-component scattering model for polarimetric SAR data. *IEEE Trans. Geosci. Remote Sens.* **1998**, *36*, 963–973. [CrossRef]
32. Wang, P.; Zhou, Z.F.; Liao, J. Study on Soil Moisture Retrieval of Tobacco Field in Karst Plateau Mountainous Area Based on Freeman Decomposition. *Geogr. Geo-Inf. Sci.* **2016**, *32*, 72–76.
33. Gherboudj, I.; Magagi, R.; Berg, A.A.; Toth, B. Characterization of the Spatial Variability of In-Situ Soil Moisture Measurements for Upscaling at the Spatial Resolution of RADARSAT-2. *IEEE J. Sel. Top. Appl. Earth Obs. Remote Sens.* **2017**, *10*, 1813–1823. [CrossRef]
34. Mei, X.; Nie, W.; Liu, J.Y. Difference Analysis of Multiply Radar Vegetation Indices Base on Radarsat-2 Full-polarization Data. *Chin. J. Agric. Resour. Reg. Plan.* **2019**, *3*, 21–28.
35. Van, Z.J.J.; Zebker, H.A.; Elachi, C. Imaging radar polarization signatures: Theory and observation. *Ratio Sci.* **1987**, *22*, 529–543.
36. Fu, Y.Z. Study on Vegetation Index of Remote Sensing and Its Applications. Master's Thesis, Fuzhou University, Fuzhou, China, 2010.
37. Zhao, X.; Wang, J.D.; Liu, S.H. Modified monitoring method of vegetation water content based on coupled radiative transfer model. *J. Infrared Millim. Wave* **2010**, *29*, 185–189. [CrossRef]
38. Wang, D.C.; Wang, J.H.; Jin, N.; Wang, Q.; Li, C.J.; Huang, J.F.; Wang, Y.; Huang, F. ANN-based wheat biomass estimation using canopy hyperspectral vegetation indices. *Trans. Chin. Soc. Agric. Eng.* **2008**, *24* (Suppl. 2), 196–201.
39. Zhao, J.H.; Zhang, B.; Li, N.; Guo, Z.W. Cooperative Inversion of Winter Wheat Covered Surface Soil Moisture Based on Sentinel-1/2 Remote Sensing Data. *J. Electron. Inf.* **2021**, *43*, 692–699.
40. Wang, Z.X.; Liu, C.; Huete, A. From AVHRR-NDVI to MODIS-EVI: Advances in Vegetation Index Research. *Acta Ecol. Sin.* **2003**, *5*, 979–987.
41. Tong, L.; Chen, Y.; Jia, M.Q. *Mechanism of Radar Remote Sensing*; Science Press: Beijing, China, 2014.
42. Fang, K.N.; Wu, J.B.; Zhu, J.P.; Xie, B.C. A Review of Technologies on Random Forests. *J. Stat. Inf.* **2011**, *26*, 32–38.
43. Chu, Q.L.; Ping, Z.D.; Yu, M.J. Prediction model of octane loss based on RBF neural network. *Internet Things Technol.* **2010**, *135*, 230–267.
44. Guo, J.; Liu, J.; Ning, J.F.; Han, W. Construction and validation of farmland surface soil moisture retrieval model based on sentinel multi-source data. *Trans. Chin. Soc. Agric. Eng.* **2019**, *35*, 71–78.
45. Brereton, R.G.; Lloyd, G.R. Support Vector Machines for classification and regression. *Analyst* **2010**, *135*, 230–267. [CrossRef]
46. Ma, Y.J.; Yun, W.X. Research progress of genetic algorithm. *Appl. Res. Comput.* **2012**, *29*, 1201–1206, 1210.
47. Li, X.L.; Zhao, H.L.; Zhao, H.L.; Wang, R.; Hao, Z. Soil Water content inversion based on extreme learning machine model. *Sci. Surv. Mapp.* **2021**, *46*, 91–97.

remote sensing

Article

A Novel Echo Separation Scheme for Space-Time Waveform-Encoding SAR Based on the Second-Order Cone Programming (SOCP) Beamformer

Shuo Han [1,2], Yunkai Deng [1], Wei Wang [1,*], Qingchao Zhao [1], Jinsong Qiu [1], Yongwei Zhang [1] and Zhen Chen [1]

[1] Department of Space Microwave Remote Sensing System, Aerospace Information Research Institute, Chinese Academy of Sciences, Beijing 100190, China
[2] School of Electronic, Electrical and Communication Engineering, University of Chinese Academy of Sciences, Beijing 100039, China
* Correspondence: wwang@mail.ie.ac.cn

Abstract: Space-time waveform-encoding (STWE)-synthetic aperture radar (SAR) is an effective way to accomplish high-resolution and wide-swath (HRWS) imaging. By designing the specific signal transmit mode, the echoes from several subswaths are received within a single receiving window and overlap each other in STWE-SAR. In order to separate the overlapped echoes, the linear-constrained minimum variance (LCMV) beamformer, a single-null beamformer, is typically used. However, the LCMV beamformer has a very narrow and unstable notch depth, which is not sufficient to accurately separate the overlapped echoes with large signal energy differences between subswaths. The issue of signal energy differences in STWE-SAR is first raised in this paper. Moreover, a novel echo separation scheme based on a second-order cone programming (SOCP) beamformer is proposed. The beam pattern generated by the SOCP beamformer allows flexible adjustment of the notch width and depth, which effectively improves the quality of separation results compared to the LCMV beamformer. The simulation results illustrate that the scheme can greatly enhance the performance of echo separation. Furthermore, the experimental results based on the X-band STWE-SAR airborne system not only demonstrate the scheme's effectiveness but also indicate that it holds great promise for future STWE-SAR missions.

Keywords: synthetic aperture radar; high-resolution and wide-swath; digital beamforming; space-time waveform-encoding; echo separation; second-order cone programming

Citation: Han, S.; Deng, Y.; Wang, W.; Zhao, Q.; Qiu, J.; Zhang, Y.; Chen, Z. A Novel Echo Separation Scheme for Space-Time Waveform-Encoding SAR Based on the Second-Order Cone Programming (SOCP) Beamformer. *Remote Sens.* **2022**, *14*, 5888. https://doi.org/10.3390/rs14225888

Academic Editors: Lan Du and Gang Xu

Received: 3 October 2022
Accepted: 16 November 2022
Published: 20 November 2022

Publisher's Note: MDPI stays neutral with regard to jurisdictional claims in published maps and institutional affiliations.

1. Introduction

Due to the increasing demand for applications, high-resolution and wide-swath (HRWS) has been a focal point for current and future research on synthetic aperture radar (SAR) systems [1–4]. The azimuth resolution and swath range of traditional SAR with a single channel cannot be enhanced simultaneously due to the minimal antenna area constraint [5]. By expanding the number of SAR channels, beamforming can overcome standard SAR system performance limitations and realize HRWS imaging. The application of multichannel reception at the azimuth [6] reduces the requirement for pulse repetition frequency (PRF) of the system and creates conditions for expanding the receiving window, allowing for a wider imaging swath. Applying the digital beamforming (DBF) technique to elevation [7] can generate a sharp receive beam with a high gain to scan the echo in real-time, hence improving the system's signal-to-noise ratio (SNR) for wide-swath imaging. Benefiting from the mature application of multichannel technology to spaceborne SAR [8], DBF has been deployed for spaceborne SAR missions, such as Radar Observing System for Europe at L-band (ROSE-L) [9], Sentinel-1 next-generation (NG) [10], Advanced Land-Observing Satellite-4 (ALOS-4) [11], Tandem-L [12], National Aeronautics and Space Administration (NASA)–Indian Space Research Organisation (ISRO) SAR (NISAR) [13],

HRWS SAR [14], etc. Additionally, the DBF technique has become an important part of the HRWS-SAR systems because of its distinctive characteristics in spatial filtering and its ability to improve system performance.

In addition to the benefits of integrating multichannel technology into the HRWS-SAR systems, the multiple degrees of freedom have come into view. For example, the innovative multidimensional waveform-encoding (MWE) [15] system can utilize the system information from multiple dimensions to improve system performance. Based on the space-time relationship of the system, the space-time waveform-encoding (STWE)-SAR can realize waveform diversity in the space-time domain [16,17]. As illustrated in Figure 1, STWE-SAR utilizes a specifically designed beam transmitting sequence that allows the subswath in the far range to be illuminated for one pulse repetition time (PRT) or several PRTs ahead of the subswath in the near range. In this way, the echoes from different subswaths can arrive at the antenna simultaneously and overlap each other within a single receiving window. STWE-SAR can gather ground feature information across a wider swath within the same receiving window and reduce the echo data volume. In order to obtain the echoes of different subswaths from the overlapped echoes, research on the separation schemes is one of the main focuses.

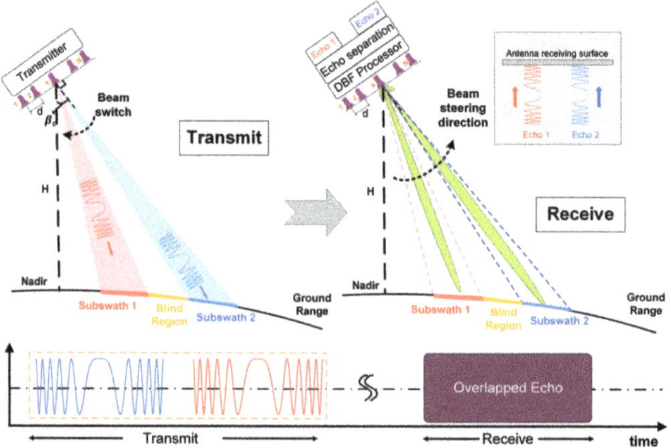

Figure 1. The principle diagram of the typical STWE-SAR system with two subswaths.

On this basis, echo separation schemes relayed on coding waveforms, such as the orthogonal frequency-division multiplexing (OFDM) chirp waveform [18], short-term shift-orthogonal (STSO) waveform [19], segmented-phase-code (SPC) waveform [20], segmented shift chirp (SSC) waveform [21], etc., have been widely and intensively studied in the multiple-input–multiple-output (MIMO)-SAR to achieve HRWS imaging. Unlike these schemes, STWE-SAR does not care about the structure of the waveform and encodes the waveform in the space-time domain. The DBF at elevation plays a crucial part in these echo separation schemes due to its spatial filtering characteristics, as the echoes of the subswaths come from distinct directions in space. Furthermore, the DBF technique and the optimization algorithm are then utilized to obtain a specific beamformer in STWE-SAR, such as the linear-constrained minimum variance (LCMV) beamformer [22], to generate a receive beam pattern with a single-null in a specific direction to achieve echo separation. In addition to the conventional LCMV beamformer, Feng et al. [16] introduced a separation strategy that incorporates the finite-impulse response (FIR) filtering process into the LCMV beamformer in order to mitigate the detrimental impacts of pulse extension loss (PEL). Zhao et al. [17] proposed an LCMV beamformer with digital scalloped beamforming (DSBF) and adaptive multiple nulls for the STWE-SAR.

However, the issue of the signal energy differences between different subswaths in real scenarios has not received significant consideration in previous research. The

LCMV beamformer and the beamformer in [16] can only suppress interference energy in a single direction. The null depth is restricted and uncontrollable. The main suppression still depends on the level of side-lobes surrounding the null. Assume a strong scatterer as interference whose signal energy exceeds the LCMV beamformer's suppression capabilities. In such a case, the remaining energy of the strong scatterer in the other subswath will degrade the separation performance. Although multiple nulls are deployed on the LCMV beamformer in [17] to increase the width of suppression, the actual suppression is not optimal, as is the case with the LCMV beamformer. Moreover, two-dimensional (2D) simulations can reflect the energy interference from azimuth directions in 2D scenes. The traces of the beam contact on the ground are circular. Targets located at different azimuth positions with varied look angles will be located on the equiphasic plane. The echoes from the targets on the trace will arrive at the antenna together with the desired signals. This complicated interference cannot be effectively suppressed by the LCMV beamformer. Consequently, the difficulties with the LCMV beamformer mentioned above could be eliminated by designing a beamformer that can generate a receive beam pattern with a flexible adjustment of the notch width and depth.

A novel echo separation scheme based on second-order cone programming (SOCP) is proposed in this paper because the null-steering constraint of the beam pattern can be summarized as a convex optimization problem [23]. The convex optimization problem can be converted to the equivalent SOCP form that can be efficiently solved by the interior point method [24]. The SOCP beamformer can constrain the side-lobes, which is meaningful for suppressing the range ambiguity to improve the image quality [12,14]. The receive pattern generated by the SOCP beamformer can achieve flexible adjustment of the notch width and depth, which ensures continuous high-intensity suppression of interference during beam steering. Even with many subswaths, the proposed SOCP beamformer can achieve high-quality separation from the overlapped echoes. Therefore, the proposed SOCP beamformer can be effectively deployed in the STWE-SAR.

The following sections are structured as follows. In Section 2, the signal energy differences in STWE-SAR are described, the deficiencies of the conventional LCMV beamformer are discussed, and the detailed design of the SOCP beamformer is given. The simulation results and the experimental separation results based on the airborne STWE-SAR data from the proposed SOCP beamformer and traditional LCMV beamformer are given in Section 3, followed by the discussion in Section 4. Section 5 draws the final conclusion of this paper.

2. Materials and Methods

In this section, the source of the signal energy differences is discussed. The deficiencies of the LCMV beamformer in separating echoes are illustrated in terms of both the depth and number of the nulls. Thus, a SOCP beamformer with flexible adjustment of the notch width and depth is proposed.

2.1. Signal Energy Differences

We assume that the antenna of STWE-SAR has N channels in elevation. The K subswaths are irradiated by K subbeams in a specific time sequence. The echoes from different subswaths arrive at the antenna within the same echo window. The angle of arrival (AoA) for the kth subswath is $\alpha_k(\tau)$. After down-conversion and range-matched filtering, the signal of the nth received channel can be approximated as:

$$s_{rn}(\eta, \tau) = \sum_{k=1}^{K} s_k(\eta, \tau) \cdot exp\left(-j\frac{2\pi d_n sin(\alpha_k(\tau) - \beta_c)}{\lambda}\right) \tag{1}$$

where η denotes the slow time, τ denotes the fast time, λ denotes the signal wavelength, d_n denotes the interval between the nth channel and the reference channel, $\theta_k(\tau) = \alpha_k(\tau) - \beta_c$ denotes the off-boresight angle of the kth subbeam, and β_c denotes the boresight angle of the antenna. $s_k(\eta, \tau)$ is the signal corresponding to the kth subswath, which can be expressed as:

$$s_k(\eta, \tau) = \sigma_k \cdot sinc\left(\tau - T_{pk} - \frac{2r_k(\eta)}{c}\right) \cdot exp\left(-j2\pi f_c T_{pk} - j4\pi \frac{r_k(\eta)}{\lambda}\right) \quad (2)$$

where σ_k denotes the backscattering coefficient corresponding to the scattering point in the kth subswath, $T_{pk} = (k-1) \cdot T_p$ denotes the transmit delay of the kth subbeam, T_p denotes the pulse duration, $r_k(\eta)$ is defined as the slope distance of the scattering point in the kth subswath, and f_c denotes the carrier frequency, and the envelope of the compressed signal in the range is $sinc(\cdot)$.

In STWE-SAR, the echoes from each subswath covered by the main lobe arrive at the antenna simultaneously. According to Equation (2), it should be noted that the signal energy from the subswath arriving at the antenna is related to the back-scattering coefficient σ_k of the scatterers in the subswath. In different scattering scenes, such as buildings and water surfaces, the energy of the signal is significantly different, as shown in Figure 2.

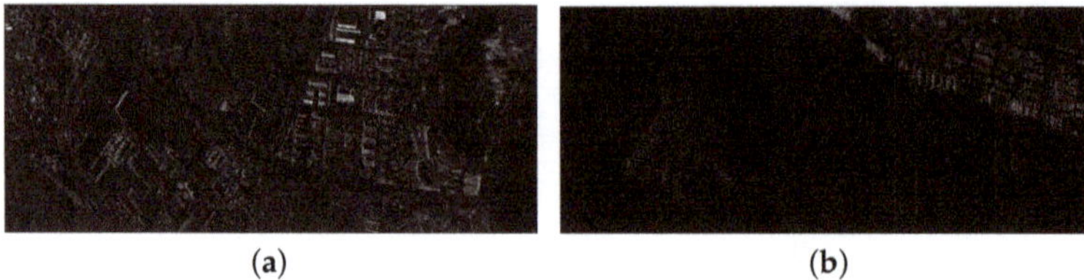

(a)   (b)

Figure 2. The SAR images with different scattering scenes. (a) Buildings. (b) Water surfaces.

2.2. Deficiencies of the LCMV Beamformer

2.2.1. LCMV Beamformer

If the received signal is expressed in the vector form, the output signal of all received channels can be written as :

$$s_r = a_s \cdot s \quad (3)$$

where

$$s_r = [s_{r1}, s_{r2}, \ldots, s_{rN}]^T, \quad (4)$$

$$s = [s_1, s_2, \ldots, s_K]^T, \quad (5)$$

$$a_s = [a_1, a_2, \ldots, a_K], \quad (6)$$

$$a_k = \left[exp\left(j\frac{2\pi d_1 sin(\theta_k(\tau))}{\lambda}\right), exp\left(j\frac{2\pi d_2 sin(\theta_k(\tau))}{\lambda}\right), \ldots, exp\left(j\frac{2\pi d_n sin(\theta_k(\tau))}{\lambda}\right)\right]^T, \quad (7)$$

$(\cdot)^T$ denotes the transpose, s denotes the receive signal matrix for all subswath, a_s denotes the receive array manifold matrix, and a_k represents the steering vector of the signal from kth subswaths (subbeams).

According to Equation (3), signals received by the antenna are the overlapped signals from different subswaths. In STWE-SAR, the interfering signals are suppressed by the single-null on the beam pattern while receiving the desired signal with a high gain. The traditional null-steering beamformer is generated based on the LCMV algorithm [22]. To obtain the vector of beamforming weighting coefficients w^H, the following constraint needs to be followed:

$$\begin{cases} \min\limits_{w} & w^H R_n w \\ s.t & w^H V = e^H \end{cases} \quad (8)$$

where $(\cdot)^H$ denotes the Hermitian transpose, $e^H = [1, 0, \ldots, 0]_{(1 \times K)}$ denotes the constraint vector, R_n denotes the spectral matrix of the thermal receiver noise, R_n can be modeled as white in the LCMV beamformer, and V represents the constraint matrix containing the one-order distortionless constraint and a $(K-1)$-order null constraint, $V \in \mathbb{C}^{N \times K}$.

$$V = [a_s^{beam}, a_s^{null}], \tag{9}$$

where a_s^{beam} denotes the receive array manifold matrix of the beam steering direction, $a_s^{beam} \in \mathbb{C}^{N \times 1}$, and a_s^{null} denotes the receive array manifold matrix of $(K-1)$-order null constraint, $a_s^{null} \in \mathbb{C}^{N \times (K-1)}$. Therefore, we can derive w^H as [22]:

$$w^H = e^H \cdot (V^H V)^{-1} \cdot V^H \tag{10}$$

We assume that the number of channels in elevation is $N = 16$ and the channel interval is $d = 0.04$ m. The direction of the desired signal is $\theta_s = 30°$. The direction of the interference signal is $\theta_j = 40°$. According to the beam pattern obtained by the LCMV beamformer in Figure 3, the LCMV beamformer can only suppress the main interfering energy at a specific angle. Because the constraint in Equation (8) does not expressly limit the depth of the null, the notch depth is uncontrollable. The single-null also leads to an insufficient notch width, which is insufficient for coping with the effects of steering deviation or stronger interference around the null.

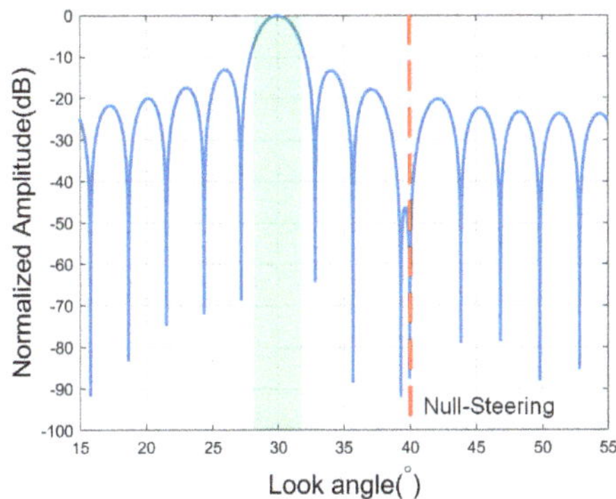

Figure 3. The single–null beam pattern based on the LCMV beamformer.

2.2.2. Restricted Notch Depth

In STWE-SAR, for a subswath, the interference signals that are overlapped can be seen as its "range ambiguity", as shown in Figure 4. The interference signals are transmitted by the main lobe of the other beam, which has much higher energy than the range ambiguity transmitted within the side-lobes. The signals to be suppressed are the superposition of interference signals and ambiguity signals, which demands the notch of the beam pattern to be deep enough at corresponding look angles. However, the notch depth of the LCMV beam pattern is uncertain, which is related to the gain of the beam pattern generated by the Scan-On-Receive (SCORE) technique at the corresponding angle of the notch on the LCMV beam pattern. The conventional SCORE beamformer and the LCMV beamformer can both be derived from the minimum variance distortionless response (MVDR) theory [22]. The difference is that the LCMV beamformer adds a null constraint. The different positions of the notch will lead to the different notch depths of the LCMV beamformer.

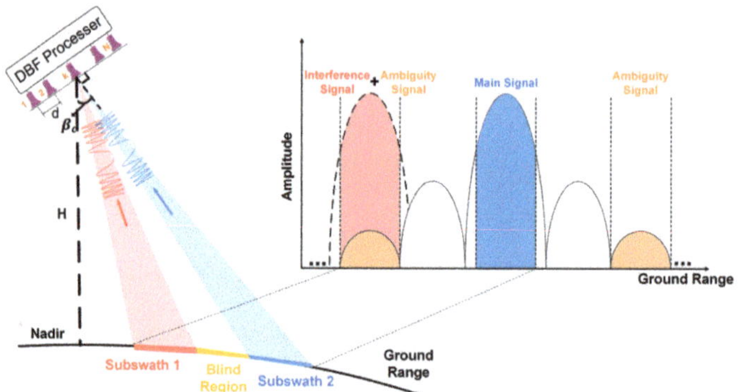

Figure 4. Diagram of the superposition of the interference signal and ambiguity signal range.

We assume that the number of channels in elevation is $N = 40$ and the channel interval is $d = 0.04$ m. Figure 5a shows the beam pattern generated by the LCMV beamformer with the null placed exactly at the notch of the beam pattern generated by SCORE. It can be seen that the LCMV beamformer can achieve an excellent suppression capability of -160 dB at this null. However, the notch depth of the other steering directions cannot be represented by this individual extreme notch depth. Only the signal in the null-steering direction is sufficiently suppressed in Figure 5a. The interference signals from the other directions are not effectively suppressed. In particular, the side-lobes near the null are still maintained at a relatively high level, which is insufficient for interference suppression. The interference signals from other subswaths in STWE-SAR will not always be located at the null. Most of the interference signals still need to be suppressed by the side-lobes. Section 2.2.3 gives a detailed analysis.

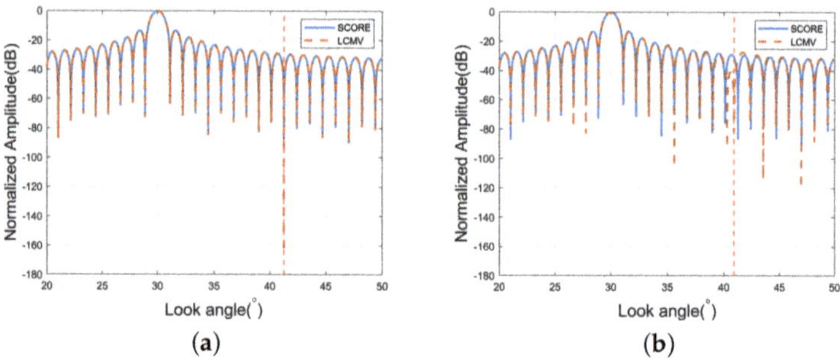

Figure 5. The beam pattern generated by the LCMV beamformer. (**a**) The null placed at the notch of the beam pattern generated by SCORE. (**b**) The null placed at the side-lobe of the beam pattern generated by SCORE.

In addition, if the null is placed on the side-lobes, as shown in Figure 5b, the suppression ability of the LCMV beamformer at the null is acceptable at -80 dB. Although the gain of the side-lobes has been reduced to -40 dB, resulting in a notch of a certain width, the overall depth of this notch does not reach -80 dB. We assume that the interference energy after being weighted by the beam pattern needs to be suppressed by the notch is still greater than 40 dB. It means that the limited depth cannot suppress the interference below the level of side-lobes when encountering huge energy differences. The residual energy will affect the final imaging results.

2.2.3. Restricted Notch Width

In addition to the limitation of the notch depth, the notch width on the beam pattern obtained by the LCMV beamformer is also an important factor limiting its capability. Since interference suppression still depends mainly on the side-lobes, the beam pattern with a high degree of suppression in a single direction, such as the beam pattern in Figure 5a, is far from sufficient. Suppose the single-null position P_{null} also exists near the strong interference target P_{inter}, as shown in Figure 6. The beam pattern in Figure 5a cannot cope with this situation well. As shown in Figure 6, the increase in notch width can assure the suppression of strong interference signals throughout a range of angles. The interfering signals are always received in the notch during the beam scanning. The separation effect can be enhanced. The increase in the notch width in Figure 5b is a suitable way to cope with it. However, it has limited notch depth.

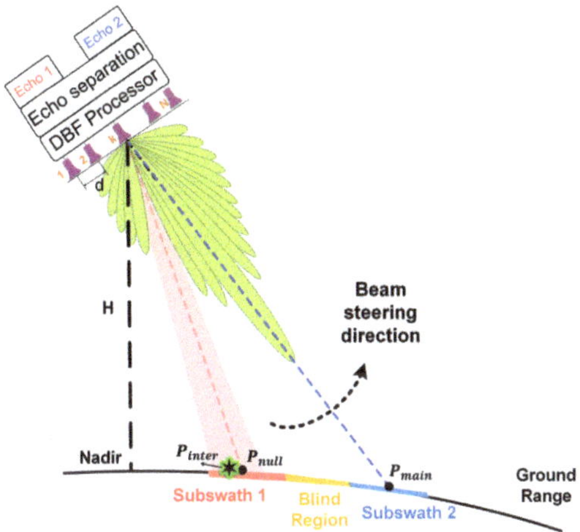

Figure 6. Diagram of the suppression of interference energy with a wide notch.

In addition, there is another case, as shown in Figure 7, where due to the particular imaging geometry of the satellite-ground, the interference signals include not only the signals of P_{null} but also the signals of targets located on the equiphasic plane (The boundary of the beam when it touches the ground is circular called the equiphasic plane) of P_{null} [25]. The signals from targets located on the equiphasic plane will simultaneously arrive at the antenna from different look angles. It is important to note that these signals cannot be handled by range migration correction and range compression in the imaging of the current subswath since the projection of the slant range in the range of these interfering signals is distinct from the center slope range of the desired subswath. The suppression of interference from these targets cannot be achieved with the single-null beam pattern. Therefore, it is necessary to widen the notch so that it can cover the angular range of these interferences in the green region, as shown in Figure 7.

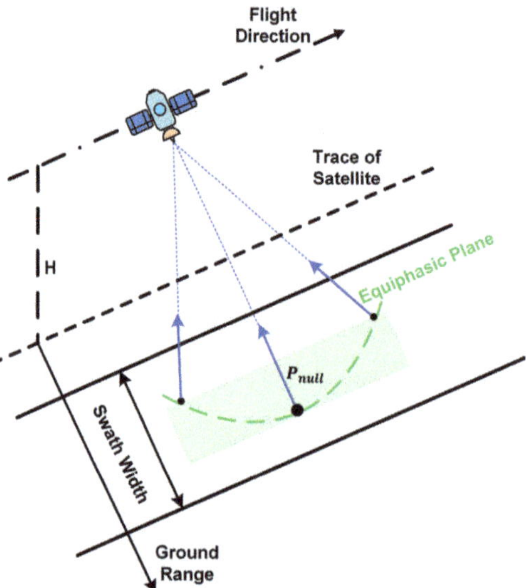

Figure 7. Diagram of the interference signals from the equiphasic plane.

2.3. Proposed SOCP Beamformer

2.3.1. SOCP Theory

SOCP is a type of convex optimization [26] and is often used for beamforming optimization [27], which is used to solve a certain linear function minimization problem of the intersection of an affine set and the product of second-order cones. The standard SOCP optimization problem can be expressed as:

$$\min_{y} \quad b^T y$$
$$s.t \quad \|A_j y + b_j\| \leq c_j^T y + d_j, \quad j = 1, 2, \cdots, J \tag{11}$$
$$Fy = g$$

where $y \in \mathbb{R}^q$ is the optimization variable, $b \in \mathbb{R}^q$, $A_j \in \mathbb{R}^{(q_j-1)\times q}$, $b_j \in \mathbb{R}^{q_j-1}$, $c_j \in \mathbb{R}^a$, $c_j^T y \in \mathbb{R}$, $d_j \in \mathbb{R}$, $b_j \in \mathbb{R}^{q_j-1}$, $F \in \mathbb{C}^{g \times a}$, $g \in \mathbb{C}^g$, $\| \bullet \|$ is the standard Euclidean norm. The inequality constraint in Equation (11) can be expressed as a second-order cone of dimension q_j:

$$\begin{bmatrix} c_j^T \\ A_j \end{bmatrix} y + \begin{bmatrix} d_j \\ b_j \end{bmatrix} \in SOC_j^{q_j} \tag{12}$$

where the q_j-dimension second-order cone $SOC_j^{q_j}$ is defined as:

$$SOC_j^{q_j} \triangleq \left\{ \begin{bmatrix} t \\ x \end{bmatrix} \middle| t \in \mathbb{R}, x \in \mathbb{C}^{q_j-1}, t \leq \|x\| \right\} \tag{13}$$

where t is a real scalar, and x is a complex $(q_j - 1)$-dimensional vector. The second-order cone is hence convex since the affine mapping does not change the convexity. The equality constraints can be expressed as zero cones:

$$g - Fy \in \{0\}^g \tag{14}$$

where the zero cone is defined as:

$$\{0\}^g \triangleq \{x \| x \in \mathbb{C}^g, x = 0\}. \tag{15}$$

Furthermore, the optimization problem in Equation (11) can be transformed into the dual standard form of the convex conic optimization problem [24]:

$$\max_{y} \quad b^T y$$
$$s.t \quad c - A^T y \in \mathcal{K} \tag{16}$$

where $\mathcal{K}$ is a symmetric cone consisting of the second-order cone, the zero cone, and the set of positive real numbers $\mathbb{R}^+$:

$$\mathcal{K} = \mathbb{R}^+ \times \{0\}^g \times SOC_1^{q_1} \times \cdots \times SOC_J^{q_J} \tag{17}$$

Each symmetric cone corresponds to a constraint.

2.3.2. Design of the Beamformer Based on SOCP Theory

In the practical application of STWE-SAR, the received beam pattern is required to form a notch with a certain width and depth in the interference direction. Therefore, it is necessary to design an optimized beam pattern to meet the requirements. The key to the design is obtaining the optimal weighting coefficients w. The comprehensive design of the optimized beam pattern includes the control of the beam steering, the constraint of the side-lobes, the constraint of the notch, and the constraint of the weighting coefficients. The optimization of the beam pattern can be expressed as:

$$\min_{w} \quad \tilde{\xi}_3$$
$$\begin{aligned}
s.t \quad & a^H(\theta_0)w = 1 \\
& |a^H(\theta_{j_{SL}})w| \leq \tilde{\xi}_1, \qquad j_{SL} = 1, \cdots, J_{SL} \\
& |a^H(\theta_{j_{NL}})w| \leq \tilde{\xi}_2, \qquad j_{NL} = 1, \cdots, J_{NL} \\
& \|w\|^2 \leq \tilde{\xi}_3
\end{aligned} \tag{18}$$

where θ_0 denotes the desired direction of the beam, $\theta_{j_{SL}}$ denotes the direction of the side-lobe area Θ_{SL}, θ_{NL} denotes the direction of the notch area Θ_{NL}, $\tilde{\xi}_1$ is the constraint of the side-lobe level, $\tilde{\xi}_2$ is the constraint of the notch depth, and $\tilde{\xi}_3$ is the constraint of weighting coefficients. It can be seen in Equation (18) that the optimized beam pattern is designed to ensure a distortionless response in the desired direction with the minimum norm of w. The side-lobe in θ_{SL} and the notch in θ_{NL} are also constrained.

According to Equation (16), the optimization in Equation (18) can be converted into the following form:

$$\min_{y} \quad y_1$$
$$\begin{aligned}
s.t \quad & a^H(\theta_0)y_4 = 1 \\
& y_1 = \sqrt{\tilde{\xi}_3} \\
& |a^H(\theta_{SL})y_4| \leq y_2, \qquad j_{SL} = 1, \cdots, J_{SL} \\
& |a^H(\theta_{NL})y_4| \leq y_3, \qquad j_{NL} = 1, \cdots, J_{NL} \\
& \|y_4\| \leq y_1
\end{aligned} \tag{19}$$

where $y_1 = \tilde{\xi}_3$, $y_2 = \tilde{\xi}_1$, $y_3 = \tilde{\xi}_2$, $y_4 = w$, and $y = [y_1, y_2, y_3, y_4^T]^T$. We define $b \triangleq [-1, 0, \cdots, 0]^T$ so that $-y_1 = b^T y$. The first two equalities can be represented as two zero-cone constraints:

$$\begin{pmatrix} \sqrt{\tilde{\xi}_3} - y_1 \\ 1 - a^H(\theta_0)y_4 \end{pmatrix} = \begin{pmatrix} \sqrt{\tilde{\xi}_3} \\ 1 \end{pmatrix} - \begin{pmatrix} 1 & 0 & 0 & \mathbf{0}^T \\ 0 & 0 & 0 & a^H(\theta_0) \end{pmatrix} y \triangleq c_1 - A_1^T y \in \{0\}^2. \tag{20}$$

The three inequalities can be expressed as SOC constraints:

$$\begin{pmatrix} y_1 \\ y_4 \end{pmatrix} = \begin{pmatrix} 0 \\ \mathbf{0} \end{pmatrix} - \begin{pmatrix} -1 & 0 & 0 & \mathbf{0}^T \\ 0 & 0 & 0 & -\mathbf{I}^T \end{pmatrix} y \triangleq c_2 - A_2^T y \in SOC^{N+1}. \tag{21}$$

$$\begin{pmatrix} y_2 \\ a^H(\theta_{j_{SL}})y_4 \end{pmatrix} = \begin{pmatrix} 0 \\ \mathbf{0} \end{pmatrix} - \begin{pmatrix} 0 & -1 & 0 & \mathbf{0}^T \\ 0 & 0 & 0 & -a^H(\theta_{j_{SL}}) \end{pmatrix} y \triangleq c_{j_{SL}+2} - A_{j_{SL}+2}^T y \in SOC^2, j_{SL} = 1, \cdots, J_{SL} \tag{22}$$

$$\begin{pmatrix} y_3 \\ a^H(\theta_{j_{NL}})y_4 \end{pmatrix} = \begin{pmatrix} 0 \\ \mathbf{0} \end{pmatrix} - \begin{pmatrix} 0 & 0 & -1 & \mathbf{0}^T \\ 0 & 0 & 0 & -a^H(\theta_{j_{NL}}) \end{pmatrix} y \triangleq c_{j_{NL}+2+J_{SL}} - A_{j_{NL}+2+J_{SL}}^T y \in SOC^2, j_{NL} = 1, \cdots, J_{NL}. \tag{23}$$

Let

$$c \triangleq [c_1, c_2, \cdots, c_{J_{SL}+2}, \cdots, c_{J_{NL}+J_{SL}+2}] \tag{24}$$

and

$$A^T \triangleq [A_1^T, A_2^T, \cdots, A_{J_{SL}+2}^T, \cdots, A_{J_{NL}+J_{SL}+2}^T] \tag{25}$$

are defined based on the above equations. The symmetric cone corresponding to the constraints can then be expressed as:

$$\mathcal{K} = \{0\}^2 \times SOC^{N+1} \times \cdots \times SOC_{J_{SL}}^2 \times \cdots \times SOC_{J_{NL}}^2. \tag{26}$$

Usually, the SOCP problem can be efficiently solved by the interior point method [26]. The optimal value of vector y can be easily obtained by the Sedumi convex optimization toolbox [24], which applies the interior point method. Then, y_4 in y is the optimal weighting coefficient of the SOCP beamformer.

We assume the channels in elevation are $N = 40$ and the beam steering angle is $\theta_0 = 30°$. The side-lobe area on the beam pattern is $\Theta_{SL} = [-60°, 28.5°] \cup [31.5°, 120°]$, and $\theta_{j_{SL}}$ is evenly distributed over the area. The notch area is placed at $\Theta_{NL} = [38°, 40°] \cup [48°, 50°]$, and $\theta_{j_{NL}}$ is also chosen with uniform spacing. $\xi_1 = 10^{-3}$ represents that the side-lobe level is required to be below -30 dB. $\xi_2 = 10^{-12}$ represents that the notch level will not exceed -120 dB. Figure 8 shows the beam pattern generated by the SOCP beamformer. The beam pattern allows the adjustment of the side-lobe level and has a sizable notch area that combines the width and depth. The beamformer designed based on SOCP theory has a strong adaptive capability and flexibility. Therefore, the SOCP beamformer is well suited for STWE-SAR to cope with the deficiencies of the LCMV beamformer for echo separation.

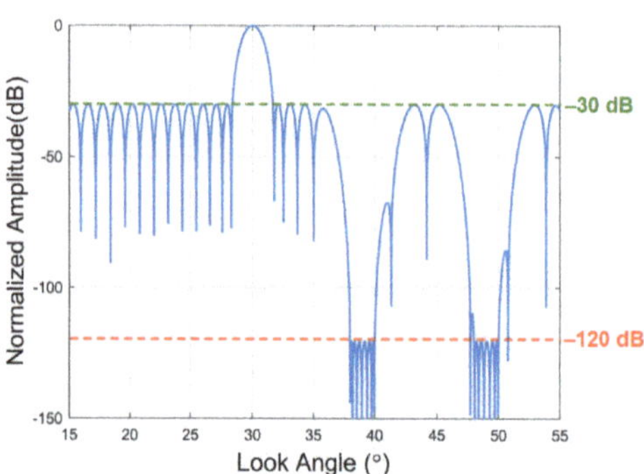

Figure 8. The beam pattern with wide and deep notches based on the SOCP beamformer.

3. Results

The spaceborne STWE-SAR was simulated to illustrate the advantages of the proposed echo separation scheme. In addition, the energy differences in the signals in different scenarios are shown based on the airborne STWE-SAR data. This is followed by the echo separation results using real data to validate the efficacy of the SOCP beamformer for echo separation.

3.1. Simulation Results

A spaceborne STWE-SAR system with three subswaths is demonstrated to evaluate the performance of the proposed method. The parameters of the simulated DBF-SAR system are given in Table 1. Three beam positions (marked in red) are illustrated in Figure 9a, which represents three subswaths. The timing diagram in Figure 9a shows the restrictions on the receive window due to transmit instances (marked in blue) and nadir echo (marked in green). The PRF and look angle values of the three beam positions are shown in Table 2. The geometric relationships among the target points are shown in Figure 9b. The energy of the target points in subswaths 1 and 2 is 40 and 20 dB higher than that in subswath 3, respectively, which is used to reflect the effect of energy differences in the echo separation results. The echoes of the three subswaths overlap in the time domain. Then, the overlapped echo is separated by the echo separation scheme.

Table 1. System Parameters of the Simulated STWE-SAR System.

Parameter	Value
Orbit height (km)	700
Platform velocity (m/s)	7474
Carrier frequency (GHz)	9.6
Signal bandwidth (MHz)	100
Pulse duration (μs)	10
Oversampling rate	1.2
Antenna height (m)	1.6
Antenna length (m)	9.6
Numbers of channels	40
PRF (Hz)	1550
Look angle of antenna normal direction (°)	30

Table 2. Information about Beam Positions.

Index	PRF (Hz)	Look Angle (°)
1	1550	[28.97, 32.92]
2	1550	[37.35, 39.91]
3	1550	[42.97, 44.82]

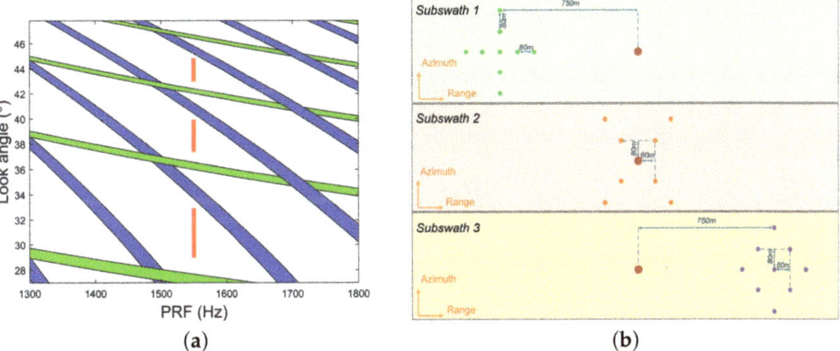

Figure 9. (**a**) The imaging diagram of the simulated system. (**b**) The imaging scenario for three subswaths.

According to the parameters in Tables 1 and 2, the beam pattern generated by the conventional LCMV beamformer can be obtained, as shown in Figure 10. It can be seen that the width of the notch in the LCMV beam pattern is small, and the depth is shallow. In contrast, the notch in the beam pattern obtained by the SOCP beamformer is wide and deep, as shown in Figure 11. The side-lobes are constrained to below −25 dB. The width of the notch is an angular pulse width [17], and the depth is below −100 dB. The gain level of the notch of the SOCP beamformer is much lower than that of the LCMV beamformer.

The final 3D and 2D separation results obtained by the LCMV beamformer and the SOCP beamformer are presented in Figure 12. Due to the limited suppression of interference by the LCMV beamformer and the lower signal energy than the interference energy, the interference energy from the other two subswaths produces obvious curved lines along the azimuth in the 2D results and the significant bulges in the 3D results, as shown in Figure 12b,c compared to Figure 12a. The interference in Figure 12a is not obvious because the interference energy levels in the other subswaths are much lower than the desired signals and can be suppressed to below the level of the side-lobes. In sharp contrast, the results obtained from the SOCP beamformer are effective for suppressing interference, as shown in Figure 12d–f. According to the results, it can be seen that the interference energy over −100 dB no longer affects the final imaging results. The 2D and 3D results illustrate the superiority of the proposed method over the conventional separation scheme based on the LCMV beamformer.

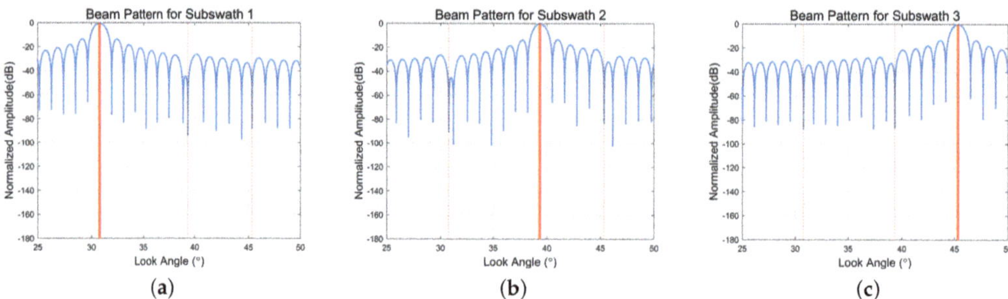

Figure 10. The beam pattern generated by the conventional LCMV beamformer. (**a**–**c**) The beam pattern for subswath 1–3.

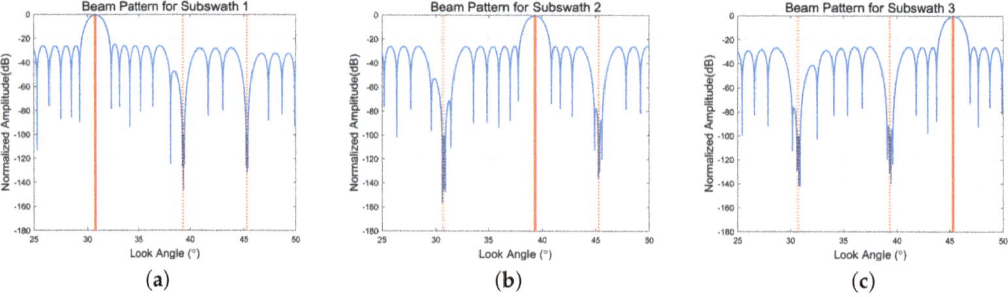

Figure 11. The beam pattern generated by the proposed SOCP beamformer. (**a**–**c**) The beam pattern for subswath 1–3.

To give a numerical comparison, the suppressing degree of the compressed interference energy of the center point is recorded in Table 3. Since the signal energy is highest in subswath 1, both LCMV and SOCP beamformers provide effective suppression of the interference energy from subswaths 2 and 3. Nevertheless, when the signal energy in the interference region is greater than the signal energy in the desired subswath, the suppression degree of the LCMV beamformer is greatly reduced compared to the SOCP beamformer. When the desired subswaths are subswaths 2 and 3, the suppression of the SOCP beamformer in subswath 1 is at least 15.2 and 39.1 dB higher than that of the LCMV beamformer, respectively. The echo separation capability of the SOCP beamformer with a wide and deep notch is significantly superior to that of the LCMV beamformer.

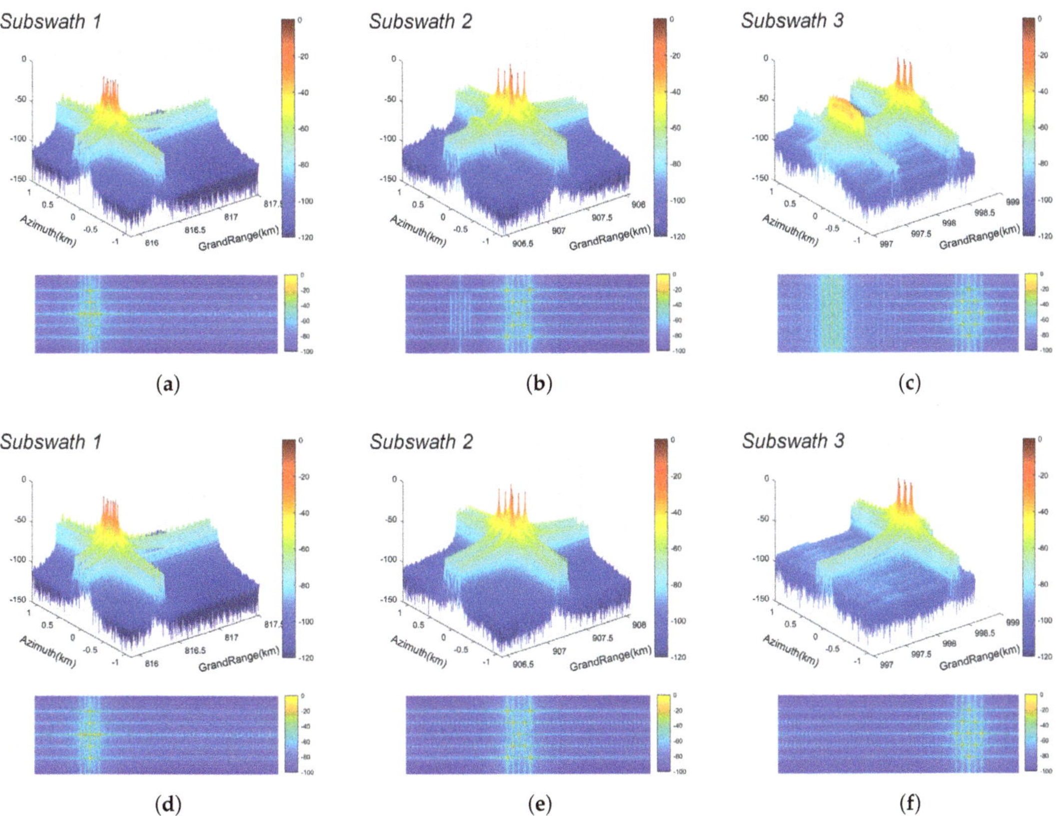

Figure 12. The final 3D and 2D separation results. (**a**–**c**) The final 3D and 2D separation results obtained by the LCMV beamformer. (**d**–**f**) The final 3D and 2D separation results obtained by the SOCP beamformer.

Table 3. Compressed Energy at the Angle of the Interference Signal for the Center Point.

Beamformer	Desired Subswath	Interference Subswath		
		1	2	3
LCMV	1	╲	−55.36 dB	−59.70 dB
	2	−47.72 dB	╲	−66.33 dB
	3	−23.33 dB	−54.10 dB	╲
SOCP	1	╲	−55.36 dB	−59.70 dB
	2	−62.96 dB	╲	−66.37 dB
	3	−62.45 dB	−58.03 dB	╲

3.2. Experimental Results

In this part, the experiment's results of the X-band airborne STWE-SAR are presented to verify the effectiveness of the proposed scheme and illustrate the effect of signal energy differences on echo separation. As shown in Figure 13, subbeams 1 and 2 are designed in the range direction, which takes inter-pulse switching to points at 60° and 70°, in turn, corresponding to subswath 1 and subswath 2, respectively. The scenes consisted primarily of ponds, hills, farming, and human-made structures. The detailed system parameters are shown in Table 4. Since it is an airborne SAR, the echoes can reach the antenna within one PRT. The echoes from the two scenes are stored on the recorder in odd and even frame formats. The sampled data from odd and even frames corresponds to the echoes of subswaths 2 and 1, respectively. The echoes of the two subbeams are artificially superimposed in the time domain as the overlapped echoes received by the STWE mode. The data processing flow is shown in Figure 14. In this flow, the error correction method in [28] is applied before the echo overlaps.

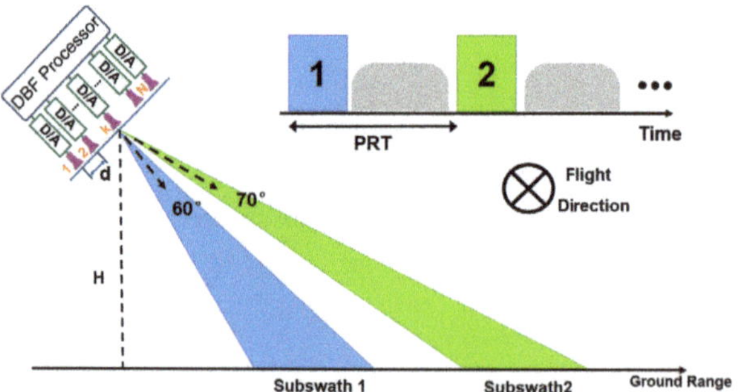

Figure 13. Real scheme.

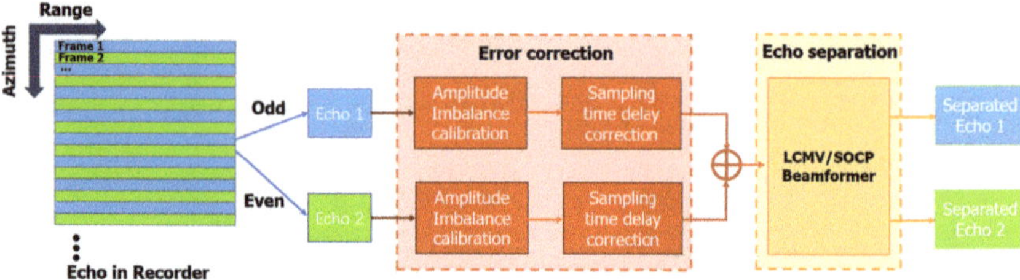

Figure 14. The data processing flow of the real data.

Table 4. Parameters of the Airborne STWE-SAR System.

Parameter	Value
Height (m)	4200
Platform velocity (m/s)	80
Carrier frequency (GHz)	9.6
Signal bandwidth (MHz)	500
Pulse duration (μs)	10
Oversampling rate	1.2
Antenna height (m)	0.32
Antenna length (m)	0.496
Numbers of channels	16
PRF (Hz)	1500
Look angle of antenna normal direction (°)	65

3.2.1. Effect of Signal Energy Differences

First, the real data are used to illustrate the effect of energy differences on the separation results. The single-channel imaging results of the two subswaths before echo separation are shown in Figure 15a,b. Figure 16a,b give the energy plots of the single-channel imaging results before overlapping the two subswaths. The scenes in subswath 1 mainly contain farmland and ponds. Except for the ponds, the overall signal energy from subswath 1 is at a high level. In subswath 2, there are a lot of human-made structures that have strong signal energy and show up brighter in the images. The imaging result is dark and the overall signal energy is low since there are fewer strong scattering points in the scene.

Furthermore, Figure 16c reflects the difference in energy between the two subswaths. The majority of subswath 2's energy is about 80 dB less than that of subswath 1, causing the separation result of subswath 2 to be affected by the energy from subswath 1. Figure 17a,b show the separation results of the LCMV beamformer in the presence of significant energy differences between two sub-beams. The residual interference energy can be observed in the separation results of both subswaths. Subswath 1, with higher overall energy, leaves a larger range of interference energy in the separation results of subswath 2, as shown in Figure 17b. Region A reflects the influences of the human-made structures in subswath 1 on subswath 2. Moreover, the energy of the strong scattering point in subswath 2 also has a clear effect on the separation results of subswath 1, as shown in regions B and C in Figure 17a. The results illustrate that the LCMV beamformer has a limited ability to suppress strong interference signals and cannot effectively separate the echoes from such a significant energy difference.

Figure 15. The single-channel imaging results before echo separation. (**a**) Subswath 1. (**b**) Subswath 2.

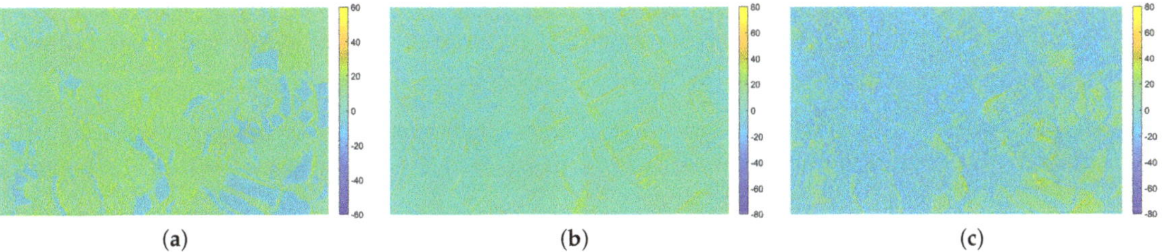

Figure 16. The single-channel energy. (**a**) Subswath 1. (**b**) Subswath 2. (**c**) Energy difference (subswath 2 − subswath 1).

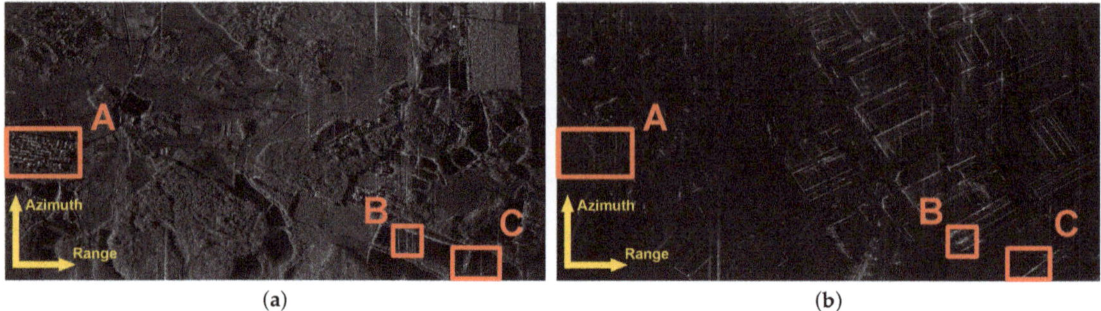

(**a**) (**b**)

Figure 17. The echo separation results of the LCMV beamformer. Region A reflects the separation results of human-made structures and roads. Region B and C compare the effect of the separation of water and human-made structures. (**a**) Subswath 1. (**b**) Subswath 2.

3.2.2. Comparison of Echo Separation Results

In order to improve the effectiveness of the traditional echo separation scheme based on the LCMV beamformer, we give the results of the echo separation scheme based on the SOCP beamformer, as shown in Figure 18. It should be noted that due to the wide beam of this airborne SAR system and the close distance between the two subswaths, the side-lobes are not constrained. The notch width is 1/10 of the angular pulse width, about $0.5°$, and the notch depth is below -80 dB. The separation results of the SOCP beamformer in Figure 18 are much improved compared to the results in Figure 17. Specifically, the special regions A, B, and C in the results are selected to reflect the difference in the separation effect of the two schemes.

Compared to Figure 17a, the interference from subswath 1 in region A has been significantly decreased, as shown in Figure 18a. Moreover, regions B and C both reflect the residual energy of the strong scattering points in subswath 2 at the weaker scatterers in subswath 1. Figure 19 gives the results of a comparison between the two schemes in region C. The residual energy in Figure 19b has been effectively lowered by the SOCP beamformer, as visible in great detail.

In brief, the results in Figures 18 and 19 illustrate that the echo separation scheme based on the SOCP beamformer can obtain better separation results with its wide and deep notch design to further suppress the interference than the LCMV beamformer. The scheme is promising for future STWE-SAR missions.

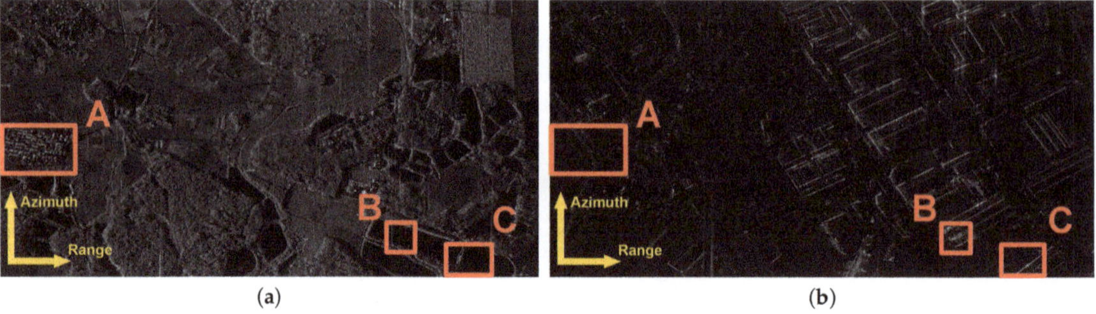

(**a**) (**b**)

Figure 18. The echo separation results of the SOCP beamformer. (**a**) Subswath 1. (**b**) Subswath 2.

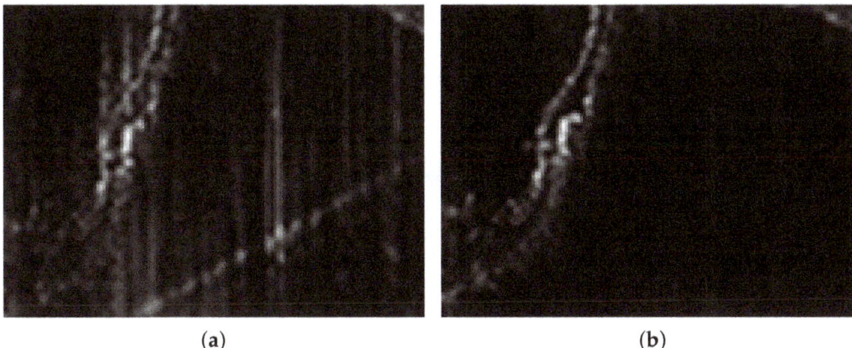

(a)   (b)

Figure 19. (**a**) Special region C selected by the red rectangles in Figure 17a. (**b**) Special region C selected by the red rectangles in Figure 18a.

4. Discussion

The traditional echo separation scheme based on the LCMV beamformer achieves separation by suppressing the interfering signal in a single direction, which will modulate the envelope of the echo. However, due to the special imaging geometry of SAR, interference signals from equiphasic planes also exist, as shown in Figure 7. The angles of these interferences are different from the angle of the null, which requires a wider width of the notch. In addition, the interfering signal is located in the range ambiguity region, and the energy to be suppressed is actually the sum of the energy of the ambiguous signal and the interfering signal, as shown in Figure 4, which requires a higher demand of the depth of the notch. As shown in Figure 3, the interference suppression capability of the side-lobes of the LCMV beamformer does not exceed −40 dB. When the level of interference energy is much higher than the desired signal, the residual energy will appear in the results after range compression. Both simulation results in Figure 12 and experimental results in Figure 17 confirm the shortcomings of the conventional LCMV beamformer for suppressing interference signals and face the problem of a finite notch width and uncontrollable notch depth.

The scheme proposed in this paper uses the second-order cone programming approach to solve the beam pattern optimization problem with a wide and deep notch. As shown in Figure 8, the beam pattern obtained by the SOCP beamformer has a settable deep notch in a specified angular range, which is designed to suppress interference signals from multiple angles to below a fixed energy level, achieving controlled and effective interference suppression. In addition, the side-lobes can be constrained to achieve improved range ambiguity performance. Compared to the LCMV beamformer, the SOCP beamformer can suppress up to 39 dB of additional interference energy in the simulation results, as shown in Figure 12. In the experimental results shown in Figure 18, the SOCP beamformer also achieves better separation results.

In STWE-SAR, the length of the echo window is constrained by the PRF, which limits the maximum continuous swath width and leads to the appearance of blind regions, as shown in Figure 1 [29]. Benefiting from the development of the variable PRF technique or Staggered-SAR [30,31], the problem of blind regions has been greatly improved. However, the echo separation in Staggered-SAR is still needed. Relying only on the space-domain filtering characteristics of DBF-SCORE to achieve echo separation is insufficient. The proposed scheme can be used to further improve the system performance of future Staggered-SAR systems.

In the dispersed SAR system based on the floating swarm concept [32], the signal is received using a small satellite as a channel at elevation and then synthesized on the ground to reap high-resolution wide-swath imaging. The STWE mode can be adopted to transmit and cover multiple consecutive subswaths based on this dispersed SAR system, which can not only achieve the elimination of blind regions but also reduce the amount of ultra-wide echo data. Therefore, the echo separation scheme proposed in this paper can also provide sufficient support for future dispersed SAR systems in the echo separation to achieve HRWS imaging.

5. Conclusions

In this study, the problem of the energy difference between the interfering signal and the desired signal in STWE-SAR is pointed out for the first time. Based on this analysis, the deficiencies of the traditional echo separation based on the LCMV beamformer in terms of notch depth and width are obtained, which affect the quality of the final separation results. To address the deficiencies of the traditional scheme, an echo separation scheme based on the SOCP beamformer was proposed, which can constrain the depth and width of the notch to improve the separation quality and can also constrain the side-lobes of the beam pattern as needed to improve the system performance. The simulation results validate that the performance of the SOCP beamformer is significantly better than that of the traditional LCMV beamformer. The experimental results also verify the superior performance of the scheme. In the future, there will be more in-depth research on the application of echo separation schemes in HRWS-SAR systems.

Author Contributions: Conceptualization, S.H. and J.Q.; methodology, S.H.; validation, S.H., Q.Z. and W.W.; formal analysis, S.H. and Y.Z.; investigation, Y.D.; resources, Y.D.; data curation, Z.C.; writing—original draft preparation, S.H. All authors have read and agreed to the published version of the manuscript.

Funding: This work was funded by the National Natural Science Foundation of China under grant numbers 61971401 and in part by the Youth Innovation Promotion Association, Chinese Academy of Sciences (CAS).

Data Availability Statement: Not applicable.

Conflicts of Interest: The authors declare no conflict of interest.

References

1. Wollstadt, S.; Lopez-Dekker, P.; De Zan, F.; Younis, M. Design principles and considerations for spaceborne ATI SAR-based observations of ocean surface velocity vectors. *IEEE Trans. Geosci. Remote Sens.* **2017**, *55*, 4500–4519. [CrossRef]
2. De Almeida, F.Q.; Younis, M.; Krieger, G.; Moreira, A. Multichannel staggered SAR azimuth processing. *IEEE Trans. Geosci. Remote Sens.* **2018**, *56*, 2772–2788. [CrossRef]
3. Baumgartner, S.V.; Krieger, G. Simultaneous high-resolution wide-swath SAR imaging and ground moving target indication: Processing approaches and system concepts. *IEEE J. Sel. Top. Appl. Earth Obs. Remote Sens.* **2015**, *8*, 5015–5029. [CrossRef]
4. Wen, Y.; Zhang, Z.; Chen, Z.; Qiu, J.; Ren, M.; Meng, X. A Novel Time-Domain Frequency Diverse Array HRWS Imaging Scheme for Spotlight SAR. *Remote Sens.* **2022**, *14*, 1085. [CrossRef]
5. Freeman, A.; Johnson, W.T.; Huneycutt, B.e.a.; Jordan, R.; Hensley, S.; Siqueira, P.; Curlander, J. The "Myth" of the minimum SAR antenna area constraint. *IEEE Trans. Geosci. Remote Sens.* **2000**, *38*, 320–324. [CrossRef]
6. Krieger, G.; Gebert, N.; Moreira, A. Unambiguous SAR signal reconstruction from nonuniform displaced phase center sampling. *IEEE Geosci. Remote Sens. Lett.* **2004**, *1*, 260–264. [CrossRef]
7. Gebert, N.; Krieger, G.; Moreira, A. Digital beamforming on receive: Techniques and optimization strategies for high-resolution wide-swath SAR imaging. *IEEE Trans. Aerosp. Electron. Syst.* **2009**, *45*, 564–592. [CrossRef]
8. Werninghaus, R.; Buckreuss, S. The TerraSAR-X mission and system design. *IEEE Trans. Geosci. Remote Sens.* **2009**, *48*, 606–614. [CrossRef]
9. Davidson, M.W.; Furnell, R. ROSE-L: Copernicus l-band SAR mission. In Proceedings of the 2021 IEEE International Geoscience and Remote Sensing Symposium IGARSS, Brussels, Belgium, 11–16 July 2021; pp. 872–873.
10. Geudtner, D.; Tossaint, M.; Davidson, M.; Torres, R. Copernicus Sentinel-1 Next Generation Mission. In Proceedings of the 2021 IEEE International Geoscience and Remote Sensing Symposium IGARSS, Brussels, Belgium, 11–16 July 2021; pp. 874–876.
11. Motohka, T.; Kankaku, Y.; Miura, S.; Suzuki, S. ALOS-4 L-band SAR observation concept and development status. In Proceedings of the IGARSS 2020–2020 IEEE International Geoscience and Remote Sensing Symposium, Waikoloa, HI, USA, 26 September–2 October 2020; pp. 3792–3794.
12. Huber, S.; de Almeida, F.Q.; Villano, M.; Younis, M.; Krieger, G.; Moreira, A. Tandem-L: A technical perspective on future spaceborne SAR sensors for Earth observation. *IEEE Trans. Geosci. Remote Sens.* **2018**, *56*, 4792–4807. [CrossRef]

13. Pinheiro, M.; Prats, P.; Villano, M.; Rodriguez-Cassola, M.; Rosen, P.A.; Hawkins, B.; Agram, P. Processing and performance analysis of NASA-ISRO SAR (NISAR) staggered data. In Proceedings of the IGARSS 2019–2019 IEEE International Geoscience and Remote Sensing Symposium, Yokohama, Japan, 28 July–2 August 2019; pp. 8374–8377.
14. Bordoni, F.; López-Dekker, P.; Krieger, G. Beam-Switch Wide-Swath Mode for Interferometrically Compatible Single-Pol and Quad-Pol SAR Products. *IEEE Geosci. Remote Sens. Lett.* **2018**, *15*, 1565–1569. [CrossRef]
15. Krieger, G.; Gebert, N.; Moreira, A. Multidimensional waveform encoding: A new digital beamforming technique for synthetic aperture radar remote sensing. *IEEE Trans. Geosci. Remote Sens.* **2007**, *46*, 31–46. [CrossRef]
16. Feng, F.; Li, S.; Yu, W.; Huang, P.; Xu, W. Echo separation in multidimensional waveform encoding SAR remote sensing using an advanced null-steering beamformer. *IEEE Trans. Geosci. Remote Sens.* **2012**, *50*, 4157–4172. [CrossRef]
17. Zhao, Q.; Zhang, Y.; Wang, W.; Deng, Y.; Yu, W.; Zhou, Y.; Wang, R. Echo separation for space-time waveform-encoding SAR with digital scalloped beamforming and adaptive multiple null-steering. *IEEE Geosci. Remote Sens. Lett.* **2020**, *18*, 92–96. [CrossRef]
18. Kim, J.H.; Younis, M.; Moreira, A.; Wiesbeck, W. A novel OFDM chirp waveform scheme for use of multiple transmitters in SAR. *IEEE Geosci. Remote Sens. Lett.* **2012**, *10*, 568–572. [CrossRef]
19. Krieger, G. MIMO-SAR: Opportunities and pitfalls. *IEEE Trans. Geosci. Remote Sens.* **2013**, *52*, 2628–2645. [CrossRef]
20. Jin, G.; Deng, Y.; Wang, W.; Wang, R.; Zhang, Y.; Long, Y. Segmented phase code waveforms: A novel radar waveform for spaceborne MIMO-SAR. *IEEE Trans. Geosci. Remote Sens.* **2020**, *59*, 5764–5779. [CrossRef]
21. Rommel, T.; Rincon, R.; Younis, M.; Krieger, G.; Moreira, A. Implementation of a MIMO SAR Imaging Mode for NASA's Next Generation Airborne L-Band SAR. In Proceedings of the EUSAR 2018; 12th European Conference on Synthetic Aperture Radar, Aachen, Germany, 4–7 June 2018; pp. 1–5.
22. Trees, H. *Optimum Array Processing: Detection, Estimation, and Modulation Theory, Ser. Detection, Estimation, and Modulation Theory*; Wiley-Interscience: Hoboken, NJ, USA, 2004. .
23. Liu, J.; Gershman, A.B.; Luo, Z.Q.; Wong, K.M. Adaptive beamforming with sidelobe control: A second-order cone programming approach. *IEEE Signal Process. Lett.* **2003**, *10*, 331–334.
24. Sturm, J.F. Using SeDuMi 1.02, a MATLAB toolbox for optimization over symmetric cones. *Optim. Methods Softw.* **1999**, *11*, 625–653. [CrossRef]
25. Cumming, I.G.; Wong, F.H. Digital processing of synthetic aperture radar data. *Artech House* **2005**, *1*, 108–110.
26. Lobo, M.S.; Vandenberghe, L.; Boyd, S.; Lebret, H. Applications of second-order cone programming. *Linear Algebra Appl.* **1998**, *284*, 193–228. [CrossRef]
27. Lorenz, R.G.; Boyd, S.P. Robust minimum variance beamforming. *IEEE Trans. Signal Process.* **2005**, *53*, 1684–1696. [CrossRef]
28. Zhou, Y.; Wang, W.; Chen, Z.; Wang, P.; Zhang, H.; Qiu, J.; Zhao, Q.; Deng, Y.; Zhang, Z.; Yu, W.; et al. Digital beamforming synthetic aperture radar (DBSAR): Experiments and performance analysis in support of 16-channel airborne X-band SAR data. *IEEE Trans. Geosci. Remote Sens.* **2020**, *59*, 6784–6798. [CrossRef]
29. Yang, T.; Lv, X.; Wang, Y.; Qian, J. Study on a novel multiple elevation beam technique for HRWS SAR system. *IEEE J. Sel. Top. Appl. Earth Obs. Remote Sens.* **2015**, *8*, 5030–5039. [CrossRef]
30. Krieger, G.; Younis, M.; Huber, S.; Bordoni, F.; Patyuchenko, A.; Kim, J.; Laskowski, P.; Villano, M.; Rommel, T.; Lopez-Dekker, P.; et al. Digital beamforming and MIMO SAR: Review and new concepts. In Proceedings of the EUSAR 2012; 9th European Conference on Synthetic Aperture Radar, Nuremberg, Germany, 23–26 April 2012; pp. 11–14.
31. Villano, M.; Krieger, G.; Moreira, A. Staggered SAR: High-resolution wide-swath imaging by continuous PRI variation. *IEEE Trans. Geosci. Remote Sens.* **2013**, *52*, 4462–4479. [CrossRef]
32. Mittermayer, J.; Krieger, G. Floating swarm concept for passive Bi-static SAR satellites. In Proceedings of the EUSAR 2018; 12th European Conference on Synthetic Aperture Radar, Aachen, Germany, 4–7 June 2018; pp. 1–6.

Article

Generation of Multiple Frames for High Resolution Video SAR Based on Time Frequency Sub-Aperture Technique

Congrui Yang [1,2], Zhen Chen [1], Yunkai Deng [1,2], Wei Wang [1], Pei Wang [1] and Fengjun Zhao [1,*]

[1] Department of Space Microwave Remote Sensing System, Aerospace Information Research Institute, Chinese Academy of Sciences, Beijing 100190, China
[2] School of Electronic, Electrical and Communication Engineering, University of Chinese Academy of Sciences, Beijing 100039, China
* Correspondence: fjzhao@mail.ie.ac.cn

Abstract: Video Synthetic Aperture Radar (ViSAR) operating in spotlight mode has received widespread attention in recent years because of its ability to form a sequence of SAR images for a region of interest (ROI). However, due to the heavy computational burden of data processing, the application of ViSAR is limited in practice. Although back projection (BP) can avoid unnecessary repetitive processing of overlapping parts between consecutive video frames, it is still time-consuming for high-resolution video-SAR data processing. In this article, in order to achieve the same or a similar effect to BP and reduce the computational burden as much as possible, a novel time-frequency sub-aperture technology (TFST) is proposed. Firstly, based on azimuth resampling and full aperture azimuth scaling, a time domain sub-aperture (TDS) processing algorithm is proposed to process ViSAR data with large coherent integration angles to ensure the continuity of ViSAR monitoring. Furthermore, through frequency domain sub-aperture (FDS) processing, multiple high-resolution video frames can be generated efficiently without sub-aperture reconstruction. In addition, TFST is based on the range migration algorithm (RMA), which can take into account the accuracy while ensuring efficiency. The results of simulation and X-band airborne SAR experimental data verify the effectiveness of the proposed method.

Keywords: video synthetic aperture radar (ViSAR); time-frequency sub-aperture technique (TFST); range migration algorithm (RMA); high resolution

Citation: Yang, C.; Chen, Z.; Deng, Y.; Wang, W.; Wang, P.; Zhao, F. Generation of Multiple Frames for High Resolution Video SAR Based on Time Frequency Sub-Aperture Technique. *Remote Sens.* **2023**, *15*, 264. https://doi.org/10.3390/rs15010264

Academic Editor: Deodato Tapete

Received: 24 November 2022
Revised: 23 December 2022
Accepted: 27 December 2022
Published: 2 January 2023

1. Introduction

As active sensors, synthetic aperture radars (SARs) provide high-resolution images under any weather conditions and at any time of day [1–4]. In recent years, many types of improved SAR systems have emerged. Among these, video-SAR (ViSAR) has attracted much attention due to its ability to provide consecutive views of a region of interest (ROI) [5–12]. Different from traditional SARs, which focus on a single ROI at a time, the update period of ViSAR images can be less than the synthetic aperture time, so that changes can be captured as they occur [5]. In general, ViSAR mainly works in spotlight mode to extend the monitoring time of the target, during which the radar platform either flies by or circles an ROI [6,8,13]. Due to the high update rate and high resolution of ViSAR images, moving objects can be tracked visually, including their direction, movement start and stop, trajectory and even speed. Moreover, the ViSAR imaging mode can also be applied to consistent and inconsistent change detection during a single pass or between passes, three-dimensional imaging and so on [6].

ViSAR requires consecutive high-resolution observation of the ROI, so there is a significant information overlap between consecutive video frames. The duplication of signal processing operations by the sub-aperture in each frame can be avoided to reduce the computational complexity. According to the analysis in [5], the back projection (BP) algorithm [14,15] is an ideal solution for video-SAR imaging because it allows highly parallel processing and avoids repeated calculation of overlapping parts. However, it is quite time-consuming. Song et al. [8] further reduced the processing complexity of ViSAR data by dividing the scene into a general region and an ROI using the fast back projection (FBP) algorithm [16–18]. Jian et al. [19] proposed a spaceborne ViSAR image algorithm based on sub-aperture extended chirp scaling processing. In that study, the frequency sub-bands of the sub-apertures were combined to achieve the required high azimuth-resolution ViSAR image. However, as the number of scenes in video increases, generating all frames still requires a lot of calculations, even at the expense of spatial resolution [8], which may result in some targets being missed. To reduce the data sampling rate while achieving similar or better imaging performance, low-rank tensor recovery was introduced into ViSAR imaging [20]. However, reducing the sampling rate more or less means loss of information.

In addition to spotlight mode ViSAR, Yamaoka et al. [21] and Kim et al. [10] proposed ViSAR imaging algorithms in stripmap mode. The results of the airborne Ku-SAR experiment in [21] showed that stripmap mode ViSAR can also offer indications of moving targets. Further, a new stripmap mode ViSAR data processing method based on a wide-angle antenna using Doppler shifting was proposed in [10]. At the same time, this method performs signal processing based on the range-Doppler algorithm (RDA). Compared to the BP algorithm, the corresponding amount of calculations is reduced. Although the processing algorithm and stripmap mode ViSAR data acquisition are relatively simple, the observation time is limited.

In this paper, we propose a novel time-frequency sub-aperture technique (TFST) for high-resolution ViSAR data processing. For airborne spotlight mode SAR, it is easy to achieve a coherent integration angle of 30°, so the SAR raw data can be divided into several large time domain sub-apertures (TDSs) and each TDS can reach 10° or more. At the same time, in practice, multiple TDSs can be processed in parallel to further improve the efficiency of ViSAR data processing. Consequently, the key is to focus the TDS accurately. The extended two-step approach (ETSA) proposed in [22] can handle a coherent integration angle of 12.5° with a full-aperture when the squint angle is 3°. However, when the squint angle is larger, the performance of ETSA decreases. Therefore, based on azimuth resampling [23] and full aperture azimuth scaling [22], an improved ETSA (IETSA), the first part of TFST, is proposed to process ViSAR data with large coherent integration angles to ensure the continuity of ViSAR monitoring.

Then, the result of TDSs processed by IETSA is converted into the Doppler domain and based on the expected frame rate and resolution of the video frame, the Doppler spectrum is then divided into multiple frequency-domain sub-apertures (FDSs). Furthermore, through FDS processing, multiple high-resolution video frames can be generated efficiently without sub-aperture reconstruction. Finally, the sequence is sorted according to the center time corresponding to each FDS to obtain the consecutive video frames.

The core of TFST is the joint processing of TDS and FDS. TDS is used to avoid unnecessary repetitive processing, while FDS is used to form consecutive video frames. Moreover, due to the TFST being based on the range migration algorithm (RMA or omega-k algorithm) [24–26] and video frames being formed by FDS processing without sub-aperture reconstruction, TFST can balance accuracy while maintaining efficiency.

Further, the novel contributions of this paper can be summarized as follows.

(1) A full-aperture processing algorithm IETSA, which can be used for squint spotlight SAR data with large coherent integration angle, is proposed.
(2) A high-resolution ViSAR data processing algorithm TFST based on TDS and FDS, which can balance imaging quality and efficiency, is proposed.
(3) A series of high-resolution video frames can be acquired without sub-aperture reconstruction.
(4) The detailed processing of accurate motion error compensation is given.

2. Geometric Model of Spotlight ViSAR and Problem Statement

The ViSAR system operating in staring spotlight-mode can illuminate the same area continuously using beam steering. It is not limited by the beam width, so it can achieve higher azimuth resolution and longer video observation times. Figure 1 shows the data acquisition geometry of staring spotlight-mode ViSAR. In this figure, the SAR sensor travels along the x-axis at an effective velocity v. R_0 is the closest range from the center of scene O to the trajectory. $R(\eta)$ is the instantaneous slant range and η_{start} and η_{end} are the starting and ending times of the recording, respectively. θ_{syn} is the relative rotation angle between η_{start} and η_{end}. The length of AD is the full synthetic aperture length L_s and BC is the sub-aperture length L_{vf} (where $L_{vf} \leq L_s$) corresponding to a video frame. The value of L_{vf} depends on the desired azimuth resolution ρ_a of a video frame. Assuming that θ_{vf} is the coherent integration angle corresponding to a video frame, then it can be expressed as [27]:

$$\theta_{vf} \approx 2arcsin\left(\frac{\lambda_c \gamma_{a,w}}{4\rho_a}\right) \tag{1}$$

where λ_c is the central wavelength and $\gamma_{a,w}$ is the mainlobe-broadening factor introduced by windowing.

Assuming that the coordinates at azimuth times η_B and η_C are (X_b, Y_b, Z_b) and (X_c, Y_c, Z_c), respectively, then L_{vf} can be approximately expressed as [28]:

$$L_{vf} = max\left(\frac{-b + \sqrt{b^2 - 4ac}}{2a}, \frac{-b - \sqrt{b^2 - 4ac}}{2a}\right) - Y_b \tag{2}$$

where

$$\begin{aligned} a &= X_b^2 cos(\theta_{vf})^2 + Y_b^2 cos(\theta_{vf})^2 - Y_b^2 \\ b &= -2X_b^2 Y_b \\ c &= X_b^2 Y_b^2 cos(\theta_{vf})^2 + X_b^4 cos(\theta_{vf})^2 - X_b^4, \end{aligned} \tag{3}$$

and $X_b = X_c$, $Z_b = Z_c$.

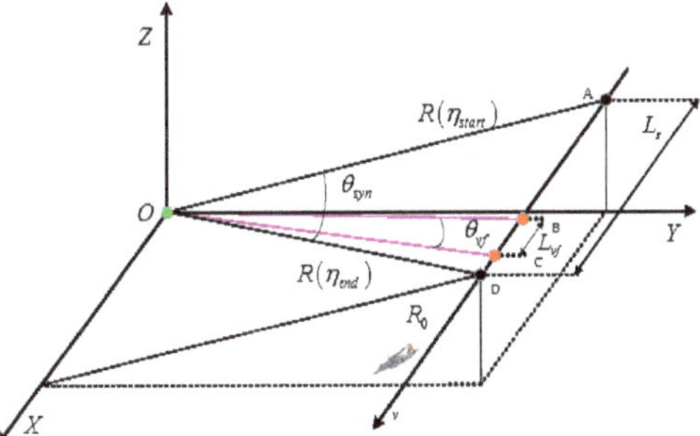

Figure 1. Geometric model of spotlight ViSAR data acquisition.

Spotlight mode has longer monitoring time than strip mode, is more practical than circular SAR and is easy to extend to spaceborne ViSAR. However, the large computational complexity limits the real-time monitoring of ViSAR. Therefore, an efficient ViSAR data processing algorithm suitable for spotlight mode is necessary.

In this paper, a novel TFST is proposed to improve the processing efficiency of high-resolution ViSAR data. In particular, TFST focuses on three main issues. First, due to the influence of atmospheric turbulence, it is necessary to compensate the motion error. In addition, in order to perform video monitoring for a long time and reduce the computational burden, it is necessary to be able to process slant spotlight SAR data with full aperture.

Finally, a series of video frames can be efficiently acquired. Moreover, TFST is validated by simulation and experimental airborne X-band data.

3. Methods

It is assumed that the radar transmits a linear frequency-modulated (LFM) signal with a duration of T_r. Then, the demodulated echoes of a TDS can be expressed as

$$ss(\tau, \eta) = w_r\left(\frac{\tau}{T_r}\right) w_a\left(\frac{\eta}{T_{\mathrm{TDS}}}\right)$$
$$\cdot exp\left(j\pi K_r\left(\tau - \frac{2R(\eta)}{c}\right)^2\right) exp\left(-j\frac{4\pi R(\eta)}{\lambda}\right) \tag{4}$$

where τ and η are the fast and slow times, c is the propagation speed of light, K_r is the frequency modulation rate of the transmitted chirp signal, λ is the wavelength, T_{TDS} is the azimuth duration corresponding to the TDS and $w_r(\cdot)$ and $w_a(\cdot)$ represent the envelope of the transmitted signal and the azimuth window function, respectively.

Then, the echo signal in the range frequency and azimuth time domains is given as follows.

$$Ss(f_r, \eta) = W_r(f_r) w_a\left(\frac{\eta}{T_{\mathrm{TDS}}}\right)$$
$$\times exp\left(-j\pi\frac{f_r^2}{K_r}\right) exp\left(-j\frac{4\pi(f_c + f_r)R(\eta)}{c}\right) \tag{5}$$

where f_r is the range frequency, f_c is the central transmitted frequency and $W_r(\cdot)$ is the Fourier transform of $w_r(\cdot)$.

3.1. Azimuth Preprocessing in TFST

Lanari et al. [29,30] proposed the two-step approach (TSA) based on azimuthal time-domain convolution to solve the azimuthal aliasing problem of spotlight/sliding spotlight mode. The reference function used for azimuth deramping in TSA is

$$s_{ref}(\eta) = exp\left(j2\pi\frac{(v\eta)^2}{R_{ref}}\frac{f_c}{c}\right) \tag{6}$$

where R_{ref} denotes the reference range and equals the closest approach R_0 of the scene center in the staring spotlight mode.

In the case of a large transmitted signal bandwidth, a significant deviation of the instantaneous range frequency from f_c will cause the Doppler bandwidth after spectral analysis to be much larger the than pulse repetition frequency (*PRF*) [31,32]. Liu et al. [31] proposed to use a range-frequency-dependent reference function to mitigate this aliasing effect:

$$S_{ref}(f_r, \eta) = exp\left(j2\pi\frac{(v\eta)^2}{R_{ref}}\frac{f_c + f_r}{c}\right). \tag{7}$$

The range-frequency-dependent Doppler centroid of $Ss(f_r, \eta)$ for the more generalized squinted SAR geometry was proposed in [22]:

$$f_{dc}(f_r) = \frac{2v\sin(\theta_0)(f_c + f_r)}{c} \tag{8}$$

where θ_0 is the squint angle of the aperture center corresponding to a certain TDS and the nominal Doppler centroid is

$$f_{dc0} = \frac{2v\sin(\theta_0)f_c}{c}. \tag{9}$$

Then, the azimuth time-domain convolution can be expressed as [22]

$$
\begin{aligned}
s_a(f_r, \eta) &= \int_{-T_{\mathrm{TDS}}/2}^{T_{\mathrm{TDS}}/2} Ss(f_r, t) \cdot exp(-j2\pi f_{dc}t) \\
&\quad \cdot S_{ref}(f_r, (\eta - t))\mathrm{d}t \\
&= S_{ref}(f_r, \eta) \cdot \int_{-T_{\mathrm{TDS}}/2}^{T_{\mathrm{TDS}}/2} Ss(f_r, t) \\
&\quad \cdot exp(-j2\pi f_{dc}t) \cdot S_{ref}(f_r, t) \\
&\quad \cdot exp(j2\pi K_{rot}\eta t)\mathrm{d}t
\end{aligned}
\tag{10}
$$

where

$$K_{rot}(f_r) = -\frac{2v^2(f_c + f_r)}{c \cdot R_{ref}}. \tag{11}$$

The operation of (10) can be broken down into deramping, Fourier transform and phase compensation. Then, Fourier transform is applied on the azimuth time η to obtain the $2-$D frequency spectrum $SS_a(f_r, f_a)$

$$SS_a(f_r, f_a) = FFT\left(s_a\left(f_r, -\frac{PRF}{N \cdot K_{rot}}m\right)\right) \tag{12}$$

where $m = -\frac{N}{2}, \ldots, \frac{N}{2} - 1$ and N must satisfy

$$N \geq \frac{PRF(B_b + B_{rot} + B_{sq})}{K_{rot0}} \tag{13}$$

where $B_b = 2v\cos(\theta_0)\theta_{bw}/\lambda_c$ is the azimuth beam bandwidth, $B_{rot} = |K_{rot0}| \cdot T_{\mathrm{TDS}}$ is the Doppler bandwidth extended by steering of the beam, $B_{sq} = 2vB_r \sin\theta_0/c$ is the squinted Doppler bandwidth and K_{rot0} is the nominal Doppler frequency modulation rate

$$K_{rot0} = -\frac{2v^2 f_c}{c \cdot R_{ref}}. \tag{14}$$

To correctly focus the signal of (12), two problems need to be solved. One is the non-uniform sampling of the azimuth spectrum due to the range-frequency-dependent K_{rot} [22]. The other is the significant range–azimuth coupling effect occurring in highly squinted SAR. The Doppler spectrum resampling proposed in [22] can solve the first problem, but cannot be used to overcome the second problem.

Although the range-azimuth coupling is weakened by (10), it is mainly aimed at the linear term of the range cell migration (RCM) and the real closest slant range of the target in the same range cell may be different; therefore, migration residuals still exist after RCM correction (RCMC). According to the principle of equivalent array in [23], in order to make the echo data received by the linear array (LA) and its equivalent array (EA) with a projection angle of θ_0 completely equivalent, the relationship between the Doppler frequency of LA and EA can be expressed as:

$$f_{a_{EA}} = \cos(\theta_0)^2 \left(f_{a_{LA}} + \frac{v}{\lambda} \sin(\theta_0) \right)$$
$$- \sin(\theta_0)\cos(\theta_0) \sqrt{\left(\frac{v}{\lambda}\right)^2 - \left(f_{a_{LA}} + \frac{v}{\lambda}\sin(\theta_0)\right)^2}, \tag{15}$$

where $f_{a_{LA}}$ and $f_{a_{EA}}$ represent the Doppler frequencies of the echo data of LA and EA, respectively. Further, for high-resolution SAR data, considering the round-trip delay, it can be further expressed in the 2−D frequency domain as

$$f_{a_{new}} = \cos(\theta_0)^2 (f_a + f_{dc}) - \sin(\theta_0)\cos(\theta_0)$$
$$\cdot \sqrt{\left(\frac{2v(f_c + f_r)}{c}\right)^2 - (f_a + f_{dc})^2}, \tag{16}$$

where $f_{a_{new}}$ is the new azimuth frequency after azimuth resampling on (12) and

$$f_a = k \cdot \Delta f_a, \quad k = (-\frac{N}{2}, \ldots, \frac{N}{2} - 1), \tag{17}$$

where

$$\Delta f_a = (\frac{|K_{rot}|}{PRF}). \tag{18}$$

Then, the equivalent new *PRF* is

$$PRF_{new} = \max(f_{a_{new}}) - \min(f_{a_{new}}). \tag{19}$$

where $\max(\cdot)$ and $\min(\cdot)$ represent the maximum and minimum operators, respectively.

After azimuth resampling, the classic broadside SAR imaging algorithm can be used to focus the squint SAR data [23]. Next, it will be explained how the azimuth resampling completes the uniform sampling of (12).

From [22], we know that in order to align the sampling grids, we need to implement the following change of variables:

$$k \cdot \Delta f_a + f_{dc} = k \cdot \Delta f_{a0} + f_{dc0} \tag{20}$$

where

$$\Delta f_{a0} = \frac{|K_{rot0}|}{PRF}. \tag{21}$$

After azimuth resampling, the squint angle θ_0 is equal to 0 [23], so $f_{dc} = f_{dc0} \equiv 0$. In addition, $f_{a_{new}}$ can be generated at equal intervals according to (16). The 2−D spectrum $SS_a(f_r, f_a)$ can then be resampled in azimuth to align the sampling grids:

$$SS_a(f_r, k \cdot \Delta f_a) \Rightarrow SS_a(f_r, k \cdot \Delta f_{a_{new}}) \tag{22}$$

where $\Delta f_{a_{new}} = (PRF_{new}/N)$.

Therefore, after azimuth resampling the 2−D spectrum is sampled uniformly and is equivalent to the spectrum of the broadside mode, as shown in Figure 2. The corresponding simulation parameters are listed in Table 1. Comparing Figure 2a,b, the signal spectrum after azimuth resampling is equivalent to the broadside SAR. At the same time, the effective velocity v is transformed into $v\cos(\theta_0)$.

Table 1. Point Target's Simulation Parameters.

Parameters	Value
Carrier frequency	9.6 GHz
Velocity	116 m/s
Height	6376.5 m
Incidence angle	54.5°
Signal bandwidth	1200 MHz
Sampling frequency	1400 MHz
Signal pulse duration	6.7 µs
Pulse repetition frequency	3000 Hz
Length of antenna	0.495 m
Scene size (Range × Azimuth)	1.0 km × 0.8 km

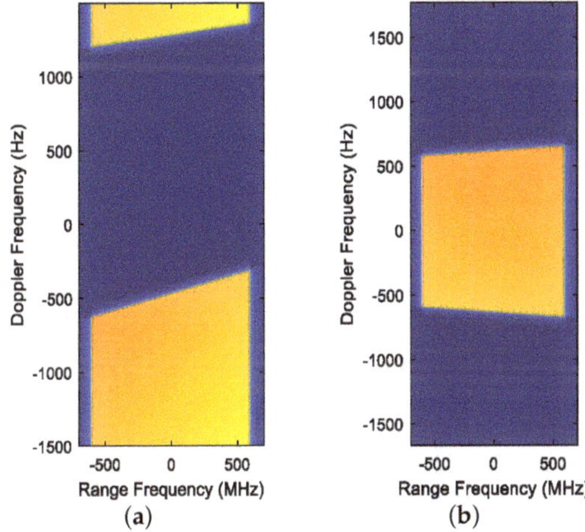

Figure 2. 2−D spectrum resulting from IETSA (**a**) before and (**b**) after azimuth resampling, with a squint angle of 15° and a coherent integration angle of 10°.

Then, the product of the 2−D spectrum $SS_a(f_r, f_{a_{new}})$ and the frequency domain phase compensation function S_{com} will be the same as the spectrum of a broadside SAR:

$$SS(f_r, f_{a_{new}}) = SS_a(f_r, f_{a_{new}}) \cdot S_{com} \tag{23}$$

where

$$S_{com} = exp\left(-j\frac{\pi \cdot (\hat{f}_{a_{new}})^2}{K_{rot}}\right) \tag{24}$$

Therefore, the existing broadside imaging algorithm can be used to process the data [23]. According to (16), $\hat{f}_{a_{new}}$ can be expressed as

$$\hat{f}_{a_{new}} = f_{a_{new}} + sin(\theta_0) \cdot \sqrt{|B - A^2|} - f_{dc} \tag{25}$$

where

$$B = \frac{f_{a_{new}}}{cos(\theta_0)}, \tag{26}$$

and

$$A = \left(\frac{2v \cdot (f_c + f_r)}{c} \right)^2.$$ (27)

Next, the classical frequency domain algorithm RMA can be used to accomplish range cell migration (RCM) correction and range compression. In this manner, we can obtain $sS_{rcmc}(R(\eta), f_{a_{new}})$ for azimuth scaling, where $R(\eta)$ is the slant range, not the closest approach defined in [22].

3.2. Modified Full-Aperture Azimuth Scaling

It is necessary to solve the problem of slow time-domain aliasing of high-resolution SAR data using TSA. Zhu et al. [22] developed a full-aperture azimuth scaling technique based on the azimuth scaling of [33–35] and combined motion compensation (MoCo) and autofocusing [34,36], which can effectively solve this problem. However, the processing here is based on the signal obtained after azimuth resampling, which is different from the Doppler Spectrum Resampling in [22], so the full-aperture azimuth scaling needs further modification.

Similar to sub-aperture azimuth scaling, full-aperture azimuth scaling can be expressed as [22]

$$\beta(R, \eta) = F^{-1}[sS(R, f_{a_{new}})] \cdot exp\left[-j\pi K_a(R_{scl}) \cdot \eta^2 \right]$$ (28)

where F^{-1} is the inverse Fourier transform, R_{scl} is the reference range and

$$\begin{aligned} sS(R, f_{a_{new}}) &= sS_{rcmc}(R, f_{a_{new}}) \\ &\cdot exp\left(j\frac{4\pi}{\lambda} \cdot R \cdot (D(f_{a_{new}}) - 1) \right) \\ &\cdot exp\left(-j\frac{\pi f_{a_{new}}^2}{K_a(R_{scl})} \right), \end{aligned}$$ (29)

and

$$D(f_{a_{new}}) = \sqrt{1 - \frac{\lambda^2 \cdot f_{a_{new}}^2}{4v^2}},$$ (30)

and

$$K_a(R_{scl}) = -\frac{2v^2}{\lambda \cdot R_{scl}}.$$ (31)

Then, as detailed in Appendix A, the dechirped signal for slow time-domain error compensation is obtained by applying the IFFT to (A4) and can be expressed as

$$\beta(R, k \cdot \Delta\eta) = IFFT(\Gamma(R, k \cdot \Delta f_{a_{new}})).$$ (32)

where $\Delta\eta = -(\Delta F_{a_{new}} / K_a(R_{scl}))$ is the sampling interval of the slow time domain.

Figure 3 shows the specific flow diagram of azimuth scaling and slow time domain error compensation and the scaling function used in the proposed algorithm is [22]

$$H_{scl}(f_{a_{new}}) = exp\left(j\frac{\pi}{K_a(R_{scl})}(\delta - 1)f_{a_{new}}^2 \right)$$ (33)

where $\delta = R_{scl}/R$, the phase perturbation function is

$$H_{php}(\eta_{new}) = exp\left(j\pi \frac{K_a(R_{scl})}{\delta}\eta_{new}^2\right)$$

(34)

where

$$\eta_{new} = \frac{k}{PRF_{new}}$$

(35)

and the inverse scaling function $H_{ins}(f_{a_{new}})$ is the conjugate of $H_{scl}(f_{a_{new}})$.

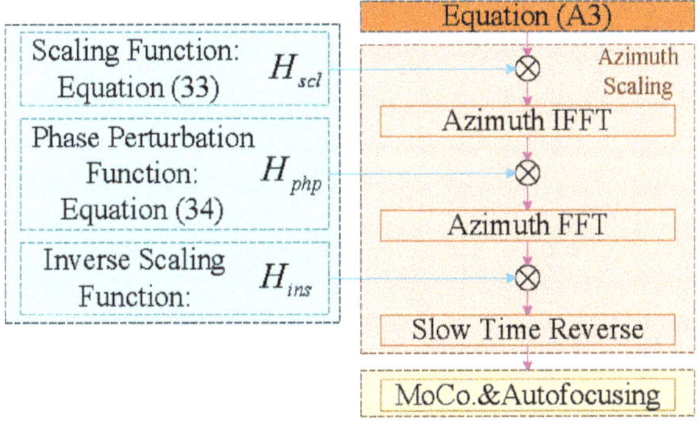

Figure 3. Flow diagram of azimuth scaling and slow time domain error compensation.

3.3. Motion Error Compensation in TFST

In addition, since airborne SAR is susceptible to atmospheric turbulence, motion compensation is essential. First, according to the motion compensation scheme proposed in [37], the phase error and envelope displacement between the real and ideal trajectories are compensated by using inertial navigation system and inertial measurement unit (INS/IMU) measurements.

The residual phase error after MOCO using INS/IMU data is

$$\Phi_e(f_r, \eta) = exp(-j\frac{4\pi}{c}(f_c + f_r)\Delta R(\eta)),$$

(36)

where $\Delta R(\eta)$ is the slope range residual error. After azimuth preprocessing, it can be expressed as [23]

$$\Phi_e(f_r, \eta_{new}) = exp(-j\frac{4\pi}{c}(f_c + f_r)\Delta R(\xi(\eta_{new}))),$$

(37)

where $\xi(\eta_{new})$ is a function of η_{new} and its analytical relationship does not need to be considered here.

Then, through the modified full-aperture azimuth scaling processing, the traditional PGA [38] can be used to estimate the residual phase error and the specific motion compensation process is shown in Figure 4.

3.4. Fundamentals of TFST

Different from traditional sub-aperture methods [6,8,29,30], in our method the TDS is required to be as large as possible, as shown in Figure 5. In addition, Figure 6 shows the corresponding relationship between TDS and FDS. It shows how to concatenate the video frame sequence formed by multiple TDSs. In the experimental data processing, the TDSs are sequentially processed by sliding in the time domain and multiple FDSs are generated in the frequency domain to form a series of video frames. In particular, TDS_n represents the nth TDS and the segment marked in yellow represents the overlapping parts (OPs) between TDS_n and its two adjacent TDSs, to ensure that video frames formed by multiple TDSs can be smoothly connected.

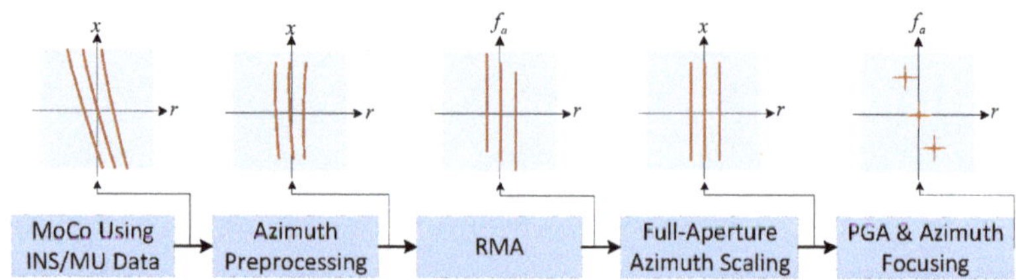

Figure 4. Flow chart of motion compensation.

Moreover, it is necessary to analyze the influence of the OPs on the amount of calculations. We performed an experiment with the parameters shown in Table 1. Figure 7 shows the ratio of OPs to TDSs with a coherent integration angle of 10° and a range resolution of $\rho_r = 0.1249$ m. Owing to the very high range resolution, even if ρ_a/ρ_r is 7, then $\rho_a = 0.8743$ m, which is still less than 1 m. Referring to Figure 7a, it can be seen that the amount of calculations incurred by the OPs is tiny. Furthermore, comparing to Figure 7a,b, it can be found that the influence of OPs on Ka-band ViSAR is almost negligible, which reveals the application potential of the proposed algorithm.

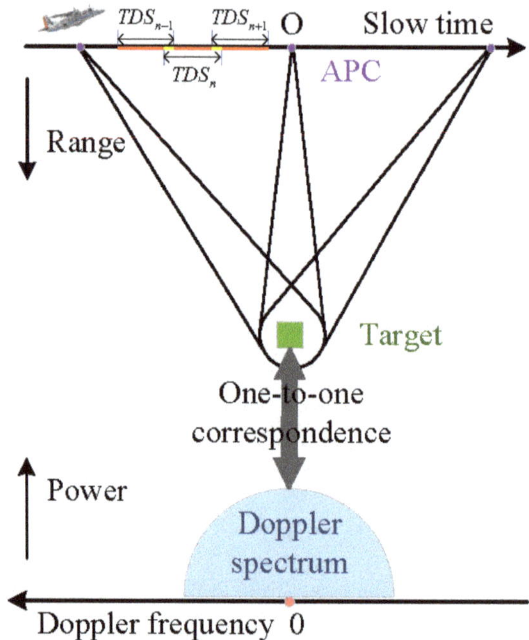

Figure 5. Schematic diagram of the formation of TDS. In addition, it can be seen that there is a one-to-one correspondence between Doppler frequency and beam direction for spotlight mode SAR. "APC" (purple dots) is the antenna phase center of transmission and reception in the slow time domain direction.

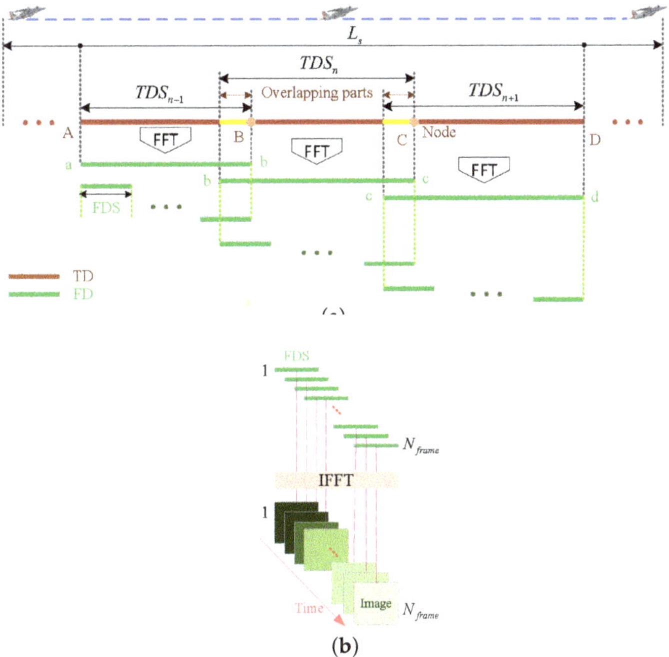

Figure 6. Correspondence between TDS and FDS (**a**). TD means 'time domain'; FD means 'frequency domain'. TDS_{n-1}, TDS_n and TDS_{n+1} correspond to those in Figure 5, respectively. A schematic diagram of the formation of multiple video frames (**b**) and N_{frame} represents the total number of FDSs.

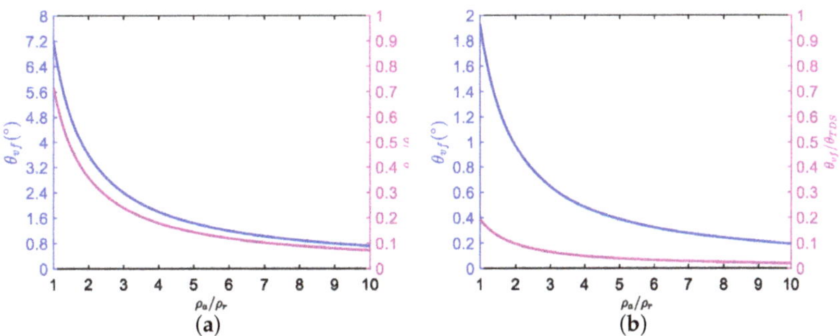

Figure 7. The ratio of OPs to TDSs with a coherent integration angle of 10°. (**a**) X band; (**b**) Ka band with carrier frequency 35.6 GHz. The other parameters are the same as in Table 1.

The complete flow diagram of high-resolution ViSAR data processing is shown in Figure 8 and it consists of two main steps: TDS processing, which corresponds to the first part of TFST and FDS processing, which corresponds to the second part of TFST. In (A), the areas marked by red, green and blue boxes, respectively, illustrate the division of TDS in a two-dimensional time domain, which corresponds to Figure 5. Next, M TDSs can be processed by IETSA in parallel and the output is shown in (B). Corresponding to (A), (C) shows the division of N FDSs. Moreover, a geometric correction is applied to remove the distortion introduced by azimuth resampling [23]. The final output is shown in (D); thus, a series of video frames for generating video are obtained.

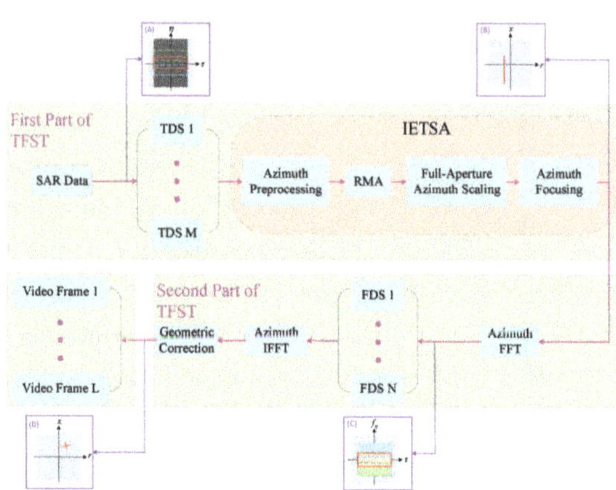

Figure 8. The complete flow diagram of high-resolution ViSAR data processing based on TFST. M and L are the total number of TDS and video frames, respectively, where $L = M \times N$. N represents the number of FDSs generated by one TDS. This implies the assumption that the number of FDSs divided by each TDS is the same. The x and r in (B) represent the azimuth and range coordinates in the image domain, respectively.

To further explain TFST, the time-frequency diagram of the formation of multiple video frames is shown in Figure 9. It can be seen that multiple video frames of desired resolution can be obtained without sub-aperture reconstruction.

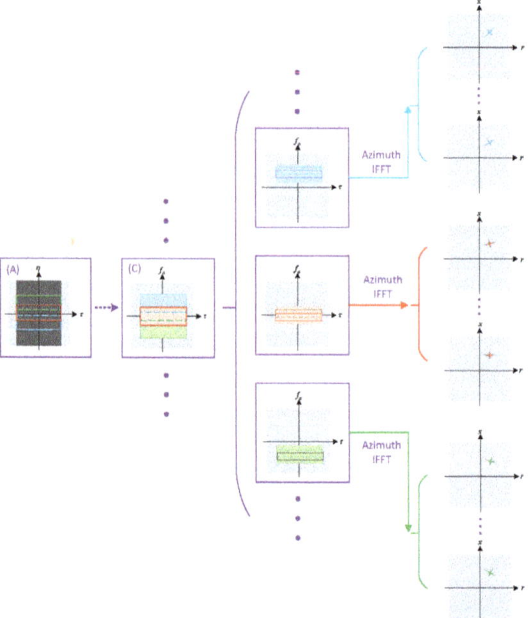

Figure 9. Time frequency diagram of generating multiple video frames. The different directions of the point targets in the image domain in the last column represent different observation times. Geometric correction is performed by default after azimuth IFFT. In particular, (A) and (C) correspond to those in Figure 8, respectively.

Next, the specific parameter design scheme is given. The synthetic aperture time corresponding to the l-th FDS is

$$T_{\text{FDS},l} = \frac{R0(tan(\theta_{start,l}) - tan(\theta_{end,l}))}{v},$$ (38)

so, the frame rate can be expressed as follows

$$F = \frac{1}{T_{\text{FDS}} \cdot (1 - \alpha)}.$$ (39)

where $l = 1, \ldots, N_{frame}$ and N_{frame} is the total number of FDSs, $\theta_{start,l}$ and $\theta_{end,l}$ are the start and end squint angles corresponding to the l-th FDS and α is the overlap rate. In practice, to obtain the minimum frame rate, we can use the maximum value of $T_{\text{FDS},l}$ for T_{FDS}.

The Doppler bandwidth corresponding to an FDS is [39]

$$B_{dop} = \frac{2v \cos \theta_{rc}}{\lambda_c} \theta_{vf}.$$ (40)

where θ_{rc} is the squint angle at the center of the aperture.

It can be seen that for a fixed frame-length ViSAR, the θ_{vf} of different FDSs is slightly different due to the change in the squint angle. In practice, the frame-length, B_{dop}, is obtained at the aperture center.

Then, the overlap rate α is obtained using (39) and the desired frame rate. Therefore, a series of FDSs can be formed in the Doppler domain, as shown in Figures 8 and 9.

In application, for airborne ViSAR, multiple different video frames need to be registered before forming video. Thus, geometric correction can be solved through registration processing, which will not increase the amount of calculation of registration, but reduce the complexity of TFST, which can further improve the efficiency.

3.5. Complexity Analysis of TFST

TFST includes two main steps: TDS processing and FDS processing, as shown in Figure 8. It is assumed that the dimensions of SAR data in range and slow time are N_r and N_a, respectively, and all operations can be divisible. Then, the size of a TDS is $N_{am} \times N_r$ and $N_{am} = N_a / M$. A complex multiplication requires a total of six floating-point operations. A 1D FFT with length N requires $5N log_2 N$ flops and an $N \times N$ 2$-$D FFT requires $10N^2 log_2 N$ flops. The interpolation operation with core length N_{ker} requires $(2N_{ker} - 1)$ flops per complex output point and N_{ker} is generally 8 or 16 in practice [39]. For convenience of analysis, let $N_{am} = N_r \equiv N$.

IETSA in TDS processing includes four main operations: complex phase multiplication, 1D FFT, 2$-$D FFT and interpolation, so the amount of operation is $\mathcal{O}(N^2) + \mathcal{O}(N log_2 N) + \mathcal{O}(N^2 log_2 N) + \mathcal{O}((2N_{ker} - 1)N^2)$. FDS processing includes two main operations: 1D FFT and interpolation, so the amount of operation is $\mathcal{O}(N log_2 N) + \mathcal{O}((2N_{ker} - 1)N^2)$. Therefore, the total complexity of TFST is $\mathcal{O}(N^2) + \mathcal{O}(N log_2 N) + \mathcal{O}(N^2 log_2 N)$. The complexity of BP is $\mathcal{O}(N^3)$, which is obviously larger than that of TFST.

To sum up, TFST is more practical in high-resolution ViSAR data processing due to its accuracy and high efficiency.

4. Results and Discussion

In this section, the proposed method is evaluated through simulations and experiments and the results are analyzed in depth. Moreover, the parameters not mentioned in the simulation are the same as those in Table 1. The data comes from the X-band airborne SAR system of the Aerospace Information Research Institute, Chinese Academy of Sciences (AIRCAS), and the transmission signal bandwidth is 1.2 GHz.

4.1. Simulation Results and Discussion

Figure 10 shows the geometric distribution of point targets in ground scene. To verify the IETSA, the case of a coherent integration angle of 10° and a squint angle of 15° is analyzed. In particular, the simulation results of the three-point targets P1, P5 and P9 in Figure 10a are analyzed, as shown in Figure 11, and the peak sidelobe ratio (PSLR), the integrated sidelobe ratio (ISLR) and the resolution achieved are summarized in Table 2. Combined with Figure 11 and Table 2, it can be concluded that the three point targets are well focused, which confirms that IETSA can effectively focus squint spotlight SAR data. This is a key step to improving the practicability of TFST.

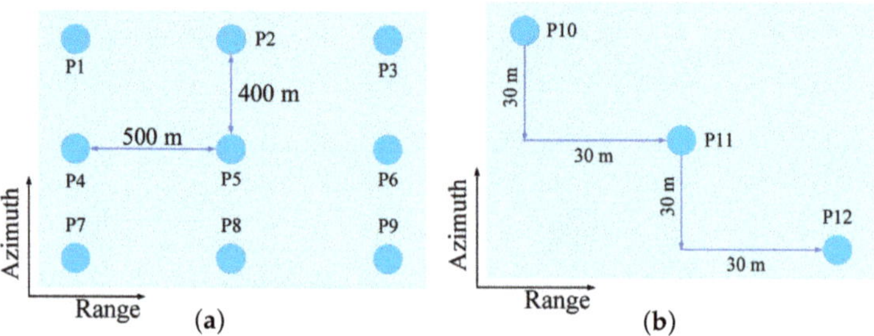

Figure 10. Geometry distribution of point targets in ground scene (**a**) for IETSA simulation, (**b**) for TFST simulation.

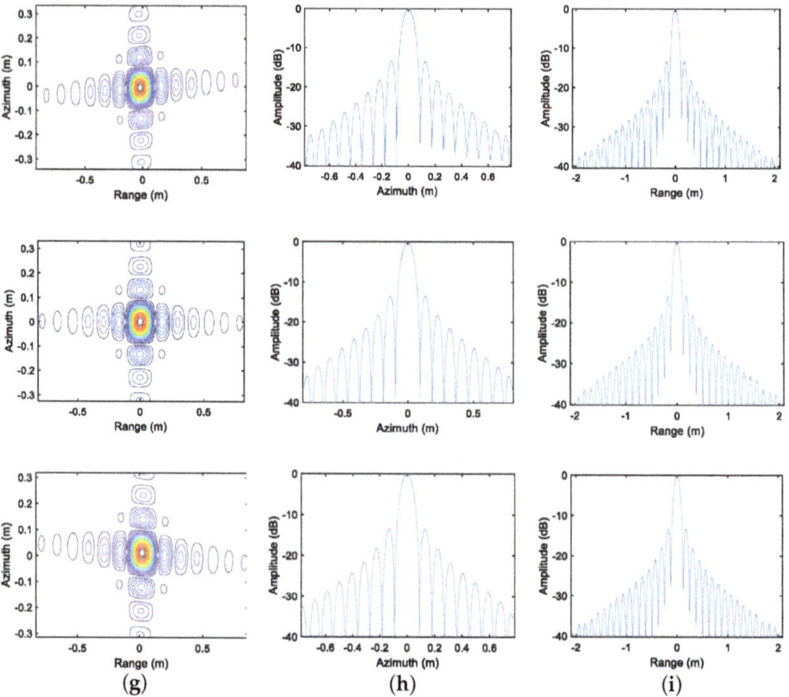

Figure 11. Contour plots (first column), azimuth (second column) and range profiles (third column) of the impulse response function for the three point targets P1 (**a–c**), P5 (**d–f**) and P9 (**g–i**).

Table 2. Evaluation of Focusing Quality In Point Target's Simulation.

Point Targets	Azimuth			Range		
	PSLR (dB)	ISLR (dB)	Resolution (m)	PSLR (dB)	ISLR (dB)	Resolution (m)
P1	−13.36	−10.99	0.0781	−13.24	−10.51	0.1110
P5	−13.38	−11.05	0.0818	−13.25	−10.51	0.1110
P9	−13.36	−11.05	0.0804	−13.23	−10.52	0.1111

In addition, the comparison results of ETSA [22] and IETSA are shown in Figure 12. The coherent integration angle and squint angle used in the simulation are 5° and 7.2°, respectively. It can be seen that for the squint mode SAR data, IETSA has better focusing performance than ETSA and the phase error caused by azimuth resampling can be compensated by TFST.

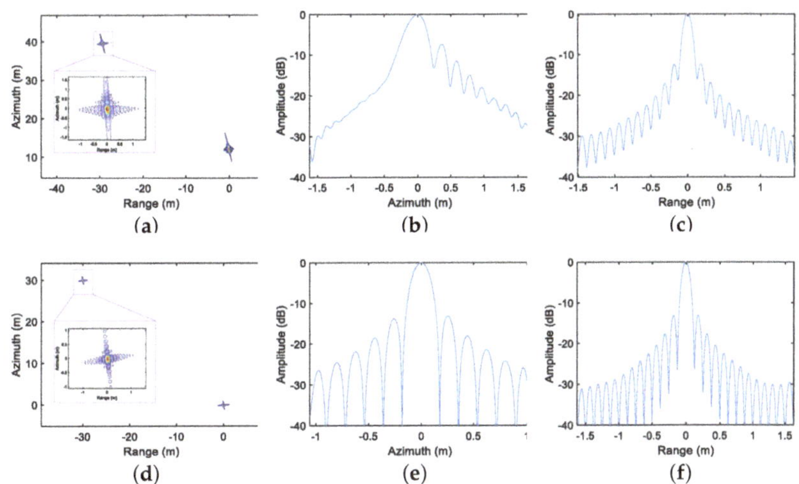

Figure 12. Contour plots (first column) for the two point targets P10 and P11. Azimuth (second column) and range profiles (third column) of the impulse response function for the point target P10. (**a–c**) and (**d–f**) are the simulation results of ETSA and IETSA, respectively.

Different from IETSA simulation, for the verification of TSFT, we designed a scene with three-point targets, as shown in Figure 10b, which includes two stationary point targets P10 and P11 and and a moving target P12 whose azimuth and range speeds are −0.4 m/s and −0.1 m/s, respectively. The parameters used for TFST verification are shown in Table 3.

Table 3. Parameters Used for TFST Verification.

Parameters	Value
Frame Rate	2.3687 Hz
Start Squint Angle	15°
End Squint Angle	−15°
Coherent Integration Time	50.78 s
Video Frame Azimuth Resolution	0.2379 m

The simulation results of TFST are shown in Figure 13 and Figure 13a is frame 49, with a squint angle of 10.17°, Figure 13b is frame 148, with a squint angle of 3.25°, Figure 13c is frame 246, with a squint angle of −3.69° and Figure 13d is frame 344, with a squint angle of $-10.40°$. The first column is the result for the entire scene, while the second and third columns are the enlarged results of P10 and P12 from the first column, respectively.

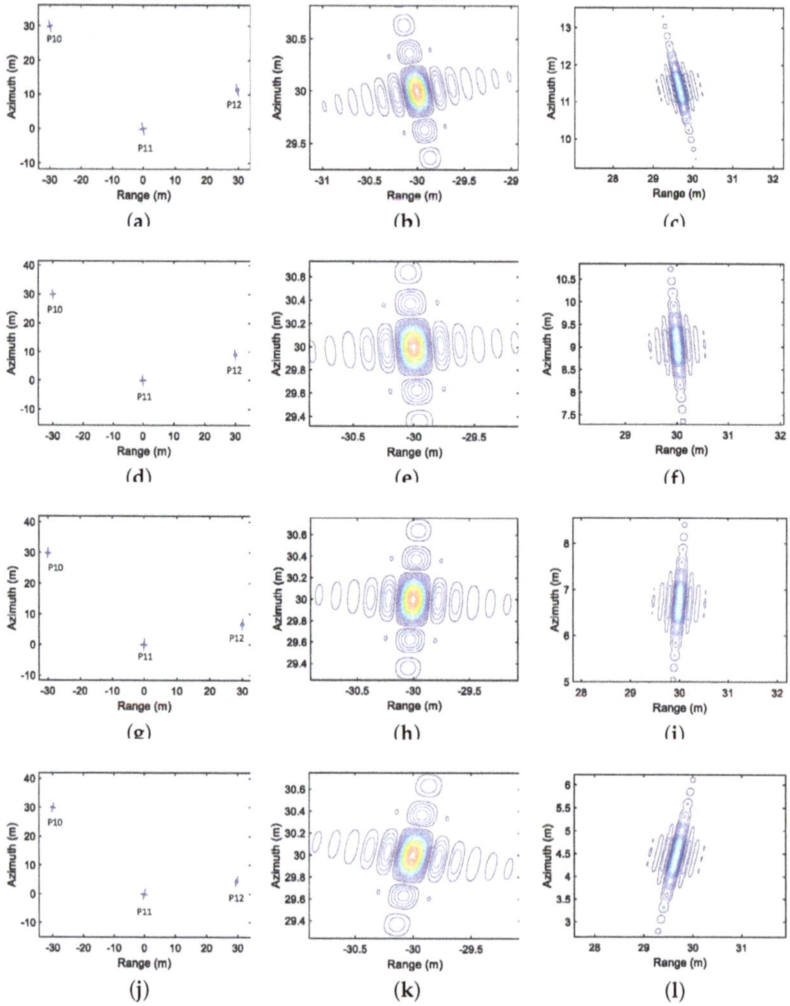

Figure 13. ViSAR simulation results of the scene shown in Figure 10b using TFST. (**a**–**c**) Frame 49, (**d**–**f**) Frame 148, (**g**–**i**) Frame 246, (**j**–**l**) Frame 344. First column: entire scene, second and third columns: enlarged results of P10 and P12 in the first column, respectively.

The second column of Figure 13 shows that the position of point target P10 in the different video frames is fixed. However, the position of moving target P12 varies among different frames, which confirms that ViSAR has the ability to discern moving targets.

To further evaluate the focusing performance of ViSAR, the point target P10 was analyzed in different frames and the PSLR, ISLR and the resolution are summarized in Table 4. It can be seen that the change in the azimuth resolution of different frames is minimal.

Table 4. Evaluation of Focusing Quality In Point Target's Simulation.

Video Frame	Azimuth			Range		
	PSLR (dB)	ISLR (dB)	Resolution (m)	PSLR (dB)	ISLR (dB)	Resolution (m)
49	−13.25	−10.87	0.2303	−13.26	−10.25	0.1107
148	−13.25	−10.87	0.2303	−13.26	−10.25	0.1107
246	−13.25	−10.87	0.2303	−13.26	−10.25	0.1107
344	−13.25	−10.87	0.2302	−13.26	−10.25	0.1107

Figure 14 shows the simulation results of the traditional SAR, with a squint angle of 10° and a coherent integration angle of 10°. Comparing the point target P12 in Figure 14 and Figure 13, it is evident that the moving target appears as a defocused line target in the traditional SAR image, which cannot show the dynamic change characteristics. However, ViSAR images can illustrate whether a target is moving, which is more helpful for moving target detection.

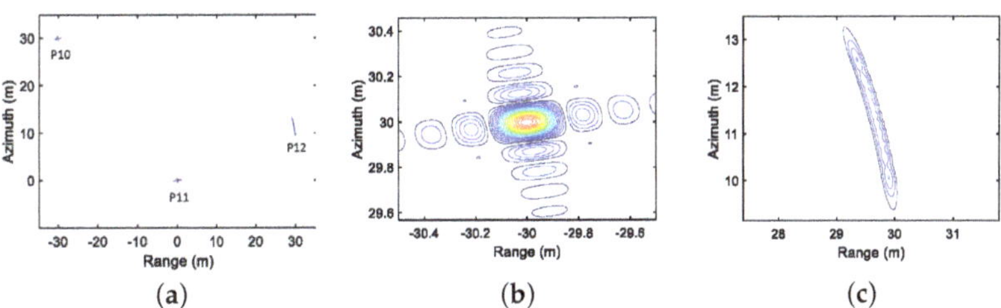

Figure 14. Traditional SAR imaging result of the scene shown in Figure 10a using IETSA. (**a**) entire scene result; (**b**,**c**) enlarged results of P10 and P12 in (**a**), respectively.

In addition, a distributed target (DT) with a size of 4×2 m is simulated, whose velocities in azimuth and range are −8 m/s and −1 m/s, respectively. The simulation results of DT are shown in Figure 15. The time interval between Figure 15a,b is about 15 s and that between Figure 15b,c is about 19 s. According to Figure 15, it can be seen that the azimuth and range position of DT both change with time and the imaging result is seriously defocused. Compared with Figure 13, the defocus and displacement of the moving target in Figure 15 are larger because of its greater velocity. Simulation of DT reveals how moving targets appear in real scenes.

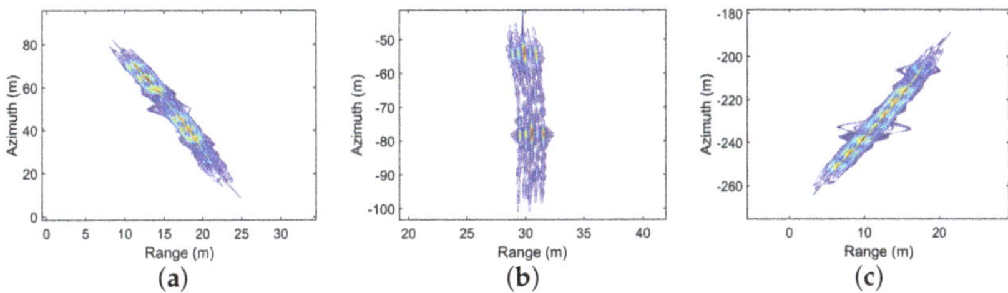

Figure 15. Imaging results of DT in different video frames. (**a**) Frame 17, (**b**) Frame 53, (**c**) Frame 81. The frame rate is 2.3687 Hz.

4.2. Airborne SAR Data Results and Discussion

A continuation of the discussion based on the X-band airborne SAR data from AIRCAS and the focused SAR image processed using IETSA is shown in Figure 16, with a squint angle of 10° and a coherent integration angle of 10°, which further confirms the effectiveness of IETSA.

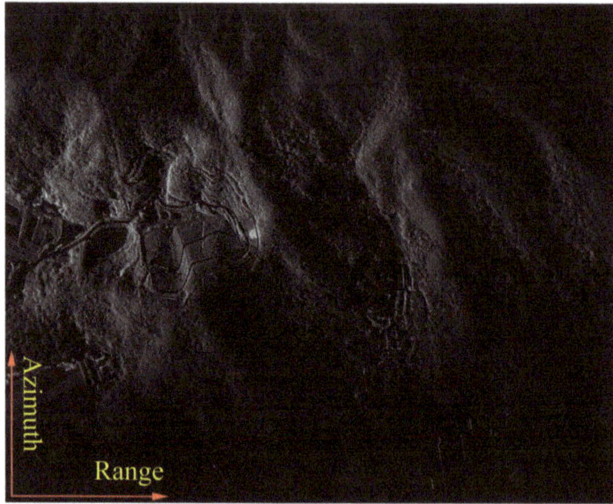

Figure 16. Focused SAR image processed using IETSA, with a squint angle of 10° and a coherent integration angle of 10°.

Moreover, Figure 17 presents the experimental data processing results of ETSA and IETSA. It is obvious that the focusing quality of IETSA is better, which is the same as the conclusion in Figure 12. At the same time, we also confirmed by the experimental data that TFST can compensate the phase error caused by azimuth resampling.

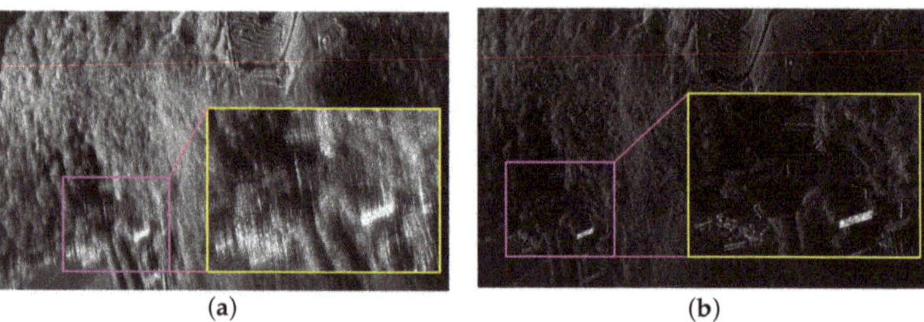

(a)   (b)

Figure 17. Comparison of data processing results of ETSA (**a**) and IETSA (**b**). Similar to the point target simulation, the squint angle and coherent integration angle of the experimental data are 7.2° and 5°, respectively.

Further, to illustrate that TFST has better performance, Figure 18 shows the comparison results of TFST and Doppler Shifting Technique (DST) proposed by [10]. In particular, Figure 18 corresponds to a video frame at an intermediate moment generated from data with a squint angle and a coherent integration angle of 11° and 8°, respectively. It can be seen that the focus quality of TFST is better compared to DST.

Then, the X-band airborne SAR data within [15°, −15°] were processed with the method shown in Figure 8 at a fixed frame rate of 2.3687 Hz. The range resolution was 0.1249 m and the azimuth resolution of the middle video frame was set to 0.2379 m. Due to FDS processing in fixed frame length mode, the resolution of different video frames will vary slightly. Nonetheless, according to the simulation results in Table 4, these changes are negligible.

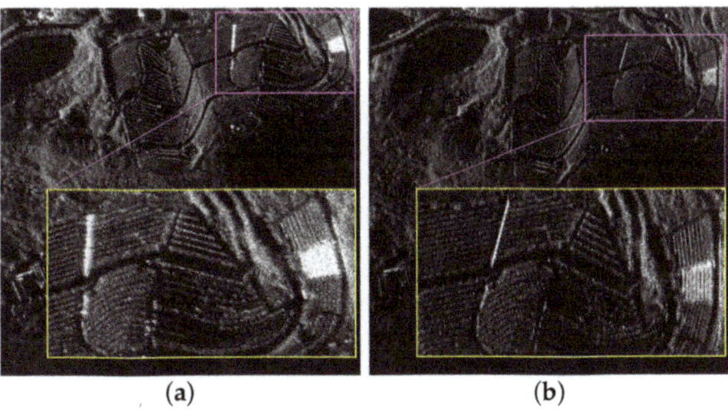

Figure 18. Comparison between TFST and DST. (**a**) DST and (**b**) TFST.

The processing results of TFST are shown in Figures 19 and 20. Figure 19 shows video frames at four different times of the area in Figure 16, from which it can be inferred that the aircraft is moving from bottom to top along the azimuth. The result shows that ViSAR can acquire images from multiple angles of the same target, which is helpful for image information interpretation. Moreover, Figure 20 is an urban scene and there is a road on the left. Comparing Figure 20a and Figure 20b, it is evident that the bright line in the area marked by the red rectangle is a moving target. Its position has shifted and the imaging result is seriously defocused. This is consistent with the simulation results of DT in Figure 15. In addition, the background of the ViSAR image is static and the position of the stationary target is constant, so the target whose position changes significantly in different video frame images should be a dynamic target. This result demonstrates the potential of ViSAR for moving target indication (MTI).

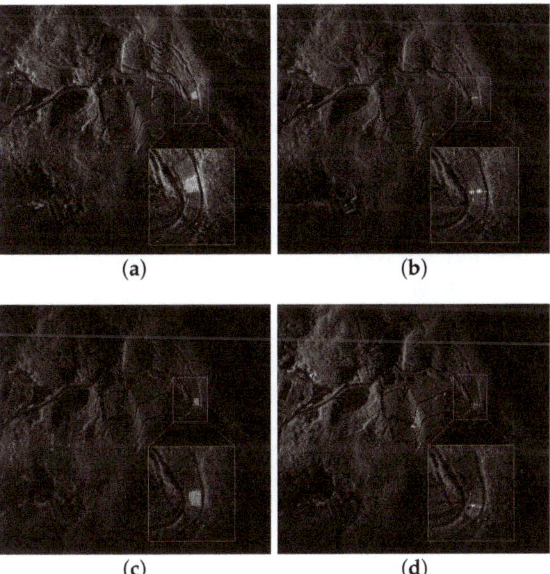

Figure 19. ViSAR image of mountain scene. (**a**) Frame 49, (**b**) Frame 148, (**c**) Frame 236 and (**d**) Frame 334.

It is worth emphasizing that, according to (39), a higher frame rate can be obtained by adjusting the overlap rate and the resolution of the video frame. In practice, these three variables need to be considered comprehensively to reduce the amount of calculations as much as possible.

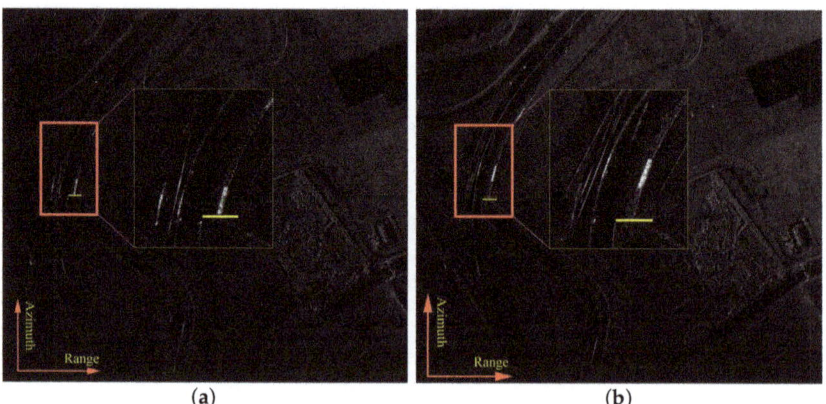

Figure 20. ViSAR image of urban scene. (**a**) Frame 17 and (**b**) Frame 59.

5. Conclusions

In this paper, we first propose an IETSA approach based on azimuth resampling and full-aperture azimuth scaling, which allows squint spotlight SAR data processing at full aperture. Then, combining TDS processing and FDS processing, a novel high-resolution ViSAR data processing technique is proposed, which is called TFST.

Moreover, since TFST is based on RMA, the image quality can be guaranteed while improving the efficiency. These advantages make the TFST a more practical choice in the high-resolution ViSAR data processing.

In addition, the effect of azimuth resampling on motion error compensation is also analyzed and the results show that azimuth resampling will not reduce the imaging quality of TFST, suggesting the high practicability of TFST.

Finally, the effectiveness of the proposed method has been confirmed by simulation and airborne SAR data processing results.

Author Contributions: Conceptualization, C.Y.; Funding acquisition, Y.D., W.W., P.W. and F.Z.; Investigation, C.Y., Z.C., W.W. and F.Z.; Methodology, C.Y.; Project administration, W.W., P.W. and F.Z.; Resources, Y.D., W.W., P.W. and F.Z.; Validation, C.Y. and W.W.; Writing—original draft, C.Y.; Writing—review & editing, Z.C., Y.D., W.W. and F.Z. All authors have read and agreed to the published version of the manuscript.

Funding: This work was supported in part by the National Science Fund under Grant 61971401 and in part by the Youth Innovation Promotion Association, Chinese Academy of Sciences (CAS).

Data Availability Statement: Not applicable.

Acknowledgments: The authors would like to thank the Department of Space Microwave Remote Sensing System, Aerospace Information Research Institute, Chinese Academy of Sciences for providing X-band high-resolution SAR data.The authors would also like to thank all colleagues who participated in this experiment.

Conflicts of Interest: The authors declare no conflict of interest.

Appendix A

The convolution in the time (frequency) domain is equal to multiplication in the frequency (time) domain [39], so (28) can be expressed as

$$\Gamma(R, f_{a_{new}}) = sS(R, f_{a_{new}}) \otimes exp\left(j\frac{\pi f_{a_{new}}^2}{K_a(R_{scl})}\right) \tag{A1}$$

Then, the convolution in (28) can be expanded as

$$
\begin{aligned}
\Gamma(R, &f_{a_{new}}) \\
&= \int_{-PRF_{new}/2}^{PRF_{new}/2} sS(R, \mu) \\
&\quad \cdot exp\left(j\frac{\pi(f_{a_{new}} - \mu)^2}{K_a(R_{scl})}\right) d\mu \\
&= exp\left(j\frac{\pi f_{a_{new}}^2}{K_a(R_{scl})}\right) \cdot \int_{-PRF_{new}/2}^{PRF_{new}/2} sS_{as}(R, \mu) \\
&\quad \cdot exp\left(-j\frac{2\pi f_{a_{new}}}{K_a(R_{scl})}\mu\right) d\mu
\end{aligned}
\tag{A2}
$$

where

$$
\begin{aligned}
sS_{as}&(R, \mu) \\
&= sS_{rcmc}(R, \mu) \cdot exp\left(j\frac{4\pi}{\lambda} \cdot R \cdot (D(\mu) - 1)\right).
\end{aligned}
\tag{A3}
$$

Expressing (A2) in discrete form, we obtain:

$$
\begin{aligned}
\Gamma(R, k \cdot \Delta f_{a_{new}}) &= exp\left(j\pi K_a(R_{scl}) \cdot \left(\frac{k}{PRF_{new}}\right)^2\right) \\
&\quad \cdot \text{IFFT}(sS_{as}(R, k \cdot \Delta f_{a_{new}})).
\end{aligned}
\tag{A4}
$$

References

1. Curlander, J.C.; McDonough, R.N. *Synthetic Aperture Radar*; Wiley: New York, NY, USA, 1991; Volume 11.
2. Soumekh, M. *Synthetic Aperture Radar Signal Processing*; Wiley: New York, NY, USA, 1999; Volume 7.
3. Stilla, U. High resolution radar imaging of urban areas. In *Proceedings of the Photogrammetric Week*; Wichmann Verlag: Heidelberg, Germany, 2007.
4. Moreira, A.; Prats-Iraola, P.; Younis, M.; Krieger, G.; Hajnsek, I.; Papathanassiou, K.P. A tutorial on synthetic aperture radar. *IEEE Geosci. Remote. Sens. Mag.* **2013**, *1*, 6–43. [CrossRef]
5. Wells, L.; Sorensen, K.; Doerry, A.; Remund, B. Developments in SAR and IFSAR systems and technologies at Sandia National Laboratories. In Proceedings of the 2003 IEEE Aerospace Conference Proceedings (Cat. No. 03TH8652), Big Sky, MT, USA, 8–15 March 2003; Volume 2, pp. 2_1085–2_1095.
6. Damini, A.; Balaji, B.; Parry, C.; Mantle, V. A videoSAR mode for the X-band wideband experimental airborne radar. In *Proceedings of the Algorithms for Synthetic Aperture Radar Imagery XVII*; International Society for Optics and Photonics: Bellingham, WA, USA, 2010; Volume 7699, p. 76990E.
7. Miller, J.; Bishop, E.; Doerry, A. An application of backprojection for video SAR image formation exploiting a sub-aperture circular shift register. In *Proceedings of the Algorithms for Synthetic Aperture Radar Imagery XX*; International Society for Optics and Photonics: bellingham, WA, USA, 2013; Volume 8746, p. 874609.
8. Song, X.; Yu, W. Processing video-SAR data with the fast backprojection method. *IEEE Trans. Aerosp. Electron. Syst.* **2016**, *52*, 2838–2848. [CrossRef]
9. Hu, R.; Min, R.; Pi, Y. A video-SAR imaging technique for aspect-dependent scattering in wide angle. *IEEE Sens. J.* **2017**, *17*, 3677–3688. [CrossRef]
10. Kim, C.K.; Azim, M.T.; Singh, A.K.; Park, S.O. Doppler Shifting Technique for Generating Multi-Frames of Video SAR via Sub-Aperture Signal Processing. *IEEE Trans. Signal Process.* **2020**, *68*, 3990–4001. [CrossRef]
11. Huang, X.; Xu, Z.; Ding, J. Video SAR image despeckling by unsupervised learning. *IEEE Trans. Geosci. Remote. Sens.* **2020**, *59*, 10151–10160. [CrossRef]
12. Hartmann, F.; Sommer, A.; Pestel-Schiller, U.; Ostermann, J. A scheme for stabilizing the image generation for VideoSAR. In Proceedings of the EUSAR 2021, 13th European Conference on Synthetic Aperture Radar, Online, 29–31 April 2021; pp. 1–5.
13. Bishop, E.; Linnehan, R.; Doerry, A. Video-SAR using higher order Taylor terms for differential range. In *Proceedings of the 2016 IEEE Radar Conference (RadarConf)*; IEEE: New York, NY, USA, 2016; pp. 1–4.

14. Soumekh, M. *Time Domain Non-Linear SAR Processing*; Technical Report; State University of New York at Buffalo Department of Electrical Engineering: Buffalo, NY, USA, 2006.
15. Frey, O.; Magnard, C.; Ruegg, M.; Meier, E. Focusing of airborne synthetic aperture radar data from highly nonlinear flight tracks. *IEEE Trans. Geosci. Remote. Sens.* **2009**, *47*, 1844–1858. [CrossRef]
16. McCorkle, J.W.; Rofheart, M. Order N^ 2 log (N) backprojector algorithm for focusing wide-angle wide-bandwidth arbitrary-motion synthetic aperture radar. In *Proceedings of the Radar Sensor Technology*; International Society for Optics and Photonics: Bellingham, WA, USA, 1996; Volume 2747, pp. 25–36.
17. Xiao, S.; Munson, D.C.; Basu, S.; Bresler, Y. An N 2 logN back-projection algorithm for SAR image formation. In Proceedings of the Conference Record of the Thirty-Fourth Asilomar Conference on Signals, Systems and Computers (Cat. No. 00CH37154), Pacific Grove, CA, USA, 29 October–1 November 2000; Volume 1, pp. 3–7.
18. Rogan, A.; Carande, R. Improving the fast back projection algorithm through massive parallelizations. In *Proceedings of the Radar Sensor Technology XIV*; International Society for Optics and Photonics: Bellingham, WA, USA, 2010; Volume 7669, p. 76690.
19. Jian, L.; Running, Z.; Lixiang, M.; Zheng, L.; Ke, J.; Dawei, W.; Zhiyun, T. An efficient image formation algorithm for spaceborne video SAR. In Proceedings of the IGARSS 2018—2018 IEEE International Geoscience and Remote Sensing Symposium, Valencia, Spain, 22–27 July 2018; pp. 3675–3678.
20. Pu, W.; Wang, X.; Wu, J.; Huang, Y.; Yang, J. Video SAR imaging based on low-rank tensor recovery. *IEEE Trans. Neural Netw. Learn. Syst.* **2020**, *32*, 188–202. [CrossRef]
21. Yamaoka, T.; Suwa, K.; Hara, T.; Nakano, Y. Radar video generated from synthetic aperture radar image. In Proceedings of the 2016 IEEE International Geoscience and Remote Sensing Symposium (IGARSS), Beijing, China, 10–15 July 2016; pp. 6509–6512.
22. Zhu, D.; Xiang, T.; Wei, W.; Ren, Z.; Yang, M.; Zhang, Y.; Zhu, Z. An extended two step approach to high-resolution airborne and spaceborne SAR full-aperture processing. *IEEE Trans. Geosci. Remote. Sens.* **2020**, *59*, 8382–8397. [CrossRef]
23. Xing, M.; Wu, Y.; Zhang, Y.D.; Sun, G.C.; Bao, Z. Azimuth resampling processing for highly squinted synthetic aperture radar imaging with several modes. *IEEE Trans. Geosci. Remote. Sens.* **2013**, *52*, 4339–4352. [CrossRef]
24. Cafforio, C.; Prati, C.; Rocca, F. SAR data focusing using seismic migration techniques. *IEEE Trans. Aerosp. Electron. Syst.* **1991**, *27*, 194–207. [CrossRef]
25. Fornaro, G.; Sansosti, E.; Lanari, R.; Tesauro, M. Role of processing geometry in SAR raw data focusing. *IEEE Trans. Aerosp. Electron. Syst.* **2002**, *38*, 441–454. [CrossRef]
26. Reigber, A.; Alivizatos, E.; Potsis, A.; Moreira, A. Extended wavenumber-domain synthetic aperture radar focusing with integrated motion compensation. *IEE Proc.-Radar Sonar Navig.* **2006**, *153*, 301–310. [CrossRef]
27. Quegan, S. Spotlight Synthetic Aperture Radar: Signal Processing Algorithms. *J. Atmos. Sol.-Terr. Phys.* **1995**, *59*, 597–598. [CrossRef]
28. Song, X. *Research on VideoSAR Parameter Design and Fast Imaging Algorithms*; Institute of Electronics, Chinese Academy of Sciences: Beijing, China, 2016.
29. Lanari, R.; Zoffoli, S.; Sansosti, E.; Fornaro, G.; Serafino, F. New approach for hybrid strip-map/spotlight SAR data focusing. *IEE Proc.-Radar Sonar Navig.* **2001**, *148*, 363–372. [CrossRef]
30. Lanari, R.; Tesauro, M.; Sansosti, E.; Fornaro, G. Spotlight SAR data focusing based on a two-step processing approach. *IEEE Trans. Geosci. Remote. Sens.* **2001**, *39*, 1993–2004. [CrossRef]
31. Liu, Y.; Xing, M.; Sun, G.; Lv, X.; Bao, Z.; Hong, W.; Wu, Y. Echo model analyses and imaging algorithm for high-resolution SAR on high-speed platform. *IEEE Trans. Geosci. Remote. Sens.* **2011**, *50*, 933–950. [CrossRef]
32. Xu, W.; Deng, Y.; Huang, P.; Wang, R. Full-aperture SAR data focusing in the spaceborne squinted sliding-spotlight mode. *IEEE Trans. Geosci. Remote. Sens.* **2013**, *52*, 4596–4607. [CrossRef]
33. Moreira, A.; Mittermayer, J.; Scheiber, R. Extended chirp scaling algorithm for air-and spaceborne SAR data processing in stripmap and ScanSAR imaging modes. *IEEE Trans. Geosci. Remote. Sens.* **1996**, *34*, 1123–1136. [CrossRef]
34. Mittermayer, J.; Moreira, A.; Loffeld, O. Spotlight SAR data processing using the frequency scaling algorithm. *IEEE Trans. Geosci. Remote. Sens.* **1999**, *37*, 2198–2214. [CrossRef]
35. Prats, P.; Scheiber, R.; Mittermayer, J.; Meta, A.; Moreira, A. Processing of sliding spotlight and TOPS SAR data using baseband azimuth scaling. *IEEE Trans. Geosci. Remote. Sens.* **2009**, *48*, 770–780. [CrossRef]
36. De Macedo, K.A.C.; Scheiber, R.; Moreira, A. An autofocus approach for residual motion errors with application to airborne repeat-pass SAR interferometry. *IEEE Trans. Geosci. Remote. Sens.* **2008**, *46*, 3151–3162. [CrossRef]
37. Chen, Z.; Zhang, Z.; Zhou, Y.; Wang, P.; Qiu, J. A Novel Motion Compensation Scheme for Airborne Very High Resolution SAR. *Remote. Sens.* **2021**, *13*, 2729. [CrossRef]
38. Wahl, D.E.; Eichel, P.; Ghiglia, D.; Jakowatz, C. Phase gradient autofocus-a robust tool for high resolution SAR phase correction. *IEEE Trans. Aerosp. Electron. Syst.* **1994**, *30*, 827–835. [CrossRef]
39. Cumming, I.G.; Wong, F.H. Digital processing of synthetic aperture radar data. *Artech House* **2005**, *1*, 108–110.

Article

Mutual Interference Mitigation of Millimeter-Wave Radar Based on Variational Mode Decomposition and Signal Reconstruction

Yanbing Li [1,*], Bo Feng [2] and Weichuan Zhang [3]

[1] School of Electronic and Information Engineering, Beijing Jiaotong University, Beijing 100044, China
[2] Beijing Institute of Radio Measurement, Beijing 100854, China
[3] Institute for Integrated and Intelligent Systems, Griffith University, Nathan, QLD 4111, Australia
* Correspondence: ybli1@bjtu.edu.cn

Abstract: As an important remote sensing technology, millimeter-wave radar is used for environmental sensing in many fields due to its advantages of all-day, all-weather operation. With the increasing use of radars, inter-radar interference becomes increasingly critical. Severe mutual interference degrades radar signal quality and affects the performance of post-processing, e.g., synthetic aperture radar (SAR) imaging and target tracking. Aiming to deal with mutual interference, we propose an interference mitigation method based on variational mode decomposition (VMD). With the characteristics that the target is a single-frequency sine wave and the interference is a broadband signal, VMD is used for decomposing the radar received signal and separating the target from the interference. Interference mitigation is then implemented in each decomposed mode, and an interference-free signal is obtained through the reconstruction process. Simulation results of multi-target scenarios demonstrate that the proposed method outperforms existing decomposition-based methods. This conclusion is also confirmed by the experimental results on real data.

Keywords: frequency modulated continuous wave; interference mitigation; millimeter-wave radar; signal reconstruction; variational mode decomposition

Citation: Li, Y.; Feng, B.; Zhang, W. Mutual Interference Mitigation of Millimeter-Wave Radar Based on Variational Mode Decomposition and Signal Reconstruction. *Remote Sens.* **2023**, *15*, 557. https://doi.org/10.3390/rs15030557

Academic Editor: Massimiliano Pieraccini

Received: 2 December 2022
Revised: 7 January 2023
Accepted: 14 January 2023
Published: 17 January 2023

1. Introduction

With the development of information technology, many intelligent applications such as autonomous driving and intelligent transportation have been aroused. As the first step in environmental perception, sensor technology has received considerable attention. In long-distance sensing applications, radar has become irreplaceable due to its all-day and all-weather work ability [1]. In recent years, the manufacturing difficulty and cost of millimeter-wave (mmWave) radar have decreased with the development of chip technology [2–4], which has led to radar being widely used in many fields [5–7]. Along with the rapid growth in deployment number, interference between radars becomes non-negligible [8]. For example, a typical radar configuration for autonomous driving contains four short-range radars deployed at each corner of a vehicle, two mid-range radars at the front and back, and one long-range radar at the front [9]. This configuration makes mutual interference inevitable, especially in congested road conditions.

Mutual interference will seriously degrade radar performance and bring security risks to important radar applications such as synthetic aperture radar (SAR) imaging [10] and multiple target tracking. Specifically, the interference can cause false targets in radar echoes and raise noise level to reduce signal-to-noise rate (SNR), which affects target detecting or imaging performance [11,12]. Therefore, an effective interference suppression solution has become an urgent need for mmWave radar applications in recent years.

Some studies on the phenomenon of radar mutual interference have been proposed. The probability density function of interference is analyzed and the statistical similarity between interference and additive white Gaussian noise is explored in [13]. The results show that interference has a white-noise-like appearance in most cases, which raises the noise floor level. The phenomenon of radar mutual interference, such as characteristics in time domain, frequency domain, and time-frequency (TF) domain for different waveforms, is analyzed in [11,14–16]. The studies mentioned above allow us to understand the characteristics of radar mutual interference and provide a basis for subsequent radar interference suppression research.

In order to mitigate the impact of interference on radar target detection, signal processing approaches are investigated. Setting the interference area to zero is a straightforward approach to suppress interference in the time domain [17], while discontinuity is induced by zeroing in signals. Adding a cosine window to the interference area can avoid the discontinuity caused by zeroing [17]; however, target echoes of interest in interfered signal parts are lost. An autoregressive (AR) model is used for predicting the target information in the interfered parts [18]. This is able to recover target information when the interference duty cycle is small, but fails to obtain good performance in large interference duty cases. To improve the loss of useful information during interference mitigation, some studies turn to signal decomposition approaches [19,20], which first decompose radar-received signals to obtain decomposed modes, then detect and suppress interference in the decomposed modes, and finally reconstruct the received signal to obtain its interference-free version. These signal decomposition methods allow the useful information in the interference-free modes to be retained.

According to the conclusion that target echo and interference can be distinguished in the TF domain [21], interference mitigation methods based on TF analysis have received attention [22]. Experimental results show that TF-domain-based methods can locate interference position more accurately than time-domain-based methods. Nevertheless, TF domain methods need a large amount of storage and computation capacity since the area of interference information extraction is expanded from one dimension of the time domain to two dimensions of the TF domain, which is not suitable to real-time application. Therefore, it is necessary to further explore methods with a better effect but low storage and calculation burden.

On the basis of the above analysis, we investigated interference mitigation based on the signal decomposition technique since it has more flexibility of interference detection than time domain methods and less storage and computational burden than TF domain methods. In this work, a mutual interference mitigation method based on variational mode decomposition (VMD) is introduced. We show that targets in different distances are separated from the broadband interference based on the role of band-pass filter bank of VMD, and target echo in interfered area of decomposed modes can be reconstructed by AR model interpolation. As a result, the improvement of signal-to-interference ratio (SIR) can be realized via replacing the interference by a reconstructed signal. The reconstructed signal has a better interference mitigation effect especially in the multi-target scene.

Compared to the existing decomposition-based methods, this work mainly contributes as follows:

1. According to the broadband frequency characteristics of the interference, the VMD method with quasi-orthogonal decomposition characteristics can effectively decompose the interference energy into different decomposed modes, thus reducing the energy of the interference in each mode and helping to improve SIR through the interference mitigation process in decomposed modes.
2. With the narrowband characteristics of VMD, the linear frequency-modulated (LFM) like interference can be decomposed into sub-band components that have a limited time-support region. This is beneficial in interference detection and signal reconstruction.
3. For multi-target scenes, targets in different ranges can be separated into different decomposed modes based on the nature of narrowband filter banks of VMD. As a result, the number of targets in each mode can be reduced and signal reconstruction can be realized by the linear prediction model with low complexity.

The remainder of this paper is structured as follows. A review of related work and the research motivation of the proposed method are given in Section 2. The signal model of the mmWave radar with mutual interference is established in Section 3; then, the interference mitigation approach based on VMD is introduced in Section 4. The performance evaluation of the proposed method is discussed by simulations and real experiments in Section 5 and Section 6, respectively. Finally, the conclusions are given in Section 7.

2. Related Work and Research Motivation

When mutual interference occurs between mmWave radars, the interference typically appears as a short time component of higher energy in a received signal of a victim radar [11]. Locating and suppressing interference in the time domain can be achieved by zeroing or adding a window to interference. However, zeroing brings discontinuity of signals, which raises the noise level and affects target detection in a range profile (RP). Adding a window can relieve the discontinuity of signals, but still loses useful information in the overlapping part of the interference [17].

Considering the limitation of direct interference suppression in the time domain, a signal-decomposition-based approach is a promising method. For instance, empirical modal decomposition (EMD) has been used for radio frequency interference (RFI) suppression in high-frequency surface-wave radar [23,24], SAR [25], and microwave radiometry [26]. These studies decompose interfered signals by EMD, then process or suppress interference in different decomposed components, and finally obtain the interference-free signal by reconstruction process. The results of these studies demonstrate the effectiveness and potential of EMD in RFI suppression. Another signal decomposition approach based on wavelet technique is proposed in [20] to decompose radar received signal into layers of different resolution, then interference mitigation is achieved by wavelet reconstruction of those layers where the interference is removed.

Although the target information can be retained during interference mitigation by signal decomposition methods such as EMD and wavelet, the frequency characteristics of EMD and wavelet need to be reconsidered in mmWave radar interference mitigation applications.

A typical frequency-modulated continuous wave (FMCW) radar workflow for an interference scenario is shown in Figure 1. The radar signal goes through a total of three phases from transmitter to receiver as follows: a signal-transmitting phase, a target scattering and interference superimposing phase, and a signal-receiving phase. The radar workflow starts from the signal transmitting phase. The radar transmitting signal is generated by a waveform generator, and then amplified by a power amplifier (PA) and radiated into space by a radar transmitter antenna. In the target scattering and interference superimposing phase, when the radar signal encounters a target, it is scattered by the target and returns to the radar. Meanwhile, if there is an interfering signal, this interfering signal will also reach the radar. In this case, the target echo and the interference are superimposed to form a radar received signal. The radar workflow then enters the signal-receiving phase. The target echo and interference are received by the radar receiver antenna and amplified by a low-noise amplifier (LNA). After mixing with the reference signal generated by the waveform generator, a beat frequency signal is obtained. Finally, the digital signal is obtained after sampling by an analog-to-digital converter (ADC) to the beat frequency signal. Here, a low-pass filter (LPF) is used before the ADC in order to avoid aliasing during signal sampling.

The majority of radars currently operate in FMCW mode and the mixer of their receiver outputs a beat frequency signal, which is the difference between the received signal and the transmitted signal [8]. The range measurement is realized in the spectrum of the beat frequency signal, which is known as RP. In automotive radar or traffic radar application scenarios, targets may appear at different distances in front of the radar. Therefore, radar detection performance at different distances will be related to the frequency characteristics of EMD and wavelet decomposition.

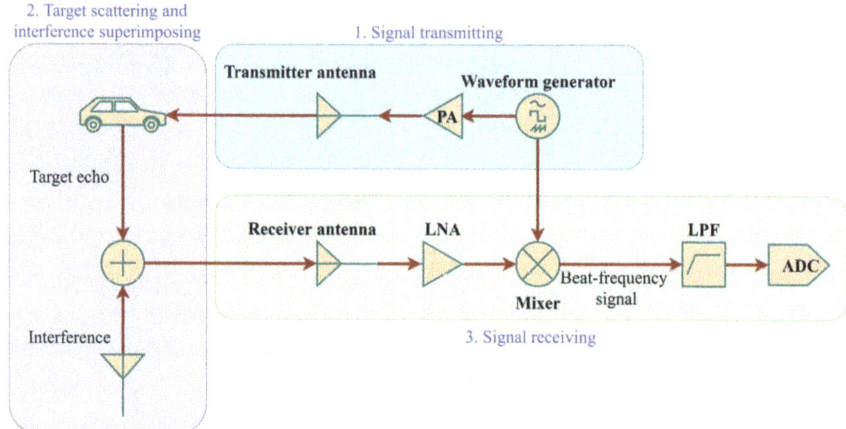

Figure 1. Flow chart of frequency-modulated continuous wave (FMCW) radar system.

The decomposition results of a real−valued broadband white noise using different methods are shown in Figure 2, and the spectrum of each decomposed mode is also given. It is worth noting that millimeter wave radars have both real−valued and complex-valued sampling structures. Currently, most radar implementations use real−valued mixers and real−valued baseband and ADC chains. Part of the reason for this type of implementation is that the number of ADCs and variable gain amplifiers does not have to be doubled, thus gaining a cost advantage [27]. Therefore, the beat frequency signals discussed in this paper are real−valued signals. It can be seen that the frequency characteristics of EMD and wavelet are similar to a set of filter banks, where each decomposed mode occupies a frequency band that is roughly the upper half of the residuals of the previous mode [28,29], as shown in Figure 2b,d. The difference is that the highest band of the filter bank corresponding to EMD is determined by the highest frequency of a signal, while the highest band of the filter bank corresponding to the wavelet is predetermined and equal to the sampling frequency of a signal. These decomposition features indicate that the filter banks corresponding to EMD and wavelet do not divide frequency bands at equal intervals in the frequency domain. In mmWave radar applications, these decomposition features correspond to a non-uniform division of the radar detection range. This means that when broadband interference is present, the SIR of target echoes at different distances are different. In typical mmWave radar detection scenarios, such as vehicle detection on roads, the use of EMD or wavelet for echo decomposition will lead to different SIRs for vehicle echoes at different distances, resulting in different radar detection performance at different distances. Thus, a decomposition method that can provide similar SIR for targets at different distances is desired.

Recently, a new signal decomposition method, named VMD, has been proposed [30]. This decomposes a signal into a collection of band-limited intrinsic mode functions (IMF) with different center frequencies that are quasi-orthogonal to each other. Based on such characteristics, the broadband signal can be uniformly divided into multiple sub-bands in the frequency domain, as shown in Figure 2f. Corresponding to the radar scenario, the received signal is uniformly decomposed into different modes by distance. This allows the targets to have similar SIRs in different IMFs, thus making interference mitigation in different modes more efficient.

In summary, VMD has potential for interference mitigation in typical mmWaveu radar applications.

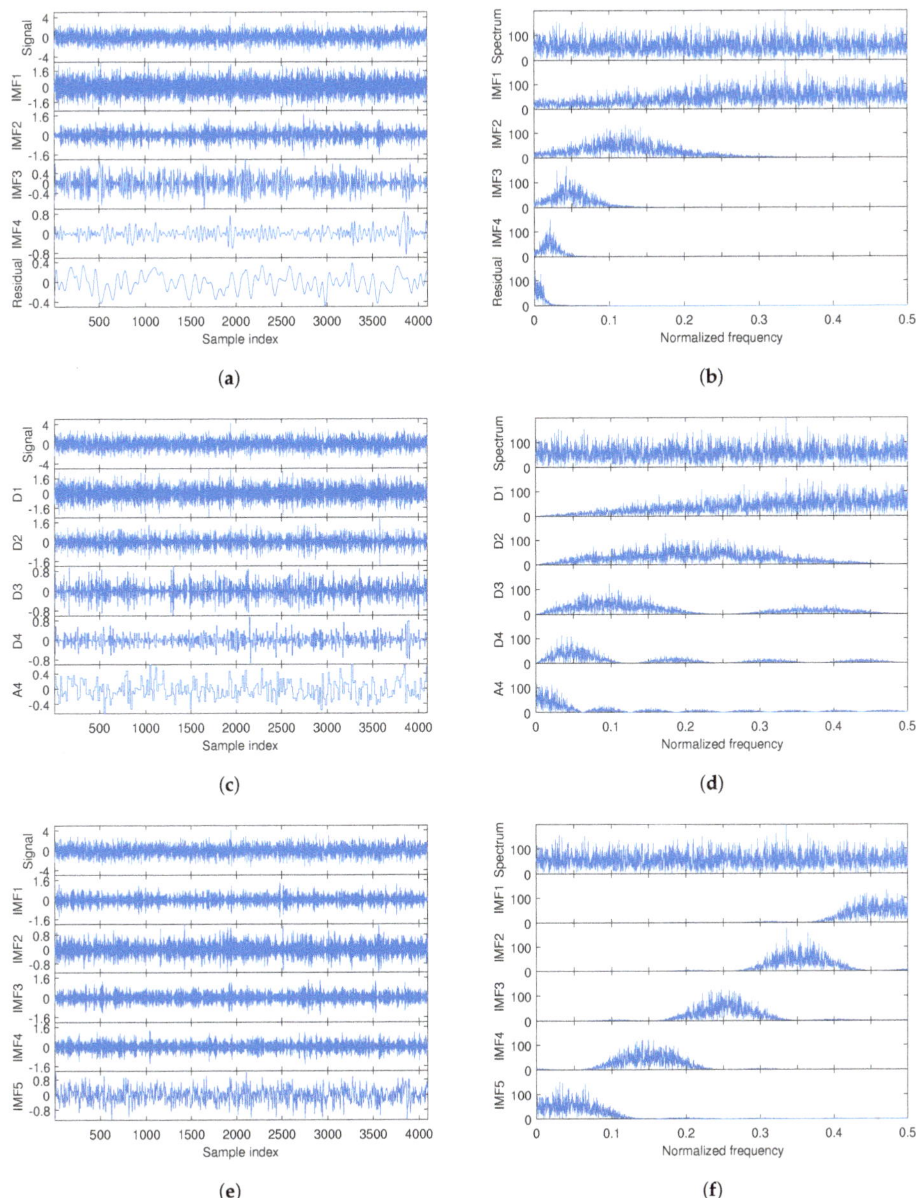

Figure 2. Decomposition results of real−valued white noise by different methods. (**a**) Results of empirical modal decomposition (EMD) in time domain, (**b**) results of EMD in frequency domain, (**c**) results of wavelet in time domain, (**d**) results of wavelet in frequency domain, (**e**) results of variational mode decomposition (VMD) in time domain, and (**f**) results of VMD in frequency domain.

3. Signal Model with Mutual Interference

The most common signal used in mmWave radar is linear FMCW [8,31], which is also called chirp signal. In this work, the radar signal and interference are modeled for the FMCW system, and the interference mitigation technique for the FMCW system is discussed. The single LFM signal transmitted by radar is

$$
\begin{aligned}
s_t(t) &= \sqrt{2P_t}\cos[2\pi\varphi(t)] \\
&= \sqrt{2P_t}\cos\left[2\pi\left(f_c + \frac{1}{2}kt\right)t\right] \\
&= \sqrt{2P_t}\cos\left[2\pi\left(f_c + \frac{1}{2}\frac{B}{T}t\right)t\right],
\end{aligned}
\tag{1}
$$

where P_t is the power of the transmitted signal, f_c is the operating frequency of the radar, k is the slope of the chirp signal, B is the chirp sweep bandwidth, and T is the chirp duration time. When a target exists, the radar-transmitted signal will be scattered by the target, and the echo signal received by the radar is

$$
s_e(t) = \sqrt{2P_e}\cos[2\pi\varphi(t - \tau)],
\tag{2}
$$

where τ is the time delay related to the range and velocity of the target and P_e is the target echo power. According to the radar equation [32], the target echo power is

$$
P_e = \frac{P_t G^2 \lambda^2 \sigma}{(4\pi)^3 R^4},
\tag{3}
$$

where λ is the wavelength, G is the antenna gain, σ is the target radar cross section (RCS), and R is the distance between the radar and the target. It should be noted that due to the two-way propagation of electromagnetic waves in space, the target echo power is inversely proportional to R^4.

When mutual interference occurs, interference signal is usually emitted by a interferer radar. A typical interference signal is

$$
\begin{aligned}
s_i(t) &= \sqrt{2P_i}\cos[2\pi\varphi_i(t - \tau_i)] \\
&= \sqrt{2P_i}\cos\left[2\pi f_{ci}(t - \tau_i) + \frac{1}{2}k_i(t - \tau_i)^2\right],
\end{aligned}
\tag{4}
$$

where P_i, f_{ci} and k_i are the power, carrier frequency and chirp slope of interference, respectively. The interference power is

$$
P_i = \frac{P_{t_i} G_i G \lambda^2}{(4\pi)^2 R_i^2},
\tag{5}
$$

where P_{t_i} and G_i are the power of the transmitted signal and the antenna gain of the interferer radar, respectively. R_i is the distance between the interferer radar and the victim radar. It should be noted that the propagation of electromagnetic waves is one-way in the interference case; thus, the interference power is inversely proportional to R_i^2. When the interferer and the target are located at the same distance, the interference power will be greater than the target echo power according to (3) and (5).

From (2) and (4), the target echo and the interference are received by radar, and the received signal is

$$
s_r(t) = s_e(t) + s_i(t) + v(t),
\tag{6}
$$

where $v(t)$ is the radar receiver noise. From (1), (2), (4), and (6), the beat frequency signal in baseband can be obtained as

$$
\begin{aligned}
s_b(t) =& s_t(t) \cdot s_r(t) \\
=& s_t(t) \cdot [s_e(t) + s_i(t) + v(t)] \\
=& \sqrt{2P_t}\cos[2\pi\varphi(t)] \\
& \cdot \left\{ \sqrt{2P_e}\cos[2\pi\varphi(t-\tau)] \right. \\
& \left. + \sqrt{2P_i}\cos[2\pi\varphi_i(t-\tau_i)] \right\} + v(t) \\
=& 2\sqrt{P_t P_e}\cos\left[2\pi(k\tau)t - 2\pi\left(\tfrac{1}{2}k\tau\tau - f_c\tau\right)\right] \\
& + 2\sqrt{P_t P_i}\cos\left[2\pi\left(f_c + \tfrac{1}{2}kt - f_{ci} + k_i\tau_i - \tfrac{1}{2}k_i t\right)t \right. \\
& \left. - 2\pi\left(\tfrac{1}{2}k_i\tau_i\tau_i - f_{ci}\tau_i\right)\right] + v(t).
\end{aligned}
\tag{7}
$$

From (7), the beat frequency corresponding to the target is $f_b = k\tau$ and the beat frequency corresponding to the interference is $f_{bi} = f_c - f_{ci} + k_i\tau_i + \frac{1}{2}(k - k_i)t$. It is worth noting that the beat frequency of interference is a LFM signal, which means the interference has a broadband spectrum in the frequency domain.

According to Figure 1, radar signal flow in an interfered condition can be described as follows: The signal is radiated into the environment through the transmitter, then scattered by the target and superimposed with the interference. At the end of the signal propagation, it reaches the radar and is received by the radar antenna. After mixing the received signal with the reference signal, a beat frequency signal can be obtained. Then, the beat frequency signal passes through an LPF before it is sampled by ADC. The function of the LPF is to prevent signal aliasing during analog-to-digital sampling. According to (7), the target echo after the dechirping process is a single-frequency sinusoid, while the interference shows LFM features after dechirping. For target echo, as long as its beat frequency is less than the cut-off frequency of LPF, the target information can be retained in the sampled digital signal. For interference, it becomes a broadband signal which occupies the whole frequency domain due to the LPF [15].

In a nutshell, after the dechirp process, the features of target echo and interference in the same distant are summarized as follows:

1. The target echo is a single-frequency and small-power sinusoid.
2. The interference is a broadband and large-power signal.

4. Interference Mitigation Method

In radar-received signals, target echo is a single-frequency component, while interference is a broadband signal. Therefore, in the spectrum of the received signal, the target is a spectral line and the interference is a broadband power signal that raises the noise floor [21]. Due to the characteristics of the target echo and the interference, we propose a interference mitigation method based on VMD. In this method, the received signal is decomposed by VMD to obtain subband signals, so that the broadband interference can be decomposed into different subbands, which reduces the interference power in each IMF, and therefore, interference mitigation can be performed in each IMF.

4.1. Introduction to Variational Mode Decomposition

VMD is an adaptive signal decomposition algorithm proposed in [30]. It is widely used in many fields due to its good adaptive decomposition characteristics [33–35]. A signal can be decomposed by VMD into an ensemble of band-limited IMFs, which have quasi-orthogonal property to each other. Essentially, VMD is a generalization of the classic Wiener filter into multiple, adaptive bands. Thus, we can obtain the decomposed IMFs for a given radar signal either exactly or in a least-squares sense. The narrow-band modes for a beat frequency signal can be obtained by solving the constrained variational problem as follows:

$$\min_{\{u_m\},\{\omega_m\}} \left\{ \sum_m \left\| \partial_t \left[\left(\delta(t) + \frac{j}{\pi t} \right) * u_m(t) \right] e^{-j\omega_m t} \right\|_2^2 \right\},$$

$$\text{s.t.} \quad \sum_m u_m(t) = s_b(t)$$

(8)

where $u_m, m = 1, ..., M$ and $\omega_m, m = 1, ..., M$ are the set of all modes and their center frequencies, respectively. There are three stages implemented for VMD in (8):

1. Firstly, for each mode u_m, compute the associated analytic signal by using the Hilbert transform.
2. Secondly, for each mode u_m, shift the mode's frequency spectrum to baseband, by mixing with an exponential tuned to the respective estimated center frequency.
3. Finally, the bandwidth is estimated through the squared L^2-norm of the gradient.

The variational problem in (8) can be converted to an unconstrained problem by introducing an augmented Lagrangian function $\mathcal{L}$ as follows:

$$\mathcal{L}(\{u_m\}, \{\omega_m\}, \beta) = \alpha \sum_m \left\| \partial_t \left[\left(\delta(t) + \frac{j}{\pi t} \right) * u_m(t) \right] e^{-j\omega_m t} \right\|_2^2$$
$$+ \left\| s_b(t) - \sum_m u_m(t) \right\|_2^2 + \left\langle \beta(t), s_b(t) - \sum_m u_m(t) \right\rangle,$$

(9)

where α represents the variance of white noise and β is the Lagrange multiplier. Then, the unconstrained problem can be solved efficiently in a classical alternate direction method of multipliers (ADMM) approach, and the decomposition modes are extracted concurrently. In detail, the solution of (9) is the saddle point of the augmented Lagrangian $\mathcal{L}$ in a sequence of iterative sub-optimizations. At step $n + 1$ of the alternating update, the mode u_m^{n+1}, the center frequency ω_m^{n+1} of the current mode, and the Lagrange multiplier β^{n+1} are:

$$\hat{u}_m^{n+1}(\omega) = \frac{\hat{s}_b(\omega) - \sum_{i \neq m} \hat{u}_i(\omega) + \frac{\hat{\beta}(\omega)}{2}}{1 + 2\alpha(\omega - \omega_m)^2},$$

(10)

$$\omega_m^{n+1} = \frac{\int_0^\infty \omega |\hat{u}_m(\omega)|^2 d\omega}{\int_0^\infty |\hat{u}_m(\omega)|^2 d\omega},$$

(11)

$$\hat{\beta}^{n+1}(\omega) \leftarrow \hat{\beta}^n(\omega) + \gamma \left(\hat{s}_b(\omega) - \sum_m \hat{u}_m^{n+1}(\omega) \right),$$

(12)

where $\hat{u}_m^{n+1}$, $\hat{s}_b$ and $\hat{\beta}^{n+1}$ are the Fourier transform corresponding to u_m^{n+1}, s_b and β^{n+1}, respectively. γ is the noise tolerance parameter. During the alternating update process, the bandwidth and centering frequency of each IMF mode are continuously updated until the iteration stop condition is satisfied as follows:

$$\sum_m \left\| \hat{u}_m^{n+1} - \hat{u}_m^n \right\|_2^2 / \|\hat{u}_m^n\|_2^2 < \epsilon,$$

(13)

where ϵ is the convergence tolerance level.

After decomposition, the reconstruction of the beat frequency signal can be obtained from the summation of the IMFs, which is

$$s_b(t) = \sum_{m=1}^M u_m(t).$$

(14)

It is worth noting that the number of decomposition modes M is an important parameter of VMD, which affects the accuracy of the decomposition results. Studies to determine the number of decomposition modes have been proposed [36,37]. In the traffic-oriented interference mitigation, the determination of M can be based on existing methods; however, the following issues need to be considered. Since the spectrum of the radar beat frequency signal represents the RP, a reasonable value of M can be determined first by existing methods, and then a further correction of M is needed to fit the traffic scenario. The goal of the correction is to divide the radar detection range into suitable range subsections to ensure that the number of targets in each range section is not too large, which helps us to simplify the order of the subsequent linear prediction model.

4.2. Interference Mitigation Realization

For the mutual interference mitigation of mmWave radars, assuming that there are multiple targets, without loss of generality, we consider that targets are evenly distributed over the radar detection range, and target echoes are the superposition of multiple sinusoids. Since the spectrum of beat frequency signal is the RP, targets with different distances are located at different frequency bins, and a high frequency corresponds to a far distance. Unlike the targets, the power of interference distributes over the entire frequency domain and behaves like broadband noise.

Based on the different characteristics of target echo and interference in the frequency domain, VMD is used for decomposing radar-received signal into narrowband modes. Due to the quasi-orthogonal property of VMD, it ensure that targets of different range are assigned to different modes, and the interference power is also divided into these narrowband modes. As a result, the number of targets reduces in each decomposed mode, and the band width of interference is limited in each mode simultaneously. In these cases, signal power is preserved while the interference power is reduced in each mode, and thus, SIR is improved. This is beneficial for performing interference suppression and signal recovery.

After decomposition by VMD, interference can be located in the time domain for each IMF mode and suppressed by zeroing; then, the target echo signal in an interfered area can be recovered by linear interpolation, which can retain more information about the target. Usually, the interpolation can be realized by a linear prediction problem defined as

$$u(t) = \sum_{i=1}^{Q} \phi_i u(t-i) + \varepsilon_t, \tag{15}$$

where Q is the model order, ϕ_i is the prediction coefficient and ε_t is the residual. The AR model is a commonly used approach to solve the linear prediction problem, and the key point of the solution is to determine its order Q. Studies have shown that when the signal is the summation of N sinusoids, the prediction in (15) can be uniquely achieved by $2N$ samples, i.e., $Q = 2N$ [38,39]. Corresponding to mmWave radar applications, the order Q of the linear model is related to the number of targets N in received signal.

It is worth noting that the order Q is not easy to determine in practice since the number of targets N is unknown, especially when the target number is large, e.g., in a congested traffic environment. In addition, an AR model of a large order will be sensitive to noise, which affects the quality of signal recovery. Therefore, it is necessary to reduce the number of targets to ensure that the AR model has lower order and is more robust to noise. From the decomposition characteristics of VMD as shown in Figure 2f, VMD has the ability to divide radar detection range into IMF modes uniformly, which will help to reduce the number of targets in each IMF. In this sense, VMD is suitable for the signal recovery in the linear prediction model as shown in (15).

In summary, the interference mitigation algorithm flow is shown in Figure 3. The algorithm steps are identified numerically in the figure, and the details of the corresponding signal processing steps are described as follows:

1. Signal input step.
   The received signal with interference is dechirped in the radar receiver to obtain the beat frequency signal, and the beat frequency signal is the input of the interference mitigation algorithm.
2. Signal decomposition step.
   VMD is used to obtain the narrowband modes of the beat frequency signal. There are total M modes.
3. Mode selection step.
   Interference mitigation is performed on each decomposed mode. For one interference mitigation process, a decomposed mode needs to be selected.
4. Interference detection and location step.
   For each mode, the interference is detected and located by means of a constant false alarm rate (CFAR) detector [40].
5. Signal recovery step.
   Based on the results of interference location, the signal at the interference points is removed and replaced by interpolation values via an AR model as shown in (15) in each mode.
6. Signal reconstruction step.
   Repeat steps 3 to 5 until all modes have been processed. Then, the beat frequency signal is reconstructed according to (14) to obtain an interference-free time domain signal.
7. Signal output step.
   The interference-free signal is output and will be used as input for subsequent radar signal processing.

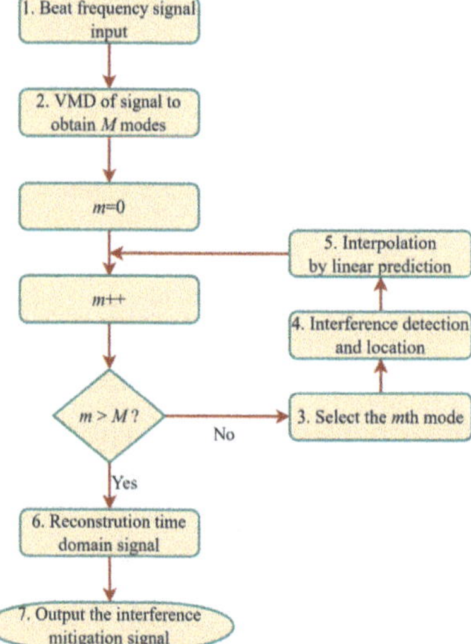

Figure 3. Interference mitigation flow chart.

5. Numerical Simulation Results

5.1. Simulation Description

A multi-target scene is simulated to evaluate the performance of the proposed method. In the multi-target scene, a total of ten targets with different ranges and velocities are simulated. Specifically, the targets are evenly distributed in the range of 90 m to 900 m at a spacing of 90 m. The targets information of the simulation are shown in Table 1.

Table 1. Target Information for Simulation.

Target Index	1	2	3	4	5	6	7	8	9	10
Range (m)	90	180	270	360	450	540	630	720	810	900
Velocity (m/s)	2.7	3.0	3.3	3.6	3.8	4.1	4.4	4.7	5.0	5.2

Considering that an automobile radar scenario is simulated in the experiment, a 77 GHz frequency band is used for simulation. Two interferer radars are configured with different modulation directions and slopes compares to the victim radar, as shown in Table 2. In this simulation, real−valued mixers and real−valued baseband and ADC chains are employed. In terms of the radar system signal flow as shown in Figure 1, the signals are analog until the ADC sampling; thus, we used a high sampling frequency, i.e., 2 GHz, to simulate the analog signal. In the simulation implementation, a complete process of radar signal transmitting, target interaction, interference superposition and signal receiving is simulated. In detail, the target echoes are generated according to (2) and (3) based on the range and speed of the targets as shown in Table 1. Meanwhile, the interference signals are generated according to (4) and (5) based on the parameters of the interferer radars as shown in Table 2. Then, the target echoes and interference signals are superposed according to (6), and the mixing process with the reference signal is performed according to (7). Finally, the sampling process of the ADC is simulated using the intermediate frequency (IF) as shown in Table 2, and the sampled beat frequency signal is obtained.

The simulated signals of targets are shown in Figure 4. Comparing the time domain signatures before and after adding interference to target echoes as shown in Figure 4a,c, the interference presents larger power at different time segments. In the range domain as shown in Figure 4b,d, the interference presents as a broadband power component. Since the power of the interference is larger than that of the target echo, the interference significantly increases the noise level in the RP, and all targets are invisible in the RP except for the first strong one. In this case, the detection performance of targets is degraded.

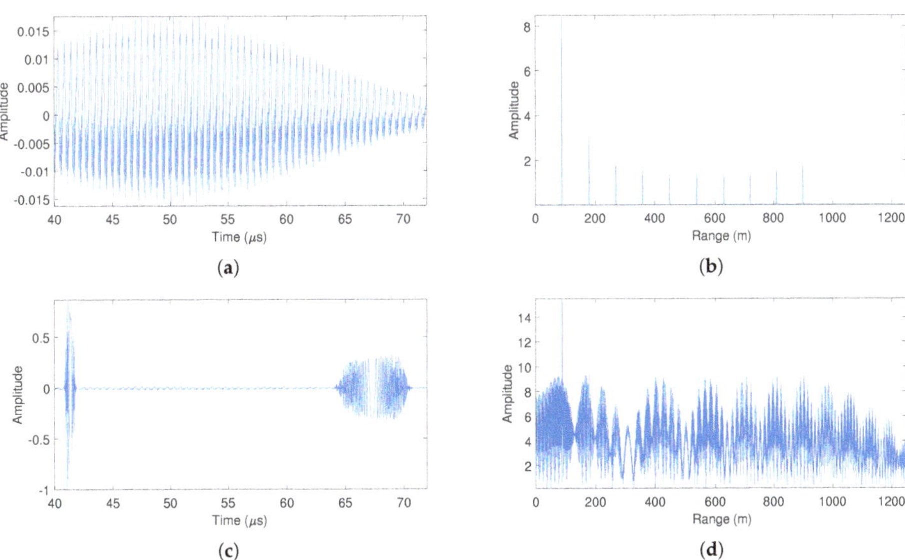

Figure 4. Real−valued beat frequency signal and correspondingrange profile (RP). (**a**) Beat frequency signal without interference, (**b**) RP without interference, (**c**) beat frequency signal with interference, and (**d**) RP with interference.

Table 2. Numerical Simulation Parameters of Radars.

Parameter	Victim	Interferer1	Interferer2
Operating frequency (GHz)	77	77	77
Sweep bandwidth (MHz)	300	600	600
Sweep time (μs)	100	10	50
Sweep direction	Up	Down	Up
IF sampling frequency (MHz)	50	-	-

5.2. Performance Evaluation Methodology

A mmWave radar operating in FMCW mode usually transmits chirp sequences for sensing the environment. In radar signal processing, the received chirp sequence is applied with two-dimensional (2D) fast Fourier transform (FFT) to obtain the corresponding range and velocity information [5]. The first FFT is the range FFT, which is performed on the sampling points of each chirp. The range FFT finds the RP, which reflects the distribution of the targets on the range domain. The second FFT is the Doppler FFT, which is based on the range FFT and is processed at different chirps. The Doppler FFT obtains the velocity profile of the targets, which reflects the velocity distribution of the targets at a specific range. After 2D FFT, the range-Doppler (RD) response of the radar signal can be obtained, and the RD response reflects the distribution of the targets in the joint range-velocity plane.

In carrying out interference mitigation performance evaluation, one-dimensional (1D) evaluation and 2D evaluation are employed in this work. In the 1D evaluation, the main purpose is to evaluate interference mitigation performance in both the time domain and range domain for a single chirp. Based on the 1D evaluation, the 2D evaluation mainly evaluates the role of interference mitigation on radar velocity measurement.

For 1D evaluation, evaluation in the time domain aims to describe the degree of interference suppression, and evaluation in the range domain focuses on the improvement of target detection ability. In order to quantify the performance, one time-domain metric and one range-domain metric are adopted. The time-domain metric is the signal to interference plus noise ratio (SINR), which is defined as

$$\text{SINR} = 10 \times \log 10 \frac{\|s_e\|_2}{\|s_e - s_{rec}\|_2}, \tag{16}$$

where s_e is the true value of target echo and s_{rec} is the reconstructed counterpart of s_e. SINR is used for measuring the suppressed degree of interference in the time domain.

The RP metric is integrated sidelobe ratio (ISLR) [41,42]. The calculation of ISLR is slightly modified in this paper in order to suit for multi-target situation. In detail, a neighborhood RP of a specific target is extracted to calculate the modified ISLR (MISLR) of the target. For a specific target, the MISLR is defined as

$$\text{MISLR} = 10 \times \log 10 \left(\frac{\sum_{m=c}^{a} F^2(m) + \sum_{m=b}^{d} F^2(m)}{\sum_{m=a}^{b} F^2(m)} \right), \tag{17}$$

where F is the spectrum of s_{rec}, the interval $[a, b]$ bounds the main lobe and the interval $[c, d]$ bounds the neighborhood of the target. A smaller MISLR value indicates a better performance of target detection, since the MISLR describes the ratio between the energy of neighborhood sidelobes with respect to the energy of the main lobe. For practical implementation, we recommend that the neighborhood can be selected as at least 5 times the width of the main lobe in order to have more adequate sample points to estimate the sidelobe level; i.e., the length of $[c, d]$ is at least 5 times the length of $[a, b]$.

5.3. Simulation Results

Two decomposition-based methods are utilized for performance comparison, including an EMD-based method [24] and wavelet-based method [20]. In this experiment, Haar basis is employed for the wavelet decomposition. In implementing interference mitigation by the proposed method, the number of decomposition modes of the VMD is set to 5, which means that the detection range of the radar of 1200 m is divided into five sub-range segments, each of which is about 240 m. In the interference detection of each mode, a cell-averaging CFAR (CA-CFAR) detector is used, where the guard cell number is set to 2 times the number of main lobe sampling points and the training cell number is set to 10 times the number of main lobe sampling points.

The decomposition results of EMD, wavelet and VMD for simulated signals are shown in Figure 5. In total, five modes are specified in the implementation of different methods. For EMD, there are four IMFs and a residual as shown in Figure 5a,b to represent the decomposition results in the time domain and RP, respectively. Correspondingly, the results of wavelet decomposition are shown in Figure 5c,d: there are four detail modes and one approximate modes. The results of VMD are shown in Figure 5e,f: there are five IMFs.

The analysis of the decomposition results in the time domain and RP are described as follows:

1. Results of EMD.

   It can be seen from the decomposition results of the RP that IMF1 contains most of the frequency components, as shown in Figure 5b. Although IMF2 occupies about half of the low-frequency band, there is still a large frequency overlap between IMF1 and IMF2. This decomposition feature makes the interference components and most of the target echoes to be contained in IMF1. The decomposition results in the time domain also show that the waveform of IMF1 is similar to the original signal. In this case, the interference mitigation based on EMD does not gain benefit in the decomposition process.

2. Results of wavelet.

   The wavelet decomposition results are similar to those of EMD, where most of the interference and target echo components are decomposed into Detail 1 and Detail 2 signals, as shown in Figure 5c,d. This also makes it impossible to obtain better interference mitigation based on this decomposition result.

3. Results of VMD.

   Based on the quasi-orthogonal and band-limited decomposition characteristics of VMD, the interfered echo signal is decomposed to obtain approximately uniform range sections in RP, as shown in Figure 5f. Such decomposition brings two benefits: the first benefit is that the interference power is uniformly decomposed into different IMFs. According to (7), the interference is characterized as an LFM signal in the beat frequency signal. When the interference is decomposed into different narrowband IMFs, it is correspondingly decomposed into different time segments in the time domain, as shown in Figure 5e. The second benefit is that targets at different distances are uniformly decomposed into different IMFs, and each target is basically decomposed into a unique IMF due to the quasi-orthogonality of VMD. As a result of the above benefits, the support area of the interference in time domain becomes smaller for each IMF, which contributes to the computational reduction in the linear prediction model. In addition, the number of targets in each IMF is reduced, which contributes to the reconstruction of target echoes by using a lower-order model.

The energy percentage of the decomposition modes for each method is counted, as shown in Figure 6. It can be seen that VMD has the most balanced decomposition results among the three methods, where each mode has a similar energy percentage. Thus, it is more suitable for the interference mitigation application of mmWave radars.

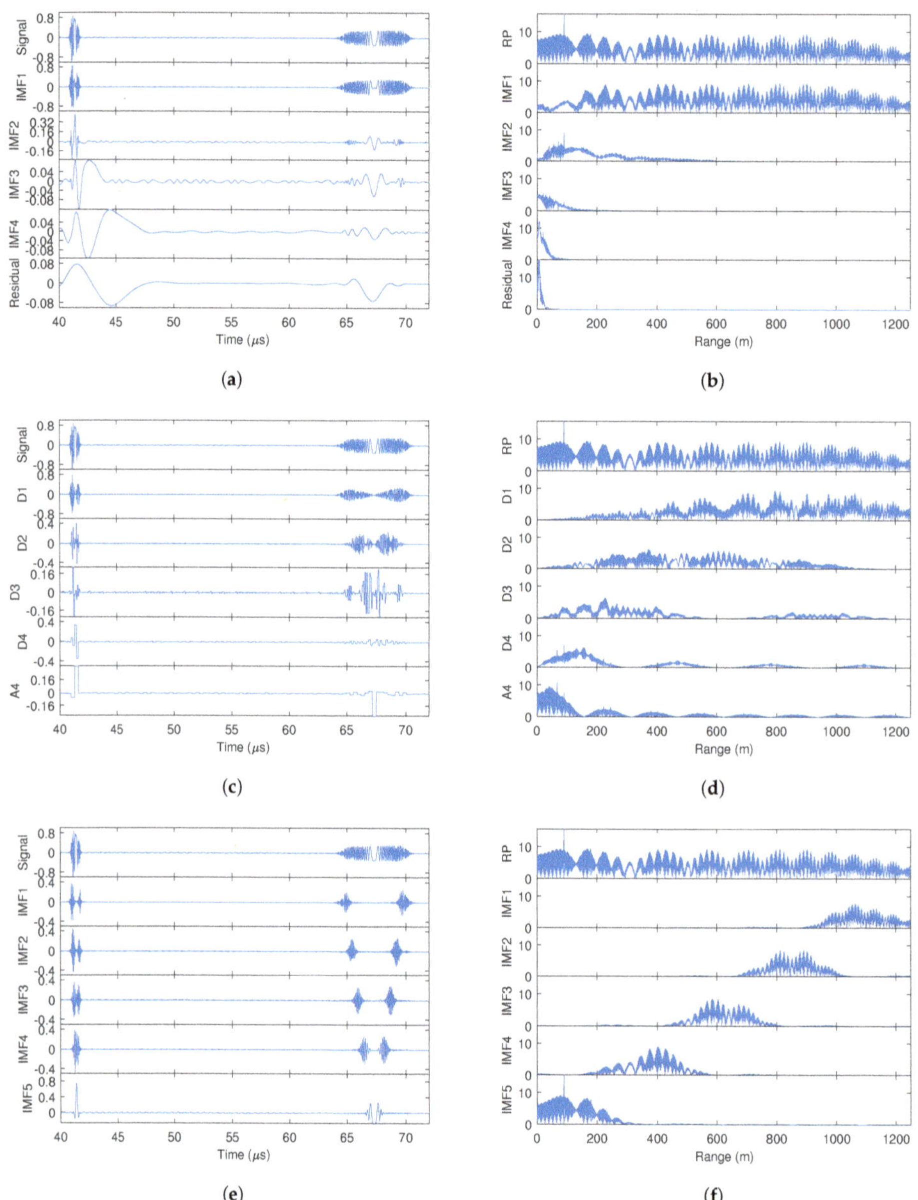

Figure 5. Decomposition results of simulated real−valued beat frequency signal by different methods. A total of five modes are specified in the implementation of different methods. For EMD, there are four intrinsic mode functions (IMF) and a residual. For wavelet, there are four detail modes and one approximate modeb. For VMD, there are five IMFs. (**a**) Results of EMD in time domain, (**b**) results of EMD in RP, (**c**) results of wavelet in time domain, (**d**) results of wavelet in RP, (**e**) results of VMD in time domain, and (**f**) results of VMD in RP.

The interference mitigation performance of all the tested methods in the time domain is shown in Figure 7. Among these results, the EMD and wavelet methods result in the interference energy being concentrated in a few decomposed modes due to the inhomogeneity of the interference decomposition. Interference mitigation in these modes is not much improved compared to the interference mitigation in the original signal. This leads to limitations in interference localization and suppression. Compared with the EMD and wavelet methods, the VMD method decomposes the interference into different time segments and spreads the interference energy into different decomposed modes, which makes the interference reduced in both the support domain and power for a mode relative to the original signal, and then the localization and suppression of interference for each mode can obtain the benefits from the decomposition. Therefore, the VMD method obtains the best interference suppression and target reconstruction effect.

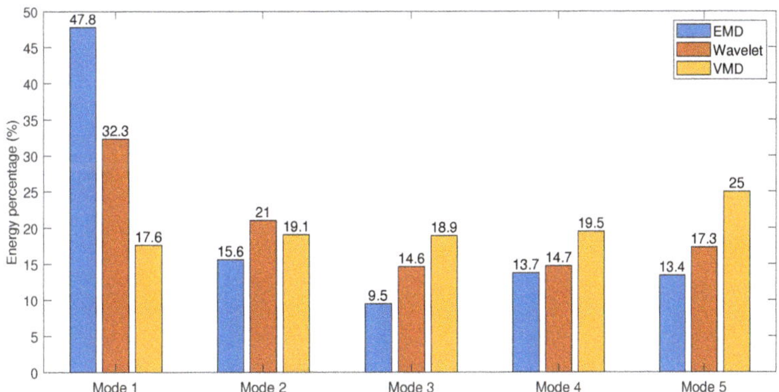

Figure 6. Energy percentage of decomposed modes for different methods.

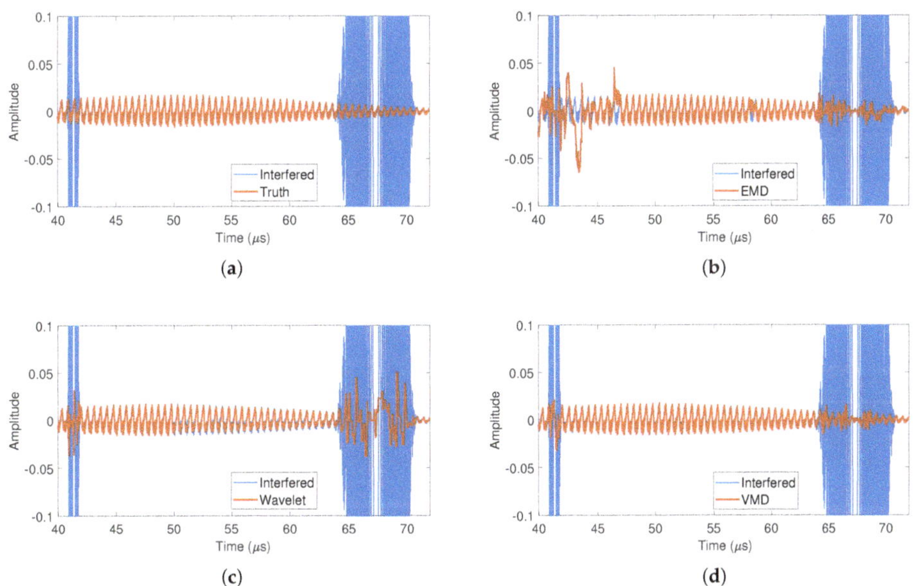

Figure 7. The interference mitigation performance of different methods in time domain for simulation experiment. (**a**) The beat frequency signal with and without interference, (**b**) the reconstructed signal by EMD, (**c**) the reconstructed signal by wavelet, and (**d**) the reconstructed signal by VMD.

The quantitative evaluation results of interference mitigation for the tested methods in the time domain are shown in Figure 8. The corresponding SINR is calculated after interference mitigation using different methods, and the original SINR level is given as a reference. All three test methods achieved interference mitigation effects, and the highest SINR level is obtained by the VMD method. Relative to the EMD and wavelet methods, the SINR of the VMD method is about 4dB higher.

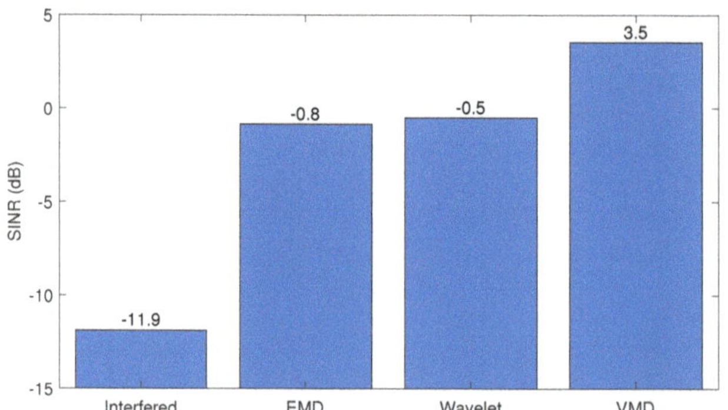

Figure 8. Signal to interference plus noise ratio (SINR) of different methods for simulation experiment.

The interference mitigation performance of all the tested methods in RP is shown in Figure 9. Before interference mitigation, only the amplitude of the first target is higher than that of the interference, and the remaining nine targets are submerged by the interference in the original signal. After interference mitigation, all three methods are able to display the ten targets completely. Among them, the VMD method obtained the lowest noise level over the whole RP. This shows the advantage of the VMD decomposition feature for interference mitigation at different distance segments.

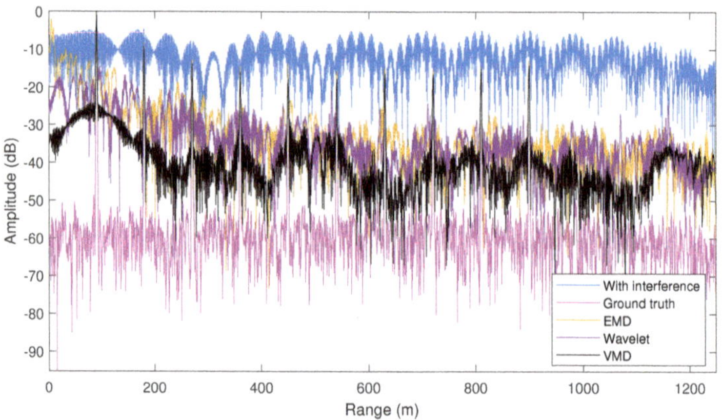

Figure 9. The interference mitigation performance of different methods in RP for simulation experiment.

The quantitative results in the range domain are shown in Figure 10, which gives the MISLR levels of the ten targets, and the original MISLR levels are given as a reference. From the statistical results, it can be seen that the VMD method obtains significant improvements in the MISLR levels of most targets compared to the other methods.

Interference mitigation performance on RD response is also evaluated. A chirp sequence is designed in the simulation experiment for obtaining velocity measurement capability. There are in total 256 chirps in one sequence, which is denoted as a frame. For each chirp in a frame, interference mitigation is performed separately, and then the processed chirps are formed into a new frame for subsequent 2D FFT processing to obtain the RD response map after interference mitigation [8].

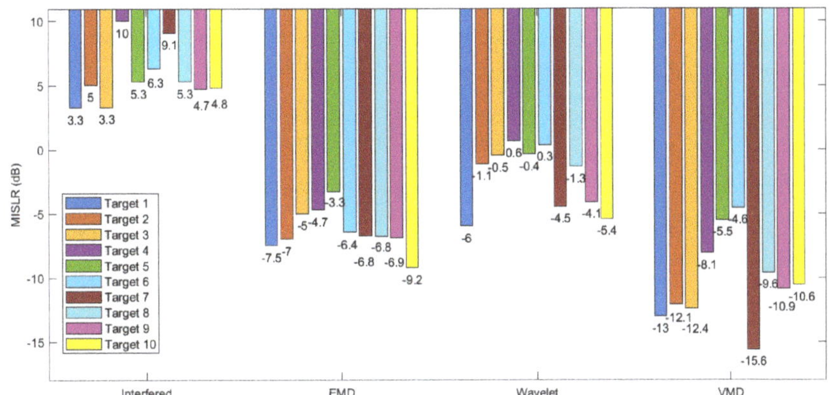

Figure 10. Modified ISLR (MISLR) of different methods for simulation experiment.

The RD maps of different methods are shown in Figure 11. Compared with the RD map with interference, all three tested methods provide effective interference mitigation. Compared with the truth, the three methods can correctly extract the RD response of the ten targets, and the VMD method obtains the best interference mitigation effect. In detail, the VMD method achieves the smallest noise level, which is beneficial to target detection performance.

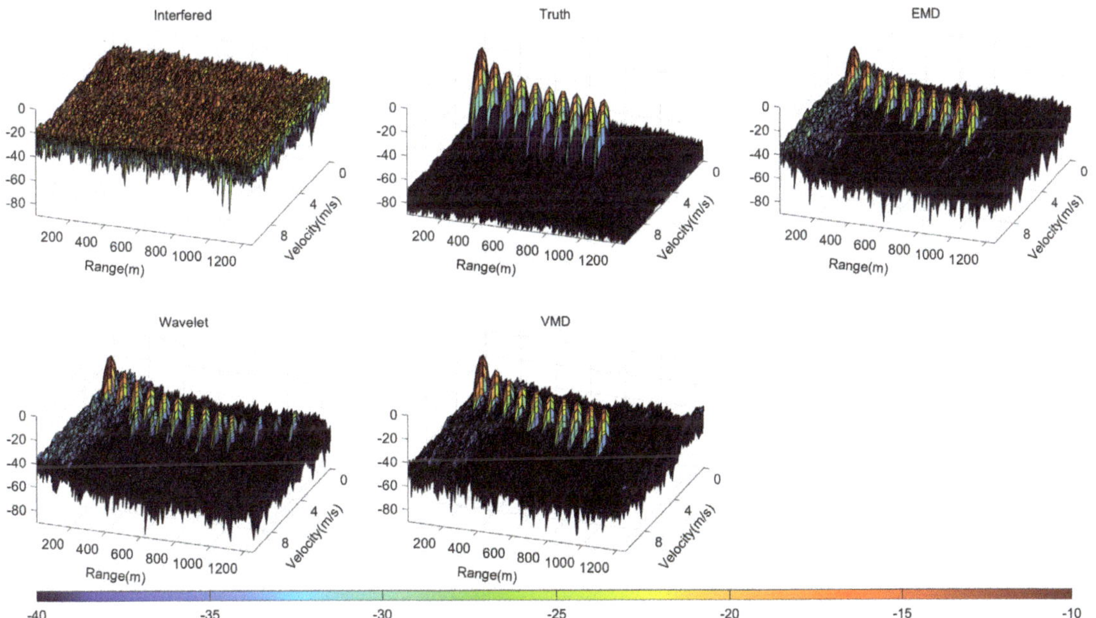

Figure 11. The interference mitigation performance of different methods in range-Doppler (RD) response for simulation experiment. Different colors represent different RD amplitudes in decibels.

In order to evaluate the computational power load of the tested methods, we use the simulation data for the running time analysis of each method. The algorithmic flow of interference mitigation is divided into two parts: signal decomposition and interference detection and mitigation. The running time of each part is counted separately and the total time consumed by each method is also given. The computer platform used for the evaluation has an AMD Ryzen 7 5800H central processing unit (CPU) and 16 GB of DDR4-3200 random access memory (RAM). The Matlab software version used in this experiment is R2021a. The running time results for the interference mitigation of a simulated chirp are shown in Table 3.

Table 3. Running Time of Different Methods for Simulation Data.

Method List	Running Time of Signal Decomposition (ms)	Running Time of Interference Detection and Mitigation (ms)	Total Running Time (ms)
EMD	47	412.9	459.9
Wavelet	38.9	28.3	67.2
VMD	192.2	749.9	942.1
VMD (parallel)	191.1	126.7	317.8

It can be seen that the proposed method has the longest running time. Specifically, for the running time of the signal decomposition part, the VMD is more time-consuming than EMD and wavelet because it is an optimization problem solving process. In the interference detection and mitigation part, the EMD and VMD methods use CFAR detector for interference detection, which has more computation and takes longer compared to the soft threshold detection of the wavelet method. In addition, the VMD method has an additional process of signal recovery using the AR model, which also increases the algorithm's time consumption. Overall, the wavelet method consumes the least amount of time due to its fast algorithm structure, and the proposed method has the longest running time. In practice, since the proposed method is performed for each mode in the interference detection and mitigation part and there is no correlation between the modes, parallel processing can be considered in this part. The running time using parallel processing is shown in the last row of Table 3, which demonstrates that the elapsed time of the proposed method is significantly reduced by parallel computing.

6. Real Experiment Results

A real scene experiment is designed to evaluate the effectiveness of the proposed method. In the experiment, we used three 77 GHz mmWave radars from Muniu Linghang Technology Company for data recording. One of these radars is used as a victim device and the other two are used as interference sources. The received data of the victim radar are recorded for performance evaluation. The locations of all devices in the experimental scene are shown in Figure 12. The two interferer radars are distributed on the left and right sides of the victim's line of sight, and the distance from the victim to the interference source 1 and source 2 is 20 m and 30 m, respectively. A corner reflector is placed at 20 m in front of the victim to simulate a typical strong target. A motorcycle rider's echo data were also recorded.

Figure 12. Data collection scenario.

The radar parameters used in the experiment are given in Table 4. All radar devices operate at 77 GHz, differing in specific signal parameters. The victim is configured as up-frequency modulation with a sweep bandwidth of 300 MHz, the interference source 1 is configured as the down-frequency modulation with a sweep bandwidth of 300 MHz, and the interference source 2 is configured as the up-frequency modulation with a sweep bandwidth of 500 MHz. The three radars are set to have different pulse repetition times (PRTs) in order to increase the probability of mutual interference.

Table 4. Radar Configurations for Real Scene Experiment

Radar Parameters	Victim	Interferer1	Interferer2
Operating frequency (GHz)	77	77	77
Sweep bandwidth (MHz)	300	300	500
Sweep time (μs)	20	20	20
Sweep direction	Up	Down	Up
Pulse repetition time (PRT) (μs)	30	43	61
Sampling frequency (MHz)	20	-	-

Similar to the system workflow in the simulation experiment, the signal flow of the victim radar in the real experiment follows Figure 1. The echoes of the targets and the two interfering radar signals are superimposed and enter the receiver of the victim radar. After mixing and low-pass filtering, a discrete version of the real−valued beat frequency signal is sampled by the ADC chip. In implementing interference mitigation by the proposed method, the number of decomposition modes of the VMD is set to 5, which is the same as the simulation experiment, and the CA-CFAR detector is also the same as in the simulation experiment.

The original beat frequency signal and its interference mitigation version by different methods are shown in Figure 13. It can be seen that two forms of interference related to the two interferer radars are observed in the received signal. The interference with a short time duration is introduced by interferer 1, and the long time duration is introduced by interferer 2. From the reconstructed waveforms of the contaminated part, the VMD method recovers the sinusoidal waveform corresponding to the corner reflector while better suppressing the interference as shown in Figure 13c. In contrast, the EMD and wavelet methods did not recover the sinusoidal signal despite the interference suppression achieved as shown in Figure 13a,b. This is due to the fact that the VMD method performs a signal reconstruction process based on the AR model after zeroing the interference in each decomposed mode, which allows the VMD method to recover the target echoes with sinusoidal characteristics.

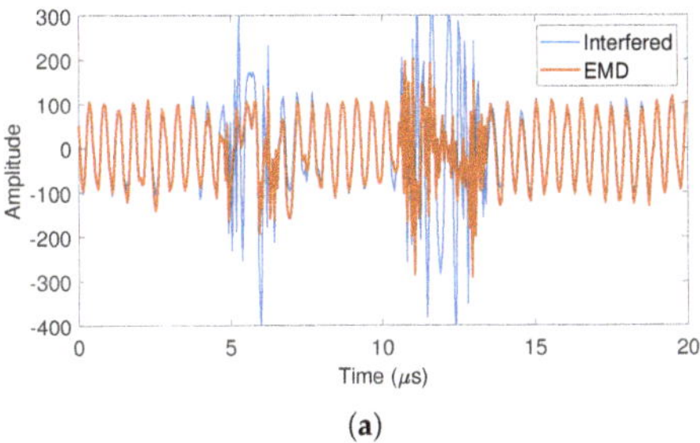
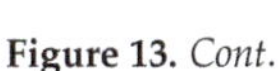

(**a**)

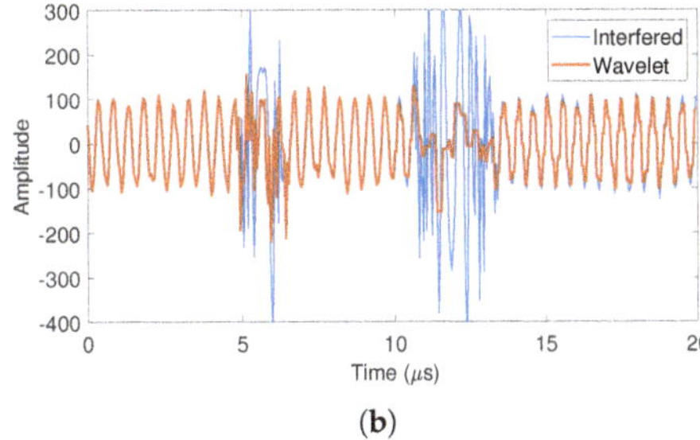

(**b**)

Figure 13. *Cont.*

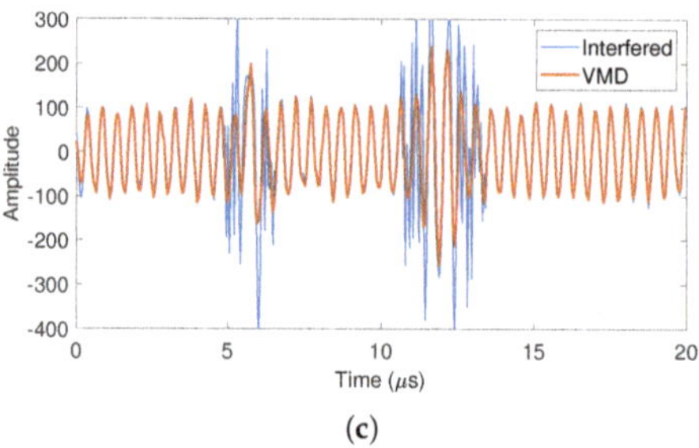

(**c**)

Figure 13. The interference mitigation performance of different methods in time domain for real scene experiment. (**a**) EMD method, (**b**) wavelet method, and (**c**) VMD method.

The RP after interference mitigation for all tested methods is shown in Figure 14. The corner reflector appears at 20 m, which coincides with the experiment setup. Overall, the proposed method has lower sidelobe levels than other methods across the entire RP. The quantitative analysis of MISLR is shown in Figure 15. It can be seen that the proposed method achieves the lowest MISLR level. These results indicate that the proposed method achieves the best interference mitigation performance and correspondingly has the best target detection performance in a real scene experiment.

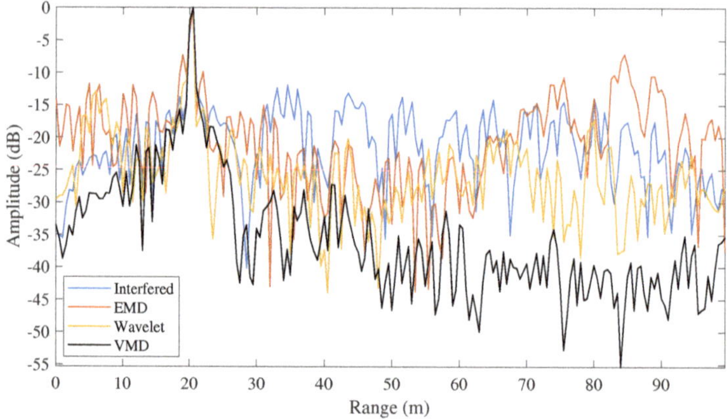

Figure 14. The interference mitigation performance of different methods in RP for real scene experiment.

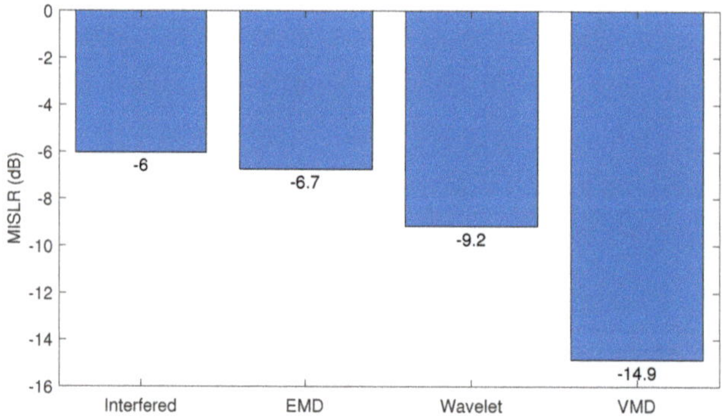

Figure 15. MISLR of different methods for real scene experiment.

The interference mitigation evaluation for a moving target scenario is implemented. The data of a motorcycle rider were recorded in the experiment. After 2D FFT processing of the received chirp sequence, the RD response can be obtained as shown in Figure 16. The RD response reflects the joint range–velocity distribution of the target, i.e., when a target locates at a specific range with a specific velocity simultaneously, a peak appears at the corresponding position in the RD map. According to the RD map of the moving target scenario in Figure 16, the motorcycle appears in the RD map at the range of 8 m and the velocity of 7.8 m/s. The corner reflector can also be observed in Figure 16, and it appears at the range of 20 m with the velocity of 0, which coincides with the experimental design of the scenario. In addition, some ground clutter components are observed in the range domain with velocity 0, which is different from the simulation experiment. As can be seen from the RD map before interference mitigation in the upper left of Figure 16, the motorcycle is barely visible in the presence of interference. When interference mitigation is performed, the motorcycle target is visible in the RD map obtained by all three tested methods. Compared to the EMD and the wavelet methods, the proposed method obtained the best interference mitigation performance. The noise level in the RD map is significantly reduced after using the proposed method for interference mitigation, as shown in the VMD result of Figure 16. It should be noted that some amounts of energy distribute in the RD cells near the range of 20 m. These amounts of energy are not introduced by the interference; in fact, they are the sidelobes of the strong corner reflector target in the velocity domain. The above experimental results show that the proposed method can also achieve good interference mitigation performance in a real-world moving target scenario.

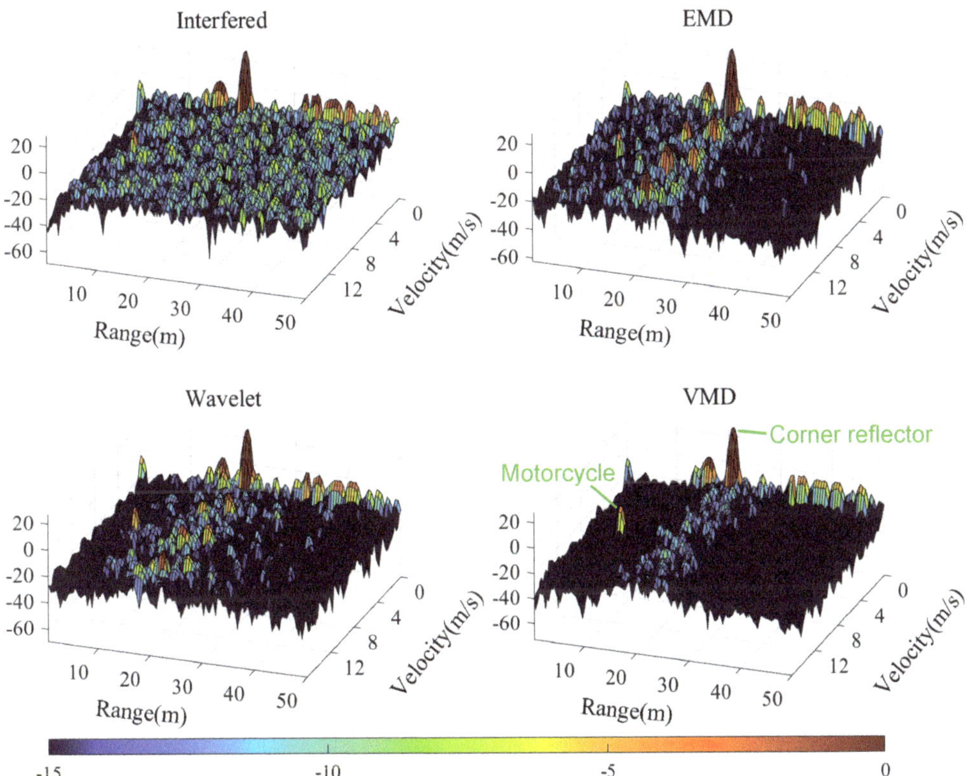

Figure 16. The interference mitigation performance of different methods in RD map for motorcycle scenario. Different colors represent different RD amplitudes in decibels.

7. Conclusions

In this paper, we analyzed the mutual interference of FMCW mmWave radars. After the dechirping process, the target echo becomes a single frequency sinusoid and the interference is a broadband signal. According to their difference, VMD is utilized to decompose the received signal into the summation of a set of narrowband modes. As a result, the interference power is broken down into each mode while the target echo is preserved. For each mode, after the interference detection, the target echo in the interfered area is restored by means of signal reconstruction, so as to achieve the effect of interference mitigation.

In the multi-target simulation scene, the modes obtained by VMD contain a reduced number of targets and reduced bandwidth interference. In this case, a good interference mitigation effect can be achieved using a simple linear prediction model. The simulation results show that the proposed method has better interference mitigation performance than the existing decomposition-based methods. The results based on the experimental data show that the proposed method also outperforms the other decomposition-based methods in a real scene.

Since the VMD implementation is based on the solution of an optimization problem, it has a computational disadvantage; however, in practice, the use of an algorithmic architecture with parallel computing can be considered to reduce the running time.

Author Contributions: Conceptualization, Y.L.; methodology, Y.L. and B.F.; software, Y.L. and W.Z.; validation, B.F.; formal analysis, B.F.; writing—original draft preparation, Y.L.; writing—review and editing, B.F. and W.Z.; supervision, Y.L.; project administration, Y.L.; funding acquisition, Y.L. All authors have read and agreed to the published version of the manuscript.

Funding: This research was funded by the Fundamental Research Funds for the Central Universities grant number 2022RC008.

Data Availability Statement: The data presented in this study are available on request from the corresponding author.

Acknowledgments: The authors would like to thank Muniu Linghang Technology Company for providing radar test equipment and data acquisition support.

Conflicts of Interest: The authors declare no conflict of interest.

References

1. Zhang, Z.; Wang, X.; Huang, D.; Fang, X.; Zhou, M.; Zhang, Y. MRPT: Millimeter-Wave Radar-based Pedestrian Trajectory-Tracking for Autonomous Urban Driving. *IEEE Trans. Instrum. Meas.* **2021**, *71*, 1–17. [CrossRef]
2. Hasch, J.; Topak, E.; Schnabel, R.; Zwick, T.; Weigel, R.; Waldschmidt, C. Millimeter-wave technology for automotive radar sensors in the 77 GHz frequency band. *IEEE Trans. Microw. Theory Tech.* **2012**, *60*, 845–860. [CrossRef]
3. Saponara, S.; Greco, M.S.; Gini, F. Radar-on-chip/in-package in autonomous driving vehicles and intelligent transport systems: Opportunities and challenges. *IEEE Signal Process. Mag.* **2019**, *36*, 71–84. [CrossRef]
4. Neri, B.; Saponara, S. Advances in technologies, architectures, and applications of highly-integrated low-power radars. *IEEE Aerosp. Electron. Syst. Mag.* **2012**, *27*, 25–36. [CrossRef]
5. Bilik, I.; Longman, O.; Villeval, S.; Tabrikian, J. The rise of radar for autonomous vehicles: Signal processing solutions and future research directions. *IEEE Signal Process. Mag.* **2019**, *36*, 20–31. [CrossRef]
6. Munoz-Ferreras, J.M.; Perez-Martinez, F.; Calvo-Gallego, J.; Asensio-Lopez, A.; Dorta-Naranjo, B.P.; Blanco-del Campo, A. Traffic Surveillance System Based on a High-Resolution Radar. *IEEE Trans. Geosci. Remote. Sens.* **2008**, *46*, 1624–1633. [CrossRef]
7. Dogru, S.; Marques, L. Pursuing Drones With Drones Using Millimeter Wave Radar. *IEEE Robot. Autom. Lett.* **2020**, *5*, 4156–4163. [CrossRef]
8. Roos, F.; Bechter, J.; Knill, C.; Schweizer, B.; Waldschmidt, C. Radar sensors for autonomous driving: Modulation schemes and interference mitigation. *IEEE Microw. Mag.* **2019**, *20*, 58–72. [CrossRef]
9. Sun, S.; Petropulu, A.P.; Poor, H.V. MIMO radar for advanced driver-assistance systems and autonomous driving: Advantages and challenges. *IEEE Signal Process. Mag.* **2020**, *37*, 98–117. [CrossRef]
10. Yamada, H.; Kobayashi, T.; Yamaguchi, Y.; Sugiyama, Y. High-resolution 2D SAR imaging by the millimeter-wave automobile radar. In Proceedings of the 2017 IEEE Conference on Antenna Measurements & Applications (CAMA), Tsukuba, Japan, 4–6 December 2017; pp. 149–150. [CrossRef]

11. Brooker, G.M. Mutual interference of millimeter-wave radar systems. *IEEE Trans. Electromagn. Compat.* **2007**, *49*, 170–181. [CrossRef]

12. Kim, G.; Mun, J.; Lee, J. A peer-to-peer interference analysis for automotive chirp sequence radars. *IEEE Trans. Veh. Technol.* **2018**, *67*, 8110–8117. [CrossRef]

13. Pirkani, A.; Norouzian, F.; Hoare, E.; Cherniakov, M.; Gashinova, M. Automotive interference statistics and their effect on radar detector. *IET Radar Sonar Navig.* **2022**, *16*, 9–21. [CrossRef]

14. Overdevest, J.; Jansen, F.; Laghezza, F.; Uysal, F.; Yarovoy, A. Uncorrelated Interference in 79 GHz FMCW and PMCW Automotive Radar. In Proceedings of the 2019 20th International Radar Symposium (IRS), Ulm, Germany 26–28 June 2019; pp. 1–8. [CrossRef]

15. Alland, S.; Stark, W.; Ali, M.; Hegde, M. Interference in Automotive Radar Systems: Characteristics, Mitigation Techniques, and Current and Future Research. *IEEE Signal Process. Mag.* **2019**, *36*, 45–59. [CrossRef]

16. Goppelt, M.; Blöcher, H.L.; Menzel, W. Analytical investigation of mutual interference between automotive FMCW radar sensors. In Proceedings of the 2011 German Microwave Conference, Darmstadt, Germany, 14–16 March, 2011; pp. 1–4.

17. Nozawa, T.; et al. An anti-collision automotive FMCW radar using time-domain interference detection and suppression. In Proceedings of the International Conference on Radar Systems, Belfast, UK, 23-26 October 2017; pp. 1–5. [CrossRef]

18. Rameez, M.; Dahl, M.; Pettersson, M.I. Autoregressive Model-Based Signal Reconstruction for Automotive Radar Interference Mitigation. *IEEE Sensors J.* **2020**, *21*, 6575–6586. [CrossRef]

19. Wu, J.; Yang, S.; Lu, W.; Liu, Z. Iterative modified threshold method based on EMD for interference suppression in FMCW radars. *IET Radar Sonar Navig.* **2020**, *14*, 1219–1228. [CrossRef]

20. Lee, S.; Lee, J.Y.; Kim, S.C. Mutual Interference Suppression Using Wavelet Denoising in Automotive FMCW Radar Systems. *IEEE Trans. Intell. Transp. Syst.* **2021**, *22*, 887–897. [CrossRef]

21. Norouzian, F.; Pirkani, A.; Hoare, E.; Cherniakov, M.; Gashinova, M. Phenomenology of automotive radar interference. *IET Radar Sonar Navig.* **2021**, *15*, 1045–1060. [CrossRef]

22. Neemat, S.; Krasnov, O.; Yarovoy, A. An interference mitigation technique for FMCW radar using beat-frequencies interpolation in the STFT domain. *IEEE Trans. Microw. Theory Tech.* **2018**, *67*, 1207–1220. [CrossRef]

23. Eslami Nazari, M.; Huang, W.; Zhao, C. Radio Frequency Interference Suppression for HF Surface Wave Radar Using CEMD and Temporal Windowing Methods. *IEEE Geosci. Remote. Sens. Lett.* **2020**, *17*, 212–216. [CrossRef]

24. Chen, Z.; Xie, F.; Zhao, C.; He, C. Radio Frequency Interference Mitigation in High-Frequency Surface Wave Radar Based on CEMD. *IEEE Geosci. Remote. Sens. Lett.* **2017**, *14*, 764–768. [CrossRef]

25. Zhou, F.; Xing, M.; Bai, X.; Sun, G.; Bao, Z. Narrow-Band Interference Suppression for SAR Based on Complex Empirical Mode Decomposition. *IEEE Geosci. Remote. Sens. Lett.* **2009**, *6*, 423–427. [CrossRef]

26. Díez-García, R.; Camps, A.; Park, H. On the Potential of Empirical Mode Decomposition for RFI Mitigation in Microwave Radiometry. *IEEE Trans. Geosci. Remote. Sens.* **2022**, *60*, 1–10. [CrossRef]

27. Ramasubramanian, K.; Instruments, T. Using a complex-baseband architecture in FMCW radar systems. *Tex. Instruments* **2017**, *19*, 1–9.

28. Flandrin, P.; Rilling, G.; Goncalves, P. Empirical mode decomposition as a filter bank. *IEEE Signal Process. Lett.* **2004**, *11*, 112–114. [CrossRef]

29. Vetterli, M.; Herley, C. Wavelets and filter banks: theory and design. *IEEE Trans. Signal Process.* **1992**, *40*, 2207–2232. [CrossRef]

30. Dragomiretskiy, K.; Zosso, D. Variational Mode Decomposition. *IEEE Trans. Signal Process.* **2014**, *62*, 531–544. [CrossRef]

31. Patole, S.M.; Torlak, M.; Wang, D.; Ali, M. Automotive radars: A review of signal processing techniques. *IEEE Signal Process. Mag.* **2017**, *34*, 22–35. [CrossRef]

32. Skolnik, M.I. *Introduction to Radar Systems*; McGraw-Hill: New York, NY, USA, 2001.

33. Upadhyay, A.; Pachori, R.B. Instantaneous voiced/non-voiced detection in speech signals based on variational mode decomposition. *J. Frankl. Inst.* **2015**, *352*, 2679–2707. [CrossRef]

34. Lahmiri, S. Intraday stock price forecasting based on variational mode decomposition. *J. Comput. Sci.* **2016**, *12*, 23–27. [CrossRef]

35. Smruthy, A.; Suchetha, M. Real-Time Classification of Healthy and Apnea Subjects Using ECG Signals With Variational Mode Decomposition. *IEEE Sensors J.* **2017**, *17*, 3092–3099. [CrossRef]

36. Zhang, X.; Miao, Q.; Zhang, H.; Wang, L. A parameter-adaptive VMD method based on grasshopper optimization algorithm to analyze vibration signals from rotating machinery. *Mech. Syst. Signal Process.* **2018**, *108*, 58–72. [CrossRef]

37. Wang, Z.; Wang, J.; Du, W. Research on Fault Diagnosis of Gearbox with Improved Variational Mode Decomposition. *Sensors* **2018**, *18*, 3510. [CrossRef]

38. Toth, M.; Meissner, P.; Melzer, A.; Witrisal, K. Performance Comparison of Mutual Automotive Radar Interference Mitigation Algorithms. In Proceedings of the 2019 IEEE Radar Conference, Boston, MA, USA, 22–26 April 2019; pp. 1–6. [CrossRef]

39. Chan, Y.; Lavoie, J.; Plant, J. A parameter estimation approach to estimation of frequencies of sinusoids. *IEEE Trans. Acoust. Speech, Signal Process.* **1981**, *29*, 214–219. [CrossRef]

40. Richards, M.A. *Fundamentals of Radar Signal Processing*; McGraw-Hill: New York, USA, 2005.

41. Davis, M.S.; Lanterman, A.D. Minimum integrated sidelobe ratio filters for MIMO radar. *IEEE Trans. Aerosp. Electron. Syst.* **2015**, *51*, 405–416. [CrossRef]
42. Chatzitheodoridi, M.E.; Taylor, A.; Rabaste, O. A Mismatched Filter for Integrated Sidelobe Level Minimization over a Continuous Doppler Shift Interval. In Proceedings of the 2020 IEEE Radar Conference, Florence, Italy, 21–25 September 2020; pp. 1–6. [CrossRef]

remote sensing 

Article

Sparsity-Based Joint Array Calibration and Ambiguity Resolving for Forward-Looking Multi-Channel SAR Imagery

Jingyue Lu [1], Xuhua Wang [1,*], Yunhe Cao [2] and Lei Zhang [3]

1. School of Computer Science and Technology, Xidian University, Xi'an 710071, China
2. National Laboratory of Radar Signal Processing, Xidian University, Xi'an 710071, China
3. School of Electronics and Communication Engineering, Sun Yat-sen University, Guangzhou 510275, China
* Correspondence: wangxuhua@xidian.edu.cn

Abstract: Forward-looking multi-channel synthetic aperture radar (FLMC-SAR) can realize two-dimension image formation in monostatic mode. This system must face the problem of left–right Doppler ambiguity. In the traditional methods, the spatial degrees of freedom of the FLMC-SAR system is expected to achieve Doppler ambiguity resolving by beamforming approaches. However, the influence of array error on beamforming cannot be ignored. In practice, the array error will lead to the mismatch of the space–time characteristic, which will reduce the performance of the Doppler ambiguity resolving method based on beamforming. This paper proposes a sparsity-based joint array calibration and ambiguity resolving method to enhance the robustness of FLMC-SAR imagery. For the FLMC-SAR system, the space–time characteristic of targets is first analyzed, based on which the observation model of FLMC-SAR Doppler ambiguity combined with array error is derived. Then, the Doppler ambiguity resolving and array error estimation are transformed into a sparse recovery problem. A modified quasi-Newton method is proposed to realize the array error estimation and Doppler ambiguity resolving of all targets in the local area. Finally, the results of the simulation and the real-data experiments verify that the proposed method can achieve FLMC-SAR Doppler ambiguity resolving and imaging.

Keywords: forward-looking multi-channel synthetic aperture radar (FLMC-SAR); Doppler ambiguity; space–time characteristic; array calibration; sparse recovery

Citation: Lu, J.; Wang, X.; Cao, Y.; Zhang, L. Sparsity-Based Joint Array Calibration and Ambiguity Resolving for Forward-Looking Multi-Channel SAR Imagery. *Remote Sens.* **2023**, *15*, 647. https://doi.org/10.3390/rs15030647

Academic Editor: Dusan Gleich

Received: 12 December 2022
Revised: 5 January 2023
Accepted: 14 January 2023
Published: 21 January 2023

1. Introduction

Synthetic aperture radar (SAR) [1–7] is an important technique for achieving two-dimensional high-resolution microwave imaging. Due to the limitation of its working mechanism, SAR has a forward-looking blind area. By optimizing the system configuration of traditional SAR, SAR is used to obtain the two-dimensional image of the forward-looking area, called forward-looking SAR. The ability to acquire forward-looking two-dimensional images makes forward-looking SAR widely applied in practical engineering, such as aircraft blind landing, remote sensing, and so on. Forward-looking SAR systems can be divided into monostatic forward-looking SAR systems and bistatic forward-looking SAR systems.

Bistatic forward-looking SAR (BFSAR) [8–13] imaging is performed by equipping an additional receiver or transmitter platform to provide forward-looking Doppler diversity to achieve high cross-range resolution. The flexible system configuration makes BFSAR widely used. However, the system configuration requirements of BFSAR may not be met in some applications (such as having only one radar platform). Therefore, this paper focuses on the monostatic forward-looking SAR imaging system. Sector imaging radar for enhanced vision (SIREV) [14,15] is the earliest monostatic forward-looking SAR imaging system with an array of real apertures in the cross-trajectory direction. SIREV forms a synthetic aperture by controlling the timing of the transmitter and receiver to obtain the forward-looking two-dimensional images, which makes the synthetic aperture length heavily dependent on the array aperture length. This means that the forward-looking resolution of the SIREV system is excessively dependent on the array aperture. However, a high-resolution forward-looking imaging requirement will cause higher system complexity, which limits the practical application of SIREV. Joint monopulse angle measurement [16] and SAR imaging is another forward-looking imaging technique. However, the accuracy of monopulse angle measurement for multiple targets in the same beam is limited, which limits the performance of the monopulse forward-looking SAR system for complex scenes. Multi-channel SAR [17–22] is used to obtain forward-looking images, known as forward-looking multi-channel SAR [23,24] (FLMC-SAR). However, Doppler ambiguity is a crucial problem that must be resolved in the imaging process. Beamforming is an excellent technique for FLMC-SAR to solve Doppler ambiguity. Unfortunately, in practical applications, the inevitable installation and other factors will introduce array errors, which is an essential factor that makes it impossible to achieve Doppler ambiguity resolving via beamforming. Rotating element electric field vector (REV) [25–27] is an effective method for array calibration. However, REV requires a set of calibration sources with known locations, which does not apply to FLMC-SAR imaging.

Sparsity-based algorithms [28–37] have shown great potential in radar imaging, such as image reconstruction from under-sampling rate image data [38–40], improving the resolution of the reconstructed image under the condition of constant sampling [41–45], and so on. At the same time, sparsity-based algorithms can also be used to estimate errors (such as motion and array) to improve imaging quality. The assumption of confirming the accuracy and success rate of image reconstruction is whether the reconstructed signal is sparse in the signal observation model, and the size of the observation matrix dimension also determines the computational complexity of the algorithm. Therefore, appropriate system resource allocation and reasonable imaging processes are key to ensure algorithm performance. In this paper, a sparsity-based joint array calibration and ambiguity resolving method is proposed to enhance the robustness of FLMC-SAR imagery. First, the space–time characteristic of FLMC-SAR is analyzed. The effect of array errors on Doppler ambiguity resolving indicates that the array errors will lead to the mismatch of space–time characteristics of the targets, causing performance degradation of Doppler ambiguity resolving. Considering the array errors, the observation model of space–time characteristic correction is derived. Then, the Doppler ambiguity resolving and array error estimation are transformed into a sparse recovery problem. Then, a modified quasi-Newton method is proposed to realize the array error estimation and Doppler ambiguity resolving in the local area. Finally, the results of the simulation and the real-data experiments verify that the proposed method can achieve FLMC-SAR Doppler ambiguity resolving and imaging. The main contribution can be summarized as follows.

(1) Both left–right Doppler ambiguity and array error estimation are considered in the observation model, which transforms the robust FLMC-SAR imaging without ambiguity into a sparse recovery problem. The modified quasi-Newton method is proposed to realizes both array error estimation and Doppler ambiguity resolving simultaneously.

(2) The sparse recovery problem is solved in the two-dimensional image domain after range–azimuth decoupling, which paves the way for sparse reconstruction in each range bin. The array error estimation and Doppler ambiguity resolving are realized in the local area, reducing the computational complexity of the proposed method. In addition, after pulse compression and azimuth focusing, the SNR of the two-dimensional image domain is significantly improved, which improves the image reconstruction accuracy.

The organizational structure of this article is as follows. In Section 2, the geometric model of FLMC-SAR is given. In Section 3, the observation model of FLMC-SAR is established. In Section 4, we propose the constrained optimization problem of array error estimation and introduce the joint Doppler ambiguity resolving and array error correction in detail. Section 5 gives the results based on simulation and real-data experiments. The results are discussed in Section 6. Section 7 summarizes the work of this paper.

2. FLMC-SAR Geometry Model

Figure 1 gives the FLMC-SAR geometry model. The Cartesian coordinate system is established. The origin O is the projection of the aperture center on the ground, the trajectory direction is the X-axis, the array distribution direction is the Y-axis, and the vertical direction up to the ground is the Z-axis.

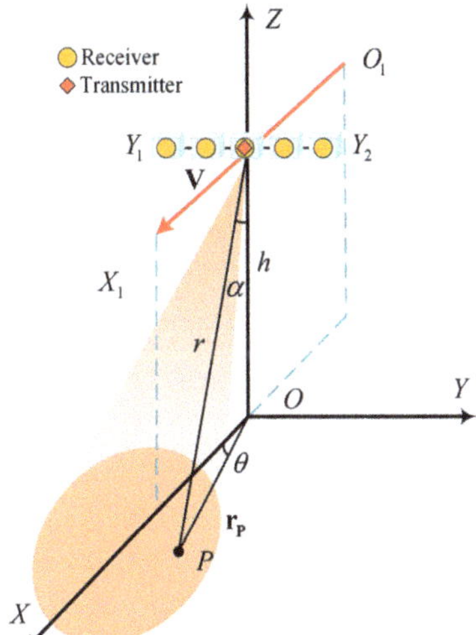

Figure 1. FLMC-SAR geometry model.

FLMC-SAR is a monopulse radar system. The radar platform moves along the trajectory O_1X_1 at a speed v and altitude h, which uses a single antenna as the transmitting antenna and a uniform linear array as the receiving antenna; D denotes the length of the real antenna, and d denotes the spacing between antenna elements. The receiver array antenna is symmetrically distributed along the Y-axis over Y_1Y_2. The imaging area is located directly in the forward-looking area; P denotes a target in the forward-looking imaging area. The radar's light of sight is defined as a vector pointing from the center of the aperture to the target. The radar's light of sight can be determined by the beam pointing pitch angle α between it and the Z-axis and the beam pointing azimuth angle θ between the projection of it on the ground and the X-axis. The reference slant range is defined as the distance between the target and the APC at the center of the aperture. With these parameters, the coordinates of the target P in the $OXYZ$ coordinate system can be expressed as

$$\mathbf{P}(x,y,z) = (r\sin\alpha\cos\theta, r\sin\alpha\sin\theta, 0) \tag{1}$$

Let t_m be the radar slow time; then the coordinate of the transmitter and receiver can be expressed as

$$\begin{cases} \mathbf{T_X}(vt_m, 0, h) \\ \mathbf{R_{Xi}}(vt_m, y, h) \end{cases} \tag{2}$$

where y denotes the Y-coordinate of the receiver array element. Then the instantaneous two-way slant range of the target P can be expressed as

$$R(t_m; r, \theta, \alpha, y) = R_R(t_m; r, \theta, \alpha, y) + R_T(t_m; r, \theta, \alpha) \tag{3}$$

where $R_T(t_m; r, \theta, \alpha)$ and $R_R(t_m; r, \theta, \alpha, y)$ denote the instantaneous slant range of the receiver and the transmitter, respectively.

$$R_R(t_m; r, \theta, \alpha, y) = \sqrt{(vt_m - r\sin\alpha\cos\theta)^2 + (y - r\sin\alpha\sin\theta)^2 + (r\cos\alpha)^2} \tag{4}$$

$$R_T(t_m; r, \theta, \alpha) = R_R(t_m; r, \theta, \alpha, 0) = \sqrt{(vt_m - r\sin\alpha\cos\theta)^2 + (r\sin\alpha\sin\theta)^2 + (r\cos\alpha)^2} \tag{5}$$

3. FLMC-SAR Observation Model

3.1. FLMC-SAR Signal Model

The radar system uses a linear frequency modulation (LFM) pulse signal, which is one of the most widely used signals in SAR system transmissions.

$$s(\tau) = \text{rect}\left(\frac{\tau}{T_p}\right) \cdot \exp\left[j2\pi\left(f_c\tau + \frac{\gamma}{2}\tau^2\right)\right] \tag{6}$$

where rect[] is the rectangular window function, τ is the radar fast time, T_p is the time width of the LFM pulse, f_c is the carrier frequency of the radar system, and γ is the chirp rate of LFM signal.

Through down-conversion, the baseband received signal can be expressed as

$$\begin{aligned}
s_P(\hat{t}, t_m; r, \theta, \alpha, y) = &A_P w_r\left[\hat{t} - \frac{R(t_m; r, \theta, \alpha, y)}{c}\right] \cdot \exp\left\{j\pi\gamma\left[\hat{t} - \frac{R(t_m; r, \theta, \alpha, y)}{c}\right]^2\right\} \\
&\cdot \exp\left[-j\frac{2\pi}{\lambda}R(t_m; r, \theta, \alpha, y)\right]
\end{aligned} \tag{7}$$

where A_p is the reflectivity coefficient of the target P, w_r is the envelope in range bin, and c is the light speed.

Through pulse compression and range cell migration correction (RCMC), the signal can be expressed as

$$s_2(\hat{t}, t_m; r, \theta, \alpha, y) = A_P\text{sinc}\left\{B\left[\hat{t} - \frac{(r - r_0)}{c}\right]\right\} \cdot \exp\left\{-j\frac{2\pi}{\lambda}[R(t_m; r, \theta, \alpha, y) - R(t_m; r, 0, \alpha, 0)]\right\} \tag{8}$$

where r_0 is the reference slant range of the center of the imaging area.

It can be seen from Equation (8) that the signal of target P has been focused on the range unit corresponding to the reference slant range r, so range–azimuth decoupling is achieved.

3.2. Doppler Ambiguity in FLMC-SAR System

To clarify the Doppler ambiguity, we rewrite Equation (8) in the following form:

$$\begin{aligned}
s_2(\hat{t}, t_m; r, \theta, \alpha, y) = &A_P\text{sinc}\left\{B\left[\hat{t} - \frac{(r - r_0)}{c}\right]\right\} \cdot \exp\left\{-j\frac{2\pi}{\lambda}[\Delta R_1(t_m; r, \theta, \alpha, y)]\right\} \\
&\cdot \exp\left\{-j\frac{2\pi}{\lambda}[\Delta R_2(t_m; r, \theta, \alpha, y)]\right\}
\end{aligned} \tag{9}$$

$$\Delta R_1(t_m; r, \theta, \alpha, y) = R(t_m; r, \theta, \alpha, y) - R(t_m; r, \theta, \alpha, 0) \tag{10}$$

$$\Delta R_2(t_m; r, \theta, \alpha) = R(t_m; r, \theta, \alpha, 0) - R(t_m; r, 0, \alpha, 0) \tag{11}$$

where ΔR_1 is the wave path difference between different channels caused by the different array element coordinate y, and ΔR_2 is the slant range history of the target, which is the same between the different channels. As the array coordinate y is much smaller than the reference slant range r, the far-field condition is satisfied. Equation (10) can be rewritten as

$$\Delta R_1(t_m; r, \theta, \alpha, y) \approx y\sin\alpha\sin\theta \tag{12}$$

Similarly, Equation (11) can be approximated as follows:

$$\Delta R_2(t_m; r, \theta, \alpha) \approx A + Bt_m \tag{13}$$

$$B = 2v \sin \alpha (1 - \cos \theta) \tag{14}$$

Then the Doppler frequency of the target can be expressed as

$$f_d(\theta, \alpha) = \frac{B}{\lambda} = \frac{2v \sin \alpha (1 - \cos \theta)}{\lambda}. \tag{15}$$

It can be seen from Equation (15) that the Doppler frequency of the target is a function related to azimuth angle θ and pitch angle α, and the Doppler frequency is an even function of azimuth angle θ. It means that targets with azimuth angles that are negative to each other have the same Doppler frequency. Traditional SAR imaging methods cannot distinguish the target with opposite azimuth angle in the same range unit, resulting in Doppler ambiguity.

As shown in Figure 2, in the imaging area, there are two targets P_1 and P_2 with the same reference slant range r, but opposite azimuth angles. In the ground grid, their coordinates can be expressed as

$$\begin{cases} P_1(r, \theta) \\ P_2(r, -\theta) \end{cases} \tag{16}$$

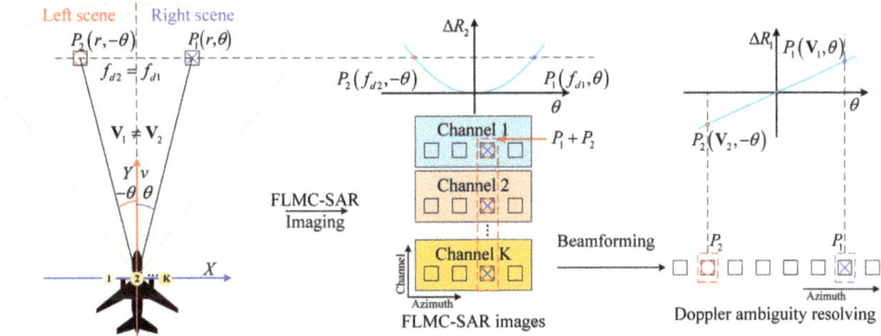

Figure 2. Doppler ambiguity in FLMC-SAR.

According to Equation (15), we can conclude that this Doppler frequency is the same. In the plane of the space–time spectrum $f_d - \theta$, there coordinates can be expressed as

$$\begin{cases} P_1(f_{d1}, \theta) \\ P_2(f_{d2}, -\theta) \end{cases} \quad f_{d1} = f_{d2} \tag{17}$$

After SAR imaging, imaging results of targets with the same Doppler frequency in the same range bin will be aliased in the same imaging unit. Therefore, it is impossible to distinguish targets with opposite azimuth angles within the same range bin only by single-channel Doppler resolution. From Equation (12), we can see that the steering vector of the target is an odd function of the azimuth angle θ. Therefore, the steering vectors of the ambiguous targets are different.

$$\mathbf{V}_1 = \mathbf{V}(\theta) = \exp\left\{ -j\frac{2\pi}{\lambda} y \sin \alpha \sin \theta \right\}$$
$$\mathbf{V}_2 = \mathbf{V}(-\theta) = \exp\left\{ -j\frac{2\pi}{\lambda} y \sin \alpha \sin(-\theta) \right\} \tag{18}$$

Then the signal model of FLMC-SAR be rewritten as

$$s_3(\hat{t}, t_m; r, \theta, \alpha, y) = s_2(\hat{t}, t_m; r, \theta, \alpha, y) + s_2(\hat{t}, t_m; r, -\theta, \alpha, y)$$
$$= \mathrm{sinc}\left\{ B\left[\hat{t} - \frac{(r - r_0)}{c} \right] \right\} \cdot \exp(-j2\pi f_d t_m) \cdot (A_{P1} \cdot \mathbf{V}_1 + A_{P2} \cdot \mathbf{V}_2). \tag{19}$$

Equation (19) has realized the range–azimuth decoupling of FLMC-SAR signals. Now, we can process the signals in each of the same range bin.

3.3. FLMC-SAR Observation Model

For any imaging unit in an FLMC-SAR image, the array's degree of freedom provides the possibility to solve the Doppler ambiguity. According to Equation (19), the following spatial observation model for each FLMC-SAR imaging unit with the same range bin can be obtained:

$$\mathbf{S}(r,\theta)_{K\times1} = \mathbf{H}_{K\times2}\mathbf{A}(r,\theta)_{2\times1} + ''_{K\times1} \tag{20}$$

where $\mathbf{S}(r,\theta)_{K\times1}$ is the imaging results for all channel for each FLMC-SAR imaging unit, $\mathbf{A}(r,\theta)_{2\times1}$ is the reflectivity coefficient vector of ambiguous targets, $''_{K\times1}$ is the additive measurement noise term, and $\mathbf{H}_{K\times2}$ is the steering vector matrix composed of ambiguous targets.

$$\mathbf{H}_{K\times2} = [h_1, h_2] \tag{21}$$

where h_i is the steering vector of the ith target.

$$h_i = \begin{bmatrix} \exp\left(-j\frac{2\pi}{\lambda}d\sin\alpha_i\sin\theta_i\right) \\ \exp\left(-j\frac{2\pi}{\lambda}2d\sin\alpha_i\sin\theta_i\right) \\ ... \\ \exp\left(-j\frac{2\pi}{\lambda}Kd\sin\alpha_i\sin\theta_i\right) \end{bmatrix}_{K\times1} \tag{22}$$

where θ_i and α_i are the azimuth angle and pitch angle of the target, respectively.

The order of the column vectors in the steering vector matrix $\mathbf{H}_{K\times2}$ corresponds to the order of the reflectivity coefficient of ambiguous targets in $\mathbf{A}(r,\theta)_{2\times1}$. It is worth noting that Doppler ambiguity is two-dimensional in the FLMC-SAR system. Therefore, the steering vector matrix $\mathbf{H}_{K\times2}$ only contains two column vectors. Based on the accuracy of the steering vector, Doppler ambiguity can be solved by spatial methods such as beamforming. However, in practical applications, the inevitable array error between channels causes the space–time characteristic mismatch of the targets and reduces the accuracy of the ambiguous targets steering vector, resulting in the degradation of the Doppler ambiguity resolving performance. To correct the array errors, the following observation model of FLMC-SAR is given in this paper.

$$\mathbf{S}(r,\theta)_{K\times1} = \boldsymbol{æ}_{K\times K}\mathbf{H}_{K\times2}\mathbf{A}(r,\theta)_{2\times1} + ''_{K\times1} \tag{23}$$

The array errors of each channel are expressed as follows:

$$\boldsymbol{æ}_{K\times K} = \begin{bmatrix} \rho_1 & & & & & \\ & \rho_2 & & & & \\ & & \ddots & & & \\ & & & \rho_k & & \\ & & & & \ddots & \\ & & & & & \rho_K \end{bmatrix} \tag{24}$$

where ρ_k denotes the steering vector of the kth channel.

4. Sparsity-Based Array Error Estimation and Doppler Ambiguity Resolving

4.1. Improved Quasi-Newton Kernel

Following the Bayesian principle, the cost function of the maximum a posteriori (MAP) estimator [46–49] for the constrained optimization problem of array error estimation can be obtained.

$$\arg\min \ J(\mathbf{A}(r,\theta), \boldsymbol{æ}) \tag{25}$$

$$J(\mathbf{A}(r,\theta), \boldsymbol{æ}) = \|\mathbf{S}(r,\theta) - \boldsymbol{æ}\mathbf{H}\mathbf{A}(r,\theta)\|_F^2 + \omega\|\mathbf{A}(r,\theta)\|_1 \tag{26}$$

where ω is the regularization parameter related to the sparsity and noise level of the signal. The cost function consists of two terms. The first term in Equation (25) is the data fidelity term of the estimator. The second term in Equation (25) incorporate the signal sparsity, which balances the reconstruction accuracy with the sparsity of the obtained solutions. As Doppler ambiguity is two-dimensional in the FLMC-SAR imaging, the results that need to be reconstructed in the observation model of Equation (23) are at most 2-sparse.

In this paper, an improved quasi-Newton method is proposed to solve the above constrained optimization problem, which uses a 2-step iterative processing. There are two steps in each iteration: image reconstruction and array error estimation Algorithm 1. The iterative process of the proposed algorithm is shown as follows.

Algorithm 1: Improved quasi-Newton

Range cycle: Traverse all range bin
 Azimuth cycle: Traverse all N azimuth bin within the same range bin
  Input: The signal of an imaging unit within the same range unit $\mathbf{S}(r,\theta)$,
  the steering vector matrix of the imaging unit $\mathbf{A}(r,\theta)$, $k=0$.
  Step 1 Image reconstruction:

$$A^{k+1}(r,\theta) = \arg\min_{A(r,\theta)} J\left(\mathbf{A}^k(r,\theta), \mathbf{æ}^k\right) \tag{27}$$

 end
 Step 2 Array error estimation:

$$\mathbf{æ}^{k+1} = \arg\min_{\mathbf{æ}} \sum_{\theta=\theta_1}^{\theta_N} J\left(\mathbf{A}^{k+1}(r,\theta), \mathbf{æ}^k\right) \tag{28}$$

 $k = k+1$ until

$$\sum_{\theta=\theta_1}^{\theta_N} \|\mathbf{S}(r,\theta) - \mathbf{æHA}(r,\theta)\|_F^2 \leq \zeta \tag{29}$$

end

The initialization array error matrix $\mathbf{æ}_0$ is set to be the identity matrix. In order to reduce the complexity of the proposed method, the array error is estimated by combining all imaging units within the same distance unit, which also reduces the influence of singularities and enhances the robustness of the algorithm.

Step 1: To achieve image reconstruction in each iteration, we first need to solve the constrained optimization problem of Equation (27). To avoid the nondifferentiability of the ℓ_1 norm in Equation (27), we use the following smooth approximation:

$$\|\mathbf{A}(r,\theta)\|_1 = \sum_{\theta=\theta_1}^{\theta_2} \left(a^2(r,\theta) + \xi\right)^{\frac{1}{2}} \tag{30}$$

where ξ is a small nonnegative constant, and $a(r,\theta)$ is the element in $\mathbf{A}(r,\theta)$.

The cost function is slightly modified as

$$J(\mathbf{A}(r,\theta), \mathbf{æ}) = \|\mathbf{S}(r,\theta) - \mathbf{æHA}(r,\theta)\|_F^2 + \omega \sum_{\theta=\theta_1}^{\theta_2} \left(a^2(r,\theta) + \xi\right)^{\frac{1}{2}}. \tag{31}$$

Then the conjugate gradient of cost function in Equation (31) over $\mathbf{A}(r,\theta)$ can be obtained:

$$\nabla J_{\mathbf{A}(r,\theta)} = 2(\mathbf{æH})^H \mathbf{æHA}(r,\theta) + \omega \mathbf{U}_{\mathbf{A}(r,\theta)} \mathbf{A}(r,\theta) - 2\mathbf{æH}^H \mathbf{S}(r,\theta) \tag{32}$$

where $\mathbf{U_A}$ is a diagonal matrix, represented as follows:

$$\mathbf{U}_{\mathbf{A}(r,\theta)} = \begin{bmatrix} \frac{1}{\sqrt{|a(r,\theta_1)|^2+\xi}} & 0 \\ 0 & \frac{1}{\sqrt{|a(r,\theta_2)|^2+\xi}} \end{bmatrix}. \tag{33}$$

Therefore, the Hessian matrix can be approximated as follows:

$$\mathbf{H}_{\mathbf{A}(r,\theta)} = 2(\text{æ}\mathbf{H})^H \text{æ}\mathbf{H} + \omega \mathbf{U}_{\mathbf{A}(r,\theta)} \tag{34}$$

In this way, we can obtain the iteration solver of image reconstruction.

$$\mathbf{A}^{k+1}(r,\theta) = \mathbf{A}^k(r,\theta) - \left[\mathbf{H}_{\mathbf{A}^k(r,\theta)}\right]^{-1} \nabla J_{\mathbf{A}(r,\theta)} \tag{35}$$

Step 2: Array error estimation is achieved by solving the constrained optimization problem of Equation (28).

The conjugate gradient of Equation (31) over æ can be obtained:

$$\nabla J_{\text{æ}} = \sum_{\theta=\theta_1}^{\theta_N} \left[2\mathbf{A}(r,\theta)^H \mathbf{H}^H \text{æ}\mathbf{H}\mathbf{A}(r,\theta) - 2\mathbf{A}(r,\theta)^H \mathbf{H}^H \mathbf{S}(r,\theta) \right]. \tag{36}$$

Let $\nabla J_{\text{æ}} = 0$, then we can obtain the iteration solver of array error estimation; æ is a diagonal matrix, so the update rule can be expressed as

$$\text{æ}^{k+1} = \text{diag}[\rho_a \exp(j\rho_b)] \tag{37}$$

where ρ_a and ρ_b is the amplitude error and phase error of the array, respectively.

$$\rho_a = \sum_{\theta=\theta_1}^{\theta_N} \{\text{abs}[\mathbf{S}(r,\theta)]./\text{abs}(\mathbf{H}\mathbf{A}(r,\theta))\} \tag{38}$$

$$\rho_b = \sum_{\theta=\theta_1}^{\theta_N} \{\text{angle}\{\mathbf{S}(r,\theta) \odot \text{conj}[\mathbf{H}\mathbf{A}(r,\theta)]\}\} \tag{39}$$

where $\odot$ is the Hadamard product. Each range cycle continues until the Equation (29) is satisfied, then æ and $\mathbf{A}$ are, respectively, the reconstructed image and the estimation of array errors within this range bin.

4.2. Computational Complexity

The above operations are performed on all range bins to achieve full-area FLMC-SAR ambiguity resolving and imaging with array error correction. The flowchart of the proposed method is shown in Figure 3. The proposed method mainly consists of two steps: FLMC-SAR imaging and joint Doppler ambiguity resolving and array error correction.

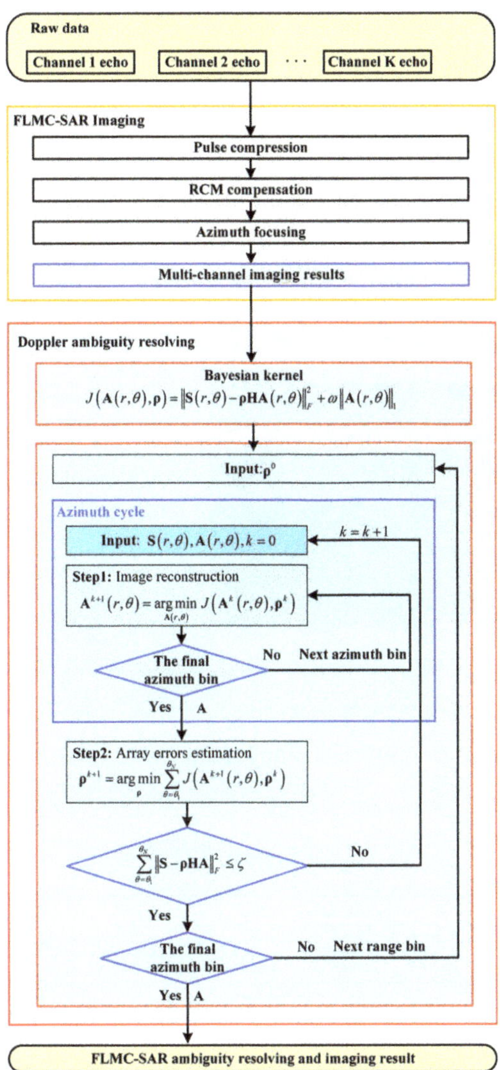

Figure 3. Flowchart of the proposed method.

Assume that the number of channels is K, the number of azimuth bins is N, and the the number of range bins is M. Then the computational complexity of the proposed method is analyzed as follows.

(1) FLMC-SAR imaging contains range pulse compression, RCMC, and azimuth focusing for multi-channels. Range pulse compression needs KN times fast Fourier transform (FFT) for the vector with the size of $M \times 1$, and KN times inverse fast Fourier transform (IFFT) for the vector with the size of $M \times 1$. RCMC needs KN times FFT for the vector with the size of $M \times 1$, KN times IFFT for the vector with the size of $M \times 1$, KM times FFT for the vector with the size of $N \times 1$, and KM times IFFT for the vector with the size of $N \times 1$. Azimuth focusing needs KM times FFT for the vector with the size of $N \times 1$. The total computation of FLMC-SAR imaging is $O(KNM(\log N + \log M))$.

(2) Joint Doppler ambiguity resolving and array error correction. Suppose the number of iterations is N_{it}. For each range bin, imaging reconstruction needs M times sparse reconstruction, which needs N_{it} times matrix inversion for the matrix with the size of $K \times K$.

Array error estimation needs $2N$ times the Hadamard product for the diagonal matrix with the size of $K \times K$. The total computation of joint Doppler ambiguity resolving and array error correction is $O\left(N_{it} NMK^3 + 2NK\right)$.

Therefore, the total computation of the proposed method is of the order $O\left(NM\left(K \log N + K \log M + N_{it} K^3\right)\right)$

5. Results

5.1. Point Target Simulation

We first give the point target simulation experiment to verify the efficacy of the proposed array error estimation and imaging reconstruction method. As shown in Figure 4, there are nine point targets in the original reference image. The red trajectory extension line in Figure 4 divides the image into the left area and the right area.

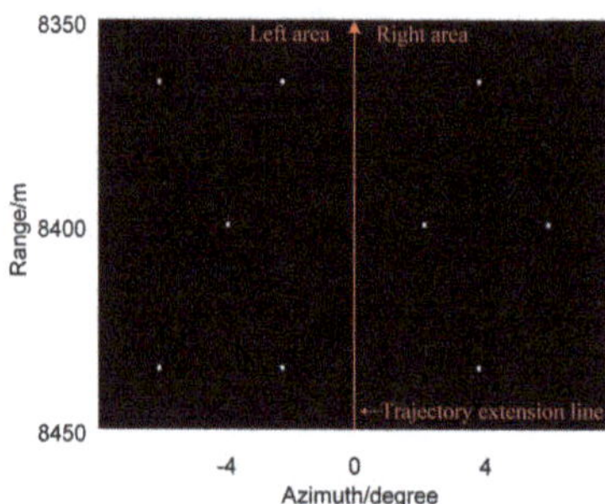

Figure 4. Original reference image.

Five point targets and four point targets are set in the left area and the right area, respectively. The coordinates of the point targets are shown in Table 1. Table 2 shows the simulation parameters. The SNR is set to 20 dB. The array error is also added to the echo.

Table 1. Point target coordinates.

		Left Area					Right Area	
P_1	P_2	P_3	P_4	P_5	P_6	P_7	P_8	P_9
$(Rs_1, -5)$	$(Rs_1, -3)$	$(Rs_2, -4)$	$(Rs_3, -5)$	$(Rs_3, -3)$	$(Rs_1, 4)$	$(Rs_2, 3)$	$(Rs_2, 5)$	$(Rs_3, 4)$

The point targets' imaging results are shown in Figure 5. Figure 5a–d indicate the ambiguous imaging results, the ambiguity resolving results of beamforming without the array calibration, the imaging result of the proposed method, and the ambiguity resolving results of beamforming with the array calibration, respectively.

Table 2. Simulation parameters.

Carrier frequency	30 GHz	Platform height	4000 m
Bandwidth	55 MHz	Platform velocity	84 m/s
Number of array element	9	Reference slant range	8400 m
PRF	2500 Hz	Synthetic aperture time	0.82 s

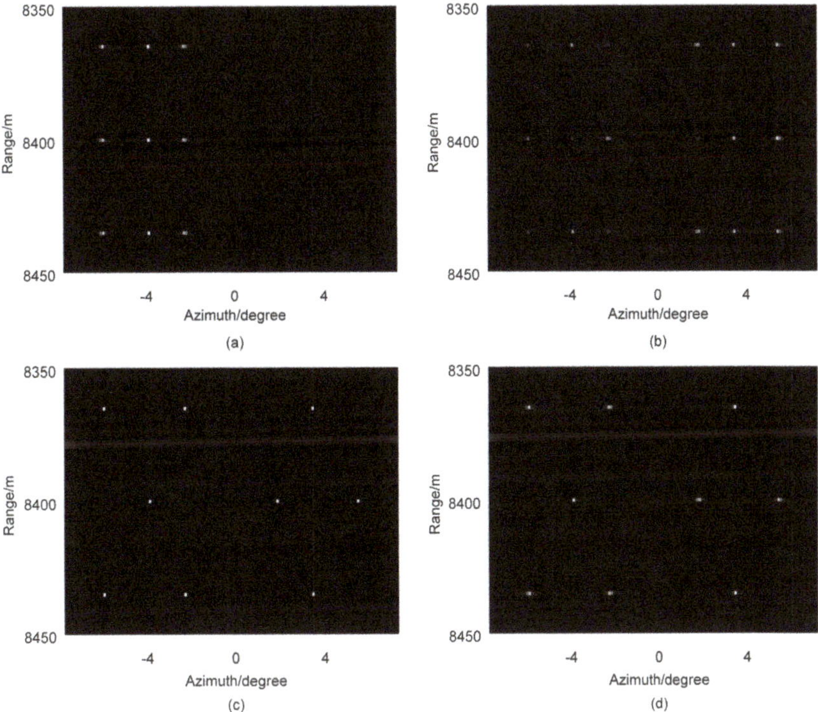

Figure 5. Imaging results of point targets: (**a**) ambiguous imaging results; (**b**) ambiguity resolving results of beamforming without the array calibration; (**c**) imaging result of the proposed method; (**d**) ambiguity resolving results of beamforming with the array calibration.

5.2. Surface Target Simulation

To verify the effectiveness of the proposed method for natural scenes, a surface target simulation is presented in this section. The original reference image is shown in Figure 6. The illuminated area in Figure 6 is divided into the left area and the right area by the red trajectory extension line. Table 2 shows the simulation parameters. The SNR is set to 20 dB. The array error is also added to the echo.

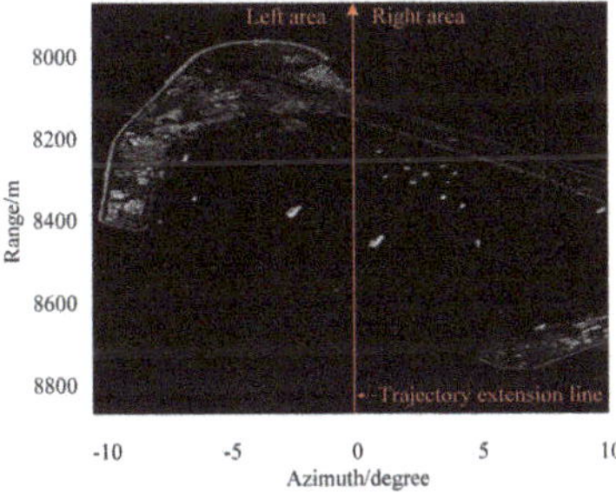

Figure 6. Original reference image.

The image results of the surface target are shown in Figure 7. Figure 7a–d indicate the ambiguous imaging results, the ambiguity resolving results of beamforming without the array calibration, the imaging result of the proposed method, and the ambiguity resolving results of beamforming with the array calibration, respectively.

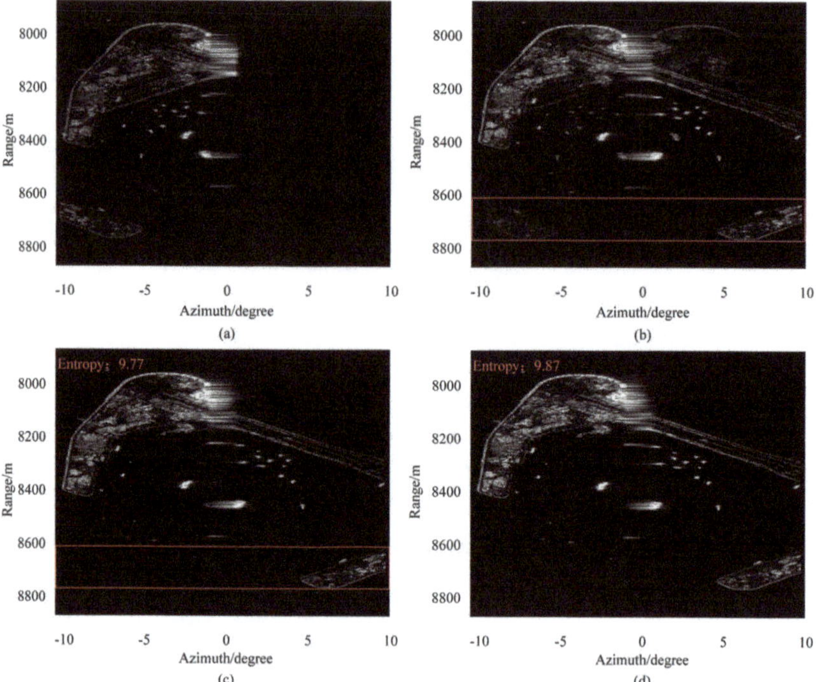

Figure 7. Imaging results of the surface target: (**a**) ambiguous imaging results; (**b**) ambiguity resolving results of beamforming without the array calibration; (**c**) imaging result of the proposed method; (**d**) ambiguity resolving results of beamforming with the array calibration.

5.3. Real Data Experiment

To demonstrate the efficacy of the proposed method in a practical application, we carried out a real-data experiment in this section. The real-data experiment was performed by a K-band FLMC-SAR. The radar is equipped with a five-channel array antenna. The radar platform is mounted on the aircraft at an altitude of 4000 m and a speed of 80 m/s. The relevant experiment parameters are shown in Table 3.

Table 3. Experiment parameters.

Carrier frequency	30 GHz	Platform height	4000 m
Bandwidth	55 MHz	Platform velocity	80 m/s
Number of array element	9	Reference slant range	8000 m
PRF	6000 Hz	Synthetic aperture time	1.3 s

Imaging results of the real-data experiment are shown in Figure 8. Figure 8a–c indicate the ambiguous imaging results, the ambiguity resolving results of beamforming, and the imaging result of the proposed method. The satellite image of the imaging area is shown in Figure 9.

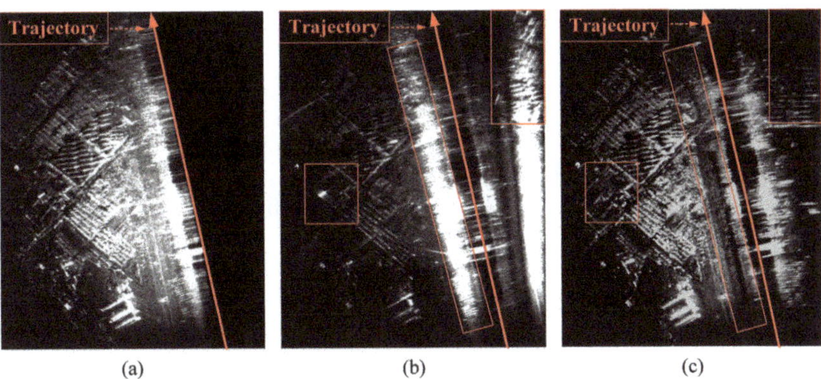

(a)   (b)   (c)

Figure 8. Imaging results of the real-data experiment: (**a**) ambiguous imaging results; (**b**) ambiguity resolving results of beamforming; (**c**) imaging result of the proposed method.

Figure 9. Satellite image of the imaging area.

6. Discussion

In point target simulation, Figure 5a shows that the targets in the left area and right area are aliased, leading to the left and right Doppler ambiguity in FLMC-SAR. In order to solve the Doppler ambiguity, a beamforming-based left–right Doppler ambiguity resolving is proposed [23]. However, due to the existence of array errors, beamforming cannot effectively solve the left–right Doppler ambiguity without array correction, as shown in Figure 5b. In order to enhance the robustness of Doppler ambiguity resolving for FLMC-SAR, we propose a sparsity-based array calibration method, which can be used for Doppler ambiguity resolving for FLMC-SAR. The azimuth ambiguity-to-signal ratio (AASR) of the nine point targets is shown in Table 4.

Table 4. AASR of point targets.

	P_1	P_6	P_2	P_8	P_3	P_7	P_4	P_9	P_5
Beamforming	3.23	6.87	2.95	9.69	2.23	3.22	3.23	6.88	2.95
Proposed method	23.42	24.79	25.03	23.72	24.71	25.11	23.77	23.56	25.41

After array calibration, the AASR of all targets is greater than 23 dB. It is concluded from the practical application that an AASR greater than 20 dB is convenient for people or computers to identify the target. Therefore, it can be seen that the proposed method is necessary for the FLMC-SAR system. To evaluate the accuracy of array error estimation of the proposed method, we define the mean square error of the array as follows:

$$\text{MSE} = \| \pmb{æ} - \hat{\pmb{æ}} \|_F^2 \tag{40}$$

where $\hat{\pmb{æ}}$ is the array error estimation. In order to reduce the complexity of the method, the array error estimation is based on the range–azimuth decoupling. Figure 10 shows the MSE of the array errors in the three range bins (Rs_1, Rs_2, and Rs_3 where targets are located) after each iteration.

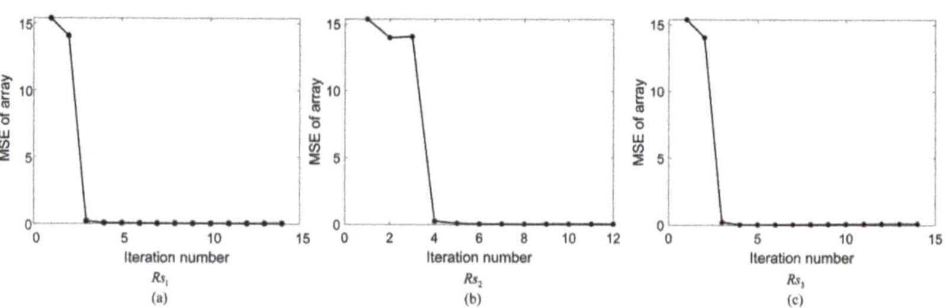

Figure 10. MSE of the array errors: (**a**) Rs_1 range bin; (**b**) Rs_2 range bin; (**c**) Rs_3 range bin.

The proposed method converges within ten iterations. Finally, the estimation of the array errors in the range bins Rs_1, Rs_2, and Rs_3 are shown in Figure 11. Compared with the actual array errors added into the echo, the estimation of array errors in each range bin are consistent with the array errors, which verifies the effectiveness of the array error estimation. Furthermore, the MSE of the array phase errors is less than $\pi/8$. After array calibration, the influence of the array errors on the Doppler ambiguity resolving can be ignored.

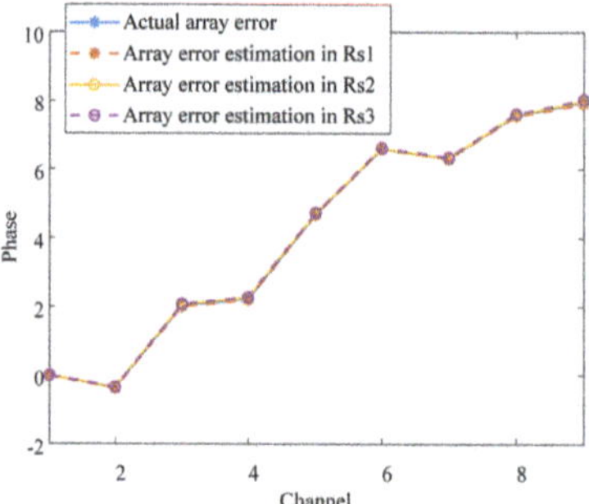

Figure 11. Estimation of the array errors.

We use the array error estimation from the proposed method for array calibration. The Doppler ambiguity resolving results via beamforming are shown in Figure 5d. Doppler ambiguity can be resolved well. The proposed method not only performs array error estimation but also achieves sparse image reconstruction. The azimuth pulse response functions of beamforming and the proposed method are shown in Figure 12. Compared with the beamforming-based Doppler ambiguity resolving, the proposed method introduces data fidelity into the image reconstruction model and combines signal sparsity, thus achieving sidelobe reduction and noise suppression.

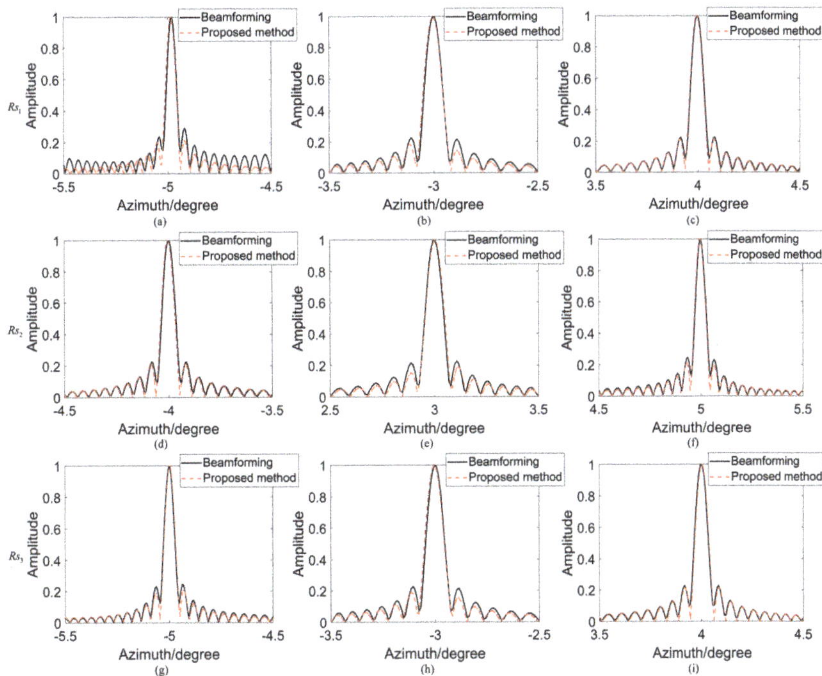

Figure 12. Azimuth pulse response functions: (**a**) P_1; (**b**) P_2; (**c**) P_3; (**d**) P_4; (**e**) P_5; (**f**) P_6; (**g**) P_7; (**h**) P_8; (**i**) P_9.

Peak sidelobe ratio (PSLR), integrated sidelobe ratio (ISLR), and impulse response width (IRW) of the point impulse responses are shown in Table 5. The values of IRWs obtained by the proposed method are basically consistent with those obtained by beamforming. However, the values of PSLR and ISLR obtained by the proposed method are significantly lower than those obtained by beamforming, which verifies that the proposed method has certain sidelobe suppression and noise suppression capabilities. Moreover, we also varied the AASR of all point targets under different SNRs, as shown in Figure 13. Thus, if SNR is greater than 10 dB, AASR is greater than 20 dB. The array error estimation and Doppler ambiguity resolving are based on the image domain. The range pulse compression and azimuth focusing can significantly improve the SNR so that the condition of SNR being greater than 10 dB can be easily met in practical applications.

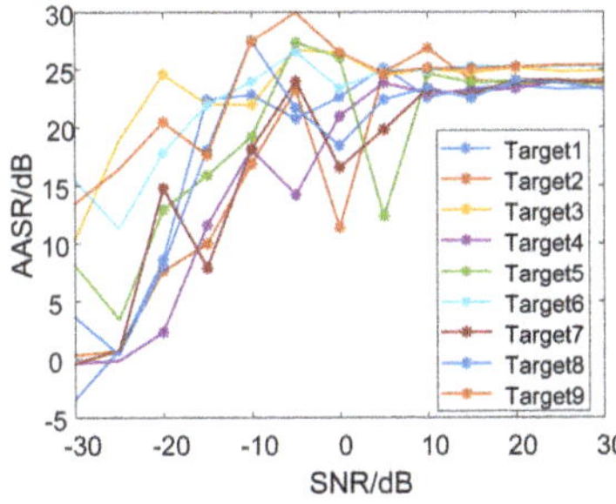

Figure 13. AASR of the point targets under different SNRs.

In surface target simulation, left–right Doppler ambiguity makes the image results of the left scene and that of the right aliased together, as shown in Figure 7a. The imaging results of beamforming-based Doppler ambiguity resolving are shown in Figure 7b. Due to the array errors, the performance of beamforming deteriorates. The average AASR of the targets in the red rectangle in Figure 7b is 6.79 dB. Clearly ambiguous residual images can be seen in Figure 7b. The proposed method was used for array error estimation and imaging reconstruction, and the image results are shown in Figure 7c. The average AASR of the targets in the red rectangle in Figure 7c is 21.81 dB. The performance of Doppler ambiguity resolving has been obviously improved.

Table 5. PSLR, ISLR, and IRW obtained by beamforming and the proposed method.

Azimuth Angle/°	Reference Slant Range	Beamforming			Proposed Method		
		PSLR/dB	ISLR/dB	IRW/m	PSLR/m	ISLR/m	IRW/m
−5	Rs_1	−12.19	−1.03	6.05	−13.46	−10.27	6.05
4	Rs_1	−12.94	−1.60	7.43	−13.39	−1.96	7.30
−1	Rs_1	−12.90	−1.75	10.03	−14.53	−12.24	10.03
5	Rs_2	−12.30	−1.06	5.89	−13.54	−10.53	5.89
−1	Rs_2	−12.91	−1.59	7.41	−13.38	−1.94	7.40
3	Rs_2	−12.89	−1.70	9.61	−14.30	−11.88	9.95
−1	Rs_3	−12.24	−1.09	5.95	−13.65	−10.79	5.95
4	Rs_3	−12.88	−1.58	7.43	−13.38	−1.92	7.30
−1	Rs_3	−12.89	−1.69	10.02	−14.29	−11.82	10.02

To evaluate the accuracy of array error estimation of the proposed method, we also give the MSE of the array errors in the three range bins (Rs_1, Rs_2, and Rs_3) after each iteration, as shown in Figure 14. The proposed method converges within ten iterations. Finally, the estimation of array errors in the three range bins (Rs_1, Rs_2, and Rs_3) are shown in Figure 15. Compared with the actual array errors added into the echo, the estimation of array errors in the three range bins are consistent with the array errors, which verifies the effectiveness of the array error estimation. The MSE of the array phase errors is less than $\pi/8$. On this basis, the Doppler ambiguity can be resolved by beamforming to get unambiguous imaging results, as shown in Figure 7d. Comparing Figure 7c,d, the imaging result of the proposed method has a lower entropy than the ambiguity resolving results of beamforming with the array calibration, which verifies that the proposed method has sidelobe suppression and noise suppression capabilities.

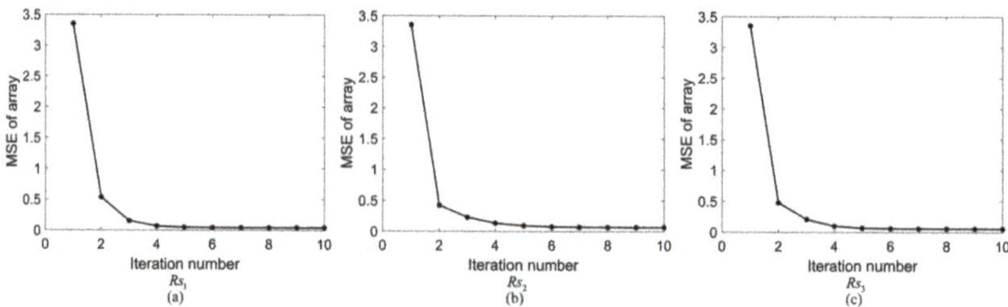

Figure 14. MSE of the array errors: (**a**) Rs_1 range bin; (**b**) Rs_2 range bin; (**c**) Rs_3 range bin.

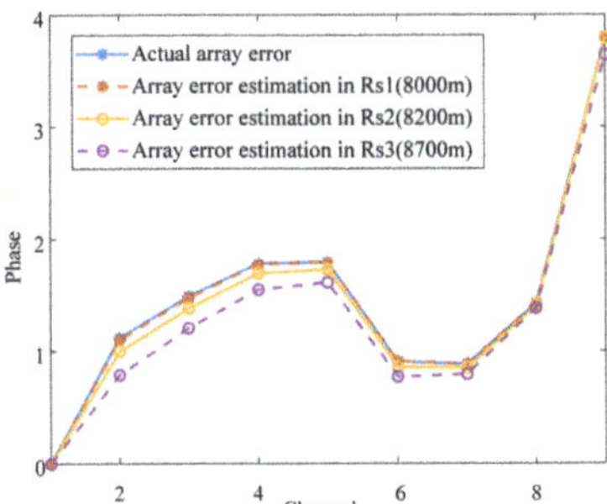

Figure 15. Estimation of the array errors.

To verify the robustness of the method under noise, we also give the image results of the surface target under different SNRs, as shown in Figure 16. Figure 16a–e are the imaging results of the surface target under SNRs of −10 dB, −10 dB, 0 dB, 10 dB, and 20 dB, respectively. Calculating the average AASR of the image results obtained by the beamforming and the proposed method, the curve of AASR changing with SNR is shown in Figure 16f. It can see that the average AASR of the image results obtained by the proposed method is more than 20 dB. The AASR improves by more than 10 dB compared to the image results obtained by the beamforming.

In the real-data experiment, the targets of the left area are aliased with that of the right area, and the imaging results are ambiguous. Based on Figure 8a, beamforming is used to solve the Doppler ambiguity, and the image results are shown in Figure 8b. One can see that the Doppler ambiguity has been resolved in some areas. However, due to the effect of array errors, most areas are still seriously ambiguous, such as the area highlighted by the red matrix. Due to Doppler ambiguity, the airstrip in Figure 16 cannot be distinguished from the ambiguous targets in Figure 8a. After the beamforming-based Doppler ambiguity resolving, due to the array error, the airstrip is still indistinct in Figure 8b. The proposed method is used to realize array error estimation and imaging reconstruction, and the airstrip is clearly distinguished, as shown in Figure 8c. Experimental results verify that the proposed method can be applied to FLMC-SAR system to realize unambiguous forward-looking imaging.

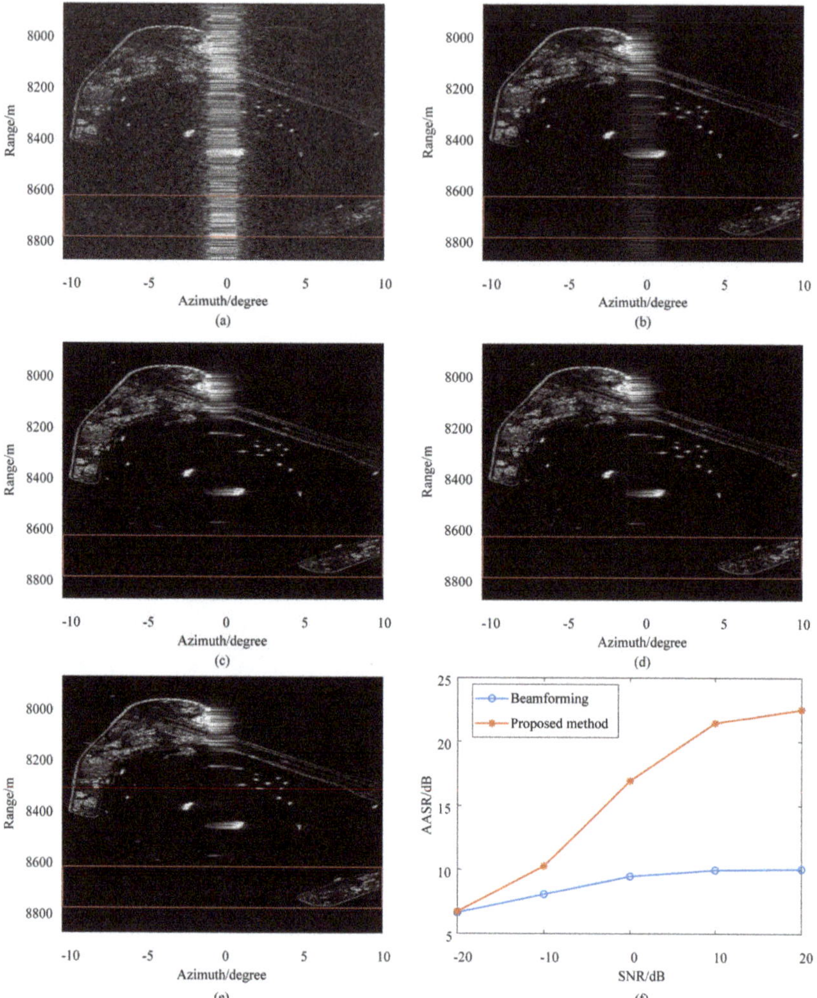

Figure 16. Imaging results of surface target under different SNRs: (**a**) imaging results under the SNR of −10 dB; (**b**) imaging results under the SNR of −10 dB; (**c**) imaging results under the SNR of 0 dB; (**d**) imaging results under the SNR of 10 dB; (**e**) imaging results under the SNR of 20 dB; (**f**) AASR of the imaging results under different SNRs.

7. Conclusions

For the FLMC-SAR system, the array error is an important factor that causes performance degradation of beamforming-based Doppler ambiguity resolving. In this paper, a sparsity-based array calibration and ambiguity resolving method is proposed for enhancing the robustness of FLMC-SAR imagery. First, the observation model of FLMC-SAR Doppler ambiguity combined with array error is derived. The model shows that array errors will lead to the mismatch of space–time characteristics of the targets, causing performance degradation of Doppler ambiguity resolving. Based on observational models, a constrained optimization problem for FLMC-SAR imaging is formulated, transforming the Doppler ambiguity resolving and array error estimation into a sparse recovery problem. Then a modified quasi-Newton method is proposed to realize both array error estimation and Doppler ambiguity resolving. Finally, simulation and real-data experiments verify the effectiveness of the proposed method.

Author Contributions: Conceptualization, J.L. and L.Z.; methodology, L.Z.; software, J.L.; validation, X.W.; investigation, X.W.; resources, X.W.; data curation, Y.C.; writing—original draft preparation, J.L.; writing—review and editing, Y.C. All authors have read and agreed to the published version of the manuscript.

Funding: This research received no external funding.

Institutional Review Board Statement: Not applicable.

Informed Consent Statement: Not applicable.

Data Availability Statement: Not applicable.

Acknowledgments: The authors would like to thank the anonymous reviewers for their valuable comments to improve the quality of this article.

Conflicts of Interest: The authors declare no conflict of interest.

References

1. Krishnan, V.; Swoboda, J.; Yarman, C.E.; Yazici, B. Multistatic Synthetic Aperture Radar Image Formation. *IEEE Trans. Image Process.* **2010**, *19*, 1290–1306. [CrossRef] [PubMed]
2. Cetin, M.; Karl,W.C. Feature-enhanced synthetic aperture radar image formation based on nonquadratic regularization. *IEEE Trans. Image Process.* **2001**, *10*, 623–631. [CrossRef] [PubMed]
3. Han, Y.;Jiao, R.; Huang, H.; Wang, Q.; Lai, T. A Framework for Distributed LEO SAR Air Moving Target 3D Imaging via Spectral Estimation. *Remote Sens.* **2022**, *14*, 5956. [CrossRef]
4. Sikaneta, I.; Gierull, C.H.; Cerutti-Maori, D. Optimum Signal Processing for Multi-channel SAR: With Application to High-Resolution Wide-Swath Imaging. *IEEE Trans. Geosci. Remote Sens.* **2014**, *52*, 6059–6109. [CrossRef]
5. Wang, Y.; Li, J.W.; Chen, J.; Xu, H.; Sun, B. A Parameter-Adjusting Polar Format Algorithm for Extremely High Squint SAR Imaging. *IEEE Trans. Geosci. Remote Sens.* **2014**, *52*, 640–650. [CrossRef]
6. Li, Z.; Wang, H.; Su, T.; Bao, Z. Generation of Wide-Swath and High-Resolution SAR Images from Multi-channel Small Spaceborne SAR Systems. *IEEE Geosci. Remote Sens. Lett.* **2005**, *2*, 82–86. [CrossRef]
7. Huang, Y.; Liao, G.; Zhang, Z.; Xiang, Y.; Li, J.; Nehorai, A. SAR Automatic Target Recognition Using Joint Low-Rank and Sparse Multiview Denoising. *IEEE Geosci. Remote Sens. Lett.* **2018**, *15*, 1570–1574. [CrossRef]
8. Yarman, C.E.; Yazici, B.; Cheney, M. Bistatic Synthetic Aperture Radar Imaging for Arbitrary Flight Trajectories. *IEEE Trans. Image Process.* **2008**, *17*, 84–93. [CrossRef]
9. Wang, L.; Yazici, B. Bistatic Synthetic Aperture Radar Imaging Using UltraNarrowband Continuous Waveforms. *IEEE Trans. Image Process.* **2012**, *21*, 3673–3686. [CrossRef]
10. Pu, W.; Wu, J.; Huang, Y.; Yang, J.; Li, W.; Yang, H. Joint Sparsity-Based Imaging and Motion Error Estimation for BFSAR. *IEEE Trans. Geosci. Remote Sens.* **2019**, *57*, 1393–1408. [CrossRef]
11. Qiu, X.; Hu, D.; Ding, C. Some reflections on bistatic SAR of forward-looking configuration. *IEEE Geosci. Remote Sens. Lett.* **2008**, *5*, 735–739. [CrossRef]
12. Pu, W.; Li, W.; Wu, J.; Huang, Y.; Yang, J.; Yang, H. An Azimuth Variant Autofocus Scheme of Bistatic Forward-Looking Synthetic Aperture Radar. *IEEE Geosci. Remote Sens. Lett.* **2017**, *14*, 689–693.
13. Liu, Z.; Ye, H.; Li, Z.; Yang, Q.; Sun, Z.; Wu, J.; Yang, J. Optimally Matched Space-Time Filtering Technique for BFSAR Nonstationary Clutter Suppression. *IEEE Trans. Geosci. Remote Sens.* **2021**, *60*, 5210617. [CrossRef]
14. Krieger, G.; Mittermayer, J.; Buckreuss, S.; Wendler, M.; Sutor, T.; Witte, F.; Moreira, A. SIREV-Sector Imaging Radar for Enhanced Vision. *Aerosp. Sci. Technol.* **2003**, *7*, 147–158. [CrossRef]
15. Krieger, G.; Mittermayer, J.; Wendler, M.; Witte, F.; Moreira, A. SIREV- Sector Imaging Radar for Enhanced Vision. *Proc. Int. Symp. ISPA.* **2001**, *3*, 377–382.
16. Soumekh, M. Moving target detection in foliage using along track monopulse synthetic aperture radar imaging. *IEEE Trans. Image Process.* **1997**, *6*, 1148–1163. [CrossRef]
17. Dai, S.; Liu, M.; Sun, Y.; Wiesbeck, W. The Latest Development of High Resolution Imaging for Forward Looking SAR with Multiple Receiving Antennas. *Proc. IEEE Int. Geosci. Remote Sens. Symp.* **2001**, *3*, 1433–1435.
18. Dai, S.; Wiesbeck, W. High Resolution Imaging for Forward Looking SAR with Multiple Receiving Antennas. *Proc. IEEE Int. Geosci. Remote Sens. Symp.* **2000**, *5*, 2254–2256.
19. Wang, W. Forward-Looking SAR Imaging with Frequency Diverse Array Antenna. *Proc. IEEE Int. Geosci. Remote Sens. Symp.* **2016**, *8*, 4191–4194.
20. Franceschetti, G.; Iodice, A.; Riccio, D. Forward-Looking Synthetic Aperture Radar (FLoSAR): The Array Approach. *IEEE Geosci. Remote Sens. Lett.* **2014**, *11*, 303–307. [CrossRef]
21. Franceschetti, G.; Iodice, A.; Riccio, D. FLoSAR: A New Concept for Synthetic Aperture Radar. *Proc. Radar Conf.* **2008**, *5*, 1–4.

22. Mahafza, B.R.; Knight, D.L.; Audeh, N.F. Forward-Looking SAR Imaging Using A Linear Array with Transverse Motion. *Proc. Southeastcon.* **1993**, *3*, 4.
23. Lu, J.; Zhang, L.; Huang, Y.; Cao, Y. High-Resolution Forward-Looking Multi-Channel SAR Imagery with Array Deviation Angle Calibration. *IEEE Trans. Geosci. Remote Sens.* **2020**, *58*, 6914–6928. [CrossRef]
24. Lu, J.; Zhang, L.; Quan, Y.; Meng, Z.; Cao, Y. Parametric Azimuth-Variant Motion Compensation for Forward-Looking Multi-channel SAR Imagery. *IEEE Trans. Geosci. Remote Sens.* **2021**, *59*, 8521–8537. [CrossRef]
25. Takemuram, N.; Deguchi, H.; Yonezawa, R.; Chiba, I. Phased Array Calibration Method with Evaluating Phase Shifter Error. *Antennas and Propag. Int. Symp.* **2001**, *3*, 259–263.
26. Mano, S.; Katagi, T. A Method for Measuring Amplitude and Phase of Each Radiating Element of A Phased Array Antenna. *Trans. IECE* **1982**, *5*, 555–560.
27. Ng, B. C.; See, C.M.S. Sensor-Array Calibration Using A Maximum-Likelihood Approach. *IEEE Trans. Antennas Propag.* **1996**, *44*, 827–835.
28. Herman, M.A.; Strohmer, T. High-resolution radar via compressed sensing. *IEEE Trans. Signal Process.* **2009**, *57*, 2275–2284. [CrossRef]
29. Varshney, K.R.; Çetin, M.; Fisher, J.W.; Willsky, A.S. Sparse Representation in Structured Dictionaries With Application to Synthetic Aperture Radar. *IEEE Trans. Signal Process.* **2008**, *56*, 3548–3561. [CrossRef]
30. Onhon, N.Ö.; Cetin, M. A Sparsity-Driven Approach for Joint SAR Imaging and Phase Error Correction. *IEEE Trans. Image Process.* **2012**, *21*, 2075–2088. [CrossRef]
31. Pu, W. Deep SAR Imaging and Motion Compensation. *IEEE Trans. Image Process.* **2021**, *30*, 2232–2247. [CrossRef]
32. Zhang, S.; Dong, G.; Kuang, G. Super Resolution Downward-Looking Linear Array Three-Dimensional SAR Imaging Based on Two-Dimensional Compressive Sensing. *IEEE J. Sel. Topics Appl. Earth Observ. Remote Sens.* **2016**, *9*, 2184–2196. [CrossRef]
33. Zhang, C.; Zhang, S.; Liu, Y.; Li, X. Joint Structured Sparsity and Least Entropy Constrained Sparse Aperture Radar Imaging and Autofocusing. *IEEE Trans. Geosci. Remote Sens.* **2020**, *58*, 6580–6593. [CrossRef]
34. Ender, J.H.G. On compressive sensing applied to radar. *Signal Process.* **2010**, *90*, 1402–1414. [CrossRef]
35. Potter, L.C.; Ertin, E.; Parker, J.T.; Cetin, M. Sparsity and compressed sensing in radar imaging. *Proc. IEEE* **2010**, *98*, 1006–1020. [CrossRef]
36. Xu, G.; Zhang, B.; Chen, J.; Hong, W. Structured Low-rank and Sparse Method for ISAR Imaging with 2-D Compressive Sampling. *IEEE Trans. Geosci. Remote Sens.* **2022**, *60*, 5239014. [CrossRef]
37. Xu, G.; Zhang, B.; Yu, H.; Chen, J. Sparse Synthetic Aperture Radar Imaging from Compressed Sensing and Machine Learning: Theories, Applications and Trends. *IEEE Geosci. Remote Sens. Magazine.* **2022**, *early access.* [CrossRef]
38. Qiu, W.; Zhou, J.; Fu, Q. Jointly Using Low-Rank and Sparsity Priors for Sparse Inverse Synthetic Aperture Radar Imaging. *IEEE Trans. Image Process.* **2020**, *29*, 100–115. [CrossRef]
39. Zhang, S.; Liu, Y.; Li, X.; Hu, D. Enhancing ISAR Image Efficiently via Convolutional Reweighted l1 Minimization. *IEEE Trans. Image Process.* **2021**, *30*, 4291–4304. [CrossRef]
40. Bi, H.; Zhang, B.; Zhu, X.X.; Hong, W.; Sun, J.; Wu, Y. L_1-Regularization-Based SAR Imaging and CFAR Detection via Complex Approximated Message Passing. *IEEE Trans. Geosci. Remote Sens.* **2017**, *55*, 3426–3440. [CrossRef]
41. Shao, S.; Zhang, L.; Wei, J.; Liu, H. Two-Dimension Joint Super-Resolution ISAR Imaging With Joint Motion Compensation and Azimuth Scaling. *IEEE Geosci. Remote Sens. Lett.* **2021**, *18*, 1411–1415. [CrossRef]
42. Ding, J.; Wang, M.; Kang, H.; Wang, Z. MIMO Radar Super-Resolution Imaging Based on Reconstruction of the Measurement Matrix of Compressed Sensing. *IEEE Geosci. Remote Sens. Lett.* **2022**, *19*, 1–5. [CrossRef]
43. Li, W.; Zhang, W.; Zhang, Q.; Zhang, Y.; Huang, Y.; Yang, J. Simultaneous Super-Resolution and Target Detection of Forward-Looking Scanning Radar via Low-Rank and Sparsity Constrained Method. *IEEE Trans. Geosci. Remote Sens.* **2020**, *58*, 7085–7095. [CrossRef]
44. Wei, Y.;Li, Y.; Ding, Z.; Wang, Y.; Zeng, T.; Long, T. SAR Parametric Super-Resolution Image Reconstruction Methods Based on ADMM and Deep Neural Network. *IEEE Trans. Geosci. Remote Sens.* **2021**, *59*, 10197–10212. [CrossRef]
45. Wu, C.; Zhang, Z.; Chen, L.; Yu, W. Super-Resolution for MIMO Array SAR 3-D Imaging Based on Compressive Sensing and Deep Neural Network. *IEEE J. Sel. Top. Appl. Earth Obs. Remote. Sens.* **2020**, *13*, 3109–3124. [CrossRef]
46. Zhang, S.; Liu, Y.; Li, X. Fast Sparse Aperture ISAR Autofocusing and Imaging via ADMM Based Sparse Bayesian Learning. *IEEE Trans. Image Process.* **2020**, *29*, 3213–3226. [CrossRef]
47. Ji, S.; Xue, Y.; Carin, L. Bayesian Compressive Sensing. *IEEE Trans. Signal Process.* **2008**, *56*, 2346–2356. [CrossRef]
48. Wu, J.; Liu, F.; Jiao, L.C.; Wang, X. Compressive Sensing SAR Image Reconstruction Based on Bayesian Framework and Evolutionary Computation. *IEEE Trans. Image Process.* **2011**, *20*, 1904–1911. [CrossRef]
49. Vogel, C.R.; Oman, M.E. Fast, robust total variation-based reconstruction of noisy, blurred images. *IEEE Trans. Image Process.* **1998**, *7*, 813–824. [CrossRef]

remote sensing

Article

A Modified 2-D Notch Filter Based on Image Segmentation for RFI Mitigation in Synthetic Aperture Radar

Zewen Fu [1,2,3], Hengrui Zhang [1,2,3], Jianhui Zhao [1,2,3], Ning Li [1,2,3,*] and Fengbin Zheng [3,4]

[1] Henan Engineering Research Center of Intelligent Technology and Application, Henan University, Kaifeng 475004, China
[2] Henan Key Laboratory of Big Data Analysis and Processing, Henan University, Kaifeng 475004, China
[3] College of Computer and Information Engineering, Henan University, Kaifeng 475004, China
[4] College of Information Engineering, Henan Kaifeng College of Science Technology and Communication, Kaifeng 475004, China
* Correspondence: hedalining@henu.edu.cn

Abstract: Synthetic aperture radar (SAR), as an active microwave sensor, can inevitably receive radio frequency interference (RFI) generated by various electromagnetic equipment. When the SAR system receives RFI, it will affect SAR imaging and limit the application of SAR images. As a kind of RFI mitigation method, notch filtering method is a classical method with high efficiency and robust performance. However, the notch filtering methods pay no attention to the protection of useful signals. This paper proposed a modified 2-D notch filter based on image segmentation for RFI mitigation with signal-protected capability. (1) The adaptive gamma correction (AGC) approach was utilized to enhance the SAR image with RFI in the range-frequency and azimuth-time domain. (2) The modified selective binary and Gaussian filtering regularized level set (SBGFRLS) model was utilized to further process the image after AGC to accurately extract the contour of the useful signals with interference, which is more conducive to protecting the useful signals without interference. (3) The Generalized Singular Value Thresholding (GSVT) based low-rank sparse decomposition (LRSD) model was utilized to separate the RFI signals and the useful signals. Then, the useful signals were restored to the raw data. The simulation experiments and measured data experiments show that the proposed method can effectively mitigate RFI and protect the useful signals whether there are RFI with single source or multiple sources.

Keywords: synthetic aperture radar; radio frequency interference; notch filter; image segmentation; low-rank sparse decomposition

Citation: Fu, Z.; Zhang, H.; Zhao, J.; Li, N.; Zheng, F. A Modified 2-D Notch Filter Based on Image Segmentation for RFI Mitigation in Synthetic Aperture Radar. *Remote Sens.* **2023**, *15*, 846. https://doi.org/10.3390/rs15030846

Academic Editors: Gang Xu, Lan Du and Haipeng Wang

Received: 12 November 2022
Revised: 14 January 2023
Accepted: 30 January 2023
Published: 2 February 2023

1. Introduction

1.1. Background

Synthetic aperture radar (SAR) is an active microwave technology that can observe the Earth all-day and during all-weather. SAR can be applied in many fields such as crop yield estimation, ground feature classification, marine environment monitoring, and military reconnaissance, etc. [1–8]. However, as an active wideband radio system, SAR can easily receive RFI signals, and these RFI signals will seriously degrade the SAR imaging quality and limit the application of SAR images. Over the past few decades, with the rapid development of electronic information field and modern radio technology, radio frequency interference (RFI) existing in SAR images has become a common phenomenon [9–13]. Figure 1 shows the common sources of RFI.

For increasingly complex RFI, it is necessary to propose some effective mitigation methods. Using scientific and effective RFI mitigation methods is beneficial to improving the survivability and practical efficiency of SAR systems in complex electromagnetic environments, and it has important practical significance. With the further development of SAR technology, researchers have proposed many mitigation methods for different types of RFI [14]. Parametric methods, semiparametric methods, and nonparametric methods are three types of RFI mitigation methods to mitigate the interference. Parametric methods and semiparametric methods can mitigate the RFI to a certain extent by adjusting the determined model and parameters. For example, Zhou et al. [15] proposed an algorithm of wideband interference suppression via instantaneous frequency estimation and regularized time-frequency filtering. Huang et al. [16–21] carried out an in-depth study on semiparametric methods and proposed a series of low rank and sparse decomposition models. Recently, they proposed an algorithm of time-varying RFI mitigation via graph Laplacian clustering techniques [22]. Braunstein et al. [23] mitigated the RFI in measured data by the parametric method. Zhang et al. [24] used wavelet transform and short-time Fourier transform to analyze the characteristics of interference in the 2-D range time–frequency domain. Yang et al. [25] proposed a postprocessing kernel, namely, the 2-D SPECtral ANalysis (2-D SPECAN) filter, to remove the RFI in SLC images. In recent years, the RFI mitigation algorithm combined with machine learning has achieved good results. Zhou et al. [26] presented a narrow-band interference and wide-band interference mitigation algorithm based on the deep residual network, and Xu et al. [27] proposed two RFI mitigation algorithms based on a modified block sparse Bayesian learning, and these algorithms have achieved a good mitigation effect. However, these above methods rely on the estimation of model parameters, and the type of interference aimed by these methods is relatively single, so the generalizability of this kind of method is relatively weak. On the other hand, nonparametric methods mitigate RFI through the characteristics of the useful signals and the RFI signals in different domains. Nonparametric methods not only have applicability and robustness, but also have high algorithm efficiency, and there is no need to establish a model of RFI. Zhou et al. [28–31] proposed a series of algorithms based on matrix decomposition theory, Yang et al. [32] proposed a generic subspace model for characterizing a variety of RFI types and designed a block subspace filter for removing RFI artifacts in SLC SAR images. In particular, the most classic nonparametric method is the notch filtering method, and the main principle of the notch filtering method is to set the RFI signal to zero and achieve the purpose of mitigating RFI. However, there is a flaw in these nonparametric methods: when the RFI signals are mitigated, part of the useful signals will be lost, so it is necessary to improve the notch method [33–38].

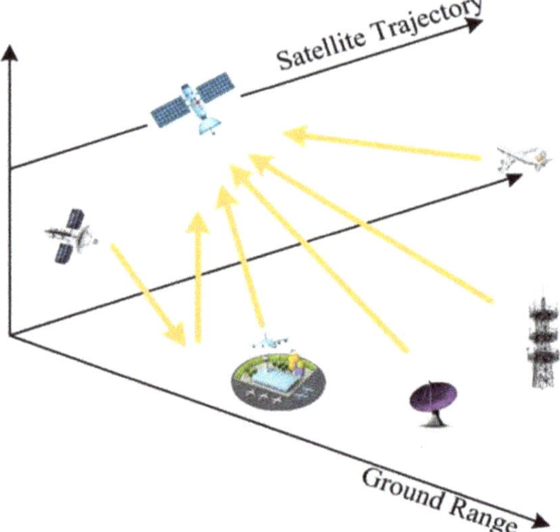

Figure 1. Common sources of RFI.

1.2. Previous Work of Notch Method

Since the 1990s, various advanced nonparametric methods have been proposed to mitigate RFI. In 1996, Cazzaniga and Guarnieri proposed the MUSIC method to estimate narrow-band interference frequencies [39]. Subsequently, in order to improve the detection probability for the peaks, Buckreuss and Horn proposed an averaging of the spectra of adjacent range lines, which was first applied to the E-SAR systems [40]. Meyer et al. designed adaptive detection to determine the RFI locations of narrow-band, wide-band, and determined the notch filter width to design a RFI mitigation process; this method was well validated in the ALOS PALSAR data [41].

Researchers have proposed a series of adaptive notch methods [42–49], which have a good compromise in terms of convergence speed, stability, computational complexity, and adaptation. The InTernational Union of Radio science (ITU-R) in its report also provides similar recommendations for the notch filter to enable its application in the Earth Exploration Satellite Service (EESS) [50]. Nabil et al. showed that there was a special near-zero RFI in the TarraSAR-X data, and proposed a modified notch filter based on the traditional notch filter to obtain two images after two mitigations, then combined them to obtain better azimuth spectrum information [51]. In [52], a sub-band spectral phase cancellation method was proposed to use the difference between adjacent spectral sub-bands to approximate the effect of the notch filter, and applied to the SAR data. The above-mentioned notch filters are mainly applied to raw data, and Reigber and Doerry proposed a new notch filtering method to eliminate the interference from the focused image and verified it on the L-band SAR data, respectively [53,54]. In [55], an azimuth-frequency domain filtering method was proposed to suppress the intermittent transmission interference in the 2-D range-time domain and azimuth-frequency domain. Li et al. proposed a time domain notch filtering (TNF) method for pulse RFI mitigation in SAR [56].

However, the drawback of the above notch filter is that when the zero-notch width of the filter exceeds 2% of the bandwidth, the spatial resolution will decrease, and the sidelobe energy will increase. Therefore, it is necessary to further improve the notch filtering method.

1.3. Main Contributions of This Paper

In order to improve the protection ability of notch filtering methods for useful signals, this paper proposed a modified 2-D notch filter based on image segmentation. First, inspired by the idea of traditional notch filtering method, the raw data were converted to the range-frequency and azimuth-time domain, the characteristics of RFI were analyzed. The image of range-frequency and azimuth-time domain was enhanced by adaptive gamma correction (AGC). Then, the modified selective binary and Gaussian filtering regularized level set (SBGFRLS) model was utilized to segment the image, and the RFI signals were extracted. Finally, the Generalized Singular Value Thresholding (GSVT) -based low-rank sparse decomposition (LRSD) model was performed on the extracted part to screen out the useful signals and the RFI signals and restore the useful signal to the initial raw data. The specific contributions of this paper are as follows:

- The method proposed combines the image segmentation technology with RFI mitigation to accurately extract the contour of the useful signals with interference, which is more conducive to protecting the useful signals without interference.
- The GSVT-based LRSD model was performed to further extract the useful signals contained in the RFI signals. The proposed method effectively improves the protection ability of the useful signals compared with the traditional notch filtering method.
- The superiority of the proposed method was verified by simulation experiments and measured data experiments. The proposed method can effectively mitigate RFI and protect the useful signals, whether there are RFI with a single source or multiple sources.

The remainder of this article is organized as follows. In Section 2, the geometric and signal models are introduced, and classical frequency domain notch filtering (FNF) is introduced. Section 3 shows the technical route of the algorithm proposed in this paper.

Section 4 shows the experimental results and performance analysis of the proposed method. Section 5 discusses the experiments in this paper. Section 6 presents our conclusions and future work.

2. Model and Related Work

2.1. Signal Model of RFI

In the SAR system, the signals exist in the 2-dimensional time domain. After quadrature demodulation and digital sampling, the raw data received by the SAR system can be written as

$$S(\tau, \eta) = X(\tau, \eta) + I(\tau, \eta) + N(\tau, \eta) \tag{1}$$

where $X(\tau, \eta)$ represents the useful signal; $I(\tau, \eta)$ represents the RFI signal; $N(\tau, \eta)$ represents the system noise; τ and η denote the range fast time and the azimuth slow time, respectively.

In general, the signal model interference can be expressed as

$$I_{NBI}(\tau, \eta) = \sum_{n=1}^{N} A_n(\eta) exp(2j\pi f_n \tau + \varphi_n) \tag{2}$$

where N represents the number of the RFI signals. $A_n(\eta)$, f_n, and φ_n represent the amplitude, frequency, and phase of the nth interference signal, respectively.

This can be divided in two terms, where ϕ_{LFM} represents the linear frequency modulation (LFM). We can obtain:

$$\phi_{LFM} = 2\pi f_n \tau + \pi K_n \tau^2 \tag{3}$$

where K_n represents the chirp rate of the nth signal, and β_n represents the modulation factor.

Normally, the SAR system transmits the LFM signal, which is expressed as

$$s_t(\tau) = rect\left(\frac{\tau}{T_\tau}\right) exp\left\{j\pi K_\tau \tau^2\right\} \tag{4}$$

where rect represents the window function; T_τ is the SAR signal receiving duration; K_τ is the frequency modulation rate of the transmitted signal; and τ_0 represents the signal transmission delay. The signal received by the SAR system after transmission delay is:

$$s_t(\tau) = rect\left(\frac{\tau - \tau_0}{T_\tau}\right) exp\left\{j\pi K_\tau (\tau - \tau_0)^2\right\} \tag{5}$$

2.2. Theory of FNF

The FNF method is a classical nonparametric interference mitigation method, which not only has applicability and robustness, but also has high algorithm efficiency. The FNF method has been widely used to solve the problem of RFI in airborne SAR and spaceborne SAR systems [57].

First, the frequency domain representation of $s_t(\tau)$ needs to be obtained because the FNF method is a frequency domain processing method. Therefore, the stationary phase method (SPM) is used, and the approximate signal spectrum of $s_t(\tau)$ can be expressed as

$$S_t(f_\tau) = C_1 rect\left(\frac{f_\tau}{B_\tau}\right) exp\left\{-j\pi \frac{f_\tau^2}{K_\tau} \pm \frac{j\pi}{4}\right\} exp\{-j2\pi f_\tau \tau_0\} \tag{6}$$

where f_τ and B_τ represent the range sampling rate and bandwidth of signal, respectively. C_1 represents the constant.

According to Equation (1), assuming that the raw data $s_t(\tau)$ contains RFI signal $s_{RFI}(\tau)$ and system noise $s_N(\tau)$, we used SPM to obtain the approximate signal spectrum of $s_t(\tau)$.

In general, the system noise can be negligible. For the sake of derivation, the constant term C_1, phase term $\pi/4$, and system noise $s_n(t)$ can be ignored. The approximate signal spectrum $S_\tau(f_\tau)$ of the signal received by the SAR system $s_t(\tau)$ can be rewritten as:

$$
\begin{aligned}
S_\tau(f_\tau) &= S_t(f_\tau) + S_{RFI}(f_\tau) \\
&= \mathrm{rect}\left(\frac{f_\tau}{B_\tau}\right) exp\left\{-j\pi\frac{f_\tau^2}{K_\tau}\right\} exp\{-j2\pi f_\tau \tau_0\} + S_{RFI}(f_\tau)
\end{aligned}
\tag{7}
$$

To show the difference between the RFI signal and the useful signal more clearly, we performed 1-dimensional range direction Fourier transform (FT) on the raw data containing RFI with single sources and multiple sources, and the 3-dimensional diagram is shown in Figure 2. Among them, Figure 2a shows the data containing RFI with a single source, Figure 2b is the data containing RFI with multiple sources, and the inside of the red ellipse represents the RFI signals.

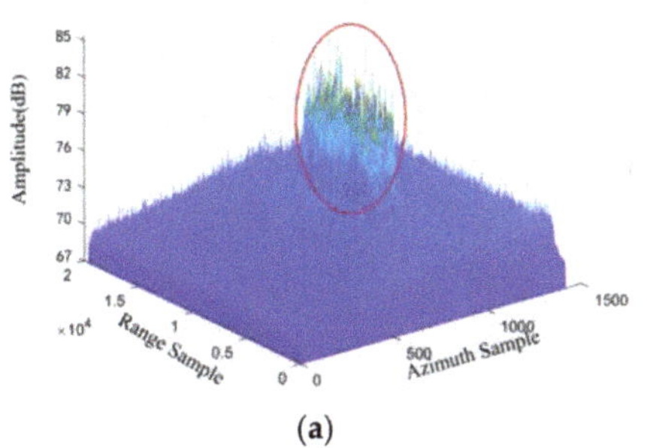
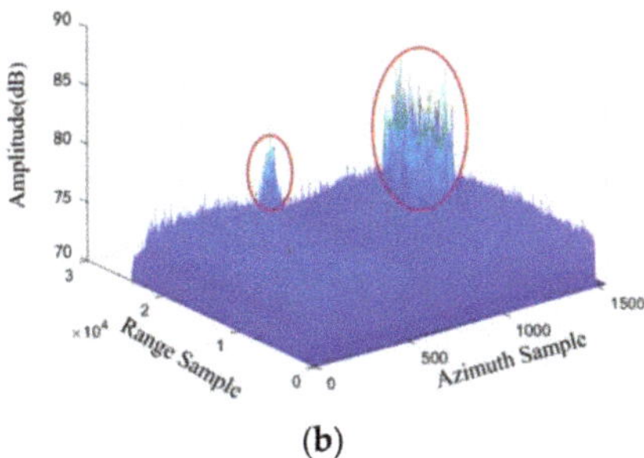

Figure 2. The 3-dimensional diagram in the range-frequency and azimuth-time domain. (**a**) The data containing RFI with a single source. (**b**) The data containing RFI with multiple sources.

We defined f_i and B_i to represent the interference frequency and bandwidth, respectively. $B_i = |K_\tau|T_\tau$, the filter can be expressed as:

$$
H_{NF}(f_\tau) = 1 - \mathrm{rect}\left(\frac{f_\tau - f_i}{B_i}\right)
\tag{8}
$$

In order to focus the signal received by the SAR system, we performed matched filtering processing, and the match filter function is

$$
H_{MF}(f_\tau) = exp\left(j\pi\frac{f_\tau^2}{K_\tau}\right)
\tag{9}
$$

At the same time, we need to multiply the signal by the notch filter $H_{NF}(f_\tau)$ in the frequency domain processing, and we can obtain:

$$
\begin{aligned}
S_{MF}(f_\tau) &= S_\tau(f_\tau)H_{MF}(f_\tau)H_{NF}(f_\tau) \\
&= (S_t(f_\tau) + S_{RFI}(f_\tau))H_{NF}(f)H(f) \\
&= S_t(f_\tau)H_{NF}(f)H(f) + S_{RFI}(f_\tau)H_{NF}(f)H(f)
\end{aligned}
\tag{10}
$$

Since the notch filter sets the frequency components of the RFI signal to zero, that is, $S_{RFI}(f_\tau)H_{NF}(f)H(f) = 0$, Equation (10) can be rewritten as:

$$
\begin{aligned}
S_{MF}(f_\tau) &= S_\tau(f_\tau)H_{MF}(f_\tau)H_{NF}(f_\tau) \\
&= \mathrm{rect}\left(\frac{f_\tau}{B_\tau}\right) exp\{-j2\pi f_\tau \tau_0\} - \mathrm{rect}\left(\frac{f_\tau - f_i}{B_i}\right) exp\{-j2\pi f_\tau \tau_0\}
\end{aligned}
\tag{11}
$$

Then, $S_{MF}(f_\tau)$ is transformed to a 2-dimensional time domain, and the impulse response of the output signal can be expressed as

$$
\begin{aligned}
s_{MF}(\tau) = &\alpha_1 sinc\{\pi B_\tau(\tau - \tau_0)\} \\
&- \alpha_2 sinc\{\pi B_i(\tau - \tau_0)\} exp\{-j2\pi f_i(\tau - \tau_0)\}
\end{aligned}
\tag{12}
$$

where α_1 and α_2 represent the amplitude response. According to the above derivation process, it can be seen that the frequency domain notch filter will eliminate the frequency components of the interference in the SAR signal, but it will be mixed with the frequency components of the useful signal. Therefore, the FNF method can suppress the RFI, but it cannot protect the useful signals, which will lead to the degradation of the image quality.

2.3. Low-Rank Characteristics of RFI

The RFI in the range-frequency domain has a relatively stable frequency in the slow time direction, and its amplitude appears as some parallel straight lines, as shown in Figure 3a. It is clear that RFI has low-rank properties in the slow time direction. For further verification, the eigenvalue decomposition of Figure 3a was performed, and the corresponding results are shown in Figure 3b, the red line represents percentage. The eigenvalues reflect the energy of different components in the SAR echo and the structural redundancy of the matrix. As can be observed, only a few large eigenvalues were related to RFI, which further illustrates the low-rank feature of RFI in the range-frequency domain.

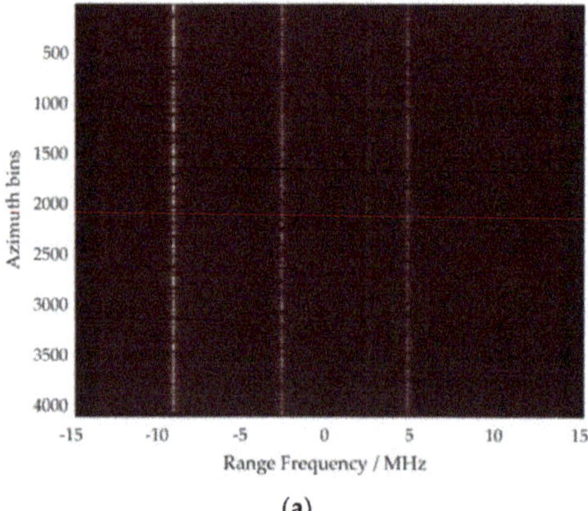

(a)

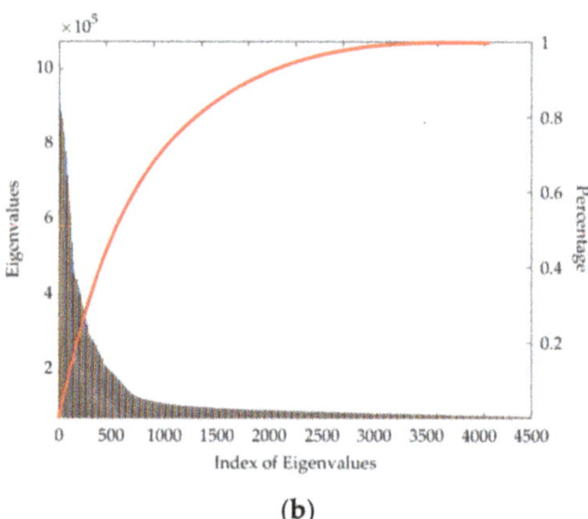

(b)

Figure 3. Structural analysis of RFI in the range-frequency domain. (**a**) Spectrogram of SAR echoes contaminated by RFI. (**b**) Eigenvalue sequences and analysis corresponding to (**a**).

3. Methodology

To solve the problem of RFI mitigation in SAR data, a modified 2-D notch filter method was proposed. The proposed method consists of three steps: enhancing the image in the range-frequency and azimuth-time domain by AGC; segmenting the edge of the RFI areas by the modified SBGFRLS model; and extracting the RFI signals to leave the useful signals by GSVT-based LRSD. The proposal can mitigate RFI robustly, and protect the useful signals effectively. The specific flowchart is shown in Figure 4.

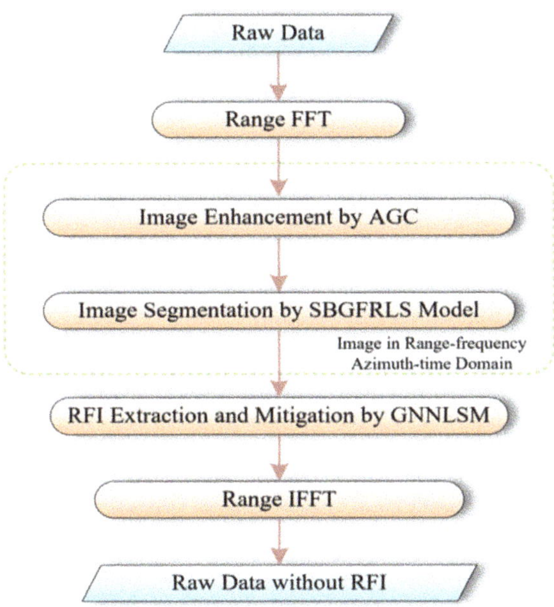

Figure 4. Flowchart of the proposed method.

3.1. Image Enhancement by AGC

AGC is an algorithm in the field of image processing, and is an improvement algorithm of gamma correction (GC) [58]. AGC can stretch the intensity level of the image and reduce the number of low intensity cells while increasing the number of high intensity cells. The AGC is used to enhance the range-frequency and azimuth-time domain image of SAR data with RFI, and this operation will enhance the discrimination between signals of RFI and useful signals.

First, we set a parameter γ; the GC algorithm adjusts the pixel value of the image by changing the size of parameter γ. The GC is formulated as follows:

$$T(l) = l_{max}(l/l_{max})^{\gamma} \tag{13}$$

where l_{max} is the maximum intensity of the input. The intensity l of each pixel in the input image is transformed as $T(l)$ after performing Equation (13).

The AGC algorithm uses a cumulative distribution function to replace the γ of the GC algorithm. The AGC is formulated as follows:

$$T(l) = l_{max}(l/l_{max})^{\gamma} = l_{max}(l/l_{max})^{1-cdf(l)} \tag{14}$$

where $cdf(l)$ represents the cumulative distribution of the intensity l. By calculating the probability density function, as shown in Equation (15):

$$cdf_w(l) = \sum_{l=0}^{l_{max}} pdf_w(l) / \sum pdf_a \tag{15}$$

where $\sum pdf_a$ represents the sum of the probability density of the whole image, and can be calculated as follows:

$$\sum pdf_a = \sum_{l=0}^{l_{max}} pdf_a(l) \tag{16}$$

where $pdf_w(l)$ refers to the probability function after l adjusts the histogram through the weighting distribution. The weighting distribution function is formulated as:

$$pdf_w(l) = pdf_{max}\left(\frac{pdf(l) - pdf_{min}}{pdf_{max} - pdf_{min}}\right)^{\alpha} \tag{17}$$

where α is the adjusted parameter; pdf_{max} is the maximum pdf of the statistical histogram; and pdf_{min} is the minimum pdf. These two parameters α and γ are empirical values, and we generally set them as 0.1–1.5. The intensity of the pixels in the image will vary with the parameters. The larger the parameter, the higher the intensity. In this paper, we compared the treatment effect through experiments. When they were in the range of 0.2–0.6, we could obtain a better result after the process.

In the image after AGC, the signals of RFI will be more obvious than the useful signals, which is equivalent to the first separation of the useful signals and RFI signals in the proposed algorithm. This operation will greatly improve the accuracy of image segmentation in the next step.

3.2. Image Segmentation by the SBGFRLS Model

The active contour model (ACM) is deformed contours that move under the force of the image and external constraints [59]. At present, ACM is generally divided into four categories: threshold based, edge based, region based, and energy functional based. According to the analysis in Section 2, the ACM based on region is more suitable for image segmentation in this paper. In the range-frequency and azimuth-time domain, the pixel intensity of RFI is usually high, and the pixel intensity in some areas is significantly higher than that of useful signals, while region-based ACM generally uses the overall intensity of the internal and external areas for segmentation. Therefore, region-based ACM is more suitable for this situation.

Among them, SBGFRLS is a kind of region-based ACM algorithm realized by selecting binary and Gaussian filter regularization, which is more suitable for image segmentation in the range-frequency and azimuth-time domain. In the range-frequency and azimuth-time domain, the strength of the RFI signals is generally significantly stronger than the strength of useful signals. However, when there are strong point targets in the region of interest such as ships at sea, corner reflectors in a certain area on land, etc., the raw data of these objects will affect image segmentation after image enhancement. Using the improved SBGFRLS algorithm can reduce the impact of these "noises" and improve the edge extraction accuracy for RFI, but the algorithm is not sensitive to the selection of the initial contour. Compared with traditional segmentation algorithms, image segmentation in this case has obvious advantages. Therefore, the modified SBGFRLS model was utilized to further process the image after AGC and accurately extract the contour of the RFI signals, and useful signals can exist, which is more conducive to protecting the signals without RFI [60]. The algorithm uses the Euclid length term to regularize the contour curve, and adds the average gray values inside and outside the curve to the SBGFRS model. The algorithm reduces the influence of other "noise" targets and is insensitive to the selection of initial contour.

Deformation energy can represent the contours of the target area, and external force on the contour, potential energy, and total energy represent the individual model functions that affect edge segmentation [59]. A suitable deformation energy $E_s(v)$ is assumed to define the contour of the target, $v(x(s), y(s))$ represents the contour, and it means the mapping from the unit parameter domain $s \in [0,1]$ to the image, while considering the external force on the contour as a differential of the potential energy $P(v)$. Then, the total energy on the contour can be defined as:

$$E(v) = E_s(v) + P(v) \tag{18}$$

with

$$E_s(v) = \int_0^1 \left(\omega_1(s)|v_s|^2 + \omega_2(s)|v_{ss}|^2\right)ds \tag{19}$$

where v represents the differential with respect to s, and an internal deformation energy of a stretchable and bendable contour is defined by $E_s(v(s))$, and the $E_s(v(s))$ consists of two parameters: the "stress" of the contour is controlled by $\omega_1(s)$, the "stiffness" of the contour is controlled by $\omega_2(s)$, these parameters control the physical behavior and local continuity of the model. In particular, assuming that $\omega_1(s_0) = \omega_2(s_0) = 0$, the discontinuous position of s_0 is allowed, and the discontinuous on tangent of the point s_0 is allowed. The external potential energy $P(v)$ can be expressed as:

$$P(v) = \int_0^1 p(v(s))\mathrm{d}s \tag{20}$$

where $P(v)$ is a scalar function defined over the entire image surface $I(x,y)$. When the external binding force is not considered, if $p(x,y) = \pm\omega_3|G_\sigma * I(x,y)|$, the contour edge will be attracted to the area of low or high intensity; if $p(x,y) = \pm\omega_3|\nabla[G_\sigma * I(x,y)]|$, the contour edge will be attracted to the edge of the region of interest. Among them, ω_3 controls the magnitude of the potential energy, and $G_\sigma * I$ represents the convolution of the image and Gaussan smoothing filter with feature density σ. The above model is conducive to unifying the target contour in a feature extraction process, and after properly initialized, it can autonomously converge to an energy minima state.

According to Equations (18)–(20), we have:

$$E(v) = \int_0^1 \left(\frac{\omega_1(s)|v_s|^2}{2} + \frac{\omega_2(s)|v_{ss}|^2}{2} + \omega_3(s)P(v) \right)\mathrm{d}s \tag{21}$$
$$= \int_0^1 F(v, v_s, v_{ss})\mathrm{d}s$$

where $E(v(s))$ represents the functionals of the $v(s)$, if $E(v)$ obtains the extreme value on a certain curve, Equation (21) satisfies the follow Equation:

$$\begin{cases} Fv - \frac{\partial}{\partial s}(Fv_s) + \frac{\partial^2}{\partial s^2}(Fv_{ss}) = 0 \\ v(s) \in [0,1] \ and \ v(0) = v_0, v'^{(0)} = v'_0, \ v(1) = v_1, v'^{(1)} = v'_1 \end{cases} \tag{22}$$

Therefore, the minimum value of the region can be obtained by solving the above Equation to obtain the edge of the target contour. The above is the process of classical ACM. Then, the modified SBGFRLS model was utilized to solve Equation (22). We assume that the internal and external average gray value are c_1 and c_2:

$$c_1(\varnothing) = \frac{\int I(x)H(\varnothing)\mathrm{d}s}{\int H(\varnothing)\mathrm{d}s} \tag{23}$$

$$c_2(\varnothing) = \frac{\int I(x)[1 - H(\varnothing)]\mathrm{d}s}{\int [1 - H(\varnothing)]\mathrm{d}s} \tag{24}$$

where I is the image after AGC; $\varnothing$ is the level set function; $H(\varnothing)$ represents the heaviside function. Then, the indicator function spf can be expressed as:

$$spf[I(x)] = \frac{I(x) - \frac{c_1 + c_2}{2}}{max\left[\left|I(x) - \frac{c_1 + c_2}{2}\right|\right]} \tag{25}$$

where $max[\cdot]$ represents the maximum in region. We can obtain the following level set Equation:

$$\frac{\partial\varnothing}{\partial t} = spf[I(x)]\beta|\nabla\varnothing| \tag{26}$$

where t represents the time; ∇ represents the gradient operation; β represents the growth power, which is used to control the contraction or expansion of the contour.

Finally, the image region can be segmented by iteratively judging whether the level set function converges according to the above equation. Although we successfully segmented the part of RFI in the range-frequency and azimuth-time domain using the modified SBGFRLS model, there were still useful signals in the extracted part. In the process of radar receiving signals, RFI was mixed with useful signals and entered the receiver at the same time. Therefore, we can understand that RFI actually covers the useful signals, but we only divided the parts with RFI, and did not protect the useful signals in the parts with RFI. In order to protect useful signals, we need to further process the area after image segmentation.

3.3. RFI Extraction by GSVT

RFI has low rank characteristics in the range-frequency and azimuth-time domain. Therefore, we can use the LRSD model to further process the signals after segmentation, and the useful signals can be effectively protected by solving the LRSD problem. The GSVT is an effective method to solve the LRSD problem. Through the analysis in Section 2, the GSVT is suitable for RFI mitigation because of the characteristics of RFI. LRSD, also known as the robust principal component analysis (RPCA) algorithm, is currently applied in many fields, for example, moving and stationary target separation in the SAR signal domain using parallel convolutional autoencoders with RPCA loss [61]. Separating the sparse matrix of moving targets from the low-rank matrix of static backgrounds by RPCA was conducted in [62], and the GSVT-based LRSD was utilized in this paper [63]. As a LRSD method, this method can better obtain the compromise factor from the observed data and solve the proposed GSVT problem through the alternating direction multiplier method (ADMM). Compared with the traditional LRSD method, this method has a better denoising effect, and the effect of protecting useful signals is better than the traditional method. The formulation of this model can be expressed as:

$$\min_{L,S} \quad rank(\mathbf{L}) + \lambda \|\mathbf{S}\|_0$$
$$s.t. \quad \mathbf{M} = \mathbf{L} + \mathbf{S} \tag{27}$$

where $rank(\cdot)$ represents the rank of the matrix; $\|\cdot\|_0$ represents the norm of the matrix, the number of non-zero elements in the matrix, and $\lambda > 0$ is a compromise factor. $\mathbf{M}$, $\mathbf{L}$, and $\mathbf{S}$ represent the SAR signals matrix, low-rank matrix, and sparse matrix, respectively.

The rank and ℓ_0 norm of the matrices can be convexly relaxed, providing a way to solve the above issues. Since the kernel norm of the matrix is the convex envelope of the rank, and the ℓ_1 norm of the matrix is the optimal convex approximation of ℓ_0, Equation (27) can be relaxed as the following convex optimization problem.

$$\min_{L,X} \quad \|\mathbf{L}\|_* + \lambda \|\mathbf{S}\|_1$$
$$s.t. \quad \mathbf{M} = \mathbf{L} + \mathbf{S} \tag{28}$$

where $\|\cdot\|_*$ is the kernel norm that can represent the sum of the singular values of the matrix; $\|\cdot\|_1$ represents the ℓ_1 norm of the matrix, that is, the sum of the absolute values of each element in the matrix.

The ability to use nonconvex surrogate functions to process non-zero singular values not only improves the accuracy of the approximate representation of low-rank matrices, but also avoids the problem of treating all singular values equally in the kernel norm. The nonconvex nonsmooth weighted nuclear norm is proposed to approximate the rank function.

$$\min_{L,X} \quad \sum_{i=1}^{n_1} g\left(\varepsilon_i\left(L^k\right)\right) + w_i^k\left(\varepsilon_i(L) - \varepsilon_i\left(L^k\right)\right) + \lambda \|\mathbf{S}\|_1$$
$$s.t. \quad \mathbf{M} = \mathbf{L} + \mathbf{S} \tag{29}$$

where $w_i^k \in \partial g\left(\varepsilon_i\left(L^k\right)\right) (i = 1, 2, \dots n_1)$; $g(\cdot) : \mathbb{R}^+ \to \mathbb{R}^+$ is a nonconvex surrogate function, which is continuous. Then, we can obtain:

$$\begin{aligned} \min_{\mathbf{L},\mathbf{S}} \quad & \sum_{i=1}^{n_1} g(\varepsilon_i(L)) + \lambda \|\mathbf{S}\|_1 \\ s.t. \quad & \mathbf{M} = \mathbf{L} + \mathbf{S} \end{aligned} \tag{30}$$

if $g(x) = x$, $\sum_i^{min(m,n)} g(\varepsilon_i(L))$ is equivalent to the nuclear norm. The generalized singular value thresholding operator $\mathbf{Prox}_g^{\varepsilon}(\cdot)$ was utilized to solve the problem of nonconvex low rank minimization, and it can be expressed as:

$$\mathbf{Prox}_g^{\varepsilon}(\mathbf{B}) = \mathop{\mathrm{arg}min}_{\mathbf{L}} \sum_{i=1}^{n_1} g(\varepsilon_i(L)) + \frac{1}{2}\|\mathbf{L} - \mathbf{B}\|_F^2 \tag{31}$$

where g is continuous, concave, and monotonically non-decreasing. Denote $\varepsilon_1(\mathbf{L}) \geq \varepsilon_2(\mathbf{L}) \geq \cdots \geq \varepsilon_{n_1}(\mathbf{L}) \geq 0$ as the singular values of $\mathbf{L}$, then, Equation (23) can be expressed as:

$$\mathop{\mathrm{arg}min}_{\varepsilon_1(\mathbf{L}) \geq \varepsilon_2(\mathbf{L}) \geq \cdots \geq \varepsilon_{n_1}(\mathbf{L}) \geq 0} \sum_{i=1}^{n_1}\left(g(\varepsilon_i(L)) + \frac{1}{2}(\varepsilon_i(L) - \varepsilon_i(B))^2 \right) \tag{32}$$

Equation (26) is equal to solve the following problem for each $b = \varepsilon_i(B)$, $i = 1, 2, \dots, n_1$.

$$\mathbf{Prox}_g(b) = \mathop{\mathrm{arg}min}_{x \geq 0} g(x) + \frac{1}{2}(x - b)^2 \tag{33}$$

The augmented Lagrangian function of the proposed problem (24) can be expressed as:

$$\begin{aligned} \mathcal{L}(\mathbf{L}, \mathbf{S}, \mathbf{Y}, \mu) = &\sum_{i=1}^{n_1} g(\varepsilon_i(L)) + \lambda\|\mathbf{S}\|_1 \\ & - \langle \mathbf{Y}, \mathbf{L} + \mathbf{S} - \mathbf{M} \rangle + \frac{\mu}{2}\|\mathbf{L} + \mathbf{S} - \mathbf{M}\|_F^2 \end{aligned} \tag{34}$$

where μ represents a variable; $\mathbf{Y}$ represents the Lagrangian multiplier; $\langle\cdot\rangle$ represents the matrix inner product. We fixed the $\mathbf{S}, \mathbf{Y}$, and update $\mathbf{L}$. In order to successfully decompose the RFI signals and the useful signals, the problem was solved via ADMM. Then, we can obtain:

$$\begin{aligned} \mathbf{L}^{k+1} &= \mathop{\mathrm{arg}min}_{\mathbf{L}} \mathcal{L}\left(\mathbf{L}, \mathbf{S}^k, \mathbf{Y}^k, \mu_k\right) \\ &= \mathop{\mathrm{arg}min}_{\mathbf{L}} \sum_{i=1}^{n_1} g(\varepsilon_i(L)) - \left\langle \mathbf{Y}^k, \mathbf{L} + \mathbf{S}^k - \mathbf{M} \right\rangle + \frac{\mu_k}{2}\|\mathbf{L} + \mathbf{S}^k - \mathbf{M}\|_F^2 \\ &= \mathop{\mathrm{arg}min}_{\mathbf{L}} \frac{1}{\mu_k}\sum_{i=1}^{n_1} g(\varepsilon_i(L)) + \frac{1}{2}\left\|\mathbf{L} - \left(\mathbf{M} - \mathbf{S}^k - \frac{\mathbf{Y}^k}{\mu_k}\right)\right\|_F^2 \end{aligned} \tag{35}$$

According to the above equation, the generalized singular value thresholding operator was used as follows:

$$\begin{aligned} \mathbf{L}^{k+1} &= \mathbf{Prox}_{\frac{g}{\mu_k}}^{\varepsilon}\left(\mathbf{M} - \mathbf{S}^k + \frac{\mathbf{Y}^k}{\mu_k}\right) \\ &= \mathbf{U}^k Diag\left\{ \mathbf{Prox}_{\frac{g}{\mu_k}}\left(\varepsilon\left(\mathbf{M} - \mathbf{S}^k + \frac{\mathbf{Y}^k}{\mu_k}\right)\right) \right\}\left(\mathbf{V}^k\right)^T \end{aligned} \tag{36}$$

where $\mathbf{Prox}_{\frac{g}{\mu_k}}^{\varepsilon}(\cdot)$ is defined as Equation (27), and $\mathbf{U}^k$, $\mathbf{V}^k$ are obtained by the matrix $\mathbf{M} - \mathbf{S}^k + \frac{\mathbf{Y}^k}{\mu_k}$.

Then, we fixed $\mathbf{L}, \mathbf{Y}$, and updated $\mathbf{S}$, and $\mathbf{S}^{k+1}$ can be expressed as:

$$
\begin{aligned}
\mathbf{S}^{k+1} &= \underset{\mathbf{S}}{\arg\min} \, \mathcal{L}\left(\mathbf{L}^{k+1}, \mathbf{S}, \mathbf{Y}^k, \mu_k\right) \\
&= \underset{\mathbf{S}}{\arg\min} \, \lambda\|\mathbf{S}\|_1 - \left\langle \mathbf{Y}^k, \mathbf{L}^{k+1} + \mathbf{S} - \mathbf{M} \right\rangle \\
&\quad + \tfrac{\mu_k}{2}\|\mathbf{L}^k + \mathbf{S} - \mathbf{M}\|_F^2 \\
&= \underset{\mathbf{S}}{\arg\min} \, \tfrac{\lambda}{\mu_k}\|\mathbf{S}\|_1 + \tfrac{1}{2}\left\|\mathbf{S} - \left(\mathbf{M} - \mathbf{L}^{k+1} + \tfrac{\mathbf{Y}^k}{\mu_k}\right)\right\|_F^2
\end{aligned}
\tag{37}
$$

The above problems can be solved via the shrinkage operator.

$$
\mathbf{S}^{k+1} = S_{\frac{\lambda}{\mu_k}}\left(\mathbf{M} - \mathbf{L}^{k+1} + \frac{\mathbf{Y}^k}{\mu_k}\right)
\tag{38}
$$

where $S_\xi(\mathbf{D}) = max(|\mathbf{D}| - \xi, 0) \cdot sign(\mathbf{D})$, $\xi > 0$ and $sign(\cdot)$ represents a function of sign. Finally, we updated the multiplier $\mathbf{Y}$, and the parameter μ can be expressed as:

$$
\mathbf{Y}^{k+1} = \mathbf{Y}^k - \mu_k\left(\mathbf{L}^{k+1} + \mathbf{S}^{k+1} - \mathbf{M}\right)
\tag{39}
$$

$$
\mu_{k+1} = min(\rho\mu_k, \mu_{max})
\tag{40}
$$

where $\rho > 1$ is the amplification factor.

To sum up, the above equations and processes can extract the RFI signals accurately, until converging, that is

$$
I_{RFI}(f_\tau, \eta) = \mathbf{L}_k
\tag{41}
$$

Removing the RFI signals in the range-frequency and azimuth-time domain to obtain the final RFI mitigation result, it can be expressed as:

$$
S\prime(f_\tau) = S(f_\tau) - I_{RFI}(f_\tau, \eta)
\tag{42}
$$

Through the above process, the RFI signals and the useful signals were successfully separated, and the final raw data after RFI mitigation were obtained. The pseudo code of the proposed algorithm is shown in Algorithm 1.

Algorithm 1. A Modified 2-D Notch Filter Based on Image Segmentation

---

Input: $\mathbf{M} = S(f_\tau)$
Initialization: $\lambda > 0$, $\mu_0 > 0$, $\mu_{max} > \mu_0$, $\rho > 1$, the starting point $\mathbf{S}^0 = 0$, $\mathbf{L}^0 = 0$, $Y_0 = \dfrac{\mathbf{M}}{max\left(\|\mathbf{M}\|_2, \sqrt{mn}\|\mathbf{M}\|_\infty\right)}$, and the iteration index
$k = 0$
Enhancement and Segmentation in Image
Update L: $\mathbf{L}^{k+1} = \mathbf{Prox}^{\varepsilon}_{\frac{g}{\mu_k}}\left(\mathbf{M} - \mathbf{S}^k + \frac{\mathbf{Y}^k}{\mu_k}\right)$
Update S: $\mathbf{S}^{k+1} = S_{\frac{\lambda}{\mu_k}}\left(\mathbf{M} - \mathbf{L}^{k+1} + \frac{\mathbf{Y}^k}{\mu_k}\right)$
Update Y: $\mathbf{Y}^{k+1} = \mathbf{Y}^k - \mu_k\left(\mathbf{L}^{k+1} + \mathbf{S}^{k+1} - \mathbf{M}\right)$
Update μ: $\mu_{k+1} = min(\rho\mu_k, \mu_{max})$
Terminate or set: $k = k + 1$ and returen to **Update L**.
Extraction of RFI: $I_{RFI}(f_\tau, \eta) = \mathbf{L}_k$
Restore the useful signals: $S\prime(f_\tau, \eta) = S(f_\tau, \eta) - I_{RFI}(f_\tau, \eta)$
Output: $S\prime(f_\tau)$

---

4. Experimental Results

The superiority of the proposed algorithm was verified through experiments in this section. The simulated SAR data were used, and the proposed algorithm was compared with the FNF method. Specifically, based on the simulated SAR data, we conducted quantitative analysis of the experimental results through root mean square error (RMSE) in the case of different signal-to-interference-noise ratio (SINR). Then, the Sentinel-1 level-0 raw data were used, two scenes with RFI and mitigation of the interference were selected, and we quantitatively analyzed the performance of the proposed method and FNF method by calculating the gray level entropy and average gradient in the measured data containing RFI with a single source or multiple sources.

4.1. Experimental Results of Simulation

The performance of the proposed method was verified by comparing the RFI mitigation effects in the case of different SINR. The parameters of the simulation experiments and measured data experiments are shown in Table 1.

Table 1. Parameters of the simulated RFI and measured SAR data.

Parameters	Values
Bandwidth of RFI	1 MHz
Carrier frequency of RFI	5.305 GHz
Pulse bandwidth	30 MHz
Pulse width	41.74 μs
Sampling frequency	32.317 MHz
Slant range	988,647 m
Efficient velocity	7000 m/s
PRF	1256.98 Hz
Carrier frequency	5.300 GHz

The original SAR image is shown in Figure 5, and the experimental results of the proposed method compared with FNF and TNF are shown in Figure 6. The first row shows the simulated data with RFI under different SINR conditions. Then, the second row shows the image after RFI mitigation by FNF, and the third row shows the image after RFI mitigation by TNF. Finally, the RFI mitigation effect of the proposed method is shown in the last row. It can be seen from Figure 6 that three kinds of RFI mitigation methods can effectively mitigate RFI in the case of different SINRs. However, FNF and TNF have limited protection capability for useful signals. The inner part of the red line is the region of interest (ROI). In the process of RFI mitigation, some useful signals will be lost because of the limited protection capability for useful signals, and the phenomenon of anomalous sidelobe effects in the range due to spectral leakage will occur. According to the ROI, the proposed method has stronger protection capability for useful signals. Figure 7 is a magnified view of the ROIs in Figure 6, where (a–d) represents the ROI after processing by FNF in the case of different SINRs, (e–h) represents the ROI after processing by TNF in the case of different SINRs, (m,n) represents the ROI after processing by the proposed method in the case of different SINRs, and we can see that the image after mitigating by the proposed method had no anomalous sidelobe effects in the range. Therefore, this method can effectively remove RFI and protect the useful signal, and the performance of the proposed method was better than the FNF method.

We conducted quantitative analysis of the experimental results through RMSE. RMSE can be defined as:

$$RMSE(\mathbf{S}, \mathbf{M}) = \frac{\|\mathbf{M} - \mathbf{S}\|_F}{\|\mathbf{M}\|_F} \tag{43}$$

According to Equation (43), we can see that RMSE represents the difference between the original SAR image and the results after RFI mitigation method processing. The smaller the RMSE, the better the effect of RFI mitigation. The RMSE results are shown in Table 2, where the proposed method had lower RMSE in the case of different SINR than the FNF method.

Figure 5. Original SAR image.

Table 2. Evaluation metrics for the three methods in the case of different SINR.

Metric	Method	FNF	TNF	Proposed Method
RMSE	SINR = 0 dB	0.2168	0.1904	0.1578
	SINR = −10 dB	0.2486	0.2089	0.2041
	SINR = −20 dB	0.2746	0.2594	0.2374
	SINR = −30 dB	0.3462	0.2896	0.2805

4.2. Experimental Results of Measured Data

We use the level-0 raw data of Sentinel-1 IW mode to verify the effectiveness of the proposed method. The performance of the proposed method and FNF method was quantitatively analyzed by calculating the gray level entropy and average gradient in the measured data containing RFI with a single source or multiple sources.

Figure 8 shows the image of the measured Sentinel-1 data, which benefits from the Terrain Observation with Progressive Scans SAR (TOPSAR) technique, which can capture three sub-swaths at once. The data were acquired on 18 December 2021. As shown in Figure 8, the SAR image was seriously polluted by RFI. Two bursts with serious RFI were chosen for analysis, and the RFI-corrupted bursts are shown in Figure 8a,b.

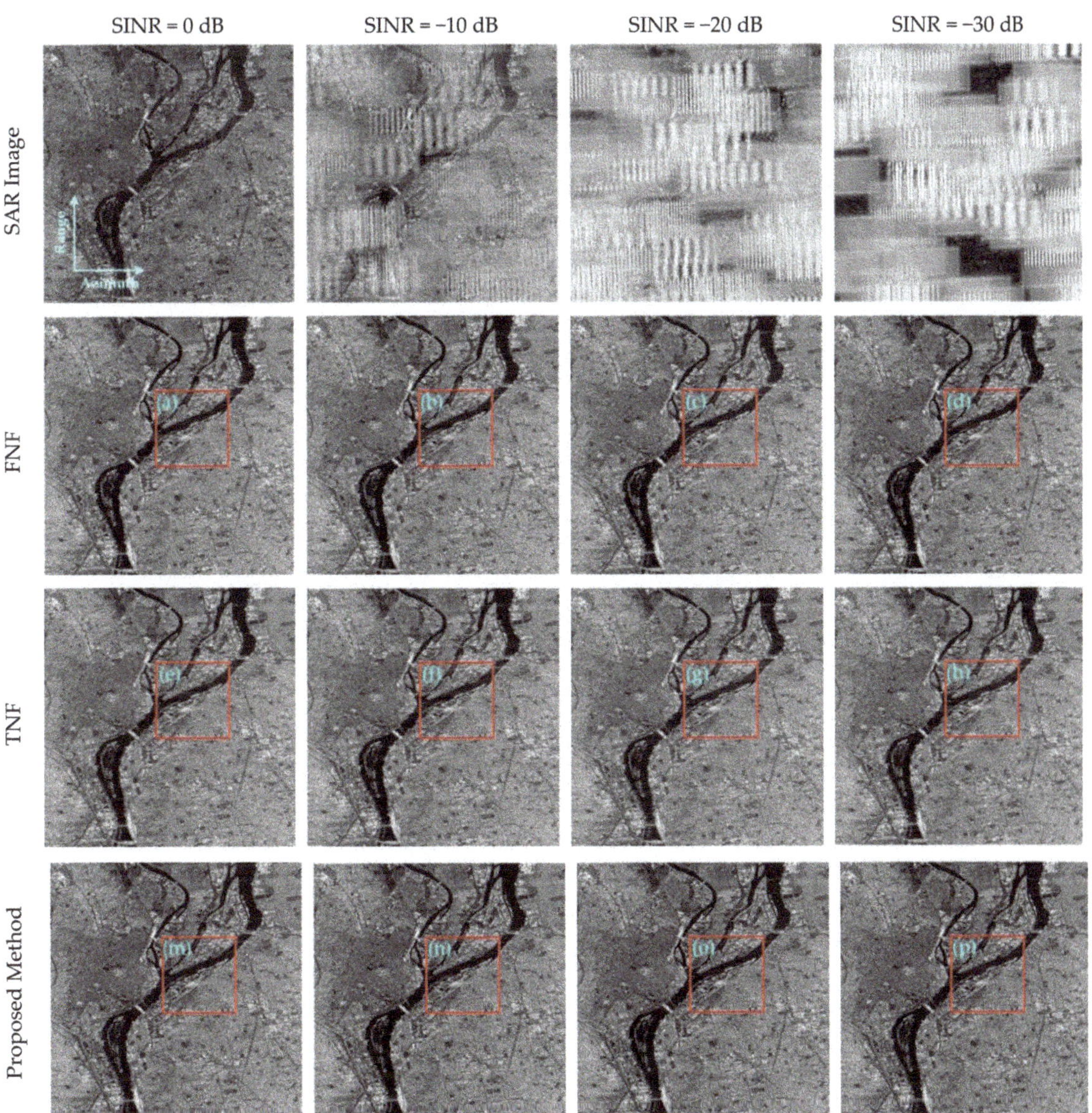

Figure 6. RFI mitigation performance for the FNF, TNF, and proposed method under different SINR conditions.

4.2.1. Experimental Results Based on Measure Data Contain RFI with Single Source

The image after AGC and the modified SBGFRLS model is shown in Figure 9, where the edge of the RFI signals was precisely segmented. The operation of image processing serves as the first layer of protection for useful signals without interference.

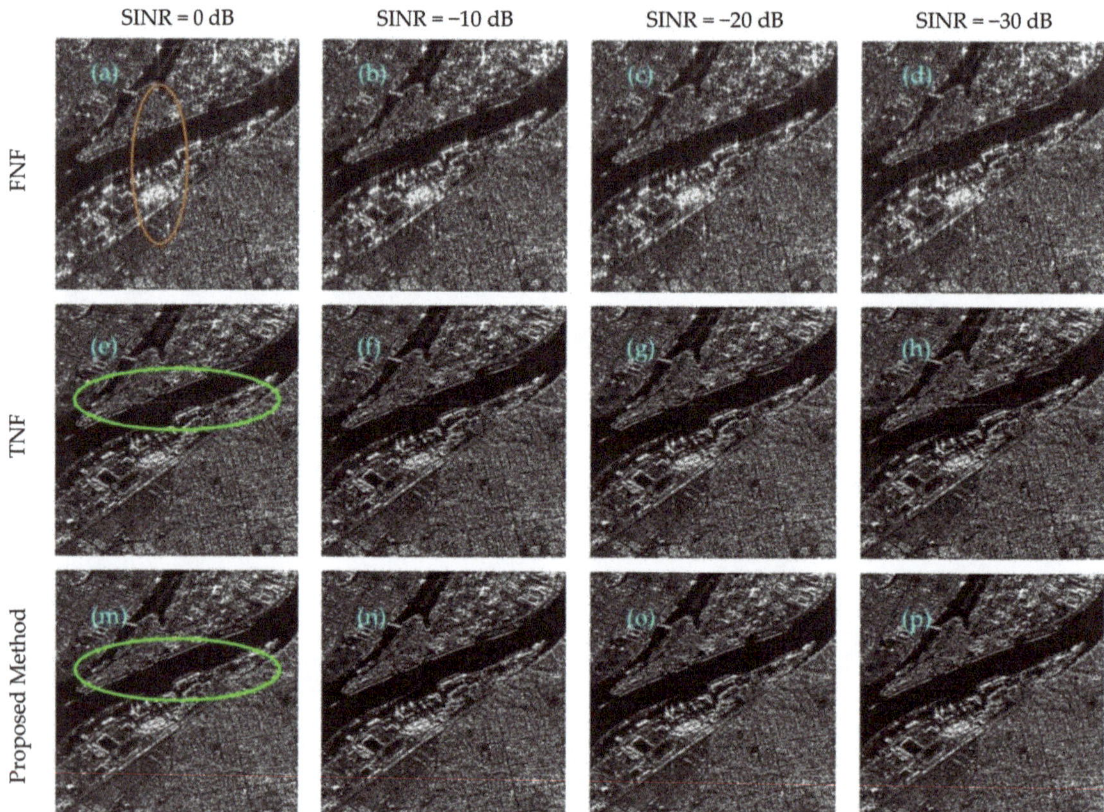

Figure 7. ROIs in Figure 6. (**a–d**) Mitigation results for the FNF method. (**e–h**) Mitigation results for the TNF method. (**m–p**) Mitigation results for the proposed method.

The three-dimensional diagram in the range-frequency and azimuth-time domain after FNF and the proposed method is shown in Figure 10, where (a) represents the RFI mitigation results by FNF, and (b) represents the RFI mitigation results by the proposed method. Compared with Figure 2a, FNF and the proposed method mitigated the RFI effectively. However, FNF failed to protect the useful signals with interference, and the signals of the inner part of the red line was set to zero. The proposed method not only mitigated the RFI, but also effectively protected the useful signals, which is conducive to the application of images.

Figure 11 shows the RFI mitigation results for the SAR data containing RFI with a single source, (a–c) represents the RFI mitigation results for FNF, TNF, and the proposed method, respectively. It can be seen from Figure 11 that the image after processing by FNF had obvious loss of useful signals, and the image after processing by TNF still had some RFI. However, the proposed algorithm had the best RFI mitigation performance, and the protection capability for useful signals was the best.

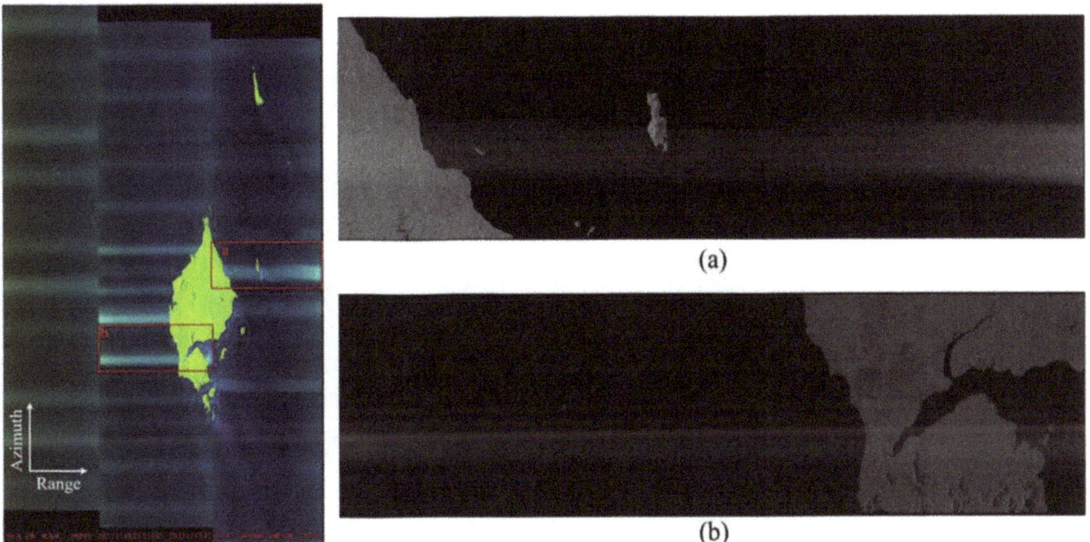

Figure 8. SAR image of the measured Sentinel-1 data. (**a**) The RFI-corrupted burst with a single source. (**b**) The RFI-corrupted burst with multiple sources.

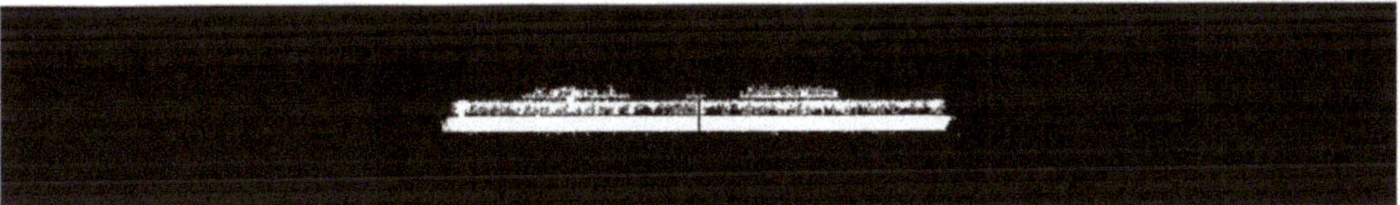

Figure 9. The image after the modified SBGFRLS model in the case of RFI with a single source.

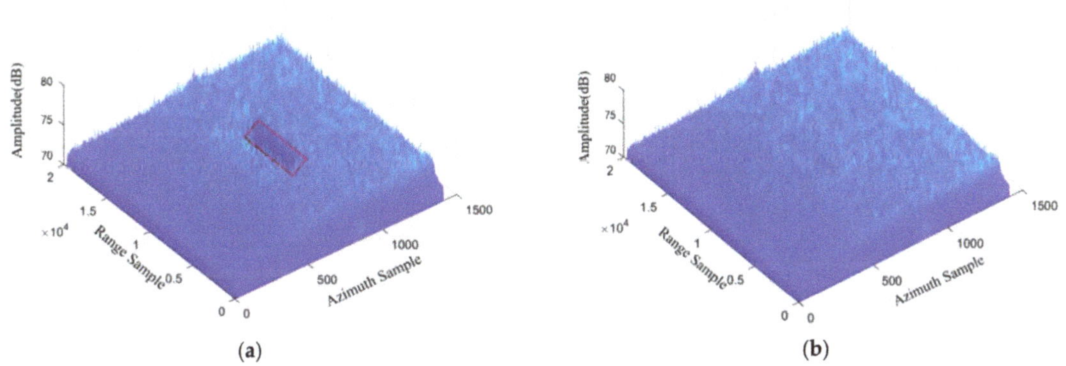

Figure 10. The 3-dimensional diagram in the range-frequency and azimuth-time domain after RFI mitigation: (**a**) FNF; (**b**) the proposed method.

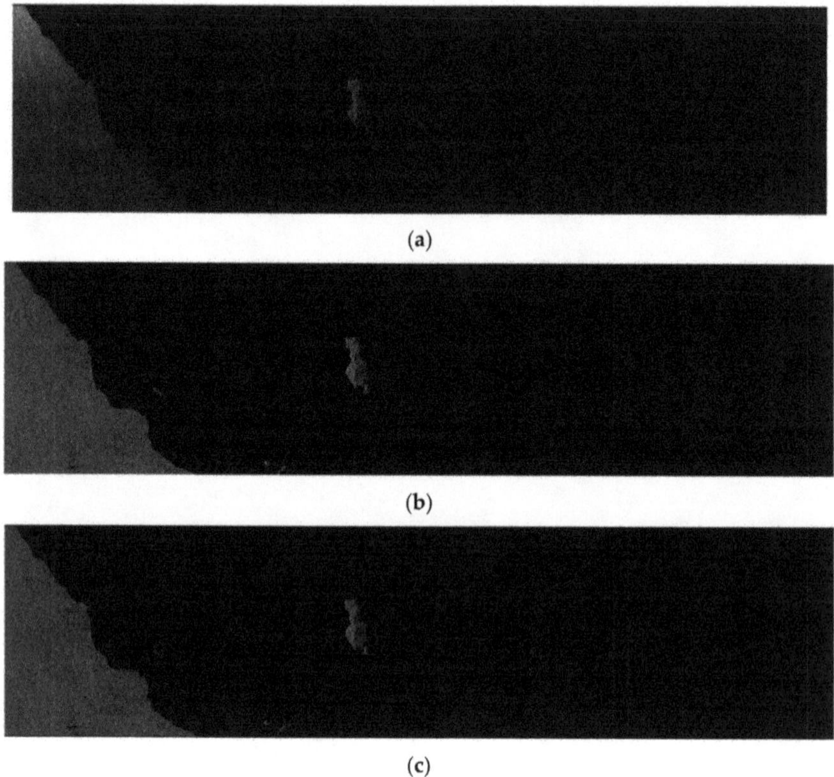

Figure 11. The RFI mitigation results for the SAR data containing RFI with a single source: (**a**) FNF; (**b**) TNF; (**c**) the proposed method.

The mitigation performance of the proposed method was analyzed by the gray level entropy and average gradients.

The gray level entropy can be expressed as:

$$E = -\sum_{k=1}^{L} P_k \log_2(P_k) \tag{44}$$

where E represents the entropy of image; L represents the total gray level; and P_k represents the probability of the occurrence of a pixel with a gray value of k. The average gradient can be expressed as:

$$AG = \frac{\sum_{\tau=1}^{N_r} \sum_{\eta=1}^{N_a} \frac{1}{4} \sqrt{\left(\frac{\partial S(\tau,\eta)}{\partial \tau}\right)^2 + \left(\frac{\partial S(\tau,\eta)}{\partial \eta}\right)^2}}{(N_r - 1)(N_a - 1)} \tag{45}$$

where $S(\tau, \eta)$ represents the position; $\partial S(\tau, \eta)/\partial \tau$ represents the grayscale gradient in the vertical direction; and $\partial S(\tau, \eta)/\partial \eta$ represents the grayscale gradient in the horizontal direction.

Through Equations (44) and (45), we could calculate the results. Then, the gray level entropy and average gradients were utilized to conduct quantitative analysis on the performance of different methods, and the advantages of the proposed method are shown in Table 3.

Table 3. Gray level entropies and average gradients after FNF, TNF, and the proposed method (single source).

Metric / Method	FNF	TNF	Proposed Method
Gray Level Entropy	1.8533	2.5791	3.0041
Average Gradient	2149.4930	2498.4109	2603.4083

4.2.2. Experimental Results Based on Measurement Data Containing RFI with Multiple Sources

The image after AGC and the modified SBGFRLS model is shown in Figure 12. Although there was RFI with multiple sources in the image, the proposed method could still accurately extract the part of RFI, and the edge of the RFI signals was precisely segmented.

Figure 12. The image after the modified SBGFRLS model in the case of RFI with multiple sources.

The 3-dimensional diagram in the case of RFI with multiple sources after FNF and the proposed method is shown in Figure 13, where (a) represents the RFI mitigation results by FNF, and (b) represents the RFI mitigation results by the proposed method. Compared with Figure 2b, although there were RFI with multiple sources, FNF and the proposed method mitigated the RFI effectively. However, FNF failed to protect the useful signals in the interference parts, and the signals of the inner parts of the red line were set to zero. The proposed method not only mitigated the RFI, but also effectively protected the useful signals, which was conducive to the application of images.

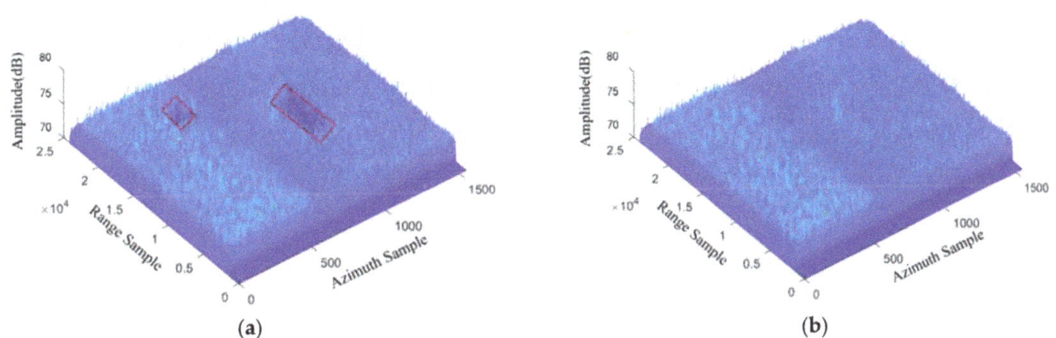

(a)   (b)

Figure 13. The 3-dimensional diagram in the range-frequency and azimuth-time domain after RFI mitigation: (**a**) FNF; (**b**) the proposed method.

Figure 14 shows the RFI mitigation results for the SAR data containing RFI with multiple sources, (a–c) represents the RFI mitigation results for FNF, TNF, and the proposed method, respectively. It can be seen from Figure 14 that, in the case of RFI with multiple sources, the mitigation effect decreased, and the image after processing by FNF had obvious loss of useful signals. The image after processing by TNF had more RFI compared with (a).

However, the proposed method still had a good inhibition effect. More importantly, Table 4 shows the gray level entropy and average gradients after mitigation by different methods, and it further proves the superiority of the proposed algorithm.

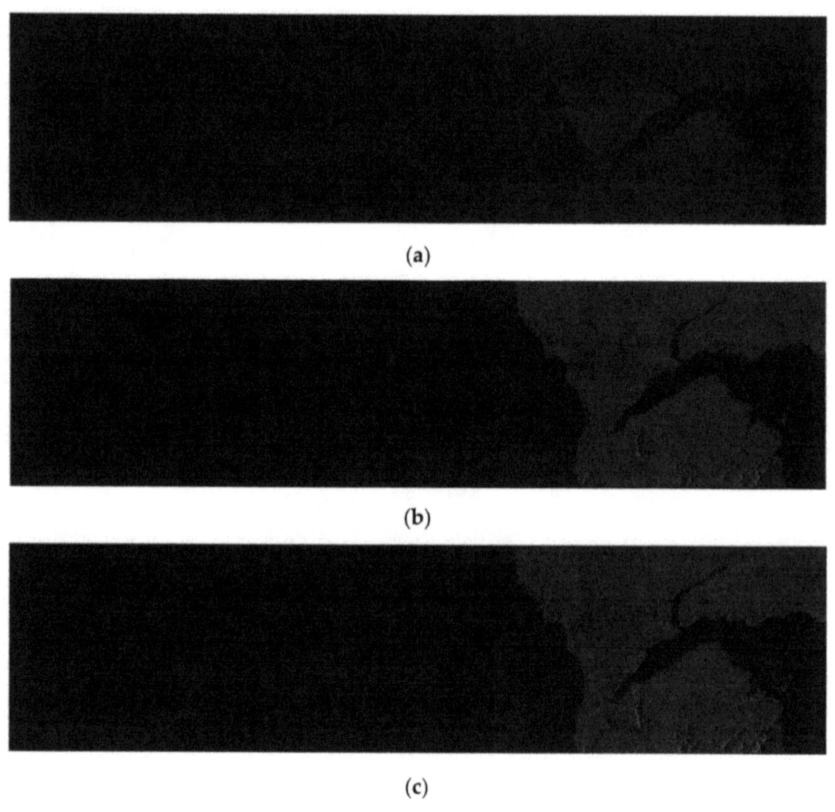

(a)

(b)

(c)

Figure 14. The RFI mitigation results for the SAR data containing RFI with multiple sources: (**a**) FNF; (**b**)TNF; (**c**) the proposed method.

Table 4. Gray level entropies and the average gradients after FNF, TNF, and the proposed method (multiple sources).

Metric \ Method	FNF	TNF	Proposed Method
Gray Level Entropy	1.7385	2.3163	2.9481
Average Gradient	1989.8290	2246.9515	2551.1923

5. Discussion

It can be seen that the proposed method had the most protective and mitigation compared with the traditional methods from the experimental results. For the experimental results based on the simulated data, the performance of the proposed method was improved by about 20% compared with the FNF algorithm. For the experimental results based on the measurement data, on one hand, the RFI mitigation effect of the proposed method was better than other notch filtering methods, and the RFI in the SAR images was obviously mitigated. On the other hand, the proposed algorithm could effectively protect useful signals, and the image quality will not be affected after RFI mitigation, which is convenient for later application. Therefore, the above experiments show that the proposed method can effectively mitigate the RFI and protect the useful signals to the maximum extent, and it is not only applicable in the case of RFI with a single source, but also had a good effect in the case of RFI with multiple sources. The other method processed image still had some interference or ambiguity. This problem is caused by the interference signals that still exist or some useful signals are missing. In comparison, the proposed method was more similar with the original image, protecting the useful signal to the maximum extent and removing the interference signal. Therefore, in terms of the RFI mitigation effect, the proposed method has obvious advantages.

Regarding the computational efficiency, although the efficiency of the proposed algorithm had some losses compared with the FNF algorithm, the efficiency of the current image segmentation algorithms has been greatly improved in the present. The target area after segmentation was smaller, which is more conducive to further processing. Therefore, compared with the low-rank sparse decomposition algorithm, the processing time of the image after segmentation by the proposed method was more efficient.

6. Conclusions

This paper proposed a modified 2-D notch filter based on image segmentation for RFI mitigation. We were inspired by the idea of a traditional notch filtering method, then converted the raw data to the range-frequency and azimuth-time domain, and analyzed the characteristics of RFI first. Then, the AGC method was utilized in the field of image processing to enhance the image. After processing by AGC, the modified SBGFRLS model was utilized to segment the image, and part of RFI was separated from the image. The useful signals also existed in part of the RFI signals, therefore, the GSVT-based LRSD model was performed on the extracted part to screen out useful signals and the RFI signals, and restore the useful signals to the initial raw data at last. The simulation experiments and measured data experiments showed that the proposed method could effectively mitigate RFI and protect the useful signals, whether there was RFI with a single source or multiple sources. More importantly, the proposed method was superior to the traditional notch method.

However, the proposed method still has some limitations. For example, the calculation efficiency of the proposed algorithm is low compared with the traditional notch filtering methods. Therefore, the CPU parallelism or GPU will be used to improve the efficiency of the proposed method in the future, so the proposed method can be more suitable for some engineering application problems.

Author Contributions: Conceptualization, Z.F. and J.Z.; Data curation, Z.F.; Formal analysis, Z.F. and N.L.; Funding acquisition, J.Z. and N.L.; Investigation, H.Z. and N.L.; Methodology, Z.F. and N.L.; Project administration, F.Z., J.Z. and N.L.; Resources, N.L.; Software, F.Z., H.Z., Z.F. and J.Z.; Supervision, F.Z. and N.L.; Validation, F.Z., J.Z. and Z.F.; Visualization, Z.F.; Writing—original draft, Z.F. and N.L. All authors have read and agreed to the published version of the manuscript.

Funding: This work was supported in part by the National Natural Science Foundation of China under grants 61871175 and 41901302, and in part by the Foundation of Key Laboratory of Radar Imaging and Microwave Photonics, Ministry of Education under grant RIMP2020003.

Institutional Review Board Statement: Not applicable.

Informed Consent Statement: Not applicable.

Data Availability Statement: Not applicable.

Acknowledgments: The authors would like to thank the anonymous reviewers for their valuable and detailed comments that are crucial in improving the quality of this paper.

Conflicts of Interest: The authors declare no conflict of interest.

References

1. Zhang, B.; Xu, G.; Zhou, R.; Zhang, H.; Hong, W. Multi-channel Back-projection Algorithm for mmWave Automotive MIMO SAR Imaging with Doppler-division Multiplexing. *IEEE J. Sel. Top. Signal Process.* **2022**, 1–13. [CrossRef]
2. Xu, G.; Zhang, B.; Yu, H.; Chen, J.; Xing, M.; Hong, W. Sparse Synthetic Aperture Radar Imaging from Compressed Sensing and Machine Learning: Theories, Applications and Trends. *IEEE Geosci. Remote Sens. Mag.* **2022**, 2–40. [CrossRef]
3. Reigber, A.; Scheiber, R.; Jager, M.; Prats-Iraola, P.; Hajnsek, I.; Jagdhuber, T.; Papathanassiou, K.P.; Nannini, M.; Aguilera, E.; Baumgartner, S. Very-High-Resolution Airborne Synthetic Aperture Radar Imaging: Signal Processing and Applications. *Proc. IEEE* **2013**, *101*, 759–783. [CrossRef]
4. Peng, B.; Peng, B.; Zhou, J.; Xie, J.; Liu, L. Scattering Model Guided Adversarial Examples for SAR Target Recognition: Attack and Defense. *IEEE Trans. Geosci. Remote Sens.* **2022**, *60*, 5236217. [CrossRef]
5. Peng, B.; Peng, B.; Zhou, J.; Xia, J.; Liu, L. Speckle-Variant Attack: Toward Transferable Adversarial Attack to SAR Target Recognition. *IEEE Geosci. Remote Sens. Lett.* **2022**, *19*, 4509805. [CrossRef]
6. Moreira, A.; Prats-Iraola, P.; Younis, M.; Krieger, G.; Hajnsek, I.; Papathanassiou, K.P. A tutorial on Synthetic Aperture Radar. *IEEE Geosci. Remote Sens. Mag.* **2013**, *1*, 6–43. [CrossRef]
7. Deng, Y.; Yu, W.; Zhang, H.; Wang, W.; Liu, D.; Wang, R. Forthcoming Spaceborne SAR Development. *J. Radars* **2020**, *9*, 1.
8. Zhou, F.; Tao, M. Research on Methods for Narrow-Band Interference Suppression in Synthetic Aperture Radar Data. *IEEE J. Sel. Top. Appl. Earth Obs. Remote Sens.* **2015**, *8*, 3476–3485. [CrossRef]
9. Li, N.; Lv, Z.; Guo, Z. Observation and Mitigation of Mutual RFI Between SAR Satellites: A Case Study Between Chinese GaoFen-3 and European Sentinel-1A. *IEEE Trans. Geosci. Remote Sens.* **2022**, *60*, 5112819. [CrossRef]
10. Li, N.; Lv, Z.; Guo, Z. Pulse RFI Mitigation in Synthetic Aperture Radar Data via a Three-Step Approach: Location, Notch, and Recovery. *IEEE Trans. Geosci. Remote Sens.* **2022**, *60*, 5225617. [CrossRef]
11. Lv, Z.; Zhang, H.; Li, N.; Guo, Z. A Two-Step Approach for Pulse RFI Detection in SAR Data. In Proceedings of the 2021 IEEE Sensors, Sydney, Australia, 3 October–3 November 2021; pp. 1–4.
12. Lv, Z.; Li, N.; Guo, Z.; Zhao, J. Detection and Mitigation of Mutual RFI in C-band SAR: A Case Study of Chinese GaoFen-3. In Proceedings of the 2021 IEEE Radar Conference (RadarConf21), Atlanta, GA, USA, 7–14 May 2021; pp. 1–5.
13. Su, J.; Tao, H.; Tao, M.; Wang, L.; Xie, J. Narrow-band Interference Suppression via RPCA-Based Signal Separation in Time–Frequency Domain. *IEEE J. Sel. Top. Appl. Earth Obs. Remote Sens.* **2017**, *10*, 5016–5025. [CrossRef]
14. Chen, B.; Lv, Z.; Lu, P.; Shu, G.; Huang, Y.; Li, N. Extension and Evaluation of SSC for Removing Wideband RFI in SLC SAR Images. *Remote Sens.* **2022**, *14*, 4294. [CrossRef]
15. Han, W.; Bai, X.; Fan, W.; Wang, L.; Zhou, F. Wideband Interference Suppression for SAR via Instantaneous Frequency Estimation and Regularized Time-Frequency Filtering. *IEEE Trans. Geosci. Remote Sens.* **2022**, *60*, 5208612. [CrossRef]
16. Huang, Y.; Wen, C.; Chen, Z.; Chen, J.; Liu, Y.; Li, J.; Hong, W. HRWS SAR Narrowband Interference Mitigation Using Low-Rank Recovery and Image-Domain Sparse Regularization. *IEEE Trans. Geosci. Remote Sens.* **2022**, *60*, 5217924. [CrossRef]
17. Huang, Y.; Liao, G.; Li, J.; Xu, J. Narrowband RFI suppression for SAR system via fast implementation of joint sparsity and low-rank property. *IEEE Trans. Geosci. Remote Sens.* **2018**, *56*, 2748–2761. [CrossRef]
18. Huang, Y.; Liao, G.; Xiang, Y.; Zhang, Z.; Li, J.; Nehorai, A. Reweighted nuclear norm and reweighted Frobenius norm minimizations for narrowband RFI suppression on SAR system. *IEEE Trans. Geosci. Remote Sens.* **2019**, *57*, 5949–5962. [CrossRef]
19. Huang, Y.; Liao, G.; Zhang, Z.; Xiang, Y.; Li, J.; Nehorai, A. Fast narrowband RFI suppression algorithms for SAR systems via matrix-factorization techniques. *IEEE Trans. Geosci. Remote Sens.* **2019**, *57*, 250–262. [CrossRef]
20. Huang, Y.; Liao, G.; Zhang, L.; Xiang, Y.; Li, J.; Nehorai, A. Efficient narrowband RFI mitigation algorithms for SAR systems with reweighted tensor structures. *IEEE Trans. Geosci. Remote Sens.* **2019**, *57*, 9396–9409. [CrossRef]
21. Huang, Y.; Liao, G.; Xu, J.; Li, J. Narrowband RFI suppression for SAR system via efficient parameter-free decomposition algorithm. *IEEE Trans. Geosci. Remote Sens.* **2018**, *56*, 3311–3322. [CrossRef]
22. Zhang, H.; Huang, Y.; Li, J.; Chen, Z.; Cai, L.; Hong, W. Time-Varying RFI Mitigation for SAR Systems via Graph Laplacian Clustering Techniques. *IEEE Geosci. Remote Sens. Lett.* **2022**, *60*, 4010805. [CrossRef]
23. Braunstein, M.; Ralston, J.; Sparrow, D. Signal Processing Approaches to Radio Frequency Interference (RFI) Suppression. In Proceedings of the SPIE 2230, Algorithms for Synthetic Aperture Radar Imagery, Orlando, FL, USA, 9 June 1994; pp. 190–208.
24. Zhang, S.; Xing, M.; Guo, R.; Zhang, L.; Bao, Z. Interference Suppression Algorithm for SAR Based on Time-Frequency Transform. *IEEE Trans. Geosci. Remote Sens.* **2011**, *49*, 3675–3779. [CrossRef]
25. Yang, H.; He, Y.; Du, Y.; Zhang, T.; Yin, J.; Yang, J. Two-Dimensional Spectral Analysis Filter for Removal of LFM Radar Interference in Spaceborne SAR Imagery. *IEEE Trans. Geosci. Remote Sens.* **2022**, *60*, 1–16. [CrossRef]
26. Fan, W.; Zhou, F.; Rong, P.; Yao, X. Interference Mitigation for Synthetic Aperture Radar Using Deep Learning. In Proceedings of the Asia-Pacific Conference on Synthetic Aperture Radar, Xiamen, China, 26–29 November 2019; pp. 1–6.
27. Lu, X.; Su, W.; Yang, J.; Gu, H.; Zhang, H.; Yu, W.; Yeo, T. Radio Frequency Interference Suppression for SAR via Block Sparse Bayesian Learning. *IEEE J. Sel. Top. Appl. Earth Obs. Remote Sens.* **2018**, *11*, 4835–4847. [CrossRef]
28. Zhou, F.; Xing, M.; Bao, Z. Narrow Band Interference Suppression for SAR Using Eigen-Subspace Based Filtering. *J. Electron. Inf. Technol.* **2005**, *27*, 767–770.
29. Zhou, F.; Tao, M.; Bai, X.; Liu, J. Narrowband interference suppression for SAR based on independent component analysis. *IEEE Trans. Geosci. Remote Sens.* **2013**, *51*, 4952–4960. [CrossRef]

30. Zhou, F.; Xing, M.; Bai, X.; Sun, G.; Bao, Z. Narrowband interference suppression for SAR based on complex empirical mode decomposition. *IEEE Geosci. Remote Sens. Lett.* **2009**, *6*, 423–427. [CrossRef]

31. Tao, M.; Zhou, F.; Liu, J.; Liu, Y.; Zhang, Z.; Bao, Z. Narrow-band interference mitigation for SAR using independent subspace analysis. *IEEE Trans. Geosci. Remote Sens.* **2014**, *52*, 5289–5301.

32. Yang, H.; Li, K.; Li, J.; Du, Y.; Yang, J. BSF: Block Subspace Filter for Removing Narrowband and Wideband Radio Interference Artifacts in Single-Look Complex SAR Images. *IEEE Trans. Geosci. Remote Sens.* **2022**, *60*, 5211916. [CrossRef]

33. Zhang, H.; Min, L.; Lu, J.; Chang, J.; Guo, Z.; Li, N. An Improved RFI Mitigation Approach for SAR Based on Low-Rank Sparse Decomposition: From the Perspective of Useful Signal Protection. *Remote Sens.* **2022**, *14*, 3278. [CrossRef]

34. Shang, R.; Lin, J.; Jiao, L.; Li, Y. SAR Image Segmentation Using Region Smoothing and Label Correction. *Remote Sens.* **2020**, *12*, 803. [CrossRef]

35. Nguyen, L.H.; Tran, T.D. Efficient and Robust RFI Extraction Via Sparse Recovery. *IEEE J. Sel. Top. Appl. Earth Obs. Remote Sens.* **2016**, *9*, 2104–2117. [CrossRef]

36. Liu, H.; Li, D. RFI Suppression Based on Sparse Frequency Estimation for SAR Imaging. *IEEE Geosci. Remote Sens. Lett.* **2016**, *13*, 63–67. [CrossRef]

37. Ding, Y.; Fan, W.; Zhang, Z.; Zhou, F.; Lu, B. Radio Frequency Interference Mitigation for Synthetic Aperture Radar Based on the Time-Frequency Constraint Joint Low-Rank and Sparsity Properties. *Remote Sens.* **2022**, *14*, 775. [CrossRef]

38. Xu, W.; Xing, W.; Fang, C.; Huang, P.; Tan, W. RFI Suppression Based on Linear Prediction in Synthetic Aperture Radar Data. *IEEE Geosci. Remote Sens. Lett.* **2021**, *18*, 2127–2131. [CrossRef]

39. Cazzaniga, G.; Guarnieri, A. Removing RF Interferences from P-band Airplane SAR Data. In Proceedings of the 1996 International Geoscience and Remote Sensing Symposium, Lincoln, NE, USA, 31 May 1996; pp. 1845–1847.

40. Buckreuss, S.; Horn, R. E-SAR P-band SAR Subsystem Design and RF-interference Suppression. In Proceedings of the 1998 IEEE International Geoscience and Remote Sensing Symposium, Seattle, WA, USA, 6–10 July 1998; pp. 466–468.

41. Meyer, F.; Nicoll, J.; Doulgeris, A. Correction and Characterization of Radio Frequency Interference Signatures in L-band Synthetic Aperture Radar Data. *IEEE Trans. Geosci. Remote Sens.* **2013**, *51*, 4961–4972. [CrossRef]

42. Koitsoudis, T.; Lovas, L. RF interference suppression in ultrawideband radar receivers. In Proceedings of the Spies Symposium on Oe/aerospace Sensing and Dual Use Photonics, Orlando, FL, USA, 5 June 1995; Volume 2487, pp. 107–118.

43. Le, C.; Hensley, S.; Chapin, E. Removal of RFI in wideband radars. International Geoscience and Remote Sensing Symposium. In Proceedings of the IEEE International Geoscience and Remote Sensing Symposium, Seattle, WA, USA, 6–10 July 1998; pp. 2032–2034.

44. Le, C.; Hensley, S.; Chapin, E. Adaptive filtering of RFI in wideband SAR signals. In Proceedings of the 7th Annual JPL AirSAR Workshop, Pasadena, CA, USA; 1998; pp. 41–50.

45. Potsis, A.; Reigber, A.; Papathanassiou, K. A phase preserving method for RF interference suppression in P-band synthetic aperture radar interferometric data. In Proceedings of the IEEE 1999 International Geoscience and Remote Sensing Symposium, Hamburg, Germany, 28 June–2 July 1999; pp. 2655–2657.

46. Luo, X.; Ulander, L.; Askne, J.; Smith, G.; Frolind, P. RFI suppression in ultra-wideband SAR systems using LMS filters in frequency domain. *Electron. Lett.* **2001**, *37*, 241–243. [CrossRef]

47. Harcke, L.; Le, C. AirMOSS P-band RF interference experience. In Proceedings of the IEEE Radar Conference, Cincinnati, OH, USA, 19–23 May 2014; pp. 761–764.

48. Vu, V.; Sjögren, T.; Pettersson, M.; Håkansson, L.; Gustavsson, A.; Ulander, L. RFI suppression in ultrawideband SAR using an adaptive line enhancer. *IEEE Geosci. Remote Sens. Lett.* **2010**, *7*, 694–698. [CrossRef]

49. Smith, T.; Hill, R.; Hayward, S.; Yates, G.; Blake, A. Filtering approaches for interference suppression in low-frequency SAR. *IEE Proc. -Radar Sonar Navig.* **2006**, *153*, 338–344. [CrossRef]

50. ITU-R Recommendation RS. Mitigation Technique to Facilitate the Use of the 1215–1300 MHz Band by the Earth Exploration-satelite Service and the Space Research Service. Available online: https://www.itu.int/rec/R-REC-RS.1749/en (accessed on 2 November 2006).

51. Nabil, H.; Chen, J.; Kamel, H. Bidirectional Notch Filter for Suppressing Pulse Modulated Radio-Frequency-Interference in SAR Data. In Proceedings of the 2014 IEEE International Geoscience and Remote Sensing Symposium, Quebec City, QC, Canada, 13–18 July 2014; pp. 1136–1139.

52. Feng, J.; Zheng, H.; Deng, Y.; Gao, D. Application of Subband Spectral Cancellation for SAR Narrow-Band Interference Suppression. *IEEE Geosci. Remote Sens. Lett.* **2012**, *9*, 190–193. [CrossRef]

53. Reigber, A.; Ferrofamil, L. Interference suppression in synthesized SAR images. *IEEE Geosci. Remote Sens. Lett.* **2005**, *2*, 45–49. [CrossRef]

54. Doerry, A. Apodized RFI Filtering of Synthetic Aperture Radar Images. Available online: https://www.osti.gov/servlets/purl/1204095 (accessed on 17 January 2023).

55. Natsuaki, R.; Motohka, T.; Watanabe, M.; Shimada, M.; Suzuki, S. An Autocorrelation-Based Radio Frequency Interference Detection and Removal Method in Azimuth-Frequency Domain for SAR Image. *IEEE J. Sel. Top. Appl. Earth Obs. Remote Sens.* **2017**, *10*, 5736–5751. [CrossRef]

56. Li, N.; Lv, Z.; Guo, Z.; Zhao, J. Time-Domain Notch Filtering Method for Pulse RFI Mitigation in Synthetic Aperture Radar. *IEEE Geosci. Remote Sens. Lett.* **2021**, *19*, 1–5. [CrossRef]

57. Wu, W.; Xiao, Y.; Lin, J.; Ma, L.; Khorasani, K. An Efficient Filter Bank Structure for Adaptive Notch Filtering and Applications. *IEEE Trans. Audio Speech Lang. Process.* **2021**, *29*, 3226–3241. [CrossRef]
58. Huang, H.; Cheng, F.; Chiu, Y. Efficient Contrast Enhancement Using Adaptive Gamma Correction with Weighting Distribution. *IEEE Trans. Image Process.* **2013**, *22*, 1032–1041. [CrossRef]
59. Kass, M.; Witkin, A.; Terzopoulous, D. Snakes: Active Contour Models. In Proceedings of the 1st International Conference on Computer Vision, London, UK, 1 October 1987; pp. 259–268.
60. Liu, Y.; Ren, M.; Zhu, L.; Hu, X. Synthetic Aperture Radar Image Segmentation Method Based on Active Contour Model. *Sci. Technol. Eng.* **2019**, *19*, 221–227.
61. Oveis, A.; Giusti, E.; Ghio, S.; Marco, M. Moving and Stationary Targets Separation in SAR Signal Domain Using Parallel Convolutional Autoencoders with RPCA Loss. In Proceedings of the IEEE Radar Conference, New York, NY, USA, 21–25 March 2022; pp. 1–6.
62. Guo, Y.; Liao, G.; Li, J.; Chen, X. A Novel Moving Target Detection Method Based on RPCA for SAR Systems. *IEEE Trans. Geosci. Remote Sens.* **2020**, *58*, 6677–6690. [CrossRef]
63. Yang, Z.; Fan, L.; Yang, Y.; Yang, Z.; Gui, G. Generalized Singular Value Thresholding Operator Based Nonconvex Low-rank and Sparse Decomposition for Moving Object Detection. *J. Frankl. Inst.* **2019**, *356*, 10138–10154. [CrossRef]

Technical Note

Recognizing the Shape and Size of Tundra Lakes in Synthetic Aperture Radar (SAR) Images Using Deep Learning Segmentation

Denis Demchev [1], Ivan Sudakow [2,*], Alexander Khodos [3], Irina Abramova [4], Dmitry Lyakhov [5] and Dominik Michels [5]

[1] Department of Space, Earth and Environment, Chalmers University of Technology, 412 96 Gothenburg, Sweden
[2] School of Mathematics and Statistics, The Open University, Milton Keynes MK7 6AA, UK
[3] The Center for Research and Invention, Veliky Novgorod 173008, Russia
[4] Arctic and Antarctic Research Institute, Saint Petersburg 199397, Russia
[5] Visual Computing Center, King Abdullah University of Science and Technology, Thuwal 23955, Saudi Arabia
* Correspondence: ivan.sudakow@open.ac.uk

Abstract: Permafrost tundra contains more than twice as much carbon as is currently in the atmosphere, and it is warming six times as fast as the global mean. Tundra lakes dynamics is a robust indicator of global climate processes, and is still not well understood. Satellite data, particularly, from synthetic aperture radar (SAR) is a suitable tool for tundra lakes recognition and monitoring of their changes. However, manual analysis of lake boundaries can be slow and inefficient; therefore, reliable automated algorithms are required. To address this issue, we propose a two-stage approach, comprising instance deep-learning-based segmentation by U-Net, followed by semantic segmentation based on a watershed algorithm for separating touching and overlapping lakes. Implementation of this concept is essential for accurate sizes and shapes estimation of an individual lake. Here, we evaluated the performance of the proposed approach on lakes, manually extracted from tens of C-band SAR images from Sentinel-1, which were collected in the Yamal Peninsula and Alaska areas in the summer months of 2015–2022. An accuracy of 0.73, in terms of the Jaccard similarity index, was achieved. The lake's perimeter, area and fractal sizes were estimated, based on the algorithm framework output from hundreds of SAR images. It was recognized as lognormal distributed. The evaluation of the results indicated the efficiency of the proposed approach for accurate automatic estimation of tundra lake shapes and sizes, and its potential to be used for further studies on tundra lake dynamics, in the context of global climate change, aimed at revealing new factors that could cause the planet to warm or cool.

Keywords: tundra lakes; synthetic aperture radar; Sentinel-1; U-Net; size distribution; Arctic; climate

Citation: Demchev, D.; Sudakow, I.; Khodos, A.; Abramova, I.; Lyakhov, D.; Michels, D. Recognizing the Shape and Size of Tundra Lakes in Synthetic Aperture Radar (SAR) Images Using Deep Learning Segmentation. *Remote Sens.* **2023**, *15*, 1298. https://doi.org/10.3390/rs15051298

Academic Editor: Lan Du

Received: 15 January 2023
Revised: 18 February 2023
Accepted: 24 February 2023
Published: 26 February 2023

1. Introduction

As permafrost thaws, some portion of its organic matter will be decomposed by microorganisms, emitting large amounts of greenhouse gases. This permafrost carbon–climate feedback is the largest terrestrial feedback to climate change, and one of the most likely to occur. It is increasingly recognized that the magnitude and timing of the permafrost carbon–climate feedback depends largely on the reorganization of hydrology (the distribution of wet and dry surfaces), and associated changes in redox conditions in permafrost landscapes [1–3]. One of the major hydrologic transitions that can occur during permafrost thaw is surface collapse due to the melting of ground ice supporting the soil profile. This abrupt, non-linear process, termed thermokarst, has proven exceedingly difficult to map, let alone predict [4–6]. In lowland landscapes, thermokarst can transform dry land to a lake.

In some areas of the permafrost zone, current or drained thermokarst lakes occupy 40–80% of the landscape surface, often forming characteristic features [7,8]. If permafrost

---

thaw continues after lake formation, loss of frozen ground supporting the water body can trigger drainage of lakes to groundwater or connected river networks, decreasing the overall number and area of large lakes, but increasing residual small lakes. Generally, high latitudes of the permafrost zone are experiencing an increase in lake area, while lower latitudes of the permafrost zone are seeing a decrease, though trends are spatiotemporally complex [9–13]. Integrating permafrost lake formation and draining into Earth system models is a high priority in the permafrost community, for improving predictions of the permafrost climate feedback [14–17].

The abundance and size distribution of tundra lakes are important spatial characteristics of the landscape, for understanding the dynamics of permafrost thawing and related processes in tundra ecosystems. In addition, the trade of the shapes of tundra lake patterns can be rigorously quantified by employing these quantities. Many studies [18–20] have recently revealed that quantification of size distribution and fractal features requires a high volume of remote sensing data on the land surface. Statistical and geometrical parameters, such as size distribution and fractal dimension, could play a role, as a metric of the accuracy of tundra lakes image segmentation when real physical parameters are limited or unavailable.

Several studies have addressed the problem of automized tundra lakes recognition from satellite optical data. Manuscript [21] reports that the traditional way is to utilize image segmentation techniques, where images are divided into regions based on some features: for example, a paper by [22] employed a cloud-removal technique, called the function of mask (Fmask) algorithm, on Landsat imagery of a tundra surface. Automatic segmentation of water bodies in satellite imagery, to estimate lake size distributions, was considered in [23]. In [17], the authors applied robust linear trends based on the Theil–Sen regression algorithm, and random forest (RF) classification technique, by feeding pre-calculated spatiotemporal trend information, to analyze the lake dynamics across Northern permafrost regions. Application of supervised classification, using a maximum likelihood classifier and digital change detection, to identify areas of high erosion or deposition of permafrost, was investigated in [24].

Although data sets of permafrost surface features based on passive-microwave satellite data do exist [25,26], their spatial resolution is relatively coarse (20–50 km), and thus only allows for estimation of inundation fractions and their variations [27,28]. For accurate monitoring of waterbodies, data with a higher spatial resolution are required, as in tens of meters in a pixel [29]. Some optical instruments acquire high-resolution images (typically 1–15 m), but they depend on cloud cover conditions and the luminance of the surface, which significantly limits their utilization, particularly in polar regions. Thus, a number of studies have focused on synthetic aperture radar (SAR) data for recognition of permafrost surface features and quantifying their dynamics [30–35], including InSAR techniques [36–38]. SAR backscatter intensity strongly depends on the dielectric constant of the surface, and is very different for water and ice [39]. Thus, thanks to the high spatial resolution and the signature differences, SAR is a suitable tool for tundra lakes recognition in many cases.

Although the manual analysis of SAR data is sensor-independent, and relatively robust against misclassification [35,40,41], it is time-consuming; therefore, automated methods are preferable. Some studies have proposed using machine learning for characterizing Arctic tundra landscapes, including lakes [30,42]. To the best of our knowledge, state-of-the-art deep learning methods have not been previously applied to the problems of automatic tundra lakes recognition from SAR, and, thus, one of *the primary aims of this study* was to investigate the potential of deep learning for accurate tundra lake shape and size recognition. C-band Sentinel-1 images, freely available under the Copernicus Program, were used to develop a machine learning methodology capable of segmenting the images into two classes: tundra lakes and background. To assess the applicability of the method, we analyzed the generated time series of segmented Sentinel-1 images, for quantification of the lake spatial features in the Arctic region.

The rest of the paper is organized as follows. In Section 2, a description of satellite SAR data processing, selected test sites, reference data preparation and tundra lakes recognition algorithm workflows are given. The results obtained in this study are presented in Section 3, followed by discussion in Section 4. The main findings and implications are summarized in the conclusions in Section 5.

2. Materials and Methods

2.1. Satellite SAR Images and Test Sites

For our study, we selected two test sites in the Arctic region, characterized by the presence of tundra lakes: the Yamal Peninsula and Alaska. The regions are located far from each other, which allowed for an improved understanding of the performance of our approach in different environmental conditions. From the two test sites, we downloaded hundreds of Sentinel-1 Level-1.5 Ground Range Detected (GRD) dual-polarized data through the Alaska SAR Facility data archive facility, for the summer months of 2015–2022, when the lakes characterized their maximum development. The spatial boundaries of the collected images are depicted on a map in Figure 1.

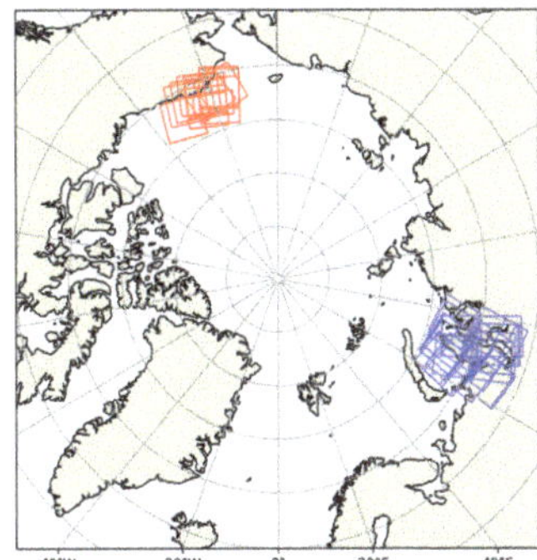

Figure 1. Sentinel-1 images collected over the two test sites: in red—Alaska test site; in blue—Yamal test site.

Then, we computed the SAR backscattering coefficient σ^0 from digital numbers DN as

$$\sigma^0 = DN^2/A^2, \tag{1}$$

where A was a calibration constant from a look-up table from auxiliary calibration files (for details, see [43]). Then, σ^0 was conventionally converted to dB: $\sigma^0_{db} = 10 \cdot log(\sigma^0)$ for both HV and HH polarizations. Finally, the data were projected onto a Polar Stereographic grid with a pixel size of 50 m. Radar backscattering σ^0 from water and ice is strongly angle-dependent, due to the specificities of SAR system geometry [39,44,45]: hence, information on the incidence angle (I_a) is very useful for SAR data analysis. Contrary to polarization and frequency, I_a is not fixed, and varies across the image. The backscatter intensity from a homogeneous surface varies with I_a, and decreases across a SAR image from near-range (low I_a) to far-range (high I_a). To compensate for this effect, most research applies a global I_a correction, using constant slope before the classification task [46–48]: however, in doing this, they neglect the known physical differences in backscatter behavior, with I_a for different surface types. Thus, to take these specificities into account, we utilized incidence angle information in our recognition framework for every corresponding pixel. The Sentinel-1 metadata contained values of I_a at certain points, which we extracted and interpolated onto the image grid by linear interpolation. Then, the values of σ^0_{HV} (Figure 2A), σ^0_{HH} (Figure 2B) and incidence angle I_a (Figure 2C) were normalized in the range 0–1 by min–max normalization ($\frac{value-min}{max-min}$), and then used to form a 3-channel image (Figure 2D). This data structure was used in both the training and segmentation stages.

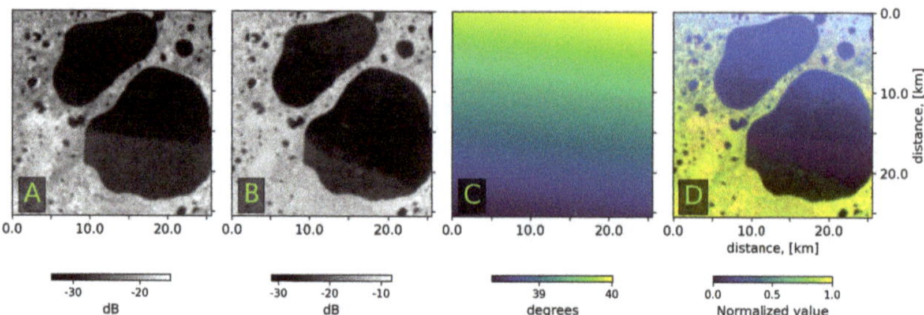

Figure 2. A pseudo-RGB image synthesized from Sentinel-1 data: (**A**) σ^0 at HV-polarization; (**B**) σ^0 at HH-polarization; (**C**) incidence angle (I_a); (**D**) synthesized 3-channel image ($\sigma^0_{HV} + \sigma^0_{HV} + I_a$) from normalized values in 0–1.

2.2. Manual Mapping of Tundra Lakes

For the training of the U-Net, we prepared binary images containing manually annotated regions of tundra lakes. The analysis was performed manually by visual expert analysis, using VGG image annotator software [49]. The lakes were mapped in the Yamal and Alaska test sites (Figure 1), from tens of images, and stored as binary tiff files, where a pixel intensity of 255 corresponded to the lake, and 0 to the background. Overall, we generated sets of 668 and 296, respectively, of 512×512 images containing labeled lakes in the Yamal and Alaska sites [50]. An example of an image generated from manual recognition of tundra lakes is shown in Figure 3.

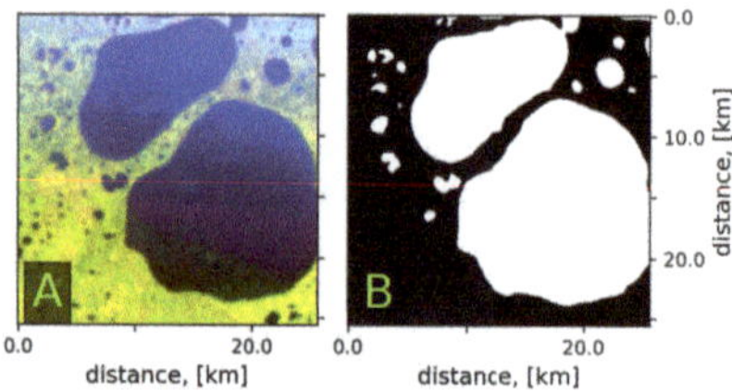

Figure 3. A pseudo-RGB Sentinel-1 image (R: HV-polarization; G: HH-polarization; B: incidence angle) taken over the Yamal Peninsula on 12 August 2015 at 12:40:52, containing tundra lakes with different shapes and sizes (**A**), and a binary image with manually extracted tundra lakes (**B**), where the white color corresponds to the tundra lake areas.

2.3. Automatic Tundra Lake Recognition from SAR Images

In this study, we propose a two-stage algorithm framework for tundra lakes recognition from SAR images. The algorithm operates with so-called semantic and instance segmentation. Semantic segmentation is the first stage, and associates every pixel of an image with a class label, such as a lake or other surface feature: it treats multiple objects of the same class as a single entity. By contrast, instance segmentation, which is the second stage, treats multiple objects of the same class as distinct individual instances. Here, we propose a combination of both methods, which is useful for separating touching or overlapping individual objects—which, in our case, were the lakes—for unambiguous estimation of their size and shape. The majority of semantic segmentation methods utilize encoder–decoder deep neural network structures, most of which are based on the U-Net architecture proposed in [51], which is an improved version of the fully convolutional network for medical images [52], and which has also been applied to the water bodies imagery [53,54]. The underlying technique of U-Net involves skipped connections, to improve gradient flow, and to allow for transferring of information between the down-sampling and up-sampling paths, by connecting each pair of encoder–decoder layers. In our algorithm framework, the U-Net took an input SAR image, classified each pixel in the image into tundra lake/background classes, and produced a binary image. The semantic segmentation by the U-Net was followed by instance segmentation with a classical watershed algorithm [55,56], which is a region-based technique that utilizes image morphology. The general scheme of our tundra lakes recognition workflow is shown in Figure 4.

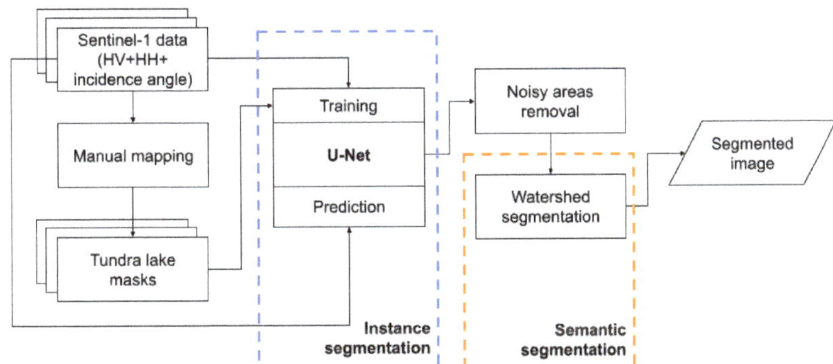

Figure 4. Tundra lakes recognition workflow.

The framework operated with Sentinel-1 data synthesized into 3-channel images, as described in Section 2.1. As a low-pass filter prior to classification and detection strongly improved the recognition results [30], we applied an anisotropic diffusion filter to suppress speckle appearance [57,58]. For training of the U-Net, the Adam optimizer [59] was employed with default parameters. A batch size of sixteen was used, and the network was trained for 200 epochs. As a loss function, the Jaccard similarity index or the Intersection over Union ($IoU = Area\ of\ Intersection\ /\ Area\ of\ Union$) was used. This metric operated with the *Area of Intersection*, which was the area where our predicted image overlapped the ground truth image, and the *Area of Union*, which combined our predicted image and the ground truth image. For the training, image patches of 512×512 pixels containing SAR data, and corresponding masks of tundra lakes labels, were used. After the U-Net training, an accuracy of 0.73, in terms of IoU, was achieved. The output of the semantic segmentation stage with U-Net was a binary image containing pixels with values of 0 (background) and 255 (tundra lakes).

Often, lakes can overlap or touch each other, therefore for accurate estimation of their shapes, special methods are required. To address this potential issue, we complemented our recognition framework with instance segmentation, which was the second stage. Here, we aimed at labeling each pixel of an image with a corresponding instance: in our case, the instance was an individual tundra lake. The watershed algorithm described in [56] was used. A chessboard metric was chosen, as it provided more robust results than other widely used metrics [60,61]. This step of our framework was implemented using a combination of several computer vision methods from the Python-based OpenCV framework [62]. In the following text, we complement the description of the processing methods with a corresponding reference to OpenCV functions.

First, we treated isolated groups of a few pixels as artificial, and discarded them using (`cv2.morphologyEx`). Then, we identified sure background areas, by convolving an image with a kernel, K, that replaced the image pixel in the anchor point position with that maximal value (`cv2.dilate`), consequently causing the tundra lake areas to extend their coverage. This increased the lake boundary to their background and, thus, we defined pixels that corresponded with certainty to the background (pixel value = 0). To find a sure foreground area, the distance transform method was used (`cv2.distanceTransform`), followed by thresholding the obtained values inside the foreground regions to their respective distances from the closest background value. The threshold of the distance transform was chosen empirically, and was 15% of its maximum value. Higher values might have resulted in the discarding of small lakes. Then, we estimated the ambiguous pixels (pixel value =

−1) by subtracting the background from the foreground. Pixel labeling was performed, using (`cv2.ConnectedComponents`) for the three categories: foreground (1); background (0); and ambiguous (−1). Finally, for defining instances that were individual lakes, watershed segmentation based on the derived markers was applied by (`cv2.watershed`). All the described steps are summarized in the pseudo-code Algorithm listing 1.

Algorithm 1: Tundra lake shapes recognition from SAR images

Prepare $\{\mathbf{S_i}\}_{i=1}^{N}$ — a set of 3-channel SAR images (HH+HV+I_a, where I_a-incidence angle values).

Label tundra lakes from a subset $\{\mathbf{S_j}\}_{j=1}^{M} \in \{\mathbf{S_i}\}$, to create set of labeled images $\{\mathbf{L_j}\}_{j=1}^{M}$.

Generate sets of 512×512 patches $\{\mathbf{P_k^S}\}_{k=1}^{K}$ and $\{\mathbf{P_k^L}\}_{k=1}^{K}$ from $\{\mathbf{S_j}\}$ and $\{\mathbf{L_j}\}$.

Train U-Net model $\mathbf{M}$ for 200 epochs, based on $\{\mathbf{P_k^S}\}_{k=1}^{K}$ and $\{\mathbf{P_k^L}\}_{k=1}^{K}$.

Generate a set of image patches $\{\mathbf{P_x^I}\}_{x=1}^{X}$ from SAR image for segmentation

for $P_x^I \in P^I$

    Segment P_x^I with $\mathbf{M}$ into binary image B_x^I, where pixel value x of 255 corresponds to the lakes, and to 0 for the background.

end

Merge $\{\mathbf{B_x^I}\}_{x=1}^{X}$ into $\mathbf{B^I}$.

Remove small noisy areas in $\mathbf{B^I}$ by morphological opening with kernel of 3×3 pixels.

Identify sure background area, by dilating $\mathbf{x_{=255}}$ a few times, to increase lake boundary to background.

Calculate Chessboard transform distance $\mathbf{P_{DT}}$ for each pixel x.

Estimate foreground pixels $\mathbf{x_{FG}}$, by thresholding $\mathbf{P_{DT}}$ as 15% of max($\mathbf{P_{DT}}$).

Estimate ambiguous pixels as $\mathbf{x_A} = \mathbf{x_{=255}} - \mathbf{x_{FG}}$.

Define pixels that correspond to a certain lake (instance) by watershed segmentation, given as $\mathbf{x_{Li}} = \mathbf{f}(\mathbf{B^I}, \mathbf{x_A})$.

2.4. Measuring Trade Features of Segmented Lakes

The formation of geological objects entails different complex natural phenomena, and it is quite remarkable that some of them have identical fractal dimensions, as was shown for clouds, in the pioneering paper by [63]. Fractal theory [64] can be used as a method for studying partially correlated (over many scales) spatial phenomena that are not differentiable but are continuous. This theory helps quantify complex shapes or boundaries, and relates them to underlying processes that may affect pattern complexity. For simple objects like circles and polygons, the perimeter P scales as the square root of the area A. However, for complex planar regions, with fractal curves as their boundaries:

$$P \sim \sqrt[D]{A}, \tag{2}$$

where the exponent D is the fractal dimension of the boundary curve.

Box Counting Method

The Box counting method is a computer method of extracting data from real pixel images, for further analysis. The methodology consists of breaking the whole image into smaller pieces, and counting the boxes with some particular feature.

In the fractal analysis, we have to count the number of boxes that cover the fractal, and to see how it changes when the size of the box is decreasing. The number of boxes should approximately follow a power law:

$$N = N_0 R^{-d},$$

where R is the size of box, and d is called Kolmogorov capacity or simply box-counting dimension.

There are other methods for counting fractal dimensions, such as the area–perimeter or divider method, which is also used in geological research [65]: in order to apply it, we need to detect boundaries with high precision, which is subject to segmentation method and resolution. Instead, the box-counting method provides a simple computer method that gives robust complexity characteristics for a mixed land–lakes object.

3. Results

3.1. Tundra Lakes Recognition by U-Net

To assess the performance of our framework (described in Section 2.3) for the lakes recognition, we applied it to Sentinel-1 Extra Wide Swath (EW) data acquired over the Yamal Peninsula on 23 August 2016 at 12:49:16. First, the data at HV- and HH-polarization mode were calibrated, and incidence angle information was extracted. Then, we projected them onto a polar stereographic grid with a grid step size of 50 m. A pseudo-RGB image was synthesized from the data, and is shown in Figure 5A. It should be noted that this image was not used in the training stage. Then, we applied a pre-trained U-Net to extract the tundra lakes from the data, and a binary image was generated (Figure 5B). Before the automatic recognition, we manually labeled the tundra lakes from the image, and used them as ground truth labels (Figure 5C). From visual inspection, in general, we saw good agreement between automatically and manually extracted lakes. However, the lake borders produced by our framework looked more smooth and more natural, compared to those manually extracted. To show the difference between the data, we subtracted the ground truth from the U-Net output, and the result is shown in Figure 5D. The blue areas correspond to lakes that were not detected by our U-Net, but were recognized by an expert, and the red, conversely, corresponded to areas which were recognized as lakes by the U-Net, but not in the manually derived image. In total, four lakes were not detected by the U-Net model (the blue blobs in Figure 5). The other differences mostly corresponded to lake border areas. As we mentioned above, the lake borders in the image generated by U-Net were more natural, and closer to what we see on the satellite image. Most of the very small lakes, the size of a few pixels, were not recognized either by U-Net or by expert analysis, because of their too-small sizes in terms of the SAR image resolution.

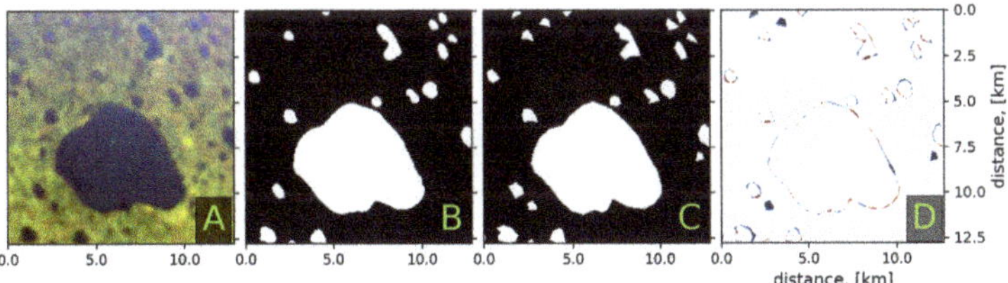

Figure 5. Results of tundra lakes recognition from Sentinel-1 data taken over the Yamal Peninsula (69.4366°E, 68.5102°N) from 23 August 2016 at 12:49:16 by our algorithm framework: (**A**) input pseudo-RGB image where R = SAR backscattering at HV-polarization, B = SAR backscattering at HH-polarization and G = incidence angle; (**B**) predicted tundra lakes (white) (**C**) ground truth image with manually mapped lakes; (**D**) the difference between ground truth and predicted data, where red indicates pixels corresponding to lakes in the framework output but not in the ground truth, blue corresponds to pixels recognized as lakes in the ground truth but not in the framework output and white stands for agreement between the data.

3.2. Instance Segmentation of the Lakes and Noise Filtering

To demonstrate the advantage of our tundra lakes recognition concept, let us consider an example of the lakes recognition products obtained from a Sentinel-1 image taken over the northern coast of Alaska (157.0301°W, 70.93083°N) on 20 October 2016 at 04:02:51. The result of segmentation by U-Net is shown in Figure 6A, after noisy areas removal in Figure 6B, and the final result obtained with instance segmentation using the watershed algorithm is in Figure 6C.

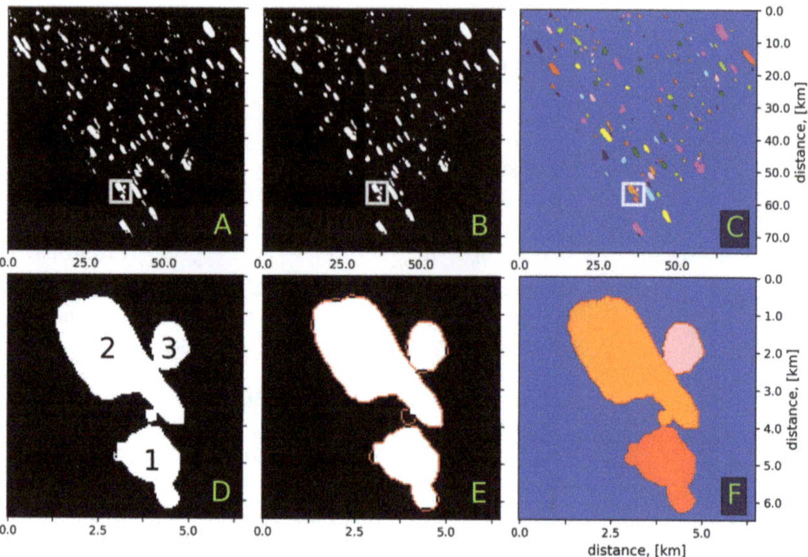

Figure 6. (**A**) segmented tundra lakes at Alaska test site (157.0301°W, 70.93083°N) from 20 October 2016 at 04:02:51, obtained by U-Net; (**B**) after filtering of small "noisy lakes"; followed by (**C**) instance segmentation, where white areas correspond to tundra lakes, and colored blobs stand for individual segmented lakes (instances). The white rectangles in (**A**–**C**) correspond to a zoomed area containing three lakes labeled as (1), (2) and (3); (**D**) raw U-Net output in the zoomed area, where lake 2 and lake 3 are overlapped (one instance); (**E**) lakes areas after morphological opening, and the obtained lake borders by the watershed segmentation, are indicated in red; (**F**) results of instance segmentation with the watershed algorithm, where each lake (instance) is depicted in different colors.

In the zoomed area in Figure 6D, we can recognize three lakes, but lake 2 and lake 3 are overlapped, which means that if we analyze their shapes they will be considered as one lake, which is not desirable, as we want to derive their spatial characteristics separately. Thanks to the instance segmentation stage of our framework, we can separate the two overlapped lakes by recognizing their borders (see Algorithm 1), and then applying the watershed algorithm. Finally, the lakes are separated and ready for further analysis, such as their shapes and sizes estimation (Figure 6F).

3.3. Fractal Dimension

Fractal dimension counted by the box-counting method led to the two following plots. It clearly showed that, except for some deviation, fractal dimension remained almost constant. Fluctuation could be explained as the computational process being vulnerable according to the choice of the size of the box, which was subject to the finite size of the pixels on the images.

3.4. Size Distribution

We calculated the size distribution of the lakes on historical maps, by means of MATLAB (R2022a, The MathWorks Inc., Natick, MA, USA) core function library skimage. The standard procedures counted the number of pixels associated with lands and lakes, which led to the plots below.

4. Discussion

Some of the most recent comprehensive comparisons between the original U-Net [51] and its modifications, such as [66,67], indicate that architectural changes in U-Net are potentially unnecessary for images formed from coherent signals and associated with increased trade and slower speed for the marginal performance gains [68]. Thus, in this study, we utilized the original architecture of U-Net. But, in future work, state-of-the-art modifications of U-Net can be assessed, to improve the performance.

As a loss function, binary cross-entropy computed by predicted values and ground truths is typically selected for segmentation tasks with U-Net. But the targets generally occupy a small part of the entire training image, so that the minimization of cross-entropy inclines to be biased towards the outside of the target (background). Thus, in our recognition framework, we utilized the Jaccard similarity index, to avoid false-positive detection, and to increase binary classification accuracy.

There are many challenges in quantifying changes in the shape and patterning of tundra lakes [69], because in comparison to mathematical fractals (like the Koch snowflake or the Sierpinski gasket) in mathematical space, we have to restrict ourselves to the finite size of pixels on images. The novel techniques of image analysis may successfully tackle these scaling challenges. Found here, time series of fractal dimensions show that the fractal features of tundra lakes are constant enough in time (see Figures 7 and 8). We did not observe an increase in shape complexity, even if a lake's size was growing. In manuscript [70], the authors analyzed Landsat images for the Western Siberia region, using region-growing segmentation techniques. While the average fractal dimension of tundra lakes lay within the interval found in our work, there were a few spikes, probably due to used technique issues. For the two sites, Alaska and Yamal, we showed that the lakes that were recognized on Sentinel-1 satellite images were lognormal distributed (see Figures 9 and 10). In manuscript [22], tundra lakes of the size of 0.002 km^2 to 50 km^2 were segmented from high-resolution Sentinel-2 satellite images covering 725,000 km^2 of the East Siberia region. The authors found that tundra lakes in the region were also log-normal distributed. The fractal dimension and size distributions exhibited universal characteristics that did not necessarily depend on the details of the driving mechanisms and associated complexities: thus, they could be used in tundra landscape modeling [71,72]. Pattern recognition, and quantifying its fractal dimension, are also important in the field of biomedical image analysis [73,74]. We believe that the methods of fractal pattern recognition developed here will be useful in other areas where the system complexity should be defined.

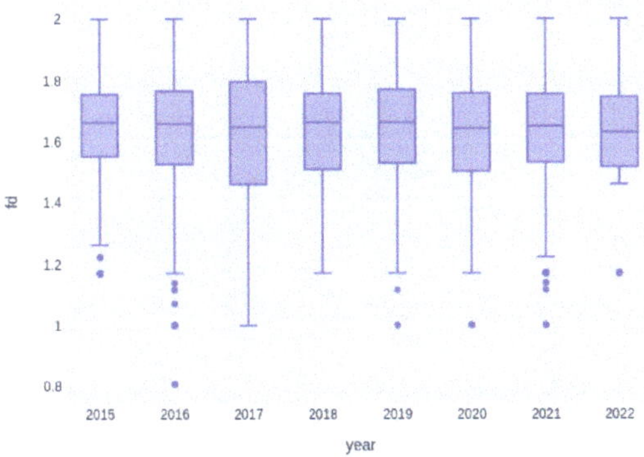

Figure 7. Box plot of the fractal dimension (fd) for the time period 2015–2022; the studied region is in Alaska.

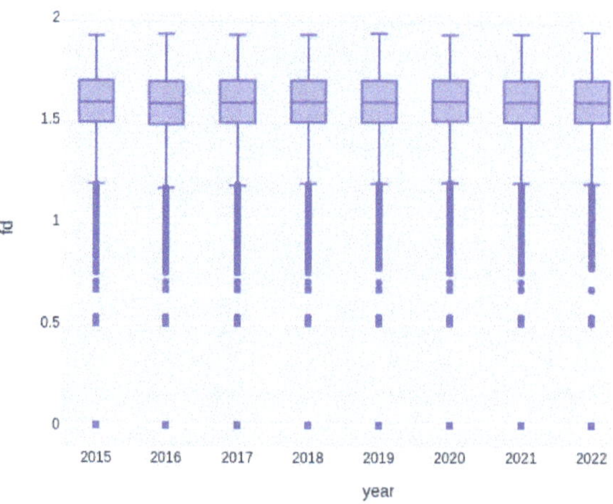

Figure 8. Box plot of the fractal dimension (fd) for the time period 2015–2022; the studied region is in Yamal.

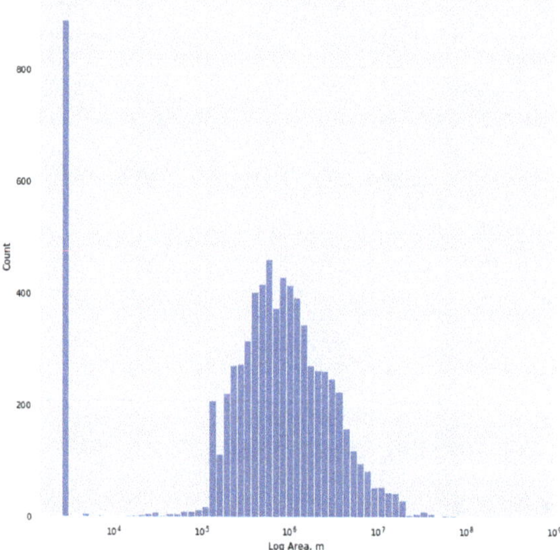

Figure 9. Histogram of the number of recognized lakes on images from Alaska vs the log of lake size (area).

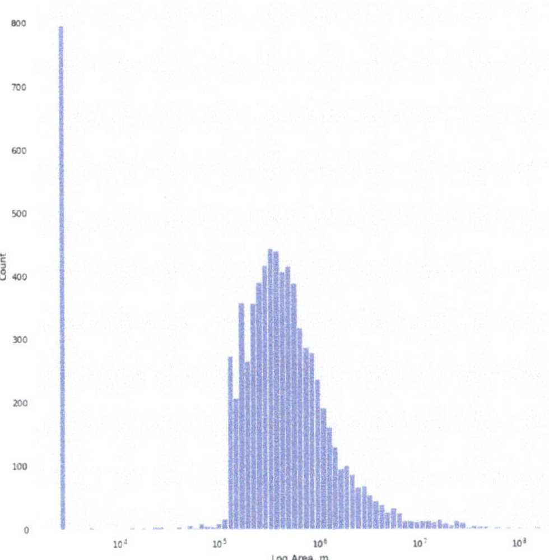

Figure 10. Histogram of the number of recognized lakes on images from Yamal vs the log of lake size (area).

5. Conclusions

This study addressed the problem of automated tundra lake recognition from SAR images for the lake shape and size calculations. We proposed a two-stage recognition framework comprising instance (1) and semantic segmentation (2), which operated with Sentinel-1 images. In the first stage, the tundra lake areas were extracted by U-Net, and a segmentation accuracy of 0.73, in terms of the Jaccard similarity, was achieved. The second stage, which was the watershed algorithm, was aimed at the separation of touching and overlapping lakes produced in the first stage, for the unambiguous estimation of the shape and size of an individual lake. The high performance of the proposed method was demonstrated, based on a generated dataset of manually labeled lakes from hundreds of images taken over the Yamal Peninsula and Alaska areas in the summer months of 2015–2022. The approach was powerful and reliable, in terms of recognition accuracy, computational efficiency and degree of automation.

The applicability of the proposed framework was assessed, based on a generated multi-year dataset of segmented binary images containing derived tundra lakes from Sentinel-1 data. Lake sizes and shapes were estimated from the framework output, and showed high potential gains for tundra lake changes analysis in the context of global climate. However, the model results for a wider range of environmental conditions and regions still need to be proven. Possible directions for improving the recognition of lakes include training over different areas and seasons, and utilizing SAR texture characteristics. State-of-the-art U-Net architectures could be incorporated into the framework, and its performance gain evaluated, but at the cost of complexity and reduced speed. Although the new approach is used here for SAR images from Sentinel-1, it has great potential for images taken from other satellite platforms, including optical imagery.

The study of the fractal characterizations of networks of tundra lakes, and their size distributions, is essential for the representation of permafrost, vegetation and landscape dynamics in climate models. The analysis of new SAR imagery of tundra lakes by a novel deep learning method showed that the average fractal dimension of recognized lake patterns was very stable over seven-year time intervals, and that all the changes fitted into the limits of accuracy. The lake sizes were lognormally distributed for every year in the studied time interval. We expect that the proposed approach to tundra lake recognition will be demanded in climate emulators that are based on conducting high-quality data analysis and revealing complex features of the systems [75].

Author Contributions: Conceptualization, D.D., I.S. and D.L.; methodology, I.S., D.M. and D.D.; software, D.D., I.S., A.K. and D.L; validation, D.D., I.S., A.K., I.A. and D.L.; data curation, D.D., A.K., I.A., I.S. and D.L.; original draft preparation, I.S., D.D. and D.L.; project administration, I.S. All authors have read and agreed to the published version of the manuscript.

Funding: This research was funded by the Russian Science Foundation (RSF), project # 21-71-10052. D.L. and D.M. were partially supported by KAUST baseline funding. I.S. was supported in part by the Microsoft AI for Earth Grant Program and by the National Science Foundation, Grant No. NSF PHY-1748958.

Data Availability Statement: The training and validation dataset, containing Sentinel-1 imagery of the tundra lakes and corresponding binary masks, is freely available through the NSF Arctic Data Center catalog [50].

Acknowledgments: The authors thank Viktoria Kharchenko for manual labeling of part of the Sentinel-1 training and validation images.

Conflicts of Interest: The authors declare no conflict of interest.

Abbreviations

The following abbreviations are used in this manuscript:

SAR	synthetic aperture radar
U-Net	full convolutional neural network architecture
InSAR	interferometric synthetic aperture radar
IoU (or Jaccard similarity index)	the Intersection of Union metric

References

1. Treat, C.C.; Frolking, S. A permafrost carbon bomb? *Nat. Clim. Chang.* **2013**, *3*, 865–867. [CrossRef]
2. Lawrence, N.S.; O'Sullivan, J.; Parslow, D.; Javaid, M.; Adams, R.C.; Chambers, C.D.; Kos, K.; Verbruggen, F. Training response inhibition to food is associated with weight loss and reduced energy intake. *Appetite* **2015**, *95*, 17–28. [CrossRef] [PubMed]
3. Schädel, C.; Bader, M.K.F.; Schuur, E.A.; Biasi, C.; Bracho, R.; Čapek, P.; De Baets, S.; Diáková, K.; Ernakovich, J.; Estop-Aragones, C.; et al. Potential carbon emissions dominated by carbon dioxide from thawed permafrost soils. *Nat. Clim. Chang.* **2016**, *6*, 950–953. [CrossRef]
4. Kokelj, S.V.; Jorgenson, M. Advances in thermokarst research. *Permafr. Periglac. Process.* **2013**, *24*, 108–119. [CrossRef]
5. Olefeldt, D.; Hovemyr, M.; Kuhn, M.A.; Bastviken, D.; Bohn, T.J.; Connolly, J.; Crill, P.; Euskirchen, E.S.; Finkelstein, S.A.; Genet, H.; et al. The Boreal–Arctic Wetland and Lake Dataset (BAWLD). *Earth Syst. Sci. Data* **2021**, *13*, 5127–5149. [CrossRef]
6. Miner, K.R.; Turetsky, M.R.; Malina, E.; Bartsch, A.; Tamminen, J.; McGuire, A.D.; Fix, A.; Sweeney, C.; Elder, C.D.; Miller, C.E. Permafrost carbon emissions in a changing Arctic. *Nat. Rev. Earth Environ.* **2022**, *3*, 55–67. [CrossRef]
7. Kirpotin, S.; Polishchuk, Y.; Zakharova, E.; Shirokova, L.; Pokrovsky, O.; Kolmakova, M.; Dupre, B. One of the possible mechanisms of thermokarst lakes drainage in West-Siberian North. *Int. J. Environ. Stud.* **2008**, *65*, 631–635. [CrossRef]
8. Jorgenson, J.C.; Jorgenson, M.T.; Boldenow, M.L.; Orndahl, K.M. Landscape change detected over a half century in the Arctic National Wildlife Refuge using high-resolution aerial imagery. *Remote Sens.* **2018**, *10*, 1305. [CrossRef]
9. Shiklomanov, N.I.; Streletskiy, D.A.; Little, J.D.; Nelson, F.E. Isotropic thaw subsidence in undisturbed permafrost landscapes. *Geophys. Res. Lett.* **2013**, *40*, 6356–6361. [CrossRef]
10. Andresen, C.G.; Lougheed, V.L. Disappearing Arctic tundra ponds: Fine-scale analysis of surface hydrology in drained thaw lake basins over a 65 year period (1948–2013). *J. Geophys. Res. Biogeosci.* **2015**, *120*, 466–479. [CrossRef]
11. Karlsson, J.M.; Jaramillo, F.; Destouni, G. Hydro-climatic and lake change patterns in Arctic permafrost and non-permafrost areas. *J. Hydrol.* **2015**, *529*, 134–145. [CrossRef]
12. Boike, J.; Grau, T.; Heim, B.; Günther, F.; Langer, M.; Muster, S.; Gouttevin, I.; Lange, S. Satellite-derived changes in the permafrost landscape of central Yakutia, 2000–2011: Wetting, drying, and fires. *Glob. Planet. Chang.* **2016**, *139*, 116–127. [CrossRef]
13. Muster, S.; Roth, K.; Langer, M.; Lange, S.; Cresto Aleina, F.; Bartsch, A.; Morgenstern, A.; Grosse, G.; Jones, B.; Sannel, A.B.K.; et al. PeRL: A circum-Arctic permafrost region pond and lake database. *Earth Syst. Sci. Data* **2017**, *9*, 317–348. [CrossRef]
14. Anthony, K.; Zimov, S.; Grosse, G.; Jones, M.C.; Anthony, P.; FS III, C.; Finlay, J.; Mack, M.; Davydov, S.; Frenzel, P.; et al. A shift of thermokarst lakes from carbon sources to sinks during the Holocene epoch. *Nature* **2014**, *511*, 452–456. [CrossRef] [PubMed]
15. Schuur, E.A.; McGuire, A.D.; Schädel, C.; Grosse, G.; Harden, J.W.; Hayes, D.J.; Hugelius, G.; Koven, C.D.; Kuhry, P.; Lawrence, D.M.; et al. Climate change and the permafrost carbon feedback. *Nature* **2015**, *520*, 171–179. [CrossRef]
16. Abbott, B.W.; Jones, J.B.; Schuur, E.A.; Chapin III, F.S.; Bowden, W.B.; Bret-Harte, M.S.; Epstein, H.E.; Flannigan, M.D.; Harms, T.K.; Hollingsworth, T.N.; et al. Biomass offsets little or none of permafrost carbon release from soils, streams, and wildfire: an expert assessment. *Environ. Res. Lett.* **2016**, *11*, 034014. [CrossRef]

17. Nitze, I.; Grosse, G.; Jones, B.M.; Arp, C.D.; Ulrich, M.; Fedorov, A.; Veremeeva, A. Landsat-based trend analysis of lake dynamics across northern permafrost regions. *Remote Sens.* **2017**, *9*, 640. [CrossRef]

18. Muster, S.; Riley, W.J.; Roth, K.; Langer, M.; Cresto Aleina, F.; Koven, C.D.; Lange, S.; Bartsch, A.; Grosse, G.; Wilson, C.J.; et al. Size Distributions of Arctic Waterbodies Reveal Consistent Relations in Their Statistical Moments in Space and Time. *Front. Earth Sci.* **2019**, *7*, 2296-6463. [CrossRef]

19. Carroll, M.L.; Loboda, T.V. Multi-Decadal Surface Water Dynamics in North American Tundra. *Remote Sens.* **2017**, *9*, 497. [CrossRef]

20. Nitze, I.; Grosse, G.; Jones, B.M.; Romanovsky, V.E.; Boike, J. Remote sensing quantifies widespread abundance of permafrost region disturbances across the Arctic and Subarctic. *Nat. Commun.* **2018**, *9*, 5423. [CrossRef]

21. Jawak, S.D.; Kulkarni, K.; Luis, A.J.; et al. A review on extraction of lakes from remotely sensed optical satellite data with a special focus on cryospheric lakes. *Adv. Remote. Sens.* **2015**, *4*, 196. [CrossRef]

22. Polishchuk, Y.M.; Bogdanov, A.N.; Polishchuk, V.Y.; Manasypov, R.M.; Shirokova, L.S.; Kirpotin, S.N.; Pokrovsky, O.S. Size distribution, surface coverage, water, carbon, and metal storage of thermokarst lakes in the permafrost zone of the Western Siberia Lowland. *Water* **2017**, *9*, 228. [CrossRef]

23. Karlsson, J.M.; Lyon, S.W.; Destouni, G. Temporal behavior of lake size-distribution in a thawing permafrost landscape in northwestern Siberia. *Remote Sens.* **2014**, *6*, 621–636. [CrossRef]

24. Payne, C.; Panda, S.; Prakash, A. Remote sensing of river erosion on the Colville River, North Slope Alaska. *Remote Sens.* **2018**, *10*, 397. [CrossRef]

25. Naeimi, V.; Bartalis, Z.; Wagner, W. ASCAT soil moisture: An assessment of the data quality and consistency with the ERS scatterometer heritage. *J. Hydrometeorol.* **2009**, *10*, 555–563. [CrossRef]

26. Kerr, Y.; Waldteufel, P.; Wigneron, J.; Martinuzzi, J.; Font, J. M. Berger: Soil moisture retrieval from space: the Soil Moisture and Ocean Salinity (SMOS) mission. *IEEE T Geosci. Remote* **2001**, *39*, 1729–1735. [CrossRef]

27. Jones, B.M.; Grosse, G.; Arp, C.; Jones, M.; Walter Anthony, K.; Romanovsky, V. Modern thermokarst lake dynamics in the continuous permafrost zone, northern Seward Peninsula, Alaska. *J. Geophys. Res. Biogeosci.* **2011**, *116*, 0148-0227. [CrossRef]

28. Schroeder, R.; Rawlins, M.; McDonald, K.; Podest, E.; Zimmermann, R.; Kueppers, M. Satellite microwave remote sensing of North Eurasian inundation dynamics: development of coarse-resolution products and comparison with high-resolution synthetic aperture radar data. *Environ. Res. Lett.* **2010**, *5*, 015003. [CrossRef]

29. Bartsch, A.; Wagner, W.; Scipal, K.; Pathe, C.; Sabel, D.; Wolski, P. Global monitoring of wetlands–the value of ENVISAT ASAR global mode. *J. Environ. Manag.* **2009**, *90*, 2226–2233. [CrossRef]

30. Sobiech, J.; Dierking, W. Observing lake-and river-ice decay with SAR: advantages and limitations of the unsupervised k-means classification approach. *Ann. Glaciol.* **2013**, *54*, 65–72. [CrossRef]

31. Widhalm, B.; Bartsch, A.; Heim, B. A novel approach for the characterization of tundra wetland regions with C-band SAR satellite data. *Int. J. Remote. Sens.* **2015**, *36*, 5537–5556. [CrossRef]

32. Hirose, T.; Kapfer, M.; Bennett, J.; Cott, P.; Manson, G.; Solomon, S. Bottomfast Ice Mapping and the Measurement of Ice Thickness on Tundra Lakes Using C-Band Synthetic Aperture Radar Remote Sensing 1. *JAWRA J. Am. Water Resour. Assoc.* **2008**, *44*, 285–292. [CrossRef]

33. Walter, K.M.; Engram, M.; Duguay, C.R.; Jeffries, M.O.; Chapin III, F. The Potential Use of Synthetic Aperture Radar for Estimating Methane Ebullition From Arctic Lakes 1. *JAWRA J. Am. Water Resour. Assoc.* **2008**, *44*, 305–315. [CrossRef]

34. Duguay, C.; Rouse, W.R.; Lafleur, P.M.; Boudreau, L.D.; Crevier, Y.; Pultz, T.J. Analysis of multi-temporal ERS-1 SAR data of subarctic tundra and forest in the northern Hudson Bay Lowland and implications for climate studies. *Can. J. Remote. Sens.* **1999**, *25*, 21–33. [CrossRef]

35. Wakabayashi, H.; Jeffries, M.; Weeks, W. C-band backscatter variation and modelling for lake ice in northern Alaska. *J. Remote. Sens. Soc. Jpn.* **1994**, *14*, 220–229.

36. Liu, L.; Schaefer, K.; Chen, A.; Gusmeroli, A.; Zebker, H.; Zhang, T. Remote sensing measurements of thermokarst subsidence using InSAR. *J. Geophys. Res. Earth Surf.* **2015**, *120*, 1935–1948. [CrossRef]

37. Liu, X.; Guo, Y.; Hu, H.; Sun, C.; Zhao, X.; Wei, C. Dynamics and controls of CO2 and CH4 emissions in the wetland of a montane permafrost region, northeast China. *Atmos. Environ.* **2015**, *122*, 454–462. [CrossRef]

38. Bartsch, A.; Leibman, M.; Strozzi, T.; Khomutov, A.; Widhalm, B.; Babkina, E.; Mullanurov, D.; Ermokhina, K.; Kroisleitner, C.; Bergstedt, H. Seasonal progression of ground displacement identified with satellite radar interferometry and the impact of unusually warm conditions on permafrost at the Yamal Peninsula in 2016. *Remote Sens.* **2019**, *11*, 1865. [CrossRef]

39. Onstott, R.G.; Carsey, F. SAR and scatterometer signatures of sea ice. *Microw. Remote. Sens. Sea Ice* **1992**, *68*, 73–104.

40. Geldsetzer, T.; Sanden, J.v.d.; Brisco, B. Monitoring lake ice during spring melt using RADARSAT-2 SAR. *Can. J. Remote. Sens.* **2010**, *36*, S391–S400. [CrossRef]

41. Jeffries, M.; Morris, K.; Weeks, W.; Wakabayashi, H. Structural and stratigraphie features and ERS 1 synthetic aperture radar backscatter characteristics of ice growing on shallow lakes in NW Alaska, winter 1991–1992. *J. Geophys. Res. Ocean.* **1994**, *99*, 22459–22471. [CrossRef]

42. Merchant, M.A.; Obadia, M.; Brisco, B.; DeVries, B.; Berg, A. Applying Machine Learning and Time-Series Analysis on Sentinel-1A SAR/InSAR for Characterizing Arctic Tundra Hydro-Ecological Conditions. *Remote Sens.* **2022**, *14*, 1123. [CrossRef]

43. Piantanida, R.; Hajduch, G.; Poullaouec, J. *Sentinel-1 Level 1 Detailed Algorithm Definition*; Techreport SEN-TN-52-7445; ESA: Paris, France, 2016.

44. Makynen, M.; Manninen, A.T.; Simila, M.; Karvonen, J.A.; Hallikainen, M.T. Incidence angle dependence of the statistical properties of C-band HH-polarization backscattering signatures of the Baltic Sea ice. *IEEE Trans. Geosci. Remote. Sens.* **2002**, *40*, 2593–2605. [CrossRef]

45. Wakabayashi, H.; Weeks, W.; Jeffries, M.O. A C-band backscatter model for lake ice in Alaska. In Proceedings of the IGARSS'93-IEEE International Geoscience and Remote Sensing Symposium, Tokyo, Japan, 18–21 August 1993; pp. 1264–1266.

46. Zakhvatkina, N.Y.; Alexandrov, V.Y.; Johannessen, O.M.; Sandven, S.; Frolov, I.Y. Classification of sea ice types in ENVISAT synthetic aperture radar images. *IEEE Trans. Geosci. Remote. Sens.* **2012**, *51*, 2587–2600. [CrossRef]

47. Karvonen, J. A sea ice concentration estimation algorithm utilizing radiometer and SAR data. *Cryosphere* **2014**, *8*, 1639–1650. [CrossRef]

48. Karvonen, J. Baltic sea ice concentration estimation using SENTINEL-1 SAR and AMSR2 microwave radiometer data. *IEEE Trans. Geosci. Remote. Sens.* **2017**, *55*, 2871–2883. [CrossRef]

49. Dutta, A.; Gupta, A.; Zissermann, A. VGG Image Annotator (VIA). 2016. Volume 2. Available online: http://www.robots.ox.ac.uk/vgg/software/via (accessed on 10 January 2023).

50. Demchev, D.; Sudakow, I. A dataset of 512 × 512 tundra lakes imagery and binary masks from Sentinel-1 in the Yamal and Alaska areas, summer, 2015-2022. *Arct. Data Cent.* **2023**. Available online: https://arcticdata.io/catalog/view/urn%3Auuid%3Aaec6b61a-7318-4680-a39c-85837fa8a5c1 (accessed on 10 January 2023).

51. Ronneberger, O.; Fischer, P.; Brox, T. U-net: Convolutional networks for biomedical image segmentation. In Proceedings of the International Conference on Medical Image Computing and Computer-Assisted Intervention, Munich, Germany, 5–9 October 2015; Springer: Berlin/Heidelberg, Germany, 2015; pp. 234–241.

52. Long, J.; Shelhamer, E.; Darrell, T. Fully convolutional networks for semantic segmentation. In Proceedings of the IEEE Conference on Computer Vision and Pattern Recognition, Boston, MA, USA, 7–12 June 2015; pp. 3431–3440.

53. Ren, Y.; Li, X.; Yang, X.; Xu, H. Development of a dual-attention U-Net model for sea ice and open water classification on SAR images. *IEEE Geosci. Remote. Sens. Lett.* **2021**, *19*, 1–5. [CrossRef]

54. Sudakow, I.; Asari, V.K.; Liu, R.; Demchev, D. MeltPondNet: A Swin Transformer U-Net for Detection of Melt Ponds on Arctic Sea Ice. *IEEE J. Sel. Top. Appl. Earth Obs. Remote. Sens.* **2022**, *15*, 8776–8784. [CrossRef]

55. Vincent, L.; Soille, P. Watersheds in digital spaces: an efficient algorithm based on immersion simulations. *IEEE Trans. Pattern Anal. Mach. Intell.* **1991**, *13*, 583–598. [CrossRef]

56. Beucher, S.; Meyer, F. The morphological approach to segmentation: the watershed transformation. In *Mathematical Morphology in Image Processing*; CRC Press: Boca Raton, FL, USA, 2018; pp. 433–481.

57. Yu, Y.; Acton, S.T. Speckle reducing anisotropic diffusion. *IEEE Trans. Image Process.* **2002**, *11*, 1260–1270.

58. Demchev, D.; Volkov, V.; Kazakov, E.; Alcantarilla, P.F.; Sandven, S.; Khmeleva, V. Sea ice drift tracking from sequential SAR images using accelerated-KAZE features. *IEEE Trans. Geosci. Remote. Sens.* **2017**, *55*, 5174–5184. [CrossRef]

59. Kingma, D.P.; Ba, J. Adam: A method for stochastic optimization. *arXiv* **2014**, arXiv:1412.6980.

60. Chen, Q.; Yang, X.; Petriu, E.M. Watershed segmentation for binary images with different distance transforms. In Proceedings of the 3rd IEEE International Workshop on Haptic, Audio and Visual Environments and Their Applications, Ottawa, ON, Canada, 2–3 October 2004; Volume 2, pp. 111–116.

61. Raju, C.N.; Raju, G.; Gottumukkala, V.V. Studies on watershed segmentation for blood cell images using different distance transforms. *IOSR J. Vlsi Signal Process. (IOSR-JVSP)* **2016**, *6*, 79–85.

62. Bradski, G. The OpenCV Library. *Dr. Dobb'S J. Softw. Tools* **2000**, *25*, 109–141.

63. Lovejoy, S. Area-perimeter relation for rain and cloud areas. *Science* **1982**, *216*, 185–187. [CrossRef]

64. Mandelbrot, B. *The Fractal Geometry of Nature.*; W. H. Freeman and Co.: San Francisco, , CA, USA, 1982; p. 460.

65. Klinkenberg, B. A Review of Methods Used to Determine the Fractal Dimension of Linear Features. *Math. Geol.* **2022**, *26*, 23–45. [CrossRef]

66. Oktay, O.; Schlemper, J.; Folgoc, L.L.; Lee, M.; Heinrich, M.; Misawa, K.; Mori, K.; McDonagh, S.; Hammerla, N.Y.; Kainz, B.; et al. Attention u-net: Learning where to look for the pancreas. *arXiv* **2018**, arXiv:1804.03999.

67. Zhou, Z.; Siddiquee, M.M.R.; Tajbakhsh, N.; Liang, J. Unet++: Redesigning skip connections to exploit multiscale features in image segmentation. *IEEE Trans. Med. Imaging* **2019**, *39*, 1856–1867. [CrossRef]

68. Kugelman, J.; Allman, J.; Read, S.A.; Vincent, S.J.; Tong, J.; Kalloniatis, M.; Chen, F.K.; Collins, M.J.; Alonso-Caneiro, D. A comparison of deep learning U-Net architectures for posterior segment OCT retinal layer segmentation. *Sci. Rep.* **2022**, *12*, 14888. [CrossRef]

69. Seekell, D.A.; Pace, M.L.; Tranvik, L.J.; Verpoorter, C. A fractal-based approach to lake size-distributions. *Geophys. Res. Lett.* **2013**, *40*, 517–521. [CrossRef]

70. Sudakov, I.; Essa, A.; Mander, L.; Gong, M.; Kariyawasam, T. The Geometry of Large Tundra Lakes Observed in Historical Maps and Satellite Images. *Remote Sens.* **2017**, *9*, 1072. [CrossRef]

71. Aleina, F.C.; Brovkin, V.; Muster, S.; Boike, J.; Kutzbach, L.; Sachs, T.; Zuyev, S. A stochastic model for the polygonal tundra based on Poisson–Voronoi diagrams. *Earth Syst. Dyn.* **2013**, *4*, 187–198. [CrossRef]

72. Polishchuk, V.Y.; Polishchuk, Y.M. Geoimitatsionnoe modelirovanie polei termokarstovykh ozer v zonakh merzloty [Geo-Simulation Modeling of Thermokarst Lakes Fields in Permafrost Zones]. *Khanty-Mansijsk Uip Yugu* **2013**, 174, 195-199.

73. Lopes, R.; Betrouni., N. Fractal and multifractal analysis: a review. *Med. Image Anal.* **2009**, *13*, 634–649. [CrossRef] [PubMed]

74. Nayak, S.R.; Mishra., J. Analysis of medical images using fractal geometry. In *The Research Anthology on Improving Medical Imaging Techniques for Analysis and Intervention*; Medical Information Science Reference; IGI Global: Hershey, PA, USA, 2023; pp. 1547–1562.

75. Sudakow, I.; Pokojovy, M.; Lyakhov, D. Statistical mechanics in climate emulation: Challenges and perspectives. *Environ. Data Sci.* **2022**, *1*, e16. [CrossRef]

Communication

A Clutter Parameter Estimation Method Based on Origin Moment Derivation

Liru Yang, Yongxiang Liu, Wei Yang, Xiaolong Su * and Qinmu Shen

College of Electronic Science and Technology, National University of Defense Technology,
Changsha 410073, China
* Correspondence: suxiaolong_nudt@163.com

Abstract: Parameter estimation is significant to prediction and estimation in the field of radar clutter characteristics. Therefore, it is necessary to study the problem of parameter estimation. The K-distribution is a commonly used model in sea clutter, which is a two-parameter model with shape parameters and scale parameters. The value of the shape parameters should be greater than 0. Moment estimation is usually used to estimate the parameters of the K-distribution. It overcomes the disadvantage of large computation compared with the maximum likelihood estimation method. However, the moment estimation usually uses two different order origin moments to solve the parameters. The joint solution of different order will cause large calculation errors, and sometimes the shape parameter is estimated to be less than 0. In the origin moment expression, the order k can be regarded as a continuous variable. By calculating the relationship between the k-order origin moment and its derivative, a parameter estimation method based on the origin moment derivative is proposed. The estimation efficiency and accuracy are compared with some moment estimation methods. Both simulation data and measured clutter data show that this method can achieve 100% estimation efficiency, can obtain higher estimation accuracy, and can also avoid the situation where the estimated value of the shape parameter is less than 0. Using the same idea to estimate the parameters in the two-parameter models, log–normal and Weibull distribution, we can also obtain the parameters with higher estimation accuracy. The experiments show that the higher-order origin moments are sensitive to the data, and the lower-order moments should be selected as far as possible. By selecting the appropriate order k, we can obtain ideal estimation parameters.

Keywords: clutter; parameter estimation; moment estimation; origin moment derivation

Citation: Yang, L.; Liu, Y.; Yang, W.; Su, X.; Shen, Q. A Clutter Parameter Estimation Method Based on Origin Moment Derivation. *Remote Sens.* **2023**, *15*, 1551. https://doi.org/10.3390/rs15061551

Academic Editor: Deodato Tapete

Received: 4 February 2023
Revised: 3 March 2023
Accepted: 9 March 2023
Published: 12 March 2023

1. Introduction

The statistical properties of radar clutter have an influence on target detection performance [1]. To improve the performance of radar constant false alarm rate (CFAR) [2] detection, it is necessary to select the optimal clutter distribution model to analyze the characteristics of clutter [3]. Radar clutter has the characteristics of random amplitude fluctuation and implies certain statistical laws. The commonly used clutter amplitude statistical models include Rayleigh distribution, log–normal distribution, Weibull distribution [4], and K-distribution [5]. The Rayleigh distribution model is suitable for describing the echo fluctuation of low-resolution radar at a high grazing angle in stationary random environments [6,7]. The log–normal distribution model is fit for describing the clutter data at a low incidence angle in complex terrain or high-resolution clutter data on flat terrain [8]. The Weibull distribution model is suitable to represent the clutter data of the high-resolution radar at a low incident angle [9]. The K-distribution is widely used in sea clutter modeling, which can match the amplitude distribution of clutter in a wide range of conditions. It can not only well represent the long tailing characteristics in the amplitude distribution, but also can describe the correlation characteristics between an echo pulse, which is suitable for describing the non-stationary sea clutter of a high-resolution radar with a low grazing angle [10,11]. When selecting the optimal clutter distribution model to fit sea or land clutter, it is necessary to estimate the distribution parameters of the models, and then select an optimal distribution model through the goodness-of-fit criterion.

When the constant false alarm rate is given, the calculation of the CFAR target detection threshold is related to the model and its parameters. Therefore, the accurate estimation of the model parameters is very important. The commonly used parameter estimation methods include the maximum likelihood estimation (MLE) method [12] and the moment estimation (ME) method [13]. The solution of the MLE is limited by the analytic form of the likelihood function. When the likelihood function is complex, it is difficult to solve or requires a large amount of calculation. ME has the characteristic of simple calculation. However, when there is noise or a small amount of data in the observation sequence, the estimation accuracy is low, and even provides no valid solution, especially in the estimation process of the K-distribution. In the ME method, the observed moment value is used to approximate the theoretical moment value, and the observed different order moment values form a simultaneous equation to solve the parameters. The calculation of these different order moments is prone to cause approximation mistakes and makes the estimation value deviate from the true value.

To overcome the shortcomings of the ME method, a parameter estimation method based on the partial derivative of the moment estimation is proposed. The proposed method only uses a single k-order origin moment and can avoid the approximate values of multiple origin moments to participate in the operation. The method can improve the efficiency and accuracy of the parameter estimation. It also can solve the problem of outliers in the K-distribution parameter estimation and be applied to the parameter estimation of log–normal and Weibull distributions. This method not only improves the calculation accuracy, but also sacrifices the calculation time. The ME methods eliminate the complex function operation through the selection of special order operations, such as first-order, second-order, and fourth-order operations. These special order operations will eliminate the gamma function or the exponential function, so the operation time is shorter. Our method uses more general operation criteria, including the operation of the gamma function or the exponential function, so the operation time is longer.

The remainder of this paper is organized as follows: In Section 2, the parameter estimation method based on the origin moment derivation is introduced through the K-distribution. The computational complexity of the proposed method is briefly explained. Both the simulated data and the measured data are used to verify the effectiveness of the proposed method. The experiments show that the proposed method has higher estimation efficiency and better estimation accuracy. The calculation time of the proposed method is also directly analyzed through experiments. In Section 3, the proposed method is also applied to the parameter estimation of log–normal and Weibull distributions. Both the simulated data and measured data experiments show that the proposed method also has higher estimation accuracy in the log–normal and Weibull distributions. Moreover, the calculation time is analyzed through experiments. Section 4 concludes the study.

2. Proposed Method

The k-order moment expression of the K-distribution [14] is defined as:

$$m_k = E[z^k] = \frac{\Gamma\left(\frac{k}{2}+1\right)\Gamma\left(\frac{k}{2}+v\right)}{\Gamma(v)}(2\sigma)^k, k \geq 0 \tag{1}$$

where $E[\bullet]$ is the mean operation, z is the clutter amplitude, $\Gamma(\bullet)$ is the gamma function, k is the order.

The origin moment of the K-distribution is found to contain the exponential function and gamma function. The derivations of the exponential function and the gamma function are related to the origin function itself. The derivation of the origin moment is based on the following Equations (2)–(4):

$$\frac{\partial \Gamma(z)}{\partial z} = \psi(z)\Gamma(z) \tag{2}$$

where $\psi(\bullet)$ is the Psi function:

$$\frac{\partial a^z}{\partial z} = a^z \ln a \tag{3}$$

where a^z is the exponential function and $\ln a$ is the logarithm of a.

D. R. Iskander et al. [15] generalized the integer moment method and proposed the fractional moment method. On this basis, the order k can be regarded as a continuous variable, and its partial derivative can be calculated.

$$\frac{\partial E[z^k]}{\partial k} = E[z^k \ln(z)] \tag{4}$$

2.1. The Origin Moment Derivation Method

This method can be expressed as:

$$\frac{\partial E[z^k]}{\partial k} = \left(\frac{1}{2}\psi\left(\frac{k}{2}+1\right) + \frac{1}{2}\psi\left(\frac{k}{2}+v\right) + \ln(2\sigma) \right) E[z^k] \tag{5}$$

Therefore, it can be obtained from Equations (4) and (5):

$$\frac{E[z^k \ln(z)]}{E[z^k]} = \frac{1}{2}\psi\left(\frac{k}{2}+1\right) + \frac{1}{2}\psi\left(\frac{k}{2}+v\right) + \ln(2\sigma) \tag{6}$$

It is found that taking the logarithm of $E[z^k]$ can also obtain the term $\ln(2\sigma)$, as follows:

$$\ln(E[z^k]) = \ln\left(\Gamma\left(\frac{k}{2}+1\right)\right) + \ln\left(\Gamma\left(\frac{k}{2}+v\right)\right) + k\ln(2\sigma) - \ln(\Gamma(v)) \tag{7}$$

By combining Equations (6) and (7), we can obtain:

$$k\frac{E[z^k \ln(z)]}{E[z^k]} - \ln(E[z^k]) = \frac{k}{2}\psi\left(\frac{k}{2}+1\right) + \frac{k}{2}\psi\left(\frac{k}{2}+v\right) - \ln\left(\Gamma\left(\frac{k}{2}+1\right)\right) - \ln\left(\Gamma\left(\frac{k}{2}+v\right)\right) + \ln(\Gamma(v)) \tag{8}$$

When the value of k is given, Equation (8) is an equation describing the shape parameter v, and can be expressed as:

$$\frac{k}{2}\psi\left(\frac{k}{2}+v\right) - \ln\left(\Gamma\left(\frac{k}{2}+v\right)\right) + \ln(\Gamma(v)) = k\frac{E[z^k \ln(z)]}{E[z^k]} - \ln(E[z^k]) - \frac{k}{2}\psi\left(\frac{k}{2}+1\right) + \ln\left(\Gamma\left(\frac{k}{2}+1\right)\right) \tag{9}$$

The left part of Formula (9) is the function of the variable v to be estimated, let:

$$f(v) = \frac{k}{2}\psi\left(\frac{k}{2}+v\right) - \ln\left(\Gamma\left(\frac{k}{2}+v\right)\right) + \ln(\Gamma(v)) \tag{10}$$

When $v > 0$, $f(v)$ is a monotone decrease function. The monotonicity is proved as follows.

The series expression of Psi function can be expressed as:

$$\psi(v) = -\gamma + \sum_{m=0}^{\infty}\left(\frac{1}{m+1} - \frac{1}{m+v}\right) \tag{11}$$

where γ is the Euler constant. There exist $\psi(z) = \Gamma'(z)/\Gamma(z)$, so the first-order derivative of $f(v)$ is:

$$\frac{df(v)}{dv} = -\sum_{m=0}^{\infty} \frac{k^2}{4(m+k/2+v)^2(m+v)} < 0 \tag{12}$$

Therefore, $f(v)$ is a monotonic decrease function.

Equation (9) is a nonlinear equation describing the shape parameter v. We use the trust-region–dogleg algorithm to solve the nonlinear equation. The estimated expression of the scale parameter σ can be obtained in Equation (13):

$$\sigma = 0.5 \left(\frac{E[z^k]\Gamma(v)}{\Gamma\left(\frac{k}{2}+1\right)\Gamma\left(\frac{k}{2}+v\right)} \right)^{\frac{1}{k}} \tag{13}$$

2.2. Complexity Analysis

The trust-region–dogleg method is a hybrid of the Gauss–Newton and steepest descent method. It is a classical trust-region technique for globalizing the Newton method. The convergence result of the dogleg is better than the Gauss–Newton. Reference [16] proves that the dogleg algorithm satisfies the first- and second-order stationary point convergence properties. Any mapping $F: X \to R^n$, differentiable at a point $x \in X \in R^n$, the $F(x_k)$ is denoted by F_k. If $1 - O(\|F_k\|) \to 1$ as $k \to \infty$, the convergence rate is quadratic [17]. Reference [16] points out that the dogleg path algorithm is reliable and easy to implement, and that it solves unconstrained optimization problems effectively. The calculation efficiency and time are assessed through the following experiments. All of the computations were executed on an Intel (R) Core (TM) i7-10750H CPU @ 2.60GHz 2.59 GHz RAM 16.0 GB. To compare the computational time, all nonlinear equations in the following experiments are solved using the trust-region–dogleg method.

2.3. Experiments and Discussion

When there is noise or a small amount of data in the K-distribution observation sequence, the shape parameter v may be estimated to be less than 0, which is not within the defined range. The estimation efficiency and estimation accuracy are verified using the data simulated from Zero-Memory Non-Linear (ZMNL) [18] and the measured IPIX data [19].

The number of invalid estimates is defined as the number of the shape parameter whose estimated value is less than 0 in the total samples. The estimation efficiency [20] ρ is defined as $\rho = (N - M)/N, M \leq N \in Z^+$. M is the number of invalid estimates and N is the number of total estimation samples. Because the simulation data may have small deviations and the parameters of the measured data are unknown, the fitted nonparametric probability density function is used to compare with the probability density function (PDF) of the estimated parameters. The Kernel smoothing density (KSdensity) [21], also known as the Parzen window, is a nonparametric estimation method. It is often used to estimate the unknown probability density in probability theory. The root-mean-square error (RMSE) [22] between the probability density obtained using the KSdensity and the parameter estimation is defined as the estimation accuracy. The smaller the RMSE, the higher the estimation accuracy. The expression of the RMSE is as follows:

$$\text{RMSE} = \sqrt{\frac{1}{N}\sum_{i=1}^{N}\left(pdf_{KSdensity} - pdf_{parmeter\ estimation}\right)^2} \tag{14}$$

The second-/fractional-order method [23], log-III estimation method [24], second-/fourth-order method [25], first-order and second-order method of log(z) [18], expectation estimation method based on z^rlog(z) [26], expectation estimation method based on zlog(z) [27], and the proposed method are used for the parameter estimation. The estimated model in references [26,27] is the power model, and the other methods are the amplitude model. The amplitude of the former is the square of the latter. When $r = 1$, the method of $z^r\log(z)$ degenerates to the method of $z\log(z)$. The method of $z\log(z)$ is equivalent to the method of log-III.

1. The simulated data experiment. The amplitude statistical models of clutter data used with different parameters are simulated using ZMNL. The scale parameter is taken randomly in the interval between 1.2 and 1.6. The shape parameter is taken in steps of 0.5 in the range of 0.5–10. The K-distribution model has 1000 groups of simulation data with a length of 256 points. When $v = 3$ and $k = 0.5$, the statistics of the estimation efficiency ρ and estimation accuracy RMSE of the seven methods are shown in Table 1.

Table 1. The statistics of estimation efficiency and estimation accuracy in simulated K-distribution ($v = 3, k = 0.5$).

Estimation Method	Second-/Fractional-Order Moment	Log-III Method	Second-/Fourth-Order Moment	Ref. [18] Method	Ref. [26] Method	Ref. [27] Method	Our Method
Invalid estimated times	2	4	2	0	0	4	0
ρ	99.8%	99.6%	99.8%	100%	100%	99.6%	100%
RMSE	0.0157	0.0155	0.0170	0.0167	0.0156	0.0155	0.0154

As can be seen from Table 1, the estimation accuracy of the second-/fourth-order ME method is the worst. The methods of references [18,26] and ours have 100% estimation efficiency, and our method has the best estimation accuracy. Here, the samples for calculating the RMSE can meet the seven methods at the same time. The number of all effective estimated samples in this experiment is 996.

The relationship between the estimated efficiency of the seven methods and the simulation parameter v is shown in Figure 1.

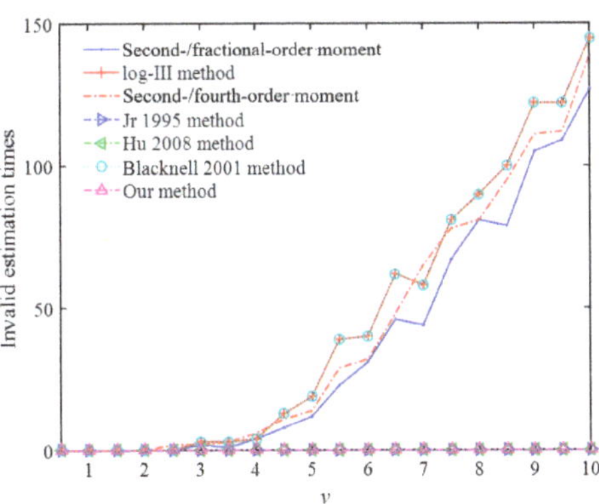

Figure 1. The relationship between v and invalid estimation times in the data of the simulated K-distribution.

As can be seen in Figure 1, when $v \leq 2$, the invalid estimation times of all methods are 0. With the increase in v, the invalid estimation times of references [18,26] and our method are still 0, and the invalid estimation times of the other methods are increasing.

When $v = 3$ and the value of k is [0.1:0.1:3], the estimation efficiency ρ of reference [26] and the proposed method can reach 100%, but the estimation accuracy of the proposed method is better. The relationship between k and the estimation accuracy of the RMSE is shown in Figure 2.

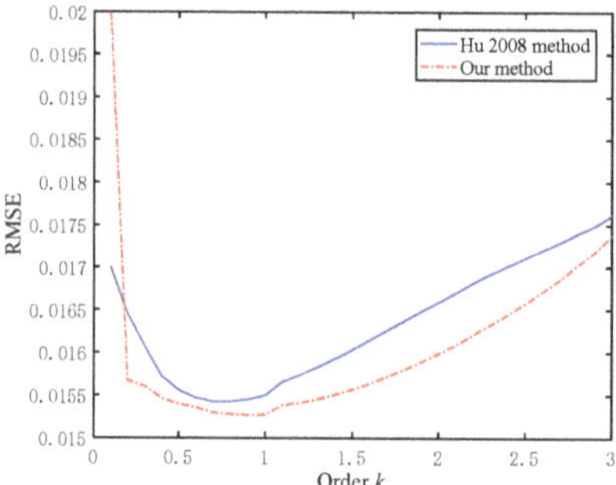

Figure 2. The relationship between the order k and the RMSE in the data of the simulated K-distribution.

When $k > 0.9$, the RMSE will also increase with the increase in k. When $k = 0.4$, some simulation samples in the method of reference [26] begin to appear with a singular value, NaN, generated by the abnormal operation. When $k = 2.9$, the singular value NaN begins to appear in our method. The k is larger and the samples with singular values are greater. With the increase in k, the estimated value of v in some samples will be relatively large, even greater than 100. When $v > 100$, Equation (1) will result in Inf/Inf, and NaN occurs. In Figure 2, the statistical result has eliminated the RMSE with a singular value. The larger the value k, the more samples will be eliminated.

The cumulative distribution function (CDF) can describe the statistical law of random variables and the tail of the K-distribution. For 1000 sets of simulation data with simulation parameters of $v = 3$, $\sigma = 1$ and a sample length of 256 points, the CDF is used to further verify the fitting of each method to the trailing part of the K-distribution. Figure 3 shows the tail fitting of each method and the simulation data. According to the CDF fitting results, except for the method in [18], which has a poor tail-fitting effect, the other methods have a good tail-fitting effect. In general, the method in [18] has the worst performance, followed by the second-/fourth-order estimation method, the method used in [26], the log-III estimation method, and the method used in [27], and the second-/fractional-moment estimation method. The method in this study has the best performance.

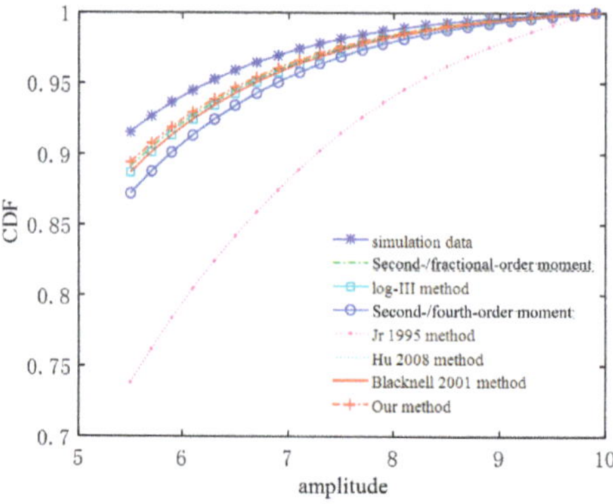

Figure 3. The tail fitting of the K-distribution.

2. The measured data experiment. In order to further verify the effectiveness of the proposed method, 160 groups of the 256 points of measured IPIX data [19] are analyzed, and the parameters of the different methods are estimated, respectively. The IPIX data were collected on the shore of Lake Ontario in Grimsby, Ontario, Canada, including lake surface echoes on different days, at different times, and under different meteorological conditions.

The measured data form is complex, including the I channel and Q channel, and the pulse repetition frequency is 1kHz.

The radar parameters are shown in Table 2.

Table 2. The IPIX radar parameters.

Parameter	Value	Parameter	Value
RF frequency (GHz)	9.39	Unambiguity velocity (m/s)	7.9872
Pulse length (ns)	200	Antenna beamwidth (°)	0.9
PRF (Hz)	1000	Antenna gain (°)	45.7
Azimuth (°)	65.0775–64.8907	Radar height (m)	20
Elevation angle (°)	359.7528–359.7693	Polarization mode	HH, HV, VV, VH
Radar latitude (°)	43.21	Radar longitude (°)	79.60

The IPIX clutter data, which has a file name of 19980205_170935_ANTSTEP.CDF, includes 28 range units, with 60,000 scanning times. The 2D and 3D amplitude of the HH polarization is shown in Figures 4 and 5, respectively.

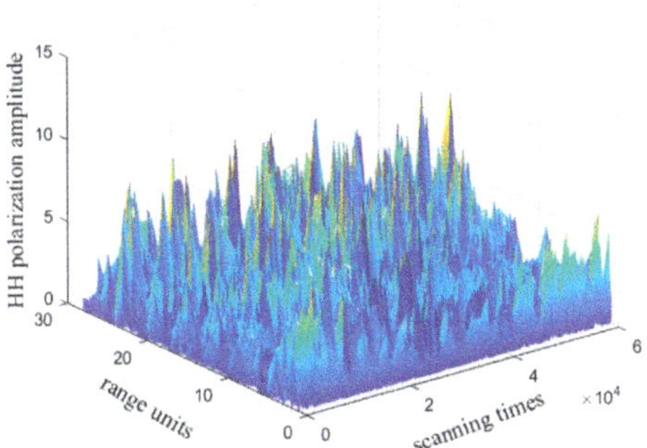

Figure 4. Partial IPIX sea clutter 3D map of HH polarization.

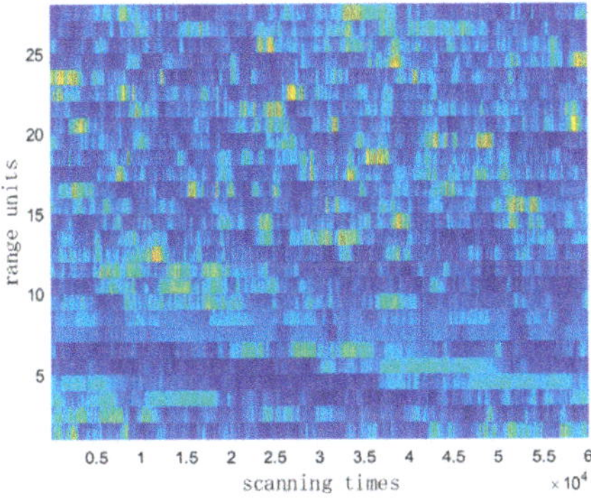

Figure 5. Partial IPIX sea clutter 2D map of HH polarization.

The estimated effective rate ρ is shown in Table 3.

Table 3. The statistics of estimation efficiency and estimation accuracy in measured K-distribution ($k = 0.5$).

Estimation Method	Second-/Fractional-Order Moment	Log-III Method	Second-/Fourth-Order Moment	Ref. [18] Method	Ref. [26] Method	Ref. [27] Method	Our Method
Invalid estimated times	0	0	0	0	0	0	0
NAN times	0	0	0	6	4	0	0
ρ	100%	100%	100%	100%	100%	100%	100%
RMSE	0.0792	0.0786	0.0815	0.0886	0.0809	0.0786	0.0780

As can be seen from Table 3, the method in reference [18] has the worst performance. The performance of our proposed method is the best. The number of all effective estimated samples in this simulation experiment is 154 because there are 6 singular sample values NaN NaN in the method of reference [18]. The reason for generating singular values NaN has been explained previously.

In the measured IPIX data, when the value of k is [0.1:0.1:5], reference [26] and the proposed method can still reach 100% estimated efficiency, but the estimation accuracy of the proposed method is higher. The relationship between k and the RMSE is shown in Figure 6.

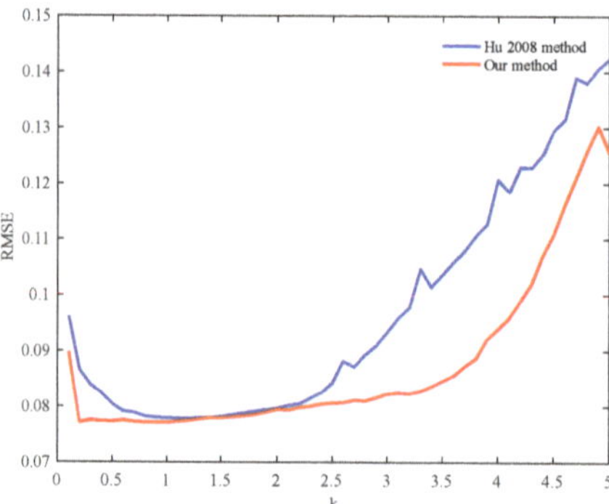

Figure 6. The relationship between k and RMSE in the data of measured K-distribution.

When $k > 0.9$, the RMSE will increase with the increase in k. When $k = 0.4$, some simulation samples in the method of reference [26] begin to appear as the singular value, NaN, generated by the abnormal operation. When $k = 4.1$, the singular value NAN begins to appear in our method. The k is larger, and the samples with singular values are greater. When $k = 5$, 59 of the 160 samples have a singular value in the method of reference [26], more than 36%. When $k = 5$, 23 of the 160 samples have a singular value in our method, more than 14%. In Figure 3, the statistical result has eliminated the RMSE with a singular value.

We analyzed the calculation time of each method to illustrate the calculation complexity. Table 4 shows that the calculation time of the second-/fractional-order method is the minimum. This is because these methods use a special origin moment so that the gamma function is eliminated. The proposed method and the methods in references [18,26] contain the operation of the gamma function, but the calculation time of the proposed method is less than that in the methods of references [18,26].

Table 4. The calculation time in simulated K-distribution ($k = 0.5$).

Estimation Method	Second-/Fractional-Order Moment	Log-III Method	Second-/Fourth-Order Moment	Ref. [18] Method	Ref. [26] Method	Ref. [27] Method	Our Method
Calculation time (s)	1.859×10^{-5}	2.779×10^{-5}	2.669×10^{-5}	6.598×10^{-3}	6.585×10^{-3}	2.586×10^{-5}	6.564×10^{-3}

Through the experiments on the simulation data and measured data, it can be found that by selecting the appropriate order k, the estimation accuracy of the K-distribution can obtain a better value. It is necessary to select an appropriate estimation method according to the actual application.

3. The Extended Application of the Proposed Method

This method also can be applied to two-parameter models, such as log–normal distribution and Weibull distribution.

3.1. Log–Normal Distribution

The k-order origin moment of the log–normal distribution is shown in Equation (15):

$$E[z^k] = \exp(k\mu + 0.5k^2\sigma^2) \tag{15}$$

The process of origin moment derivation is as follows:

$$\frac{\partial E[z^k]}{\partial k} = \left(\mu + k\sigma^2\right)E[z^k] \tag{16}$$

Then, Equation (17) can be obtained from Equations (4) and (16).

$$\frac{E[z^k \ln(z)]}{E[z^k]} = \mu + k\sigma^2 \tag{17}$$

Because:

$$\ln E[z^k] = k\mu + 0.5k^2\sigma^2 \tag{18}$$

By combining Equations (18) and (19), we can obtain:

$$k\frac{E[z^k \ln(z)]}{E[z^k]} - \ln E[z^k] = 0.5k^2\sigma^2 \tag{19}$$

Equation (19) can be further sorted into Equation (20).

$$\hat{\sigma} = \left(\frac{2}{k^2}\left(k\frac{E[z^k \ln(z)]}{E[z^k]} - \ln E[z^k]\right)\right)^{1/2} \tag{20}$$

Substituting the estimated value of the shape parameter $\hat{\sigma}$ into Equation (17), the estimated expression of the scale parameter $\hat{\mu}$ can be obtained in Equation (21).

$$\hat{\mu} = \frac{1}{k}\left(2\ln E[z^k] - k\frac{E[z^k \ln(z)]}{E[z^k]}\right) \tag{21}$$

1. **The simulated data experiment.** The scale parameter is taken randomly in the interval between 0.5 and 0.9. The shape parameter is taken randomly in the range of -1.5 and -0.1. The MLE method [28], ME method [28], estimation method based on sample expectation and variance [29], and the proposed method are used for parameter estimation. The statistics of estimation accuracy of the RMSE are shown in Table 5.

Table 5. The statistics of estimation accuracy in simulated log–normal distribution ($k = 0.5$).

Estimation Method	MLE	ME	Ref. [29]	Our Method
RMSE	0.0943	0.1000	0.1000	0.0943

Table 5 shows that the MLE and the proposed method have the highest estimation accuracy. In the simulation data, when the value of k is [0.1:0.1:3], the relationship between k and the RMSE is shown in Figure 7.

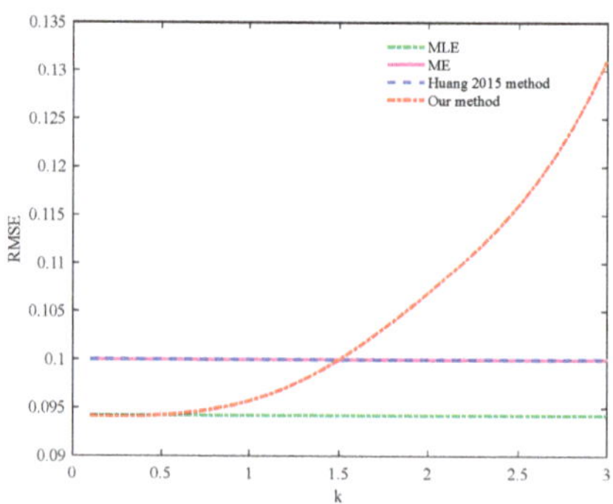

Figure 7. The relationship between k and RMSE in the data of simulated log–normal distribution.

When $k = 0.1{\sim}0.4$, the RMSE of the proposed method is smaller than that in MLE; When $k = 0.5$, the RMSE of the proposed method is the same as that in MLE, both of which are 0.0943. When $k = 1.5$, the RMSE of the proposed method is the same as that of the method in reference [29], both of which are 0.1000. When $k > 0.4$, with the increase in k, the RMSE will also increase. When $k = 1.6$, the RMSE exceeds the method in reference [29].

2. The measured data experiment. In order to further verify the effectiveness of the proposed method, 107 groups of measured IPIX data with a length of 256 points are modeled and analyzed, and the parameters of the different methods are estimated, respectively. The estimation accuracy of the RMSE is shown in Table 6.

Table 6. The statistics of estimation accuracy in measured log–normal distribution ($k = 0.5$).

Estimation Method	MLE	ME	Ref. [29]	Our Method
RMSE	0.0668	0.0723	0.0723	0.0648

Table 6 shows that the proposed method has the highest estimation accuracy. In the measured data, when the value of k is [0.1:0.1:3], the relationship between k and the RMSE is shown in Figure 8.

When $k = 0.1{\sim}1$, the RMSE of the proposed method is smaller than that in MLE; When $k = 0.5$ and 0.6, the RMSE has the minimum value of 0.0648. When $k > 0.6$, the RMSE will also increase with the increase in k. When $k = 1.1$, the RMSE of the proposed method exceeds the method in MLE.

We analyzed the calculation time of each method to illustrate the calculation complexity. Table 7 shows that the calculation time of the ME method is the minimum. The calculation time of the proposed method is longer than the other methods, but the calculation time is acceptable.

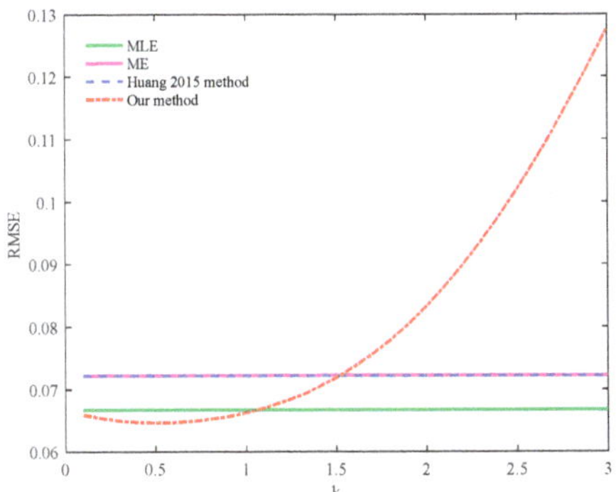

Figure 8. The relationship between k and RMSE in the data of measured log–normal distribution.

Table 7. The calculation time in simulated log–normal distribution ($k = 0.5$).

Estimation Method	MLE	ME	Ref. [29]	Our Method
Calculation time (s)	2.0749×10^{-5}	1.3723×10^{-5}	2.1330×10^{-5}	4.9372×10^{-5}

Through the experiments, it can be found that when the value of k is 0.1~1, the estimation accuracy of the log–normal distribution can obtain a better value.

3.2. Weibull Distribution

The k-order origin moment of the Weibull distribution is shown in Equation (22):

$$E[z^k] = q^k \Gamma\left(1 + \frac{k}{p}\right) \tag{22}$$

The process of the origin moment derivation is as follows:

$$\frac{\partial E[z^k]}{\partial k} = \left(\ln q + \frac{1}{p}\psi\left(1 + \frac{k}{p}\right)\right)E[z^k] \tag{23}$$

Equation (24) can then be obtained from Equations (4) and (23):

$$\frac{E[z^k \ln(z)]}{E[z^k]} = \ln q + \frac{1}{p}\psi\left(1 + \frac{k}{p}\right) \tag{24}$$

It is found that taking the logarithm of $E[z^k]$ can also obtain the term $\ln q$, that is:

$$\ln E[z^k] = k \ln q + \ln \Gamma\left(1 + \frac{k}{p}\right) \tag{25}$$

By combining Equations (24) and (25), we can obtain:

$$k\frac{E[z^k \ln(z)]}{E[z^k]} - \ln E[z^k] = \frac{k}{p}\psi\left(1 + \frac{k}{p}\right) - \ln \Gamma\left(1 + \frac{k}{p}\right) \tag{26}$$

Equation (26) is a nonlinear equation describing the shape parameter p. The estimated value of the shape parameter p can be obtained by solving Equation (26). We use the trust-region–dogleg [16] algorithm to solve the nonlinear equation. The complexity of the solution technology is provided in the following theoretical and experimental analysis.

The estimated value of the shape parameter p is substituted into Equation (24), and the estimated expression of the scale parameter q can be obtained in Equation (27).

$$q = \left(E[z^k] / \Gamma\left(1 + \frac{k}{p}\right) \right)^{1/k} \tag{27}$$

1. **The simulated data experiment.** The scale parameter is taken randomly in the interval between 0.3 and 3.2. The shape parameter is taken randomly in the range of 2.1–7.7. The MLE method [30], ME method [30], MENON estimation method [31], and the proposed methods are used for parameter estimation. The statistics of estimation accuracy of the RMSE are shown in Table 8.

Table 8. The statistics of estimation accuracy in simulated Weibull distribution ($k = 0.5$).

Estimation Method	MLE	ME	MENON Method	Our Method
RMSE	0.0727	1.0434	0.0796	0.0688

Table 8 shows that the estimation accuracy of the proposed method is the highest. When the value of k is [0.3:0.1:4], the relationship between k and the RMSE is shown in Figure 9.

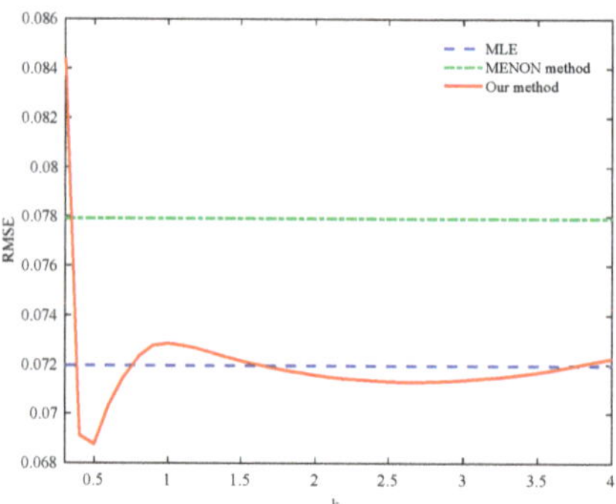

Figure 9. The relationship between k and RMSE in the data of simulated Weibull distribution.

Through our analysis, when $k = 0.4\sim0.7$ and $k = 1.7\sim3.7$, the RMSE of the proposed method is slightly smaller than the MLE method. When $k = 0.5$, the RMSE has the minimum value of 0.0688.

2. **The measured data experiment.** In order to further verify the effectiveness of the proposed method, 100 groups of IPIX data with a length of 256 points are modeled and analyzed, and the parameters of the different methods are estimated, respectively. The estimation accuracy of the RMSE is shown in Table 9.

Table 9. The statistics of estimation accuracy in measured Weibull distribution ($k = 1$).

Estimation Method	MLE	ME	MENON Method	Our Method
RMSE	0.1517	2.0937	0.1664	0.1510

Table 9 shows that the estimation accuracy of our method is the best. In the measured data, when the value of k is [0.3:0.1:4], the relationship between the RMSE and k is shown in Figure 10.

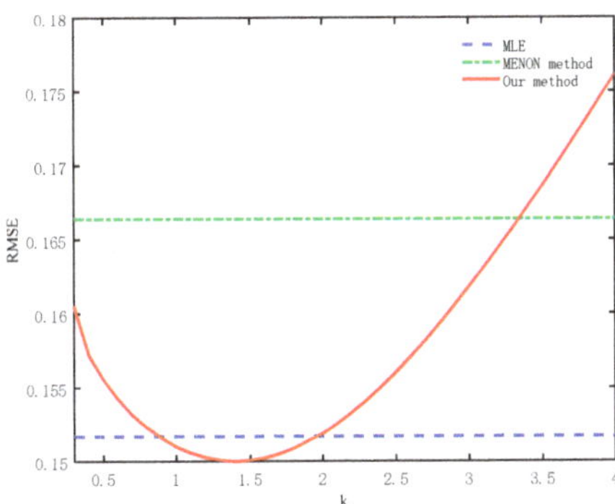

Figure 10. The relationship between k and RMSE in the data of measured Weibull distribution.

When $k = 0.9{\sim}1.9$, the RMSE of the proposed method is smaller than that of the MLE method. When $k = 1.4$, the RMSE has the minimum value of 0.1500. When $k > 1.4$, the RMSE will increase with the increase in k.

We analyzed the calculation time of each method to illustrate the calculation complexity. Table 10 shows that the calculation time of the MENON method is minimum. The calculation time of the proposed method is similar to that of the MLE method.

Table 10. The calculation time in simulated Weibull distribution ($k = 0.5$).

Estimation Method	MLE	ME	MENON Method	Our Method
Calculation time (s)	6.5399×10^{-3}	8.6569×10^{-3}	2.9555×10^{-5}	5.8162×10^{-3}

Through the experiments using the simulation data and measured data, it can be found that the estimation accuracy of the Weibull distribution can obtain a better value by selecting an appropriate value of k.

4. Conclusions

Both the simulation and measured data experiments indicate that when the order k is an appropriate value, such as a smaller value, the origin moment derivation method can obtain an ideal estimation result and accuracy with a high probability. This conclusion is the same as that in reference [32], which points out that the high-order moment in the parameter estimation is sensitive to data, and the low-order moment should be selected as much as possible. The commonly used ME methods have the use of high-order moments, such as first-order, second-order, and fourth-order moments. These methods use different orders of k to solve simultaneous equations for convenience. The data are amplified using different orders, and the joint operation of different orders introduces different multiples of calculation errors. The derivation of the proposed method is based on a single order k; it will not introduce the error of different order operations. Thus, the calculation errors are smaller while the order k is an appropriate value. When k is given, the computational complexity of the proposed method is related to the PDF. The more complex the PDF, the greater the computational complexity, e.g., K-distribution and Weibull distribution contain complex gamma function and Psi function operations. The log–normal distribution has the operation of the square root. The proposed method is suitable for solving the parameter estimation problem in the form of the exponential function and the gamma function.

Author Contributions: Conceptualization, L.Y. and W.Y.; methodology, L.Y. and Y.L.; software, L.Y. and X.S.; validation, L.Y.; data curation, L.Y.; writing—review and editing, L.Y. and X.S.; supervision, Q.S.; All authors have read and agreed to the published version of the manuscript.

Funding: This work was supported by the National Natural Science Foundation of China (61921001, 61871384).

Data Availability Statement: Not applicable.

Conflicts of Interest: The authors declare there is no conflict of interest.

References

1. Gong, M.; Cao, Y.; Wu, Q. A neighborhood-based ratio approach for change detection in SAR images. *IEEE Geosci. Remote Sens. Lett.* **2011**, *9*, 307–311. [CrossRef]
2. Hakim, W.L.; Achmad, A.R.; Eom, J. Land subsidence measurement of Jakarta coastal area using time series interferometry with Sentinel-1 SAR data. *J. Coast. Res.* **2020**, *102*, 75–81. [CrossRef]
3. Kang, M.S.; Baek, J.M. Efficient SAR imaging integrated with autofocus via compressive sensing. *IEEE Geosci. Remote Sens.* **2022**, *19*, 4514905. [CrossRef]
4. Sayama, S.; Sekine, M. Amplitude statistics of ground clutter included town using a millimeter wave radar. *IEICE Trans. Commun.* **2003**, *26*, 829–836.
5. Sayama, S.; Sekine, M. Log–Normal, Log–Weibull and K–distributed sea clutter. *IEICE Trans. Commun.* **2002**, *E85–B*, 1375–1381.
6. Ozgun, O.; Kuzuoglu, M. Physics-based modeling of sea clutter phenomenon by a full-wave numerical solver. *Wave Motion Vol.* **2022**, *109*, 102872. [CrossRef]
7. Shao, Z.; Ji, W.; Qian, C.; Yang, Y. Ship detection for SAR images with sea clutter models estimation. In Proceedings of the 2021 2nd China International SAR Symposium, Shanghai, China, 3–5 November 2021; pp. 1–4.
8. Greco, M.; Bordoni, F.; Gini, F. X–band sea–clutter nonstationarity: Influence of long waves. *IEEE J. Ocean. Eng.* **2004**, *29*, 269–283. [CrossRef]
9. Chan, H.C. Radar sea—Clutter at low grazing angles. *Radar Signal Process. IEEE Proc.* **1990**, *137*, 102–112. [CrossRef]
10. Walker, D. Doppler modelling of radar sea clutter. *IEEE Proc.—Radar Sonar Navig.* **2001**, *148*, 73–80. [CrossRef]
11. Watts, S. Radar detection prediction in K–distributed sea clutter and thermal noise. *Aerosp. Electron. Syst.* **1987**, *23*, 40–45. [CrossRef]
12. Lauritzen, S.; Uhler, C.; Zwiernik, P. Maximum likelihood estimation in Gaussian models under total positivity. *Ann. Stat.* **2019**, *47*, 1835–1863. [CrossRef]
13. Du, X.; Liao, K.; Shen, X. Secondary radar signal processing based on deep residual separable neural network. In Proceedings of the 2020 IEEE International Conference on Power, Intelligent Computing and Systems, Shenyang, China, 28–30 July 2020; pp. 12–16.
14. Abraham, D.A.; Lyons, A.P. Novel physical interpretations of K–distributed reverberation. *Ocean. Eng. IEEE J.* **2002**, *27*, 800–813. [CrossRef]
15. Iskander, D.R.; Zoubir, A.M. Estimation of the parameters of the K-distribution using higher order and fractional moments. *IEEE Transcations Aerosp. Electron. Syst.* **1999**, *35*, 1453–1457. [CrossRef]
16. Zhang, J.Z.; Xu, C.X. Trust region dogleg path algorithms for unconstrained minimization. *Ann. Oper. Res.* **1999**, *87*, 407–418. [CrossRef]
17. Bellavia, S.; Macconi, M.; Pieraccini, S. Constrained Dogleg methods for nonlinear systems with simple bounds. *Kluwer Acad. Publ.* **2012**, *53*, 771–794. [CrossRef]
18. Marier, L., Jr. Correlated K-Distributed clutter generation for radar detection and track. *IEEE Trans. Aerosp. Electron. Syst.* **1995**, *31*, 568–580. [CrossRef]
19. Xu, X.H. Entropy metrics of radar signatures of sea surface scattering for distinguishing targets. *Remote Sens.* **2021**, *13*, 3950.
20. Xu, W.; Chen, Y.S. An estimation method for K–distribution clutter parameters. *Shipboard Electron. Count. Meas.* **2013**, *36*, 82–84. (In Chinese).
21. Tang, W.; He, H.; Gunzler, D. Kernel smoothing density estimation when group membership is subject to missing. *J. Stat. Plan. Inference* **2012**, *142*, 685–694. [CrossRef] [PubMed]
22. Zhang, W.; Wen, A.; Wang, Q.; Li, Y.Y. Simple and flexible photonic microwave waveform generation with low RMSE of square waveform. *IEEE Photonics Technol. Lett.* **2019**, *31*, 829–832. [CrossRef]
23. Ossant, F.; Patat, F.; Lebertre, M.; Teriierooiterai, M.L.; Pourcelot, L. Effective density estimators based on the K-distribution: Interest of low and fractional order moments. *Ultrason. Imaging* **1988**, *20*, 243–259. [CrossRef]
24. Deng, Z.H. Statistical Modeling of Sea Clutter Based on Measured Data. Ph.D. Thesis, Xidian University, Xi'an, China, 2014. (In Chinese).
25. Abraham, D.A.; Lyons, A.P. Reliable methods for estimating the K-distribution shape parameter. *IEEE J. Ocean. Eng.* **2010**, *35*, 288–302. [CrossRef]
26. Hu, W.L.; Wang, Y.L.; Wang, S.Y. Estimation of the parameters of K–distribution based on zrlog(z) Expectation. *J. Electron. Inf. Technol.* **2008**, *30*, 203–205. (In Chinese) [CrossRef]
27. Blacknell, D.; Tough, R. Parameter estimation for the K–distribution based on [zlog(z)]. *IEEE Proc Radar Sonar Navig.* **2001**, *148*, 309–312. [CrossRef]

28. Bílková, D. Lognormal distribution and using L-moment method for estimating its parameters. *Int. J. Math. Model. Methods Appl. Sci.* **2012**, *6*, 30–44.
29. Huang, C. Parameter estimation of the lognormal distribution. *Stud. Coll. Math.* **2015**, *18*, 19–20. (In Chinese)
30. Bhattacharya, P.; Bhattacharjee, R. A study on Weibull distribution for estimating the parameters. *J. Appl. Quant. Methods* **2010**, *5*, 234–241. [CrossRef]
31. Menon, M.V. Estimation of the shape and scale parameters of the Weibull distribution. *Technometrics* **1963**, *15*, 175–182. [CrossRef]
32. Li, Q.L.; Yin, Z.Y.; Zhu, X.Q. *Measurement and Modeling of Radar Clutter from Land and Sea*; National Defense Industry Press: Beijing, China, 2017; pp. 262–265. (In Chinese)

 *remote sensing*

Article

An Efficient Channel Imbalance Estimation Method Based on Subadditivity of Linear Normed Space of Sub-Band Spectrum for Azimuth Multichannel SAR

Zongxiang Xu [1,2], Pingping Lu [1,2,*], Yonghua Cai [1,2], Junfeng Li [1,2], Tianyuan Yang [1], Yirong Wu [1,2] and Robert Wang [1,2]

[1] National Key Laboratory of Microwave Imaging Technology, Aerospace Information Research Institute, Chinese Academy of Sciences, Beijing 100190, China
[2] School of Electronic, Electrical and Communication Engineering, University of Chinese Academy of Sciences, Beijing 100049, China
* Correspondence: lupp@aircas.ac.cn

Abstract: Azimuth multichannel (AMC) technology is one of the mainstream technical approaches to realize high-resolution wide-swath (HRWS) imaging. It has been successfully applied to several synthetic aperture radar (SAR) satellites in orbit. However, the inevitable imbalance between channels can seriously affect the azimuth reconstruction spectrum, introducing ghost targets into the final imaging results and degrading the SAR image quality. In order to address this issue, this paper proposes a channel imbalance estimation method based on minimizing the sum of the sub-band norm (MSSBN) for the reconstructed azimuth spectrum. First, the amplitude imbalance is calibrated in the range-Doppler domain. Then, the echo in each channel with phase imbalances is reconstructed by filters separately and converted to the range-Doppler domain. Finally, the global optimization algorithm is used to find the phase error of each channel so that the reconstructed postcompensation spectrum has the smallest sub-band spectrum norm sum. By two-dimensional blocking, this method can also estimate the space-varying phase imbalance in the range dimension and the time-varying phase imbalance in the azimuth dimension. Experimental results using simulated and actual AMC SAR data from the GF-3 system validate the proposed algorithm's high estimation accuracy and excellent computational efficiency.

Keywords: azimuth multichannel (AMC) synthetic aperture radar (SAR); high-resolution wide-swath (HRWS); channel imbalance estimation; minimizing the sum of sub-band norm (MSSBN)

Citation: Xu, Z.; Lu, P.; Cai, Y.; Li, J.; Yang, T.; Wu, Y.; Wang, R. An Efficient Channel Imbalance Estimation Method Based on Subadditivity of Linear Normed Space of Sub-Band Spectrum for Azimuth Multichannel SAR. *Remote Sens.* **2023**, *15*, 1561. https://doi.org/10.3390/rs15061561

Academic Editors: Gang Xu, Lan Du and Haipeng Wang

Received: 6 January 2023
Revised: 27 February 2023
Accepted: 9 March 2023
Published: 13 March 2023

1. Introduction

Synthetic aperture radar (SAR) is an all-day, all-weather active microwave remote sensing system, which has become an indispensable means for Earth observation [1]. High resolution can provide finer feature information, and wide imaging can provide a broader observation scene. Therefore, high-resolution wide-swath (HRWS) imaging is a significant development direction for a modern SAR system [2–8]. However, the conventional SAR system cannot realize both azimuth high-resolution and range wide-swath imaging due to the minimum antenna area constraint [4,9–12]. At present, the basic working modes of SAR systems, such as ScanSAR [13], Spotlight SAR [14], TOPS [15], and Mosaic [16], achieve a compromise between azimuth resolution and range swath by allocating and adjusting the synthetic illumination time of ground scene. However, they cannot fundamentally solve the contradiction between imaging swath and resolution. The displaced phase center multiple azimuth beams (DPC-MAB) SAR system [9,10,12,17,18] arranges multiple phase centers along the azimuth direction with equal spacing to receive scene echoes, which equivalently increases the sampling rate in time dimension by increasing sampling rates in the spatial dimension. Then, the alias-free full azimuth spectrum is reconstructed by the digital beamforming (DBF) technique to alleviate constraints of azimuth resolution and range swath on pulse repetition frequency (PRF) [2–4,6,12,17]. The DPC-MAB technique has become one of the effective means for realizing HRWS SAR imaging and has been successfully deployed on TerraSAR-X [19], RADARSAT-2 [20], ALOS-2 [21], GF-3 [22,23], LT-1 [24], and other azimuth multichannel (AMC) SAR systems.

Theoretically, the HRWS SAR image can be well focused after AMC signal reconstruction. However, due to error factors, such as temperature variations, antenna radiation pattern difference, and undesirable receiver components, there are often inconsistencies in amplitude, phase, range sampling time, and antenna phase center locations between channels. These imbalances among channels, especially amplitude and phase inconsistencies, can significantly reduce the suppression effect of spectrum ambiguity in subsequent reconstruction processing algorithms, which, in turn, introduces virtual targets on the final image and seriously degrades the quality of SAR imaging. Therefore, channel imbalance consistency calibration is a crucial step in AMC SAR data processing [19,25]. Amplitude error can be accurately corrected by the channel balancing technique [26]. Generally speaking, range sampling time imbalance (RSTI) can be accurately measured and compensated by an internal calibration system. Meanwhile, the azimuth time-domain cross-correlation (ATC) method can also effectively estimate the RSTI from echo data [27]. For the spaceborne SAR system, the position error of the antenna phase center is tiny, the position error along the track can be ignored, and the radial position error can be regarded as a tiny phase error [28–30]. Thus, the main task of channel consistency calibration is the efficient estimation and calibration of phase imbalance between channels.

For the phase imbalance problem between channels, the available channel error calibration methods are divided into two categories in this paper: processing for signal data and processing in the image domain, both of which have various advantages and disadvantages. The approaches based on signal data processing are mainly the following: The ATC method is operationally efficient in estimating the RSTI and phase error [27]. However, the system Doppler center frequency accuracy affects phase error estimation results in more. Although the signal subspace (SSP) method has more accurate estimation results [31], it needs eigen decomposition of the covariance matrix to obtain signal subspace and noise subspace, so it has much computation and is affected by the SNR of echo data. The minimum variance distortionless response (MVDR) method is also an accurate phase imbalance estimation method [32] but is affected by Doppler center frequency estimation accuracy. However, it has a lower computational complexity compared with the SSP method. Then, methods that perform processing in the image domain, including the image weighted minimum entropy (WME) method [33,34], the image least L^1-norm (LLN) method [35], and the maximum normalized image sharpness (MNIS) method [30], are both characterized by high estimation accuracy and robustness, but generally have high computational complexity.

Although existing methods have made significant progress, reducing computational complexity while estimating channel phase errors accurately and robustly remains a challenge due to the complexity of a scene and the need for fast processing in real time. In order to address this challenge, a novel method of minimizing the sum of sub-band norms (MSSBN) for reconstruction azimuth spectrums is proposed to estimate phase errors of the AMC SAR system. Specifically, the reconstructed azimuth spectrum by the DBF technique has the smallest sum of sub-band spectrum norm when channel phases are consistent. In this paper, by modeling phase imbalance estimation as an optimization problem, typical global optimization algorithms, such as optima quest of nonlinear program (OQNLP) [36] and particle swarm algorithms (PSO) [36] can be used to find phase imbalances by the MSSBN method. Meanwhile, the proposed algorithm limits the influence of range samples by range pulse compression, and range-variation phase errors can be fitted by estimating the phase error at different sampling locations in the range direction. The MSSBN method does not require more spatial degrees of freedom for phase error estimation and has a broader range of application scenarios than the SSP method. Meanwhile, the MSSBN method has a higher estimation accuracy than the ATC method and is unaffected by the

Doppler center frequency estimation accuracy. Since there is no need to image the reconstructed signal data of each channel, the MSSBN method has higher computational efficiency than the LLN method. The phase information of all scenes is converted to the same Doppler bandwidth using pulse compression and azimuth fast Fourier transform (FFT), and computational efficiency can be further improved by downsampling for azimuth spectrum without degrading estimation accuracy.

The remainder of this paper is organized as follows: Section 2 introduces the signal model and reconstruction algorithm of the AMC SAR system, then analyzes the effect of channel imbalance and performs precompensation for phase error estimation. Section 3 details the derivation of the proposed algorithm. Section 4 reports the simulation and actual experimental results with a discussion. Then, conclusions are provided in Section 5.

2. Material

2.1. Signal Model

The AMC SAR system sets the azimuth antenna as a transmitting subaperture and multiple receiving subapertures. Multiple echo signals can be obtained by transmitting a chirp signal once in a pulse repetition time (PRT) [3]. The echo signals received by subchannels simultaneously can be reconstructed to an azimuth signal sampled sequentially in the time dimension after reconstruction by a filter. The multichannel technique reduces the PRF required by azimuth resolution to times of channel number. It compensates for the lack of sampling in the time dimension by increasing the spatial dimension, thus extending the swath while ensuring that azimuth resolution remains unchanged. Figure 1a shows the imaging geometry of the AMC SAR system, with red boxes indicating transmit and receive subapertures, all other boxes indicating receive subapertures, and a circle indicating the effective phase center (EPC).

For an AMC SAR system, as shown in Figure 1a, the separation between adjacent two channels is d. According to the principle, the distance between the mth and m_{ref}th channels (reference channel) is $\Delta x_m = (m - m_{ref})d$, where $m = 1, 2, \ldots, M$ denotes the number of channels. V_s, V_r, and η represent the orbital flight speed, equivalent radar speed, and azimuth slow time of the AMC SAR platform. As shown in Figure 1b, taking the first channel of the antenna as the transmitting channel, the distance from the ground target to the mth EPC and propagation history of the echo signal received by the mth channel can be expressed, respectively, as

$$R_{em}(\eta) = \sqrt{R_0^2 + \left(V_r\eta - \frac{\Delta x_m}{2}\right)^2} \tag{1}$$

$$R_m(\eta) = R_{rm}(\eta) + R_{tm}(\eta) = \sqrt{R_0^2 + (V_r\eta - \Delta x_m)^2} + \sqrt{R_0^2 + (V_r\eta)^2} \tag{2}$$

where R_0 donates the nearest slant range between the platform and ground target. According to Equation (1), the azimuth time dimension echo signal of the mth EPC in self-transmit self-received mode can be expressed as

$$s_{em}(\eta) = w_a(\eta - \eta_c) \times \exp\left\{-j\frac{4\pi}{\lambda}R_{em}(\eta)\right\} \tag{3}$$

where η_c, λ, and w_a denote the azimuth center time, wavelength, and azimuth signal envelop, respectively. The difference between propagation histories of echo signals received by the mth channel and the mth EPC channel is $-\Delta x_m^2/4R_0$ after Taylor expansion of Equations (1) and (2), respectively. Therefore, the azimuth echo signal of the mth channel can be expressed as

$$s_m(\eta) = s_{em}(\eta) \times \exp\left\{-j\frac{\pi\Delta x_m^2}{2\lambda R_0}\right\} \tag{4}$$

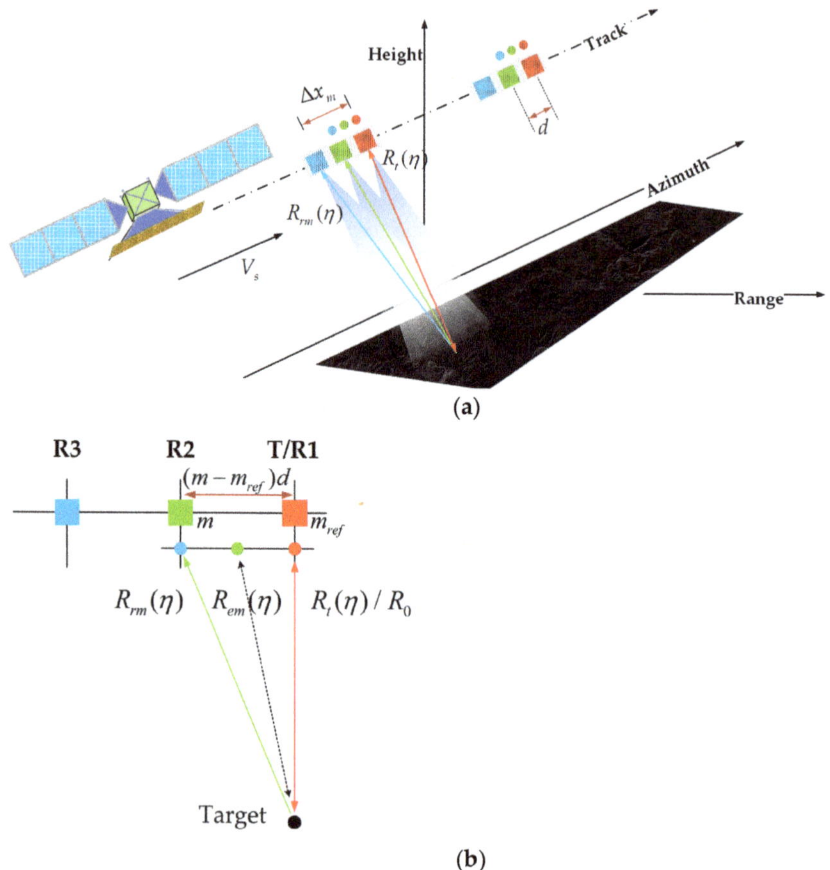

Figure 1. Illustration of the AMC SAR system. (**a**) Imaging geometry of AMC SAR; (**b**) EPC distribution.

The azimuth spectrum of the reference channel is denoted by $S_{ref}(f_\eta)$. Assuming that its spectrum is band-limited, the PRF design of an azimuth multichannel SAR system satisfies $(M-1)\text{PRF} \leq B_d \leq M\text{PRF}$, where B_d is the bandwidth of the azimuth unambiguity spectrum. Therefore, according to the "stop-go-stop" hypothesis, the echo signal is sampled in the azimuth direction at a PRF much smaller than the Doppler bandwidth, and the number of sampling points in the azimuth direction of each channel cannot satisfy the Nyquist sampling theorem, so it works in the sub-Nyquist sampling state. The signal sampled at each frequency point is the superposition of a signal with the frequency f_η with signals whose frequency is apart from f_η with an integer multiple of PRF interval. The alias-free full azimuth spectrum of the reference channel is denoted by S_0, whose sampling frequency is $N\text{PRF}$, and $N(N \leq M)$ represents the Doppler ambiguity numbers. The azimuth spectrum is equally divided into N successive sub-bands of the length PRF. Therefore, the signal of the reference channel in the range-Doppler domain is aliased as

$$S_{ref}(f_\eta) = \sum_{n=1}^{N} S_0(f'_\eta + (n-1)\text{PRF}) \tag{5}$$

where $f_\eta \in [-\text{PRF}/2, \text{PRF}/2]$ denotes the azimuth sampling frequency range, $f'_\eta \in [-N\text{PRF}/2, -N\text{PRF}/2 + \text{PRF}]$ belongs to the first sub-band, and for convenience of subsequent discussion, $f'_{\eta,n}$ is used to represent the frequency variable of the nth sub-band. The EPC of the reference channel coincides with the reference channel phase center. Therefore, by combining Equations (4) and (5), the range-Doppler domain aliased signal of each channel is derived as

$$
\begin{aligned}
S_m(f_\eta) &= \sum_{n=1}^{N} S_0(f'_\eta + (n-1)\mathrm{PRF}) \\
&\times \exp\left\{-j2\pi\Delta x_m(f'_\eta + (n-1)\mathrm{PRF})\right\} \\
&\times \exp\left\{-j\Delta\varphi_m\right\}
\end{aligned}
\tag{6}
$$

where $\Delta\eta_m$ and $\Delta\varphi_m$, respectively, represent an azimuth time delay and a constant phase shift of the mth channel relative to the reference channel, which can be derived as

$$
\Delta\eta_m = \frac{1}{2}\frac{\Delta x_m}{V_r}
\tag{7}
$$

$$
\Delta\varphi_m = \frac{1}{2}\frac{\pi\Delta x_m^2}{\lambda R_0}
\tag{8}
$$

Therefore, signal reconstruction must recover the complete unaliased spectrum for the subsequent SAR imaging operation from the aliased azimuth signals.

2.2. Signal Reconstruction

A multichannel data reconstruction theory is mathematically given for the problem of nonuniform sampling of band-limited signals [3]. A signal with the bandwidth B_d is passed through N pre-filters and sampled at a sampling rate of $\mathrm{PRF} \geq B_d/N$. Then the N channel sampled signal can then be recovered to the original lowpass signal by designing a suitable reconstruction filter.

Comparing Equations (5) and (6), the azimuth spectrum of the mth channel can be obtained from a single-channel SAR spectrum by time-shifting and multiplying by a constant phase term. The last two terms of Equation (6) are transfer functions of the multichannel system, which can be recovered from the original equivalent single-channel signal by building a digital reconstruction filter. In a matrix form, the sampled signals of N channels can be reformulated as

$$
\mathbf{S} = \mathbf{\Lambda H S_0}
\tag{9}
$$

with

$$
\mathbf{S} = \left[S_1(f_\eta), \ldots, S_N(f_\eta)\right]^T
\tag{10}
$$

$$
\mathbf{\Lambda} = \mathrm{diag}\left[\exp\left\{-j2\pi f'_\eta\Delta\eta_1\right\}, \ldots, \exp\left\{-j2\pi f'_\eta\Delta\eta_N\right\}\right]
\tag{11}
$$

$$
\mathbf{H} = [\mathbf{h}_1, \ldots, \mathbf{h}_n, \ldots, \mathbf{h}_N]
\tag{12}
$$

$$
\begin{aligned}
\mathbf{h}_n &= \exp\left\{-j2\pi(n-1)PRF \times [\Delta\eta_1, \ldots, \Delta\eta_N]^T\right\} \\
&\times \exp\left\{-j[\Delta\varphi_1, \ldots, \Delta\varphi_N]^T\right\}
\end{aligned}
\tag{13}
$$

$$
\mathbf{S_0} = \left[S_0\left(f'_{\eta,1}\right), \ldots, S_0\left(f'_{\eta,N}\right)\right]^T
\tag{14}
$$

where the superscript $\{\cdot\}^T$ represents the matrix transpose operation, and $\mathrm{diag}\{\cdot\}$ denotes the operator that rewrites a vector into a diagonal matrix with the vector as a diagonal element. $\mathbf{\Lambda}$ is a diagonal matrix dependent on the Doppler frequency, and $\mathbf{H}$ is a constant matrix independent of the Doppler frequency. $\mathbf{\Lambda H}$ forms the transfer function of the multi-channel SAR system.

From Equation (12), it can be seen that the main diagonal elements of the matrix $\mathbf{\Lambda}$ are in complex exponential form. Hence, the matrix $\mathbf{\Lambda}$ is invertible, and its inverse matrix $\mathbf{\Lambda}^{-1}$ is its conjugate transpose the matrix $\mathbf{\Lambda}^H$. Both sides of Equation (10) are multiplied by $\mathbf{\Lambda}^H$ simultaneously on the left to obtain a new system transfer equation as

$$
\mathbf{\Lambda}^H\mathbf{S} = \mathbf{H S_0}
\tag{15}
$$

The signal reconstruction process is an inverse process shown in the above equation, where $\mathbf{H}$ can be considered as a new system transfer function independent of the Doppler frequency, and let $\mathbf{S_1}=\mathbf{\Lambda}^H\mathbf{S}$; then Equation (15) can be expressed as

$$\mathbf{S_1}=\mathbf{H}\mathbf{S_0} \tag{16}$$

From the $\mathbf{H}$ matrix, it is known that its inverse the matrix $\mathbf{P}$ is also a constant matrix independent of the Doppler frequency. By left multiplying the matrix $\mathbf{P}$ on both sides of Equation (16) simultaneously, the data reconstruction process of the multichannel SAR system can be expressed as

$$\mathbf{S_0}=\mathbf{P}\mathbf{S_1} \tag{17}$$

The aliased spectrum $\mathbf{S_1}$ of each channel is recovered to the unaliased spectrum $\mathbf{S_0}$ through a reconstruction filter $\mathbf{P}$. After rearranging each sub-band spectrum in the order of frequency, a full unaliased spectrum is obtained. The AMC SAR system lowers system PRF to achieve extensive bandwidth imaging in the range direction, uses signal reconstruction and spectrum rearrangement algorithms to increase PRF equivalently to solve the azimuth ambiguity problem, and then realizes HRWS imaging.

In order to introduce the phase imbalance estimation method proposed in this paper by the MSSBN, this section analyzes the effect of channel imbalance on the Doppler spectrum and precalibrates the amplitude imbalance and RSTI.

2.3. Channel Imbalance Analysis

Ideally, the effect of channel imbalance should not be considered in the reconstruction of multichannel signals. However, in an actual operation of a spaceborne AMC SAR system, under the influence of multiple factors [37], such as manufacturing process, system implementation mode, and operating environment, relative errors will inevitably appear between echoes of each channel. The change in antenna radiation characteristics will lead to other deviations in amplitude and phase of echo signals received by each channel. The change in the position of the antenna subaperture along the track will lead to additional phase imbalances. Moreover, the RSTI is caused by the inconsistent transmission characteristics of each channel receiver. Therefore, the channel mismatch of the system needs to be preprocessed before signal reconstruction. First, the influence caused by various imbalances is analyzed, and the channel imbalance model is established. Since the attitude error of the antenna [38] varies with azimuth time, the actual phase imbalance is time varying in azimuth and space varying in range. However, in an image, the temporal and spatial variation of the phase imbalance is minimal, which has little impact on the product quality. This paper considers the phase imbalance of the channel to be a constant in an image. Consequently, the channel imbalance is as

$$s_{me}(\tau,\eta) = \Delta a_m s_m(\tau - \Delta\tau_m, \eta - \frac{\Delta d_m}{2V_r})\exp\{j\Delta\theta_m\} \tag{18}$$

where τ, Δa_m, $\Delta\tau_m$, Δd_m, and $\Delta\theta_m$ represent the range fast time, amplitude imbalance, RSTI, antenna location imbalance, and phase imbalance of other channels relative to the reference channel, respectively.

As shown in Section 2.2, the signal reconstruction filter matrix of the AMC SAR system is designed by inverting the transfer function matrix of the system. However, when the imbalances mentioned above are introduced into the echo signal of each channel, the reconstruction filter becomes mismatched with the echo signal. Using mismatched reconstruction filters to process the spectrum of each channel will introduce other sub-band spectrum components in the reconstruction process of each sub-band spectrum, resulting in aliasing of the reconstructed spectrum and, ultimately, leading to ghosts in the imaging results. Therefore, mismatch of the echo signal between channels is equivalent to mismatch of the reconstruction filter, which can be analyzed by studying the effect of channel errors on the reconstructed filter. The parameters of the AMC SAR system are shown in Table 1, and the influence of channel errors will be analyzed through simulation.

Table 1. Simulation system parameters.

Parameter	Symbol	Value	Unit
Platform velocity	V_r	7563	m/s
Carrier frequency	f_0	5.4	GHz
Signal bandwidth	B_r	300	MHz
Signal pulse duration	T_p	2.5	μs
Nearest slant range	R_0	900	km
Azimuth antenna length	L_{az}	3.75×3	m
Range sampling frequency	F_r	360	MHz
Azimuth sampling frequency	F_a	1429	Hz
Number of channels	M	3	\

2.3.1. Influence of Amplitude Imbalance

The simulation parameters in Figure 2 are shown in Tables 1 and 2. Figure 2 gives the effect of amplitude imbalance on the signal reconstruction of the AMC SAR system through simulation. From Figure 2a, it can be seen that when there is no amplitude imbalance, the reconstruction filter forms a deep null at the ambiguous frequency component, which can effectively suppress the ambiguous frequency component. When the amplitude imbalance exists in the channel, the null position of the reconstruction filter is shifted to both sides of the ±PRF frequency point. At the same time, the shallower null depth introduces ambiguous spectrum energy, leading to the degradation of the reconstruction filter performance, which is consistent with the result in Figure 2b that the spectrum energy of reconstruction with error is higher than that of reconstruction without error.

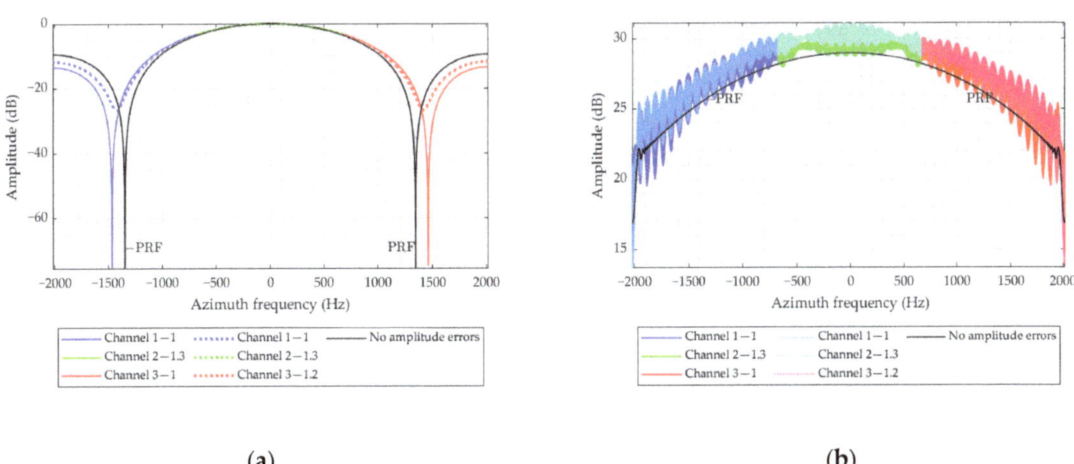

(**a**) (**b**)

Figure 2. Effect of channel amplitude imbalance. (**a**) Effect of channel amplitude imbalance on the reconstructed filter. The solid black line indicates the response of the reconstruction filter without amplitude error. The colored solid line indicates the reconstruction filter response with channel 2 set to 1.3 times the reference channel amplitude. The colored dashed lines indicate the reconstructed filter responses for channels 2 and 3 with 1.3 and 1.2 times the reference channel amplitude, respectively. (**b**) Effect of channel amplitude imbalance on the reconstructed spectrum.

Table 2. Amplitude imbalance.

	Channel 1	Channel 2	Channel 3
No amplitude errors	1	1	1
Amplitude errors 1	1	1.3	1
Amplitude errors 2	1	1.3	1.2

2.3.2. Influence of Phase Imbalance

The simulation parameters in Figure 3 are shown in Tables 1 and 3. Figure 3a shows that when the channel phase imbalance is consistent, the filter can completely pass the spectrum at the target frequency while forming a very deep null at the aliasing frequency, thus effectively suppressing the ambiguous spectrum. When there is a phase error, although the frequency at the target spectrum can still be filtered, a phase error of just 0.2 rad causes the notch of the reconstructed filter at the ambiguous frequencies to become shallow rapidly. At the same time, the null position is related to the phase imbalance. With the same effect as the amplitude error, the reconstructed azimuth spectrum illustrated in Figure 3b appears to have severe aliasing, but the difference is that the degree of aliasing of each sub-band spectrum is different.

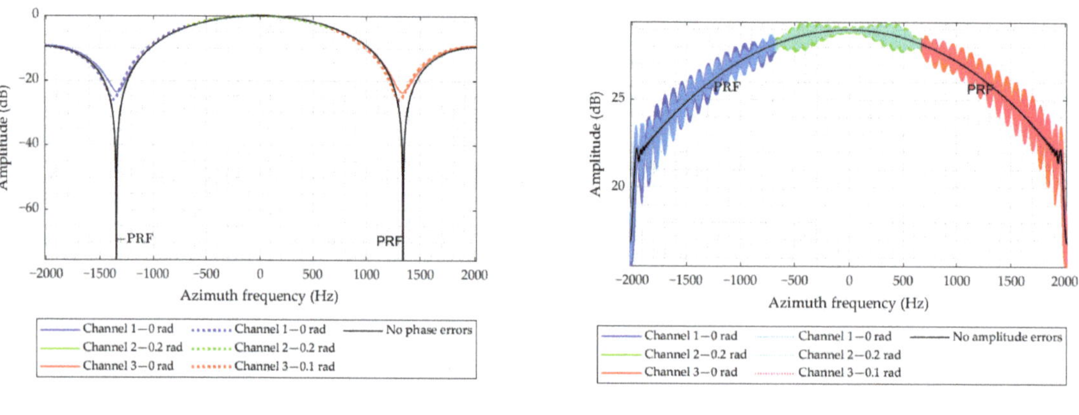

(a)   (b)

Figure 3. Effect of channel phase imbalance. (**a**) Effect of channel phase imbalance on the reconstructed filter. Solid black line: reconstruction filter response without phase error. The colored solid line shows the reconstruction filter response with 0.2 radians set for channel 2. Colored dashed lines represent the reconstruction filter response for channels 2 and 3 with 0.2 and 0.1 radians, respectively. (**b**) Effect of channel phase imbalance on the reconstructed spectrum.

Table 3. Phase imbalance.

	Channel 1	Channel 2	Channel 3
No phase errors	0 rad	0 rad	0 rad
Phase errors 1	0 rad	0.2 rad	0 rad
Phase errors 2	0 rad	0.2 rad	0.1 rad

2.3.3. Influence of RSTI

The RSTI may exist between channels due to imperfect alignment of electronics and satellite attitude. Due to the presence of the RSTI, range migration correction (RMC) cannot be performed perfectly, so the energy of a point target will be spread over two or more range bins, resulting in degraded imaging performance. There are two methods of the RSTI estimation and calibration: the accurate determination and calibration of the RSTI by an internal calibration system; the other uses the ATC method to effectively estimate the RSTI from echo data. Theoretically, the RSTI should not exceed half a sampling interval. For example, the RSTI of the GF-3 azimuth dual-channel SAR system is about 0.2 ns, which is much smaller than its sampling interval of 7.5 ns, and thus has a negligible impact on imaging [39].

2.3.4. Influence of Antenna Position Imbalance

The antenna position error due to antenna attitude can be divided into two parts: along-track antenna position error and radial antenna position error. The radial antenna position error can be equated to phase error and eliminated through phase error correction. As shown in Equation (19) [32], the phase variation caused by the position error of the along-track antenna is limited by the multichannel SAR azimuth resolution:

$$\frac{2\pi f_\eta \Delta d_m}{V_r} \leq \frac{\pi \mathrm{PRF} \Delta d_m}{V_r} \approx \frac{\pi \Delta d_m}{N \delta_{az}} \tag{19}$$

where Δd_m and δ_{az} represent the antenna position error along track and azimuth resolution, and Δd_m is much smaller than the azimuthal resolution δ_{az}. This makes the phase change caused by baseline error tend to be close to zero and can be neglected. Taking the GF-3 azimuth dual-channel SAR system as an example, the estimated along-track position error of the antenna is about 8 cm, much smaller than the azimuth subaperture antenna length of 3.75 m, and its effect on the image is also negligible.

Therefore, the AMC SAR system is mainly affected by amplitude error and phase error. Furthermore, there are already efficient amplitude error and RSTI correction methods, so this paper divides channel imbalance calibration into two independent parts: precorrection of amplitude error and RSTI, followed by the estimation and calibration of phase error by the MSSBN method proposed in this paper.

2.4. Precalibration Processing

This section calibrates the amplitude imbalance and RSTI between channels first. Generally, the first channel is used as the reference channel, and the amplitude imbalance is calibrated in the range-Doppler domain using the channel balancing technique [26]. The amplitude imbalance is estimated as

$$\Delta \hat{a}_m = \frac{E_\tau \left\{ E_\eta \left\{ |S_m(\tau, f_\eta)| \right\} \right\}}{E_\tau \left\{ E_\eta \left\{ |S_1(\tau, f_\eta)| \right\} \right\}} \tag{20}$$

where $E_\eta\{\cdot\}$ and $E_\tau\{\cdot\}$ denote the statistical averaging operations in the azimuth and range, respectively. Then, the signal after amplitude calibration is expressed by the following equation:

$$S'_m(\tau, f_\eta) = \frac{S_m(\tau, f_\eta)}{\Delta \hat{a}_m} \tag{21}$$

After the channels are calibrated for amplitude errors, the RSTI is estimated according to the ATC technique. As seen in Equation (22), inconsistent phase information is obtained by the intercorrelation operation of each channel echo data with the reference channel echo data:

$$E_\eta \left\{ S^*_{ref}(f_\tau, \eta) S_m(f_\tau, \eta) \right\}$$
$$= r_{ant}(f_\tau, \Delta \eta_m) \exp\{ j2\pi f_{dc} \Delta \eta_m - j2\pi f_\tau \Delta \tau_m + j\Delta \theta_m \} \tag{22}$$

where $r_{ant}(f_\tau, \Delta \eta_m)$ denotes the azimuth inverse fast Fourier transform (IFFT) of the antenna's two-way power pattern [27]. Thus, it can be seen that the phase is a linear function of the range frequency f_τ with the slope $2\pi \Delta \tau_m$. Therefore, the RSTI can be obtained by differentiating the phase of Equation (22) with respect to f_τ:

$$\Delta \tau_m = -\frac{1}{2\pi} \frac{d(\arg(E_\eta \{ S^*_{ref}(f_\tau, \eta) S_m(f_\tau, \eta) \}))}{d f_\tau} \tag{23}$$

where $\arg(\cdot)$ denotes the operation of finding the phase of a complex number. Finally, the echo signal of each channel after RSTI calibration is represented as

$$S''_m(\tau, f_\eta) = S''_m(\tau, f_\eta) \exp\{ -j2\pi f_\tau \Delta \tau_m \} \tag{24}$$

3. Method

When the phase between channels is consistent, the sum of the norm of reconstructed spectrum sub-bands has a minimum value, so the proposed method will be proven and derived based on the subadditivity of the linear normed space [40] in this section. According to the analysis in Section 2, the phase imbalance is treated as a constant error in this paper. As is known, the phase imbalance between channels is a relative quantity, so the first channel is set as the reference channel, and the phase error of the mth channel is set as $\Delta\theta_m, m = 1, \cdots M$, where $\Delta\theta_1 = 0 \ rad$. Therefore, the phase error matrix is shown as

$$\boldsymbol{\Gamma} = \text{diag}\{\alpha_1, \cdots, \alpha_m, \cdots, \alpha_M\} = \text{diag}\{\exp\{j[\Delta\theta_1, \cdots, \Delta\theta_m, \cdots, \Delta\theta_M]\}\} \tag{25}$$

According to the Krieger reconstruction matrix [3], the echo signal only with a phase error of each channel is reconstructed as

$$\begin{bmatrix} S_{re}(f'_{\eta,1}) \\ \vdots \\ S_{re}(f'_{\eta,M}) \end{bmatrix} = \mathbf{P}\boldsymbol{\Lambda}^H \begin{bmatrix} S_{1e}(f_\eta) \\ \vdots \\ S_{Me}(f_\eta) \end{bmatrix} = \mathbf{P} \begin{bmatrix} \alpha_1 & \cdots & 0 \\ \vdots & \ddots & \vdots \\ 0 & \cdots & \alpha_M \end{bmatrix} \mathbf{H} \begin{bmatrix} S_0(f'_{\eta,1}) \\ \vdots \\ S_0(f'_{\eta,M}) \end{bmatrix} \tag{26}$$

where $S_{re}(f'_{\eta,.}), S_{.e}(f_\eta)$, and $S_0(f'_{\eta,.})$ denote the reconstructed spectrum with phase error, the aliased echo with phase error for each channel, and the full alias-free spectrum, respectively. As shown in the second equation of Equation (26), the phase error can be considered a part of the filter matrix or echo signal. The matrix $\boldsymbol{\Lambda}^H$ related to the Doppler frequency does not have an effect on the analysis.

The reconstructed filter matrix $\mathbf{P}$ and transmission matrix $\mathbf{H}$ generate a weighted matrix $\mathbf{W}$, and the complete aliasing-free spectrum is mapped to the reconstructed spectrum by $\mathbf{W}$. At the same time, it can be seen from Equation (26) that $\mathbf{W}$ has nothing to do with the Doppler frequency but is only related to the reconstruction filter, the system transmission matrix, and channel errors. Therefore, this paper first analyzes the weighted matrix. The phase errors from $-\pi$ to π were assigned to channel 2 and channel 3, respectively, to obtain the reconstruction filter $\mathbf{P}_e$ with errors, and the weighted matrix $\mathbf{W}_e$ with errors was obtained by matrix multiplication.

The simulation parameters in Figure 4 are from Table 1, which shows the schematic diagram of each component of the weighted matrix affected by the channel phase error. As can be seen from the figure, when the phase error of the channel is consistent with that of the reference channel, that is, when the phase error in the figure is simultaneously 0, the diagonal elements of the weighted matrix all reach the maximum value 1, and other elements reach the minimum value 0. However, the reconstruction filter is known in the actual channel phase error estimation. In contrast, the phase error is an unknown quantity present in the echo, and the phase error cannot be estimated directly by the filter matrix. According to Perceval's theorem [1], when the power of the ambiguous spectrum is minimized, an alias-free reconstructed spectrum can be obtained [41]. However, the ambiguous spectrum cannot be separated from the signal spectrum, so the ambiguous spectrum is considered together with the signal spectrum in this paper. Fortunately, the sub-band of the reconstructed spectrum is obtained by the weighted summation of the sub-band of the aliasing-free full spectrum, so it is necessary to add and analyze the corresponding elements of the weight matrix.

The simulation parameters in Figure 5 are shown in Table 1. It can be seen from Figure 5 that the shift in the position of the reconstructed filter null caused by phase imbalance leads to a change in the amplitude of each spectrum. However, a minimum value exists for their sum of the norm, and the minimum value is obtained when there is no channel phase imbalance. The spectrum of different sub-bands is different, so this paper will mathematically derive the relationship between the sub-band spectrum norm (SBN) and the channel phase imbalance.

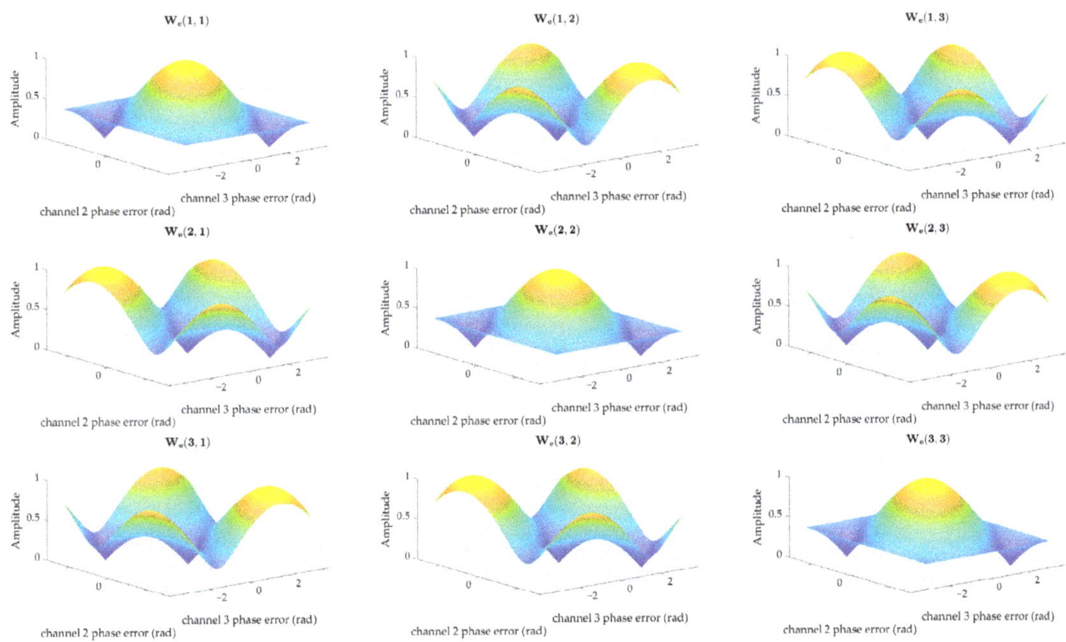

Figure 4. Weighted matrices affected by different phase errors.

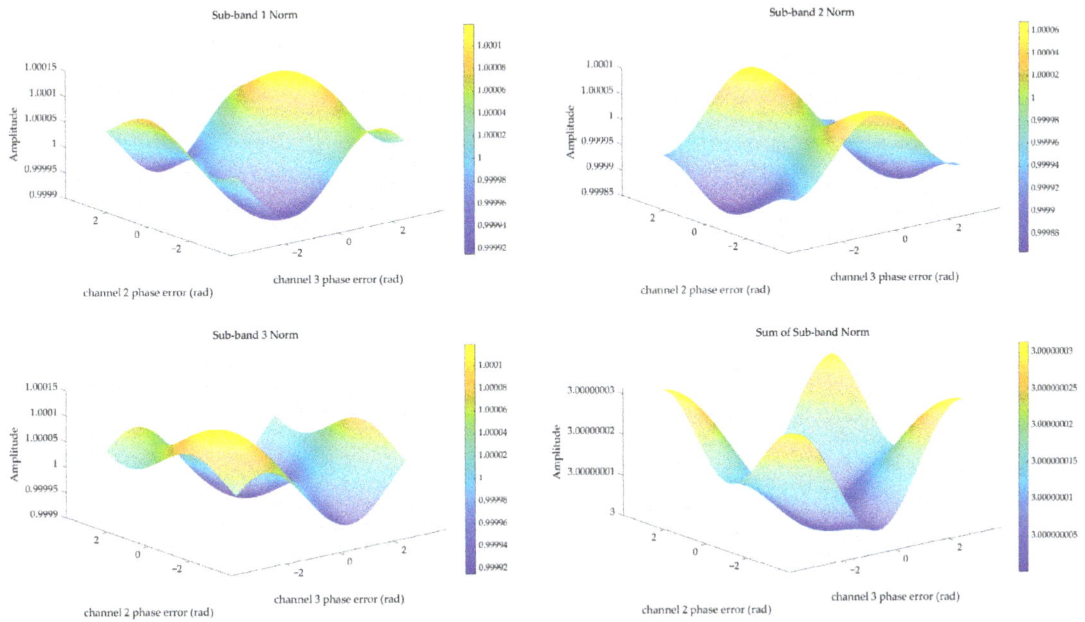

Figure 5. The norm of the row's sum of the weight matrix and the sum of the norm.

According to the subadditivity of linear normed space, the SBN satisfies the following equation:

$$\left|S_{re}(f'_{\eta,1})\right| + \cdots + \left|S_{re}(f'_{\eta,M})\right| \geq \left|S_{re}(f'_{\eta,1}) + \cdots + S_{re}(f'_{\eta,M})\right| = \left|S_0(f'_{\eta,1}) + \cdots + S_0(f'_{\eta,M})\right| = C \qquad (27)$$

From the above equation, it can be seen that there is a constant lower bound for the sum of the sub-band norm (SSBN), but the minimum cannot be reached because the argument of the sub-band spectral cannot be the same. Again, according to the subadditivity of linear normed space, SBN with error can be written as

$$\left|S_{re}(f'_{\eta,1})\right| \geq \left|\left|\alpha_1 S_0(f'_{\eta,1})\right| - C\right| \qquad (28)$$

where C denotes a constant number; the detailed duction can be found in Appendix A. There exists $\left|S_{re}(f'_{\eta,1})\right| = \left|S_0(f'_{\eta,1})\right|$ when, and only when, the phase error between channels is consistent, at which time the norm of the sub-band 1 spectrum reaches a minimum value [40]. All sub-band spectral norms satisfy the above equation, so the SSBN with phase errors can be written as

$$\left|S_{re}(f'_{\eta,1})\right| + \cdots + \left|S_{re}(f'_{\eta,M})\right| \geq \left|S(f'_{\eta,1})\right| + \cdots + \left|S(f'_{\eta,M})\right| = C \qquad (29)$$

Next, the above equation is verified by simulation experiments of three channels. The simulation parameters of Figure 6 are shown in Table 1. As shown in Figure 6a, the sum of three sub-band spectral norms reaches a minimum value, and the minimum value is equal to the SSBN without phase errors when the phase errors of channels 2 and 3 are 0 at the same time, i.e., when they are in phase with channel 1. As shown in Figure 6b, the norm of the sum of sub-band spectral (NSSB) with errors is also a constant and is less than the SSBN.

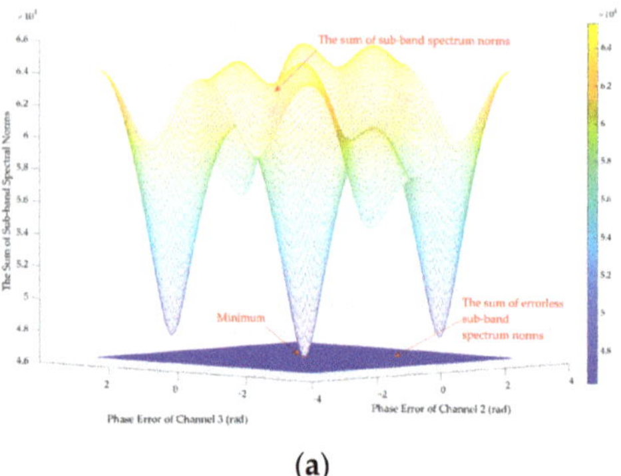

(a)

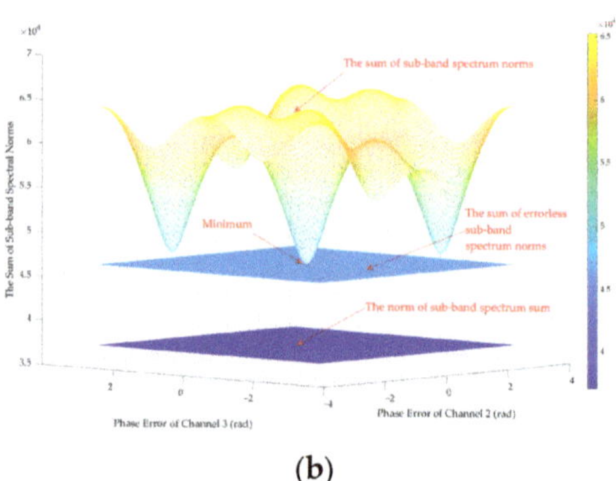

(b)

Figure 6. SSBN of the reconstructed spectrum with different phase errors. (**a**) SSBN has a minimum value when channel phase error is consistent. (**b**) The NSSB is smaller than the SSBN without phase error.

Now the optimization function is established to estimate channel phase error by the MSSBN. The phase error estimates are shown as

$$\hat{\boldsymbol{\Gamma}} = \mathrm{diag}\{\hat{\alpha}_1, \cdots, \hat{\alpha}_m, \cdots, \hat{\alpha}_M\} = \mathrm{diag}\{\exp\{j[\Delta\hat{\theta}_1, \cdots, \Delta\hat{\theta}_m, \cdots, \Delta\hat{\theta}_M]\}\} \qquad (30)$$

According to Equation (26), the reconstructed spectrum after channel phase error compensation is shown as

$$
\begin{bmatrix} \hat{S}(f'_{\eta,1}) \\ \vdots \\ \hat{S}(f'_{\eta,M}) \end{bmatrix} = \mathbf{P} \begin{bmatrix} \hat{\alpha}_1 & \cdots & 0 \\ \vdots & \ddots & \vdots \\ 0 & \cdots & \hat{\alpha}_M \end{bmatrix} \begin{bmatrix} S_{1e}(f_\eta) \\ \vdots \\ S_{Me}(f_\eta) \end{bmatrix}
$$
$$
= \begin{bmatrix} p_{11} \\ \vdots \\ p_{N1} \end{bmatrix} \hat{\alpha}_1 S_{1e}(f_\eta) + \cdots + \begin{bmatrix} p_{1N} \\ \vdots \\ p_{NN} \end{bmatrix} \hat{\alpha}_N S_{Ne}(f_\eta) \tag{31}
$$

where $\mathbf{P} = [\mathbf{P}_1, \cdots, \mathbf{P_m}, \cdots, \mathbf{P_N}], \mathbf{P_m} = [p_{1m}, \cdots, p_{Nm}]^T$. Then the norm of the $n-$th sub-band spectrum can be expressed as

$$
\left| \hat{S}(f'_{\eta,n}) \right| = \left| \sum_{m=1}^{M} p_{nm} \hat{\alpha}_m S_{me}(f_\eta) \right| = \left| \sum_{m=1}^{M} \hat{S}'_m(f'_{\eta,n}) \right|, n = 1, \cdots, M \tag{32}
$$

Since the number of channels M is equal to the number of spectrum ambiguities N, the SSBN can be further written in the following form:

$$
\sum_{n=1}^{N} \left| \hat{S}(f'_{\eta,n}) \right| = \sum_{n=1}^{N} \left| \sum_{m=1}^{N} p_{nm} \hat{\alpha}_m S_{me}(f_\eta) \right| = \sum_{n=1}^{N} \left| \sum_{m=1}^{N} p_{nm} \exp(-j\Delta\hat{\theta}_m) S_{me}(f_\eta) \right| \tag{33}
$$

In this case, the phase error can be estimated by the MSSBN. The estimation model for the MSSBN is developed as

$$
\Delta\hat{\theta} = \underset{\Delta\hat{\theta}}{\operatorname{argmin}} \left\{ \sum_{n=1}^{N} \left| \sum_{m=1}^{N} p_{nm} \exp(-j\Delta\hat{\theta}_m) S_{me}(f_\eta) \right| \right\}
$$
$$
s.t. \left| \Delta\hat{\theta} \right| \le \pi \tag{34}
$$

where $\Delta\hat{\theta} = [\Delta\hat{\theta}_1, \cdots, \Delta\hat{\theta}_m, \cdots, \Delta\hat{\theta}_N]$. Many mature global optimization algorithms, such as OQNLP and PSO, can solve the optimization problem in Equation (34).

The estimation algorithm proposed in this paper is performed in the range-Doppler domain, where the phase information of all sampling points is transformed to the same Doppler frequency band by the azimuth FFT and range pulse compression operations. The phase information of all sampling points is then converted to the same Doppler band. Furthermore, after the range pulse compression, the Doppler spectrum is aggregated into a few range bins, which can significantly reduce the influence of the range dimension on the computational complexity. In order to improve the computational efficiency, downsampling must be uniformly sampled in the whole Doppler band so as not to affect the estimation results.

After estimating the channel phase error, channel error compensation is carried out for the multichannel signal with error reconstruction, and the compensation results are as

$$
S_r(f'_{\eta,n}) = \sum_{m=1}^{M} \hat{S}'_m(f'_{\eta,n}) \operatorname{conj}\left\{ \exp\left\{ j\Delta\hat{\theta}_m \right\} \right\} \tag{35}
$$

The complete process of the algorithm proposed in this paper is shown in Figure 7. The combined flow chart summarizes the main steps of the channel error estimation and calibration process as follows:

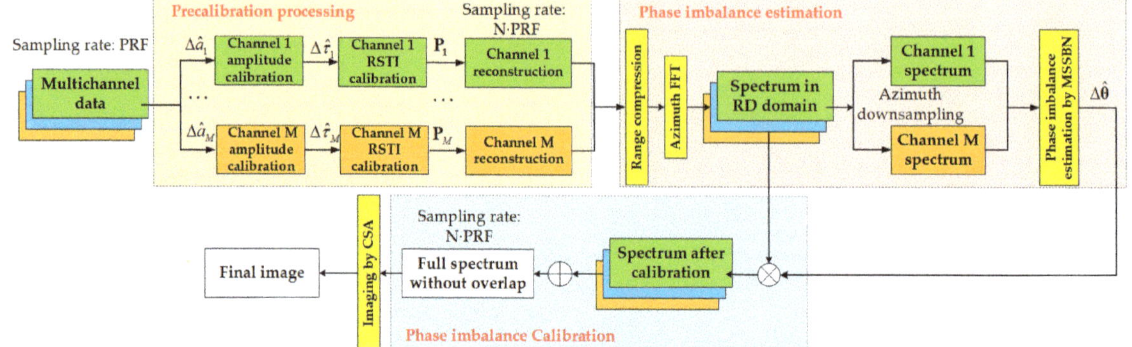

Figure 7. The processing flowchart of the proposed method.

Step 1: Estimate and calibrate the amplitude imbalance Δa_m between channels using the channel balancing method.

Step 2: Estimate and calibrate the RSTI between channels by using the ATC technique.

Step 3: Perform the spectrum reconstruction of each channel's echo signal by the reconstruction filters **P**.

Step 4: Transform each channel's constructed signal into the range-Doppler domain utilizing range pulse compression and azimuth FFT.

Step 5: Uniformly sample the reconstructed azimuth spectrum of each channel echo signal and search for phase error $\Delta\theta_m$ by the MNSSB.

Step 6: The reconstructed azimuth spectrum of each channel echo signal is calibrated for phase error and then summed to obtain the alias-free full azimuth spectrum.

Step 7: Image the reconstructed echo signal after imbalance calibration.

4. Results and Discussions

In this section, the simulation and measured data will be processed and analyzed, and the proposed method's effectiveness, accuracy, and robustness will be verified by analyzing the processing results. Phase errors are artificially added to different channel echo data in the simulation. By comparing the estimated results with the actual values, the method proposed in this paper can be proved effective. The actual channel imbalance is unknown when processing the accurate data of GF-3, so the method's effectiveness can be illustrated by comparing the SAR images before and after processing.

4.1. Simulation Experiment and Results

This section uses a three-channel AMC SAR system to generate simulated echo data for simulation experiments. The parameters of the AMC SAR system are shown in Table 1. The platform's velocity is 7563 m/s, and the nearest distance of the scene center from the radar platform is 900 km. The system transmits a chirp signal with a bandwidth of 300 MHz and a pulse width of 2.5 us. The azimuth antenna length is 11.25 m, with three antenna subapertures uniformly distributed along the radar motion track, and the reference channel is the first. From the transmitting antenna length, the Doppler bandwidth of the system is 3574 Hz, and since the PRF is only 1429 Hz, the ambiguity number of the system is three, which is equal to the number of channels. The two-way antenna pattern weights the azimuth echo signal, and the weighting function is denoted by $\text{sinc}^2(0.886\theta/\theta_{bw})$, where θ and θ_{bw} denote the angle measured from boresight in the slant range plane and azimuth beamwidth, respectively.

In order to generate a simulated echo signal, 9-point targets were placed in the imaging scene to form a 3×3 matrix with azimuth and range intervals of 400 and 100 m between points, respectively. The point target's signal-to-noise ratio (SNR) was set to 20 dB, and the phase errors between channels are shown in Table 4. The top panel of Figure 8a shows the imaging results in the presence of channel errors, and the bottom panel of Figure 8a shows the imaging results after phase error calibration. Figure 8b shows the ambiguity suppression effect after phase calibration, from which it can be seen that the method proposed in this paper effectively suppresses azimuth ambiguity.

Table 4. Channel phase imbalance estimation results.

Method	Channel 1	Channel 2	Channel 3	SNR	Execution Time
Initial error	0.00°	50.00°	−100.00°	20 dB	\
ATC [27]	0.00°	49.88°	−100.34°	20 dB	0.65 s
MVDR [32]	0.00°	49.90°	−99.95°	20 dB	0.54 s
LLN [35]	0.00°	50.00°	−99.99°	20 dB	17.04 s
MSSBN-1	0.00°	50.01°	−100.00°	20 dB	1.93 s
MSSBN-2	0.00°	49.96°	−99.97°	20 dB	0.41 s
MSSBN-3	0.00°	49.95°	−100.00°	20 dB	0.15 s
MSSBN-2	0.00°	50.12°	−100.17°	0 dB	0.39 s
MSSBN-3	0.00°	49.91°	−100.67°	0 dB	0.17 s

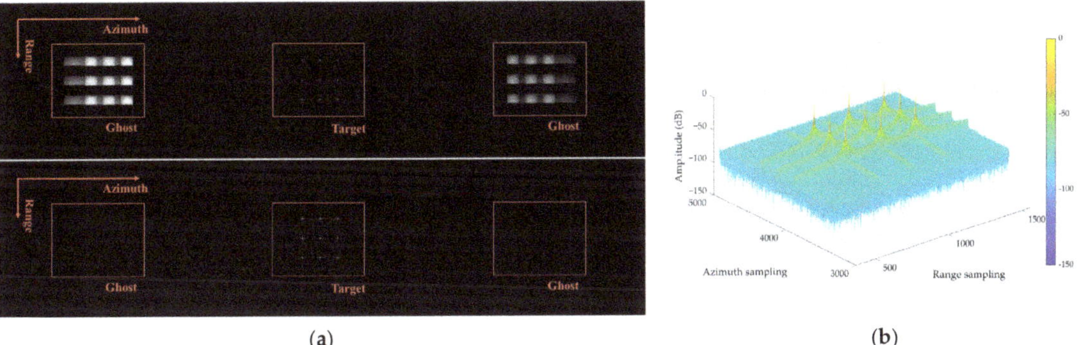

Figure 8. Imaging results after phase error calibration. (**a**) Phase errors introduce false targets, and ambiguity is suppressed when phase errors are calibrated. (**b**) Ambiguity suppression effect for point targets.

As shown in Table 4, comparing the estimation accuracy of the method in this paper with that of ATC, MVDR, and LLN, it can be seen that the method proposed in this paper has high estimation accuracy. Its accuracy is consistent with that of the image-domain method. The computational effort can be significantly reduced by uniform sampling operation of the sub-band spectrum without affecting the estimation accuracy. MSSBN-1, MSSBN-2, and MSSBN-3 were processed without sampling operation, 10 times sampling, and 100 times sampling for the azimuth sub-band spectrum, respectively. The method proposed in this paper has high computational efficiency regarding computing time.

The algorithm proposed in this paper finds the global minimum by iteration, so the convergence speed needs to be tested. Figure 9 shows the iterative process of the MSSBN-3 experiment, and simulation results show that the method proposed in this paper has a good convergence effect.

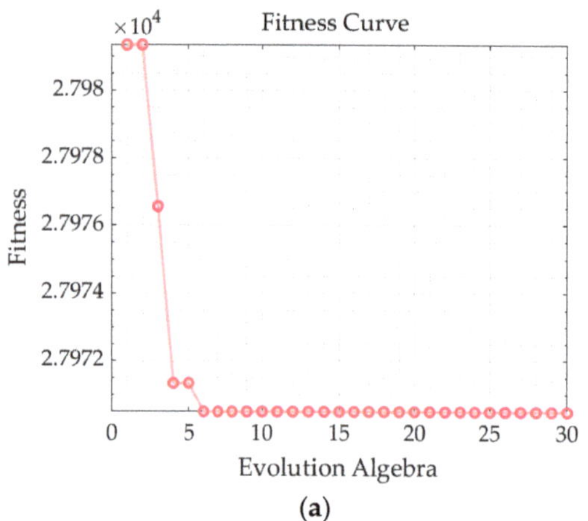

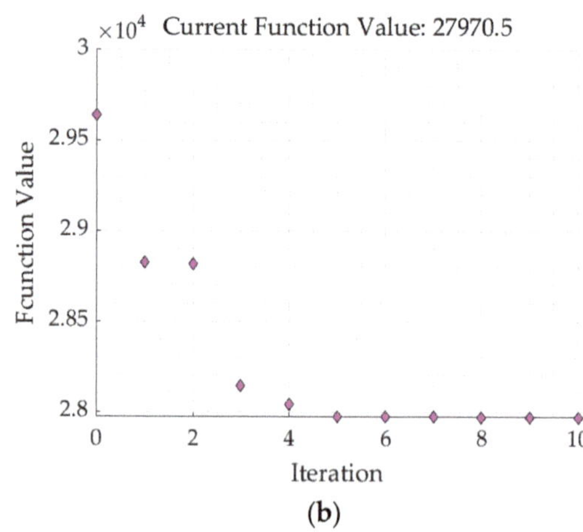

Figure 9. Iterative process of the MSSBN-3 experiment. (**a**) PSO. (**b**) OQNLP.

Monte Carlo experiments test the robustness of the algorithm proposed in this paper. Simulation data are generated using the parameters in Table 1, and the effect of SNR on the estimation accuracy is first considered. An SNR interval of -10 to 20 dB is chosen for evaluation, and the phase error of each channel is set as a random amount from $[-\pi, \pi]$. SNR is varied in steps of 5 dB, and 30 Monte Carlo experiments are performed for each SNR. Average root mean square error (ARMSE) is used to evaluate the estimation accuracy of phase errors, which is defined as

$$\text{ARMSE} = \frac{1}{N} \sum_{n=1}^{N} \sqrt{\frac{1}{M} \sum_{m=1}^{M} \left(\Delta \hat{\theta}_{nm} - \Delta \theta_{nm} \right)^2} \tag{36}$$

where N and M represent the number of experiments and the number of channels, respectively. The estimation accuracy of the proposed method is compared with that of the ATC, MVDR, and LLN methods. Figure 10a represents the relationship between ARMSE and SNR, and it is clear that the ARMSE of the proposed method is the smallest among them.

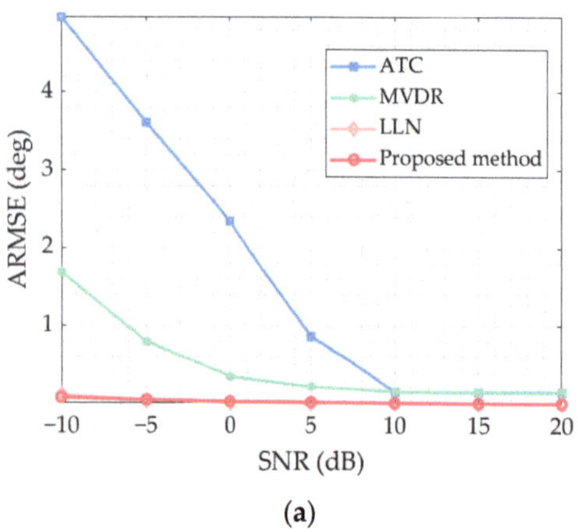

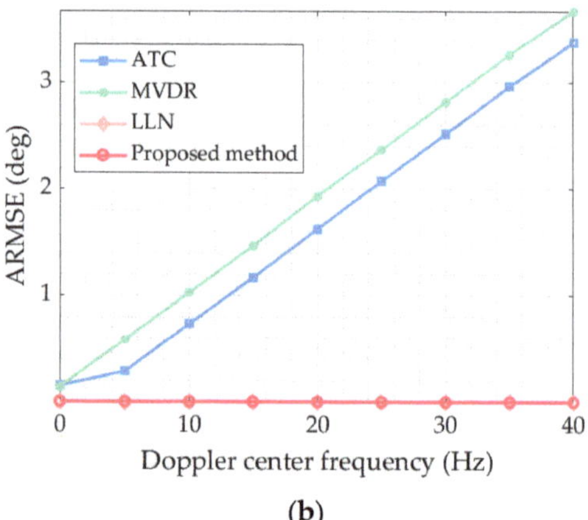

Figure 10. Comparison of stability of different estimation methods. (**a**) ARMSE of the phase error versus SNR. (**b**) ARMSE of the phase error versus Doppler center frequency.

Affected by the antenna attitude, the Doppler frequency in the echo signal is not always zero, and the Doppler center estimation error also impacts the estimation accuracy of the phase error. Therefore, another Monte Carlo experiment is used to test the sensitivity of the proposed algorithm to the Doppler center error in this paper. The Doppler center frequency is set to 0 Hz in the echo data generation. As in the previous experiment, the phase error of each channel is set as a random variable. Furthermore, in the data processing, the estimated error is added to the Doppler center frequency, which is varied from 0 to 40 Hz in steps of 5 Hz. Figure 10b shows the ARMSE of the phase error versus the Doppler center frequency. It can be seen from the figure that the sensitivity of the algorithm proposed in this paper to the Doppler center error is equally low.

In the method proposed in this paper, the phase error estimation results are only related to the Doppler spectrum and are independent of the range direction when RCM is excluded. In order to exclude the effect of range sampling, the proposed method estimates the phase error in the range-Doppler domain. After the range pulse compression, the azimuth spectrum is clustered in a few range bins. Therefore, the phase error at the selected range bin can be accurately estimated by chunking the range direction using the algorithm in this paper. First, the multichannel echo signal is generated according to the phase error settings in Table 5. Then the range direction is chunked into five range bins, and the phase error of each range bin is estimated by the proposed method. The estimation results are shown in Figure 11.

Table 5. Linear phase error setting of range.

	Channel 1	Channel 2	Channel 3
Range space variation	0	$0.5 + (R - R_c)/\delta_r \pi/6$	$-1 + (R - R_c)/\delta_r \pi/4$

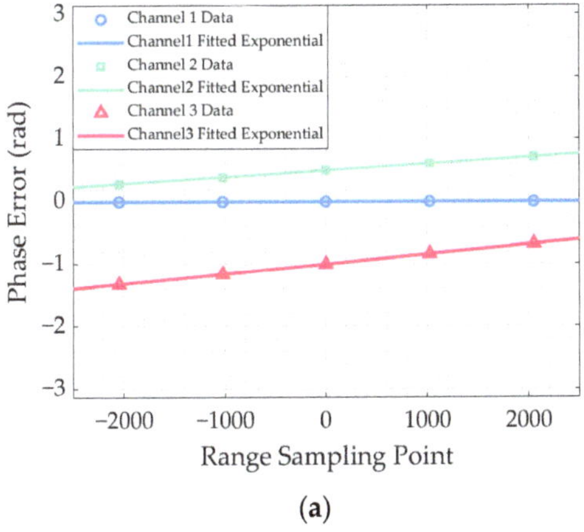

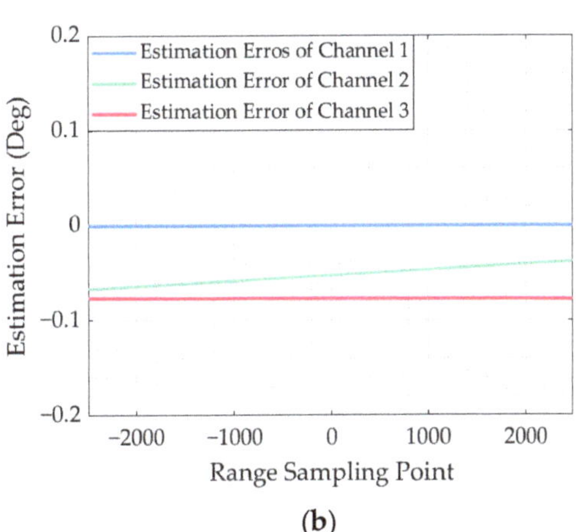

Figure 11. Range space-variant phase error estimation results. (**a**) Range chunking estimation and least-squares fitting results. (**b**) Estimation error of range linear space-variant phase.

Figure 11a represents the phase error results using least squares fitting, and Figure 11b represents the difference between the estimated and actual phase errors. The figure shows that the algorithm proposed in this paper can effectively estimate the range space-variation linear phase error, and the estimation error is within 0.1 Deg.

When the echo signal covers the complete Doppler spectrum, the algorithm proposed in this paper can accurately estimate the phase error. Therefore, when estimating the azimuth time-varying linear phase error, the azimuth blocks need to cover the entire synthetic aperture time. The multichannel echo signal is generated according to the phase error settings in Table 6. The phase error of each azimuth block is estimated by the proposed method.

Table 6. Linear phase error setting of azimuth.

	Channel 1	Channel 2	Channel 3
Azimuth time variation	0	$0.5 + (\eta - \eta_c)/\mathrm{PRF}\pi/6$	$-1 + (\eta - \eta_c)/\mathrm{PRF}\pi/4$

Figure 12a shows the phase error results using least squares fitting, and Figure 12b indicates the difference between the estimated and actual values of the azimuth time-varying linear phase error. The figure shows that the algorithm proposed in this paper can effectively estimate the phase error of the azimuth time-varying linear phase error. Although the estimation accuracy is lower compared with the range space-varying phase error estimation results, the estimated error is within 0.8 Deg.

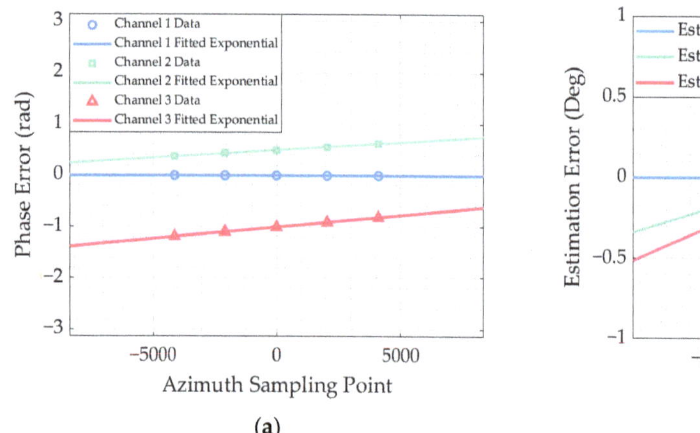

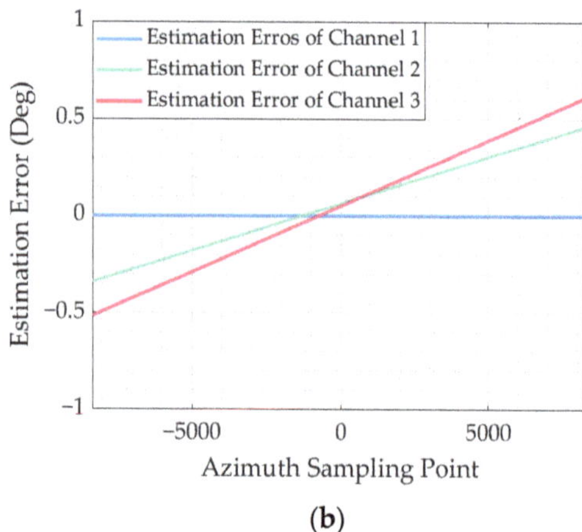

Figure 12. Azimuth time-variant phase error estimation results. (**a**) Azimuth chunking estimation and least-squares fitting results (**b**) Estimation error of azimuth linear time-variant phase.

4.2. Experimental Results of GF-3 Measured Data

This experiment uses satellite-based dual-channel SAR data from the GF-3 system acquired over Lake Inawashiro, the fourth lake northwest of Koriyama, Fukushima, Japan, on 13 March 2018. The main system parameters are listed in Table 7. Figure 13 shows imaging results after the same amplitude but different phase imbalance calibrations. Figure 13a shows imaging results without imbalance calibration, which is severely aliased. The village that should be located up the lake appears in the center of the lake with its ghost's energy significantly higher than the energy of the actual target. Figure 13b shows the imaging results after compensating for the phase error by the MVDR method, from which it can be seen that the ambiguity is significantly suppressed but still exists. Figure 13c shows the image after compensating for the phase error by the ATC method, the ambiguity is almost invisible, and some ghosting can be seen after magnification. Figure 13d shows the imaging results after phase error correction by the LLN method, and it can be seen that the figure shows that ambiguity is further suppressed compared with the other methods. Figure 13e shows the imaging results after the phase error correction by the proposed method, which has the best ambiguity suppression effect. The compensated phase error in Figure 13f is obtained through 40 times uniform downsampling of the azimuth reconstruction signal, and its value is consistent with the estimated result without downsampling.

To quantitatively analyze the effectiveness of different methods, the ghost–target energy ratio (GTER) is employed to measure the suppression of ghosts in the final focused image [30], which is defined as

$$\text{GTER} = 20 \log_{10} \frac{\max(|I_G|)}{\max(|I_T|)} \tag{37}$$

where $\max(|I_G|)$ denotes the maximum amplitude value of the pixels of the ghost, and $\max(|I_T|)$ denotes the maximum value of the target. The more accurate the calibration of phase imbalance between channels, the smaller the value of the GTER. The target and ghost, marked by a box in the image, are selected to calculate the GTER, where the ghost in the lake is GTER1, and the ghost on land is GTER2. The measured results of different images calibrated by different methods are listed in Table 8. GTER1 shows that the best estimation results are obtained using the method proposed in this paper to estimate range space-variant linear phase imbalance. From GTER2, the best estimation results are obtained using the proposed method to estimate the constant phase imbalance. The GTER shows that range space-variant linear phase error estimation results are similar to constant phase error estimation results.

Table 7. GF-3 system parameters.

Parameter	Symbol	Value	Unit
Platform velocity	V_r	7563	m/s
Carrier frequency	f_0	5.4	GHz
Slant angle	θ_{rc}	0	deg
Nearest slant range	R_0	900	km
Azimuth antenna length	L_{az}	3.75*2	m
Pulse repetition frequency	PRF	1994	Hz
Number of channels	M	2	/

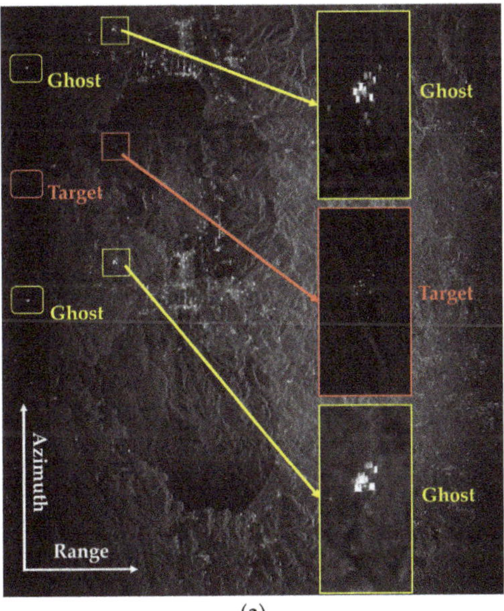

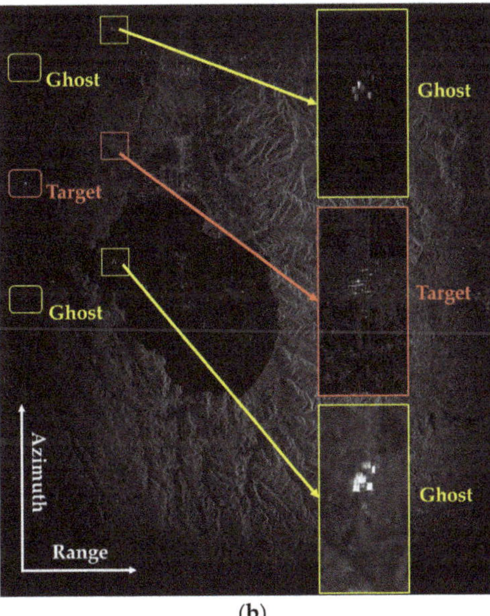

Figure 13. *Cont.*

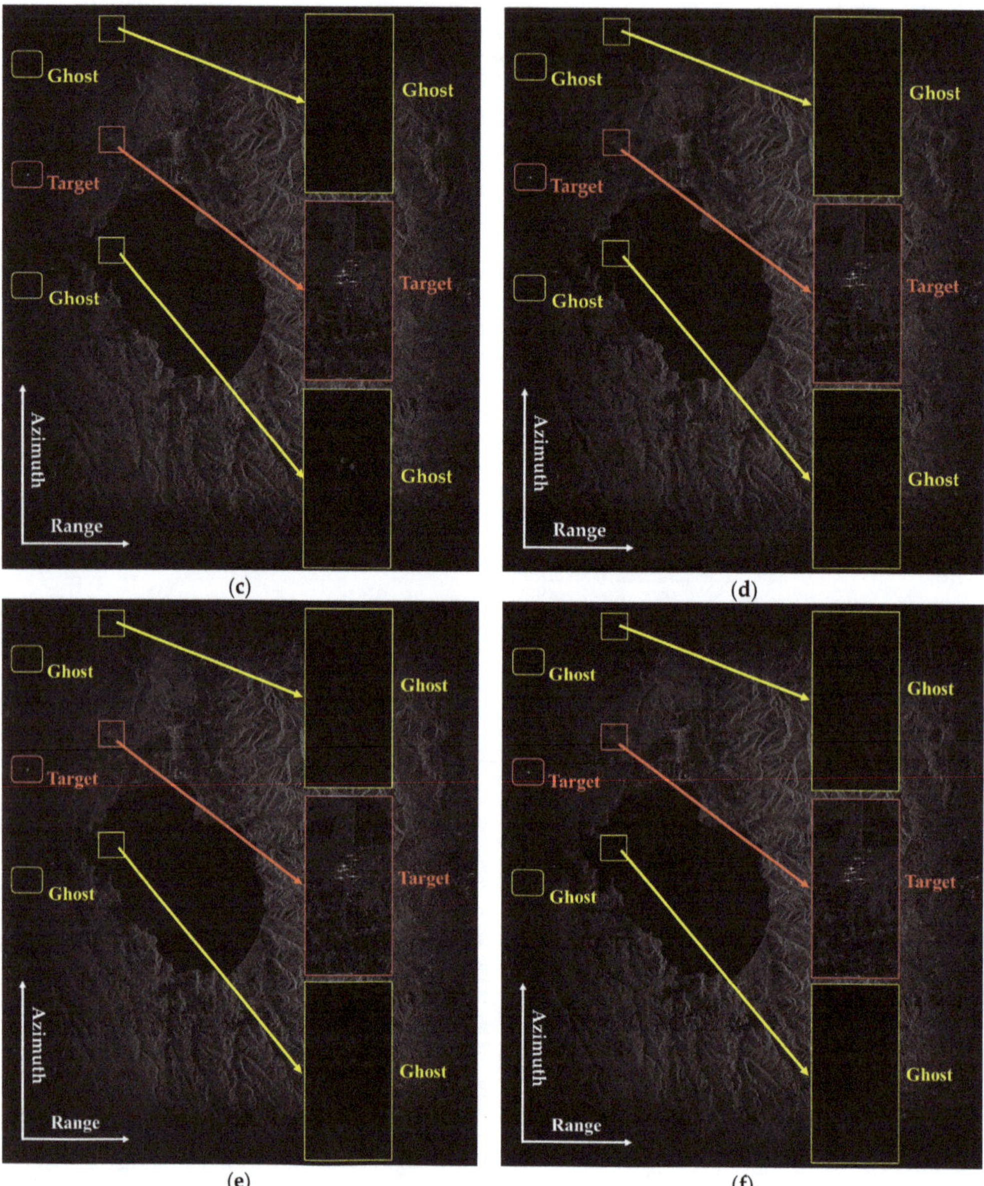

Figure 13. GF-3 two-channel signal imaging results. (**a**) imaging results without phase calibration, (**b**) imaging results after phase calibration using the MVDR method, phase error = 1.96 rad, (**c**) imaging results after phase calibration using the ATC method, phase error = 2.95 rad, (**d**) imaging results after phase calibration using the LLN method, phase error = 2.79 rad, (**e**) imaging results after phase calibration using the method proposed by this paper, phase error = 2.79 rad, and (**f**) imaging results after phase calibration using the method proposed by this paper with 40 times downsampling for reconstructed azimuth spectrum, phase error = 2.79 rad.

Table 8. GTER of different methods.

Method	GTER1	GTER2	Execution Time
Without calibration	−11.45 dB	−11.05 dB	\
ATC [27]	−45.71 dB	−43.91 dB	33.93 s
MVDR [32]	−34.08 dB	−32.82 dB	41.12 s
LLN [35]	−49.38 dB	−44.71 dB	318.63 s
MSSBN	−50.75 dB	−44.75 dB	49.15 s
Range space variation Imbalance of MSSBN	−50.82 dB	−44.72 dB	337.04 s
Azimuth time variation Imbalance of MSSBN	−50.19 dB	−44.76 dB	118.19 s

In contrast, azimuth time-variant phase error estimation results are different from the constant phase error estimation results, which shows that the constant phase error estimation can achieve a good compensation effect. The method proposed in this paper improves the computational efficiency by reducing the sample size, and its computational time is mainly spent on range pulse compression and azimuth FFT processing. The global optimization only takes about 2 s. A comprehensive comparison of the above methods shows that the method proposed in this paper has the highest estimation accuracy while having high computational efficiency.

5. Conclusions

This paper proposes a new method for estimating the channel inconsistency of the AMC SAR system, and the influence of channel inconsistency is analyzed. The method mainly calibrates the phase imbalance of channels. Analysis and experiments show that when channel phase imbalances are consistent, the SSBN reaches its minimum value, the norm sum of the unambiguous sub-band spectrum. According to the analysis in Section 3, this minimum value is only affected by the consistency of channel phase imbalance, so phase imbalance can still be accurately estimated even when amplitude imbalance is inconsistent. In this paper, the mathematical model between channel phase imbalance and SSBN is established, and phase imbalance estimation is transformed into the problem of finding the minimum sub-band spectral norm. Channel phase imbalance can be obtained effectively using various global optimal search algorithms. By dividing signal data into two dimensions, the proposed method can also estimate space-varying phase imbalance of range and time-varying phase imbalance of azimuth. The performance of the proposed method is verified by simulation data and actual spaceborne multichannel SAR data and compared with other methods. Results show that this method ensures the accuracy and robustness of estimation and has higher computational efficiency.

Author Contributions: Conceptualization, Z.X.; methodology, Z.X, J.L. and Y.C.; funding acquisition, P.L., T.Y. and R.W.; projection administration, P.L. and R.W.; writing—original draft, Z.X.; writing—review and editing, Z.X., Y.C., R.W. and Y.W. All authors have read and agreed to the published version of the manuscript.

Funding: This research was funded by the National Science Fund for Distinguished Young Scholars, China, grant number No. 61825106.

Data Availability Statement: Not applicable.

Conflicts of Interest: The authors declare no conflict of interest.

Appendix A

Expand Equation (26), and according to the subadditivity of linear normed space, the spectrum recovered by the channel 1 can be written as follows:

$$\begin{aligned}
\left| S_{0e}\left(f'_{\eta,1}\right) \right| &= \left| (\alpha_1 p_{11} h_{11} + \cdots + \alpha_N p_{1N} h_{N1}) S_0\left(f'_{\eta,1}\right) + \cdots + (\alpha_1 p_{11} h_{1N} + \cdots + \alpha_N p_{1N} h_{NN}) S_0\left(f'_{\eta,N}\right) \right| \\
&= \left| \begin{array}{l} \alpha_1 S_0\left(f'_{\eta,1}\right) - (\alpha_1 - \alpha_2) p_{12} h_{21} S_0\left(f'_{\eta,1}\right) - \cdots - (\alpha_1 - \alpha_N) p_{1N} h_{N1}\, S_0\left(f'_{\eta,1}\right) - \\ \cdots - (\alpha_1 - \alpha_2) p_{12} h_{2N} S_0\left(f'_{\eta,N}\right) - \cdots - (\alpha_1 - \alpha_N) p_{1N} h_{NN}\, S_0\left(f'_{\eta,N}\right) \end{array} \right| \\
&\geq \left| \left| \alpha_1 S_0(f'_{\eta,1}) \right| - \left| \begin{array}{l} (\alpha_1 - \alpha_2) p_{12} h_{21} S_0(f'_{\eta,1}) + \cdots + (\alpha_1 - \alpha_N) p_{1N} h_{N1} S_0(f'_{\eta,1}) + \\ \cdots + (\alpha_1 - \alpha_2) p_{12} h_{2N} S_0(f'_{\eta,N}) + \cdots + (\alpha_1 - \alpha_N) p_{1N} h_{NN} S_0(f'_{\eta,N}) \end{array} \right| \right|
\end{aligned} \tag{A1}$$

When the channel phase error is consistent, the minimum value of the reconstructed signal spectrum norm is obtained. This formula holds for the reconstructed signal spectrum recovered by each channel.

References

1. Cumming, I.G.; Wong, F.H. *Digital Processing of Synthetic Aperture Radar Data: Algorithms and Implementation*; Artech House: Boston, MA, USA, 2005; Volume 1, pp. 108–110.
2. Li, Z.; Wang, H.; Su, T.; Bao, Z. Generation of Wide-Swath and High-Resolution SAR Images From Multichannel Small Spaceborne SAR Systems. *IEEE Geosci. Remote Sens. Lett.* **2005**, *2*, 82–86. [CrossRef]
3. Krieger, G.; Gebert, N.; Moreira, A. Unambiguous SAR Signal Reconstruction From Nonuniform Displaced Phase Center Sampling. *IEEE Geosci. Remote Sens. Lett.* **2004**, *1*, 260–264. [CrossRef]
4. Krieger, G.; Gebert, N.; Moreira, A. Multidimensional waveform encoding: A new digital beamforming technique for synthetic aperture radar remote sensing. *IEEE Trans. Geosci. Remote Sens.* **2007**, *46*, 31–46. [CrossRef]
5. Jing, W.; Xing, M.; Qiu, C.-W.; Bao, Z.; Yeo, T.-S. Unambiguous reconstruction and high-resolution imaging for multiple-channel SAR and airborne experiment results. *IEEE Geosci. Remote Sens. Lett.* **2008**, *6*, 102–106. [CrossRef]
6. Krieger, G.; Gebert, N.; Younis, M.; Bordoni, F.; Patyuchenko, A.; Moreira, A. Advanced concepts for ultra-wide-swath SAR imaging. In Proceedings of the 7th European Conference on Synthetic Aperture Radar, Friedrichshafen, Germany, 2–5 June 2008; pp. 1–4.
7. Yunkai, D.; Weidong, Y.; Heng, Z.; Wei, W.; Dacheng, L.; Robert, W. Forthcoming spaceborne SAR development. *J. Radars* **2020**, *9*, 1–33.
8. Fu, Z.; Zhang, H.; Zhao, J.; Li, N.; Zheng, F. A Modified 2-D Notch Filter Based on Image Segmentation for RFI Mitigation in Synthetic Aperture Radar. *Remote Sens.* **2023**, *15*, 846. [CrossRef]
9. Currie, A.; Brown, M.A. Wide-swath SAR. *IEE Proc. F* **1992**, *139*, 122–135. [CrossRef]
10. Callaghan, G.; Longstaff, I. Wide-swath space-borne SAR using a quad-element array. *IEE Proc.-Radar Sonar Navig.* **1999**, *146*, 159–165. [CrossRef]
11. Ender, J.H.; Klare, J. System architectures and algorithms for radar imaging by MIMO-SAR. In Proceedings of the 2009 IEEE Radar Conference, Pasadena, CA, USA, 4–8 May 2009; pp. 1–6.
12. Süß, M.; Grafmüller, B.; Zahn, R. A novel high resolution, wide swath SAR system. In Proceedings of the IGARSS 2001. Scanning the Present and Resolving the Future. In Proceedings of the IEEE 2001 International Geoscience and Remote Sensing Symposium (Cat. No. 01CH37217), Sydney, Australia, 9–13 July 2001; pp. 1013–1015.
13. Moore, R.K.; Claassen, J.P.; Lin, Y. Scanning spaceborne synthetic aperture radar with integrated radiometer. *IEEE Trans. Aerosp. Electron. Syst.* **1981**, *AES-17*, 410–421. [CrossRef]
14. Soumekh, M. *Synthetic Aperture Radar Signal Processing*; Wiley: New York, NY, USA, 1999; Volume 7.
15. De Zan, F.; Guarnieri, A.M. TOPSAR: Terrain observation by progressive scans. *IEEE Trans. Geosci. Remote Sens.* **2006**, *44*, 2352–2360. [CrossRef]
16. Naftaly, U.; Levy-Nathansohn, R. Overview of the TECSAR satellite hardware and mosaic mode. *IEEE Geosci. Remote Sens. Lett.* **2008**, *5*, 423–426. [CrossRef]
17. Younis, M.; Fischer, C.; Wiesbeck, W. Digital beamforming in SAR systems. *IEEE Trans. Geosci. Remote Sens.* **2003**, *41*, 1735–1739. [CrossRef]
18. Yingjie, W.; Robert, W.; Weidong, Y.; Qingchao, Z.; Kaiyu, L.; Dacheng, L.; Yunkai, D.; Naiming, O.; Xiaoxue, J.; Heng, Z.; et al. See-Earth: SAR Constellation with Dense Time-SEries for Multi-dimensional Environmental Monitoring of the Earth. *J. Radars* **2021**, *10*, 842–864.
19. Kim, J.-H.; Younis, M.; Prats-Iraola, P.; Gabele, M.; Krieger, G. First spaceborne demonstration of digital beamforming for azimuth ambiguity suppression. *IEEE Trans. Geosci. Remote Sens.* **2012**, *51*, 579–590. [CrossRef]
20. Luscombe, A. Image quality and calibration of RADARSAT-2. In Proceedings of the 2009 IEEE International Geoscience and Remote Sensing Symposium, Cape Town, South Africa, 12–17 July 2009; pp. II757–II760.
21. Shimada, M. ALOS-2 science program. In Proceedings of the 2013 IEEE International Geoscience and Remote Sensing Symposium-IGARSS, Melbourne, Australia, 21–26 July 2013; pp. 2400–2403.

22. Qingjun, Z. System design and key technologies of the GF-3 satellite. *Acta Geod. Et Cartogr. Sin.* **2017**, *46*, 269.
23. Sun, J.; Yu, W.; Deng, Y. The SAR payload design and performance for the GF-3 mission. *Sensors* **2017**, *17*, 2419. [CrossRef]
24. Lin, H.; Deng, Y.; Zhang, H.; Liu, D.; Liang, D.; Fang, T.; Wang, R. On the Processing of Dual-Channel Receiving Signals of the LuTan-1 SAR System. *Remote Sens.* **2022**, *14*, 515. [CrossRef]
25. Gebert, N.; de Almeida, F.Q.; Krieger, G. Airborne demonstration of multichannel SAR imaging. *IEEE Geosci. Remote Sens. Lett.* **2011**, *8*, 963–967. [CrossRef]
26. Yang, T.; Li, Z.; Liu, Y.; Bao, Z. Channel error estimation methods for multichannel SAR systems in azimuth. *IEEE Geosci. Remote Sens. Lett.* **2012**, *10*, 548–552. [CrossRef]
27. Feng, J.; Gao, C.; Zhang, Y.; Wang, R. Phase Mismatch Calibration of the Multichannel SAR Based on Azimuth Cross Correlation. *IEEE Geosci. Remote Sens. Lett.* **2013**, *10*, 903–907. [CrossRef]
28. Zhang, S.-X.; Xing, M.-D.; Xia, X.-G.; Zhang, L.; Guo, R.; Liao, Y.; Bao, Z. Multichannel HRWS SAR imaging based on range-variant channel calibration and multi-Doppler-direction restriction ambiguity suppression. *IEEE Trans. Geosci. Remote Sens.* **2013**, *52*, 4306–4327. [CrossRef]
29. Shang, M.; Qiu, X.; Han, B.; Yang, J.; Zhong, L.; Ding, C.; Hu, Y. The space-time variation of phase imbalance for GF-3 azimuth multichannel mode. *IEEE J. Sel. Top. Appl. Earth Obs. Remote Sens.* **2020**, *13*, 4774–4788. [CrossRef]
30. Yang, W.; Guo, J.; Chen, J.; Liu, W.; Deng, J.; Wang, Y.; Zeng, H. A Novel Channel Inconsistency Estimation Method for Azimuth Multi-channel SAR Based on Maximum Normalized Image Sharpness. *IEEE Trans. Geosci. Remote Sens.* **2022**, *60*, 1–16.
31. Li, Z.; Bao, Z.; Wang, H.; Liao, G. Performance improvement for constellation SAR using signal processing techniques. *IEEE Trans. Aerosp. Electron. Syst.* **2006**, *42*, 436–452.
32. Zhang, L.; Gao, Y.; Liu, X. Robust channel phase error calibration algorithm for multichannel high-resolution and wide-swath SAR imaging. *IEEE Geosci. Remote Sens. Lett.* **2017**, *14*, 649–653. [CrossRef]
33. Xiang, J.; Ding, X.; Sun, G.-C.; Zhang, Z.; Xing, M.; Liu, W. An Efficient Multichannel SAR Channel Phase Error Calibration Method Based on Fine-Focused HRWS SAR Image Entropy. *IEEE J. Sel. Top. Appl. Earth Obs. Remote Sens.* **2022**, *15*, 7873–7885. [CrossRef]
34. Zhang, S.-X.; Xing, M.-D.; Xia, X.-G.; Liu, Y.-Y.; Guo, R.; Bao, Z. A robust channel-calibration algorithm for multi-channel in azimuth HRWS SAR imaging based on local maximum-likelihood weighted minimum entropy. *IEEE Trans. Image Process.* **2013**, *22*, 5294–5305. [CrossRef] [PubMed]
35. Cai, Y.; Deng, Y.; Zhang, H.; Wang, R.; Wu, Y.; Cheng, S. An Image-Domain Least L1-Norm Method for Channel Error Effect Analysis and Calibration of Azimuth Multi-Channel SAR. *IEEE Trans. Geosci. Remote Sens.* **2022**, *60*, 1–14.
36. Ugray, Z.; Lasdon, L.; Plummer, J.; Glover, F.; Kelly, J.; Martí, R. Scatter search and local NLP solvers: A multistart framework for global optimization. *INFORMS J. Comput.* **2007**, *19*, 328–340. [CrossRef]
37. Lun, M.; Guisheng, L.; Zhenfang, L. An approach for Multi-channel SAR Array Error Compension and Its Verification by Measured Data. *J. Electron. Inf. Technol.* **2009**, *31*, 1305–1309.
38. Laskowski, P.; Bordoni, F.; Younis, M. Antenna pattern compensation in multi-channel azimuth reconstruction algorithm. In Proceedings of the Advanced RF Sensors and Remote Sensing Instruments (ARSI), Noordwijk, The Netherlands, 13–15 September 2011; pp. 1–10.
39. Shang, M.; Qiu, X.; Han, B.; Ding, C.; Hu, Y. Channel Imbalances and Along-Track Baseline Estimation for the GF-3 Azimuth Multichannel Mode. *Remote Sens.* **2019**, *11*, 1297. [CrossRef]
40. Maligranda, L. Some remarks on the triangle inequality for norms. *Banach J. Math. Anal.* **2008**, *2*, 31–41. [CrossRef]
41. Zhang, Y.; Wang, W.; Deng, Y.; Wang, R. Signal reconstruction algorithm for azimuth multichannel SAR system based on a multiobjective optimization model. *IEEE Trans. Geosci. Remote Sens.* **2020**, *58*, 3881–3893. [CrossRef]

 *remote sensing*

Article

A Novel Multistage Back Projection Fast Imaging Algorithm for Terahertz Video Synthetic Aperture Radar

Qibin Zheng [1,2], Shuangli Shang [1], Yinwei Li [2,*] and Yiming Zhu [2]

[1] School of Health Science and Engineering, University of Shanghai for Science and Technology, Shanghai 200093, China
[2] Terahertz Technology Innovation Research Institute, University of Shanghai for Science and Technology, Shanghai 200093, China
* Correspondence: liyw@usst.edu.cn

Abstract: Terahertz video synthetic aperture radar (THz-ViSAR) has tremendous research and application value due to its high resolution and high frame rate imaging benefits. However, it requires more efficient imaging algorithms. Thus, a novel multistage back projection fast imaging algorithm for the THz-ViSAR system is proposed in this paper to enable continuous playback of images like video. The radar echo data of the entire aperture is first divided into multiple sub-apertures, as with the fast-factorized back projection algorithm (FFBP). However, there are two improvements in sub-aperture imaging. On the one hand, the back projection algorithm (BPA) is replaced by the polar format algorithm (PFA) to improve the sub-aperture imaging efficiency. The imaging process, on the other hand, uses the global Cartesian coordinate system rather than the local polar coordinate system, and the wavenumber domain data of the full aperture are obtained step by step through simple splicing and fusion, avoiding the amount of two-dimensional (2D) interpolation operations required for local polar coordinate system transformation in FFBP. Finally, 2D interpolation for full-resolution images is carried out to image the ground object targets in the same coordinate system due to the geometric distortion caused by linear phase error (LPE) and the mismatch of coordinate systems in different imaging frames. The simulation experiments of point targets and surface targets both verify the effectiveness and superiority of the proposed algorithm. Under the same conditions, the running time of the proposed algorithm is only about 6% of FFBP, while the imaging quality is guaranteed.

Keywords: THz-ViSAR; multistage back projection; PFA; global Cartesian coordinate; wavenumber domain fusion

Citation: Zheng, Q.; Shang, S.; Li, Y.; Zhu, Y. A Novel Multistage Back Projection Fast Imaging Algorithm for Terahertz Video Synthetic Aperture Radar. *Remote Sens.* **2023**, *15*, 2602. https://doi.org/10.3390/rs15102602

Academic Editor: Timo Balz

Received: 15 April 2023
Revised: 12 May 2023
Accepted: 15 May 2023
Published: 16 May 2023

1. Introduction

Synthetic aperture radar (SAR) is a high-resolution imaging radar that can operate from a long distance, in all types of weather, and throughout the day [1–4]. However, due to the low image frame rate, the conventional SAR can only obtain static target images, but not information on the moving targets, and cannot even distinguish between dynamic and static targets. Video synthetic aperture radar (Video SAR) [5] is an extension of the traditional SAR. It allows continuous video observation of the region of interest (ROI), generates a series of images during the flight of the radar platform [6], and broadcasts them in video form. Inheriting the advantages of traditional SAR, video SAR overcomes the disadvantages of traditional SAR with a low frame rate and inability to monitor in real time. Video SAR has the capability to continuously observe, and it has great potential in fields such as tracking of ground moving target indication (GMTI) and 3D imaging [7–9]. The frame rate usually exceeds 5 Hz to track moving targets and obtain information such as velocity and direction [10]. Furthermore, for a given resolution, the frame rate is proportional to the frequency [11], so a higher operating frequency results in a higher frame rate. With recent research on terahertz waves [12–17], terahertz video SAR (THz-ViSAR), which operates in the terahertz band, has attracted extensive attention because of its unique advantages. In contrast to traditional microwave video SAR, it can easily accomplish a high frame rate and high-resolution imaging. However, the required real-time performance will be much higher, so more efficient and applicable imaging algorithms [18] are needed.

There are a variety of imaging techniques. References [19–21] approximately modeled the video SAR imaging problem as tensor analysis and tensor recovery and greatly reduced the amount of required data samples during video SAR echo collection. However, in practical applications, the actual value of the tensor cannot be obtained, the rank value will also affect the error or calculation cost of the algorithms, and the dimension of vectors and matrices also imposes memory requirements. Therefore, the methods mentioned in [19–21] are usually suitable for specific scenarios.

For video SAR imaging in general scenes, two commonly employed approaches are the polar format algorithm (PFA) and the back projection algorithm (BPA) [22]. PFA can significantly mitigate the effects of distance migration by storing data in the polar format and has higher computational efficiency. Conventional PFA leads to the presence of residual phase errors due to the plane wave assumption, where linear phase error (LPE) leads to geometric distortion of the image [23] and quadratic phase error (QPE) results in defocusing. Its effective imaging scene radius is inversely proportional to the open square of the wavelength. Therefore, the algorithm is not applicable when dealing with large scenes and high-resolution cases.

BPA is a typical time-domain imaging algorithm introduced by computer tomography (CT) technology. It is suitable for arbitrary trajectories in any imaging mode, with accurate motion compensation capability and without assumptions and approximations. For an image with $N \times N$ pixels, its operation complexity for coherent accumulation within N pulses is N^3. Such a large computation load greatly restricts the broad application of BPA. To this end, academics have undertaken extensive studies and proposed several methods [24–31] to reduce the computational burden of the traditional BPA and applied them to different modes [32–36]. The most representative ones are the fast back projection algorithm (FBP) and the fast factorized back projection algorithm (FFBP). FBP was formally proposed by A. F. Yegulalp [24] at Lincoln Laboratory, thus laying the foundation for the fast time-domain algorithm. FBP performs sub-aperture division and reconstructs the sub-images in local polar coordinates. Then, based on the geometric relationship between the sub-aperture and the scene center, the sub-images are transformed into the Cartesian coordinate system and coherently summed to obtain a full-resolution image. Although FBP is computationally efficient, it reduces image quality. FFBP, proposed by Ulander et al., has higher operational efficiency [26]. It adopts the same processing as FBP in the initial stage, but the difference is that the final image is obtained through fusion step-by-step. This process mainly realizes the mapping between coordinates using 2D interpolation operations [36], which inevitably leads to the introduction and accumulation of interpolation errors. Additionally, it is difficult for FFBP to balance efficiency and image quality and to reach its theoretical calculation amount in practical applications.

To address the contradiction between efficiency and image quality in traditional FFBP, and further improve the efficiency of the traditional imaging algorithms, this paper proposes a novel multistage back projection algorithm for THz-ViSAR fast imaging. The main technical contributions are as follows:

(1) FFBP uses inefficient BP integration to obtain the sub-images in the initial stage. The calculation amount increases with the sub-aperture length and is close to BPA. Differently, the proposed algorithm uses more efficient PFA to process the sub-aperture data, reducing the number of interpolations in this stage.

(2) FFBP is based on local polar coordinates, and the fusion stage requires many 2D interpolations. Differently, the proposed algorithm is based on the global Cartesian coordinate with a simpler geometric configuration, and the fusion stage can be realized by wavenumber domain splicing, preventing the introduction and accumulation of interpolation errors in FFBP. Through the above improvements, the efficiency is significantly improved.

(3) Aiming at the geometric distortion caused by linear phase error (LPE) and considering the image rotation caused by different flight trajectories, the proposed algorithm carries out 2D resampling to correct the geometric distortion and rotate the images into the same ground Cartesian coordinate.

The rest of this paper is organized as follows. Section 2 introduces the radar echo model and the FFBP algorithm. Section 3 proposes a novel imaging algorithm to address the inconsistencies existing in traditional imaging algorithms. In Section 4, the proposed algorithm is simulated and compared with the imaging result of FFBP to verify the superiority of the proposed algorithm. Finally, the conclusions are drawn in Section 5.

2. Materials

2.1. Radar Echo Model

In a typical THz-ViSAR imaging mode, the radar platform is in linear spotlight mode during the flight of each frame. In this mode, SAR can continuously monitor the region of interest (ROI) while obtaining images to generate video. The imaging geometry configuration is shown in Figure 1. The top view is shown in (a), where green, black, and blue lines correspond to the flight trajectory of the $k-1$, k, and $k+1$ frames, respectively, and each frame is in linear spotlight mode. Taking frame k as an example, its 3D view is shown in Figure 1b. During the data collection, the platform flies at a constant elevation angle φ. The shaded area is the irradiation scene, and the coordinate origin O is the scene's center. The radar platform makes a uniform linear motion with velocity V_a, and its instantaneous position is $(V_a t_a, Y_s, H)$. The radar assumes transmit linear frequency modulation (LFM) signal, as shown in (1).

$$s_t(t_a, t_r) = rect\left(\frac{t_r}{T_r}\right) exp\left\{ j2\pi\left(f_c t_r + \frac{1}{2}\gamma t_r^2 \right) \right\}, \tag{1}$$

where t_a is the azimuth slow time, t_r is the range fast time, T_r is the pulse width, f_c is the center frequency, modulation frequency $\gamma = B/T_r$, B is the bandwidth, and $rect(\cdot)$ is rectangular window function.

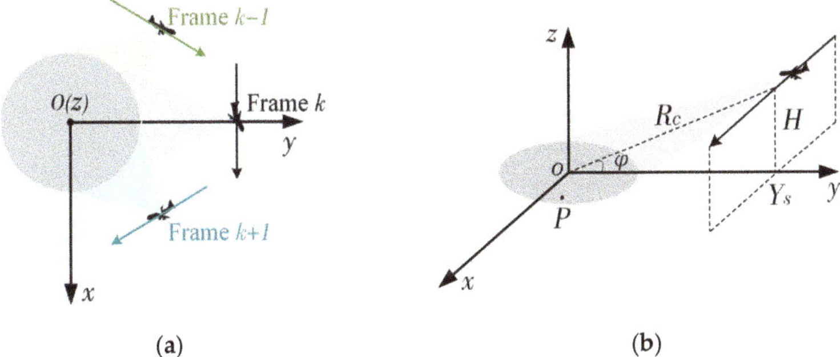

Figure 1. ViSAR imaging geometry: (**a**) Top view; (**b**) 3D view.

For any point target P $(x, y, 0)$ in the scene, the instantaneous slant range from the radar platform to point P is:

$$R_p = \sqrt{(V_a t_a - x)^2 + (Y_s - y)^2 + H^2}, \tag{2}$$

Therefore, the echo signal of the target is expressed as:

$$s_r(t_a, t_r) = rect\left(\frac{t_r - \tau}{T_r}\right) exp\left\{ j2\pi\left(f_c(t_r - \tau) + \frac{1}{2}\gamma(t_r - \tau)^2 \right) \right\}, \tag{3}$$

where time delay $\tau = 2R_p/c$.

Since the radar operates in the THz band, dechirping is frequently used to receive the echo, lowering the frequency bandwidth of the signal and, thus, reducing the stress on the hardware. Combining the imaging geometry shown in Figure 1, the slant range from the scene center O to the radar platform is selected as the reference range, i.e.,

$$R_{ref} = \sqrt{(V_a t_a)^2 + Y_s^2 + H^2},\qquad (4)$$

Then, the reference signal is shown as:

$$s_{ref}(t_a, t_r) = rect\left(\frac{t_r - \tau_{ref}}{T_{ref}}\right) exp\left\{ j2\pi\left(f_c\left(t_r - \tau_{ref}\right) + \frac{1}{2}\gamma\left(t_r - \tau_{ref}\right)^2 \right)\right\},\qquad (5)$$

where time delay $\tau_{ref} = 2R_{ref}/c$, and T_{ref} is the pulse width of the reference signal, which is larger than T_r. The differential frequency signal, as shown in (6), is obtained by mixing the reference signal (5) with the echo (3), and the differential range $\Delta R = R_p - R_{ref}$.

$$s_{if}(t_a, t_r) = s_r(t_a, t_r) \cdot s_{ref}^*(t_a, t_r) = rect\left(\frac{t_r - \tau}{T_r}\right) exp\left(-j\frac{4\pi f_c}{c}\Delta R\right) \cdot exp\left[-j\frac{4\pi}{c}\gamma\left(t_r - \tau_{ref}\right)\Delta R\right] exp\left(-j\frac{4\pi}{c^2}\gamma\Delta R^2\right),\qquad (6)$$

2.2. Review of FFBP

FFBP is a classical time-domain imaging algorithm that significantly improves the operational efficiency of the traditional back projection algorithm. Reference [26] notes that FFBP requires the least interpolations when the factorization factor is the natural logarithm e. Practically, only integers can be taken, so 2 or 4 are often taken as the base. The flow chart of the algorithm with base two is shown in Figure 2.

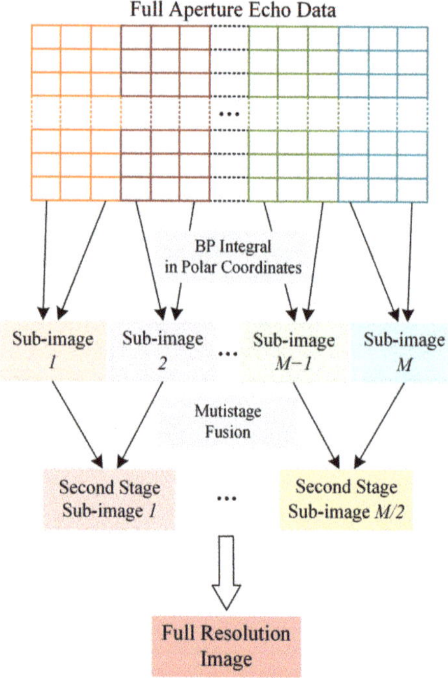

Figure 2. Flow chart of FFBP.

Firstly, FFBP divides the whole synthetic aperture into several shorter sub-apertures and performs back projection in local polar coordinates to obtain sub-images. Each sub-image has a full resolution in range and a lower resolution in azimuth. Then, two adjacent sub-images are coherently fused to obtain an image with higher azimuthal resolution in a new polar coordinate system. As the recursive fusion proceeds, the azimuthal resolution of sub-images increases. Finally, the polar coordinate image is transformed into the Cartesian coordinate system, and the full-resolution image is obtained.

However, there are still many drawbacks to this algorithm. The algorithm still uses inefficient BPA to reconstruct sub-images. Moreover, the mapping between different coordinate systems during image fusion requires several 2D interpolations, which will inevitably introduce interpolation errors. As the fusion proceeds gradually, the errors will be accumulated and amplified continuously, eventually lowering the image quality. The image quality can be improved by lengthening the sub-aperture or using a more accurate interpolation kernel, which will sacrifice the algorithm's efficiency [37,38]. Therefore, it is difficult for FFBP to reach the theoretical computing capacity in practical applications, and it is usually challenging to achieve high image quality with less computational burden. Notably, the algorithm does not consider the rotation of the images due to different flight trajectories. Therefore, it is urgent to improve the imaging algorithm for the THz-ViSAR system.

3. Methods

The efficiency of the imaging algorithm is crucial to the ViSAR system, which is related to whether ViSAR can realize real-time observation of ROI. Unfortunately, the traditional FFBP requires numerous 2D interpolations to achieve image fusion, significantly affecting the algorithm's efficiency and even the image quality. This paper proposes a novel multistage back projection algorithm to address these problems.

3.1. Principle of the Proposed Algorithm

For the above imaging geometry, the principle and processing flow of the proposed algorithm is described as follows. In the initial stage, the full aperture (length of L_a) is divided. If the number of sub-apertures is M, the sub-aperture length is $l = L_a/M$, and the global Cartesian coordinate system with the center of the full aperture as the origin is established. Figure 3 depicts the sub-apertures' division and the coordinate system's establishment. The direction of flight is taken as the positive x-axis, the direction perpendicular to the flight path and pointing to the center of the scene is taken as the positive y-axis, and the center of the full aperture is taken as the origin. Figure 3 is illustrated in a two-dimensional coordinate system to simplify the geometric configuration.

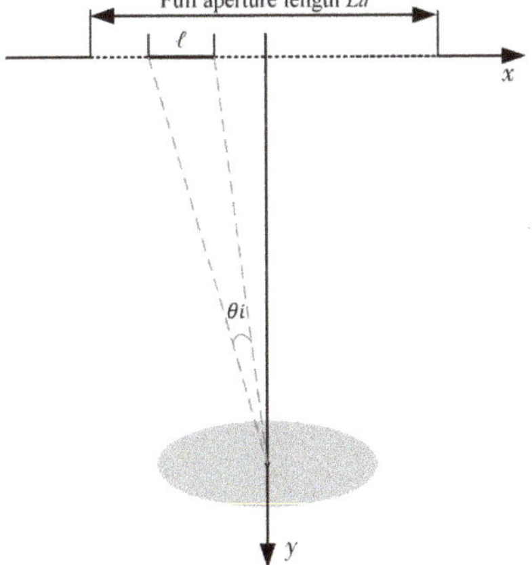

Figure 3. The division of sub-apertures and the establishment of Cartesian coordinate.

In this case, the differential frequency signal is expressed as:

$$s_{if}\left(t_a^{(i)}, t_r\right) = rect\left(\frac{t_r - \tau}{T_r}\right)\exp\left(-j\frac{4\pi f_c}{c}\Delta R\right)\cdot exp\left[-j\frac{4\pi}{c}\gamma\left(t_r - \tau_{ref}\right)\Delta R\right]exp\left(-j\frac{4\pi}{c^2}\gamma\Delta R^2\right), \tag{7}$$

where time delay $\tau = 2R_p/c$, the instantaneous slant range $R_p\left(t_a^{(i)}\right)$ from target P to the sub-aperture $i(i = 1,2\ldots M)$, and the azimuth slow time $t_a^{(i)}$ of the sub-aperture i are shown in (8) and (9), respectively.

$$R_p\left(t_a^{(i)}\right) = \sqrt{\left(V_a t_a^{(i)} - x\right)^2 + (Y_s - y)^2 + H^2}, \tag{8}$$

$$t_a^{(i)} \in \left[-\frac{L_a}{2V_a} + \frac{(i-1)l}{V_a}, -\frac{L_a}{2V_a} + \frac{il}{V_a}\right), \tag{9}$$

The differential frequency signal (7) is converted to the range frequency domain by range fast Fourier transform (FFT), as shown in (10).

$$s_{if}\left(t_a^{(i)}, f_r\right) = T_r sinc[T_r(f_r + \frac{2\gamma}{c}\Delta R)]exp\left(-j\frac{4\pi f_c}{c}\Delta R\right)exp\left(-j\frac{4\pi f_r}{c}\Delta R\right)exp\left(-j\frac{4\pi\gamma}{c^2}\Delta R^2\right), \tag{10}$$

where f_r is the range frequency.

The last two terms are the oblique item and the remaining video phase (RVP), respectively, which need to be removed by multiplying with the compensation function (11).

$$H_{RVP} = \exp\left(-j\frac{\pi f_r^2}{\gamma}\right), \tag{11}$$

Then, range fast inverse Fourier transform (IFFT) is performed and the result is shown in (12).

$$s_i\left(t_a^{(i)}, t_r\right) = rect\left(\frac{t_r - \tau}{T_r}\right)exp[-jK_r\Delta R], \tag{12}$$

where the wave number $K_r = \frac{4\pi}{c}(f_c + f_r)$.

If the plane wave-front hypothesis is adopted, the differential distance $\Delta R = R_p - R_{ref}$ can be expanded by the Taylor series as:

$$\Delta R = -x\frac{V_a t_a^{(i)}}{\sqrt{\left(V_a t_a^{(i)}\right)^2 + Y_s^2 + H^2}} - y\frac{Y_s}{\sqrt{\left(V_a t_a^{(i)}\right)^2 + Y_s^2 + H^2}} + \xi(x,y) + o(x,y), \tag{13}$$

where $\xi(x,y)$ is a second-order Taylor series expansion, and $o(x,y)$ is a third-order and higher-order term. This equation can be simplified to (14).

$$\Delta R \approx -x sin\theta cos\varphi - y cos\theta cos\varphi, \tag{14}$$

where φ is the elevation angle, θ is the azimuth accumulation angle, and

$$\begin{cases} cos\varphi = \dfrac{\sqrt{\left(V_a t_a^{(i)}\right)^2 + Y_s^2}}{\sqrt{\left(V_a t_a^{(i)}\right)^2 + Y_s^2 + H^2}} \\[4mm] sin\theta = \dfrac{V_a t_a^{(i)}}{\sqrt{\left(V_a t_a^{(i)}\right)^2 + Y_s^2}} \\[4mm] cos\theta = \dfrac{Y_s}{\sqrt{\left(V_a t_a^{(i)}\right)^2 + Y_s^2}} \end{cases}, \tag{15}$$

Then, (12) can be written as:

$$s_i(K_x, K_y) = rect\left(\frac{t_r - \tau}{T_r}\right) exp\left[j(xK_x + yK_y)\right],\tag{16}$$

where the azimuth and range wavenumber are, respectively:

$$\begin{cases} K_x = K_r sin\theta cos\varphi \\ K_y = K_r cos\theta cos\varphi \end{cases},\tag{17}$$

At this point, the data in polar format are converted to the Cartesian format by range interpolation and azimuth interpolation, respectively. Then, since there is no need to output sub-images, the multistage fusion is performed in the wavenumber domain (shown in Figure 4).

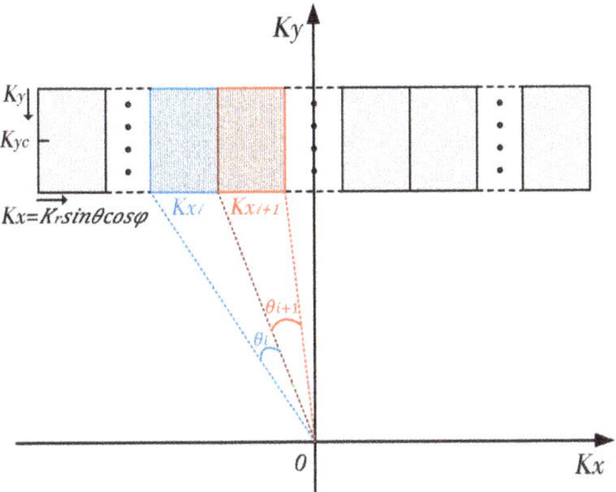

Figure 4. Wavenumber domain fusion step by step.

If the total number of stages is G, for stage $g(g = 1, 2 \ldots G)$, the number of sub-apertures is represented by K_g. For example, the number of sub-apertures in the initial stage, $K_1 = 2^{G-1}$. $s_i^g(K_x, K_y)$, is used to represent the wavenumber domain data of stage g and sub-aperture $i(i = 1, 2 \ldots K_g)$. Since the proposed algorithm adopts the global Cartesian coordinate system with the center of the full aperture as the origin, wavenumber domain splicing is adopted in this paper to achieve the fusion stage, and the process can be expressed as (18).

$$s_i^g(K_x, K_y) = \left[\left(s_{2i-1}^{g-1}(K_x, K_y)\right); \left(s_{2i}^{g-1}(K_x, K_y)\right)\right],\tag{18}$$

The wavenumber domain data of two adjacent sub-apertures are spliced to obtain the wavenumber domain data of the larger sub-apertures, and the azimuth resolution is increased two times. The procedure is repeated until the full-aperture wavenumber domain data (i.e., $s(K_x, K_y)$) are obtained. Then, 2D IFFT is performed for $s(K_x, K_y)$ to obtain the full-resolution image, described as:

$$I(x, y) = \iint s(K_x, K_y) exp\left[-j(xK_x + yK_y)\right] dxdy,\tag{19}$$

3.2. Geometric Distortion Analysis and Correction

Since the proposed algorithm extends ΔR to (14) based on the far-field assumption, leading to linear, quadratic, and higher-order phase errors, the linear phase error will cause image distortion, and the quadratic phase error will defocus the image. As shown in (12), its phase can be expressed as:

$$\Phi = K_r \Delta R = K_r \left(R_p - R_{ref} \right), \tag{20}$$

The phase Φ can be rewritten as a Taylor expansion at the center $(K_x = 0, K_y = K_{yc})$ of 2D wavenumber domain [39]:

$$\Phi = a_0 + a_x K_x + a_y \left(K_y - K_{yc} \right) + a_{xx} K_x^2 + a_{yy} \left(K_y - K_{yc} \right)^2 + a_{xy} K_x \left(K_y - K_{yc} \right) + \dots, \tag{21}$$

After formula derivations, the first-order and second-order coefficients a_x, a_y, a_{xx}, and a_{yy} can be calculated as:

$$a_x = \left. \frac{\partial \Phi}{\partial K_x} \right|_{K_x=0, K_y=K_{yc}} = \frac{-x Y_s}{\alpha \cos\varphi}, \tag{22}$$

$$a_y = \left. \frac{\partial \Phi}{\partial K_y} \right|_{K_x=0, K_y=K_{yc}} = \frac{\alpha - R_c}{\cos\varphi}, \tag{23}$$

$$a_{xx} = \left. \frac{\partial^2 \Phi}{\partial K_x^2} \right|_{K_x=0, K_y=K_{yc}} = \frac{\alpha - R_c}{K_{yc} \cos\varphi} + \frac{Y_s^2}{K_{yc} \cos\varphi} \left(\frac{-x^2}{\alpha^3} + \frac{1}{\alpha} - \frac{1}{R_c} \right), \tag{24}$$

$$a_{yy} = \left. \frac{\partial^2 \Phi}{\partial K_y^2} \right|_{K_x=0, K_y=K_{yc}} = 0, \tag{25}$$

where $\alpha = \sqrt{x^2 + (Y_s - y)^2 + H^2}$, and $R_c = \sqrt{Y_s^2 + H^2}$.

The first-order terms a_x, a_y cause geometric distortion. Additionally, the offset coordinate of the target (x, y) in the image is

$$\begin{cases} \widetilde{x} = \frac{x Y_s}{\alpha \cos\varphi} \\ \widetilde{y} = \frac{\alpha - R_c}{\cos\varphi} \end{cases}, \tag{26}$$

where $\widetilde{x}$ and $\widetilde{y}$ are the distorted azimuth and range coordinates, respectively. Therefore, the target position offset in each sub-image in the initial stage is consistent, and then as the fusion stage continues, the geometric distortion in each image in each step is also consistent. In other words, the target position offset in the full-resolution image also conforms to (26).

To describe the image distortion more directly, r_e is defined as the range error between the true position (x, y) and the offset position $\left(\widetilde{x}, \widetilde{y} \right)$ of the point targets.

$$r_e = \sqrt{\left(x - \widetilde{x} \right)^2 + \left(y - \widetilde{y} \right)^2}, \tag{27}$$

Figure 5 shows the range error in the whole scene, in which the solid red line is the contour with a range error of 1 m, and the solid green line is the contour with a distance error of 0.1 m. It is easy to find that the closer the point target is to the center, the smaller the range error is, and vice versa. For example, the range error of the point at (50, 50) is larger than 1 m, which is almost equivalent to ten resolution units (resolution of 0.12 m). The ViSAR system usually requires the target to be positioned at the actual location in the image, so geometric distortion correction [40] of the image is essential.

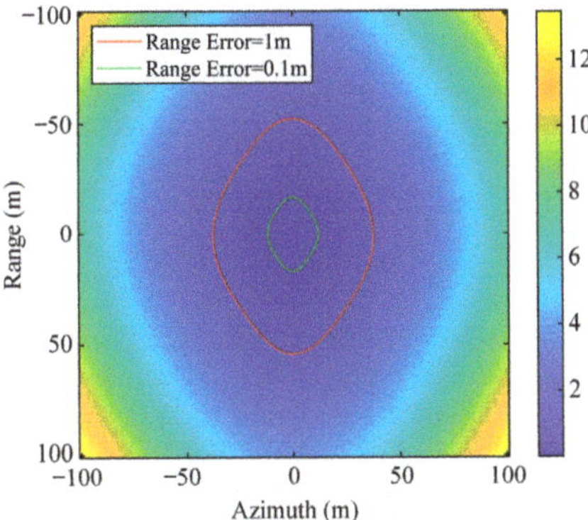

Figure 5. Range error in 220 GHz ViSAR image.

It should be noted that when there is an angle ϑ between the flight trajectories of different frames (see Figure 1a), the proposed algorithm reconstructs the image in a Cartesian coordinate centered on the full aperture, which results in a corresponding angular rotation of the full-resolution image. The rotation is shown in (28).

$$\begin{bmatrix} x' \\ y' \end{bmatrix} = \begin{bmatrix} \cos\vartheta & -\sin\vartheta \\ \sin\vartheta & \cos\vartheta \end{bmatrix} \begin{bmatrix} \tilde{x} \\ \tilde{y} \end{bmatrix}, \tag{28}$$

However, each frame should be built in the coordinate system with the scene center as the origin. Therefore, each point (x', y') in the image needs to be corrected and rotated into the ground coordinate system, and its principle is shown in Figure 6.

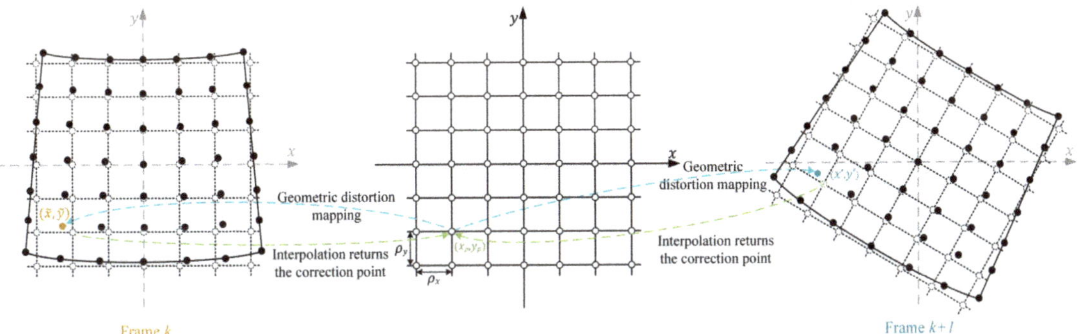

Figure 6. Schematic diagram of the image correction to the same coordinate system.

As can be seen from (26) and (28), the corresponding offset position $\left(\tilde{x}, \tilde{y}\right)$ or (x', y') in the full-resolution image can be obtained according to the real position coordinates (x, y) of the ground point target. Furthermore, the mapping relationship between them is known so that the correction can be realized by image domain resampling. Firstly, the correction area is selected, and the correction grid is divided, as shown in Figure 6. The correction points are evenly distributed in the xy coordinate system, and the horizontal and vertical intervals of adjacent correction points are ρ_x, ρ_y, respectively. For each correction point (x_p, y_p), its coordinates $\left(\tilde{x}, \tilde{y}\right)$ or (x', y') in the full-resolution image are calculated, and then the correction point is returned from this coordinate in the full-resolution image by bilinear interpolation so that the correction of a single correction pixel is achieved. Then, the above process is repeated for all pixels of the full-resolution image. While realizing geometric distortion correction, the images are rotated into the same ground coordinate system.

The quadratic terms a_{xx}, a_{yy} are space-variable functions [41], which will defocus the image. Moreover, the farther the target is from the scene's center, the more severe the defocusing is. Therefore, the QPE leads to the scene size limitation, and the quantitative expressions for the maximum allowable scene radius are given in (29).

$$r_{\pi/4} = \rho_a \sqrt{\frac{R_c}{\lambda_c}},$$

(29)

where $r_{\pi/4}$ is the scene radius using $\pi/4$ as the allowable quadratic phase error. From (29), it is known that the allowed scene radius is inversely proportional to the square root of the wavelength λ_c, or proportional to the square root of the center frequency. Once the imaging region does not satisfy this limitation, the image requires phase error correction. Generally, it is challenging to correct QPE with spatial post-processing [42,43]. Fortunately, this paper considers the characteristics of the small imaging region of THz-ViSAR and the large allowable radius of QPE. This paper considers that QPE can be ignored, i.e., all targets in the whole scene have good focusing performance.

3.3. Algorithm Processing Flow

The proposed algorithm flow chart is shown in Figure 7, and the key steps are as follows. Firstly, the aperture division is performed, and the below processes are carried out separately in the global Cartesian coordinate: removal of RVP and oblique items, range and azimuth interpolation. Then, two adjacent sub-apertures are spliced in the wavenumber domain, which is repeated until the full aperture wavenumber domain is obtained. Furthermore, the full-resolution image is obtained using 2D IFFT. Finally, correction is performed to obtain the final image.

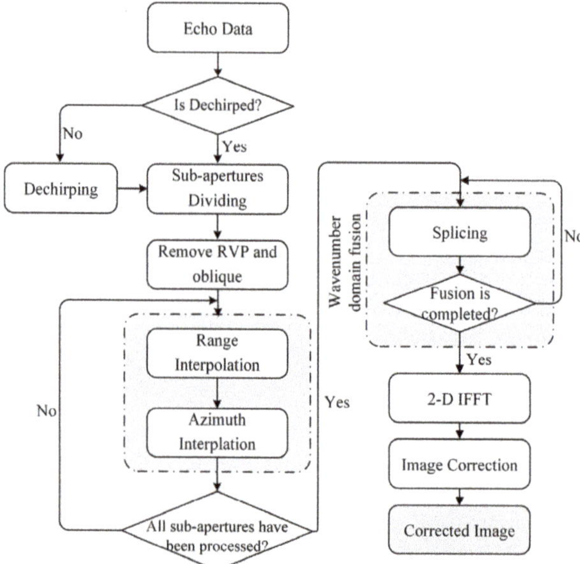

Figure 7. Flow chart of the proposed algorithm.

Compared with traditional FFBP, the proposed algorithm has the following advantages. Firstly, the reconstruction of sub-images abandons the inefficient BPA and only requires fewer interpolation operations [44–47]. Secondly, the global Cartesian coordinate is used to replace local polar coordinates. Compared with the latter, the former has a simpler geometric structure and is easier to implement programmatically. Moreover, simple wavenumber domain splicing can realize the fusion stage, thus avoiding introduction and accumulation of interpolation errors. Therefore, the proposed algorithm further reduces the processing complexity of the traditional imaging algorithm. Last, the geometric distortion and rotation between different frames are considered and corrected in this paper.

3.4. Computing Load Analysis

Assume that the number of pulses contained in the full aperture is N, and the full-resolution image contains $N \times N$ pixels. The full aperture is divided into N/n sub-apertures containing n pulses. According to the algorithm processing flow, the calculation amount is analyzed as follows:

The following processing for each sub-aperture data is performed separately.

(1) Removal of RVP and oblique items requires the following operations: range FFT ($Nn\log_2\sqrt{N}$), multiplied by the compensation function (Nn) and range IFFT ($Nn\log_2\sqrt{N}$).

(2) Range interpolation is realized via range FFT ($Nn\log_2\sqrt{N}$), 8-fold up-sampling, and range IFFT ($8Nn\log_2\sqrt{8N}$).

(3) Azimuth interpolation is realized via azimuth FFT ($Nn\log_2\sqrt{n}$), 8-fold up-sampling, and azimuth IFFT ($8Nn\log_2\sqrt{8n}$).

Thus, the number of complex multiplication operations required for each sub-aperture processing at the initial stage is $25Nn + 11Nn\log_2\sqrt{N} + 9Nn\log_2\sqrt{n}$. Therefore, the number of complex multiplication operations dealing with N/n sub-apertures data is $N/n\left(25Nn + 11Nn\log_2\sqrt{N} + 9Nn\log_2\sqrt{n}\right)$, i.e., $25N^2 + 11N^2\log_2\sqrt{N} + 9N^2\log_2\sqrt{n}$.

Then, the wavenumber domain fusion stage is realized via splicing, and the number of multiplications in this stage can be ignored. Additionally, the multiplication times of 2D IFFT to get the full-resolution image is $2N^2\log_2\sqrt{N}$. Finally, the image correction is achieved via bilinear interpolation ($4N^2$).

Summing up the above, the total multiplications of the algorithm in this paper are $29N^2 + 13N^2\log_2\sqrt{N} + 9N^2\log_2\sqrt{n}$. Additionally, through calculation, the total multiplications required by FFBP under the same conditions are $8nN^2 + 16N^2\log_2 N/n$.

Set $N = 2048$ to compare the computation burden of the proposed algorithm and FFBP, as shown in Figure 8. It is not difficult to find that the computation amount of FFBP increases with the increase of the number of sub-aperture pulses n, and the trend is significant. However, the changing trend of the proposed algorithm is much slower and lower than FFBP. Therefore, the proposed algorithm has higher efficiency. The area in the red box is enlarged as shown in the figure. It can be seen that the computing load of the proposed algorithm increases monotonically with the number of sub-aperture pulses n. In other words, the larger the n, the longer the sub-aperture length, and the larger the arithmetic amount. However, n cannot be infinitesimally small, resulting in a low azimuth resolution of the sub-images. Therefore, the value of n should be selected according to the practical application.

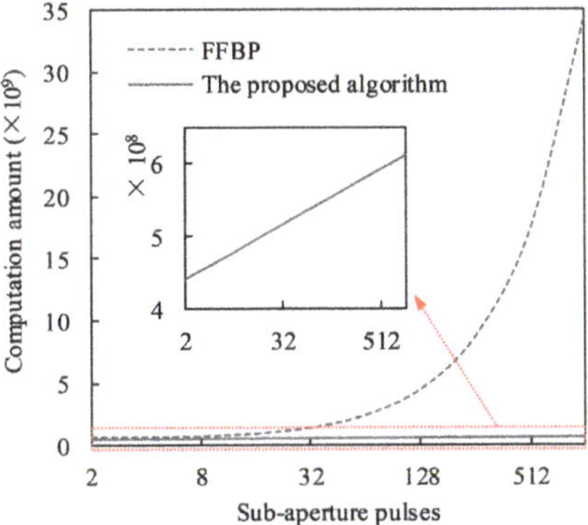

Figure 8. Comparison of the computation amount of the proposed algorithm and FFBP. The red arrow points to the enlarge of the area selected by the red dotted line.

4. Results

4.1. Point Targets Simulation Results

The simulation experiments were conducted in this paper to verify the proposed algorithm's effectiveness and superiority. The main radar system parameters for the simulation experiment are shown in Table 1, which were used to simulate the echo data. There are 11 × 11 point targets evenly distributed in the scene (see Figure 9). The final imaging grid's pixel points are 2048 × 2048 (Range × Azimuth). The theoretical value of 2D resolution is about 0.12 m × 0.12 m (Range × Azimuth). Here, the whole synthetic aperture is divided into eight sub-apertures, and each sub-aperture contains 256 pulses. Both algorithms adopt two as the radix for fusion in this simulation experiment.

Table 1. Radar system parameters.

Parameters	Explain	Value
f_c	center frequency	220 GHz
B	bandwidth	1.2 GHz
T_r	pulse width	50 μs
φ	elevation angle	45°
R_c	slant range of scene center	1 km
V_a	flight speed	50 m/s
r_i	radius of the imaging area	60 m
ρ_r	range resolution	0.12 m
ρ_a	azimuth resolution	0.12 m

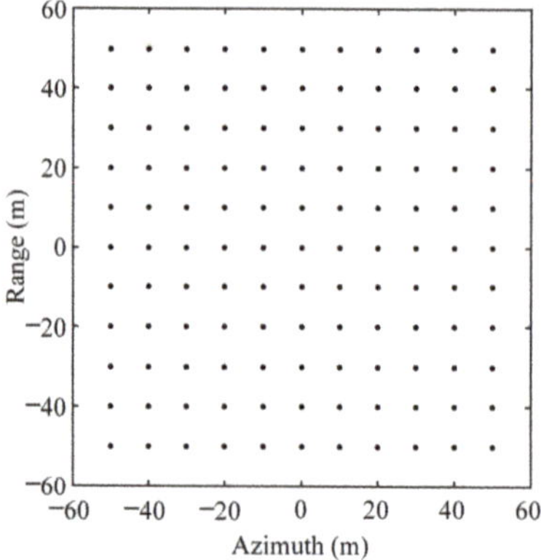

Figure 9. Point targets distribution.

Some images of each stage and their geometric distortion are given. For example, eight sub-apertures are divided in this paper; both algorithms have four steps. Firstly, eight sub-images can be obtained in the initial stage, and some of them were selected for illustration, as shown in Figure 10. In these images, the red dots indicate the theoretical point target locations and the white dots indicate the distribution during imaging. It can be seen that the target position offset in each sub-image in the initial stage is the same and follows the theoretical situation.

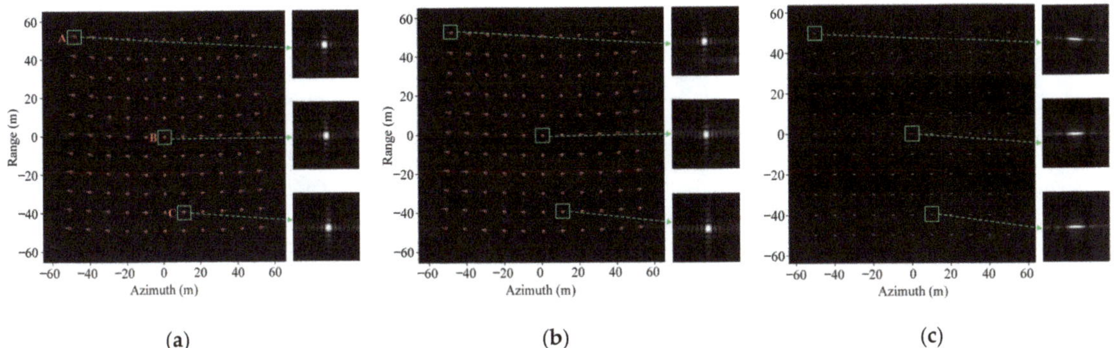

Figure 10. Sub-images in the initial step: (**a**,**b**) Sub-images 1, 2 of the proposed algorithm; (**c**) Sub-image 1 of FFBP.

Three targets, A, B, and C, were enlarged, and their focusing effects were analyzed. The results of impulse response width (IRW), peak sidelobe ratio (PSLR), and integral sidelobe ratio (ISLR) are shown in Table 2. It can be seen that the azimuth resolution of the sub-image at this stage is about eight times the theoretical value, and the range resolution is close to the theoretical value. That is, the sub-images in this stage have the characteristics of lower resolution in azimuth and full resolution in range. However, the azimuth resolution of the proposed algorithm is slightly lower than FFBP due to no assumptions and approximations in the initial stage of FFBP processing. In addition, the results of PSLR and ISLR indicate that the focusing effect is good.

Table 2. IRW, PSLR, and ISLR values of point targets A/B/C in the initial step sub-images.

		Sub-Image 1 of the Proposed Algorithm (A/B/C)	Sub-Image 2 of the Proposed Algorithm (A/B/C)	Sub-Image 1 of FFBP (A/B/C)
IRW (m)	Range	0.16/0.15/0.16	0.16/0.16/0.15	0.16/0.15/0.15
	Azimuth	1.03/0.99/1.04	1.01/0.97/1.02	0.93/0.97/0.98
PSLR (dB)	Range	−12.90/−13.31/−13.16	−12.95/−13.22/−13.04	−13.30/−13.43/−13.26
	Azimuth	−12.38/−12.97/−12.88	−12.33/−11.69/−12.53	−17.80/−11.79/−12.70
ISLR (dB)	Azimuth	−28.49/−29.21/−31.64	−28.43/−30.35/−28.63	−26.70/−27.91/−25.82
	Range	−26.81/−25.75/−29.48	−27.66/−23.92/−22.39	−45.21/−26.27/−29.89

In the second step, four sub-images can be obtained, some shown in Figure 11. Since the geometric distortion of each sub-image in the initial stage is consistent and in line with the theoretical derivation (see (26)), the distortion will not change after splicing two adjacent sub-aperture wavenumber domains. Then, the third step can obtain two images (see Figure 12) with the same position offset satisfying the previous analysis. The same analysis of the three targets A, B, and C in each image shows that the fusion of two adjacent sub-images doubles the azimuth resolution, and the focusing effect is good. This section is omitted to save article space.

The full-resolution images are shown in Figure 13. Similarly, the geometric distortion in Figure 13a still conforms to the previous analysis, i.e., it also satisfies (26). The result of the correction is shown in Figure 13b. It can be seen from the image before correction shown in Figure 13a that the point targets that should be uniformly distributed in a rectangle are imaged with a certain degree of distortion due to wave-front bending error. After the geometric distortion correction processing, the distortion of the image is improved, as shown in Figure 13b.

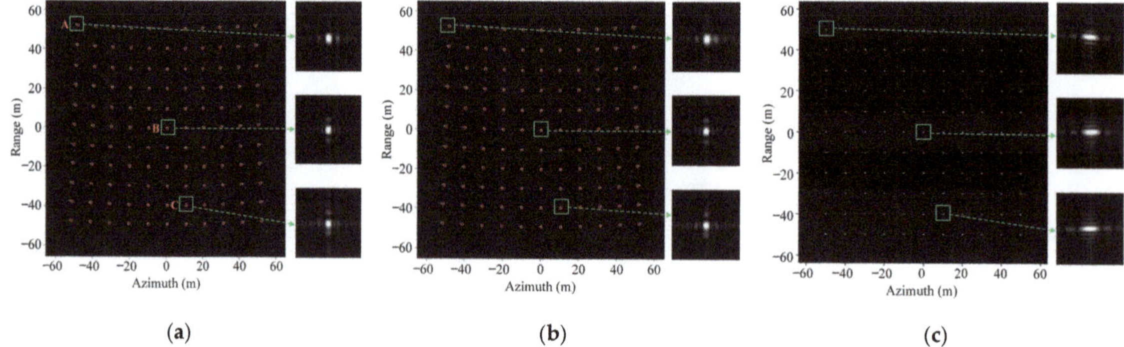

Figure 11. Sub-images in the second step: (**a**,**b**) Sub-images 1, 2 of the proposed algorithm; (**c**) Sub-image 1 of FFBP.

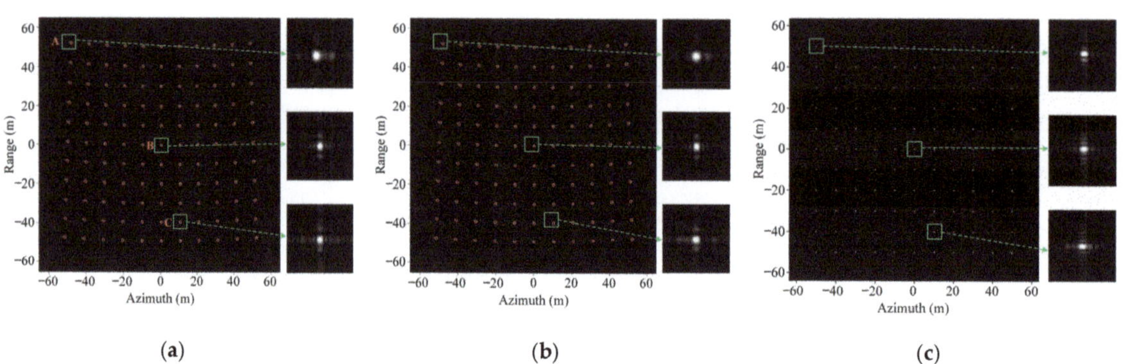

Figure 12. Sub-images in the third step: (**a**,**b**) Sub-images 1, 2 of the proposed algorithm; (**c**) Sub-image 1 of FFBP.

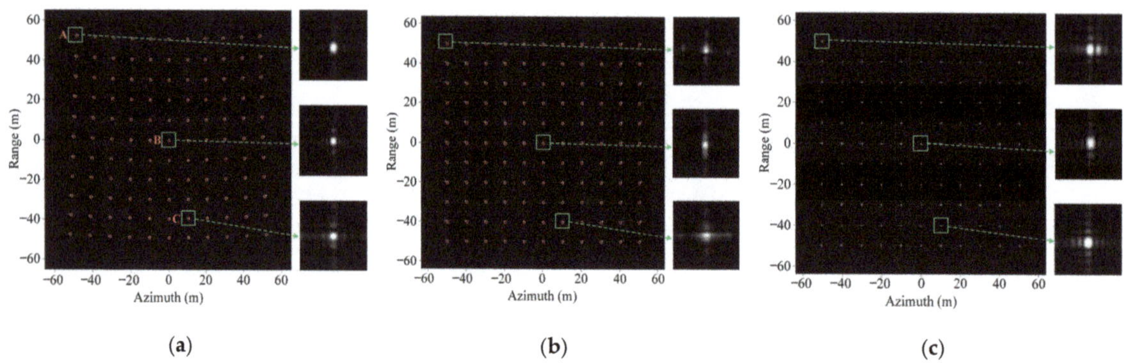

Figure 13. Full-resolution images: (**a**,**b**) Images before and after correction of the proposed algorithm; (**c**) Image of FFBP.

Selecting representative points A, B, and C to analyze, and the 2D profiles of these points are, respectively, given in Figure 14. It is not difficult to see that the sidelobes of the 2D profiles in this proposed algorithm are some lower. The results of IRW, PSLR, and ISLR are shown in Table 3. The PSLR and ISLR of the proposed algorithm are lower than those of the FFBP, indicating that the image obtained by the proposed algorithm is better focused. Since there are assumptions in the proposed algorithm, the azimuth resolution of the algorithm is slightly lower than that of FFBP. However, it is still close to the theoretical value (0.12 m), so this situation is acceptable. Here, the image has full resolutions in azimuth and range. In summary, FFBP imaging accuracy has a certain loss, and more accurate interpolation methods or higher interpolation multipliers are needed to improve imaging accuracy, but this will inevitably increase the complexity of FFBP operations. In contrast, the proposed algorithm has higher imaging accuracy than FFBP, thus confirming the advantages of this algorithm.

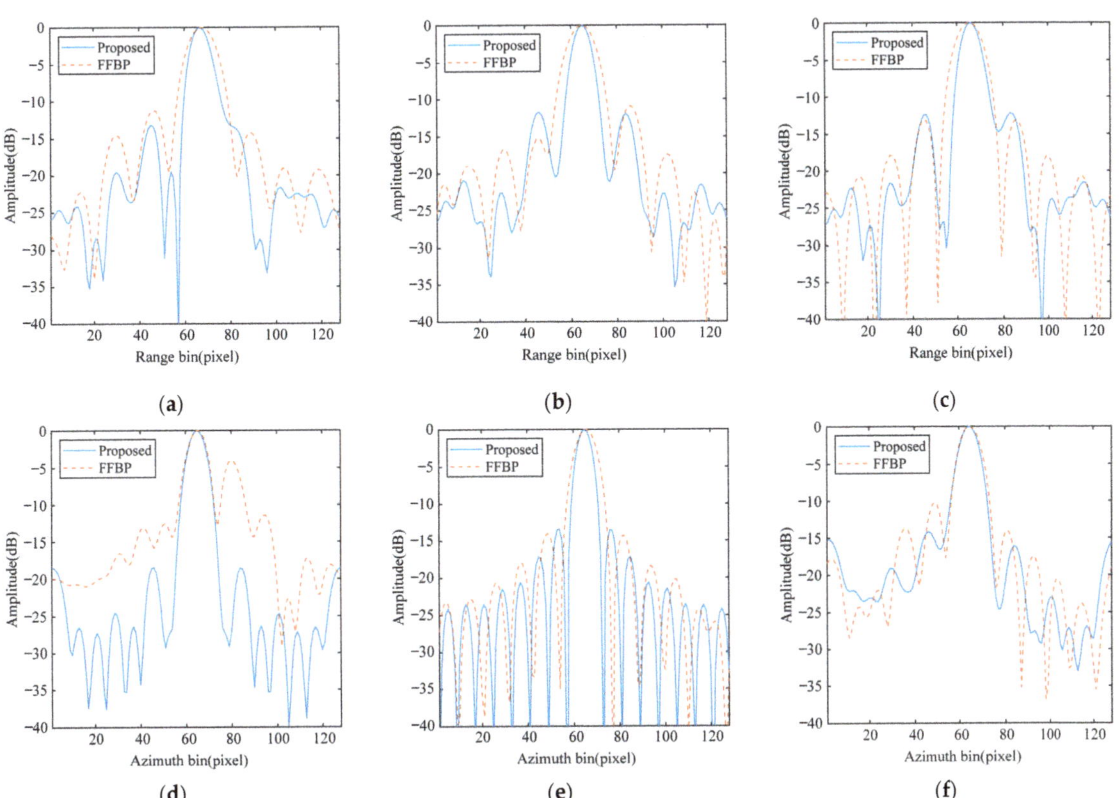

Figure 14. 2D profiles of point targets A/B/C in the full-resolution images: (**a–c**), respectively, are the range profiles of A/B/C; (**d–f**), respectively, are the azimuth profiles of A/B/C.

Table 3. IRW, PSLR, and ISLR values of point targets A/B/C in the full-resolution images.

		The Proposed Algorithm (A/B/C)	FFBP (A/B/C)
IRW (m)	Range	0.15/0.15/0.15	0.16/0.16/0.15
	Azimuth	0.14/0.12/0.15	0.11/0.12/0.12
PSLR (dB)	Range	$-11.89/-11.69/-12.23$	$-11.18/-10.88/-13.10$
	Azimuth	$-18.35/-13.42/-14.17$	$-3.95/-14.03/-10.20$
ISLR (dB)	Range	$-$Inf/-Inf/-Inf	$-22.53/-23.51/-25.57$
	Azimuth	$-$Inf/-25.08/-Inf	$-15.85/-25.90/-19.91$

The azimuth direction of the target in Figure 10c is widened by comparing the zoomed point target images at each stage. As the fusion stage continues, the azimuth is further compressed. This is because the initial stage of FFBP uses BP for processing with a pre-established imaging grid and a fixed sampling interval. The number of pulses accumulates and the azimuthal resolution increases continuously as the fusion continues. However, the proposed algorithm does not need to establish an imaging grid in the initial stage, which means that the sampling interval decreases with the processing of each stage, the azimuthal resolution increases, and the full-resolution image is finally obtained.

Table 4 shows the actual position and the coordinate comparison before and after correction to display the correction effect visually. Here, the same three targets, A, B, and C, are chosen. It is easy to see from the table that the imaging coordinates obtained after geometric distortion correction are very close to the actual position. Although there are still errors, they are within acceptable limits.

Table 4. Geometric distortion correction results of the proposed algorithm.

	Point A	Point B	Point C
Real Position	$(-50, 50)$	$(0, 0)$	$(10, -40)$
Before Correction	$(-48.9, 52.6)$	$(0.1, 0.1)$	$(10.6, -39.4)$
After Correction	$(-50.4, 50.2)$	$(0.1, 0.1)$	$(10.3, -39.8)$

When the flight trajectory of Frame $k + 1$ and Frame k are at an angle $\vartheta = 30°$, the imaging results of the two algorithms are shown in Figure 15. The images before and after correction in this paper are shown in Figure 15a,b, where the red dots represent the theoretical distribution. It is not difficult to see that the full-resolution image undergoes a rotation of the corresponding angle. The corrected image Figure 15b is established in the Cartesian coordinate system with the scene's center as the origin. However, the traditional FFBP does not consider the image rotation correction caused by the different flight trajectories.

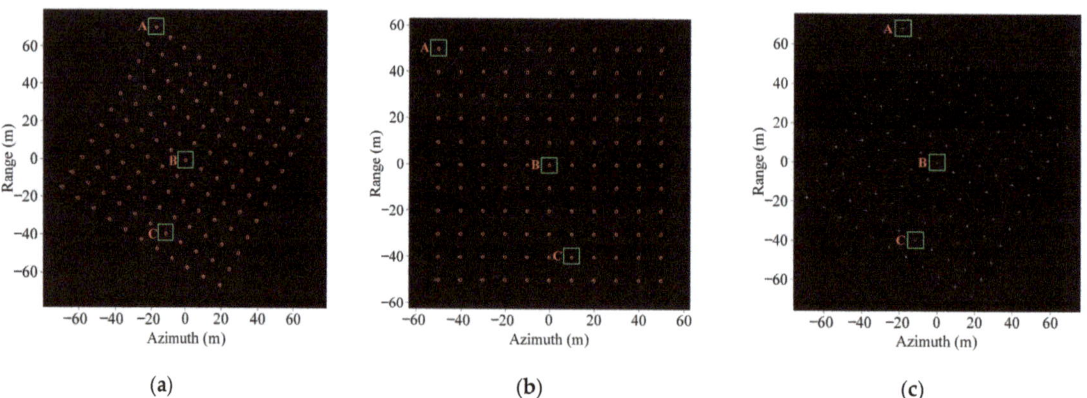

(a)　　　　　(b)　　　　　(c)

Figure 15. Imaging results of point targets with $\vartheta = 30°$: (**a,b**) Images before and after correction of the proposed algorithm; (**c**) Image of FFBP.

Three targets, A, B and C, are also selected, and their focusing effect analysis and position comparison are shown in Tables 5 and 6. It can be seen that the focusing effect of the proposed algorithm is slightly better than FFBP, and the rotation and geometric distortion of the image are substantially corrected. Therefore, this algorithm is still suitable for arbitrary linear trajectory SAR.

Table 5. IRW, PSLR, and ISLR values of point targets A/B/C in the images with $\vartheta = 30°$.

		The Proposed Algorithm (A/B/C)	FFBP (A/B/C)
IRW (m)	Range	0.15/0.15/0.16	0.17/0.16/0.15
	Azimuth	0.16/0.12/0.14	0.11/0.12/0.12
PSLR (dB)	Range	$-12.65/-11.77/-14.61$	$-13.02/-10.79/-13.03$
	Azimuth	$-12.36/-13.30/-16.68$	$-9.43/-13.73/-9.77$
ISLR (dB)	Range	$-34.70/$-Inf$/-34.38$	$-27.51/-23.52/-25.92$
	Azimuth	-Inf$/-24.84/$-Inf	$-27.57/-25.75/-25.63$

Table 6. Correction results of geometric distortion and rotation of the proposed algorithm.

	Point A	Point B	Point C
Real Position	$(-50, 50)$	$(0, 0)$	$(10, -40)$
Before Correction	$(-17.6, 70.1)$	$(0.1, -0.1)$	$(-11.81, -39.1)$
After Correction	$(-50.5, 50)$	$(-0.1, 0)$	$(9.6, -40)$

4.2. Surface Target Simulation Results

The surface target simulation was also performed to verify the proposed algorithm's effectiveness. The input image of the simulation experiment, shown in Figure 16, is a SAR image of the stationary ground scene, which was acquired in Mianyang City, Sichuan Province, China, on June 2011 by the X-band airborne dual-antenna SAR system, developed by the Institute of Electronics, Chinese Academy of Sciences [48].

Figure 16. Input SAR image of the surface target simulation.

The system parameters, shown in Table 1, were also used in the surface target simulation. Then, the radar echo data in the linear spotlight mode were generated [49–51]. Figure 17a,b show the imaging results with $\vartheta = 0°$ for the proposed algorithm. It can be seen from the image before correction that LPE causes the image to be distorted into a fan-shaped image. Compared with Figure 17a, Figure 17b shows that the geometric distortion has been effectively corrected. Additionally, the imaging result of FFBP is shown in Figure 17c for comparison.

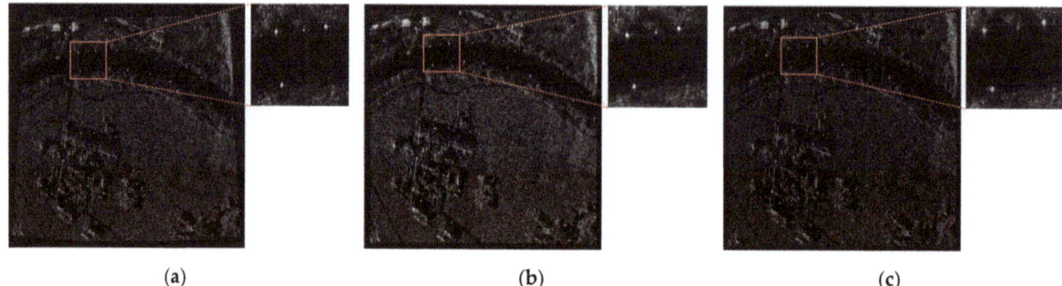

(a)   (b)   (c)

Figure 17. Imaging results of the surface target with $\vartheta = 0°$: (**a**,**b**) Images before and after correction of the proposed algorithm; (**c**) Image of FFBP.

For a more visual illustration, Table 7 shows the entropy, normalized root means square error (NRMSE), peak signal-to-noise ratio (PSNR), and running time of the two imaging algorithms. Both algorithms were run on the same computer processor with "Intel(R) Core(TM) i5-10400 CPU @ 2.90 GHz", 16 GB RAM, and 12 threads. As can be seen from the table, the entropy, NRMSE and PSNR of the images are very close, so the imaging quality of both algorithms is comparable. However, the time consumption of the proposed algorithm is about 6% of FFBP, indicating that the proposed algorithm is much more efficient. The experimental results prove that the proposed algorithm can obtain images with an excellent focusing effect in the THz band, and is more conducive to ViSAR fast imaging.

Table 7. Image quality and time consumption of different algorithms.

	Entropy	NRMSE	PSNR (dB)	Running Time (min)
Input image	12.88	0	∞	/
Imaging of the proposed algorithm	13.74	0.18	31.05	2.12
Imaging of FFBP	13.52	0.22	30.14	35.27

In addition, this section considers the angle $\vartheta = 30°$ between the flight trajectories of different frames, and the imaging results are shown in Figure 18. It is not difficult to see from Figure 18a that geometric distortion and corresponding rotation occur in the full-resolution image. After correction, as shown in Figure 18b, the image is established in the ground coordinate system with the scene's center as the origin. Both the geometric distortion and the rotation are corrected effectively. However, FFBP does not consider the correction of image rotation (see Figure 18c). This further confirms that the proposed algorithm still applies to the non-ideal linear trajectory.

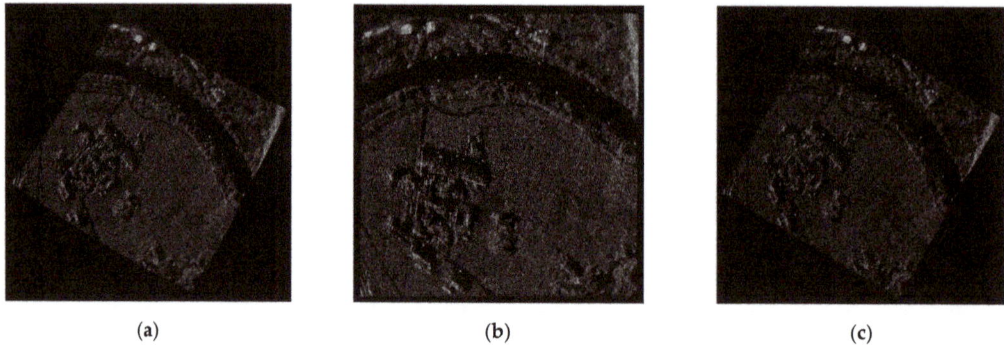

(a)   (b)   (c)

Figure 18. Imaging results of the surface target with $\vartheta = 30°$: (**a**,**b**) Images before and after correction of the proposed algorithm; (**c**) Image of FFBP.

5. Conclusions

The high frame rate and high-resolution imaging of THz-ViSAR require higher efficiency of the imaging algorithm. BPA is applicable to arbitrary modes and can reconstruct images on arbitrary imaging grids. However, its computational complexity is too high to meet the system requirements. Although FFBP has improved its efficiency significantly, it is difficult to reach its theoretical computational power in practical applications and to balance image quality and computational efficiency. To solve these problems and further improve the efficiency of the imaging algorithm, a novel multistage back projection algorithm is proposed in this paper. The processing, based on PFA and global Cartesian coordinate, greatly avoids interpolation times and errors in the traditional algorithm. Moreover, the geometric distortion and rotation existing in the images are analyzed and corrected to ensure uniformity between frame images. The results of this study are as follows:

(1)   By analyzing the number of complex multiplications, it is shown that the computational effort of the proposed algorithm is significantly lower than that of FFBP.

(2)   The point target simulation experiment analyses the IRW, PSLR, and ISLR at each stage, confirming that the proposed algorithm and the FFBP focusing effect is comparable. Analyzing the positions of point targets before and after image correction confirms that the proposed algorithm can effectively complete the correction of geometric distortion and rotation.

(3)   The surface target simulation experiment analyses the entropy, NRMSE, PSNR, and the running time of different algorithms, indicating that the proposed algorithm is more efficient and ensures the quality and uniformity of images.

Experimental results show that the proposed algorithm can obtain images with a good focusing effect more efficiently, which is more conducive to ViSAR fast imaging. Relevant research results are of great significance to the development of video SAR imaging technology. However, the measured data contain many motion errors, so the implementation of the proposed algorithm in the theoretical case is only considered in this paper. In future studies, the proposed algorithm will be combined with motion error compensation for continuous and rapid imaging of THz-ViSAR and extended to other research fields, such as ship imaging.

Author Contributions: Conceptualization, Q.Z., S.S. and Y.L.; methodology, S.S.; validation, Q.Z. and S.S.; formal analysis, S.S. and Y.L.; writing—original draft preparation, Q.Z. and S.S.; writing—review and editing, Q.Z., S.S., Y.L. and Y.Z. All authors have read and agreed to the published version of the manuscript.

Funding: This research was funded by the National Natural Science Foundation of China (12105177), the National Natural Science Foundation of China (61988102), the Natural Science Foundation of Shanghai (21ZR1444300), and the Opened Foundation of Hongque Innovation Center (HQ202204002).

Data Availability Statement: Not applicable.

Acknowledgments: The authors wish to acknowledge MaoSheng Xiang at Aerospace Information Research Institute, Chinese Academy of Sciences, for his helpful discussions and data support.

Conflicts of Interest: The authors declare no conflict of interest.

References

1.   Song, X.; Yu, W. Processing video-SAR data with the fast backprojection method. *IEEE Trans. Aerosp. Electron. Syst.* **2017**, *52*, 2838–2848. [CrossRef]

2.   Xu, G.; Zhang, B.; Yu, H.; Chen, J.; Xing, M.; Hong, W. Sparse synthetic aperture radar imaging from compressed sensing and machine learning: Theories, applications, and trends. *IEEE Geosci. Remote Sens. Mag.* **2022**, *10*, 32–69. [CrossRef]

3.   Zhang, B.; Xu, G.; Zhou, R.; Zhang, H.; Hong, W. Multi-channel back-projection algorithm for mmwave automotive MIMO SAR imaging with Doppler-division multiplexing. *IEEE J. Sel. Top. Signal Process.* **2022**, 1–13. [CrossRef]

4.   Shi, J.; Zhou, Y.; Xie, Z.; Yang, X.; Guo, W.; Wu, F.; Li, C.; Zhang, X. Joint autofocus and registration for video-SAR by using sub-aperture point cloud. *Int. J. Appl. Earth Obs. Geoinf.* **2023**, *118*, 103295. [CrossRef]

5. Defense Advanced Research Projects Agency. Broad Agency Announcement: Video Synthetic Aperture Radar (Visar) System Design and Development. 2012. Available online: https://govtribe.com/project/videosynthetic-aperture-radarvisar-system-design-and-development (accessed on 2 March 2022).

6. Zuo, F.; Li, J.; Hu, R.; Pi, Y. Unified Coordinate System Algorithm for Terahertz Video-SAR Image Formation. *IEEE Trans. Terahertz Sci. Technol.* **2018**, *8*, 725–735. [CrossRef]

7. Zhao, B.; Han, Y.; Wang, H.; Tang, L.; Liu, X.; Wang, T. Robust shadow tracking for video SAR. *IEEE Geosci. Remote Sens. Lett.* **2021**, *18*, 821–825. [CrossRef]

8. Zhang, Z.; Shen, W.; Xia, L.; Lin, Y.; Shang, S.; Hong, W. Video SAR Moving Target Shadow Detection Based on Intensity Information and Neighborhood Similarity. *Remote Sens.* **2023**, *15*, 1859. [CrossRef]

9. Yang, C.; Chen, Z.; Deng, Y.; Wang, W.; Wang, P.; Zhao, F. Generation of Multiple Frames for High Resolution Video SAR Based on Time Frequency Sub-Aperture Technique. *Remote Sens.* **2023**, *15*, 264. [CrossRef]

10. Miller, J.; Bishop, E.; Doerry, A. An application of backprojection for video SAR image formation exploiting a subaperture circular shift register. In Proceedings of the Algorithms for Synthetic Aperture Radar Imagery XX, Baltimore, MD, USA, 23 May 2013; pp. 874609.1–874609.14.

11. Wallace, H.B. Development of a video SAR for FMV through clouds. In Proceedings of the Open Architecture/Open Business Model Net-Centric Systems and Defense Transformation, Baltimore, MD, USA, 21 May 2015; pp. 64–65.

12. Langdon, R.M.; Handerek, V.; Harrison, P.; Eisele, H.; Stringer, M.; Tae, C.F.; Dunn, M.H. Military applications of terahertz imaging. In Proceedings of the 1st EMRS DTC Technical Conference, Edinburgh, UK, 20–21 May 2004.

13. Li, Y.; Wu, Q.; Jiang, J.; Ding, X.; Zheng, Q.; Zhu, Y. A High-Frequency Vibration Error Compensation Method for Terahertz SAR Imaging Based on Short-Time Fourier Transform. *Appl. Sci.* **2021**, *11*, 10862. [CrossRef]

14. Tonouchi, M. Cutting-edge terahertz technology. *Nat. Photonics* **2007**, *1*, 97–105. [CrossRef]

15. Li, Y.; Ding, L.; Zheng, Q.; Zhu, Y.; Sheng, J. A Novel High-Frequency Vibration Error Estimation and Compensation Algorithm for THz-SAR Imaging Based on Local FrFT. *Sensors* **2020**, *20*, 2669. [CrossRef] [PubMed]

16. Appleby, R.; Anderton, R.N. Millimeter-Wave and Submillimeter-Wave Imaging for Security and Surveillance. *Proc. IEEE* **2007**, *95*, 1683–1690. [CrossRef]

17. Li, Y.; Wu, Q.; Wu, J.; Li, P.; Ding, L. Estimation of High-Frequency Vibration Parameters for Terahertz SAR Imaging Based on FrFT with Combination of QML and RANSAC. *IEEE Access* **2021**, *9*, 5485–5496. [CrossRef]

18. Jiang, J.; Li, Y.; Zheng, Q. A THz Video SAR Imaging Algorithm Based on Chirp Scaling. In Proceedings of the 2021 CIE International Conference on Radar, Haikou, China, 15–19 December 2021; pp. 656–660.

19. Pu, W.; Wang, X.; Wu, J.; Huang, Y.; Yang, J. Video SAR Imaging Based on Low-Rank Tensor Recovery. *IEEE Trans. Neural Netw. Learn. Syst.* **2021**, *32*, 188–202. [CrossRef] [PubMed]

20. An, H.; Wu, J.; Teh, K.C.; Sun, Z.; Li, Z.; Yang, J. Joint Low-Rank and Sparse Tensors Recovery for Video Synthetic Aperture Radar Imaging. *IEEE Trans. Geosci. Remote Sens.* **2022**, *60*, 5214913. [CrossRef]

21. Moradikia, M.; Samadi, S.; Hashempour, H.R.; Cetin, M. Video-SAR Imaging of Dynamic Scenes Using Low-Rank and Sparse Decomposition. *IEEE Trans. Comput. Imaging* **2021**, *7*, 384–398. [CrossRef]

22. Gorham, L.; Moore, R.J. SAR image formation toolbox for MATLAB. In Proceedings of the Algorithms for Synthetic Aperture Radar Imagery XVII, Orlando, FL, USA, 8–9 April 2010; Volume 7699, pp. 223–263.

23. Musgrove, C. *Polar Format Algorithm: Survey of Assumptions and Approximations*; Sandia National Laboratories (SNL): Albuquerque, NM, USA; Livermore, CA, USA, 2012.

24. Yegulalp, A.F. Fast backprojection algorithm for synthetic aperture radar. In Proceedings of the 1999 IEEE Radar Conference. Radar into the Next Millennium (Cat. No. 99CH36249), Waltham, MA, USA, 22–22 April 1999.

25. Basu, S.K.; Bresler, Y. O(N2log2N) filtered backprojection reconstruction algorithm for tomography. *IEEE Trans. Image Process.* **2000**, *9*, 1760–1773. [CrossRef]

26. Ulander, L.; Hellsten, H.; Stenstrom, G. Synthetic-aperture radar processing using fast factorized back-projection. *IEEE Trans. Aerosp. Electron. Syst.* **2003**, *39*, 760–776. [CrossRef]

27. Wahl, D.E.; Yocky, D.A.; Jakowatz, C.V., Jr.; Zelnio, E.G.; Garber, F.D. An implementation of a fast backprojection image formation algorithm for spotlight-mode SAR. *Proc. Spie* **2008**, *6970*, 8.

28. Yang, Z.M.; Sun, G.C.; Xing, M. A new fast Back-Projection Algorithm using Polar Format Algorithm. In Proceedings of the Synthetic Aperture Radar (APSAR), Tsukuba, Japan, 23–27 September 2013; pp. 373–376.

29. Lei, Z.; Li, H.L.; Qiao, Z.J.; Xu, Z.W. A Fast BP Algorithm With Wavenumber Spectrum Fusion for High-Resolution Spotlight SAR Imaging. *IEEE Geosci. Remote Sens. Lett.* **2014**, *11*, 1460–1464.

30. Yang, Z.; Xing, M.; Zhang, L.; Bao, Z. A coordinate-transform based FFBP algorithm for high-resolution spotlight SAR imaging. *Sci. China Inf. Sci.* **2015**, *2*, 11. [CrossRef]

31. Gorham, L.; Majumder, U.K.; Buxa, P.; Backues, M.J.; Lindgren, A.C. Implementation and analysis of a fast backprojection algorithm. In Proceedings of the Algorithms for Synthetic Aperture Radar Imagery XIII, Orlando, FL, USA, 17–21 April 2006; Volume 6237.

32. Rodriguez-Cassola, M.; Prats, P.; Krieger, G.; Moreira, A. Efficient Time-Domain Image Formation with Precise Topography Accommodation for General Bistatic SAR Configurations. *Aerosp. Electron. Syst. IEEE Trans.* **2011**, *47*, 2949–2966. [CrossRef]

33. Yang, L.; Zhao, L.; Zhou, S.; Bi, G.; Yang, H. Spectrum-Oriented FFBP Algorithm in Quasi-Polar Grid for SAR Imaging on Maneuvering Platform. *IEEE Geosci. Remote Sens. Lett.* **2017**, *14*, 724–728. [CrossRef]
34. Xie, H.; Shi, S.; An, D.; Wang, G.; Wang, G.; Hui, X.; Huang, X.; Zhou, Z.; Chao, X.; Feng, W. Fast Factorized Backprojection Algorithm for One-Stationary Bistatic Spotlight Circular SAR Image Formation. *IEEE J. Sel. Top. Appl. Earth Obs. Remote Sens.* **2017**, *10*, 1494–1510. [CrossRef]
35. Zhang, L.; Li, H.; Xu, Z.; Wang, H.; Yang, L.; Bao, Z. Application of fast factorized back-projection algorithm for high-resolution highly squinted airborne SAR imaging. *Sci. China Inf. Sci.* **2017**, *60*, 1–17. [CrossRef]
36. Frölind, P.O.; Ulander, L. Evaluation of angular interpolation kernels in fast back-projection SAR processing. *IEE Proc.-Radar Sonar Navig.* **2006**, *153*, 243–249. [CrossRef]
37. Hanssen, R.; Bamler, R. Evaluation of Interpolation Kernels for SAR Interferometry. *IEEE Trans. Geosci. Remote Sens.* **1999**, *37*, 318–321. [CrossRef]
38. Selva, J.; Lopez-Sanchez, J.M. Efficient Interpolation of SAR Images for Coregistration in SAR Interferometry. *IEEE Geosci. Remote Sens. Lett.* **2007**, *4*, 411–415. [CrossRef]
39. Garber, W.L.; Hawley, R.W. Extensions to polar formatting with spatially variant post-filtering. In Proceedings of the Algorithms for Synthetic Aperture Radar Imagery XVIII, Orlando, FL, USA, 25–29 April 2011; p. 8051.
40. Mao, D.; Rigling, B.D. Distortion correction and scene size limits for SAR bistatic polar format algorithm. In Proceedings of the 2017 IEEE Radar Conference (RadarConf), Seattle, WA, USA, 8–12 May 2017; pp. 1103–1108.
41. Rigling, B.D.; Moses, R.L. Taylor expansion of the differential range for monostatic SAR. *IEEE Trans. Aerosp. Electron. Syst.* **2008**, *41*, 60–64. [CrossRef]
42. Jakowatz, C.V.; Wahl, D.E.; Thompson, P.A.; Doren, N.E. Space-variant filtering for correction of wavefront curvature effects in spotlight-mode SAR imagery formed via polar formatting. In Proceedings of the Algorithms for Synthetic Aperture Radar Imagery IV, Orlando, FL, USA, 21–25 April 1997; pp. 33–42.
43. Doerry, A.W. *Wavefront Curvature Limitations and Compensation to Polar Format Processing for Synthetic Aperture Radar Images*; Sandia National Laboratories (SNL): Albuquerque, NM, USA; Livermore, CA, USA, 2007.
44. Zhu, D.; Zhu, Z. Range resampling in the polar format algorithm for spotlight SAR image formation using the chirp z-transform. *IEEE Trans. Signal Process.* **2007**, *55*, 1011–1023. [CrossRef]
45. Yu, T.; Xing, M.; Zheng, B. The Polar Format Imaging Algorithm Based on Double Chirp-Z Transforms. *IEEE Geosci. Remote Sens. Lett.* **2008**, *5*, 610–614.
46. Zuo, F.; Li, J. A ViSAR Imaging Method for Terahertz Band Using Chirp Z-Transform. In Proceedings of the Communications, Signal Processing, and Systems: Proceedings of the 2018 CSPS Volume II: Signal Processing 7th, Dalian, China, 14–16 July 2018; pp. 796–804.
47. Cumming, I.G.; Wong, F.H. Digital processing of synthetic aperture radar data. *Artech House* **2005**, *1*, 108–110.
48. Mao, Y.; Wang, X.; Xiang, M. Joint Three-dimensional Location Algorithm for Airborne Interferometric SAR System. *J. Radars* **2013**, *2*, 60–67. [CrossRef]
49. Franceschetti, G.; Migliaccio, M.; Riccio, D.; Schirinzi, G. SARAS: A synthetic aperture radar(SAR) raw signal simulator. *IEEE Trans. Geosci. Remote Sens.* **1992**, *30*, 110–123. [CrossRef]
50. Shoalehvar, A. Synthetic Aperture Radar (SAR) Raw Signal Simulation. Master's Thesis, California Polytechnic State University, San Luis Obispo, CA, USA, 2012.
51. Zhang, S.-S.; Zeng, T.; Long, T.; Chen, J. Research on echo simulation of space-borne bistatic SAR. In Proceedings of the 2006 CIE International Conference on Radar, Shanghai, China, 16–19 October 2006; pp. 1–4.

Article

Performance Analysis of Channel Imbalance Control and Azimuth Ambiguity Suppression in Azimuth Dual Receiving Antenna Mode of LT-1 Spaceborne SAR System

Zongxiang Xu [1,2], Pingping Lu [1,2,*], Yonghua Cai [1,2], Yirong Wu [1,2] and Robert Wang [1,2]

[1]   National Key Laboratory of Microwave Imaging Technology, Aerospace Information Research Institute, Chinese Academy of Sciences, Beijing 100190, China

[2]   School of Electronic, Electrical and Communication Engineering, University of Chinese Academy of Sciences, Beijing 100049, China

*   Correspondence: lupp@aircas.ac.cn

Abstract: The LuTan-1(LT-1), known as the L-band differential interferometric synthetic aperture radar (SAR) satellite system, is an essential piece of civil infrastructure in China, providing extensive applications such as surface deformation monitoring and topographic mapping. To achieve high-resolution and wide-swath (HRWS) observation abilities, the LT-1 takes the dual receiving antenna (DRA) imaging mode as its working mode. However, amplitude and phase errors between channels lead to a mismatch between the reconstruction filter and the multichannel echo signal, worsen the reconstructed azimuth spectrum, and introduce ambiguity targets in the final imaging results, seriously affecting the final imaging quality. In order to better evaluate the channel error and azimuth ambiguity performance of the LT-1 system, this paper proposed an advanced channel consistency correction method and conducted many measured data experiments. The experimental results show that the proposed method is effective, and the LT-1 system has excellent channel error control and azimuth ambiguity performance.

Keywords: synthetic aperture radar (SAR); azimuth ambiguity suppression; high-resolution and wide-swath (HRWS); dual receive antenna (DRA); channel error estimation

Citation: Xu, Z.; Lu, P.; Cai, Y.; Wu, Y.; Wang, R. Performance Analysis of Channel Imbalance Control and Azimuth Ambiguity Suppression in Azimuth Dual Receiving Antenna Mode of LT-1 Spaceborne SAR System. *Remote Sens.* **2023**, *15*, 2765. https://doi.org/10.3390/rs15112765

Academic Editor: Piotr Samczynski

Received: 20 February 2023
Revised: 3 April 2023
Accepted: 8 April 2023
Published: 26 May 2023

1. Introduction

Synthetic aperture radar (SAR) is an all-day and all-weather active microwave remote sensing tool with powerful Earth observation capability [1]. High resolution can provide detailed feature information of observation scenes, and wide-swath imaging can provide extensive scene information. Therefore, high-resolution and wide-swath (HRWS) imaging is a significant trend in modern SAR systems [2–9]. However, it is an irreconcilable contradiction to improve azimuth resolution and range swath. High-resolution imaging requires higher pulse frequency repetition (PRF) to satisfy the Nyquist sampling theorem, while wide-swath imaging requires lower PRF to ensure a sufficient echo-receiving window [10,11]. It is hard for the traditional SAR system to realize HRWS imaging simultaneously. Fortunately, by reconstruction algorithm, digital beamforming (DBF) technology can recover a full alias-free spectrum from the aliasing spectrum and suppress azimuth ambiguity caused by low PRF [3,12,13]. Therefore, an azimuth multichannel SAR system can realize HRWS imaging with DBF technology. Azimuth multichannel (AMC) technology was first applied in the TerraSAR-X satellite launched by Germany in 2007 [12]. The experimental results show that DBF has good azimuth ambiguity suppression ability. Then, the ALOS-2 satellite launched by Japan in 2014 first used the dual-antenna receiving (DRA) mode as a working mode [14]. In 2016, China's first dual-channel spaceborne SAR sensor, the Gaofen-3 satellite, was launched [15,16]. In recent years, AMC technology has been widely used in spaceborne SAR.

LuTan-1 (LT-1) [17], a vital component of the "China National Civil Space Infrastructure Long-Term Development Plan (2015–2025)", is an innovative spaceborne Earth observation constellation. It consists of two identical satellites carrying advanced, fully polarized L-band SAR capable of obtaining SAR images. The LT-1's primary aim is to acquire global surface deformation measurement and digital elevation model (DEM). LT-1 data can also serve multiple industries such as land, earthquake, surveying and mapping, environment, disaster mitigation, and forestry. Achieving tasks such as surface deformation monitoring and topographic mapping in a given area requires extensive coverage capabilities, and higher spatial resolution is essential for disaster monitoring. Therefore, to achieve HRWS imaging, LT-1 uses DRA mode as its operating mode. The duration of its mission is divided into two phases. In phase I, two satellites fly in a formation with a variable baseline. The bistatic InSAR strip map mode is utilized to acquire global digital elevation and terrain models with high accuracy and spatial resolution [18]. In phase II, two satellites shall share the standard reference orbit with a 180° orbital phasing difference. LT-1 will continue to provide high-quality observation data throughout its future life cycle, providing new possibilities, advances, and relevant information for dynamic monitoring on land.

Although AMC SAR has many advantages, using multiple channels to receive echo signals also brings a variety of error sources, including phase synchronization error, baseline error, channel error, and attitude error [19–21]. These errors lead to inconsistent amplitude and phase characteristics of each channel's echo, seriously restricting the performance of the AMC SAR system. To meet the requirements of interferometric altimetry and deformation measurement, LT-1 built a high-precision spaceborne internal calibration system consisting of internal calibrators, antenna calibration networks, and interconnection cables. The system includes three calibration loops: no delay calibration loop, delay calibration loop, and synchronous calibration loop. It is difficult for the internal calibration system to achieve high-precision calibration by relying on onboard self-service. Then, LT-1 compensates through ground processing, including the temperature curve data, to compensate for amplitude and phase errors. The LT-1 can obtain an amplitude-temperature curve and phase-temperature curve by setting temperature points reasonably in the antenna calibration network or internal calibrator through temperature experiments in the ground test. When in orbit, it can use the measured temperature value and temperature curve to compensate for the amplitude-phase calibration results of the system during ground data processing. Using temperature compensation measures, LT-1 can achieve the internal calibration index requirements of the amplitude of no more than 0.4 dB and phase of no more than 3°. The two main factors of channel amplitude-phase inconsistency are the phase center deviation error of receiving antenna and amplitude-phase error caused by channel characteristics. Internal calibration technology can only measure the amplitude and phase errors of the channel itself, but not the amplitude and phase characteristics of the passive part of the antenna. Therefore, the channel errors caused by antenna phase center errors and other residual channel errors must be corrected before imaging by other channel consistency correction techniques [12,22].

Channel consistency correction methods mainly include external calibration and self-calibration methods based on echo data. The external calibration technology provides end-to-end absolute measurement and calibration of SAR system parameters and their changes through the standard calibration equipment deployed in the ground calibration field. It covers the whole channel link and has high calibration accuracy. However, the external calibration technology needs to lay out the calibration equipment in the scene, so the flexibility needs to be improved. In contrast, the self-calibration method estimates the channel error by processing the original data and does not need additional calibration equipment. At the same time, the estimation accuracy of the self-correction method has been dramatically improved through constant updating and iteration. Currently, the commonly used self-calibration methods include the signal subspace method (SSP) [23], azimuth time domain cross-correlation method (ATC) [24], minimum variance distortionless response method (MVDR) [25], the image weighted minimum entropy method (WME) [26,27], the image least L1-norm method (LLN) [19], maximum normalized image sharpness method (MNIS) [20], and the minimizing the sum of the sub-band norm method (MSSBN) [28]. Considering the estimation accuracy and computational efficiency, this paper proposes a phase error estimation method that maximizes the L1 norm of the sum (MLNS) of multichannel 2-Dimension (2D) frequency domain echo signals based on the MSSBN method. This paper uses this new method to estimate and correct the channel amplitude and phase errors of LT-1.

LT-1 has been in orbit for nearly a year, accumulating much observation data from different scenarios under different operating modes. This paper aims to evaluate the channel error control capability of the LT-1 system by estimating the channel amplitude-phase error of the DRA receive data and to study its performance by focusing on the azimuth ambiguity of the image. Experiments include different incident angles and representative scene types. The experimental results demonstrate the excellent channel error control and azimuth ambiguity suppression capability of LT-1.

The paper is organized as follows. The second part of this paper briefly introduces the channel error model of a dual-channel SAR system. The third parts show the influence of channel error on imaging through simulation experiments. Section 4 proposes a new phase error estimation and correction method, and the measured data results of LT-1 are given and analyzed in detail. Finally, Section 5 summarizes the system performance of LT-1 and discusses its future application prospects.

2. LT-1 DRA Mode and Channel Error Model

2.1. LT-1 Signal Model and Channel Constant Amplitude-Phase Error

Figure 1 is a brief illustration of the imaging geometry of the LT-1 DRA mode. The LT-1 normal orbit flight posture is on the right, and there is a need to adjust to the left view mode. The antenna phase center (APC) settings of LT-1 are shown in Figure 1, the antenna transmits chirp signals at the Tx channel, and two separate channels (Rx 1 and Rx 2) in azimuth receive echoes simultaneously. The aperture size is set to $d_{rx,az}$, and the distance between two receive channels is 4.9 m. The echo history of the signal received by the mth channel and the echo history of the signal received by the mth equivalent phase center in self-transmit and self-received can be expressed, respectively, as

$$
\begin{aligned}
R_m(\eta) &= R_{rx,m}(\eta) + R_{tx,m}(\eta) = \sqrt{R_0^2 + V_s^2\left(\eta - \tfrac{\Delta x_m}{V_s}\right)^2} + \sqrt{R_0^2 + V_s^2\eta^2} \\
&\approx 2R_0 + \frac{V_s^2\left(\eta - \tfrac{\Delta x_m}{2V_s}\right)^2}{R_0} + \frac{\Delta x_m^2}{4R_0}
\end{aligned}
\tag{1}
$$

$$
R_{eq,m}(\eta) = 2\sqrt{R_0^2 + V_s^2\left(\eta - \frac{\Delta x_m}{2V_s}\right)^2} \approx 2R_0 + \frac{V_s^2\left(\eta - \tfrac{\Delta x_m}{2V_s}\right)^2}{R_0}
\tag{2}
$$

where $R_{rx,m}$, $R_{tx,m}\eta$, R_0, V_s donate the range history from target to receive channel, range history from target to transmission channel, azimuth slow time, nearest slant range, and satellite speed of the LT-1 SAR system, respectively. $\Delta x_m = (m - m_{ref})d_{rx,az}$ donates the distance between the mth channel and the reference channel. According to Equation (1), the 2-D time domain echo signal of the mth equivalent phase center (EPC) can be expressed as

$$
\begin{aligned}
s_{eq,m}(\tau, \eta) &= w_r\left(\tau - \frac{R_{eq,m}(\eta)}{c}\right)w_a(\eta - \eta_c) \\
&\times \exp\left\{-j\tfrac{4\pi}{\lambda}R_{eq,m}(\eta)\right\}\exp\left\{j\pi K_r\left(\tau - \tfrac{2R_{eq,m}(\eta)}{c}\right)^2\right\}
\end{aligned}
\tag{3}
$$

where τ, w_r, w_a, η_c, λ, K_r and c denote the range fast time, range signal envelop, azimuth signal envelop, azimuth center time, wavelength, chirp rate, and velocity of light, respectively. The difference between $R_m(\eta)$ and $R_{eq,m}(\eta)$ is $-\Delta x_m^2/4R_0$ after Taylor expansion is shown in Equations (1) and (2). Therefore, the 2-D time domain echo signal of the *m*th channel can be expressed as

$$s_m(\tau, \eta) = s_{eq,m}(\tau, \eta) \times \exp\left\{ -j\frac{\pi \Delta x_m^2}{2\lambda R_0} \right\} \tag{4}$$

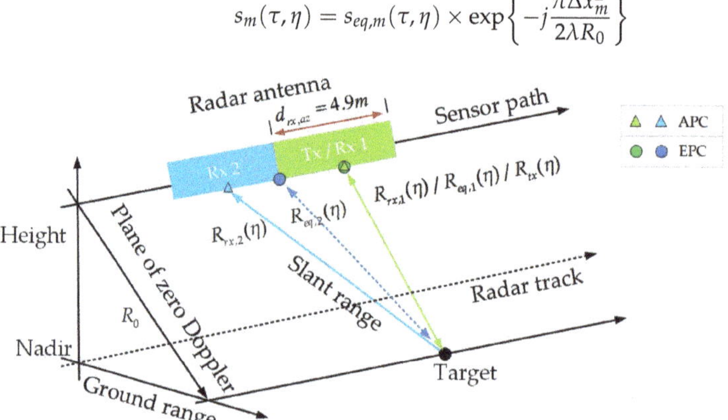

Figure 1. Imaging geometry of LT-1 DRA mode.

AMC SAR system adopts multiple channels for sampling. It uses the increase in spatial dimension sampling to exchange for the decrease in temporal dimension sampling [3], which makes the performance of an HRWS realized. Multichannel systems require multi-channel echo signal reconstruction processing, and the key to reconstruction is to ensure that the reconstruction filter is consistent with the echo signal. That is, the echo signal amplitude and phase between channels are consistent. However, in actual operation, due to various factors such as electronic equipment, antenna arrays, and satellite platforms [23] in each channel, amplitude, and phase errors inevitably exist between the echo signals in each channel. Final imaging quality will be seriously affected if the above errors are uncompensated. The satellite platform operates very stably for spaceborne SAR systems during startup time. Then the attitude, velocity, and orbit curvature errors involved in the satellite platform can be ignored [29,30]. Radar electronic equipment can introduce amplitude, phase, and range sampling time errors in the echo signal due to processing technology, device aging, and temperature changes. However, since the LT-1 SAR system operates in the L-band, the impact of range sampling time error can be ignored. At the same time, channel errors caused by radar electronics are generally considered stable during a startup operation [30]. Therefore, considering only the errors introduced by radar electronics, error models can be established as

$$s_{me}(\tau, \eta) = \alpha_m s_m(\tau, \eta) \exp(j\psi_m) \tag{5}$$

where α_m and ψ_m are amplitude error and phase error, respectively, the error takes the channel of the first antenna as a reference, then $\alpha_{1r} = 1, \psi_{1r} = 0$.

2.2. APC Position Error

In addition to the amplitude error, phase error, and range sampling time error caused by the electronic equipment of the SAR system, APC position error also introduces significant phase errors. APC position error is affected by antenna thermal deformation, installation error, satellite attitude error, and other factors. At the same time, phase error caused by APC position error is affected by the incident angle and radar operating wavelength. Therefore, this section will quantitatively analyze the APC position error of the dual-channel SAR system. Figure 2 shows the geometric schematic diagram of the APC position error.

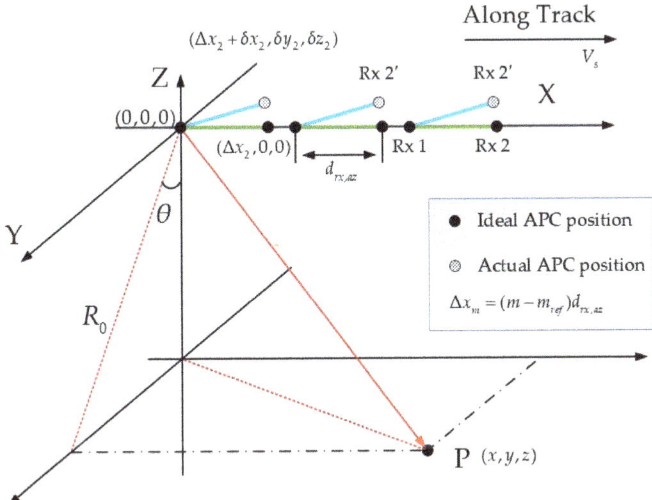

Figure 2. Schematic diagram of APC in a dual-channel SAR system.

Firstly, the spatial coordinate system of the AMC SAR system is defined as:

- The X-axis direction is the satellite track direction.
- The Y-axis direction is perpendicular to the X-axis direction.
- The Z-axis faces away from the center of the Earth.
- The P point is the target point.

The first antenna coordinate is (0,0,0), and the solid circles represent the ideal position of the APC, all on the X-axis. In contrast, the hollow circles represent the actual APC position, which deviates from the ideal position. Set the APC position deviation of the mth antenna as $(\delta x_m, \delta y_m, \delta z_m)$, and take the first antenna as the reference channel, then its position error is $(\delta x_1, \delta y_1, \delta z_1) = (0,0,0)$. The APC position error will change the distance between the APC and the target.

When there is no APC position error, the phase center coordinate of the mth antenna is $(\Delta x_m + V_s\eta, 0, 0)$. If the coordinate of the point target is (x, y, z), then the distance from the APC to the point target can be expressed as

$$R_{rx,m} = \sqrt{(x - \Delta x_m - V_s\eta)^2 + (y)^2 + (z)^2} \tag{6}$$

When the APC position error exists, the APC coordinate of the mth antenna is $(\Delta x_m + \delta x_m + V_s\eta, \delta y_m, \delta z_m)$, then the distance between the antenna and the target is expressed as

$$R'_{rx,m} = \sqrt{(x - \Delta x_m - \delta x_m - V_s\eta)^2 + (y - \delta y_m)^2 + (z - \delta z_m)^2} \tag{7}$$

Then, by combining Equations (3)–(5) and (7), the echo signal with APC position error can be expressed as

$$s_{me}(\tau, \eta) = \alpha_m w_r(\tau - \frac{R'_{eq,m}(\eta)}{c}) w_a(\eta - \eta_c)$$
$$\times \exp\left\{-j\frac{4\pi}{\lambda}R'_{eq,m}(\eta)\right\} \exp\left\{j\pi K_r(\tau - \frac{2R'_{eq,m}(\eta)}{c})^2\right\} \exp\{j\psi_m\} \tag{8}$$

The Taylor expansion is carried out on Equation (7), and the square term of the error minor term and the term of third or higher order are ignored. Then the distance history can be approximated as follows

$$R'_{rx,m} = R_{rx,m} - \frac{x - \Delta x_m - V_s\eta}{R_{rx,m}}\delta x_m - \frac{y}{R_{rx,m}}\delta y_m - \frac{z}{R_{rx,m}}\delta z_m \tag{9}$$

According to Equation (9), the phase error generated by the APC position deviation can be expressed as

$$\delta\varphi_m = \frac{4\pi}{\lambda}\frac{(R_{rx,m} - R'_{rx,m})}{2} = \frac{2\pi}{\lambda}\frac{x - \Delta x_m - V_s\eta}{R_{rx,m}}\delta x_m - \frac{2\pi}{\lambda}\frac{y}{R_{rx,m}}\delta y_m - \frac{2\pi}{\lambda}\frac{z}{R_{rx,m}}\delta z_m \tag{10}$$

Since $x \ll y, x \ll z$, the above equation can be simplified as

$$\delta\varphi_{m_x} \approx \frac{2\pi}{\lambda}\frac{x - \Delta x_m - V_s\eta}{\sqrt{y^2 + z^2}}\delta x_m \tag{11}$$

$$\delta\varphi_{m_y} \approx \frac{2\pi}{\lambda}\frac{y}{\sqrt{y^2 + z^2}}\delta y_m \tag{12}$$

$$\delta\varphi_{m_z} \approx \frac{2\pi}{\lambda}\frac{z}{\sqrt{y^2 + z^2}}\delta z_m \tag{13}$$

where $\delta\varphi_{m_x}, \delta\varphi_{m_y}, \delta\varphi_{m_z}$ are the phase errors caused by the deviation δx_m along the X-axis, δy_m along the Y-axis, and δz_m along the Z-axis, respectively. Among them, $\delta\varphi_{m_y}$ and $\delta\varphi_{m_z}$ are independent of azimuth time but related to the slant range of the echo, which is named the space-variant phase error in this paper. $\delta\varphi_{m_x}$ changes with azimuth slow time, referred to as azimuth time-variant phase error in this paper. Due to the Doppler effect, the azimuth slow time domain is also a linear frequency-modulated signal. The relation between azimuth slow time and Doppler frequency is as

$$\eta = \frac{x - \Delta x_m}{V_s} - \frac{\lambda\sqrt{y^2 + z^2}}{2V_s^2}f_\eta \tag{14}$$

Substituting the above equation into Equation (11), the frequency domain form of time-variant phase error can be expressed as

$$\delta\varphi_{m_x} = \frac{2\pi f_\eta}{V_s}\delta x_m \tag{15}$$

According to the geometric diagram in Figure 2, Equations (12) and (13) can be written as expressions for the angle of view

$$\delta\varphi_{m_y} = \frac{2\pi}{\lambda}\delta y_m \sin\theta \tag{16}$$

$$\delta\varphi_{m_z} = \frac{2\pi}{\lambda}\delta z_m \cos\theta \tag{17}$$

where θ is the angle of view, thus, the space-variant phase error caused by the APC position error can be written as

$$\delta\varphi_{myz} = \frac{2\pi}{\lambda}(\delta y_m \sin\theta + \delta z_m \cos\theta) \tag{18}$$

From the above analysis, it can be seen that the channel phase error caused by APC position error is related to not only wavelength and a function of view of angle. Therefore, the impact of APC position error on phase error is studied through simple simulation.

Generally, the APC position error does not exceed 5 mm [30], while the APC position error caused by satellite attitude errors is less than 1 mm and can be ignored. Therefore, select a uniform distribution of δy_2 and δz_2 between -5 mm and 5 mm, $\theta = 40°$, and compare the relationship between APC position error and phase error at different wavelengths. Secondly, compare the phase errors generated by APC position error at different antenna beams in the L-band. The relationship between the space-variation error, wavelength, and angle of view is shown in Figure 3.

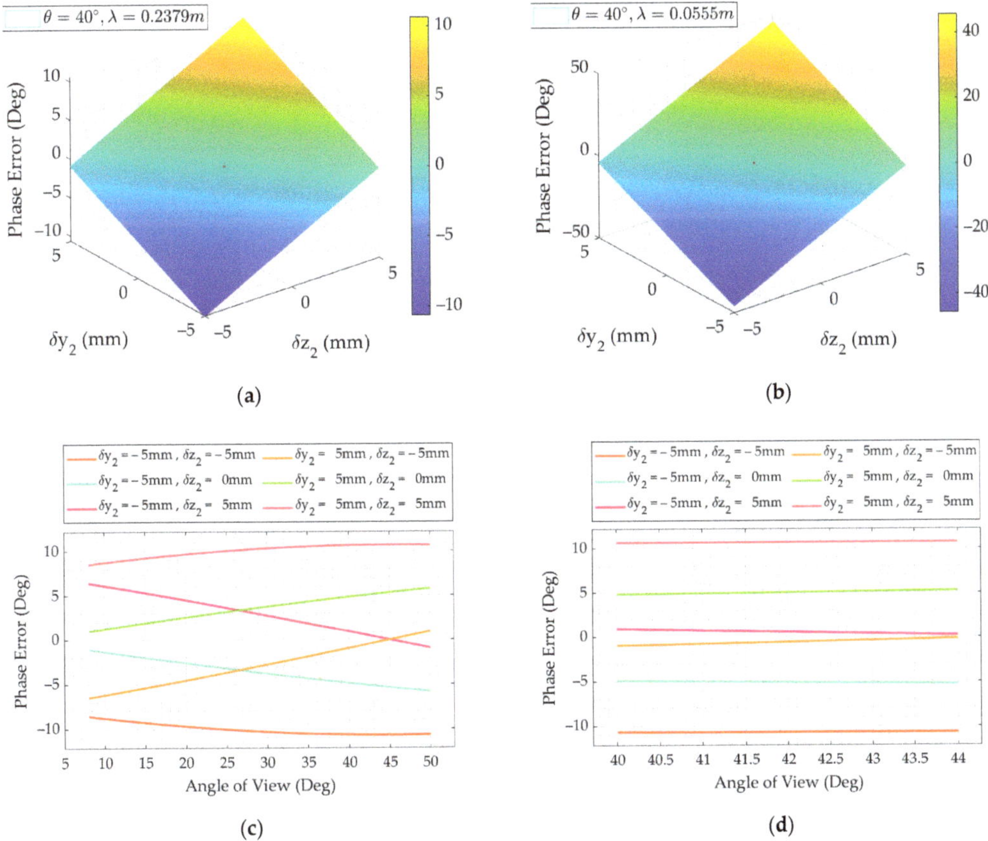

Figure 3. Phase error caused by APC position error. (**a**) L-band phase error, (**b**) C-band phase error, (**c**) the phase error caused by APC position error and the different antenna beam, and (**d**) the phase error caused by APC position error and the same antenna beam.

From Figure 3a,b, the space-varying phase error caused by APC decreases with the increased wavelength. Therefore, unlike other bands, the L-band SAR satellite has a natural advantage in space-varying phase error control. From Figure 3c, the spatial variability of the phase is maximum when the $|\delta y_m - \delta z_m|$ maximum. As shown in Figure 3d, the phase error caused by APC position error can be considered approximately constant at the same antenna beam. Therefore, the error model shown in Equation (5) can be used when considering channel errors within a scene. In contrast, when considering channel errors over a wider swath, it is necessary to consider space-variant phase errors.

3. Influence of Channel Error on Imaging Performance

The influence of channel amplitude-phase characteristic error and APC position error on imaging is analyzed for the dual-channel model. The simulation parameters are shown in Table 1. By analyzing peak sidelobe ratio (PSLR), integral sidelobe ratio (ISLR), and impulse response width (IRW), the effects of various errors on the system imaging were evaluated quantitatively.

Table 1. AMC SAR system point-like target simulation parameters.

Parameter	Symbol	Value	Unit
Platform velocity	V_s	7635	m/s
Carrier frequency	f_0	1.26	GHz
Nearest slant range	R_0	817	km
Azimuth antenna length	L_{az}	2×4.9	m
Doppler bandwidth	B_a	2761	Hz
Azimuth sampling frequency	F_a	1795	Hz
Number of channels	M	2	\

According to the analysis in Section 2, the amplitude and phase errors of the channel can be decomposed into invariant amplitude errors, invariant phase errors, space-variant phase errors, and time-variant phase errors caused by APC position errors. Regarding the parameters in Table 1, various errors were added for the point target simulation.

The following figure shows the azimuth profile of a point-like target obtained without error and by adding 1.3 dB amplitude error, 0.5 rad phase error, and $\delta x_2 = 0.005$ mm, $\delta y_2 = 0.005$ mm, $\delta z_2 = 0.005$ mm APC position error, respectively.

As can be seen from Figure 4b,c, for the azimuth dual-channel system, channel amplitude-phase characteristic errors cause severe false targets in the azimuth direction, and channel phase errors have a more profound impact on imaging quality than amplitude errors. L-band multichannel SAR system has the advantage of long wavelength, and the space–time variation error caused by APC position error is small. Meanwhile, it can be seen from Figure 4d that the effect of time-variant phase error caused by APC position error on imaging is almost negligible. However, in Figure 4e, the space-variant phase error caused by APC position error seriously impacts imaging. Therefore, the channel correction of the L-band multichannel SAR system must focus on the correction of invariant channel errors and space-variant phase errors caused by APC position errors.

Then, the specific effects of invariant amplitude error, invariant phase error, time-variant phase error, and space-variant phase error on imaging quality will be studied through simulation. Three indexes of integral sidelobe ratio (ISLR), peak sidelobe ratio (PSLR), and resolution broadening factor were studied.

The first is the influence of invariant amplitude error on imaging quality. Set the invariant amplitude error of channel 2 relatives to the reference channel to be 0 dB to 2 dB and the step size to be 0.2 dB.

As seen from Figure 5, PSLR and ISLR are linear with invariant amplitude error, and both increase with amplitude error. However, PSLR-R and PSLR-L are different because the amplitude error shifts the filter reconstruction matrix to one side [28], resulting in a different amplitude of the first side lobe on either side of the peak in the final image result.

The second is the influence of channel invariant phase error on imaging quality. Set the invariant phase error of channel 2 relatives to the reference channel to be 0 deg to 40 deg and the step size to be 4 deg.

As can be seen from Figure 6, PSLR and ISLR have a nonlinear relationship with the invariant phase error and increase rapidly with the increase in the invariant phase error. At the same time, PSLR-R and PSLR-L are almost the same because the phase error will cause the same change in the filter reconstruction matrix.

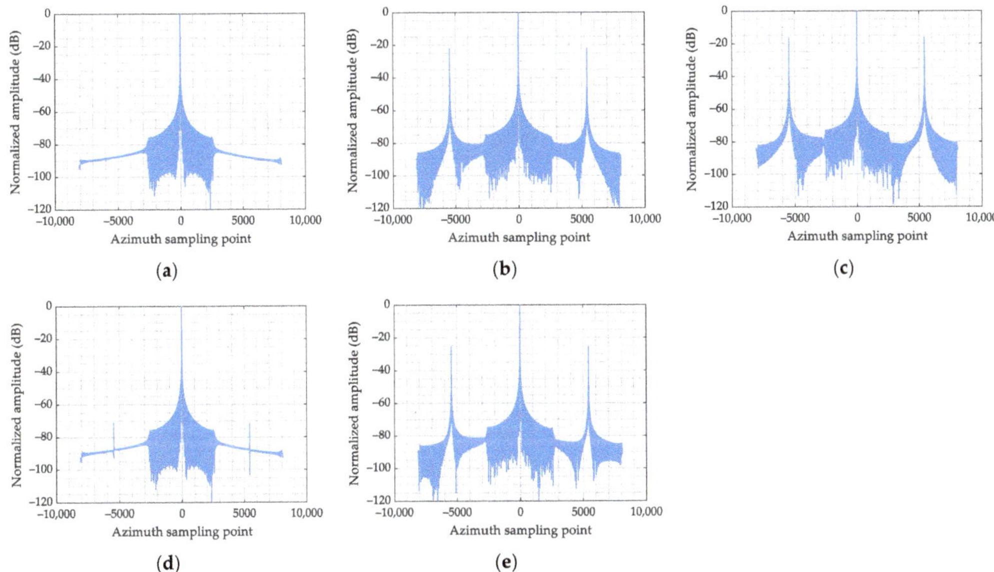

Figure 4. Azimuth profile of point-like target. (**a**) No channel error, (**b**) 1.3 dB amplitude error, (**c**) 0.5 rad phase error, (**d**) time-variant phase error caused by APC position error, and (**e**) space-variant phase error caused by APC position error.

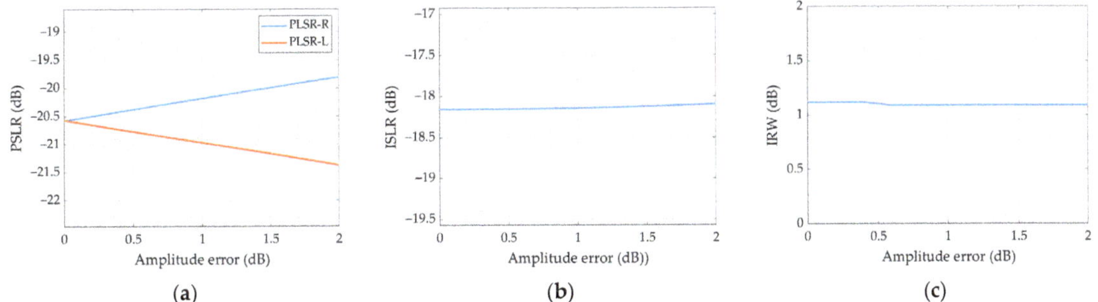

Figure 5. Effect of invariant amplitude error on point-like target imaging. (**a**) PSLR, (**b**) ISLR, and (**c**) IRW.

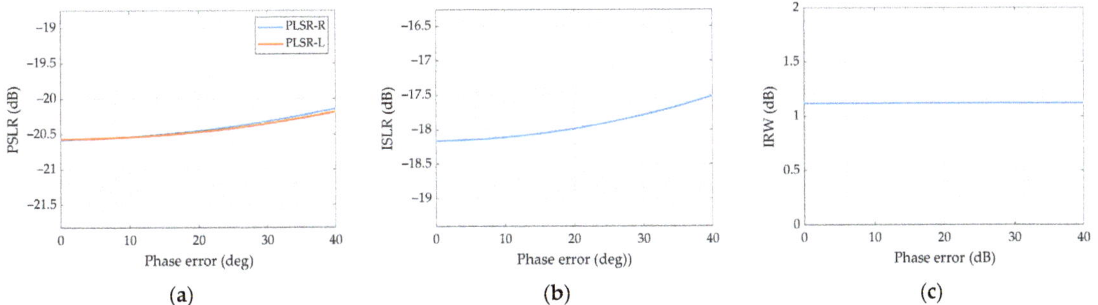

Figure 6. Effect of invariant phase error on point imaging. (**a**) PSLR, (**b**) ISLR, and (**c**) IRW.

The third is the influence of azimuth APC position error on imaging quality. Set the APC error of channel 2 relatives to the reference channel to be 0 mm to 10 mm and the step size to be 1 mm.

The influence of the azimuth APC position error is small and almost negligible. The last is the influence of range APC position error on imaging quality. Set the APC position error of channel 2 relatives to the reference channel to be $\delta y_2 = 0\,mm$ to 10 mm and $\delta z_2 = 0$ mm to 10 mm, respectively, and the step size to be 2 mm.

The effect of range APC position error on image quality is consistent with the invariant channel phase error. In general, the influence of channel error on image mainly includes two aspects: the appearance of ghost targets on both sides of real targets along the azimuth direction (as shown in Figure 4) and defocusing of real targets (as shown in Figures 5–8). The reason is that with the existence of channel errors, the ambiguous components are not suppressed completely and the full Doppler spectrum cannot be reconstructed well. As can be seen from the above figures, the PSLR and ISLR are easily affected by phase errors, but the IRW does not change significantly. This is because channel errors lead to imaging defocusing. Among the four errors, constant phase error and space-variant phase error caused by APC position error have the most severe effect on imaging quality.

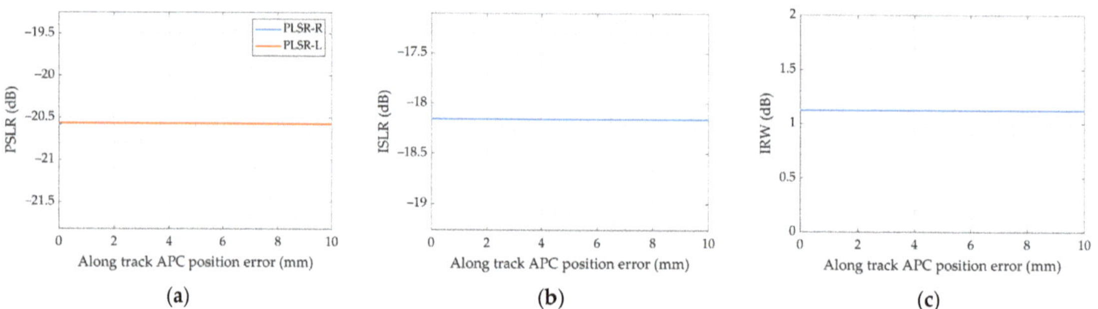

Figure 7. Effect of azimuth APC position error on point imaging. (**a**) PSLR affected by azimuth APC error, (**b**) ISLR affected by azimuth APC error, and (**c**) IRW affected by azimuth APC error.

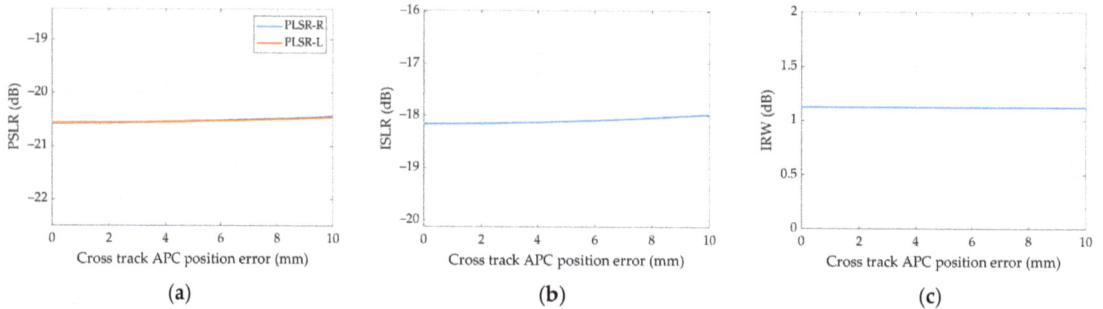

Figure 8. Effect of range APC position error on point imaging. (**a**) PSLR, (**b**) ISLR, and (**c**) IRW.

4. LT-1 SAR Performance Analysis

As shown in Figure 9, the adequate aperture size of the LT-1 antenna array is 9.8 m × 3.4 m (azimuth × range). In order to realize HRWS imaging, AMC technology is adopted. Namely, the azimuth is divided into two independent sub-apertures on the antenna +X wing and −X wing, respectively [31]. Due to the DRA, the LT-1 strip map mode can achieve approximately 3 m azimuth resolution and a 50km swath imaging area with a PRF of less than 2000 Hz. It is difficult to achieve such high spatial resolution for traditional strip map SAR using the same PRF.

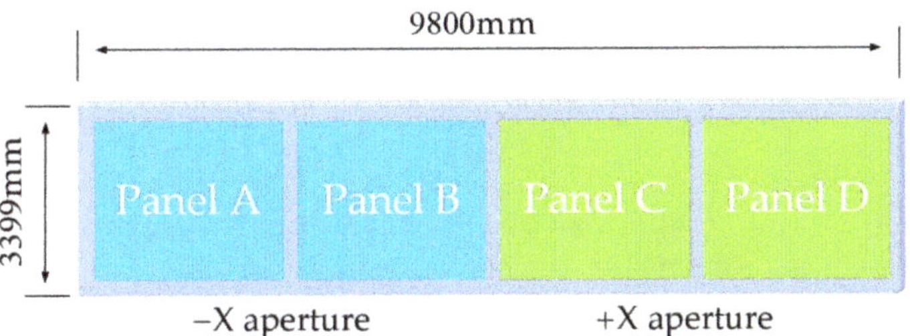

Figure 9. SAR antenna architecture diagram.

4.1. Consistency Calibration Method for Channel Error

This paper proposes a novel phase error estimation method based on the MLNS method to evaluate the phase error of LT-1 SAR system DRA mode. It has fast calculation speed, high estimation accuracy, and strong estimation stability. Moreover, according to the estimation characteristics of this method, many blocks can be divided in the range direction without adding a calculation amount, so the space-variation phase error caused by APC position error can be estimated.

According to the subadditivity of normed linear spaces [32], two complex numbers, a and b, satisfy the following triangular inequalities:

$$||a| - |b|| \leq |a \pm b| \leq |a| + |b| \tag{19}$$

If, and only if, the angles of the complex numbers a and b are the same, the equal can be taken on both sides of the above equation, and the absolute values of the sum of two complex numbers can obtain the maximum value. For a dual-channel system, the echo signals of both channels are complex and can be expressed as a set of normed linear spaces. The SAR system's time domain echo signal phase satisfies a random distribution and needs to be transformed into a two-dimensional frequency domain. In order to estimate the channel phase error, based on the feature selection characteristics of the L1 norm, an L1 norm model of the sum of frequency domain echo signals of dual channels SAR system can be established as

$$F(\Delta\varphi) = \sum_{i=1}^{N} |S_1(f_{\eta,i}) + \exp\{\Delta\varphi\}S_2(f_{\eta,i})| \tag{20}$$

$S_1(f_{\eta,i})$ and $S_2(f_{\eta,i})$ are aliased spectra obtained by performing azimuth Fast Fourier Transform (FFT) on $s_1(\eta)$ and $s_2(\eta)$, respectively, where $f_\eta \in [-\text{PRF}/2, \text{PRF}/2]$. According to spectrum reconstruction theory [3], the aliased spectrum of each channel can be represented as the product of a pre-filter matrix $\mathbf{H}$ [28], and the original aliased-free frequency domain signal

$$\begin{bmatrix} S_1(f_\eta) \\ S_2(f_\eta) \end{bmatrix} = \mathbf{H} \cdot \begin{bmatrix} S_0(f'_{\eta,1}) \\ S_0(f'_{\eta,2}) \end{bmatrix} = \begin{bmatrix} \exp\left\{-j\pi\frac{\Delta x_1}{V_s}f'_{\eta,1}\right\} & \exp\left\{-j\pi\frac{\Delta x_1}{V_s}f'_{\eta,2}\right\} \\ \exp\left\{-j\pi\frac{\Delta x_2}{V_s}f'_{\eta,1}\right\} & \exp\left\{-j\pi\frac{\Delta x_2}{V_s}f'_{\eta,2}\right\} \end{bmatrix} \cdot \begin{bmatrix} S_0(f'_{\eta,1}) \\ S_0(f'_{\eta,2}) \end{bmatrix} \tag{21}$$

$S_0(f'_{\eta,1})$ and $S_0(f'_{\eta,2})$ are two sub-band spectra of the complete aliased-free spectrum, where $f'_{\eta,i} \in [-\text{PRF}, 0] + (i-1)\text{PRF}$. According to Equation (21), $S_1(f_{\eta,i})$ and $S_2(f_{\eta,i})$ can be expressed as

$$S_1(f_{\eta,i}) = S(f'_{\eta,1,i}) + S(f'_{\eta,2,i}) \tag{22}$$

$$\begin{aligned} S_2(f_{\eta,i}) &= \exp\left\{-j\pi\frac{d_{az,rx}}{V_s}f'_{\eta,1,i}\right\}S(f'_{\eta,1,i}) + \exp\left\{-j\pi\frac{d_{az,rx}}{V_s}f'_{\eta,2,i}\right\}S(f'_{\eta,2,i}) \\ &= \exp\left\{-j\pi\frac{d_{az,rx}}{V_s}f'_{\eta,1,i}\right\}\left(S(f'_{\eta,1,i}) + \exp\left\{-j\pi\frac{d_{az,rx}}{V_s}\text{PRF}\right\}S(f'_{\eta,2,i})\right) \end{aligned} \tag{23}$$

Compensate one phase term $\exp\{j\pi d_{az,rx}f_{\eta,1,i}/V_s\}$ for Equation (23) to obtain its simplified form:

$$S'_2(f_{\eta,i}) = S(f'_{\eta,1,i}) + \exp\left\{-j\pi\frac{d_{az,rx}}{V_s}\mathrm{PRF}\right\}S(f'_{\eta,2,i}) \tag{24}$$

After compensating for the channel phase error for channel 2, Equation (20) can be expressed as

$$\begin{aligned}
F(\Delta\varphi - \Delta\hat{\varphi}) &= \sum_{i=1}^{N}\left|S_1(f_{\eta,i}) + \exp\{-j\Delta\hat{\varphi}\}\exp\{j\Delta\varphi\}S'_2(f_{\eta,i})\right| \\
&\leq \sum_{i=1}^{N}\left|S_1(f_{\eta,i})\right| + \sum_{i=1}^{N}\left|\exp\{-j\Delta\hat{\varphi}\}\exp\{j\Delta\varphi\}S'_2(f_{\eta,i})\right|
\end{aligned} \tag{25}$$

Fortunately, in the actual operation process, the phase difference between $S_1(f_{\eta,i})$ and $S'_2(f_{\eta,i})$ is symmetrically distributed with respect to the channel phase error. According to the parameters in Table 1, we verify it through simulation and measured data. As shown in Figure 10a,b, the differences between argument angles of $S_1(f_{\eta,i})$ and $S'_2(f_{\eta,i})$ concentrate near the phase error and distribute symmetrically along the phase error. As can be seen from Figure 10c,d, when compensating for channel errors, $F(\Delta\varphi - \Delta\hat{\varphi})$ has a maximum value. When compensating for an error inversely related to the phase error, $F(\Delta\varphi - \Delta\hat{\varphi})$ has a minimum value, and the maximum and minimum results are consistent with Equation (25).

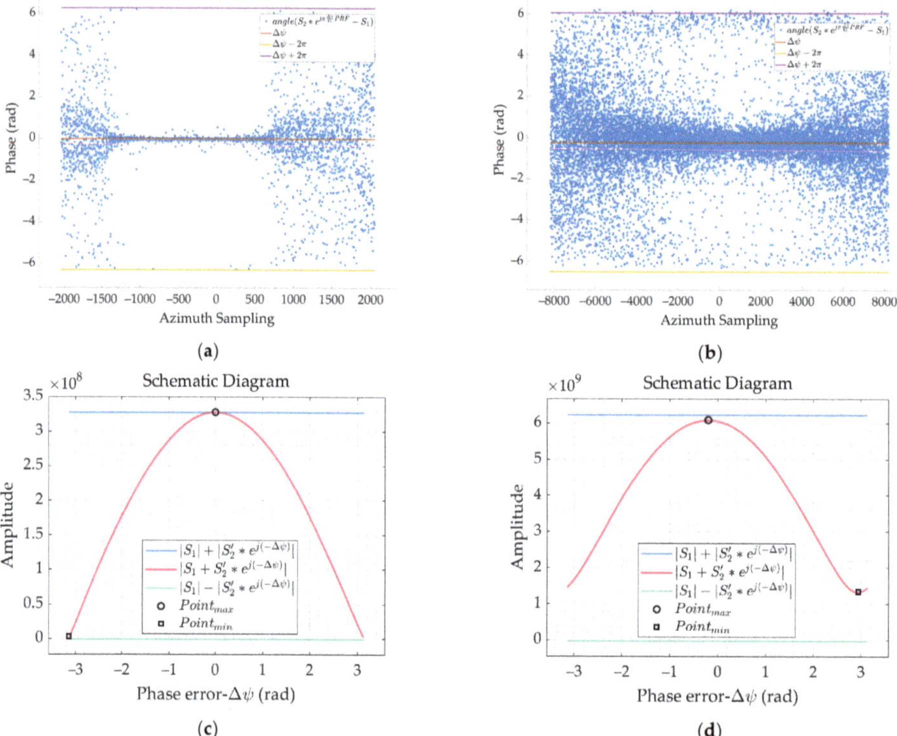

Figure 10. The phase difference and L1 norm of dual-channel frequency domain signal. (**a**) The phase difference of simulation results, (**b**) the phase difference of LT-1 measured data results, (**c**) the L1 norm of simulation results, and (**d**) the L1 norm of LT-1 measured data results.

According to the subadditivity of normed linear space, when the estimated phase error and the channel error are the same, there is a maximum value for $F(\Delta\varphi - \Delta\hat{\varphi})$. When the estimated phase and channel errors are inverted, there is a minimum value for $F(\Delta\varphi - \Delta\hat{\varphi})$. Therefore, the MLNS optimization model is established as

$$\Delta\hat{\varphi} = \underset{\Delta\hat{\varphi}}{\mathrm{argmax}}\{F(\Delta\varphi - \Delta\hat{\varphi})\}$$
$$s.t.|\Delta\hat{\varphi}| \leq \pi \tag{26}$$

To facilitate the use of optimization algorithms in MSSBN, Equation (26) is rewritten as

$$\Delta\hat{\varphi} = \underset{\Delta\hat{\varphi}}{\mathrm{argmin}}\left\{\frac{1}{F(\Delta\varphi - \Delta\hat{\varphi})}\right\}$$
$$s.t.|\Delta\hat{\varphi}| \leq \pi \tag{27}$$

Compared to the MSSBN method, the MLNS method can directly estimate channel phase error using two-dimensional frequency domain echo signals and has higher computational efficiency with consistent accuracy.

4.2. Calibration Point Imaging Results

SAR resolution and sidelobe ratios are essential parameters for quantifying the quality of focused SAR images. A point-like target analysis from a corner reflector's two-dimensional impulse response function can determine them. The peak-to-sidelobe ratio (PSLR) defines the ratio between the prominent lobe power peak and the highest sidelobe power peak. It represents the ability of a SAR system to identify a weak target from a nearby strong one and gives information about the system's sensitivity. The integrated sidelobe ratio (ISLR) is the ratio of the energy in the main lobe to the total energy in all sidelobes. It represents the ability of the SAR system to detect weak targets in the neighborhood of bright targets. Therefore, the imaging capability of LT-1 will be evaluated first by imaging the calibration point.

As shown in Figure 11a, this paper takes the Hami South Gobi area in Xinjiang as the research area, and its geographic location is 90°15′E–93°35′E and 40°05′–42°25′N. The study area has a typical temperate continental arid climate with low soil humidity and no vegetation coverage [33]. It is an ideal natural calibration field for microwave radiation calibration. LT-1 acquired the selected data on July 7 in the DRA mode. The system parameters and coordinates of the point-like target experiment are shown in Tables 2 and 3. The incident angle was 39°, and the initial resolution of the SAR imagery was 3 m.

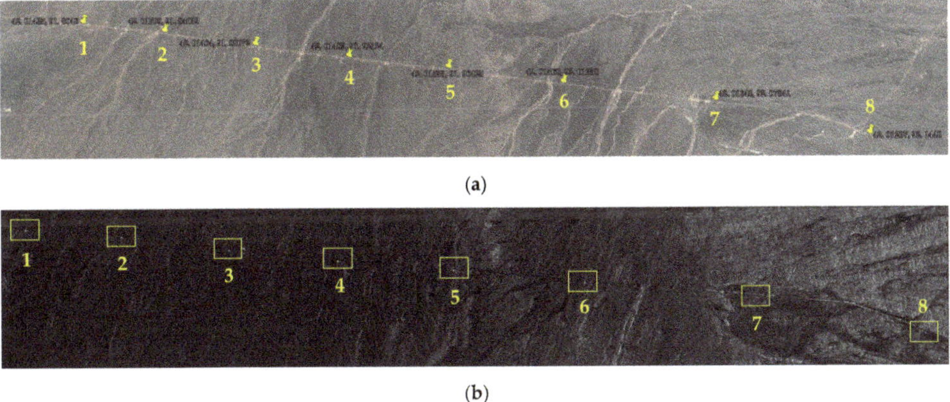

(a)

(b)

Figure 11. Location of the test sites and study area. (**a**) optical image of the test site, (**b**) SAR image of the test site.

Table 2. Imaging parameters.

Parameter	Symbol	Value	Unit
Platform velocity	V_s	7635	m/s
Carrier frequency	f_0	1.26	GHz
Signal bandwidth	B_r	80	MHz
Signal pulse duration	T_p	70	μs
Nearest slant range	R_0	817	km
Azimuth antenna length	L_{az}	4.9×2	m
Rang sampling frequency	F_r	90	MHz
Azimuth sampling frequency	F_a	1742	Hz
Number of channels	M	2	\

Table 3. Calibration point coordinates.

Calibration Point	1	2	3	4	5	6	7	8
Latitude	42.01435	42.01508	42.01464	42.01483	42.01553	42.01568	42.01648	42.01237
Longitude	91.80460	91.84039	91.88072	91.92124	91.96425	92.01358	92.07841	92.14480

According to the amplitude error estimation, the amplitude error of channel 2 compared with the reference channel is 0.316 dB, which is within the internal calibration precision (0.4 dB) of LT-1. Figure 12a,b show the azimuth time-varying phase error and range space-varying phase error estimated by the MSSBN channel error estimation method. Compared with the reference channel, the azimuth time-varying phase error and range space-varying error are linearly related to the sampling interval, and the absolute value of the phase error relative to the reference channel is within 10 deg. As seen in Figure 6, the phase error within 10 degrees has a minimal impact on the imaging quality and can usually be ignored. This result is consistent with the azimuth spectrum in Figure 12c,d. When there is a significant phase error for uniform land scenes, there is a significant jump in the reconstructed spectral sub-bands at the junction. Figure 12c,d show that the reconstructed spectrum with phase error and the reconstructed spectrum after compensating for phase error are both very smooth. Although the phase error in this scene has significant time–space varying due to other factors, it does not affect the spectrum reconstruction results.

After channel error correction and two-channel signal reconstruction, the calibration field scene was imaged. As can be seen from Figures 13 and 14, calibration points are well-focused. Figure 14 shows the analysis results of range and azimuth slices of calibration points. It can be seen from the figure that ISLR indexes of all calibration points are less than -16 dB, satisfying the system design indexes of LT-1 (ISLR ≤ -13 dB). Furthermore, the PSLR of all calibration points is close to the system design indexes of LT-1 (PSLR ≤ -20 dB). It can be further reduced later by adding Windows.

4.3. Scenarios Imaging Results

Then, DAR echo data of different scenarios will be processed in this paper. The estimation of channel errors, azimuth reconstruction spectrum, imaging results, and AASR evaluated the performance of the LT-1 DRA imaging mode.

4.3.1. Wilderness Scenarios

The data collected by LT-1 on 19 October 2022 were processed. Other imaging parameters were consistent with Table 2 except that the R_0 was 692 km. The estimated amplitude error of this scenario is 0.3677 dB, which meets the design requirements. It can be seen from Figure 15a,b that the phase error of space–time variation is very tiny, and the phase error is approximately a constant error within 10 deg. Therefore, the azimuth spectrum is recovered ideally after channel error correction in Figure 15c,d. As can be seen from the imaging results in Figure 15e,f, the channel error within 10 degrees has very little influence on the imaging quality.

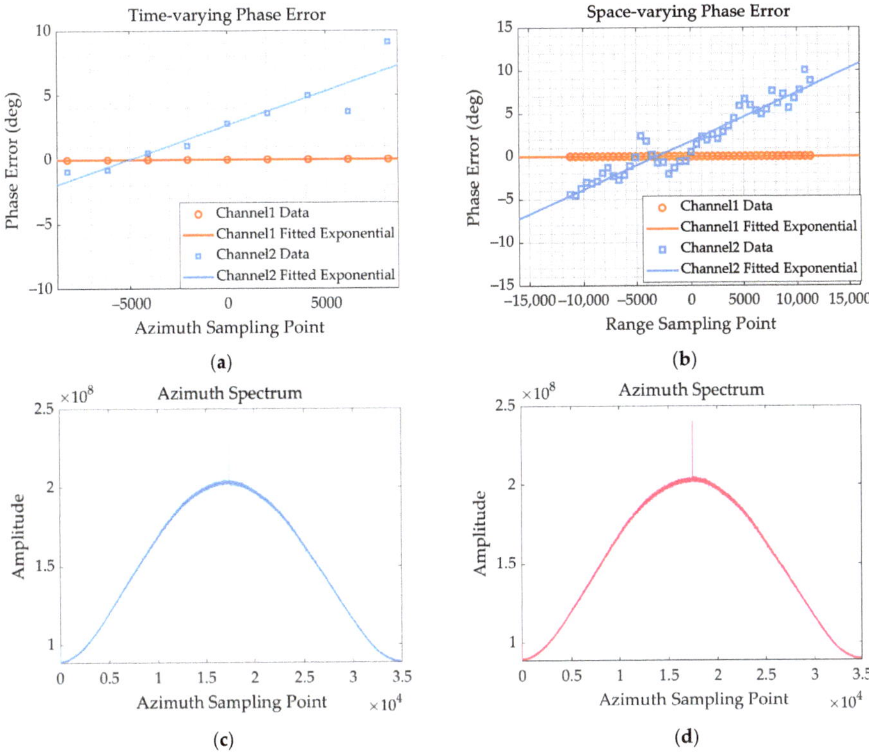

Figure 12. Phase error estimation and reconstruction spectrum of calibration field scene. (**a**) Azimuth time-varying phase error estimation, (**b**) range space-varying phase error estimation, (**c**) reconstruction spectrum without channel error calibration, and (**d**) reconstruction spectrum with channel error calibration.

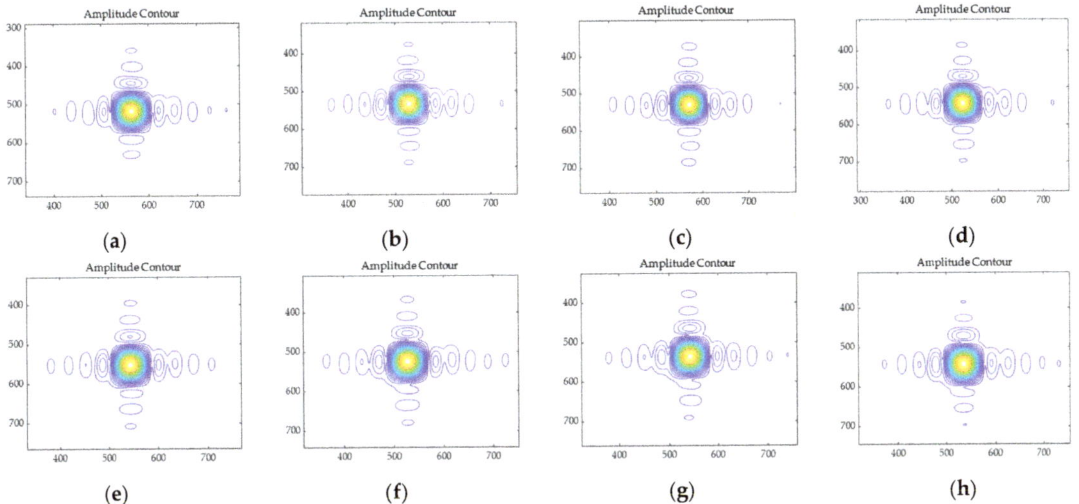

Figure 13. Contour of calibration points. (**a**) Point 1, (**b**) point 2, (**c**) point 3, (**d**) point 4, (**e**) point 5, (**f**) point 6, (**g**) point 7, and (**h**) point 8.

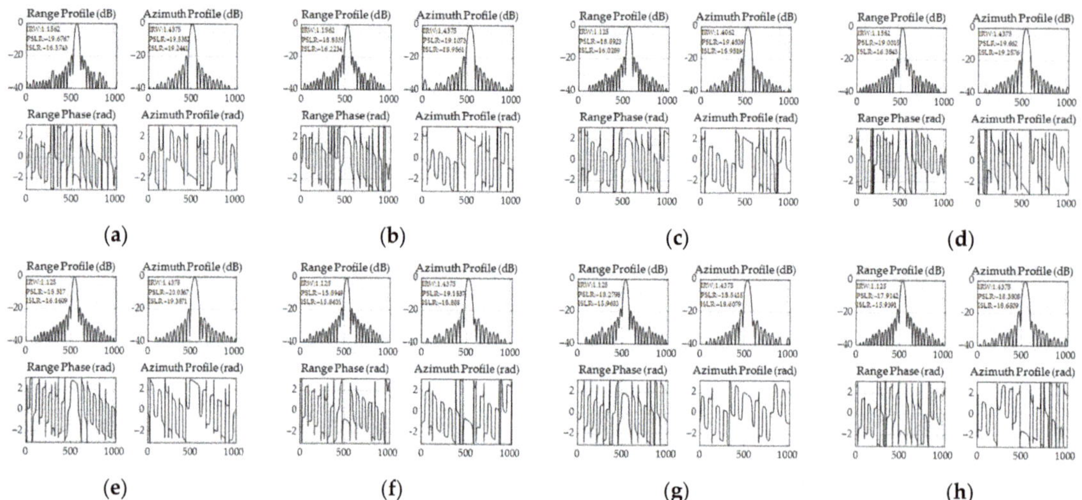

Figure 14. Profile of calibration points. (**a**) Point 1, (**b**) point 2, (**c**) point 3, (**d**) point 4, (**e**) point 5, (**f**) point 6, (**g**) point 7, and (**h**) point 8.

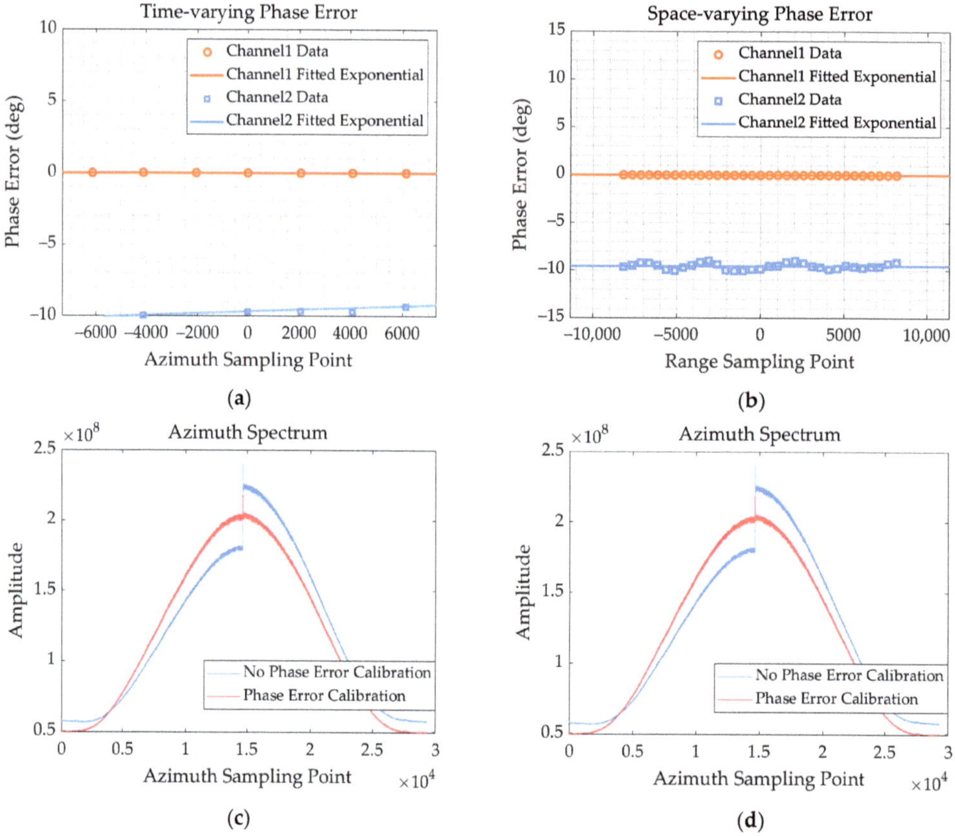

Figure 15. *Cont.*

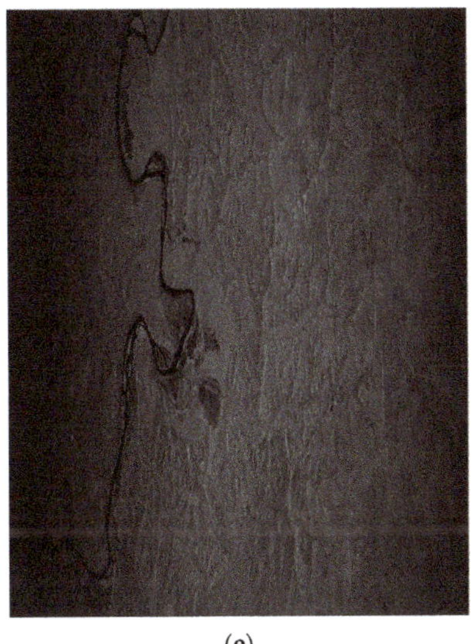 

(e)   (f)

Figure 15. Imaging results of wilderness scenarios: (**a**) azimuth time-varying phase error estimation results, (**b**) range space-varying phase error estimation, (**c**) reconstruction spectrum without channel error calibration, (**d**) reconstruction spectrum with channel error calibration, (**e**) focused image without channel error calibration, and (**f**) focused image with channel error calibration.

4.3.2. City Scenarios

The city scenarios data collected by LT-1 on 29 August 2022 were processed. Other imaging parameters were consistent with Table 2 except that the R_0 was 692 km and V_s was 7640 m/s, and the estimated amplitude error of this scenario is 0.3469 dB.

The LT-1 system always maintains good channel error control ability during operation. From Figure 16a,b, it can be seen that the time-varying phase error of the channel is controlled within 5 deg, and the space-varying error is controlled within 10 deg. It can be seen from the reconstructed spectrum that the reconstructed spectrum without channel error correction is consistent with the reconstructed spectrum after channel error correction. The imaging results in Figure 16e,f show that there is no apparent azimuth ambiguity in the images.

4.3.3. Land and Sea Interface Scenarios

In ambiguity azimuth analysis, other targets in the land scene often interfere with the amplitude of the false target. In the case of superimposed clutter interference, the amplitude of the false target even exceeds the target itself, so AASR cannot be accurately estimated. Therefore, this paper chooses a land and sea interface scene for processing. The scene of this experiment is the dual-antenna receiving data measured by LT-1 in the sea off Ningbo City on 9 February 2022. The R_0 was 767 km and V_s was 7277 m/s. The estimated amplitude error of this scenario is 0.3890 dB. Due to the excellent channel error control performance of LT-1, the dual-channel echo data of LT-1 has excellent azimuth fuzzy suppression performance. By comparing Figure 17a without channel error correction with Figure 17b focusing image after channel error correction, it can be found that the azimuth fuzzy of the focusing image directly reconstructed by echo data is almost invisible.

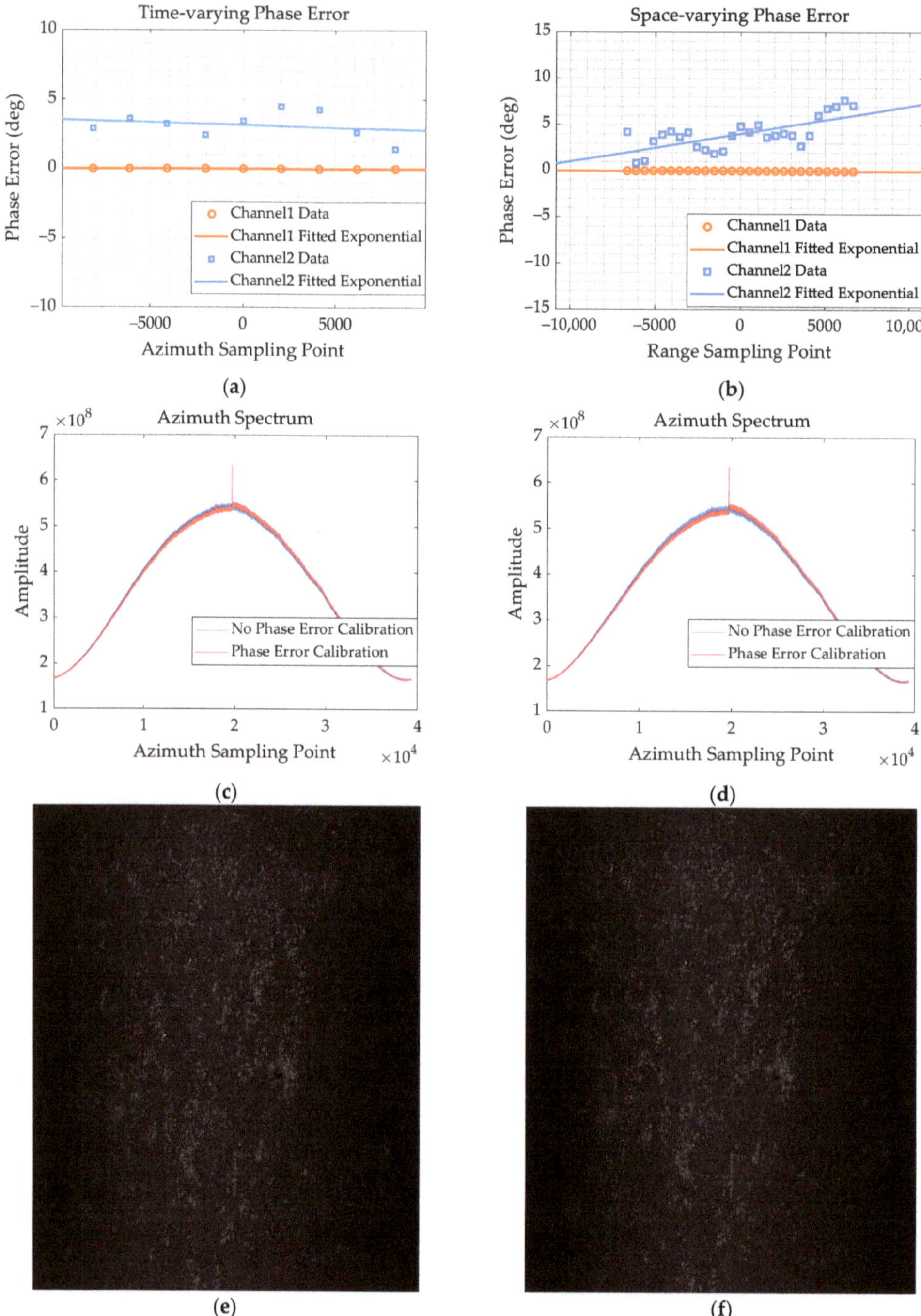

Figure 16. Imaging results of city scenarios: (**a**) azimuth time-varying phase error estimation results, (**b**) range space-varying phase error estimation, (**c**) reconstruction spectrum without channel error calibration, (**d**) reconstruction spectrum with channel error calibration (**e**) focused image without channel error calibration, and (**f**) focused image with channel error calibration.

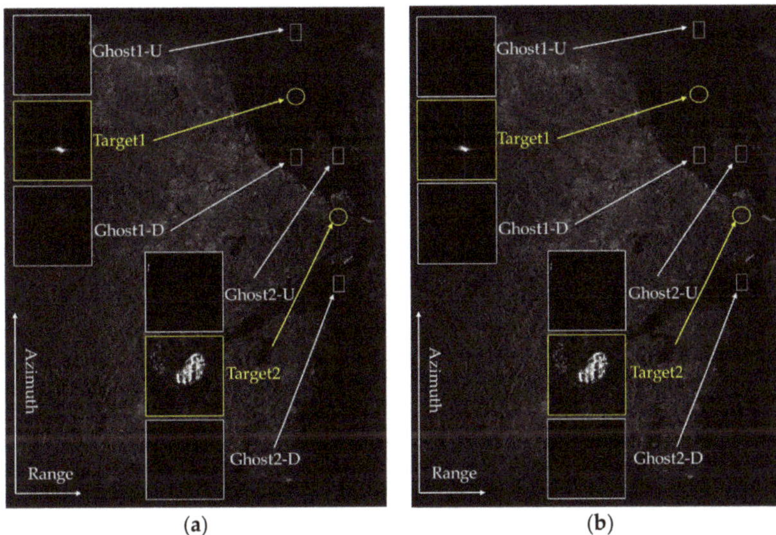

Figure 17. Imaging results of land and sea interface scenarios: (**a**) focused image without channel error calibration, and (**b**) focused image with channel error calibration.

The displacement in azimuth and slant range of the first-order ambiguity relative to the primary response is approximately given by [34,35]

$$\Delta_{az,1} \approx \frac{\lambda \mathrm{PRF} R_0}{2V_s} \tag{28}$$

$$\Delta_{rg,1} \approx \sqrt{R_0^2 - \Delta_{az,1}^2} - R_0 \tag{29}$$

According to Equations (28) and (29), the distance between the first-order ambiguity and the target and the ambiguity dispersed distance is 18.255 km and 217.2085 m. It can be seen from Figure 18 that the azimuth direction of the first-order ambiguity and the range direction of the first-order ambiguity are consistent with the calculated results. Due to the long wavelength of the L-band, the imaging results are mainly affected by the first-order ambiguity, and the first-order ambiguity discretization is serious.

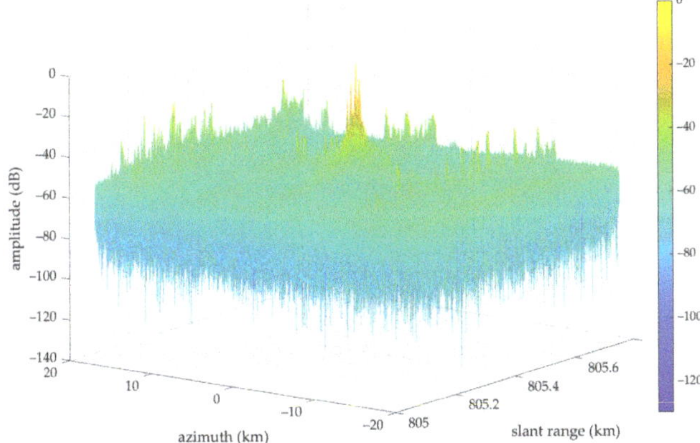

Figure 18. Imaging results of Target 1 and Ghost 1.

To quantitatively analyze the image quality, the ghost-real target energy ratio (GTER) [20] is employed to measure the suppression of the ghost target in the final focused image without phase error calibration, which is defined as

$$\mathrm{GTER} = 20 \log_{10} \frac{\max(|I_G|)}{\max(|I_R|)} \tag{30}$$

where $\max(|I_G|)$ denotes the maximum amplitude value of pixels of ghost targets, and $\max(|I_R|)$ denotes the maximum value of pixels of real targets. The GTER result of Target 1 is -52 dB, which shows the LT-1's sound azimuth ambiguity performance.

4.4. Discussion

The experimental results in Sections 3 and 4 demonstrate an effective suppression of azimuth ambiguities for different scene targets with the LT-1 DRA mode. The simulation experiment in Section 3 shows that, compared with other SAR systems with shorter wavelength segments, LT-1 adopts an L-band, which can significantly reduce the influence of antenna installation on the channel phase error. At the same time, thanks to the excellent system design of LT-1, the channel phase error of LT-1 is always controlled within 10 deg during the on-orbit operation, which enables the dual antennas of LT-1 to receive data and carry out signal reconstruction and focusing imaging operations without going through the channel error correction. Therefore, in the face of sudden disasters and other emergencies, HRWS SAR imaging can save time for channel error correction. In Section 4, this paper deals with the measured data from different scenarios. Point-like target analysis is performed by analyzing the system's response to the reflector in the Xinjiang calibration field. The experimental results thoroughly verify the ambiguity suppression capability of the LT-1 system.

5. Conclusions

This paper established the channel error model of LT-1 DRA mode, and the influence of channel error on imaging quality was analyzed. The channel phase error of LT-1 was efficiently estimated using the MLNS method. Experiments showed that the LT-1 system can suppress azimuth ambiguity successfully by using DRA mode. The LT-1 benefits from an excellent SAR system design that allows amplitude-phase errors to be controlled within the system design requirements, requiring little additional channel error calibration operations. The LT-1 has shown the expected performance and resolution during in-orbit operation, and the parameters related to image quality are excellent.

Author Contributions: Conceptualization, Z.X.; Methodology, Z.X. and Y.C.; Funding acquisition, P.L. and R.W.; Projection administration, P.L. and R.W.; Writing—original draft, Z.X.; Writing—review and editing, Z.X., Y.C., R.W. and Y.W. All authors have read and agreed to the published version of the manuscript.

Funding: This research was funded by the National Science Fund for Distinguished Young Scholars, China, grant number No. 61825106.

Data Availability Statement: Not applicable.

Conflicts of Interest: The authors declare no conflict of interest.

References

1. Cumming, I.G.; Wong, F.H. Digital processing of synthetic aperture radar data. *Artech House* **2005**, *1*, 108–110.
2. Li, Z.; Wang, H.; Su, T.; Bao, Z. Generation of wide-swath and high-resolution SAR images from multichannel small spaceborne SAR systems. *IEEE Geosci. Remote Sens. Lett.* **2005**, *2*, 82–86. [CrossRef]
3. Krieger, G.; Gebert, N.; Moreira, A. Unambiguous SAR signal reconstruction from nonuniform displaced phase center sampling. *IEEE Geosci. Remote Sens. Lett.* **2004**, *1*, 260–264. [CrossRef]
4. Krieger, G.; Gebert, N.; Moreira, A. Multidimensional waveform encoding: A new digital beamforming technique for synthetic aperture radar remote sensing. *IEEE Trans. Geosci. Remote Sens.* **2007**, *46*, 31–46. [CrossRef]

5. Jing, W.; Xing, M.; Qiu, C.-W.; Bao, Z.; Yeo, T.-S. Unambiguous reconstruction and high-resolution imaging for multiple-channel SAR and airborne experiment results. *IEEE Geosci. Remote Sens. Lett.* **2008**, *6*, 102–106. [CrossRef]
6. Krieger, G.; Gebert, N.; Younis, M.; Bordoni, F.; Patyuchenko, A.; Moreira, A. Advanced concepts for ultra-wide-swath SAR imaging. In Proceedings of the 7th European Conference on Synthetic Aperture Radar, Friedrichshafen, Germany, 2–5 June 2008; pp. 1–4.
7. Yunkai, D.; Weidong, Y.; Heng, Z.; Wei, W.; Dacheng, L.; Robert, W. Forthcoming spaceborne SAR development. *J. Radars* **2020**, *9*, 1–33.
8. Yingjie, W.; Robert, W.; Weidong, Y.; Qingchao, Z.; Kaiyu, L.; Dacheng, L.; Yunkai, D.; Naiming, O.; Xiaoxue, J.; Heng, Z.; et al. See-Earth: SAR Constellation with Dense Time-SEries for Multi-dimensional Environmental Monitoring of the Earth. *J. Radars* **2021**, *10*, 842–864. [CrossRef]
9. Fu, Z.; Zhang, H.; Zhao, J.; Li, N.; Zheng, F. A Modified 2-D Notch Filter Based on Image Segmentation for RFI Mitigation in Synthetic Aperture Radar. *Remote Sens.* **2023**, *15*, 846. [CrossRef]
10. Currie, A.; Brown, M.A. Wide-swath SAR. In *IEE Proceedings F (Radar and Signal Processing)*; IET: Wales, UK, 1992; pp. 122–135.
11. Süß, M.; Grafmüller, B.; Zahn, R. A novel high resolution, wide swath SAR system. In Proceedings of the IGARSS 2001. Scanning the Present and Resolving the Future. IEEE 2001 International Geoscience and Remote Sensing Symposium (Cat. No. 01CH37217), Sydney, NSW, Australia, 9–13 July 2001; pp. 1013–1015.
12. Kim, J.-H.; Younis, M.; Gabele, M.; Prats, P.; Krieger, G. First spaceborne experiment of digital beam forming with TerraSAR-X dual receive antenna mode. In Proceedings of the 2011 8th European Radar Conference, Manchester, UK, 12–14 October 2011; pp. 41–44.
13. Younis, M.; De Almeida, F.Q.; Bordoni, F.; López-Dekker, P.; Krieger, G. Digital beamforming techniques for multi-channel synthetic aperture radar. In Proceedings of the 2016 IEEE International Geoscience and Remote Sensing Symposium (IGARSS), Beijing, China, 10–15 July 2016; pp. 1412–1415.
14. Hatooka, Y.; Kankaku, Y.; Arikawa, Y.; Suzuki, S. First result from ALOS-2 operation. In Proceedings of the Earth Observing Missions and Sensors: Development, Implementation, and Characterization III, San Francisco, CA, USA, 18 December 2014; pp. 138–147.
15. Jin, T.; Qiu, X.; Hu, D.; Ding, C. Unambiguous imaging of static scenes and moving targets with the first Chinese dual-channel spaceborne SAR sensor. *Sensors* **2017**, *17*, 1709. [CrossRef]
16. Qingjun, Z. System design and key technologies of the GF-3 satellite. *Acta Geod. Cartogr. Sin.* **2017**, *46*, 269.
17. Shi, J.; Lü, D.; Wang, Y.; Du, Y.; Pang, Y.; Yang, D.; Wang, X.; Dong, X.; Yang, X. Recent Progress of Earth Science Satellite Missions in China. *Chin. J. Space Sci.* **2022**, *42*, 712–723. [CrossRef]
18. Cai, Y.; Wang, R.; Yu, W.; Liang, D.; Liu, K.; Zhang, H.; Chen, Y. An Advanced Approach to Improve Synchronization Phase Accuracy with Compressive Sensing for LT-1 Bistatic Spaceborne SAR. *Remote Sens.* **2022**, *14*, 4621. [CrossRef]
19. Cai, Y.; Deng, Y.; Zhang, H.; Wang, R.; Wu, Y.; Cheng, S. An Image-Domain Least L 1-Norm Method for Channel Error Effect Analysis and Calibration of Azimuth Multi-Channel SAR. *IEEE Trans. Geosci. Remote Sens.* **2022**, *60*, 1–14. [CrossRef]
20. Yang, W.; Guo, J.; Chen, J.; Liu, W.; Deng, J.; Wang, Y.; Zeng, H. A Novel Channel Inconsistency Estimation Method for Azimuth Multichannel SAR Based on Maximum Normalized Image Sharpness. *IEEE Trans. Geosci. Remote Sens.* **2022**, *60*, 1–16. [CrossRef]
21. Shang, M.; Qiu, X.; Han, B.; Ding, C.; Hu, Y. Channel Imbalances and Along-Track Baseline Estimation for the GF-3 Azimuth Multichannel Mode. *Remote Sens.* **2019**, *11*, 1297. [CrossRef]
22. Gebert, N.; de Almeida, F.Q.; Krieger, G. Airborne demonstration of multichannel SAR imaging. *IEEE Geosci. Remote Sens. Lett.* **2011**, *8*, 963–967. [CrossRef]
23. Li, Z.; Bao, Z.; Wang, H.; Liao, G. Performance improvement for constellation SAR using signal processing techniques. *IEEE Trans. Aerosp. Electron. Syst.* **2006**, *42*, 436–452.
24. Feng, J.; Gao, C.; Zhang, Y.; Wang, R. Phase mismatch calibration of the multichannel SAR based on azimuth cross correlation. *IEEE Geosci. Remote Sens. Lett.* **2012**, *10*, 903–907. [CrossRef]
25. Zhang, L.; Gao, Y.; Liu, X. Robust channel phase error calibration algorithm for multichannel high-resolution and wide-swath SAR imaging. *IEEE Geosci. Remote Sens. Lett.* **2017**, *14*, 649–653. [CrossRef]
26. Xiang, J.; Ding, X.; Sun, G.-C.; Zhang, Z.; Xing, M.; Liu, W. An Efficient Multichannel SAR Channel Phase Error Calibration Method Based on Fine-Focused HRWS SAR Image Entropy. *IEEE J. Sel. Top. Appl. Earth Obs. Remote Sens.* **2022**, *15*, 7873–7885. [CrossRef]
27. Zhang, S.-X.; Xing, M.-D.; Xia, X.-G.; Liu, Y.-Y.; Guo, R.; Bao, Z. A robust channel-calibration algorithm for multi-channel in azimuth HRWS SAR imaging based on local maximum-likelihood weighted minimum entropy. *IEEE Trans. Image Process.* **2013**, *22*, 5294–5305. [CrossRef] [PubMed]
28. Xu, Z.; Lu, P.; Cai, Y.; Li, J.; Yang, T.; Wu, Y.; Wang, R. An Efficient Channel Imbalance Estimation Method Based on Subadditivity of Linear Normed Space of Sub-Band Spectrum for Azimuth Multichannel SAR. *Remote Sens.* **2023**, *15*, 1561. [CrossRef]
29. Laskowski, P.; Bordoni, F.; Younis, M. Antenna pattern compensation in multi-channel azimuth reconstruction algorithm. In Proceedings of the Advanced RF Sensors and Remote Sensing Instruments (ARSI), Noordwijk, The Netherlands, 15 September 2011; pp. 1–10.
30. Shang, M.; Qiu, X.; Han, B.; Yang, J.; Zhong, L.; Ding, C.; Hu, Y. The space-time variation of phase imbalance for GF-3 azimuth multichannel mode. *IEEE J. Sel. Top. Appl. Earth Obs. Remote Sens.* **2020**, *13*, 4774–4788. [CrossRef]

31. Yu, W.; Wang, K.; Wu, J.; Li, S.; Xie, W.; Sun, H.; Ou, N. The LuTan-1 SAR Antenna System. In Proceedings of the EUSAR 2022, 14th European Conference on Synthetic Aperture Radar, Leipzig, Germany, 25–27 July 2022; pp. 1–4.
32. Maligranda, L. Some remarks on the triangle inequality for norms. *Banach J. Math. Anal.* **2008**, *2*, 31–41. [CrossRef]
33. Mou, J.; Hong, J.; Wang, Y.; Du, S.; Xing, K.; Qiu, T. LT-1 Baseline Calibration Method Based on Improved Baseline Calibration Model. In Proceedings of the IGARSS 2022–2022 IEEE International Geoscience and Remote Sensing Symposium, Kuala Lumpur, Malaysia, 17–22 July 2022; pp. 7452–7455.
34. Villano, M.; Krieger, G. Spectral-based estimation of the local azimuth ambiguity-to-signal ratio in SAR images. *IEEE Trans. Geosci. Remote Sens.* **2013**, *52*, 2304–2313. [CrossRef]
35. Li, F.K.; Johnson, W.T.K. Ambiguities in Spacebornene Synthetic Aperture Radar Systems. *IEEE Trans. Aerosp. Electron. Syst.* **1983**, *AES-19*, 389–397. [CrossRef]

 *remote sensing*

Article

SCM: A Searched Convolutional Metaformer for SAR Ship Classification

Hairui Zhu, Shanhong Guo *, Weixing Sheng and Lei Xiao

School of Electronic and Optical Engineering, Nanjing University of Science and Technology, Nanjing 210094, China; hairui.zhu@njust.edu.cn (H.Z.); shengwx@njust.edu.cn (W.S.); leixiao@njust.edu.cn (L.X.)
* Correspondence: guosh@njust.edu.cn

Abstract: Ship classification technology using synthetic aperture radar (SAR) has become a research hotspot. Many deep-learning-based methods have been proposed with handcrafted models or using transplanted computer vision networks. However, most of these methods are designed for graphics processing unit (GPU) platforms, leading to limited scope for application. This paper proposes a novel mini-size searched convolutional Metaformer (SCM) for classifying SAR ships. Firstly, a network architecture searching (NAS) algorithm with progressive data augmentation is proposed to find an efficient baseline convolutional network. Then, a transformer classifier is employed to improve the spatial awareness capability. Moreover, a ConvFormer cell is proposed by filling the searched normal convolutional cell into a Metaformer block. This novel cell architecture further improves the feature-extracting capability. Experimental results obtained show that the proposed SCM provides the best accuracy with only 0.46×10^6 weights, achieving a good trade-off between performance and model size.

Keywords: synthetic aperture radar; ship classification; deep learning; transformer; network architecture searching

Citation: Zhu, H.; Guo, S.; Sheng, W.; Xiao, L. SCM: A Searched Convolutional Metaformer for SAR Ship Classification. *Remote Sens.* **2023**, *15*, 2904. https://doi.org/10.3390/rs15112904

Academic Editor: Gerardo Di Martino

Received: 29 April 2023
Revised: 28 May 2023
Accepted: 30 May 2023
Published: 2 June 2023

1. Introduction

With the increasing number of ships on the oceans, efficient classification of ships has become vital [1,2]. Synthetic aperture radar (SAR) is a popular type of positive detection equipment used in modern ocean surveillance. SAR can work in all weathers and at all times [3]; thus, ship classification with SAR images has become a research hotspot in remote sensing.

Traditional SAR ship classification methods depend on feature extraction and classification using manual design, including support vector machine (SVM) [4], decision tree [5], random forest [6], Bayesian classifier [7], Adaboost [8], etc. However, these methods require large amounts of time and labor regarding their design, testing, analysis, and verification. The robustness and accuracy of these methods may significantly decrease in actual application scenarios with complex ship shapes, high sea conditions, and low image quality.

Recently, deep learning has achieved many excellent results in several fields of computer vision, including image processing [9,10] and pattern recognition [11,12]. Many studies have been conducted on applying deep learning to SAR ship classification. Compared to conventional methods, convolutional neural network (CNN)-based methods result in a remarkable boost in accuracy. Based on advanced training technologies, suitable natural optical image object classification models can operate SAR ship classification tasks [13–15]. For high accuracy, another more common method is to manually design networks for SAR data [16,17]. Furthermore, traditional features, such as histograms of oriented gradients (HOGs) and polarization coherence, can be integrated into neural networks to further improve performance [18,19].

Review of existing SAR ship classification networks suggests that most of the networks that are transplanted from computer vision or that are handcrafted have unsatisfactory

---

accuracies on SAR data or have an unfeasible model size for embedded devices. To further improve efficiency, a novel searched convolutional Metaformer (SCM) is proposed for SAR ship classification in this paper. The design of the SCM starts from a baseline network searched in a small SAR ship dataset. We improve the network architecture searching (NAS) algorithm. Considering the mobile and embedded platforms used in SAR tasks that require a high prediction efficiency, we improve the baseline network with transformer encoders and efficient convolutional cells. The SCM achieves a good trade-off between performance and computational complexity with a very small size.

Partial channel connection differentiable architecture searching (PC-DARTS) [20] treats the search as a bi-level optimization problem and updates the target network architecture by gradient descent [21]. The target network produced by the original PC-DARTS contains too many weight-free operations, which leads to low feature extraction capability. After an analysis of the involved SAR ship dataset, the small number of available samples, the poor quality of the SAR images, and the high similarity of the different ship categories result in the super net learning little about the features. Weight-free operations have a very high probability to be chosen at the early stage of searching. This problem is unsolvable automatically even at the end of searching, which leads to a target network with poor performance. Hence, progressive data augmentation partial channel connection differentiable architecture searching (PDA-PC-DARTS) is proposed to obtain a target network with strong feature excretion capability on a small size SAR ship dataset with poor image quality. This paper employs data augmentation and progressive learning to divide the whole search into three stages. A different data augmentation policy is assigned to a searching stage, so the target architecture is optimized progressively. The data augmentation can expand sample diversity. Progressive learning enhances the gradient learning of the super-net and architecture parameters. Our target network searching in the low-quality small-size SAR ship dataset shows improved performance.

Obviously, the performance of the target network has a great deal of room for further improvement. Networking scaling is the most common way to improve the performance of a searched network architecture. However, network scaling substantially increases the number of weights and computational operations. Considering the application scenario of SAR ship classification using mobile or embedded devices, we have to give up network scaling and choose an efficient way to improve performance. The convolutional operations and classifier in the target network are updated. The basic blocks of convolutional operations used in NAS are switched from depth-wise separable convolution (DSCONV) to the EfficientNet version inverted residual block (MBCONV) [22], which can improve the efficiency and accuracy. A transformer classifier is connected to the searched cells, which contributes to a searched convolutional transformer (SCT). The searched cells play the role of tokenizers instead of common simple tokenizers, such as a single fully connected layer [23] or stacked vanilla convolution layers [24]. The searched CNN has the advantage of processing two-dimensional data and the transformer classifier provides better learning capability. The SCT can combine the advantages of both parts and improve the performance of ship classification.

A transformer has a higher computational complexity than a CNN. Usually, transformer–CNN mixed networks in computer vision employ many transformer-based operations as the main computational module, which results in high computational complexity. In our approach, the performance of the CNN part on SAR ship data is guaranteed by NAS. Hence, the CNN is the main part in the proposed network. We only use two transformer encoders as an additional module and maintain the computational complexity of the whole network at a low level.

Then, following the concept of a Metaformer [25] and modern CNNs, a ConvFormer cell is proposed as the novel cell architecture. Studies have found that transformers have advantages for both self-attention and architecture. Hence, we convert a searched convolutional cell into a Metaformer block by referring to the transformer encoder architecture.

The token mixer in the ConvFormer cell is searched from the SAR ship dataset. Hence, the ConvFormer cell has a strong feature extraction ability on SAR ships.

SCM is a combination of SCT and ConvFormer cells. The experimental results on the OpenSARShip dataset show that SCM achieves 82.06% accuracy with 81.46×10^6 multiply-accumulate operations (MAdds). Compared with existing handcrafted learning-based methods and networks from computer vision, SCM outperforms them by 3% in accuracy and reaches a state-of-the-art level. To verify the generalization capability of the searched architecture, we train the proposed network on a more challenging classification task from the FUSARship dataset without prior research, resulting in a good performance of 63.90%. SCM only has 0.46×10^6 weights. In addition, the searching and training of SCM can be operated with only a single RTX3060.

We summarize the main contributions of the paper as follows:

1. A novel mixture network, SCM, is proposed for SAR ship classification. This network is built with the help of NAS and a transformer, which have the characteristics of high accuracy and small model size.
2. PDA-PC-DARTS is proposed to explore a high-performance target network in the poor quality, small sample number SAR ship dataset.
3. For searching, DSCONV is replaced with MBCONV to achieve better efficiency.
4. SCT is proposed by combining NAS, CNN, and a transformer, which improves the classification accuracy.
5. A novel cell architecture is proposed to further enhance the accuracy by filling a searched node into a Metaformer block.

The rest of this paper is organized as follows: Section 2 reviews related work. In Section 3, the SCM is described in detail. Section 4 describes the experiments. The results and ablation studies are described in Section 5. A discussion is provided in Section 6. Section 7 concludes the paper.

2. Related Work

2.1. CNN-Based SAR Ship Classification

In contrast to natural optical image object classification, samples of SAR ship classification show many special features, such as single channel, low quality, and simple background. CNN-based SAR ship classification methods have two main aspects: improving computer vision networks on SAR ship data and designing networks for SAR ship data. Table 1 summarizes some published CNN-based SAR ship classification methods.

Directly applying computer vision models to SAR ship data can give unsatisfactory results. Improving the adaptability of computer vision models on SAR data is the key to the first approach. A SAR ship classification method based on transfer learning was presented, where some layers of the visual geometry group network (VGGNet) pretrained on the ImageNet dataset, which is a popular public natural optical image object classification dataset, were retrained on SAR data [13]. VGGNet has a large mode size and high computational complexity because it was designed for optical object classification, which is a very difficult task in computer vision. A hybrid channel feature loss was designed to improve the training of ship SAR images [14]. This dual-polarization classification method also employed VGGNet and achieved good accuracy. The model size and computational complexity of this method are much larger than many methods using networks designed for SAR ship data. Thanks to the development of deep learning, a dual-polarization classification method based on mini-hourglass region extraction and a dual-channel efficient fusion network has achieved good accuracy with moderate computational complexity [15]. Many advanced technologies based on deep learning have been integrated into this method, which has boosted its efficiency. However, the core of this method is EfficientNet, whose target platform is a desktop. Compared with some networks designed for SAR aiming at mobile or embedded platforms, this method has relatively higher computational complexity.

Networks designed for SAR ship data are different to networks designed for computer vision. A sequential CNN for high-resolution SAR ship images has been designed, including ten convolutional layers and three fully connected layers [16]. Experimental results on the dataset obtained from the Gaofen-3 satellite showed that this CNN had good accuracy over many ship categories. However, this method has very higher computational complexity compared to VGGNet-based methods. Furthermore, the number of weights in this method is not suitable for embedded devices. Traditionally handcrafted feature extraction can be integrated into neural networks. A combination of HOGs, principal component analysis (PCA), spatial attention, and a CNN was proposed and referred to as the network with HOG feature fusion (HOG-ShipCLSNet) [18]. This method contains two data flows. For one data flow, the features in HOGs are extracted by PCA. For the other data flow, convolutional layers with spatial attention are used to extract features from the original SAR image at different scales. Concatenated feature vectors from the two flows are fed to a linear classifier. Compared with other CNN-based SAR ship classification methods, HOG-ShipCLSNet has low computational complexity and high performance. Two disadvantages of this method are the large model size and the need for many additional computations outside of the neural network. A dual-polarization SAR ship classification network was constructed with a channel-wise attention fusion module, a squeeze-and-excitation (SE) module, and a Laplacian feature pyramid [19]. This squeeze-and-excitation Laplacian pyramid network was dubbed "SE-LPN-DPFF". The dual-polarization SAR images and the polarization coherence are the inputs. This method shows good performance and depends largely on dual-polarization with a large prediction latency. Many SAR products do not provide dual-polarization. Single-polarization-based methods can work on dual polarization data with simple fusion. Hence, compared to single-polarization-based methods, SE-LPN-DPFF has limited scope for application.

Table 1. Summary of CNN-based SAR ship classification methods.

Method	Use Traditional Feature	Network Architecture Designed for SAR Ships	Model Size	Computational Complexity	Dual-Polarization Only
Finetuned VGGNet [13]	×	×	Large	High	×
VGGNet With Hybrid Channel Feature Loss [14]	×	×	Large	High	√
Mini Hourglass Region Extraction and Dual-Channel Efficient Fusion Network [15]	×	×	Moderate	Moderate	√
Plain CNN [16]	×	√	Large	High	×
HOG-ShipCLSNet [18]	√	√	Large	Low	×
SE-LPN-DPFF [19]	√	√			√

2.2. Network Architecture Searching

NAS is a kind of algorithm that automatically designs network architectures for predefined purposes. A manually designed network architecture always requires a long development period with low labor efficiency. Furthermore, unnecessary interventions from involved researchers may damage these handcrafted networks.

Hence, to improve upon manually designed networks, NAS has become one of the most important research areas in computer science. NAS algorithms can be divided into three main categories reflecting their different approaches to exploring the searching space: reinforcement-learning-based searching, evolutionary-based searching, and gradient-based searching.

Reinforcement-learning-based searching is the most important area to date and has had a remarkable impact on all NAS algorithms. An outside meta-controller is defined to generate the network architectures. The controller is taught to act following predefined goals through the use of reinforcement learning. Application of the concept of a node cell prevents unfinishable search with an exaggerated large searching space for the whole network [26]. The target network can be built with several stacked searched cells sharing the same prediction graph, which means the output of searching changes from a whole network to a few cells, significantly reducing the computational complexity. Weight-sharing searching significantly reduces the time consumption for searching [27]. Clearing trained weights is forbidden, which eliminates the time used to retrain from scratch at the evaluation stage.

Evolutionary-based searching focuses on applying evolutionary algorithms and genetic operations on network architectures with predefined targets. With the help of encoding technology, neural networks can be presented as architecture gene queries [28]. Furthermore, some network architectures for special uses, such as resisting noise [29], can be found with this approach.

Gradient-based searching was established based on converting architecture searching into an optimization problem and achieves excellent searching efficiency. Differentiable architecture searching (DARTS) was the first gradient-based searching algorithm [21]. Searching is defined as a bi-level optimization problem which finds the target network by applying gradient descent. In the meantime, both node cell searching and weights sharing are utilized. All candidate operations are packaged into a super-net and can be trained together, which significantly decreases the search time. However, DARTS still has server-level hardware requirements and can be criticized for filling many weight-free operations in target networks. PC-DARTS is a personal-computer-level NAS [20]. Randomly sampled channels can effectively reduce the computational complexity and memory occupation, while the hit rates of weight-free operations with constant output are decreased to a moderate state.

2.3. Efficient Convolutional Blocks

Due to popular modern CNNs being built with convolutional blocks, many researchers are working on modifying convolutional blocks to improve the efficiency of CNNs. Related studies are being conducted on pure convolutional blocks and additional module-equipped blocks.

An efficient pure convolutional block usually contains one or several convolutional layers, activations, normalizations, and shortcut connections. DSCONV is a well-known efficient convolutional block, where a normal convolution is replaced by a depth-wise convolution and a point-wise convolution [30]. DSCONV has an advantage in having very low computational complexity and has become a pioneer for the implementation of CNNs in mobile devices. In depth-wise convolution, each kernel is only connected to an input channel. Point-wise convolution can be considered as a normal convolution whose kernel size is set as 1. However, when the number of input channels is small, the performance of DSCONV is significantly reduced. MobileNetv2 version MBCONV has an inverted residual architecture to alleviate this problem [31]. This architecture first expands the dimension of the feature maps with a point-wise convolution. After computing the depth-wise convolution, the dimension is projected back with another point-wise convolution.

The second approach towards creating efficient convolutional blocks is by inserting additional modules into blocks to improve the performance with a small cost in computational complexity. Inserting the SE module into the residual blocks can significantly improve accuracy [32]. An SE module contains two fully connected layers and one global average pooling layer. After processing of an SE module, the key channels are enhanced, which can be considered as a kind of channel-level attention. Based on MobileNetv2, an improved version of MBCONV, that contains the SE module and uses Swish for activation, is the basic block of EfficientNet [22]. With the help of NAS, EfficientNet achieves a good trade-off between computational complexity and performance. However, the model size and computational complexity of the smallest member of the EfficientNet family is unaffordable for most embedded devices because EfficientNet is designed for middle- and large-sized devices.

2.4. Networks Related to Transformers

After the great success of transformer-based models in natural optical image classification and detection, researchers are increasingly working with transformers on two-dimensional data. The three main types of networks related to transformers can be summarized as transformer-based networks, CNN–transformer mixed networks, and Metaformer-based networks.

Transformer-based networks use self-attention to process features. Vision transformer (ViT) is one of the pioneers in this approach [23]. An image is divided into 286 (14×14) patches of 16×16 pixels. After tokenization and positional embedding, tokens are fed into stacked transformer encoders. The Hierarchical Vision transformer using Shifted Windows (Swin) uses cross-attention to achieve CNN-like multi-level feature extraction and provides multi-scale feature maps [33]. Self-attention involves processing the features within its receptive field and sensing the whole feature map with slicing windows, which decreases the computational complexity.

Some studies on integrating CNNs and transformers have revealed that mixture networks exhibit advantages from both sides. CoAtNet-equipped MBCONV blocks and transformer encoders exhibit more powerful performance on some small-size datasets, indicating that CNNs and transformers exhibit differences in learning capability and generalization capability [34].

The compact convolutional transformer (CCT) is aimed at lightweight model applications [24]. A classic CNN with several normal convolutional layers and max pooling layers is used as a tokenizer, which can efficiently extract features and reduce some of the requirements for transformer parts regarding feature extraction. The CCT outperforms most transformer-based networks and other mixed networks in model size and computational complexity.

Metaformer-based networks focus on the characteristics of the transformer encoder architecture and use other operations to act as token-mixers instead of self-attention. Poolformer uses average pooling, which has no weights, as the token-mixer [25]. Compared to transformer-based networks with multi-head self-attention, Poolformer has much lower computational complexity. Another Metaformer uses multilayer perceptron (MLP) as the token-mixer and achieves acceptable results on image classification tasks [35]. Although a gap in performance between this method and the most advanced CNNs can be observed, its special architecture can significantly help future investigations into interactions in a neural network. Figure 1 displays the general concept of the Metaformer block and the architecture of the transformer encoder and poolformer block.

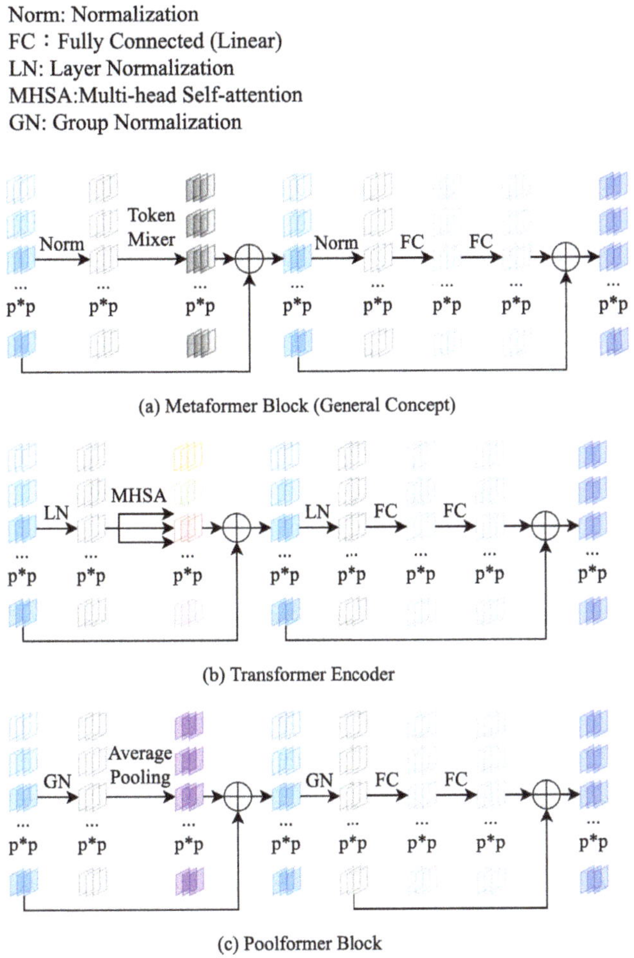

Norm: Normalization
FC：Fully Connected (Linear)
LN: Layer Normalization
MHSA:Multi-head Self-attention
GN: Group Normalization

(a) Metaformer Block (General Concept)

(b) Transformer Encoder

(c) Poolformer Block

Figure 1. Architecture of a Metaformer block and its specific applications.

2.5. SAR Image

Radars transmit modulated radio signals and receive the reflected signals from objects to detect targets, which is very different from cameras, which use natural light or infrared radiation [36]. The wavelength of the radio frequency signal and the polarization processes of radar can affect the SAR images produced. In addition, SAR technology itself and the ocean also present problems for the SAR imaging of ships.

Radar detection waves are able to penetrate, which allows SAR to operate in all weathers. Different wavelengths have different penetration distances, which could affect SAR images. The X-band has a wavelength of three centimeters (cm), the C-band has a wavelength of 6 cm, the L-band has a wavelength of 24 cm, and the P-band has a wavelength of 65 cm. Usually, a signal in the C-band can only penetrate the upper levels of a forest's canopy and can show much roughness scattering combined with some volume scattering. A signal in the P-band or L-band could result in more volume scattering and double-bounce scattering [37]. The C-band has been widely applied in many SAR-equipped satellites, such as Sentinel-1 and Gaofen-3.

Different polarization processes show different features in the resultant SAR images. When describing radar polarization processes, the letters H and V stand for horizontal and vertical polarization, respectively. The transmitter and receiver of a radar may employ polarization processes; hence, the complete polarization process of an SAR image can be described in two letters. VV polarization means both the transmitter and receiver use vertical polarization. This polarization can be more easily influenced by sea clutter. If the transmitter and receiver are set with different polarization processes, VH or HV will be used. These polarizations are susceptible to volume scattering. HH polarization has a stronger double-bounce [37].

SAR images of the sea do not show distance distortion, which is common in optical images, but other problems remain. Geometric distortion can be generated in SAR images by deviation of the detecting geometry, refraction, turbulence, etc. Many SAR detectors are present in high-speed aircraft or where the target is moving rapidly [38]. The relative speed leads to range migration [39]. In addition, speckle noise is generated by the SAR imaging technology itself and is unavoidable in SAR images [40].

3. Searched Convolutional Metaformer

Convolutional cells and a transformer classifier are integrated together in the proposed network for SAR ship classification. The overall architecture of the SCM is shown in Figure 2 and Table 2. The SCM contains four searched convolutional cells, two proposed ConvFormer cells, and one transformer classifier. To achieve a small model size and low computational complexity, a very small number of initial channels is selected and the sizes of the feature maps are reduced many times in the SCM.

In this section, convolutional cells are introduced, with details covered including cell architectures, updated basic blocks, and the proposed searching algorithm which is designed for SAR data. Then, the transformer classifier and proposed ConvFormer cell are described.

Table 2. Details of the SCM network architecture.

Name	Type	Input(s)	Input Shape(s)	Output Shape
Stem	Normal Convolution	SAR Image	$1 \times 100 \times 100$	$12 \times 48 \times 48$
Cell1	Normal Cell	#1 Stem #2 Stem	#1 $12 \times 48 \times 48$ #2 $12 \times 48 \times 48$	$16 \times 48 \times 48$
Cell2	Reduction Cell	#1 Stem #2 Cell1	#1 $12 \times 48 \times 48$ #2 $16 \times 48 \times 48$	$32 \times 24 \times 24$
Cell3	Reduction Cell	#1 Cell1 #2 Cell2	#1 $16 \times 48 \times 48$ #2 $32 \times 24 \times 24$	$64 \times 12 \times 12$
Cell4	Normal Cell	#1 Cell2 #2 Cell3	#1 $32 \times 24 \times 24$ #2 $64 \times 12 \times 12$	$64 \times 12 \times 12$
Cell5	ConvFormer Cell	#1 Cell3 #2 Cell4	#1 $64 \times 12 \times 12$ #2 $64 \times 12 \times 12$	$64 \times 12 \times 12$
Cell6	ConvFormer Cell	#1 Cell4 #2 Cell5	#1 $64 \times 12 \times 12$ #2 $64 \times 12 \times 12$	$64 \times 12 \times 12$
Transformer Classifier	Compact Transformer Classifier	Cell6	$64 \times 12 \times 12$	3

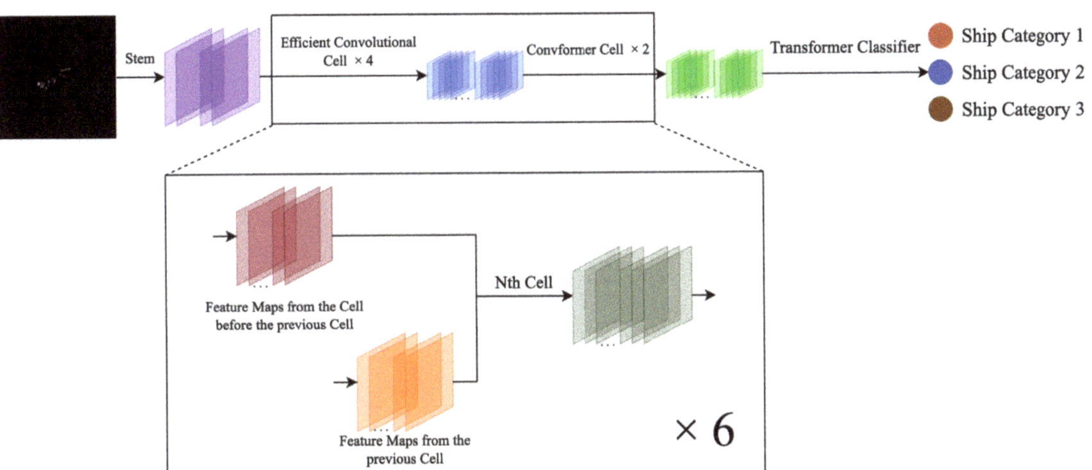

Figure 2. Overall network architecture of the SCM.

3.1. Efficient Convolutional Cells

The efficient convolutional cells in this paper include normal cells and reduction cells, which are shown in Figure 3. A normal cell is constructed with five efficient 5×5 convolutional operations, two efficient 5×5 dilated convolutional operations, and one efficient 3×3 convolutional operation. A reduction cell contains four efficient 5×5 convolutional operations, two efficient reduction 5×5 convolutional operations, one efficient 3×3 convolutional operation, and one efficient reduction 3×3 convolutional operation. All the efficient operations use MBCONV as the basic block instead of DSCONV from DARTS and PC-DARTS. The type and the connections of each operation in our efficient convolutional cells are determined by the proposed PDA-PC-DARTS.

Progressive Network Architecture Searching

Inspired by studies about progressive learning and data augmentation [41], PDA-PC-DARTS is proposed based on PC-DARTS. The search is divided into three stages with different data augmentation policies, which makes the super-net learn better with low-quality SAR ship data. The convolutional cells in the SCM are obtained with the help of PDA-PC-DARTS.

Original PC-DARTS divides the training set into two parts equally. One part is used to train the super-net and the other is used to optimize the architecture parameters. Meanwhile, only a moderate data augmentation policy is employed during the entire search. As a result, weight-free operations, such as pooling, are highly likely to be selected in SAR ship data. The target network is unable to produce a satisfactory performance.

In this paper, more complex data augmentation policies are used. In addition, both super-net training and architecture parameter searching are conducted on the whole training set, which can improve the data utilization rate and the search performance. Following the concept of progressive learning, PDA-PC-DARTS lets the super-net learn small-size images with weak transformations at the beginning stage. Then, large images with a strong data augmentation policy are used in training. Considering the characteristics of SAR images, the search progress contains three stages where data augmentation technologies are used to change the difficulty of learning and increase the variety of data. Hence, the super-net in PDA-PC-DARTS can learn more about samples and produces a target network with better performance. The data augmentation policy in each stage is shown below:

In the first stage, the size of the input images is 76 × 76, obtained by center cropping and Bicubic. The first stage consists of 12 epochs.

In the second stage, the size of the input images is 84 × 84, obtained by random cropping and use of Bicubic. Additional sample transformation methods include random horizontal flipping and cutout [42]. The second stage consists of 12 epochs.

In the third stage, the size of the input images is 100 × 100, obtained by random cropping. Additional sample transformation methods include random horizontal flipping, vertical flipping, rotation, and cutout. The third stage consists of 36 epochs.

Figure 4 demonstrates the different outputs with all three data augmentation policies used in the different searching stages. Notice that human eyes are not sensitive to dark pixels in the original SAR images. For a better viewing experience, a row of enlarged images with increased lightness is shown in Figure 4.

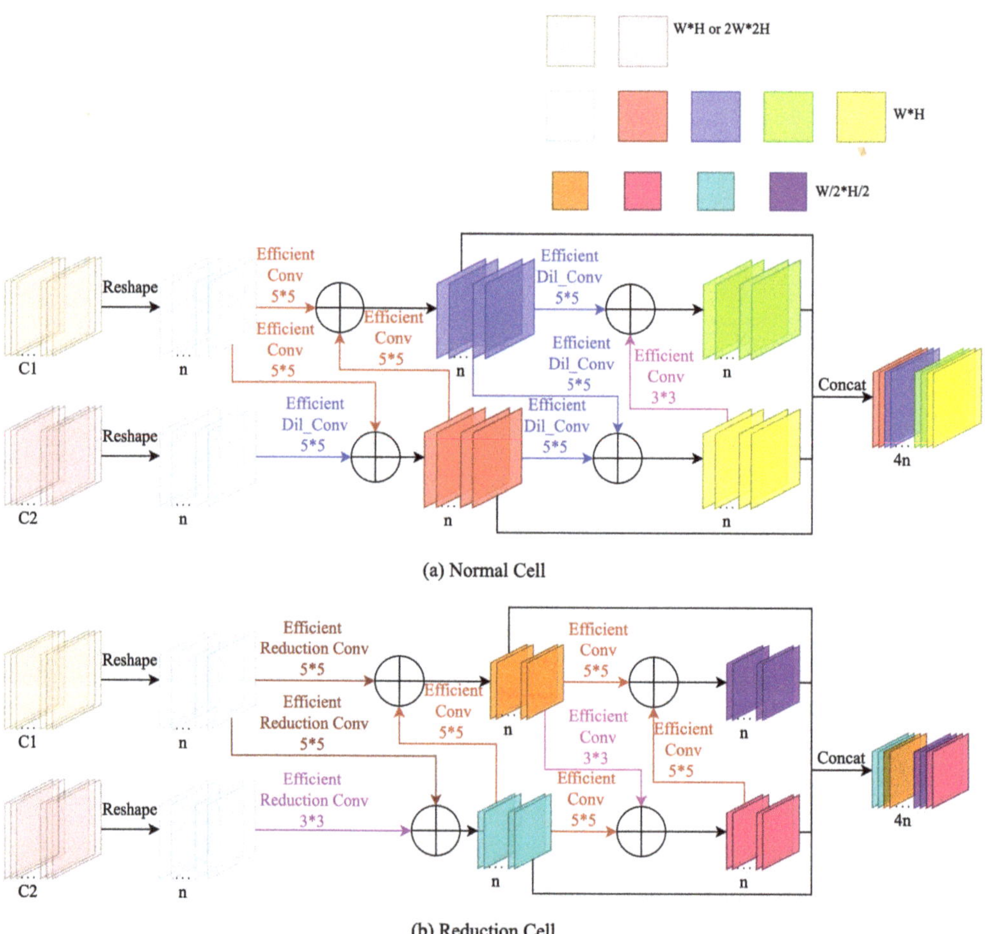

Figure 3. Architectures of searched cells. Efficient convolutional and dilated convolutional operations do not change the shape of feature maps. Efficient reduction convolutional operations reduce the height and width of feature maps by half.

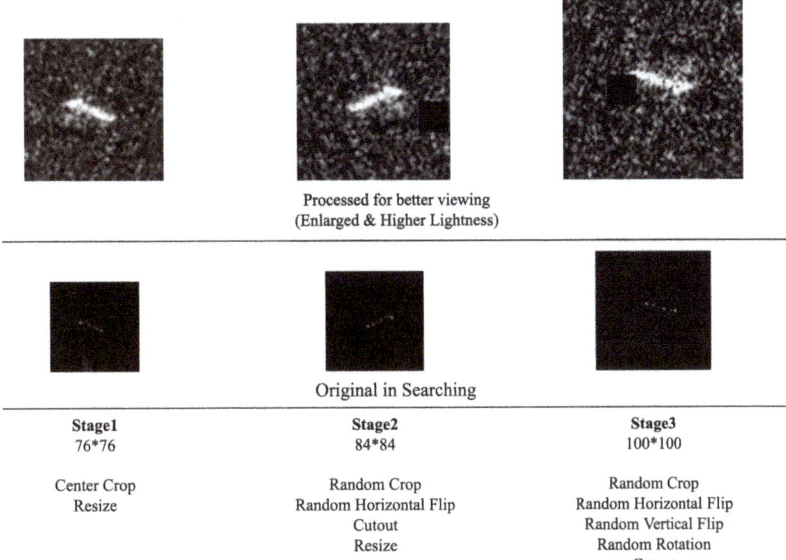

Figure 4. Demonstration of different data augmentation policies.

The mixture operation used in PDA-PC-DARTS is the same as the original random partial connection mixture operation in PC-DARTS, which ensures very low computational complexity and high memory efficiency. Taking the information propagating from node i to node j as an example, the mixture operation $f_{i,j}(.)$ is shown below:

$$f_{i,j}(x_i; S_{i,j}) = \sum_{o \in O} \frac{\exp\{\alpha_{i,j}^o\}}{\sum_{o' \in O} \exp\{\alpha_{i,j}^{o'}\}} o(S_{i,j} \times x_i) + (1 - S_{i,j}) \times x_i \tag{1}$$

where O is the predefined searching space, $o(.)$ is the candidate operation, x_i is the output of node i, $\alpha_{i,j}^o$ is the weight for choosing the candidate operation $o(.)$, and $S_{i,j}$ are randomly sampled channels.

Random sampling can directly reduce the consumption of computational resources and memory, but the search progress is highly likely to be unstable. Similarly, we apply the solution from PC-DARTS. Edge normalization is used to stabilize searching. The output of node j in the search with edge normalization can be described as:

$$x_j = \sum_{i<j} \frac{\exp\{\beta_{i,j}\}}{\sum_{i'<j} \exp\{\beta_{i',j}\}} f_{i,j(x_i; S_{i,j})} \tag{2}$$

where $\beta_{i,j}$ is the edge normalization coefficient.

3.2. Efficient Convolutional Operations

DSCONV has the advantage of low computational complexity, but low accuracy performance can be observed when a small channel number is set. Recently, research on efficient networks has shown the EfficientNet version MBCONV has strong feature extraction and generalization capability [22,24]. To achieve better performance, we replace DSCONV with the EfficientNet version MBCONV as the basic block of candidate operations. The architecture of the EfficientNet version MBCONV block is represented in Figure 5.

PW CONV:Point-wise Convolution
DW CONV: Depth-wise Convolution
GAP: Global Average Pooling
FC: Fully Connected (Linear)

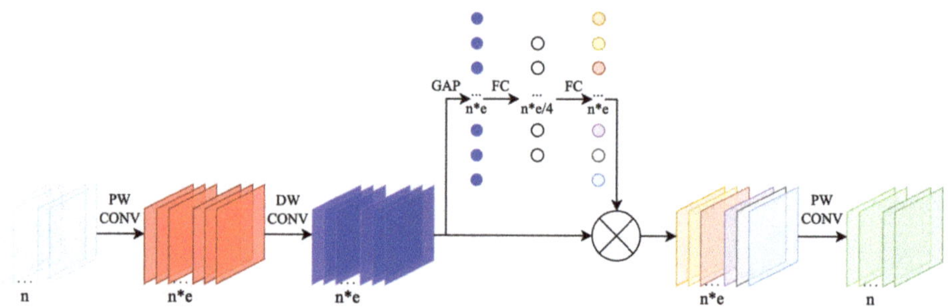

Figure 5. Architecture of EfficientNet version MBCONV block.

Each operation in the SCM consists of one or two stacked MBCONV blocks. To ensure the search space meets the requirements, our replacement keeps the features of the original candidate operations. The hyperparameters of the first depth-wise convolution in each of our efficient operations, such as the kernel size and dilation, are kept the same as the original operations. The details of the updated operation are shown in Table 3.

Table 3. Configuration of efficient convolutional operations.

Original Operation	Efficient Operation	Original Block(s)	Updated Block(s)	Hyperparameters of Depth-Wise Convolution(s)
Conv 3×3	Efficient Conv 3×3	$2\times$ DSCONV	$2\times$MBCONV	#1 Kernel 3×3, Stride 1, Dilation 1 #2 Kernel 3×3, Stride 1, Dilation 1
Conv 5×5	Efficient Conv 5×5	$2\times$ DSCONV	$2\times$MBCONV	#1 Kernel 5×5, Stride 1, Dilation 1 #2 Kernel 5×5, Stride 1, Dilation 1
Dil_Conv 3×3	Efficient Dil_Conv 3×3	DSCONV	MBCONV	Kernel 3×3, Stride 1, Dilation 2
Dil_Conv 5×5	Efficient Dil_Conv 5×5	DSCONV	MBCONV	Kernel 5×5, Stride 1, Dilation 2
Reduction Conv 3×3	Efficient Reduction Conv 3×3	$2\times$ DSCONV	$2\times$MBCONV	#1 Kernel 3×3, Stride 2, Dilation 1 #2 Kernel 3×3, Stride 1, Dilation 1
Reduction Conv 5×5	Efficient Reduction Conv 5×5	$2\times$ DSCONV	$2\times$MBCONV	#1 Kernel 5×5, Stride 2, Dilation 1 #2 Kernel 5×5, Stride 1, Dilation 1
Reduction Dil_Conv 3×3	Efficient Reduction Dil_Conv 3×3	DSCONV	MBCONV	Kernel 3×3, Stride 2, Dilation 2
Reduction Dil_Conv 5×5	Efficient Reduction Dil_Conv 5×5	DSCONV	MBCONV	Kernel 5×5, Stride 2, Dilation 2

Similar to DSCONV, point-wise convolution and depth-wise convolution play important roles in MBCONV. Point-wise convolution is a normal convolution with a kernel size of 1. Furthermore, depth-wise convolution is a modified convolution where each channel is connected to only one input channel. The equations of normal convolution and depth-wise convolution are shown below:

$$CONV_{C_{out}}(In) = B_{C_{out}} + \sum_{k=0}^{C_{in}-1} W_{C_{out}}^k \bigotimes In_k \tag{3}$$

$$DWCONV_{C_{in}}(In) = B_{C_{in}} + W_{C_{in}} \bigotimes In_{C_{in}} \tag{4}$$

where W is the weight, B is the bias, In is the input, and C indicates the number of channels.

A novel activation, Swish, is equipped in the EfficientNet version MBCONV. This function is obtained with the help of exhaustive and reinforcement-learning-based automatic search techniques and is characterized by smoothing, the first derivative, and non-monotonicity [43]. The computation of Swish is as follows:

$$Swish = In \times Sigmoid(In) \tag{5}$$

3.3. Transformer Classifier

Some studies have shown that combining a transformer and a CNN together is beneficial to tasks with two-dimensional data. To enhance the learning capability, the compact transformer classifier from CCT [24] is used as the classifier in the SCM.

The transformer classifier of the SCM only contains two transformer encoders. Firstly, convolution shows more advantages than transformer blocks in a small-size two-dimensional dataset. Secondly, the transformer encoder suffers more from overfitting and usually has higher computational complexity [34]. Lastly, the searched convolutional cells already have strong feature extraction capability in SAR ship images, and adding a small number of self-attention-based modules can help obtain improved learning capability. We believe that the ratio of convolutional operations to transformer blocks in the SCM leads to a better trade-off between accuracy and computational complexity. The classifier used in the SCM is shown in Figure 6, including one trainable positional embedding, two staked transformer encoders, one sequence pooling, and one fully connected layer.

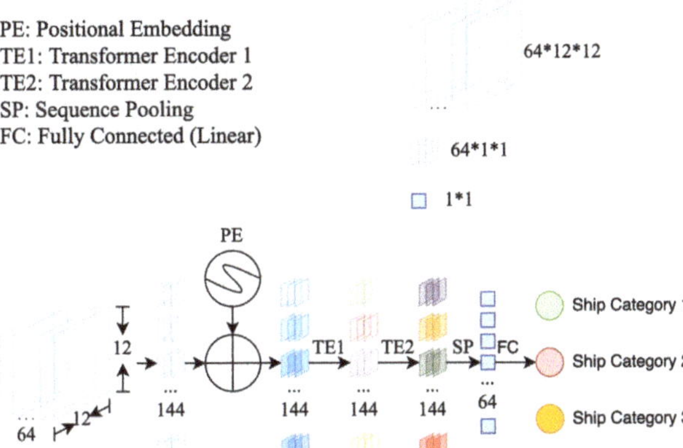

Figure 6. Architecture of the transformer classifier.

Our transformer classifier assigns a positional vector to each pixel of the input feature maps and processes information at the spatial level with transformer encoders. One multi-head self-attention mechanism, two normalizations, two residual connections, and two fully connected layers can be found in a transformer encoder. The multi-head self-attention mechanism can be thought of as the parallel computing of several self-attentions with a concatenated and transformed output, which can be calculated as:

$$MultiHead(Q, K, V) = [head_1, head_2, \ldots, head_h]W_0 \tag{6}$$

$$head_l = Softmax(\frac{Q_l K_l{}^T}{\sqrt{d_k}})V_l \tag{7}$$

$$Q_l = In W_l^Q \tag{8}$$

$$K_l = In W_l^K \tag{9}$$

$$V_l = InW_l^V \tag{10}$$

3.4. ConvFormer

Inspired by recent studies on network architecture [25], a novel cell, the ConvFormer cell, is proposed in this paper. The searched convolutional cells show higher power feature extraction capability than some simple layers or weight-free operations, such as MLP and pooling. Following the concept of a Metaformer, enhancement of the node cell architecture can further improve the performance of the SCM.

The ConvFormer Cell is designed by filling a Metaformer block. The searched efficient convolutional cell is used as the token mixer, which can combine the advantages of both NAS and Metaformer. The architecture of our proposed ConvFormer cell is shown in Figure 7. This cell adds two residual connections, two group normalization layers [44], and two fully connected layers activated by GeLu [45]. Notice that the residual connection requires that both inputs share the same shape. Hence, the token mixer in a ConvFormer cell is only a normal cell.

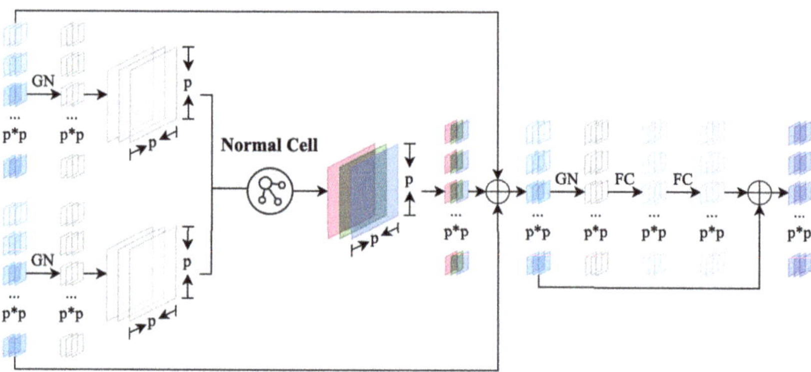

Figure 7. Architecture of the ConvFormer cell.

Metaformer networks with simple token mixers usually require a large number of Metaformer blocks to ensure satisfactory performance, which leads to a large model size. The performance of the SCM is guaranteed by searched convolutional cells. A very small number of ConvFormer cells is used in the SCM to obtain a small model. Furthermore, the SCM does not require an additional patch embedding module, the function of which can be replaced by convolutional operations. As a result, considering the trade-off between computational complexity and performance, we build an SCM with six cells by appending two ConvFormer cells to four stacked efficient convolutional cells. The channel number and type of each efficient convolutional cell are set according to the rules of DARTS.

4. Experiments

A series of experiments were conducted on the OpenSARShip dataset and FUSARShip to verify the superiority of the proposed SCM. A personal computer with an Intel core i5-11600 CPU, 16 G memory, and only a single RTX 3060 was used. The main software environment was an Ubuntu20.04 LTS operating system and the Pytorch framework [46] with the compute unified device architecture (CUDA) toolkit.

4.1. Datasets

4.1.1. OpenSARShip

OpenSARShip is an open-access dataset that has been widely used in many SAR studies [47]. A total of 11,346 SAR ship images from the ground range detected mode or the single look complex mode are contained in OpenSARShip. The size of a category in this dataset is distributed over a wide range. The single look complex mode provides both VV and VH data for a target with good range resolution. Bulk carriers, container ships, and tankers are selected to construct the classification task. The involved categories include approximately 80% of international routes [48].

To avoid the long tail problem, we follow the rule of HOG-ShipCLSNet [18] to take 70% of the least sample number in all three categories as the uniform number of the training targets for each category. Furthermore, the remainder of the targets are put into the test set. VV and VH data of the same target are treated as two samples and put in the same set. The sample number of each category is shown in Table 4; some samples are shown in Figure 8.

Table 4. Sample numbers of each category in the three-category classification task from OpenSARShip.

Ship Category	VH Samples	VV Samples	Training Samples	Test Samples
Bulk Carrier	333	333	338	328
Container Ship	573	573	338	808
Tanker	242	242	338	146

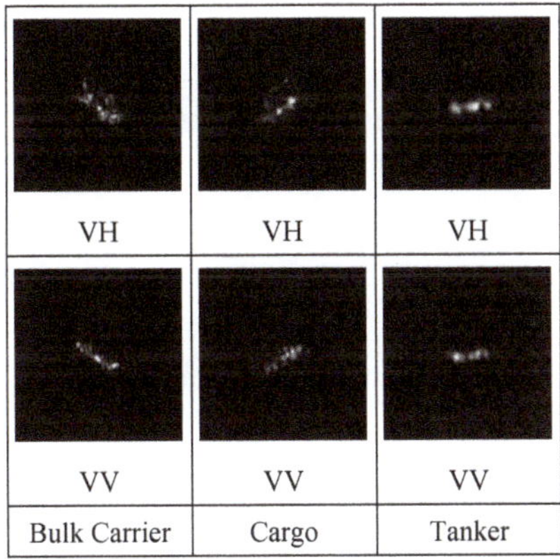

Figure 8. SAR ship samples in the three-category classification task from OpenSARShip. The targets in a column are the same.

4.1.2. FUSARShip

FUSARShip is another open-access dataset for SAR research and is built with the Gaofen-3 civil C-band fully polarimetric spaceborne SAR and related ship automatic identification systems [16]. Ships, non-ship targets, land, sea clutters, and false alarms are contained in this dataset. Although the number of ship categories in FUSARShip is large, the data sizes of the categories are much more unbalanced. A classification task is built with the bulk carriers, fishing ships, cargoes, and tankers, including both onshore and offshore scenarios.

Similarly, 70% of the least sample number in all four categories is set as the uniform number of training targets for each category. Moreover, the sample numbers of the cargo and fishing categories are much larger than the other categories. The testing sample number of each category is set to be the same as the third sample number in the involved categories. The sample number of each category is shown in Table 5; some samples are shown in Figure 9.

Table 5. Sample numbers of each category in the four-category classification task from FUSARShip.

Ship Category	Total Samples	Training Samples	Test Samples
Bulk Carrier	274	174	100
Fishing	787	174	100
Cargo	1735	174	100
Tanker	248	174	74

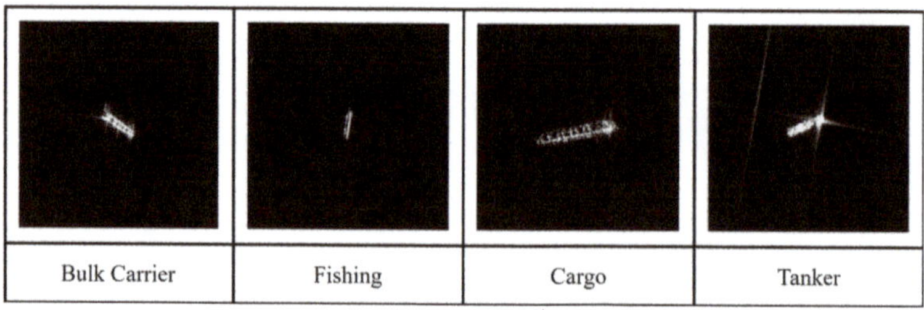

| Bulk Carrier | Fishing | Cargo | Tanker |

Figure 9. SAR ship samples in the four-category classification task from FUSARShip.

4.2. Data Preprocessing

OpenSARShip provides processed Uint8 images with different image shapes. To reduce the distortion effect of resizing, samples over 128×128 are center-cropped into 112×112. Meanwhile, samples smaller than 112×112 are resized to 112×112 by Bicubic. Different sample shapes are made uniform to 100×100 with cropping. In detail, random and center cropping are used in training and testing, respectively.

The samples from FUSARShip share the same image shape of 512×512. Hence, samples are first resized to 128×128. Then, cropping is used to create images with a size of 100×100.

4.3. Architecture Searching

The convolutional cells in the SCM are searched in the three-category classification task from OpenSARShip with the proposed PDA-PC-DARTS. The search is operated with a batch size of 32. The SGD optimizer [49] is used to update the super-net and architecture parameters. The learning rate is set with an initial value of 0.04 and a final value of 0.001. The cosine annealing scheduler changes the learning rate during training following the equation shown below:

$$\eta_t = \eta_{min} + \frac{1}{2}(\eta_{max} - \eta_{min})(1 + cos(\frac{T_{cur}}{T_{max}}))$$ (11)

where η_t is the current learning rate, η_{max} is the max learning rate, η_{min} is the final learning rate, and T_{cur} and T_{max} indicate the current epoch and final epoch, respectively.

4.4. Network Training

To improve learning with a small number of available samples, the SCM is trained with data augmentation and multi-task learning technologies.

Data augmentation can significantly increase the variety of samples and has been proven to improve the learning of SAR data [18]. The applied transformations in the SCM training policy include random cropping, flipping, and cutout.

Usually, transformer encoders require a large amount of training data to obtain satisfactory learning, which has led to it being referred to as "data hungry". To reduce this phenomenon in training with the small SAR dataset, this paper takes an approach based on multitask learning to improve performance [50]. The loss in training the SCM is the sum of the distance loss and cross-entropy. To calculate the distance loss, a pair of patches from the last transformer encoder are randomly picked up. An MLP is used to output the distance between the patches. The L1 loss of the outputted value and the real value is the result of the distance loss. The equation for the distance loss is shown below:

$$L_{dr} = \sum_{s \in B} E_{(p_{i1,j1}, p_{i2,j2})} G_s [|(t_{dh}, t_{dv})^T - (d_{dh}, d_{dv})^T|_1] \tag{12}$$

where B is the mini-batch with N samples, s is a sample in B, $(p_{i1,j1}, p_{i2,j2})$ is a pair of patches randomly picked from the last transformer encoder, d_{dh} and d_{dv} indicate the predicated distance horizontally and vertically, respectively, and t_{dh} and t_{dv} are the true distances.

Cross-entropy is the most common classification loss and can be calculated as follows:

$$L_{ce} = -\frac{1}{N} \sum_{s \in B} log \frac{exp(y_{(s,label_s)})}{\sum_m^M exp(y_{s,m})} \tag{13}$$

where M is the total number of categories, N is the total number of samples, $label_s$ is the true category of sample s, and $y_{s,m}$ indicates the probability of a category produced by the network.

The training of the SCM consists of 600 epochs with a batch size of 32. The weights in the SCM are updated by the SGD optimizer. The cosine annealing scheduler with an initial value of 0.0025 and a final value of 0 is employed during training.

5. Results

5.1. Comparison with CNN-Based SAR Ship Classification Methods

The results of the SCM on the three-category classification task from OpenSARShip are shown in Table 6. For comparison, four other SAR ship classification methods are listed together, including the finetuned VGGNet [13], the plain CNN [16], the group squeeze excitation sparsely connected CNN (GSESCNNs) [17], and HOG-ShipCLSNet [18]. Additional computations outside of the neural networks are necessary in HOG-ShipCLSNet.

Table 6. Results of the three-category classification task from OpenSARShip.

Method	Precision	Accuracy	MAdds	Weights
Finetuned VGGNet [13]	58.72%	69.27%	13.84×10^9	15.52×10^6
Plain CNN [16]	69.44%	67.41%	2.17×10^9	47.44×10^6
GSESCNNs [17]	69.56%	74.98%	—	—
HOG-ShipCLSNet [18]	72.42%	78.15%	89.46×10^6 (Not including HOG and PCA)	65.11×10^6
SCM (ours)	77.74%	82.06%	81.46×10^6	0.46×10^6

As shown in Table 6, our proposed SCM achieves the best accuracy, 82.06%, and outperforms all listed single-polarization and dual-polarization methods. Moreover, the SCM has the smallest model size and lowest computational complexity. The results of the SCM regarding the number of weights and MAdds are 0.46×10^6 and 81.46×10^6, respectively, which means the SCM is very efficient in computation and storage.

Table 7 displays the classification confusion matrix for the SCM on the three-category classification task from OpenSARShip, indicating that the majority of samples can be correctly predicted. However, the recall of each category is imbalanced; tankers and bulk carriers have the highest and lowest values, respectively. Many other CNN-based SAR networks have reported similar results [15,16,18,19]. The main confusion can be found between container ships and bulk carriers. These two categories have many similarities, and the quality of samples is poor. Furthermore, the number of samples in the different categories in the test set is imbalanced.

Table 7. Confusion matrix of SCM on the three-category classification task from OpenSARShip.

	Bulk Carrier	Container Ship	Tanker
Bulk Carrier	75.61%	20.73%	3.66%
Container Ship	12.50%	82.80%	4.70%
Tanker	2.05%	5.48%	92.47%

5.2. Comparison with Efficient Networks from Computer Vision

To further verify the efficiency of the SCM, four famous light computation models from computer vision are used for comparison. The models used for comparison are the 18-layer residual net(ResNet-18) [51], MobileNetv2 [31], ViT_tiny [23], and CCT_722 [24]. ResNet-18 is a popular small-size general network that has been widely applied in many fields. MobileNetv2 is designed for mobile devices and is widely applied in light computation tasks. ViT_tiny is the first model ranked by computational complexity in the vision transformer family. CCT_772 is a tiny mixture network dedicated to datasets with a small sample size.

As the results of the three-category classification task from OpenSARShip depicted in Table 8 show, the SCM outperforms MobileNetv2 by over 15%. Furthermore, the computational complexity of the SCM is slightly higher than MobileNetv2. Moreover, the SCM requires far fewer weights than the other listed models. In detail, the storage space of the proposed SCM is less than a quarter of MobileNetv2, which is the second-smallest model in the list.

Table 8. Comparison with light computation networks on the three-category classification task from OpenSARShip.

Method	Precision	Accuracy	MAdds	Weights
ResNet-18 [51]	69.40%	74.64%	333.10×10^6	11.18×10^6
MobileNetv2 [31]	61.80%	65.83%	56.65×10^6	2.23×10^6
ViT_tiny [23]	46.10%	64.35%	208.03×10^6	5.49×10^6
CCT_722 [24]	72.29%	75.66%	274.07×10^6	4.51×10^6
SCM (ours)	77.74%	82.06%	81.46×10^6	0.46×10^6

5.3. Results on Dual-Polarization

The SCM can also operate dual-polarization classification with high accuracy. The results of the SCM with decision fusion on the three category dual-polarization classification task from OpenSARShip dataset are shown in Table 9. Three CNN-based dual-polarization SAR ship classification methods are listed together for comparison, including the VGGNet with hybrid channel feature loss [14], the Mini Hourglass Region Extraction and Dual-Channel Efficient Fusion Network [15], and SE-LPN-DPFF [19]. SE-LPN-DPFF requires additional computation outside of the neural networks. Dual-polarization classification takes both VV and VH data together as inputs to predict a target. The decision fusion used in the SCM is very simple without additional weights. The SCM outputs two results based on VV and VH, respectively. If the results are different, the result with a higher confidence level will be the final output of the SCM.

Table 9. Results of the three-category dual-polarization classification task from OpenSARShip.

Method	Precision	Accuracy	MAdds	Weights
VGGNet With Hybrid Channel Feature Loss [14]	74.05%	77.41%	27.68×10^9	15.01×10^6
Mini Hourglass Region Extraction and Dual-Channel Efficient Fusion Network [15]	71.50%	75.44%	$\geq 369.93 \times 10^6$ (Dynamic)	7.45×10^6
SE-LPN-DPFF [19]	76.45%	79.25%	—	—
SCM with decision fusion (ours)	80.36%	83.78%	162.92×10^6	0.46×10^6

5.4. Results of Generalization Capability

The computer vision networks compared in Section 5.2 are designed with ImageNet or CIFAR datasets and are well known for their strong generalization capabilities. To verify the generalization capability of the proposed method, the SCM is trained on the four-category classification task from FUSARShip with the architecture seached in the three-category classification from OpenSARShip. Considering the application requirement for computational complexity, the results of the compared computer vision networks are obtained using the same data processing of the SCM and are shown in Table 10. Although the task difficulty has increased, the SCM still has the best accuracy in the list.

Table 10. Comparison of light computation networks on the four-category classification task from FUSARShip.

Method	Precision	Accuracy	MAdds	Weights
ResNet-18 [51]	64.12%	60.16%	333.10×10^6	11.18×10^6
MobileNetv2 [31]	59.35%	60.43%	56.65×10^6	2.23×10^6
ViT_tiny [23]	60.76%	58.29%	208.03×10^6	5.49×10^6
CCT_722 [24]	57.31%	57.75%	274.07×10^6	4.51×10^6
SCM(ours)	61.95%	63.90%	81.46×10^6	0.46×10^6

Table 11 displays the classification confusion matrix for the SCM on the four-category classification task from FUSARShip. SCM can effectively classify bulk carriers and fishing and cargo ships. The tanker category is identified with a low recall and is easily misclassified as another category. We analyzed the confusion matrices of the tested computer vision networks and obtained similar results. The following reason may explain the poor performance in this category. Data processing may remove much information on the tanker category. This work focuses on network architecture; hence, we do not apply additional training methods for small training sets or imbalanced test sets. Many onshore images of tankers can be found in the dataset, which may further decrease the learning. Similarly, no background processing is employed in our method. Lastly, the tanker category in FUSARShip is built with over ten subcategories, including asphalt bitumen, gas, chemical, and crude oil tankers, among others. Some subcategories have strong differences.

Table 11. Confusion matrix of the SCM on the four-category classification task from FUSARShip.

	Bulk Carrier	Fishing	Cargo	Tanker
Bulk Carrier	68.00%	7.00%	8.00%	17.00%
Fishing	3.00%	70.00%	13.00%	14.00%
Cargo	2.00%	10.00%	76.00%	12.00%
Tanker	21.62%	21.62%	22.97%	33.78%

5.5. Ablation Study

A series of ablation experiments were conducted with the three-category classification task from OpenSARShip to confirm the positive effect of each proposed improvement in the SCM. For fairness, all networks used were trained with the same script covering data augmentation, the optimizer, and the learning rate. Table 12 displays the overall results—the conclusion can be drawn that each improvement results in an accuracy increase. In the cell type column, N and R indicate a searched normal cell and a reduction cell, respectively, and C represents the proposed ConvFormer cell. The details of each improvement can be seen in Sections 5.5.1–5.5.4.

Table 12. Results of each improvement in the SCM.

Method	Cell Type	Accuracy
PC-DARTS [20] (Baseline)	NNRNRN	75.66%
+PDA	NNRNRN	80.11%
+PDA +MBCONV	NNRNRN	80.50%
+PDA +MBCONV +Transformer Classifier	NNRNRN	81.59%
+PDA +MBCONV +Transformer Classifier +ConvFormer Cell	NRRNCC	82.06%

5.5.1. Progressing Network Architecture Searching

Table 13 shows the ablation study on the proposed PDA-PC-DARTS. In this experiment, other improvements, the efficient blocks, the transformer classifier, and the ConvFormer cell are used in a target network searched by the original PC-DARTS. As can be seen, employing PDA-PC-DARTS for searching results in higher accuracy. This is because the original PC-DARTS fills its target network with a lot of weight-free operations, leading to weak feature extraction capability.

Table 13. Effectiveness of progressive searching.

Searching Algorithm	Accuracy
PDA-PC-DARTS	82.06%
PC-DARTS [20]	80.81%

5.5.2. MBCONV Block

Table 14 shows the ablation study on replacing DSCONV blocks with MBCONV blocks. A target network based on DSCONV is searched by PDA-PC-DATRS. Then, the transformer classifier and ConvFormer cells are appended under the rule mentioned in Section 3. MBCONV has a stronger feature extraction capability and produces higher accuracy.

Table 14. Comparison of DSCONV and MBCONV.

Basic Block	Accuracy
MBCONV	82.06%
DSCONV	79.41%

5.5.3. Compact Transformer Classifier

Table 15 shows the results of the ablation study on the transformer classifier. A popular linear classifier with global average pooling is listed for comparison. The spatial awareness and learning capability of the transformer classifier boost the performance.

Table 15. Comparison of classifiers.

Classifier	Accuracy
Compact transformer classifier	82.06%
Linear classifier	78.39%

5.5.4. ConvFormer

Table 16 shows the results of the ablation study on the proposed ConvFormer Cell. Firstly, we replace the ConvFormer cells with normal cells to build an SCT. As discussed in Section 3.4, the SCM does not follow the rule of DARTS to assign the type of each cell. Hence, another SCT is built with the DARTS rule. From Table 16, changing the assignment of cells has a negative effect. However, the ConvFormer cell has more advantages to make up for this negative effect drop and achieves the best accuracy in the table.

Table 16. Comparison of different arrangement of cells.

Network	Cell Type	Accuracy
SCM	NRRNCC	82.06%
SCT_1	NRRNNN	81.27%
SCT_2	NNRNRN	81.59%

6. Discussion

6.1. Practical Application Scenario

To meet the increasing requirements for ocean surveillance, increasingly more SAR-equipped small aircraft, such as satellites, balloons and drones, are deployed for large area detection and monitoring. Considering security and their energy consumption, application-specific integrated circuit (ASIC) solutions are more suitable for these small aircraft. However, ASIC platforms have many strict requirements for neural networks, especially for model size. Using external memory leads to a large latency in data input/output. The proposed network has the advantages of small model size with good trade-off between accuracy and computational complexity. ASIC implementation can be more flexible than for a CUDA graphics processing unit (GPU), which means ASIC is recommended for the complex prediction graph of the proposed network.

6.2. Trade-Off between Accuracy, Number of Weights, and Computational Complexity

Compared with many state-of-the-art SAR classification methods, the proposed SCM has relatively much smaller computational complexity. The proposed network achieves a good trade-off between accuracy and computational complexity. However, we admit that the computational complexity of SCM is slightly higher than MobileNetv2. The efficiency operations, node style prediction graph, ConvFormer cells, and transformer classifier mean that SCM has better accuracy with many small operators. Each small operator is computed with the total feature maps, which leads to increased complexity. To outperform networks for mobile devices both on computational complexity and performance, the operations and cells in the SCM should be further optimized with advanced methods, such as fusion of operators.

6.3. Adaptability of the ConvFormer Cell

As we described, only normal cells can be used to build ConvFormer cells which are appended to the DARTS cells. The ablation study showed that breaking the DARTS rule to assign convolutional cells is not recommended. It would be meaningful to design novel ConvFormer cells which contain both normal-type and reduction-type cells. In addition, replacing all DARTS cells with ConvFormer cells should be considered.

6.4. Imbalance of Performance over Different Categories

The proposed method has high average accuracy, but the performance of each category is imbalanced. Furthermore, the performance changes in different radar production and application scenarios. Offshore scenarios have higher accuracy than onshore scenarios. This work focuses on network architecture and does not consider additional methods for small training sets and unbalanced test sets. Meanwhile, preprocessing the background may improve the performance of some categories of ships. To obtain better performance of the SCM over more categories and application scenarios, some special preprocessing and training methods should be studied in future work.

7. Conclusions

In this paper, a searched convolutional Metaformer, SCM, is proposed to classify SAR ship images. Firstly, PDA-PC-DARTS, which is designed for SAR datasets with a small data size, is proposed. Cells with strong feature extraction capability can be searched by PDA-PC-DARTS. Secondly, the basic block of operations used in NAS was changed from DSCONV to MBCONV, which results in better accuracy. At the same time, NAS, a CNN, and the transformer are integrated to further improve the learning capability by employing the transformer classifier. In addition, a ConvFormer cell is proposed to further increase the accuracy. The experimental results show that the SCM has many advantages. On the performance side, the SCM achieves state-of-the-art accuracy. Moreover, both the computational complexity and the number of weights of the SCM are very low.

Author Contributions: Conceptualization, H.Z., W.S. and S.G.; methodology, H.Z.; software, H.Z.; validation, H.Z., L.X., S.G. and W.S.; formal analysis, H.Z.; investigation, H.Z.; data curation, H.Z.; writing—original draft preparation, H.Z.; writing—review and editing, H.Z., L.X., S.G. and W.S.; visualization, H.Z.; supervision, S.G.; project administration, W.S.; funding, S.G. All authors have read and agreed to the published version of the manuscript.

Funding: This work was supported in part by the National Natural Science Foundation of China under grants 62001227, 62001232, and 61971224.

Data Availability Statement: Data are contained within this article.

Acknowledgments: The authors thank the reviewers for their help with the article during the review process.

Conflicts of Interest: The authors declare no conflict of interest.

References

1. Petit, M.; Stretta, J.M.; Farrugio, H.; Wadsworth, A. Synthetic aperture radar imaging of sea surface life and fishing activities. *IEEE Trans. Geosci. Remote Sens.* **1992**, *30*, 1085–1089. [CrossRef]
2. Park, J.; Lee, J.; Seto, K.; Hochberg, T.; Wong, B.A.; Miller, N.A.; Takasaki, K.; Kubota, H.; Oozeki, Y.; Doshi, S.; et al. Illuminating dark fishing fleets in North Korea. *Sci. Adv.* **2020**, *6*, eabb1197. [CrossRef] [PubMed]
3. Brusch, S.; Lehner, S.; Fritz, T.; Soccorsi, M.; Soloviev, A.; van Schie, B. Ship surveillance with TerraSAR-X. *IEEE Trans. Geosci. Remote Sens.* **2010**, *49*, 1092–1103. [CrossRef]
4. Cortes, C.; Vapnik, V. Support-vector networks. *Mach. Learn.* **1995**, *20*, 273–297. [CrossRef]
5. Quinlan, J.R. Induction of decision trees. *Mach. Learn.* **1986**, *1*, 81–106. [CrossRef]
6. Ho, T.K. The random subspace method for constructing decision forests. *IEEE Trans. Pattern Anal. Mach. Intell.* **1998**, *20*, 832–844.
7. Friedman, N.; Geiger, D.; Goldszmidt, M. Bayesian network classifiers. *Mach. Learn.* **1997**, *29*, 131–163. [CrossRef]
8. Freund, Y.; Schapire, R.E. A decision-theoretic generalization of on-line learning and an application to boosting. *J. Comput. Syst. Sci.* **1997**, *55*, 119–139. [CrossRef]
9. Abdal, R.; Qin, Y.; Wonka, P. Image2stylegan: How to embed images into the stylegan latent space? In Proceedings of the IEEE/CVF International Conference on Computer Vision, Seoul, Republic of Korea, 27 October–2 November 2019; pp. 4432–4441.
10. Yasarla, R.; Sindagi, V.A.; Patel, V.M. Syn2real transfer learning for image deraining using gaussian processes. In Proceedings of the IEEE/CVF Conference on Computer Vision and Pattern Recognition, Seattle, WA, USA, 14–19 June 2020; pp. 2726–2736.
11. Zhong, Y.; Deng, W.; Wang, M.; Hu, J.; Peng, J.; Tao, X.; Huang, Y. Unequal-training for deep face recognition with long-tailed noisy data. In Proceedings of the IEEE/CVF Conference on Computer Vision and Pattern Recognition, Long Beach, CA, USA, 15–20 June 2019; pp. 7812–7821.
12. Krizhevsky, A.; Sutskever, I.; Hinton, G.E. Imagenet classification with deep convolutional neural networks. *Adv. Neural Inf. Process. Syst.* **2012**, *25*, 1097–1105 [CrossRef]
13. Wang, Y.; Wang, C.; Zhang, H. Ship classification in high-resolution SAR images using deep learning of small datasets. *Sensors* **2018**, *18*, 2929. [CrossRef]
14. Zeng, L.; Zhu, Q.; Lu, D.; Zhang, T.; Wang, H.; Yin, J.; Yang, J. Dual-polarized SAR ship grained classification based on CNN with hybrid channel feature loss. *IEEE Geosci. Remote Sens. Lett.* **2021**, *19*, 1–5. [CrossRef]
15. Xiong, G.; Xi, Y.; Chen, D.; Yu, W. Dual-polarization SAR ship target recognition based on mini hourglass region extraction and dual-channel efficient fusion network. *IEEE Access* **2021**, *9*, 29078–29089. [CrossRef]

16. Hou, X.; Ao, W.; Song, Q.; Lai, J.; Wang, H.; Xu, F. FUSAR-Ship: Building a high-resolution SAR-AIS matchup dataset of Gaofen-3 for ship detection and recognition. *Sci. China Inf. Sci.* **2020**, *63*, 1–19. [CrossRef]

17. Huang, G.; Liu, X.; Hui, J.; Wang, Z.; Zhang, Z. A novel group squeeze excitation sparsely connected convolutional networks for SAR target classification. *Int. J. Remote Sens.* **2019**, *40*, 4346–4360. [CrossRef]

18. Zhang, T.; Zhang, X.; Ke, X.; Liu, C.; Xu, X.; Zhan, X.; Wang, C.; Ahmad, I.; Zhou, Y.; Pan, D.; et al. HOG-ShipCLSNet: A novel deep learning network with hog feature fusion for SAR ship classification. *IEEE Trans. Geosci. Remote Sens.* **2021**, *60*, 1–22. [CrossRef]

19. Zhang, T.; Zhang, X. Squeeze-and-excitation Laplacian pyramid network with dual-polarization feature fusion for ship classification in sar images. *IEEE Geosci. Remote Sens. Lett.* **2021**, *19*, 1–5. [CrossRef]

20. Xu, Y.; Xie, L.; Zhang, X.; Chen, X.; Qi, G.J.; Tian, Q.; Xiong, H. Pc-darts: Partial channel connections for memory-efficient architecture search. *arXiv* **2019**, arXiv:1907.05737.

21. Liu, H.; Simonyan, K.; Yang, Y. Darts: Differentiable architecture search. *arXiv* **2018**, arXiv:1806.09055.

22. Tan, M.; Le, Q. Efficientnet: Rethinking model scaling for convolutional neural networks. In Proceedings of the International Conference on Machine Learning, Long Beach, CA, USA, 9–15 June 2019; pp. 6105–6114.

23. Dosovitskiy, A.; Beyer, L.; Kolesnikov, A.; Weissenborn, D.; Zhai, X.; Unterthiner, T.; Dehghani, M.; Minderer, M.; Heigold, G.; Gelly, S.; et al. An image is worth 16x16 words: Transformers for image recognition at scale. *arXiv* **2010**, arXiv:2010.11929.

24. Hassani, A.; Walton, S.; Shah, N.; Abuduweili, A.; Li, J.; Shi, H. Escaping the big data paradigm with compact transformers. *arXiv* **2021**, arXiv:2104.05704.

25. Yu, W.; Luo, M.; Zhou, P.; Si, C.; Zhou, Y.; Wang, X.; Feng, J.; Yan, S. Metaformer is actually what you need for vision. In Proceedings of the IEEE/CVF Conference on Computer Vision and Pattern Recognition, New Orleans, LA, USA, 18–24 June 2022; pp. 10819–10829.

26. Zoph, B.; Vasudevan, V.; Shlens, J.; Le, Q.V. Learning transferable architectures for scalable image recognition. In Proceedings of the IEEE Conference on Computer Vision and Pattern Recognition, Salt Lake City, UT, USA, 18–20 June 2018; pp. 8697–8710.

27. Pham, H.; Guan, M.; Zoph, B.; Le, Q.; Dean, J. Efficient neural architecture search via parameters sharing. In Proceedings of the International Conference on Machine Learning, Stockholm, Sweden, 10–15 July 2018; pp. 4095–4104.

28. Xie, L.; Yuille, A. Genetic cnn. In Proceedings of the IEEE International Conference on Computer Vision, Venice, Italy, 22–29 October 2017; pp. 1379–1388.

29. Real, E.; Aggarwal, A.; Huang, Y.; Le, Q.V. Regularized evolution for image classifier architecture search. In Proceedings of the AAAI Conference on Artificial Intelligence, Honolulu, HI, USA, 27 January–1 February 2019; Volume 33, pp. 4780–4789.

30. Howard, A.G.; Zhu, M.; Chen, B.; Kalenichenko, D.; Wang, W.; Weyand, T.; Andreetto, M.; Adam, H. Mobilenets: Efficient convolutional neural networks for mobile vision applications. *arXiv* **2017**, arXiv:1704.04861.

31. Sandler, M.; Howard, A.; Zhu, M.; Zhmoginov, A.; Chen, L.C. Mobilenetv2: Inverted residuals and linear bottlenecks. In Proceedings of the IEEE Conference on Computer Vision and Pattern Recognition, Salt Lake City, UT, USA, 18–20 June 2018; pp. 4510–4520.

32. Hu, J.; Shen, L.; Sun, G. Squeeze-and-excitation networks. In Proceedings of the IEEE Conference on Computer Vision and Pattern Recognition, Salt Lake City, UT, USA, 18–20 June 2018; pp. 7132–7141.

33. Liu, Z.; Lin, Y.; Cao, Y.; Hu, H.; Wei, Y.; Zhang, Z.; Lin, S.; Guo, B. Swin transformer: Hierarchical vision transformer using shifted windows. In Proceedings of the IEEE/CVF International Conference on Computer Vision, Virtual, 11–17 October 2021; pp. 10012–10022.

34. Dai, Z.; Liu, H.; Le, Q.V.; Tan, M. Coatnet: Marrying convolution and attention for all data sizes. *Adv. Neural Inf. Process. Syst.* **2021**, *34*, 3965–3977.

35. Touvron, H.; Bojanowski, P.; Caron, M.; Cord, M.; El-Nouby, A.; Grave, E.; Izacard, G.; Joulin, A.; Synnaeve, G.; Verbeek, J.; et al. Resmlp: Feedforward networks for image classification with data-efficient training. *IEEE Trans. Pattern Anal. Mach. Intell.* **2022**, *45*, 5314–5321. [CrossRef]

36. Singh, P.; Shree, R.; Diwakar, M. A new SAR image despeckling using correlation based fusion and method noise thresholding. *J. King Saud Univ.-Comput. Inf. Sci.* **2021**, *33*, 313–328. [CrossRef]

37. Singh, P.; Shankar, A.; Diwakar, M. Review on nontraditional perspectives of synthetic aperture radar image despeckling. *J. Electron. Imaging* **2023**, *32*, 021609. [CrossRef]

38. Toutin, T. Geometric processing of remote sensing images: Models, algorithms and methods. *Int. J. Remote Sens.* **2004**, *25*, 1893–1924. [CrossRef]

39. Choo, A.; Chan, Y.; Koo, V.; Electromagnet, A. Geometric correction on SAR imagery. In Proceedings of the Progress in Electromagnetics Research Symposium Proceedings, Kuala Lumpur, Malaysia, 27–30 March 2012.

40. Moreira, A.; Prats-Iraola, P.; Younis, M.; Krieger, G.; Hajnsek, I.; Papathanassiou, K.P. A tutorial on synthetic aperture radar. *IEEE Geosci. Remote Sens. Mag.* **2013**, *1*, 6–43. [CrossRef]

41. Tan, M.; Le, Q. Efficientnetv2: Smaller models and faster training. In Proceedings of the International Conference on Machine Learning, Virtual, 18–24 July 2021; pp. 10096–10106.

42. DeVries, T.; Taylor, G.W. Improved regularization of convolutional neural networks with cutout. *arXiv* **2017**, arXiv:1708.04552.

43. Ramachandran, P.; Zoph, B.; Le, Q.V. Searching for activation functions. *arXiv* **2017**, arXiv:1710.05941.

44. Wu, Y.; He, K. Group normalization. In Proceedings of the European Conference on Computer Vision (ECCV), Munich, Germany, 8–14 September 2018; pp. 3–19.
45. Hendrycks, D.; Gimpel, K. Gaussian error linear units (gelus). *arXiv* **2016**, arXiv:1606.08415.
46. Paszke, A.; Gross, S.; Massa, F.; Lerer, A.; Bradbury, J.; Chanan, G.; Killeen, T.; Lin, Z.; Gimelshein, N.; Antiga, L.; et al. Pytorch: An imperative style, high-performance deep learning library. *Adv. Neural Inf. Process. Syst.* **2019**, *32*, 8026–8037
47. Huang, L.; Liu, B.; Li, B.; Guo, W.; Yu, W.; Zhang, Z.; Yu, W. OpenSARShip: A dataset dedicated to Sentinel-1 ship interpretation. *IEEE J. Sel. Top. Appl. Earth Obs. Remote Sens.* **2017**, *11*, 195–208. [CrossRef]
48. Zhang, H.; Tian, X.; Wang, C.; Wu, F.; Zhang, B. Merchant vessel classification based on scattering component analysis for COSMO-SkyMed SAR images. *IEEE Geosci. Remote Sens. Lett.* **2013**, *10*, 1275–1279. [CrossRef]
49. Robbins, H.; Monro, S. A stochastic approximation method. *Ann. Math. Stat.* **1951**, *22*, 400–407. [CrossRef]
50. Liu, Y.; Sangineto, E.; Bi, W.; Sebe, N.; Lepri, B.; Nadai, M. Efficient training of visual transformers with small datasets. *Adv. Neural Inf. Process. Syst.* **2021**, *34*, 23818–23830.
51. He, K.; Zhang, X.; Ren, S.; Sun, J. Deep residual learning for image recognition. In Proceedings of the IEEE Conference on Computer Vision and Pattern Recognition, Las Vegas, NV, USA, 27–30 June 2016; pp. 770–778.

Article

A SAR Image-Despeckling Method Based on HOSVD Using Tensor Patches

Jing Fang [1], Taiyong Mao [1], Fuyu Bo [1], Bomeng Hao [1], Nan Zhang [1], Shaohai Hu [2], Wenfeng Lu [3] and Xiaofeng Wang [4,*]

[1] School of Physics and Electronics, Shandong Normal University, Jinan 250014, China; fangjing@sdnu.edu.cn (J.F.); 2021317025@stu.sdnu.edu.cn (T.M.); 2020317005@stu.sdnu.edu.cn (F.B.); 202109020335@stu.sdnu.edu.cn (B.H.); 2021317022@stu.sdnu.edu.cn (N.Z.)
[2] Institute of Information Science, Beijing Jiaotong University, Beijing 100044, China; shhu@bjtu.edu.cn
[3] School of Management Engineering, Shandong Jianzhu University, Jinan 250101, China; luwenfeng@sdjzu.edu.cn
[4] School of Geography and Environment, Shandong Normal University, Jinan 250358, China
* Correspondence: wangxiaofeng@sdnu.edu.cn

Abstract: Coherent imaging systems, such as synthetic aperture radar (SAR), often suffer from granular speckle noise due to inherent defects, which can make interpretation challenging. Although numerous despeckling methods have been proposed in the past three decades, SAR image despeckling remains a challenging task. With the extensive use of non-local self-similarity, despeckling methods under the non-local framework have become increasingly mature. However, effectively utilizing patch similarities remains a key problem in SAR image despeckling. This paper proposes a three-dimensional (3D) SAR image despeckling method based on searching for similar patches and applying the high-order singular value decomposition (HOSVD) theory to better utilize the high-dimensional information of similar patches. Specifically, the proposed method extends two-dimensional (2D) to 3D for SAR image despeckling using tensor patches. A new, non-local similar patch-searching measure criterion is used to classify the patches, and similar patches are stacked into 3D tensors. Lastly, the iterative adaptive weighted tensor cyclic approximation is used for SAR image despeckling based on the HOSVD method. Experimental results demonstrate that the proposed method not only effectively reduces speckle noise but also preserves fine details.

Keywords: synthetic aperture radar (SAR); despeckling; tensor patches; higher-order singular value decomposition (HOSVD)

Citation: Fang, J.; Mao, T.; Bo, F.; Hao, B.; Zhang, N.; Hu, S.; Lu, W.; Wang, X. A SAR Image-Despeckling Method Based on HOSVD Using Tensor Patches. *Remote Sens.* **2023**, *15*, 3118. https://doi.org/10.3390/rs15123118

Academic Editor: Gerardo Di Martino

Received: 28 April 2023
Revised: 6 June 2023
Accepted: 13 June 2023
Published: 14 June 2023

1. Introduction

Synthetic aperture radar (SAR) is an earth observation system that can generate high-resolution remote sensing images, allowing for daylong and all-weather observations. Hence, the SAR system offers unique benefits in various fields, including disaster monitoring, resource exploration, and military applications [1]. However, SAR images are often degraded by speckle noise, which is a natural consequence of coherent scattering phenomena. Unfortunately, speckle noise cannot be eliminated from SAR images and can significantly hinder interpretation, such as target recognition and image segmentation [2]. In contrast to additive white Gaussian noise (AWGN) typically found in optical images, the speckle noise in SAR images is usually multiplicative.

Speckle noise significantly impacts the understanding and interpretation of SAR images. Therefore, despeckling SAR images can considerably improve the subsequent application performance [3–6]. Owing to the limitations of early SAR image technology, traditional spatial domain despeckling approaches primarily relied on spatial filtering algorithms, such as the Lee filter [7], the Frost filter [8], and the Kuan filter [9], to remove the speckles. These methods tend to over-smooth the image while removing speckle noise, resulting in the loss of textural details. Subsequently, transform domain methods have been proposed for despeckling SAR images. Currently, the most commonly used transform domain despeckling methods include wavelet transform [10,11], contourlet transform [12], and shearlet transform [13]. Despite being more effective than spatial domain despeckling methods, transform domain filtering demands substantial computational resources, can cause image blurring, and introduce artifacts, which adversely affect subsequent image processing. Regularization-based despeckling is another popular approach to SAR images [14,15], which transforms the despeckling problem into a minimization energy function. Regularization models based on the total variance model can effectively suppress speckle noise and maintain texture. In fact, they may create an unpleasant staircase effect. In recent years, sparse representation has emerged as a captivating field of research [16]. Sparse representation aims to represent the original image with as few atoms as possible in an overcomplete dictionary, leveraging natural prior knowledge as a clean image can be represented sparsely while noise cannot. Furthermore, some methods increase adaptability by iteratively and alternately updating the atoms and sparse solutions in the dictionary using a learning strategy [17]. Unfortunately, these methods have limitations in terms of noise separation and computational efficiency. Deep learning has shown significant potential in various research areas and has also made outstanding achievements in the despeckling of SAR images. For instance, the SAR-convolution neural network (CNN) combines homomorphic transformation SAR images with deep CNN [18–20]. The development of deep learning has significantly improved the performance and speed of SAR image despeckling by leveraging the powerful feature extraction and adaptive parameter adjustment capabilities of deep learning [21,22]. To achieve optimal despeckling performance, deep learning models heavily rely on training with a large volume of clean real SAR images. Nevertheless, acquiring such datasets can be challenging. As a result, models trained solely on simulated SAR images are often limited to specific scenes or scenarios.

The non-local means (NLM) denoising algorithm uses similar features, including non-adjacent pixels [23]. The fundamental concept is the self-similarity of natural images, which involves estimating clean pixels by searching for similar patches in the entire image or a large window. Therefore, the NLM has achieved promising denoising results and has been widely used in the field of SAR image despeckling. The probabilistic patch-based (PPB) filter [24] replaces the Euclidean distance with a statistical similarity criterion and utilizes an iterative method to update weights. Several algorithms have combined NLM with multiscale geometric transformations and achieved good despeckling results. For instance, SAR-block-matching 3D (SAR-BM3D) filter [25] combined 3D block matching and collaborative filtering. Subsequently, Cozzolino et al. proposed a fast adaptive nonlocal SAR (FANS) filter [26] and enhanced the despeckling efficiency of SAR-BM3D. NLM has also been employed in total variation regularization [27], low-rank approximation [28], and sparse representation [29].

The NLM algorithm is effective in suppressing speckle noise. However, it can also generate artificial texture due to the influence of the blocking criterion and patch-matching accuracy, which can impede image interpretation. Various strategies, such as multiscale [30] and the gray theory [31], have been employed to improve the selection of patches with similar properties. Liang et al. [32] introduced gradient information into the similarity measure by constructing the gradient orientation map and accelerated despeckling using the fast Fourier transform algorithm. Giampaolo et al. [33] proposed an independent model-free NLM despeckling framework that provided a generic solution for despeckling a variety of SAR products. However, most of the above-mentioned despeckling methods rearrange SAR images into vectors or matrices and rely on vector or matrix calculation methods for despeckling. Unfortunately, the vectorization operation destroys the topology between similar patches, resulting in suboptimal despeckling performance. Tensors [34] offer a natural representation for multilinear data, serving as a high-order generalization of vectors and matrices. Liu et al. [35] proposed a method for estimating missing values in tensor visual data by generalizing low-rank matrices to low-rank tensors. Tensor completion techniques have demonstrated significant potential for applications such as image inpainting, video compression, and bidirectional reflectance distribution function (BRDF) data estimation. Zhang et al. [36] introduced the tensor singular value decomposition (t-SVD) for denoising and completion of multilinear data, which can handle a broader range of multilinear data as long as they are compressible in the t-SVD-based representation. Xue et al. [37] developed a method for reducing noise in hyperspectral images (HSI) using CANDECOMP/PARAFAC (CP) decomposition modeling with tensor and rank automatic determination. This algorithm significantly improves the denoising performance of HSI in various quality assessments. This paper proposes a three-dimensional (3D) SAR image-despeckling method based on patch matching and the tensor patch higher-order singular value decomposition (HOSVD) theory to effectively capture the potential information shared among similar patches and better exploit the correlation between different dimensions of SAR images. By using the non-local framework, the classified two-dimensional (2D) similar patches are stacked to construct third-order non-local tensor patches, which can better utilize the similarity of patches. It should be noted that this is the first time higher-order tensor decomposition has been applied to the field of SAR image despeckling. The proposed method effectively achieves high-dimensional SAR image despeckling by utilizing the low-rank tensor approximation technique to exploit the potential correlation and low-rank structure of SAR image data. The effectiveness of the proposed method is evaluated and compared with the existing advanced despeckling algorithms.

The remainder of this paper is structured as follows. Section 2 presents a detailed description of the material and methodology used for despeckling. Section 3 provides an analysis of the experimental results. In Section 4, a comprehensive discussion is presented to provide deeper insights. Finally, Section 5 outlines the conclusions drawn from the method. The framework of the proposed method is illustrated in Figure 1.

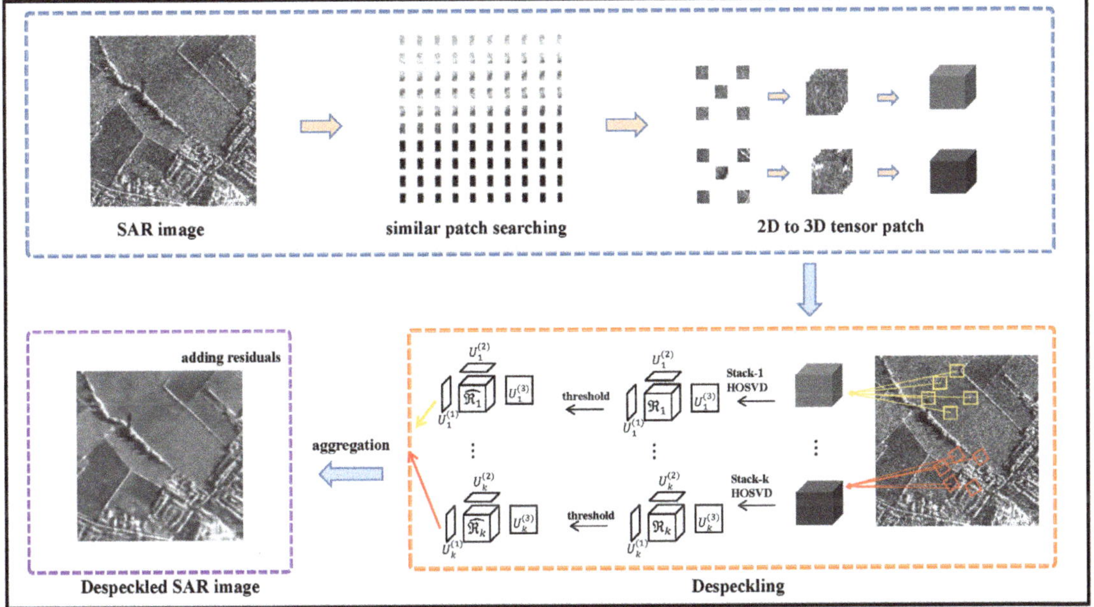

Figure 1. The framework of the proposed method.

2. Materials and Methods

This section presents the proposed method for SAR image despeckling. Prior to the proposed method, the multiplicative noise is converted to an additive model using a logarithmic transformation. Then, the entire image is divided into overlapping patches and similar patches within a local search window are selected for each reference patch. Firstly, the gradient of the image is calculated and used to identify similar patches. Next, these similar patches are combined into third-order tensor patches. Subsequently, the iterative low-rank tensor patch approximation is applied to recover the clean tensor patches. Finally, the ultimate results are obtained by exponential transformation and aggregation of all patch estimates.

2.1. Statistics of Log-Transformed Speckle

Considering an intensity SAR image, the intensity Y is related to the backscatter return X and speckle noise B by the following multiplicative model:

$$Y = BX \tag{1}$$

Assuming that the speckle noise is fully developed, Goodman's model indicates that it follows a gamma distribution with a probability density function (PDF), expressed as follows:

$$p(B) = \frac{L^L B^{L-1}}{\Gamma(L)} \exp(-LB), B \geq 0 \tag{2}$$

where $\Gamma(\cdot)$ denotes the gamma function. Logarithmic transformation is often used to convert multiplicative noise into additive ones. Using the logarithmic operator applied on Equation (1):

$$\widetilde{Y} = \ln(Y) = \ln(B) + \ln(X) = \widetilde{B} + \widetilde{X} \tag{3}$$

The random variable $\widetilde{B}$ follows the Fisher–Tippett distribution, defined as follows:

$$p(\widetilde{B}) = \frac{L^L e^{\widetilde{B}L}}{\Gamma(L)} \exp(-Le^{\widetilde{B}}) \tag{4}$$

The mean and variance of $\widetilde{B}$ can be computed as follows:

$$E[\widetilde{B}] = \Psi(L) - \ln(L), \mathrm{Var}[\widetilde{B}] = \Psi(1, L) \tag{5}$$

where $\Psi(\cdot)$ is the digamma function, and $\Psi(\cdot, L)$ is the polygamma function of order L. Formula (5) shows that the noise has a non-zero mean. Therefore, a de-biasing step is required after inverse logarithmic operations.

2.2. Searching Similar Patches of SAR Images

2.2.1. Measure for Non-Local Similarity

In the presence of noise, a similar patch-searching algorithm may select irrelevant candidates, resulting in undesired artifacts due to averaging the selected patches with irrelevant data values. Conversely, it can be difficult to find enough similar patches in regions with edges or unique structures, leading to the incorrect despeckling of pixels in such areas. To address these weaknesses in the similar patch-searching algorithm, the similarity criterion proposed in [38] is utilized in this study. The similarity criterion uses the gradient information to calculate the similarity between patch x and patch y in the SAR images, as follows:

$$S(x,y) = \|x - y\|_{Ga}^2 + \rho(\overline{E_x}) \|E_x - E_y\|_{Ga}^2 \tag{6}$$

where $\| \bullet \|_{G_a}$ represents the Gaussian norm [39] and ρ is the continuous increasing bounded function. The basic idea behind the similarity criterion is to consider the gradient information, especially in regions near singular points. The gradient vector $|\nabla x(i,j)|$ of the patch x in the position (i,j) is calculated and expressed as $(E_x)_{ij} = |\nabla x(i,j)|$. The matrix with entries is denoted by E_x, while $\overline{E_x}$ represents the average value of the elements of matrix E_x.

2.2.2. Computation of Gradient

This paper utilizes the constrained least squares method [40] to estimate gradient information from SAR images. Specifically, given a pixel point $\theta \in \mathbb{R}^2$ in a patch, the vector of data values is denoted by $\mathbf{f} := [f(\theta_v) : v = 1, \ldots, V]^T$. Let $\{Z_u : u = 1, \ldots, U\}, U < V$ be a basis for Π_U, which represents the space of algebraic polynomials of degree $< U$, and put

$$\Phi := \left[z_j(\theta_i) : i = 1, \ldots, V, j = 1, \ldots, U \right], Z^{(\varphi)}(\theta) := \left[z_j^{(\varphi)}(\theta) : j = 1, \ldots, U \right]^T.$$

The approximation of the derivative $f^{(\varphi)}$ can be expressed as follows:

$$f^{(\varphi)}(\theta) \approx \mathrm{K}^{(\varphi)}(\theta)^T \mathbf{f} \tag{7}$$

where $\mathrm{K}^{(\varphi)}(\theta) := \left[K_v^{(\varphi)}(\theta) : v = 1, \ldots, V \right]^T$ is the appropriate coefficient vector that can be obtained by solving the following convex optimization problem:

$$\min_{\mathrm{K}^{(\varphi)}(\theta)} \sum_{v=1}^{V} K_v^{(\varphi)}(\theta)^2 \delta(\theta - \theta_v) \quad s.t. \quad \Phi^T \cdot \mathrm{K}^{(\varphi)}(\theta) = Z^{(\varphi)}(\theta) \tag{8}$$

The penalty function δ is typically chosen to be a smooth function that increases rapidly. A typical example is as follows:

$$\delta(\theta) = \exp\left(\|\theta\|^2 / 2 \right) \tag{9}$$

The matrix form $\mathrm{K}^{(\varphi)}(\theta) = \mathrm{M}^{-1} \Phi \left(\Phi^T \mathrm{M}^{-1} \Phi \right)^{-1} z^{(\varphi)}(\theta)$ can be used to represent the solution to the optimization problem (8), where matrix M is defined as follows:

$$\mathrm{M} := 2 \operatorname{diag}(\delta(\theta - \theta_v) : v = 1, \ldots, V).$$

2.3. Tensor and Third-Order Tensor Decomposition

2.3.1. Definition of Tensor

To better exploit the self-similarity of similar patches, 2D patches are converted into third-order tensors for SAR image despeckling. The tensors are equivalent to multidimensional vector arrays, which are vectors and matrices extended to multiple dimensions. Specifically, scalars, vectors, and matrices can be viewed as zero-order, first-order, and second-order tensors, respectively. To further explain the third-order tensor decomposition, a few basic definitions of tensors are provided as follows. The mode-n vector of the third-order tensor is shown in Figure 2.

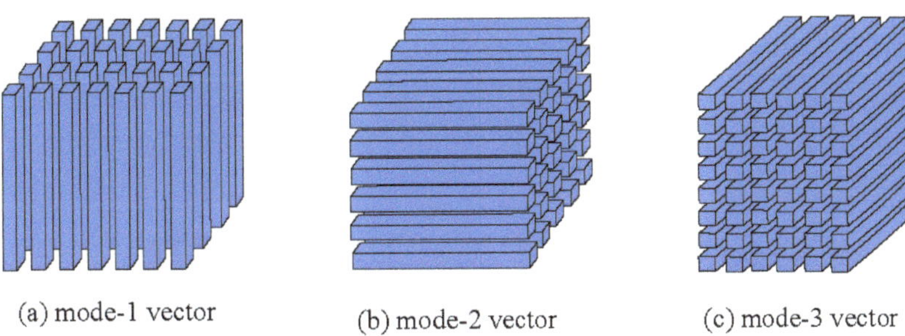

<table>
<tr><td>(a) mode-1 vector</td><td>(b) mode-2 vector</td><td>(c) mode-3 vector</td></tr>
</table>

Figure 2. The mode-n vector of the third-order tensor.

Definition 1. *Given two tensors $A \in R^{I_1 \times I_2 \times I_3 \cdots \times I_N}$ and $B \in R^{I_1 \times I_2 \times I_3 \cdots \times I_N}$ of the same dimension, their inner products can be calculated as follows:*

$$\langle A, B \rangle = \sum_{i1} \sum_{i2} \sum_{i3} \cdots \sum_{iN} a_{i1i2i3\ldots iN} \cdot b_{i1i2i3\ldots iN} \tag{10}$$

the Frobenius norm of tensor $A \in R^{I_1 \times I_2 \times I_3 \cdots \times I_N}$ can be obtained as follows:

$$\|A\|_F = \sqrt{\langle A, A \rangle} = \sqrt{\sum_{i1,i2,\cdots,iN} |a_{i1i2\ldots iN}|^2} \tag{11}$$

Definition 2. *The elements of tensor A are mapped to the elements of the mode-n matrix $A_{(n)} \in R^{I_n \times (I_1 \times \cdots \times I_{n-1} \times I_{n+1} \times \cdots \times I_N)}$, and a mode-n expansion of a tensor $A \in R^{I_1 \times I_2 \times I_3 \cdots \times I_N}$ of order n can be obtained. All elements of the tensor are expanded using modulo and then rearranged into a two-dimensional matrix.*

Definition 3. *The mode-n rank of $A \in R^{I_1 \times I_2 \times I_3 \cdots \times I_N}$ for a given tensor A of order N is the rank of the tensor A mode-n expanded matrix $A_{(n)}$, it can signify $\mathrm{rank}_n(A)$.*

The corresponding mode-n expansion of the third-order tensor A is shown in Figure 3.

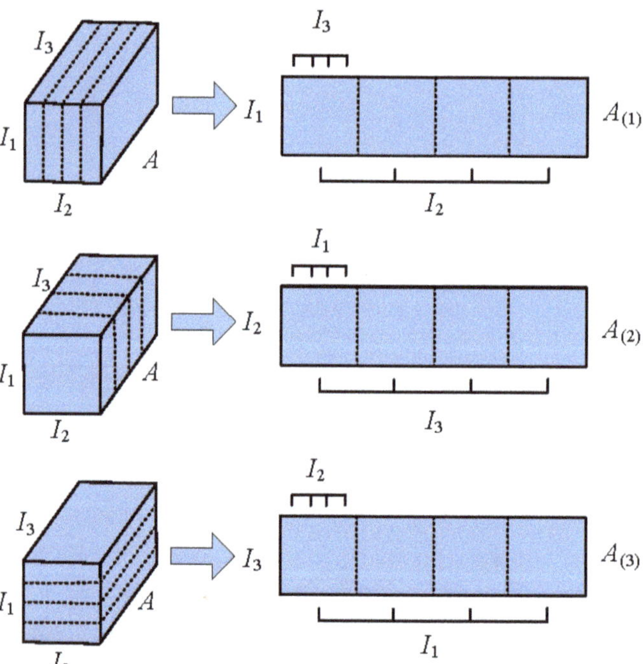

Figure 3. Mode-n expansion of the third-order tensor A.

2.3.2. Higher-Order Singular Value Decomposition

The CP decomposition and Tucker decomposition are two commonly used methods for tensor decomposition. The CP decomposition expresses a given observation tensor as a combination of a sequence of rank-one quantities. The Tucker decomposition can be regarded as the extension of the matrix component analysis to higher dimensions. The HOSVD is a special form of the Tucker decomposition and can be obtained by generalizing the mode-n product of the matrix singular value decomposition (SVD) to the nth-order tensor $A \in R^{I_1 \times I_2 \times I_3 \cdots \times I_N}$ [41].

$$A = \Re \times_1 U^{(1)} \times_2 U^{(2)} \times_3 U^{(3)} \times \cdots \times_N U^{(N)} \tag{12}$$

where $U^{(n)} \in R^{I_n \times I_n}, n = 1, 2, 3, \cdots, N$ is the orthogonal basis matrix and $\Re$ is the kernel tensor obtained by factorization. Unlike SVD for 2D matrices, the kernel tensor $\Re$ coefficient distribution for HOSVD does not have a diagonal structure and is not necessarily non-negative. $\Re$ can be expressed as follows:

$$\Re = A \times_1 U^{(1)^T} \times_2 U^{(2)^T} \times_3 U^{(3)^T} \times \cdots \times_N U^{(N)^T} \tag{13}$$

The HOSVD of the third-order tensor $A \in R^{I_1 \times I_2 \times I_3}$ is expressed as follows:

$$A = \Re \times_1 U^{(1)} \times_2 U^{(2)} \times_3 U^{(3)} \tag{14}$$

Figure 4 shows the schematic diagram of the HOSVD of the third-order observation tensor.

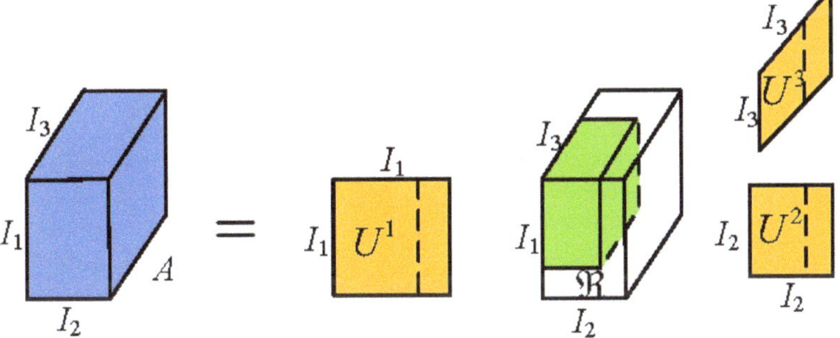

Figure 4. Schematic diagram of the HOSVD of the third-order tensor.

where $U^{(n)} \in R^{I_n \times I_n} (n = 1, 2, 3)$ is the orthogonal basis matrix and $\Re \in R^{I_1 \times I_2 \times I_3}$ is the kernel tensor obtained by factorization. Since the third-order tensor A is obtained by stacking similar patches, the traditional HOSVD denoising algorithm assumes that the core tensor is sparse. The kernel tensor $\Re$ coefficient of the HOSVD decomposition is shrunk by a hard threshold as follows:

$$\widehat{\Re} = T_\tau(\Re) \tag{15}$$

where T_τ is hard-thresholding and is defined as follows:

$$T_\tau(\Re) = \begin{cases} \Re(i, j, k), & \Re(i, j, k) \geq \tau \\ 0, & \Re(i, j, k) < \tau \end{cases} \tag{16}$$

The coefficient of the kernel tensor that is less than the threshold τ is set to zero by Equation (16), and then the estimated value $\widehat{A}$ of the kernel tensor intercepted by the hard threshold can be obtained by the inverse HOSVD transformation.

$$\widehat{A} = \widehat{\Re} \times_1 U^{(1)} \times_2 U^{(2)} \times_3 U^{(3)} \tag{17}$$

Therefore, the HOSVD is used for image denoising, mainly by decomposing the high-order structure of the image tensor to extract useful information. Subsequently, the tensor is reconstructed to achieve the denoising effect. For denoising, appropriate thresholds can be selected to remove the noise from the image tensor, and the remaining parts can be recombined to form a denoised image. Let us explore the application of SVD in image denoising. The self-similarity of an image refers to the presence of a high degree of similarity between similar patches. If there is no noise, the similar patch matrix should have a low-rank property when converted to column vectors. Based on this observation, image denoising has been formulated as a low-rank matrix approximation problem. Let

$Y \in R^{m \times n}$ be the observation matrix that comprises similar blocks of noisy observation images converted to column vectors, and $X \in R^{m \times n}$ be the corresponding noiseless matrix to be estimated, which has a low-rank property. Typically, the denoising model [28] based on the 2D low rank is approximated as follows:

$$\widehat{X} = \arg\min_{X} \frac{1}{2}\|Y - X\|_F^2 \quad s.t. \ \mathrm{rank}(X) \leq r \tag{18}$$

The above equation can be solved by performing SVD on the noisy observation matrix $Y \in R^{m \times n}$. In the SVD domain, Y can be decomposed into

$$Y = U\Lambda V^T \tag{19}$$

the following is

$$Y_r = U\Lambda_r V^T \tag{20}$$

including,

$$\Lambda_r = \mathrm{diag}(\lambda_1, \lambda_2, \cdots, \lambda_r, 0, \cdots, 0) \tag{21}$$

That is, the r singular values before the singular value decomposition of Y are retained, and Y_r is the solution of Equation (18). However, the above rank constraint is an NP problem. Determining the shrinkage of rank r for singular values is the key. However, it is difficult to determine a reasonable threshold in despeckling based on SVD. Building on these observations, a new methodology for approximating low-rank tensor patches is proposed in this study. The proposed approach involves penalizing the adaptive weighted singular values of the core tensor, which is obtained via HOSVD, to achieve the low-rank tensor patch approximation.

2.4. SAR Image Despeckling Based on the Iterative Low-Rank Tensor Patch Approximation Algorithm

The basis of HOSVD is different from the fixed basis used in the SAR-BM3D wavelet transform and is obtained by decomposing the 3D observation tensor. The objective function for the low-rank tensor estimation of SAR image despeckling is expressed as follows:

$$\arg\min_{X} \frac{1}{2}\|Y - X\|_F^2 + \mathcal{W}\|F(X)\|_*$$

where Y and X denote the noisy and latent clean images, respectively; $\| \bullet \|_F$ represents the Frobenius norm, which can be calculated using Equation (11). $\mathcal{W}$ denotes adaptive weights; $\|F(X)\|_*$ denotes the proposed low-rank tensor patch nuclear norm, and $F(X)$ represents HOSVD, which is used to obtain the low-tensor patches from X.

In the field of SAR image processing, natural scenes are often represented by SAR images that exhibit low-rank features. Generally, ℓ_1 norm regularization or nuclear norm regularization is frequently employed to approximate the rank function of the image. While these regularization methods can yield sparse solutions, the convexity in both approaches may introduce significant estimation bias [42]. In contrast, non-convex penalties, such as $\ell_q (0 \leq q < 1)$, smoothly clipped absolute deviation (SCAD), or minimax concave (MC) penalties can improve the deviation problem. Non-convex penalty regularization has demonstrated many advantages over convex penalty regularization in numerous applications. In recent years, there has been significant interest in non-convex regularization methods for sparse and low-rank recovery, driven by advancements in non-convex and non-smooth optimization algorithm theories. Building upon these developments, this paper proposes an iterative adaptive weight scheme for the regularization term, the adaptive weights with ℓ_1-norm can be seen as types of non-convex penalties applied to the core tensor [43]. The proposed scheme involves assigning varying penalty weights to the singular values of the tensor to achieve an optimal approximation of the low-rank tensor patches. Furthermore, soft threshold operators are used to solve the non-convex objective function.

For each tensor patch $A^k, k = 1 \cdots K$, the count of reference patches is represented by K, and the minimization problem is further expressed as follows:

$$\min_{X} w^k \left\| A^k \right\|_* = \sum_i w_j^k \left| \lambda_j^k \right| \tag{22}$$

the adaptive weight is denoted by w_j^k, where λ_j^k is the elemental value in the core tensor $\mathfrak{R}$. Due to the varying weights w_j^k, the adaptive weights with the ℓ_1-norm penalized optimization problem are non-convex. However, if the weights w_j^k are assigned in a non-increasing manner to the increasing absolute values $\left| \lambda_j^k \right|$ of the core tensor $\mathfrak{R}^k$, the penalized optimization problem becomes a convex problem [35].

With the above analysis, the third-order tensor A is directly despeckled. The objective function can be expressed as follows:

$$\arg\min_{A_x^t} \frac{1}{2} \left\| A_y^t - A_x^t \right\|_F^2 + \sum_j w_j^t \left| \lambda_j^t \right|, t = 1, \ldots, T \tag{23}$$

the kth reference tensor patch of noisy image Y is denoted by A_y^t, where A_x^t is the potential clean tensor patch corresponding to A_y^t, and λ_j^t is the value of the core tensor $\mathfrak{R}$. The adaptive weight is denoted by w_j^t, which is assigned to $\left| \lambda_j^t \right|$. After obtaining the despeckling estimate A_x^t, the despeckled tensor patches are put back into their original positions to form a clean image, which requires an aggregation procedure. Up to this point, the despeckling of the nth image is completed and the despeckled SAR image $\hat{X}_n$ is obtained.

2.5. Soft-Thresholding Proximal Operator

Compared to traditional convex penalty functions, the non-convex penalty functions may be more difficult to solve. Therefore, this paper uses the proximity operator to solve the non-convex penalty function. By solving the proximity problem for the objective function at each iteration, the proximity operator can provide an approximation to the optimal solution of the objective function. This approach effectively overcomes the difficulty of solving non-convex penalty functions and facilitates efficient optimization in high-dimensional spaces. Given a proper and lower semi-continuous penalty function $P_\lambda(\cdot)$ and a threshold parameter $\lambda > 0$, the scalar proximal projection can be defined as follows:

$$\text{proxp}_\lambda(h) = \arg\min_{x} \left\{ P_\lambda(x) + \frac{1}{2}(x - h)^2 \right\} \tag{24}$$

the $\text{proxp}_\lambda(h)$ represents the proximity operator of a vector $h = [h_1, \ldots, h_n] \in R^n$, which can be expressed as follows:

$$\text{proxp}_\lambda(h) = \left[\text{proxp}_\lambda(h_1), \ldots, \text{proxp}_\lambda(h_n) \right]^\mathrm{T} \tag{25}$$

For commonly used proximity operators, the following soft-thresholding proximal operator is used in this paper:

$$\text{proxp}_\lambda(\mathrm{h}) = \text{sign}(h) \max\{|h| - \lambda, 0\} \tag{26}$$

Given a patch tensor A, $\widetilde{P_\lambda}(\cdot)$ denotes a generalized penalty on the elemental values λ_i of the core tensor $\mathfrak{R}$, which can be expressed as follows:

$$\widetilde{P_\lambda}(A) = \sum_i P_\lambda(\lambda_i) \tag{27}$$

2.6. Residual Iteration and Adaptive Weight Setting to w_j^t

To further enhance the efficacy of SAR image despeckling, the residual image is utilized and iteratively processed. The residual image is effective in capturing the differences between noise and signal. The iterative process facilitates continuous optimization of despeckling outcomes, resulting in improved despeckling effects and image quality. The key element of the iterative algorithm is to add the residual $\left(Y - \hat{X}_n\right)$ from the nth iteration to refine the $(n+1)$th step despeckled image Y_{n+1}.

$$Y_{n+1} = \hat{X}_n^t + \mu\left(Y - \hat{X}_n^t\right) \tag{28}$$

where n stands for the number of iterations; and the relaxation parameter is denoted by μ. To incorporate the residual information into the despeckled image, the remaining noise variance should be estimated, which is expressed as follows:

$$\sigma_n = \eta\sqrt{\sigma^2 - \frac{1}{q \times r}\|Y - Y_{n-1}\|_F^2} \tag{29}$$

where η represents the scaling factor, which is used to control the re-estimation of noise variance. The noise variance of Y is indicated by σ, and the number of pixels in Y is denoted by $q \times r$.

The iterative adaptive weights w_j^t related to each λ_j are assigned as follows:

$$w_j^t = \frac{2\sqrt{2}\sqrt{N}\sigma_n^2}{|\lambda_j| + \varrho} \tag{30}$$

where N denotes the number of patches per tensor A_y^t; the small positive parameter is represented by $\varrho > 0$, which is chosen to prevent division by zero, and σ_n can be computed using Equation (29). The optimization problem (22) of the proximal soft-threshold operator (26) is solved using an iterative adaptive weight w_j^t. The adaptive thresholding retains the large values while filtering out the small values, thereby preserving important structural information in the image. Consequently, solution τ_j^t to the jth element of A_x^t can be expressed as follows:

$$\tau_j^k = \text{sign}(\lambda_j)\max\left\{|\lambda_j| - w_j^t, 0\right\} \tag{31}$$

2.7. Aggregation of Despeckled Tensor Patches

When the patches are reassembled into the image, a weight is assigned to each tensor patch based on its level of noise. Specifically, the weight is defined as follows:

$$\omega = \begin{cases} \dfrac{r^2 \times N}{r^2 \times N + C}, & \text{if } C \geq 1 \\ 0, & \text{otherwise} \end{cases} \tag{32}$$

where r denotes the patch size; C represents the number of thresholded elements in the core tensor and N is the count of patches in the current tensor. The resulting image can be obtained as follows:

$$\hat{X}(x,y) = \frac{\sum_{i \in N_{x,y}} \sum_{j \in J(i)_{x,y}} \omega_{i,j} \Omega_{i,j}}{\sum_{i \in N_{x,y}} \sum_{j \in J(i)_{x,y}} \omega_{i,j}} \tag{33}$$

where $N_{x,y}$ represents all tensor patches overlapping in position (x, y); $J(i)_{x,y}$ represents all tensors in the ith tensor that overlap at position (x, y); and $\Omega_{i,j}$ represents a pixel located at position (x, y). The flow of the tensor despeckling algorithm is shown in Algorithm 1.

Algorithm 1 Iterative low-rank tensor patch approximation algorithm for SAR image despeckling

---

Input: SAR image Y, the ENL L, the number of reference patches K, and iteration F
Output: despeckled SAR image X
  1: Initialization:
    Initialize $X_0 = Y$, $Y^0 = Y$, SAR image patch tensor A
  2: Iteration:
  ① Outer loop: for $n = 1{:}F$ do
    (I) Re-estimate Y^n by (28)
    (II) Re-estimate noise variance σ_n by (28)
  ② Inner loop: for $t = 1{:}T$ do
    (I) Compute $U^{(1)}, U^{(2)}, U^{(3)}$ and core tensor $\mathfrak{R}_y^{(t)}$ of $A_y^{(t)}$ by HOSVD via Equation (14)
    (II) For each λ_j in core $\mathfrak{R}_y^{(t)}$ calculate the w_j^t via Equation (22)
    (III) Apply threshold w_j^t to λ_j in $\mathfrak{R}_y^{(t)}$ via Equation (27)
    (IV) Estimate despeckling patches tensor by (23)
      End for
  ③ Obtain the nth step despeckled SAR image $\hat{X}_n$ via Equation (33)
      End for
  ④ Obtain the despeckled SAR image X

---

3. Results

This section presents the despeckling experiments conducted on simulated multiplicative noise images and real SAR images. The experimental results verify the effectiveness of the proposed method. The proposed method is compared with the current state-of-the-art filters, including PPB [24], SAR-BM3D [25], FANS [26], and Mulog [44]. The executable codes for these methods can be downloaded from the authors' websites, accessed on 29 May 2023. (https://www.charles-deledalle.fr/pages/ppb.php; http://www.grip.unina.it/download/prog/SAR-BM3D/; http://www.grip.unina.it/download/prog/FANS/; https://www.charles-deledalle.fr/pages/mulog.php) To better measure the despeckling performance of the proposed method, three objective evaluation metrics, i.e., peak signal-to-noise ratio (PSNR), feature similarity index (FSIM) [45], and structural similarity index (SSIM) [46], are used for the simulated multiplicative noise images. A higher PSNR indicates better image quality, while a higher FSIM suggests improved preservation of image structure information, which is represented as a number between 0 and 1. For real SAR images, the equivalent number of looks (ENL) and the mean of ratio image (MoR) are used to measure the superiority of the proposed method. Moreover, the ratio image is calculated to objectively evaluate the real SAR image-despeckling performance. To facilitate a well-informed selection of parameters within the proposed algorithm, this paper conducted an analysis of their influence using PSNR and SSIM metrics, as presented in Table 1. The PSNR and SSIM values were computed by averaging across multiple images. In the proposed algorithm, the patch size was set to 7×7 and the search window size was set to 30×30. Additionally, the parameter patch number and patch stack number were set to 300 and 60, respectively.

Table 1. PSNR and SSIM performance comparison with different patch sizes, patch numbers, patch stack numbers, and search windows. The best results are highlighted in bold.

Patch size	5×5	6×6	7×7	8×8	9×9
PSNR	23.11	23.14	**23.26**	23.24	23.19
SSIM	0.679	0.681	**0.687**	0.680	0.679
Patch number	100	150	200	250	300
PSNR	23.23	23.27	23.33	23.31	**23.38**
SSIM	0.683	0.687	0.691	0.693	**0.695**
Patch stack number	30	40	50	60	70
PSNR	23.07	23.25	23.40	**23.42**	23.37
SSIM	0.683	0.687	0.691	0.693	**0.695**
Search window	10×10	15×15	20×20	25×25	30×30
PSNR	23.21	23.34	23.41	23.43	**23.45**
SSIM	0.686	0.691	0.697	0.696	**0.698**

3.1. Experiments on Simulated Multiplicative Noise Images

In this study, initial evaluations were conducted on three simulated multiplicative noise images, i.e., house, monarch, and Napoli. These three images are commonly used as test images in SAR image despeckling. To simulate realistic conditions, multiplicative speckle noise with different appearance levels (L = 1, 2, 4, 8) was added to each noise-free image. Reference images and corresponding noisy images (with L = 2) are presented in Figure 5.

Figure 5. Test of synthetic multiplicative noise images (top line, from left to right—house, monarch, and Napoli are the clean optical images. Bottom line, from left to right—house, monarch, and Napoli are the simulated SAR images, L = 2).

Figure 6 shows the despeckling results of each filter on the house image, along with the relevant quantitative indexes reported in Table 2. Visually, compared to the clear images, the PPB filter smooths the image excessively, resulting in the loss of detail and structural information, such as the window details highlighted by the red box. FANS effectively suppresses the noise while preserving the structural information of the image, but it has poor performance in retaining details. Although both SAR-BM3D and Mulog can preserve the details and structural information of the image well and have good despeckling performance in homogeneous regions, some unwanted artifacts are introduced during the despeckling process. Overall, the proposed method effectively despeckles the SAR images while preserving more details and structural information. Regarding the quantitative evaluation, Table 2 shows that Mulog performs well in terms of SSIM and PSNR, while FANS performs well in terms of ENL. The proposed method achieves the highest PSNR when L = 1, and consistently achieves higher FSIM and PSNR than the PPB and SAR-BM3D filters for $L > 2$.

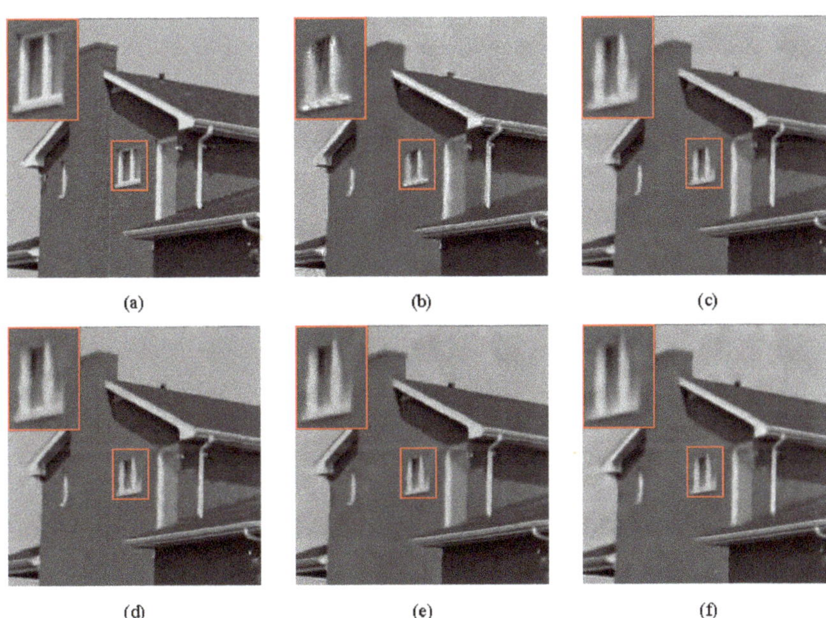

Figure 6. The results of different filters on the house ($L = 4$). The red box highlights the zoomed region of interest. (**a**) Clean image; (**b**) PPB; (**c**) FANS; (**d**) SAR-BM3D; (**e**) Mulog; (**f**) proposed method.

Table 2. The PSNR, FSIM, and SSIM of different filters on the simulated multiplicative noise images ($L = 1, 2, 4, 8$). The best results are highlighted in bold.

	Methods	$L = 1$ PSNR	FSIM	SSIM	$L = 2$ PSNR	FSIM	SSIM	$L = 4$ PSNR	FSIM	SSIM	$L = 8$ PSNR	FSIM	SSIM
	Noisy image	12.16	0.427	0.096	14.71	0.504	0.152	17.32	0.584	0.225	20.04	0.663	0.316
	PPB	25.13	0.786	0.642	27.16	0.840	0.724	29.02	0.877	0.786	30.39	0.893	0.823
House	FANS	25.34	0.804	0.757	**28.66**	0.854	0.811	**31.17**	0.883	0.842	32.95	0.903	0.860
	SAR-BM3D	24.62	**0.836**	0.772	28.14	**0.877**	0.816	30.90	**0.905**	0.845	32.12	**0.922**	**0.863**
	Mulog	25.01	0.835	**0.783**	28.36	0.870	**0.822**	31.15	0.894	**0.847**	**32.97**	0.914	0.862
	Proposed	**25.54**	0.813	0.733	28.28	0.857	0.797	30.76	0.888	0.836	32.38	0.904	0.854
	Noisy image	13.47	0.536	0.258	16.04	0.614	0.349	18.75	0.691	0.444	21.52	0.762	0.546
	PPB	23.00	0.826	0.716	24.72	0.866	0.790	25.99	0.919	0.835	27.63	0.916	0.873
Monarch	FANS	**24.21**	0.856	0.805	**26.57**	0.895	0.864	**28.52**	0.919	0.900	30.23	0.938	0.924
	SAR-BM3D	23.61	0.853	0.800	26.13	0.890	0.856	28.13	0.915	0.893	29.84	0.933	0.919
	Mulog	23.80	**0.866**	**0.813**	26.24	**0.901**	**0.867**	28.43	**0.924**	**0.904**	30.30	0.942	**0.929**
	Proposed	23.51	0.843	0.750	26.04	0.890	0.825	28.28	0.921	0.890	30.30	**0.943**	0.920
	Noisy image	14.64	0.606	0.229	17.27	0.628	0.337	20.08	0.759	0.463	22.97	0.826	0.593
	PPB	21.74	0.713	0.561	23.23	0.783	0.659	24.94	0.845	0.741	26.40	0.885	0.800
Napoli	FANS	22.26	0.710	0.598	24.24	0.804	0.703	26.24	0.869	0.784	**28.12**	0.910	0.848
	SAR-BM3D	**22.64**	0.760	**0.639**	**24.42**	0.825	**0.724**	**26.36**	0.880	**0.802**	**28.12**	**0.916**	**0.865**
	Mulog	22.43	0.735	0.628	24.08	0.801	0.704	26.02	0.861	0.778	27.96	0.908	0.843
	Proposed	22.33	**0.780**	0.600	24.15	**0.826**	0.690	26.14	**0.881**	0.775	28.04	0.915	0.842

Figure 7 shows the results of the Monarch image, while the relevant quantitative indexes are reported in Table 2. It can be observed that, although the PPB filter can suppress speckle well, it removes a significant amount of textural details. FANS preserves more details than PPB, but the solution appears slightly over-smoothed. The SAR-BM3D can preserve details and textures well, but it has poor despeckling performance in homogeneous regions, which is opposite to Mulog. Overall, the proposed method can effectively balance despeckling and detail preservation, albeit with some minor artifacts. Regarding the quantitative evaluation, Mulog shows the best results in terms of SSIM and FSIM, while the highest values of PSNR are achieved by FANS. The proposed method consistently outperforms PPB, and when $L = 8$, it achieves the best performance in both FSIM and PSNR.

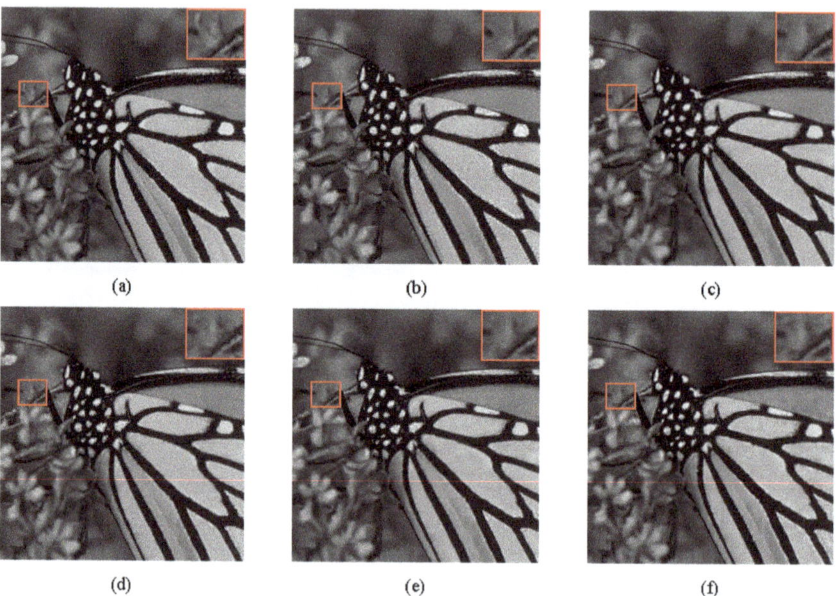

Figure 7. The results of different filters on monarch ($L = 4$). The red box highlights the zoomed region of interest. (**a**) Clean image; (**b**) PPB; (**c**) FANS; (**d**) SAR-BM3D; (**e**) Mulog; (**f**) proposed method.

Figure 8 shows the results of the Napoli image, with the corresponding quantitative indexes reported in Table 2. It can be seen from Figure 8 that PPB over-smooth the image, resulting in a significant loss of details and texture. FANS and Mulog preserve more details than the PPB, but the resulting solution is slightly over-smoothed. The SAR-BM3D can preserve details and textures well, but its despeckling performance in homogeneous regions is suboptimal. The despeckling ability of the proposed method needs to be strengthened in the homogeneous region, but the texture and details are well preserved, which is better conducive to image interpretation. In terms of quantitative assessment, SAR-BM3D shows the best results in terms of SSIM and PSNR, while the proposed method achieves the highest values of SSIM. It can be observed from Table 2 that the proposed method consistently outperforms the PPB. All of the above objective results show that our proposed method can suppress speckle noise in homogeneous regions and preserve image details well in edge regions.

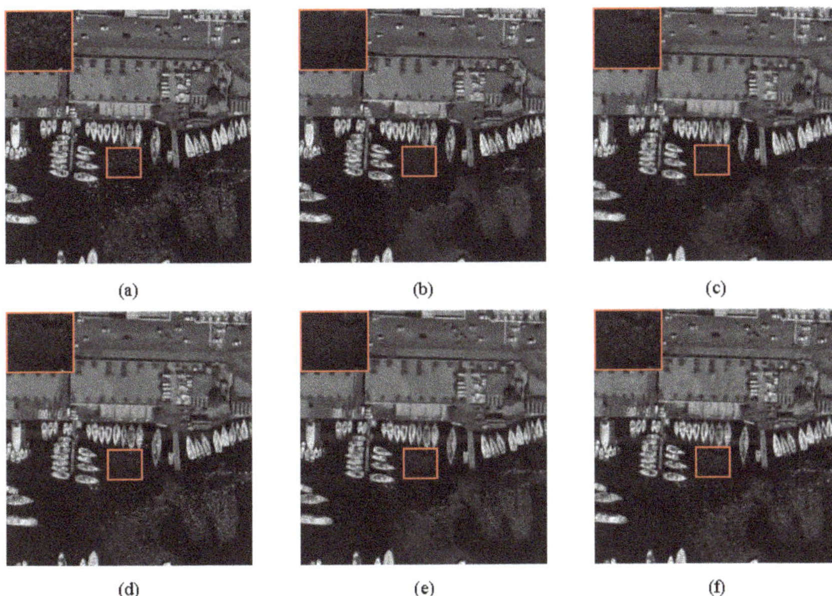

Figure 8. The results of different filters on Napoli ($L = 4$). The red box highlights the zoomed region of interest. (**a**) Clean image; (**b**) PPB; (**c**) FANS; (**d**) SAR-BM3D; (**e**) Mulog; (**f**) proposed method.

3.2. Experiments on Real SAR Images

To assess the despeckling performance of the proposed method on real SAR images, three remote-sensing images from different sensors were selected. Figure 9a is a 2-look SAR image, denoted as R1, acquired by the British DRA SAR. Figure 10a is a 1-look SAR image, denoted as R2, acquired by the German TerraSAR-X satellite. The 8-look SAR image, denoted as R3, as shown in Figure 11a, was acquired by Sentinel-1. In the experiment, the ENL values were measured in the blue rectangular areas. The specific results are shown in Table 3. A larger ENL generally indicates better despeckling ability. Furthermore, ratio images were computed to compare the residual speckle noise. As per the findings of [47], an ideal filter's ratio image (the ratio between noisy and despeckled images) should solely preserve the speckle. Hence, based on the multiplicative noise model, the MoR of the three images should be 1. To facilitate a comprehensive assessment of the proposed despeckling method's efficiency in real-world scenarios, a comparative analysis of computational times was performed for different filters. The results are presented in Table 3, which highlights the differences in processing times among the different filters.

Table 3. The ENL, MoR, and times of different filters on real SAR images. The best results are highlighted in bold.

	R1				R2				R3			
Methods	ENL1	ENL2	MoR	Time	ENL1	ENL2	MoR	Time	ENL1	ENL2	MoR	Time
Noisy image	2.90	2.62	-	-	2.20	2.62	-	-	22.47	12.78	-	-
PPB	**55.58**	**366.65**	**1.00**	23.08	**1345.02**	**340.70**	0.99	22.36	1327.92	**1181.27**	**1.00**	23.11
FANS	32.65	74.58	1.02	**1.62**	100.75	98.03	1.11	**1.74**	525.66	770.60	1.01	**1.58**
SAR-BM3D	19.63	22.27	0.99	24.31	127.05	77.99	**0.99**	24.54	**3837.38**	111.76	0.99	24.09
Mulog	29.74	69.61	1.07	10.25	216.92	144.60	1.38	10.02	878.76	859.67	1.01	11.36
Proposed	38.04	151.58	0.96	50.63	137.74	151.58	**0.99**	49.47	401.67	951.58	0.98	49.33

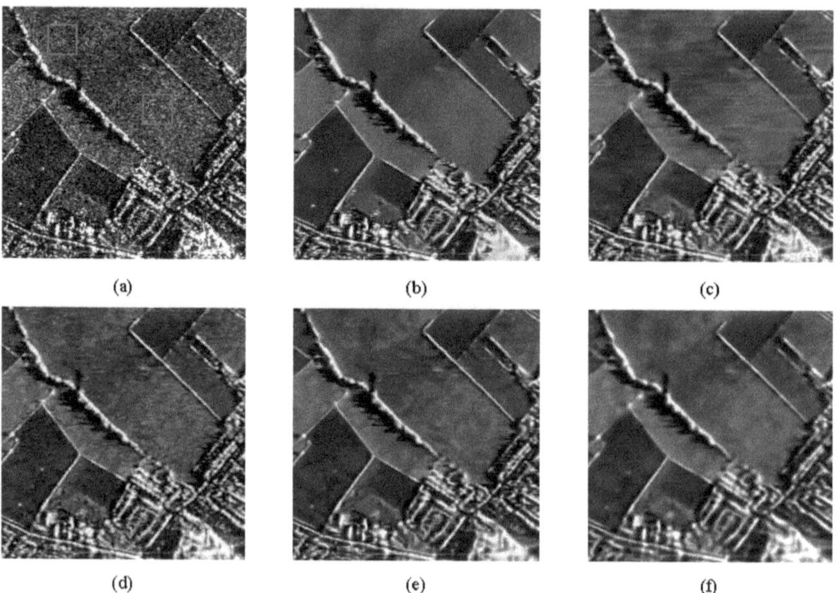

Figure 9. The results of different filters on R1 ($L = 2$). The blue boxes highlight the regions of interest used to compute the equivalent number of looks (ENL). (**a**) Noisy image; (**b**) PPB; (**c**) FANS; (**d**) SAR-BM3D; (**e**) Mulog; (**f**) proposed method.

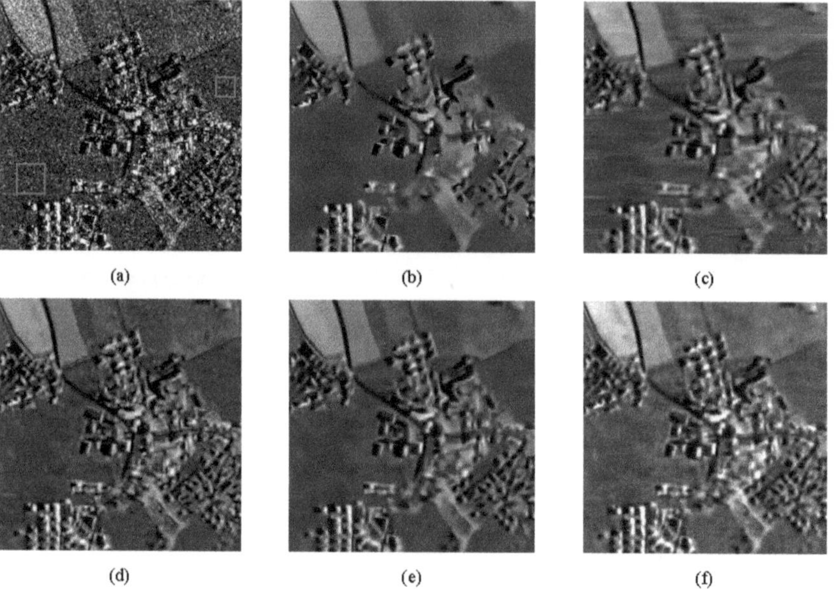

Figure 10. The results of different filters on R2 ($L = 1$). The blue boxes highlight the regions of interest used to compute the equivalent number of looks (ENL). (**a**) Noisy image; (**b**) PPB; (**c**) FANS; (**d**) SAR-BM3D; (**e**) Mulog; (**f**) proposed method.

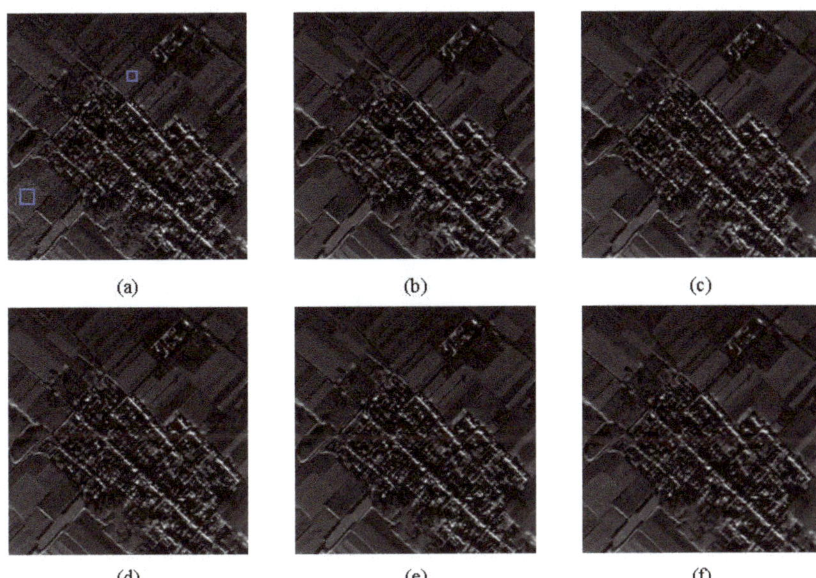

Figure 11. The results of different filters on R3 ($L = 8$). The blue boxes highlight the regions of interest used to compute the equivalent number of looks (ENL). (**a**) Noisy image; (**b**) PPB; (**c**) FANS; (**d**) SAR-BM3D; (**e**) Mulog; (**f**) proposed method.

Figures 9–11 display the despeckling results of real-world SAR images. Compared to the other filters, PPB leads to excessive smoothing, resulting in the loss of a considerable amount of detail and texture information, as shown in Figures 9b and 10b. The SAR-BM3D can preserve details and structural information well but it has poor noise removal performance in homogeneous areas, as shown in Figure 10d. Both FANS and Mulog can reduce noise while preserving details and structural information well; however, they may introduce some unnecessary artifacts, as shown in Figures 9c,e and 10c,e. Since the proposed method is an edge-preserving filter, it focuses on preserving edges and does not over-smooth details. Therefore, it does not over-smooth the homogeneous regions, such as the PPB filter, resulting in lower ENL values. Table 3 also shows that the proposed method can better use the similarity of the image to denoise and retain the detailed information. Figures 12–14 show the ratio image results. The despeckled SAR images show that both PPB and FANS tend to remove excessive details during despeckling, especially in SAR images with more structural information, as shown in Figure 13a,b. This excessive despeckling of structural and edge information adversely affects the subsequent SAR image processing. On the other hand, the SAR-BM3D effectively despeckles homogeneous regions while preserving edge regions and structural information, as seen in the ratio image results in Figures 12c and 13c. However, the despeckled SAR images appear distorted, resulting in the blurring of many details. While Mulog preserves details and edges information well, it introduces unnecessary artifacts in homogeneous regions, as shown in Figures 12d and 13d. In contrast, the proposed method achieves perfect homogeneous despeckling while maintaining details and structural information to the greatest extent. To summarize, the experimental comparative analysis of SAR image despeckling shows that the proposed method achieves advanced visual quality, structure preservation, and speckle reduction results. However, a comparison of the filtering time results in Table 3 reveals that the despeckling time for the PPB and SAR-BM3D methods is approximately 23–24 s, while the Mulog method takes 10–11 s, and the FANS method requires only 1–2 s. In contrast, the proposed method exhibits a longer despeckling time of around 49–50 s. This indicates that there is still considerable room for improvement in achieving real-time despeckling with the proposed method. The primary contributing factor is the computationally demanding task of gradient-based similar patch classification, which necessitates significant computational resources. Additionally, the 3D tensor-based despeckling process involves the stacking of numerous 2D similar patches and relies on HOSVD, both of which further contribute to the overall computation time. Nevertheless, in practical applications of SAR despeckling, if real-time processing is not a strict requirement, the preservation of texture and detailed information is of utmost importance, and the proposed method remains a good choice.

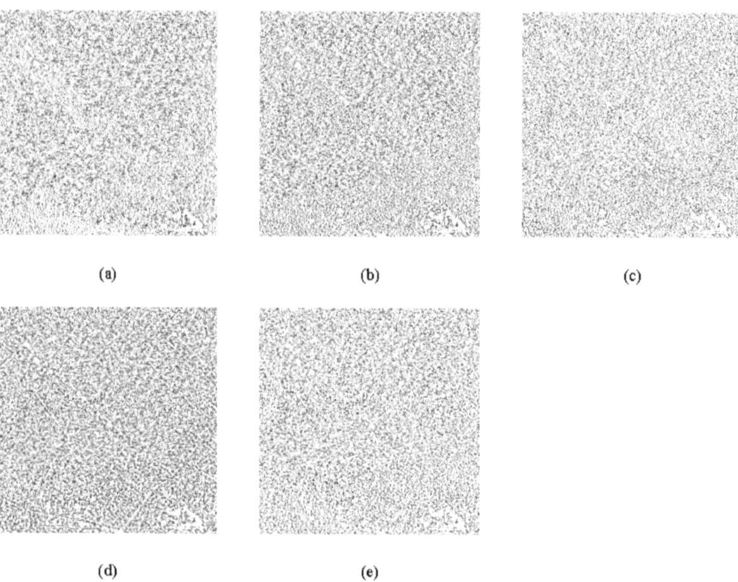

Figure 12. The corresponding ratio images for R1. (**a**) PPB; (**b**) FANS; (**c**) SAR-BM3D; (**d**) Mulog; (**e**) proposed method.

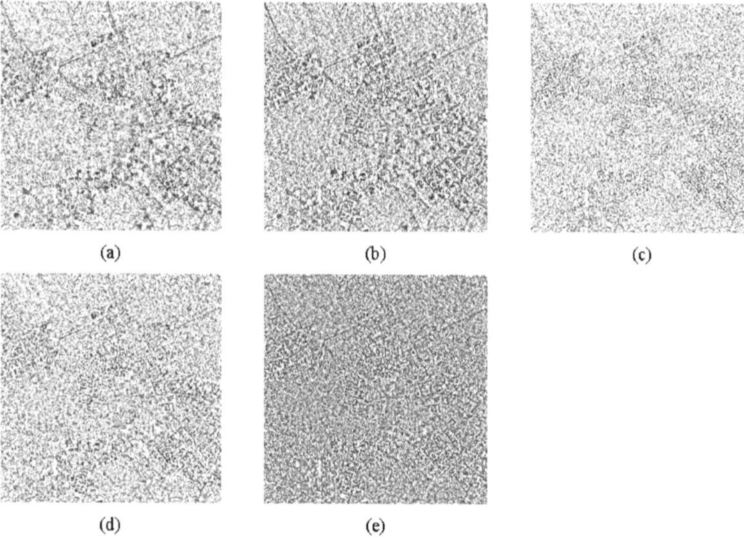

Figure 13. The corresponding ratio images for R2. (**a**) PPB; (**b**) FANS; (**c**) SAR-BM3D; (**d**) Mulog; (**e**) proposed method.

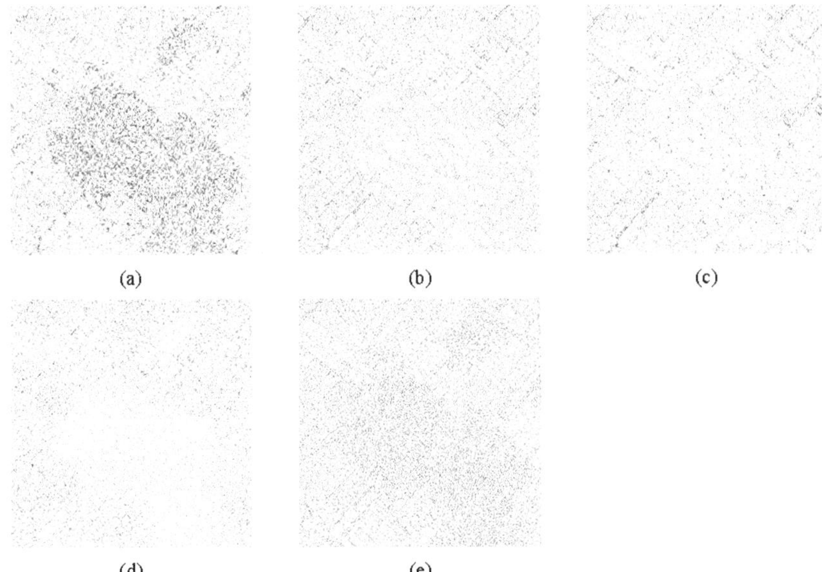

Figure 14. The corresponding ratio images for R3. (**a**) PPB; (**b**) FANS; (**c**) SAR-BM3D; (**d**) Mulog; (**e**) proposed method.

4. Discussion

In this paper, a 3D despeckling method is proposed that stacks similar patches into tensor patches and uses the HOSVD method for low-rank tensor approximation despeckling. The proposed approach extends the application of image self-similarity to 3D and better utilizes the spatial geometric structure information of similar patches. The effectiveness of the proposed method is comprehensively evaluated and compared with the state-of-the-art despeckling methods through despeckling experiments on both simulated and real SAR images.

The experimental results demonstrate that the FSIM metric consistently outperforms FANS and PPB, particularly in the case of SAR images with more complex structural information, such as the Napoli image. Despite not having the highest PSNR and SSIM scores, the proposed method consistently outperforms PPB in terms of despeckling value across all view numbers. Visually, the proposed method achieves excellent speckle reduction while preserving image details. Although the ENL value is not the highest, the MoR values demonstrate that the proposed method effectively preserves edge information. Overall, the experimental results demonstrate and validate that the proposed method achieves an optimal balance between preserving an image's structural and texture information and achieving effective despeckling.

However, it is noted that the proposed method may result in some artifacts while despeckling homogeneous areas. To minimize these artifacts, future work will focus on improving the accuracy of patch classification. Additionally, the process of classifying similar patches based on gradients poses significant computational demands, necessitating substantial computational resources. Moreover, the 3D tensor-based despeckling procedure involves the aggregation of multiple 2D similar patches and relies on HOSVD, both of which contribute to the overall computation time. This complexity is crucial for achieving improved despeckling performance and preserving essential image details. In future research, the authors intend to utilize approximate methods or efficient algorithms to minimize the computational complexity and reduce the despeckling time as much as possible.

5. Conclusions

This paper proposes an SAR image-despeckling method that utilizes high-dimensional information from similar patches by employing a search for similar patches and applying the tensor patch HOSVD theory. Specifically, the method extends 2D to 3D for SAR image despeckling using tensor patches for the first time. The proposed effectively achieves high-dimensional SAR image despeckling using a low-rank tensor approximation technique to exploit the potential correlation and low-rank structure of SAR image data. Previous image low-rank algorithms tended to destroy the topological structures of image patches by pulling similar patches into column vectors, which is detrimental to preserving image details. As a high-order generalization of vectors and matrices, tensors can stack 2D similar patches into 3D tensor patches, enabling HOSVD to directly despeckle the 3D tensor patches and leverage the high-dimensional structural information of similar patches. Extensive experiments demonstrate and validate the superior performance of the proposed method in synthetic and real SAR image despeckling compared with the existing advanced filters in the subjective visual evaluation and objective evaluation indices. However, the computational complexity of classifying similar patches and performing high-order singular value decomposition to approximate 3D tensor patches makes the proposed method time-consuming for despeckling. In the future, the focus will be on optimizing the mathematical algorithm and extending 3D despeckling to higher dimensions. This will leverage the self-similarity information of similar patches for SAR despeckling.

Author Contributions: Conceptualization, J.F., T.M., and F.B.; methodology, J.F. and T.M.; validation, X.W., W.L., and N.Z.; investigation, J.F., T.M., and B.H.; writing—original draft preparation, T.M.; writing—review and editing, J.F. and S.H. All authors have read and agreed to the published version of the manuscript.

Funding: This work was supported by the Natural Science Foundation of China under grant nos. 62002208 and 62172030 and the Natural Science Foundation of Shandong Province under grant nos. ZR2020MA082 and ZR2020MF119.

Data Availability Statement: The images used in this paper are sourced from the technical library.

Conflicts of Interest: The authors declare no conflict of interest.

References

1. Moreira, A.; Prats-Iraola, P.; Younis, M.; Krieger, G.; Hajnsek, I.; Papathanassiou, K.P. A tutorial on synthetic aperture radar. *IEEE Geosci. Remote Sens. Mag.* **2013**, *1*, 6–43. [CrossRef]
2. Ren, H.; Yu, X.; Zou, L.; Zhou, Y.; Wang, X.; Bruzzone, L. Extended convolutional capsule network with application on SAR automatic target recognition. *Signal Process.* **2021**, *183*, 108021. [CrossRef]
3. Baraha, S.; Sahoo, A.K.; Modalavalasa, S. A systematic review on recent developments in nonlocal and variational methods for SAR image despeckling. *Signal Process.* **2022**, *196*, 108521. [CrossRef]
4. Ponmani, E.; Saravanan, P. Image denoising and despeckling methods for SAR images to improve image enhancement performance: A survey. *Multimed. Tools Appl.* **2021**, *80*, 26547–26569. [CrossRef]
5. Wang, G.; Bo, F.; Chen, X.; Lu, W.; Hu, S.; Fang, J. A collaborative despeckling method for SAR images based on texture classification. *Remote Sens.* **2022**, *14*, 1465. [CrossRef]
6. Bo, F.; Lu, W.; Wang, G.; Zhou, M.; Wang, Q.; Fang, J. A Blind SAR Image Despeckling Method Based on Improved Weighted Nuclear Norm Minimization. *IEEE Geosci. Remote Sens. Lett.* **2022**, *19*, 1–5. [CrossRef]
7. Lee, J.S. Digital image enhancement and noise filtering by use of local statistics. *IEEE Trans. Pattern Anal. Mach. Intell.* **1980**, *PAMI-2*, 165–168. [CrossRef]
8. Frost, V.S.; Stiles, J.A.; Shanmugan, K.S.; Holtzman, J.C. A model for radar images and its application to adaptive digital filtering of multiplicative noise. *IEEE Trans. Pattern Anal. Mach. Intell.* **1982**, *PAMI-4*, 157–166. [CrossRef]
9. Kuan, D.T.; Sawchuk, A.A.; Strand, T.C.; Chavel, P. Adaptive noise smoothing filter for images with signal-dependent noise. *IEEE Trans. Pattern Anal. Mach. Intell.* **1985**, *PAMI-7*, 165–177. [CrossRef]
10. Ranjani, J.J.; Thiruvengadam, S. Dual-tree complex wavelet transform based SAR despeckling using interscale dependence. *IEEE Trans. Geosci. Remote Sens.* **2010**, *48*, 2723–2731. [CrossRef]
11. Bianchi, T.; Argenti, F.; Alparone, L. Segmentation-based MAP despeckling of SAR images in the undecimated wavelet domain. *IEEE Trans. Geosci. Remote Sens.* **2008**, *46*, 2728–2742. [CrossRef]

12. Tao, R.; Wan, H.; Wang, Y. Artifact-free despeckling of SAR images using contourlet. *IEEE Geosci. Remote Sens. Lett.* **2012**, *9*, 980–984. [CrossRef]

13. Hou, B.; Zhang, X.; Bu, X.; Feng, H. SAR image despeckling based on nonsubsampled shearlet transform. *IEEE J. Sel. Top. Appl. Earth Obs. Remote Sens.* **2012**, *5*, 809–823. [CrossRef]

14. Sun, Y.; Lei, L.; Guan, D.; Li, X.; Kuang, G. SAR image speckle reduction based on nonconvex hybrid total variation model. *IEEE Trans. Geosci. Remote Sens.* **2020**, *59*, 1231–1249. [CrossRef]

15. Maji, S.K.; Thakur, R.K.; Yahia, H.M. SAR image denoising based on multifractal feature analysis and TV regularisation. *IET Image Process.* **2020**, *14*, 4158–4167. [CrossRef]

16. Zhang, Z.; Xu, Y.; Yang, J.; Li, X.; Zhang, D. A survey of sparse representation: Algorithms and applications. *IEEE Access* **2015**, *3*, 490–530. [CrossRef]

17. Jiang, J.; Jiang, L.; Sang, N. Non-local sparse models for SAR image despeckling. In Proceedings of the 2012 International Conference on Computer Vision in Remote Sensing, Xiamen, China, 16–18 December 2012; pp. 230–236. [CrossRef]

18. Wang, P.; Zhang, H.; Patel, V.M. SAR image despeckling using a convolutional neural network. *IEEE Signal Process. Lett.* **2017**, *24*, 1763–1767. [CrossRef]

19. Lattari, F.; Gonzalez Leon, B.; Asaro, F.; Rucci, A.; Prati, C.; Matteucci, M. Deep learning for SAR image despeckling. *Remote Sens.* **2019**, *11*, 1532. [CrossRef]

20. Liu, S.; Gao, L.; Lei, Y.; Wang, M.; Hu, Q.; Ma, X.; Zhang, Y.D. SAR speckle removal using hybrid frequency modulations. *IEEE Trans. Geosci. Remote Sens.* **2020**, *59*, 3956–3966. [CrossRef]

21. Vitale, S.; Ferraioli, G.; Pascazio, V. Multi-objective CNN-based algorithm for SAR despeckling. *IEEE Trans. Geosci. Remote Sens.* **2020**, *59*, 9336–9349. [CrossRef]

22. Liu, Z.; Lai, R.; Guan, J. Spatial and transform domain CNN for SAR image despeckling. *IEEE Geosci. Remote Sens. Lett.* **2020**, *19*, 1–5. [CrossRef]

23. Buades, A.; Coll, B.; Morel, J.M. A non-local algorithm for image denoising. In Proceedings of the 2005 IEEE Computer Society Conference on Computer Vision and Pattern Recognition (CVPR'05), San Diego, CA, USA, 20–25 June 2005; Volume 2, pp. 60–65. [CrossRef]

24. Deledalle, C.A.; Denis, L.; Tupin, F. Iterative weighted maximum likelihood denoising with probabilistic patch-based weights. *IEEE Trans. Image Process.* **2009**, *18*, 2661–2672. [CrossRef] [PubMed]

25. Parrilli, S.; Poderico, M.; Angelino, C.V.; Verdoliva, L. A nonlocal SAR image denoising algorithm based on LLMMSE wavelet shrinkage. *IEEE Trans. Geosci. Remote Sens.* **2011**, *50*, 606–616. [CrossRef]

26. Cozzolino, D.; Parrilli, S.; Scarpa, G.; Poggi, G.; Verdoliva, L. Fast adaptive nonlocal SAR despeckling. *IEEE Geosci. Remote Sens. Lett.* **2013**, *11*, 524–528. [CrossRef]

27. Chen, G.; Li, G.; Liu, Y.; Zhang, X.P.; Zhang, L. SAR image despeckling based on combination of fractional-order total variation and nonlocal low rank regularization. *IEEE Trans. Geosci. Remote Sens.* **2019**, *58*, 2056–2070. [CrossRef]

28. Guo, Q.; Zhang, C.; Zhang, Y.; Liu, H. An efficient SVD-based method for image denoising. *IEEE Trans. Circuits Syst. Video Technol.* **2015**, *26*, 868–880. [CrossRef]

29. Ozcan, C.; Sen, B.; Nar, F. Sparsity-driven despeckling for SAR images. *IEEE Geosci. Remote Sens. Lett.* **2015**, *13*, 115–119. [CrossRef]

30. Zhou, X.; Yang, C.; Zhao, H.; Yu, W. Low-rank modeling and its applications in image analysis. *ACM Comput. Surv. (CSUR)* **2014**, *47*, 1–33. [CrossRef]

31. Zhao, Y.Q.; Yang, J. Hyperspectral image denoising via sparse representation and low-rank constraint. *IEEE Trans. Geosci. Remote Sens.* **2014**, *53*, 296–308. [CrossRef]

32. Liang, D.; Jiang, M.; Ding, J. Fast patchwise nonlocal SAR image despeckling using joint intensity and structure measures. *IEEE J. Sel. Top. Appl. Earth Obs. Remote Sens.* **2022**, *15*, 6283–6293. [CrossRef]

33. Aghababaei, H.; Ferraioli, G.; Vitale, S.; Zamani, R.; Schirinzi, G.; Pascazio, V. Nonlocal model-free denoising algorithm for single-and multichannel SAR data. *IEEE Trans. Geosci. Remote Sens.* **2021**, *60*, 1–15. [CrossRef]

34. Kolda, T.G.; Bader, B.W. Tensor decompositions and applications. *SIAM Rev.* **2009**, *51*, 455–500. [CrossRef]

35. Liu, J.; Musialski, P.; Wonka, P.; Ye, J. Tensor completion for estimating missing values in visual data. *IEEE Trans. Pattern Anal. Mach. Intell.* **2012**, *35*, 208–220. [CrossRef] [PubMed]

36. Zhang, Z.; Ely, G.; Aeron, S.; Hao, N.; Kilmer, M. Novel methods for multilinear data completion and de-noising based on tensor-SVD. In Proceedings of the IEEE Conference on Computer Vision and Pattern Recognition, Washington, DC; USA, 23–28 June 2014; pp. 3842–3849.

37. Xue, J.; Zhao, Y.; Liao, W.; Chan, J.C.W. Nonlocal low-rank regularized tensor decomposition for hyperspectral image denoising. *IEEE Trans. Geosci. Remote Sens.* **2019**, *57*, 5174–5189. [CrossRef]

38. Yang, H.; Park, Y.; Yoon, J.; Jeong, B. An improved weighted nuclear norm minimization method for image denoising. *IEEE Access* **2019**, *7*, 97919–97927. [CrossRef]

39. Buades, A.; Coll, B.; Morel, J.M. A review of image denoising algorithms, with a new one. *Multiscale Model. Simul.* **2005**, *4*, 490–530. [CrossRef]

40. Jang, S.; Nam, H.; Lee, Y.J.; Jeong, B.; Lee, R.; Yoon, J. Data-adapted moving least squares method for 3-D image interpolation. *Phys. Med. Biol.* **2013**, *58*, 8401. [CrossRef]

41. De Lathauwer, L.; De Moor, B.; Vandewalle, J. A multilinear singular value decomposition. *SIAM J. Matrix Anal. Appl.* **2000**, *21*, 1253–1278. [CrossRef]

42. Gandy, S.; Recht, B.; Yamada, I. Tensor completion and low-n-rank tensor recovery via convex optimization. *Inverse Probl.* **2011**, *27*, 025010. [CrossRef]

43. Shen, Z.; Sun, H. Iterative Adaptive Nonconvex Low-Rank Tensor Approximation to Image Restoration Based on ADMM. *J. Math. Imaging Vis.* **2019**, *61*, 627–642. [CrossRef]

44. Deledalle, C.A.; Denis, L.; Tabti, S.; Tupin, F. MuLoG, or how to apply Gaussian denoisers to multi-channel SAR speckle reduction? *IEEE Trans. Image Process.* **2017**, *26*, 4389–4403. [CrossRef] [PubMed]

45. Zhang, L.; Zhang, L.; Mou, X.; Zhang, D. FSIM: A feature similarity index for image quality assessment. *IEEE Trans. Image Process.* **2011**, *20*, 2378–2386. [CrossRef] [PubMed]

46. Wang, Z.; Bovik, A.C.; Sheikh, H.R.; Simoncelli, E.P. Image quality assessment: From error visibility to structural similarity. *IEEE Trans. Image Process.* **2004**, *13*, 600–612. [CrossRef] [PubMed]

47. Di Martino, G.; Poderico, M.; Poggi, G.; Riccio, D.; Verdoliva, L. Benchmarking framework for SAR despeckling. *IEEE Trans. Geosci. Remote Sens.* **2013**, *52*, 1596–1615. [CrossRef]

Article

Refocusing Swing Ships in SAR Imagery Based on Spatial-Variant Defocusing Property

Jin Wang [1], Xiangguang Leng [1,*], Zhongzhen Sun [1], Xi Zhang [2] and Kefeng Ji [1]

[1] College of Electronic Science and Technology, National University of Defense Technology, Changsha 410073, China; wangjinwj123@nudt.edu.cn (J.W.); sunzhongzhen14@nudt.edu.cn (Z.S.); jikefeng@nudt.edu.cn (K.J.)
[2] First Institute of Oceanography, Ministry of Natural Resources, Qingdao 266061, China; xi.zhang@fio.org.cn
* Correspondence: guang@nudt.edu.cn; Tel.: +86-15507487405

Abstract: Synthetic aperture radar (SAR) is an essential tool for maritime surveillance in all weather conditions and at night. Ships are often affected by sea breezes and waves, generating a three-dimensional (3D) swinging motion. The 3D swing ship can thereby become severely defocused in SAR images, making it extremely difficult to recognize them. However, refocusing 3D swing ships in SAR imagery is challenging with traditional approaches due to different phase errors at each scattering point on the ship. In order to solve this problem, a novel method for refocusing swing ships in SAR imagery based on the spatial-variant defocusing property is proposed in this paper. Firstly, the spatial-variant defocusing property of a 3D swing ship is derived according to the SAR imaging mechanism. Secondly, considering the spatial-variant defocusing property, each azimuth line of the SAR 3D swing ship image is modeled as a multi-component linear frequency modulation (MC-LFM) signal. Thirdly, Fractional Autocorrelation (FrAc) is implemented in order to quickly calculate the optimal rotation order set for each azimuth line. Thereafter, Fractional Fourier Transform (FrFT) is performed on the azimuth lines to refocus their linear frequency modulation (LFM) components one by one. Finally, the original azimuth lines are replaced in the SAR image with their focused signals to generate the refocused SAR image. The experimental results from a large amount of simulated data and real Gaofen-3 data show that the proposed algorithm can overcome the spatial-variant defocusing of 3D swing ships. Compared with state-of-the-art algorithms, our approach reduces the image entropy by an order of magnitude, leading to a visible improvement in image quality, which makes it possible to recognize swing ships in SAR images.

Keywords: synthetic aperture radar (SAR); spatial-variant defocusing; Fractional Fourier Transform (FrFT); Fractional Autocorrelation (FrAc); refocusing

Citation: Wang, J.; Leng, X.; Sun, Z.; Zhang, X.; Ji, K. Refocusing Swing Ships in SAR Imagery Based on Spatial-Variant Defocusing Property. *Remote Sens.* **2023**, *15*, 3159. https://doi.org/10.3390/rs15123159

Academic Editor: Dusan Gleich

Received: 26 April 2023
Revised: 14 June 2023
Accepted: 15 June 2023
Published: 17 June 2023

1. Introduction

Synthetic aperture radar (SAR) is capable of obtaining high-resolution images of the sea surface, even in adverse weather conditions or at night, which is not possible with passive sensors, such as optical cameras [1–3]. Ships are critical targets for maritime surveillance and their detection and identification in SAR images hold significant value for economic and military interests [4–6]. Researchers have proposed numerous algorithms for ship detection in SAR images, with satisfactory results [7–19]. However, it is very difficult to further recognize ships, because ships in motion at sea exhibit azimuthal defocusing in SAR imagery [20–29].

In ideal conditions, SAR emits electromagnetic waves in a uniform linear motion to illuminate a stationary target on the ground. The reflected signal from the target is then collected and compressed in the range and azimuth direction to generate a two-dimensional image. In practice, the trajectory of the SAR platform is non-ideal, due to the effects of turbulence and wind. In order to achieve high resolution imaging, researchers have proposed numerous autofocus algorithms [30]. However, these autofocus algorithms only compensate for phase errors caused by the non-ideal motion of the SAR platform, and not those caused by the ship's motion. As a result, the moving ship remains defocused in the SAR image.

In order to improve the recognition performance, we need to refocus SAR for moving ships to accurately obtain the structure of the ship. Since the motion of ships can be equated to the non-ideal motion of the SAR platform, SAR autofocus methods are also applicable to SAR moving ship refocusing. Classical SAR autofocus methods assume that each scattering point has the same phase error. There are two main types of autofocus algorithms. One class of methods for autofocusing is based on phase error function, such as map drift (MD) [31,32] and phase gradient autofocus algorithms (PGAs) [33,34]. Another class of methods is based on SAR image quality, such as fast minimum entropy phase compensation (FMEPC) algorithms [35–38] and maximum contrast algorithms [39–41]. In calm sea conditions, ships primarily move in linear motion through their own propulsion, and the phase error of each scattering point on the translational ship is the same. Therefore, the above methods can improve the image quality of SAR-translated ships. However, ships can swing as they are affected by wind and waves in rough seas [42–44]. When this occurs, the velocity of each scattering point on the ship is different, resulting in varying phase error for each part of the ship. Therefore, the above traditional methods cannot achieve well-focused performance on the SAR images of swing ships.

There are currently four types of methods with which to refocus 3D swing ships in SAR images. The first class of methods is based on the idea of splitting the image into subimages. Aron Sommer et al. [45] first split an SAR ship image into subimages and used a maximum contrast algorithm to estimate the phase error of each subimage. A second-order fit was then performed with the Levenberg–Marquardt algorithm in order to obtain the phase error of the entire SAR image, thus refocusing the SAR 3D swing ship image. Based on this, Li et al. [46] proposed a hybrid coordinate system and improved the calculation of the phase error for the subimages, which ultimately improved the speed of the algorithm. Yu et al. [47] divided the ship image into subimages with different sizes, according to the difference in the image entropy of a ship's different parts, and then used the PGA algorithm to compensate for each subimage. However, this type of approach must ensure that each subblock is small enough that the phase error within the subimage can be considered spatial-invariant phase error. Moreover, reducing the size of the subimage can lead to a decrease in the accuracy of the phase error estimation. In contrast, if the subimage is large, the phase error may still contain spatial-variant phase error.

The second class of methods aims to establish a fine spatial-variant phase error model. Brian et al. [48] constructed a range-variant phase error model that solved phase error through image quality optimization and used it to refocus the SAR images of rotating targets. Huang et al. [49] considered the azimuth-variant phase error and proposed a two-dimensional spatial-variant phase error compensation method based on maximum image contrast for refocusing ISAR images. However, the phase error model constructed using these methods may not represent the true phase error of swing ships.

The third class of methods uses ISAR techniques to select the time period when the ship's motion is stable. Based on this idea, Jia et al. [50] mapped the SAR defocused ship images into the ISAR equivalent echo domain. The phase error caused by the ship translation is first compensated. Thereafter, the optimal imaging time was selected using a maximum image contrast search method. Finally, a high-resolution image was obtained using an iterative adaptive method. Cao et al. [51] used the range-instantaneous Doppler (RID) algorithm in the equivalent ISAR echo domain to obtain a clear SAR ship image. However, this type of approach reduces the resolution of the SAR image, as it selects only a part of the time.

The final class of methods uses deep learning techniques to learn complex mapping relationships between defocused and focused SAR ship images. Lu et al. [52] improved the original Unet network based on SAR imaging characteristics. The trained network was able to refocus simulated moving targets, demonstrating the effectiveness of deep learning techniques in SAR moving target refocusing. Liu et al. [53] proposed an autofocus method based on ensemble learning, constructing a convolutional extreme learning machine (CELM) with which to estimate phase errors. This method achieved a good balance between focusing quality and processing speed. Hua et al. [54] considered this problem as a regression problem in the field of deep learning. A complex-valued neural network was then constructed and trained using simulated data to refocus SAR 3D swing ship images. However, deep learning techniques require a large dataset, and obtaining real samples can be difficult.

In fact, the main reason for SAR moving ship defocusing is the erroneous matched filtering in the azimuth direction, which results in residual linear frequency-modulated (LFM) signals in this direction of the SAR image. The LFM signal exhibits excellent energy aggregation after Fractional Fourier Transform (FrFT) at its optimal rotation order [55]. The application of FrFT to SAR moving targets has been extensively studied in recent years [56–63]. These methods are all specific to SAR echo data, while Pelich et al. [64] used this property to perform FrFT on each azimuth line of the single-look complex (SLC) SAR image, which can refocus the SAR translational ship very well. However, each azimuth line of the 3D swing ship is not a single LFM signal, as there are multiple scattering points with different velocities at each range cell. In this paper, the phenomenon that scattering points at different locations on a 3D swing ship exhibit different degrees of defocusing is defined as the spatial-variant defocusing property. Based on the spatial-variant defocusing property, a new SAR 3D swing ship image refocusing algorithm is proposed in this paper. The contribution of this paper mainly includes the following aspects:

1. By investigating the SAR imaging characteristics of 3D swing ships, this paper constructs a SAR imaging model for such ships and further reveals the spatial-variant defocusing property of 3D swing ships in SAR imaging.
2. The authors of this paper propose to model each azimuth line of the SAR 3D swing ship image as an MC-LFM signal to address the spatial-variant defocusing problem. This approach simplifies the task of refocusing the 3D swing ship by converting it into the task of refocusing the MC-LFM signal.
3. In order to refocus each azimuth line, the authors of this paper first use Fractional Autocorrelation (FrAc) to accelerate the precise estimate for the optimal rotation order set of the MC-LFM signal [65]. The azimuth line is then refocused by performing FrFT on each LFM component at its optimal rotation order. This achieves a good balance between focusing quality and processing speed.
4. We conduct sufficient experiments using simulated data and Gaofen-3 SAR images. The experimental results show that the proposed algorithm can overcome the spatial-variant defocusing property of 3D swing ships. Compared with state-of-the-art algorithms, our method reduces image entropy by an order of magnitude, which can improve the recognition of SAR ships. These experimental data are published at https://github.com/RefocusWang/SAR_Swing_Ship (accessed on 1 June 2023). Scholars interested in this issue can further study this data.

The remainder of this paper is organized as follows. Section 2 establishes the SAR imaging signal model for a 3D swing ship and analyzes its spatial-variant defocusing property. Section 3 introduces the method that the authors of this paper propose. Section 4 validates the performance of the proposed method using simulated SAR images and real high-resolution SAR images from Gaofen-3.

2. Geometry and Signal Model

The 3D swing of a ship is the pendulum motion of the ship around its own longitudinal, lateral, and vertical axes [27], which are referred to as roll, pitch, and yaw, respectively, as shown in Figure 1. The geometric relationship between the 3D swing ship and the SAR platform is shown in Figure 2. Firstly, a fixed space coordinate system $o - xyz$ is established to describe the absolute motion of the ship. Assume that the coordinate of the ship's center of gravity O in the fixed-space coordinate system is represented by $(x_0, y_0, 0)$. Taking the ship's center of gravity O as the origin, the bow direction as the axis X, a direction perpendicular to the bow direction and parallel to the plane of the ship's body as the axis Y, and a direction perpendicular to the plane of the ship's body as the axis Z, a ship-fixed coordinate system $O - XYZ$, which describes the positions of a ship's scattering points relative to the ship's center of gravity O, is established.

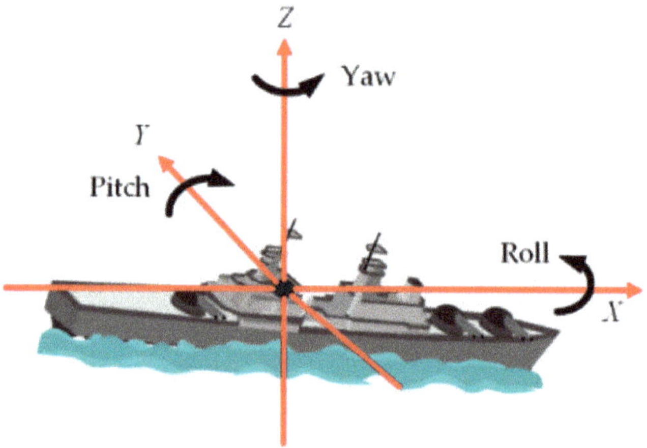

Figure 1. Schematic of the ship's 3D swing.

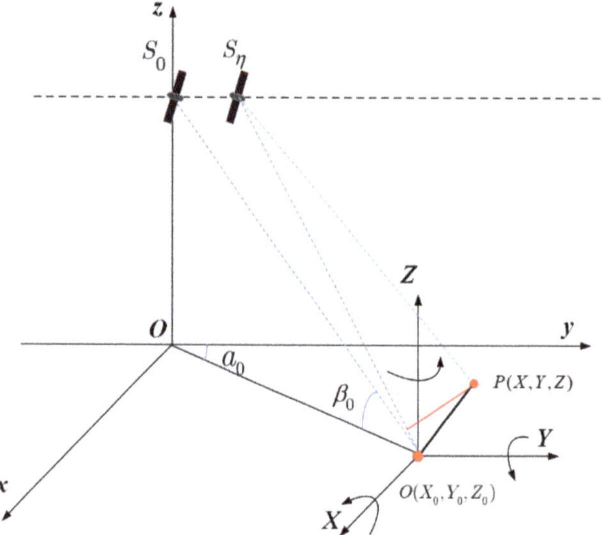

Figure 2. Geometric relationship between the 3D swing ship and the SAR platform.

Assume that at the initial moment, the ship-fixed coordinate system is parallel to the fixed-space coordinate system. At time η, the roll, pitch, and yaw angles of a certain scattering point $P(X_p, Y_p, Z_p)$ on the ship-fixed coordinate system are represented by θ_x, θ_y, and θ_z respectively.

$$\begin{cases} \theta_x = A_x \sin(w_x \eta + \varphi_x) \\ \theta_y = A_y \sin(w_y \eta + \varphi_y) \\ \theta_z = A_z \sin(w_z \eta + \varphi_z) \end{cases} \tag{1}$$

where A_x, A_y, A_z represent the maximum swing amplitude of roll, pitch, and yaw motions, w_x, w_y, w_z represent the angular frequencies of swing motions, and φ_x, φ_y, φ_z represent the values at the initial phase. Thereafter, the coordinates of point P in the fixed-space coordinate system can be expressed through the coordinate rotation matrix $\mathrm{Rot}(\theta_x, \theta_y, \theta_z)$ as

$$\begin{bmatrix} x_p(\eta) \\ y_p(\eta) \\ z_p(\eta) \end{bmatrix} = \mathrm{Rot}(\theta_x, \theta_y, \theta_z) \begin{bmatrix} X_p \\ Y_p \\ Z_p \end{bmatrix} + \begin{bmatrix} x_0 \\ y_0 \\ 0 \end{bmatrix} \tag{2}$$

$$\mathrm{Rot}(\theta_x, \theta_y, \theta_z) = \begin{bmatrix} 1 & 0 & 0 \\ 0 & \cos\theta_x & -\sin\theta_x \\ 0 & \sin\theta_x & \cos\theta_x \end{bmatrix} \begin{bmatrix} \cos\theta_y & 0 & \sin\theta_y \\ 0 & 1 & 0 \\ -\sin\theta_y & 0 & \cos\theta_y \end{bmatrix} \begin{bmatrix} \cos\theta_z & -\sin\theta_z & 0 \\ \sin\theta_z & \cos\theta_z & 0 \\ 0 & 0 & 1 \end{bmatrix} \tag{3}$$

The distance from the SAR platform to point P can be approximately expressed as

$$\begin{aligned} R(\eta) &\approx |OS_\eta| - |OP|\cos\langle OS_\eta, OP\rangle \approx |OS_\eta| - |OP|\cos\langle OS_0, OP\rangle \\ &= \sqrt{x_0^2 + (v_{sar}\eta - y_0)^2 + H^2 -} \\ &\quad (-\sin a_0 \cos \beta_0, -\cos a_0 \cos \beta_0, \sin \beta_0) * Rot(\theta_x, \theta_y, \theta_z)(X_p, Y_p, Z_p)^T \end{aligned} \tag{4}$$

From the above equation, it can be seen that an additional phase is created when the ship swings, compared with a stationary target. To simplify the analysis, when the ship is only yawing (i.e., $\theta_x = \theta_y = 0$), the additional phase error can be expressed as follows:

$$\begin{aligned} \varphi_{m-yaw} &= -\tfrac{4\pi}{\lambda}\left(\cos \beta_0 \sqrt{X_p^2 + Y_p^2} * \sin(\theta_z + a_0 + \arg(X_p + jY_p)) - Z_p \sin \beta_0\right) \\ &= -\tfrac{4\pi}{\lambda}\left(L_{zeff} * \sin(\sin(w_z t + \theta_z) + \theta_{z0rff}) - Z_p \sin \beta_0\right) \end{aligned} \tag{5}$$

where $L_{zeff} = \cos \beta_0 \sqrt{X_p^2 + Y_p^2}$ and $\theta_{z0eff} = a_0 + \arg(X_p + jY)$.

The above equation shows that the phase error caused by yaw is a sinusoidal function. As the ship swings slowly, over a period of more than ten seconds, the phase error can be approximated as a polynomial of no higher than third order using Taylor expansion within the synthetic aperture time (one second) of the space-borne SAR.

However, compared with the ship's translational motion, the ship's 3D swing is not a homogeneous motion, resulting in spatial-variant phase error. In the above equation, $\sqrt{X_p^2 + Y_p^2}$ represents the distance from the scattering point to the swing axis Z. This shows that the phase error of each scattering point is related to its distance from the swing axis when the ship yaws. Therefore, the 3D swing of the ship causes the phase error of each scattering point to be closely related to its spatial position, resulting in spatial-variant phase error. The spatial-variant phase error caused by the ship's 3D swing renders traditional autofocus algorithms ineffective. Consequently, refocusing 3D swing ship images is a very challenging problem.

3. Proposed Method

In this section, the proposed method for refocusing swing ships in SAR imagery based on the spatial-variant defocusing property is described. The flowchart of the proposed method is shown in Figure 3.

3.1. Selecting a Ship's Azimuth Lines

SAR usually obtains large-area images of the sea, and in order to refocus SAR moving ships, it is first necessary to obtain a subimage of the ship through ship detection algorithms. However, there are still sea surface backgrounds in the subimage, which need to be further removed in order to obtain the azimuth line set of the ship target. Only the signals in the azimuth line set need to be processed, which can effectively reduce the computational load.

Ships are typically made of metal, and their backscattering coefficients are strong, so a ship's azimuth lines can be selected based on the energy difference between the ship target and the sea surface background.

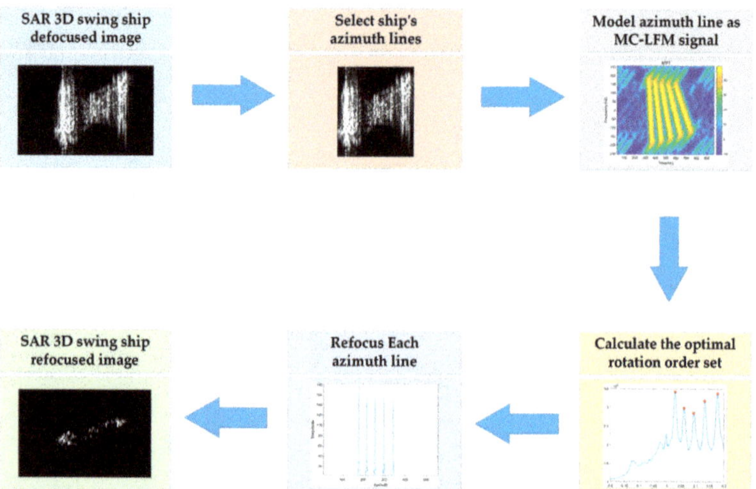

Figure 3. The flowchart of the proposed method.

For a SAR ship image $g(m,n)$ with a size of $M \times N$, the energy of the azimuth line at the n'th range cell $E(n)$ and the mean energy of all azimuth lines $\overline{E}$ can be defined as

$$E(n) = \sum_{m=1}^{M} |g(m,n)|^2 \tag{6}$$

$$\overline{E} = \frac{1}{N} \sum_{m=1}^{M} \sum_{n=1}^{N} |g(m,n)|^2 \tag{7}$$

where M represents the number of azimuth cells, N represents the number of range cells, and $g(m,n)$ is a two-dimensional complex matrix representing a single-look complex (SLC) SAR image.

As the energy of a ship's azimuth lines on the SAR image is much greater than the energy of the azimuth lines on the sea surface, the azimuth lines with energy greater than the average energy $\overline{E}$ are selected as the ship's azimuth line set S.

$$S = [g(:,n_1), g(:,n_2), \cdots, g(:,n_k)] \tag{8}$$

3.2. Modeling Each Azimuth Line as an MC-LFM Signal

From the analysis in Section 2, it is clear that the phase error of scattering points at different positions on the 3D swing ship in SAR imaging is not the same. The Equation (5) can be further derived as

$$\begin{aligned}\varphi_{m-yaw} &= -\tfrac{4\pi}{\lambda}\left(L_{zeff} * \sin(\sin(w_z t + \theta_z) + \theta_{z0rff}) - Z_p \sin\beta_0\right) \\ &= -\tfrac{4\pi}{\lambda}\left(L_{zeff} * (\sin(\sin(w_z t + \theta_z)) \cos\theta_{z0rff}) + \cos(\sin(w_z t + \theta_z)) \sin\theta_{z0rff} - Z_p \sin\beta_0\right)\end{aligned} \tag{9}$$

Since the swing period of the ship is much longer than the synthetic aperture time of space-borne SAR, the above equation can be approximated using the Taylor expansion as

$$\sin(A_z \sin(w_z t + \theta_z)) \approx A_z w_z t \tag{10}$$

$$\cos(A_z \sin(w_z t + \theta_z)) = 1 - \frac{A_z^2}{2} w^2 t^2 \tag{11}$$

$$\varphi_{m-yaw} = -\frac{4\pi}{\lambda} L_{zeff}\left(A_z \cos\theta_{z0rff} w_z t - \frac{A_z^2}{2}\sin\theta_{z0rff} w^2 t^2 + \sin\theta_{z0rff} - Z_p \sin\beta_0\right) \tag{12}$$

As shown in Equation (12), the phase error of a scattering point on the 3D swing ship is an LFM signal, where L_{zeff} is the distance between the scattering point and the swing axis. Therefore, the phase error of each scattering point on the 3D swing ship is the LFM signal with different modulation frequencies. After azimuth compression, each scattering point on the ship corresponds to a residual LFM signal in the SAR image due to the erroneous matched filtering. Based on the spatial-variant defocusing property of the SAR 3D swing ship image, each range cell in the SAR image can be modeled as an MC-LFM signal. Thus, the azimuth line $g(:, n_j)$ in the ship's azimuth line set S can be represented as

$$g(:, n_j) = \sum_{i=1}^{N} l_i(m) = \sum_{i=1}^{N} e^{j2\pi(\frac{1}{2}a_i m^2 + b_i m)} \tag{13}$$

where $l_i(m)$ is the i'th component of $g(:, n_j)$, with a modulation frequency of a_i and a center frequency of b_i.

3.3. Calculating the Optimal Rotation Order Set

FrFT decomposes a signal into the space of orthogonal basis functions consisting of LFM functions and is a generalized form of the Fourier Transform. The FrFT of a signal $x(t)$ can be expressed as

$$X_a(u) = \int_{-\infty}^{+\infty} x(t) K_a(t, u) dt \tag{14}$$

where the kernel function $K_a(t, u)$ is defined as

$$K_a(t, u) = \begin{cases} \sqrt{\frac{1 - j\cot a}{2\pi}} e^{j(\frac{1}{2}t^2 \cot a - ut \csc a + \frac{1}{2}u^2 \cot a)}, & a \neq n\pi \\ \delta(t + u), & a = 2n\pi \\ \delta(t - u), & a = (2n+1)\pi \end{cases} \tag{15}$$

where n is an integer, $a = p\pi/2$ is the rotation angle, and p is the rotation order of the FrFT.

In FrFT, the modulation frequency $\cot(-a)$ and initial frequency $u \csc a$ of the LFM basis function change as the rotation angle a changes from 0 to 2π. As a result, the signal is projected on different LFM basis functions as the rotation angle changes, and the magnitude of the projection value reflects the similarity of the signal to the different LFM basis functions. Therefore, by selecting the optimal rotation order in which to perform FrFT on the LFM signal, the LFM signal after FrFT will exhibit energy aggregation and become an impulse signal. In order to obtain the optimal rotation order in practical calculation, it is necessary to perform FrFT on the LFM signal at different rotation orders. The maximum value of the transformed signal energy in the two-dimensional parameter space (a, u) corresponds to the optimal rotation order of the LFM signal.

$$(a_{opt}, u_{opt}) = \underset{a,u}{\operatorname{argmax}} |X_a(u)|^2 \tag{16}$$

Based on the excellent aggregation capability of FrFT on the LFM signal, Pelich performed FrFT for each azimuth line on the SAR image at its optimal rotation order to obtain a clear SAR image of the moving ship. However, the azimuth line of a 3D swing ship is an MC-LFM signal. At the optimal rotation order, only one LFM component of the azimuth line can be refocused using FrFT, but the rest of the LFM components cannot be refocused. In order to refocus the MC-LFM signal, the optimal rotation order of each

LFM component must be obtained. The optimal rotational order for each component of the MC-LFM signal is combined as its optimal rotation order set in this paper. Fractional autocorrelation (FrAc) can effectively distinguish LFM signals with different modulation frequencies. Therefore, the authors of this paper use FrAc to calculate the optimal rotation order set of the MC-LFM signal.

The FrAc of the signal $x(t)$ can be expressed as

$$R_\beta(\rho) = e^{j\pi\rho^2 \cos\beta \sin\beta} \int x(t)\overline{x}(t - \rho\cos\beta)e^{-j2\pi t\rho \sin\beta} dt \tag{17}$$

where $\overline{x}$ represents the complex conjugate, ρ represents the delay factor, and β represents the rotation angle. The FrAc of the signal can be calculated using FrFT and inverse Fourier Transform.

$$R_\beta(\rho) = F_{u\to r}^{-1}\left\{\left|X^{\beta+\frac{\pi}{2}}(u)\right|^2\right\} \tag{18}$$

where F^{-1} represents the inverse Fourier Transform and $X^{\beta+\frac{\pi}{2}}(u)$ represents the signal $x(t)$ after FrFT at the rotation angle $\beta + 0.5\pi$.

The FrAc for one azimuth line $g(:, n_j)$ of the SAR 3D swing ship image $g(m, n)$ can be expressed as

$$R_\beta(\rho) = e^{j\pi\rho^2 \cos\beta \sin\beta} \sum_{i,i'=1}^{N} F_{t\to\rho \sin\beta}\{l_i(t)l_{i'}^*(m - \rho\cos\beta)\} \tag{19}$$

For $i = i'$, the auto-term of $R_\beta(\rho)$ can be expressed as

$$R_\beta(\rho)_{auto} = \sum_{i=1}^{N} e^{j2\pi b_i\rho \cos\beta}\delta(\rho\sin\beta - a_i\rho\cos\beta) \tag{20}$$

For $i \neq i'$, the cross-term of $R_\beta(\rho)$ can be expressed as

$$
\begin{aligned}
R_\beta(\rho)_{cross} = &\sum_{\substack{i,i'=1, i<i' \\ a_i = a_{i'}}} e^{j\pi(b_i+b_{i'})\rho\cos\beta} \times [\delta(\rho\sin\beta - a_i\rho\cos\beta - b_{ii'}) + \delta(\rho\sin\beta - a_i\rho\cos\beta + b_{ii'})] \\
&+ \sum_{\substack{i,i'=1:1i<i' \\ a_i \neq a_{i'}}} \frac{2}{\sqrt{a_{ii'}}}e^{j\frac{2\pi\rho}{a_{ii'}}[b_i(\sin\beta - a_{i'}\cos\beta) - b_{i'}(\sin\beta - a_i\cos\beta)]} \\
&\times \cos\left(\frac{\pi}{a_{ii'}}\left[\rho^2(\sin\beta - a_i\cos\beta)(\sin\beta - a_{i'}\cos\beta) + b_{ii'}^2 - \frac{a_{ii'}}{4}\right]\right)
\end{aligned}
\tag{21}
$$

From the expressions of the auto-term and cross-term, it can be seen that the energy of the auto-term is concentrated on the straight lines passing through the origin, where the slopes of these lines represent the different modulation frequencies of the LFM component $l_i(m)$. On the other hand, most of the cross-term energy is concentrated on straight lines that do not pass through the origin, and cross-term energy with different modulation frequencies can be ignored, compared with the auto-term. The energy of the signal after FrAc is defined as

$$E_\rho(\beta) = \int_{-\infty}^{+\infty} |R(\beta)(\rho)|^2 d\rho \tag{22}$$

When the rotation order β matches the modulation frequency of one LFM component, the energy of the signal after FrAc has a peak, and the optimal rotation order corresponding to this component in FrFT is $\alpha = \beta + 0.5\pi$. Therefore, the energy distribution of the signal after FrAc at different rotation angles can be used to calculate the parameters of each component for the MC-LFM signal.

The scattering points on the 3D swing ship have different velocities, but the velocities are relatively close in magnitude, which places the optimal rotation order of each LFM component within a certain range. Therefore, the optimal rotation order set is calculated using a two-step method and Algorithm 1 illustrates the specific implementation. First, the energy $E_\rho(\beta)$ of the MC-LFM signal after FrAc with a greater step size Δ in the interval $\beta_1 = linspace(-\frac{\pi}{2}, \frac{\pi}{2}, \Delta)$ is calculated, and the rotation order β_{coarse} in which the position of the maximum value of $E_\rho(\beta)$ is obtained. Since there are multiple optimal rotational orders of the MC-LFM signal near the rotation order β_{coarse}, the energy $E_\rho(\beta)$ of the MC-LFM signal after FrAc is further calculated using a smaller step Δ' in the subinterval $\beta_2 = linspace(\beta_{coarse} - \Delta, \beta_{coarse} + \Delta, \Delta')$. As a result, the peaks of the energy $E_\rho(\beta)$ correspond to the precise optimal rotation order of each LFM component. Finally, the corresponding rotation orders are sorted according to the magnitude of the peak to obtain the optimal rotation order set α_{fine} of the MC-LFM signal.

$$\beta_{coarse} = \underset{\beta}{\mathrm{argmax}}(E_\rho(\beta)) \tag{23}$$

$$\alpha_{fine} = [\alpha_1, \alpha_2, \cdots, \alpha_q] \tag{24}$$

Algorithm 1 The algorithm of calculating the optimal rotation order set

---

    Input: A ship's azimuth line $g(:, n_j)$
    Output: The optimal rotation order set $\alpha_{fine} = [\alpha_1, \alpha_2, \cdots \alpha_q]$
    $\beta_1 = linspace(-0.5\pi, 0.5\pi, \Delta)$
    for β in β_1
       $E_\rho(\beta) = \int_{-\infty}^{+\infty} abs(FrAc(g(:, n_j), \beta))^2 d\rho$
    end for
    $\beta_{coarse} = \underset{\beta}{\mathrm{argmax}}(E_\rho(\beta))$
    $\beta_2 = linspace(\beta_{coarse} - \Delta, \beta_{coarse} + \Delta, \Delta')$
    for β in β_2
        $E_\rho(\beta) = \int_{-\infty}^{+\infty} abs(FrAc(g(:, n_j), \beta))^2 d\rho$
    end for
    $[\beta_1, \beta_2, \cdots \beta_q] = Sort(Findpeaks(E_p(\beta)))$
    $\alpha_{fine} = [\alpha_1, \alpha_2, \cdots, \alpha_q] = [\beta_1 + 0.5\pi, \beta_2 + 0.5\pi, \cdots, \beta_q + 0.5\pi]$

3.4. Refocusing Each Azimuth Line

The components of the MC-LFM signal are difficult to distinguish in the time domain, but after performing FrFT at each component's optimal rotation order, each component will become an impulse signal. Therefore, after obtaining the optimal rotation order set for the azimuth line, the azimuth line can be refocused using FrFT and Algorithm 2 illustrates the specific implementation.

First, FrFT is performed at the optimal rotation order α_1 on the azimuth line $g(:, n_j)$, and the LFM component corresponding to this optimal rotation order α_1 will be refocused. All peaks of the signal $g(:, n_j)_{FrFT}$ are detected, and if the peak magnitude of the signal $g(:, n_j)_{FrFT}$ is close to its maximum value, the peak is the refocused signal of the corresponding LFM component. In the fractional-order domain, each refocused signal is filtered out with a narrowband filter and accumulated to obtain the refocused signal $g(:, n_j)_{refocus\_1}$ of the corresponding LFM component at optimal rotational order α_1. The residual signal $g(:, n_j)_{left}$ is further transformed to the time domain using inverse FrFT.

$$g(:, n_j)_{FrFT} = FrFT(g(:, n_j), \alpha_i) \tag{25}$$

$$[p_1, p_2, \cdots p_K] = FindPeak(abs(g(:, n_j)_{FrFT}) > 0.7 * max(abs(g(:, n_j)_{FrFT}))) \tag{26}$$

$$Win(p_k) = \begin{cases} 1, & p_k - 1/2 < p_k < p_k + 1/2 \\ 0, & otherwise \end{cases} \tag{27}$$

$$g(:,n_j)_{refocus\_i} = \sum_{k=1}^{K} g(:,n_j)_{FrFT} * Win(p_k) \tag{28}$$

$$g(:,n_j)_{left} = FrFT(g(:,n_j)_{FrFT} - g(:,n_j)_{refocus\_i'} - \alpha_i) \tag{29}$$

Thereafter, FrFT is performed on the residual signal $g(:,n_j)_{left}$ at the second optimal rotation order α_2. If the maximum value of the refocused signal after the second transformation is much less than that of the first transformation, the loop is terminated. Otherwise, the refocused signal $g(:,n_j)_{refocus\_2}$ is continually filtered. By repeating this operation, each refocused signal obtained is accumulated to obtain the entire refocused signal $g(:,n_j)_{refocus}$ from the MC-LFM signal.

$$g(:,n_j)_{refocus} = g(:,n_j)_{refocus\_1} + g(:,n_j)_{refocus\_2} + \cdots + g(:,n_j)_{i-1} \tag{30}$$

If the energy of the refocused signal $g(:,n_j)_{refocus}$ is less than a certain proportion of the energy of the original signal $g(:,n_j)$, this indicates that there are still LFM components outside of the current subinterval. In this case, the subinterval is further expanded, the optimal rotation order set is recalculated, and the refocusing process is repeated as described above until the energy of the refocused signal meets the predetermined threshold. Finally, by replacing the corresponding original azimuth lines in the SAR image with the refocused azimuth lines, a clear SAR image of the 3D swing ship can be obtained.

Algorithm 2 The specific implementation of refocusing one azimuth line

---

Input: Optimal rotation order set $\alpha_{fine} = [\alpha_1, \alpha_2, \cdots \alpha_q]$ and azimuth line $g(:,n_j)$
Output: Refocused azimuth line $g(:,n_j)_{refocuse}$
for $i = 1,2,\cdots,q$
    $g(:,n_j)_{FrFT} = FrFT(g(:,n_j),\alpha_i)$
    $[p_1,p_2,\cdots p_K] = FindPeak(abs(g(:,n_j)_{FrFT}) > 0.7 * max(abs(g(:,n_j)_{FrFT})))$
    $g(:,n_j)_{refocus\_i} = \sum_{k=1}^{K} g(:,n_j)_{FrFT} * Win(p_k)$
    if $max(abs(g(:,n_j)_{refocus\_i})) < 0.7 * max(abs(g(:,n_j)_{refocus\_1}))$
        break
    else
        $g(:,n_j)_{left} = FrFT(g(:,n_j)_{FrFT} - g(:,n_j)_{refocus\_i'} - \alpha_i)$
        $g(:,n_j) = g(:,n_j)_{left}$
    end if
end for
$g(:,n_j)_{refocus} = g(:,n_j)_{refocus\_1} + g(:,n_j)_{refocus\_2} + \cdots + g(:,n_j)_{i-1}$

3.5. Numerical Analysis of Computional Burden

Suppose the size of a SAR 3D swing ship image is $M \times N$, where M is the number of range cells and N is the number of azimuth cells. The time complexity, for both the inverse FFT and the Discrete FrFT, for a signal of length N is $O(N \log_2 N)$. Therefore, the time complexity required to calculate the energy of an azimuth line after FrAc at each rotation order is $2O(N \log_2 N)$. If the coarse step is a_1 and the fine step is a_2, then the time complexity required to compute the optimal rotation order set for the azimuth line of the i'th range cell is C_1.

$$C_1 = 2 * (\frac{2}{a_1} + \frac{2a_1}{a_2})O(N \log_2 N) \tag{31}$$

If the azimuth line consists of K_i LFM signals with different tuning frequencies, the time complexity required to refocus the azimuth line is C_2.

$$C_2 = C_1 + K_i O(N \log_2 N) = (K_i + 4\frac{a_2 + a_1^2}{a_1 a_2})O(N \log_2 N) \tag{32}$$

Thus, the time complexity C_3 required to refocus the entire SAR image can be expressed as

$$C_3 = M * C_2 = (\sum_{i=1}^{M} K_i + 4M\frac{a_2 + a_1^2}{a_1 a_2})O(N \log_2 N) \tag{33}$$

If the optimal rotation order set of the azimuth line is calculated using the method in [56], the required time complexity C_4 can be expressed as

$$C_4 = K_i * (\frac{2}{a_1} + \frac{2a_1}{a_2})O(N \log_2 N) \tag{34}$$

For a 3D swing ship, there are multiple scattering points at each range cell, and K_i is much greater than 2. Therefore, the proposed algorithm can reduce the time complexity.

4. Experiments

4.1. Refocusing the MC-LFM Signal of One Azimuth Line

For SAR 3D swing ship images, there are several different LFM signals in each range cell. Therefore, in order to refocus a SAR 3D swing ship image, it is first necessary to refocus the MC-LFM signal of each range cell. The authors of this paper conduct a simulation experiment in order to validate the refocusing performance of the algorithm for MC-LFM signals. Table 1 shows the simulation parameters for the radar system, and Table 2 shows the azimuth velocity value of each scattering point in the same range cell.

Table 1. The SAR system parameters of the simulation experiment.

Parameter	Value
Carrier Frequency	3 GHZ
Pulse Repetition Frequency	188 HZ
Band Width	150 MHZ
Platform Height	3000 m
Antenna Length	2 m
Platform Velocity	150 m/s
Pulse Width	1.5 μs

Table 2. The azimuth velocity value of each scattering point.

Scattering Point	P1	P2	P3	P4	P5
Azimuth Velocity	5 m/s	10 m/s	15 m/s	20 m/s	25 m/s

The SAR images of these five scattering points, as they move individually, are shown in Figure 4. It can be observed that, due to the various azimuth velocities, each scattering point exhibits a different degree of defocusing. Specifically, the greater the velocity of the scattering point, the more severe the defocus. The SAR image of these five scattering points when stationary is shown in Figure 5a, and the SAR image of these five scattering points as they move together is shown in Figure 5b. The azimuthal profile of the range cell where these five scattering points are located is shown in Figure 6. It can be seen that the signals of the five scattering points are gathered together, so it is difficult to distinguish the signal of each scattering point. As a result, the problem of azimuth-variant defocusing is difficult to overcome using the split-block method.

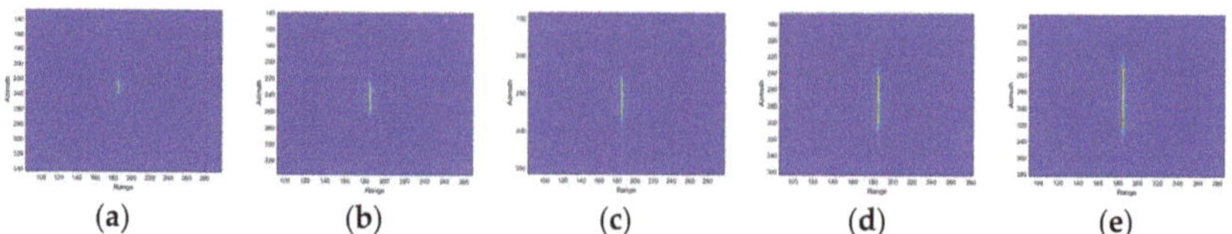

(a) (b) (c) (d) (e)

Figure 4. SAR images of different azimuth velocity scattering points: (**a**) SAR image of point P1. (**b**) SAR image of point P2. (**c**) SAR image of point P3. (**d**) SAR image of point P4. (**e**) SAR image of point P5.

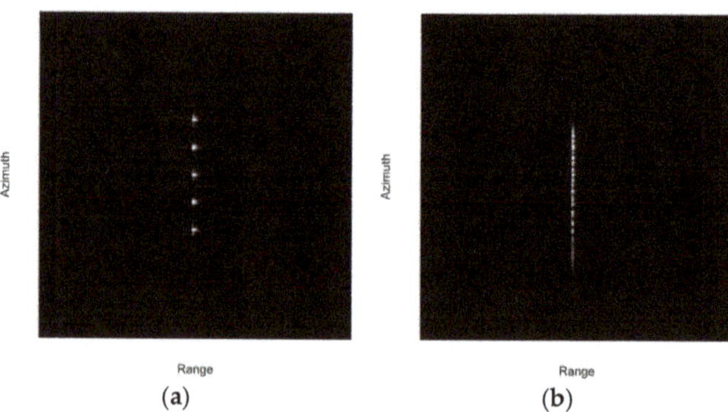

(a) (b)

Figure 5. SAR image of five points: (**a**) Five points when stationary. (**b**) Five points when in motion.

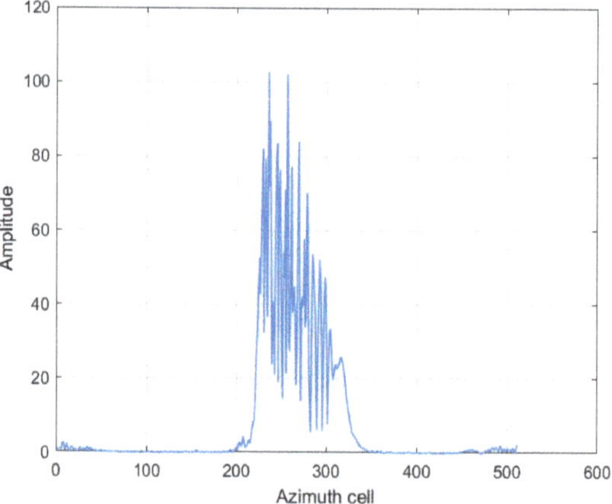

Figure 6. Azimuth profile of the range cell where the five points are located.

The two-dimensional distribution of the time–frequency spectrum obtained using FrFT on the azimuth line where these five scattering points are located is shown in Figure 7a, and its three-dimensional distribution is shown in Figure 7b. It can be observed that there are five energy concentration points in this figure, and the scattering points, each with a different velocity, are well distinguished from one another after FrFT. Moreover, the defocused signal of each scattering point becomes the impulse signal at its respective optimal rotation order, which is equivalent to being re-matched. The correlation between the signal energy after FrAc and the rotation order when a coarse step is used is illustrated in Figure 8a. Similarly, this relationship is shown in Figure 8b when a fine step is used. Five distinct peaks can be seen in Figure 8b, and the positions of these peaks correspond to the optimal rotation order set. The result of performing FrFT at one of the optimal rotation orders on the azimuth line is shown in Figure 9, where a focused impulse signal appears.

Thus, by performing FrFT on the MC-LFM signal in its optimal rotational order set, and then accumulating the refocused signal of each LFM component, the original azimuth line can be refocused. This is the main idea of the algorithm proposed in this paper, and the specific details were explained in the previous section. The results of the defocused SAR image in Figure 5b after processing with PGA, FMEPC, and Pelich's method, and the proposed algorithm are shown in Figure 10. It can be seen that only one scattering point is refocused when using PGA, FMEPC, and Pelich's method, while the proposed algorithm can refocus all five scattering points. This demonstrates the effectiveness of the proposed algorithm in refocusing the MC-LFM signal. For a SAR image of a 3D swing ship, a clear image can be obtained by refocusing each of its azimuth lines. The following experiments show the performance of the proposed algorithm in refocusing SAR 3D swing ship images.

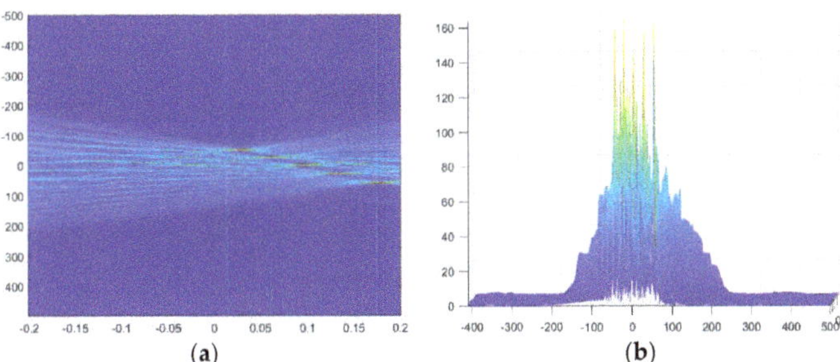

Figure 7. Time−frequency diagram of defocusing signal after FrFT: (**a**) Two-dimensional distributed time−frequency graph. (**b**) Three-dimensional distributed time–frequency graph.

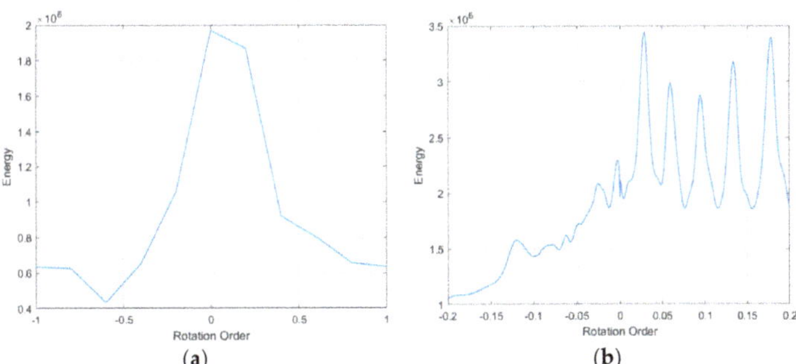

Figure 8. The relationship between the energy of the signal after FrAc and the rotation order. (**a**) Coarse search. (**b**) Fine search.

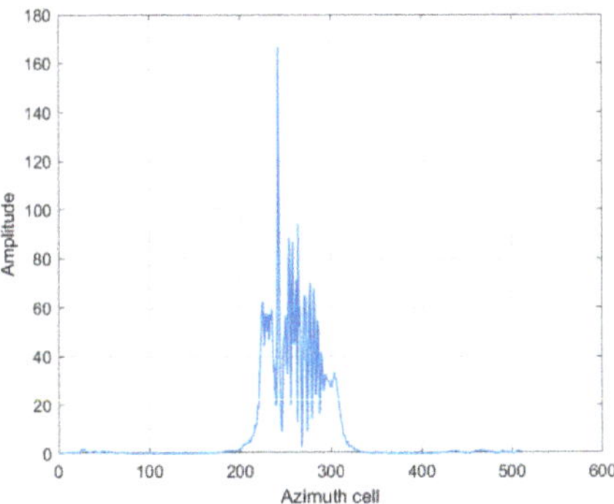

Figure 9. The result of performing FrFT on the signal at one optimal rotation order.

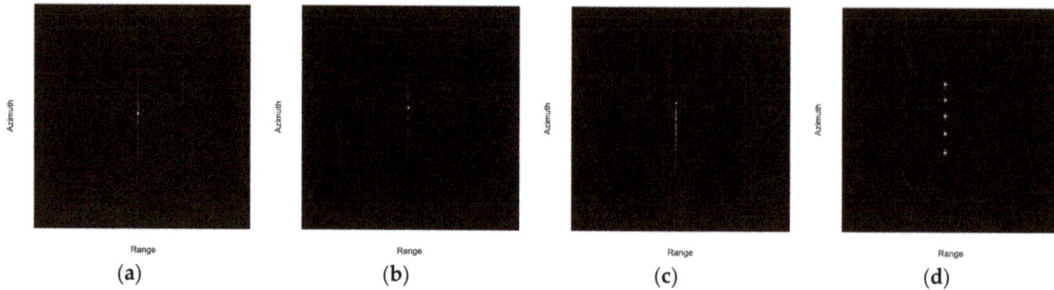

Figure 10. The performance of different methods on the simulated SAR image. (**a**) PGA. (**b**) FMEPC. (**c**) Pelich's method. (**d**) The proposed method.

4.2. Refocusing Simulated SAR Images of 3D Swing Ship

In order to simulate the SAR image of a 3D swing ship, the 3D swing parameters of the ship in this experiment are shown in Table 3, with an initial phase of 0.5π. Simulated SAR images of the rolling, pitching and yawing ship, as well as the corresponding SAR images when the ship is stationary, are shown in Figure 11. It can be seen that, compared with those of a stationary ship, SAR 3D swing ship images are severely defocused, making it difficult to distinguish its scattering points. Although each scattering point on the ship has the same amplitude and frequency of swing, the velocity of each scattering point is also related to its distance from the swing axis, due to the particular nature of the swing motion. As a result, the velocity of each scattering point on the ship is no longer the same, resulting in spatial-variant defocusing. As can be seen from Figure 11, the further the point on the ship is from the swing axis, the more severe the defocus appears, and the final SAR imaging results form an 'X' shape.

Table 3. The 3D swing parameters of a simulated ship.

	Roll	Pitch	Yaw
Double Amplitude (deg)	17.2	3.4	38
Average Period (s)	12.2	6.7	14.2

The results of refocusing simulated SAR images using the PGA, FMEPC, and Pelich's methods, and the algorithm proposed in this paper are shown in Figures 12–14. It can be seen that the PGA and FMEPC methods cannot refocus the SAR image of a 3D swing ship. This is because the phase error of each part of the 3D swing ship is different, while the PGA and FMEPC methods use a unified phase error to compensate. Pelich's method can overcome range-variant defocusing because it processes each azimuth line of the ship separately. When the ship pitches, the phase error of each scattering point on the ship is related to its proximity to the transverse axis, exhibiting only range-variant defocusing. Thus, Pelich's method is able to refocus the SAR image of a pitching ship. However, when the ship experiences yaw or roll, both azimuth-variant and range-variant defocusing exist. In such situations, Pelich's method is only capable of refocusing certain scattering points, while others remain defocused. The proposed method in this paper models each azimuth line as an MC-LFM signal and processes each azimuth line of the ship separately. As a result, it can overcome both azimuth-variant and range-variant defocusing. The severely defocused SAR images of the 3D swing ship are successfully refocused to 57 scattering points, as shown in Figures 12–14. The experimental results from simulated images show that the proposed method effectively refocuses SAR 3D swing ship imagery with high accuracy. It is worth noting that the refocused SAR ship images obtained using the proposed method still exhibit some deformation compared with the corresponding stationary ship. This is due to the fact that, in SAR imaging, velocity along the range direction causes targets to shift additionally in the azimuth direction. When the ship undergoes 3D swing, different velocities along the range direction of each scattering point result in different azimuth shifts, leading to distortions in the final SAR imaging of the ship. Table 4 shows the image entropy of simulated 3D swing ships after refocusing with different algorithms. From the average entropies presented in the table, it can be seen that the average entropy of the proposed algorithm is much lower than those of other algorithms.

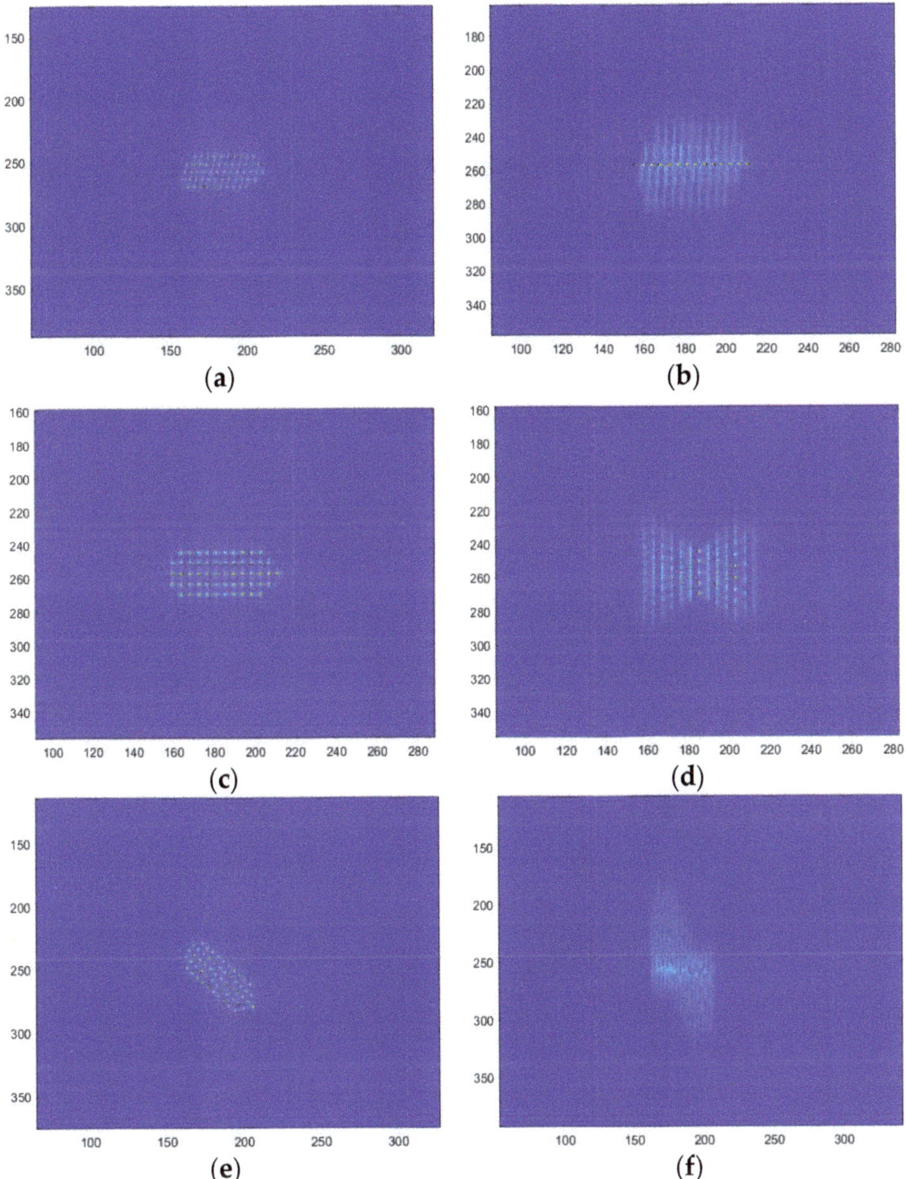

Figure 11. Simulated SAR swing ship images: (**a**) The simulated SAR image of a stationary ship target corresponding to the roll motion. (**b**) The simulated SAR image of a rolling ship. (**c**) The simulated SAR image of a stationary ship target corresponding to the pitch motion. (**d**) The simulated SAR image of a pitching ship. (**e**) The simulated SAR image of a stationary ship target corresponding to the yaw motion. (**f**) The simulated SAR image of a yawing ship.

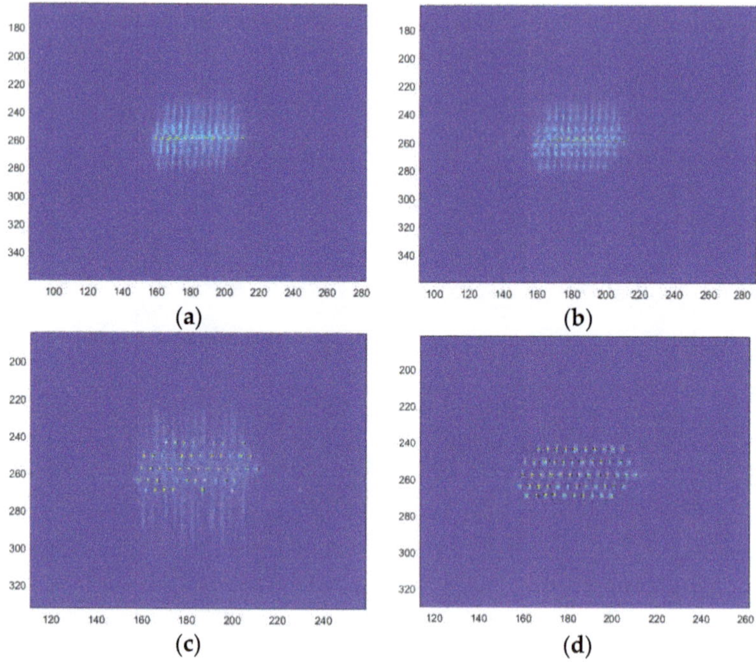

Figure 12. The performance of different methods on the simulated SAR rolling ship image: (**a**) PGA. (**b**) FMEPC. (**c**) Pelich's method. (**d**) The proposed method.

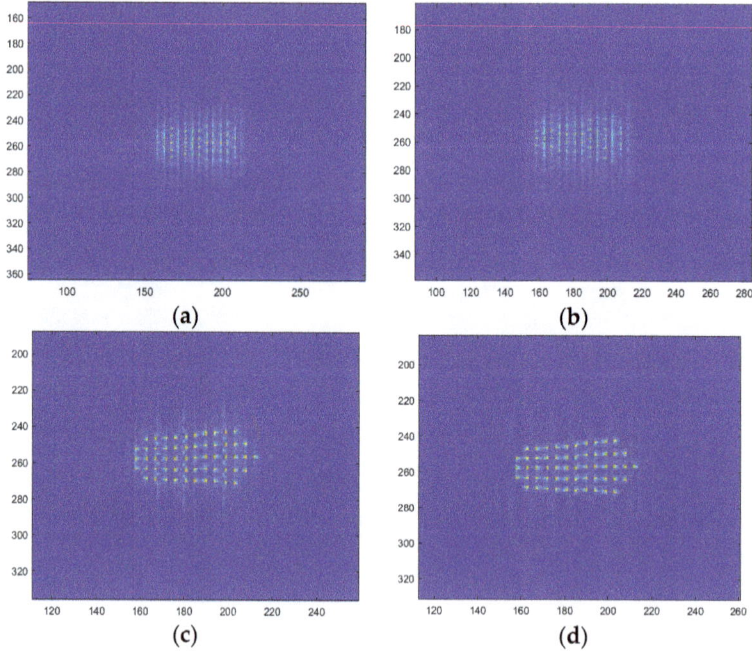

Figure 13. The performance of different methods on the simulated SAR pitching ship image: (**a**) PGA. (**b**) FMEPC. (**c**) Pelich's method. (**d**) The proposed method.

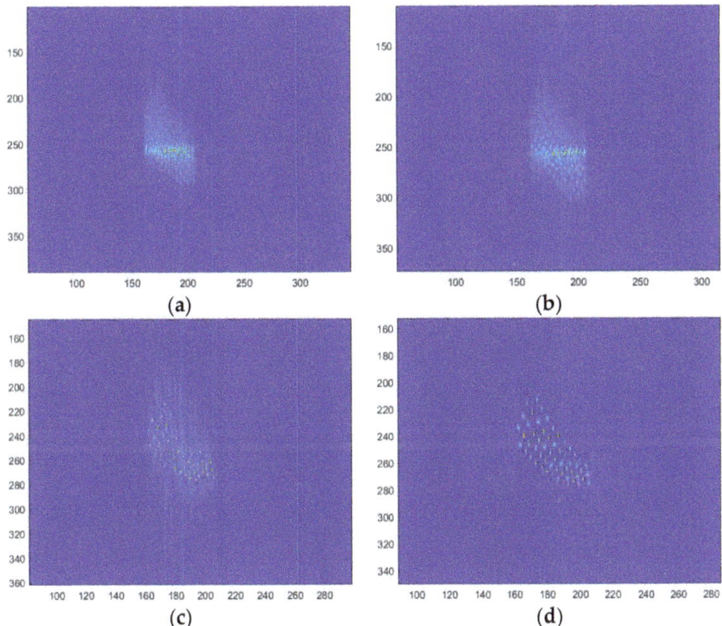

Figure 14. The performance of different methods on the simulated SAR yawing ship image: (**a**) PGA. (**b**) FMEPC. (**c**) Pelich's method. (**d**) The proposed method.

Table 4. The image entropies of different algorithms on simulated SAR swing ship images.

Simulated Image	Rolling Ship	Pitching Ship	Yawing Ship	Average
Original Image	7.05	6.57	7.67	7.10
PGA	7.03	6.63	7.45	7.04
FMEPC	6.90	6.49	7.32	6.90
Pelich's Method	6.40	6.04	6.62	6.35
Proposed Method	5.50	5.95	5.57	5.67

4.3. Refocusing Gaofen-3 SAR Images of a 3D Swing Ship

Gaofen-3 is a Chinese C-band polarimetric synthetic aperture radar satellite launched in August 2016. It includes multiple imaging modes, with a resolution of one meter in sliding spotlight mode. However, severe defocusing occurs when imaging moving ships in this mode. In order to further validate the performance of the proposed algorithm, the authors of this paper also conduct experiments on seven defocused SAR swing ship images taken using Gaofen-3 in sliding spotlight mode. Each SAR image is processed separately using the PGA, FMEPC, and Pelich's methods, and the proposed algorithm. Table 5 shows the system parameters of Gaofen-3 in sliding spotlight (SL) mode. The experimental results are shown in Figures 15–21. It can be seen that each ship has a different degree of defocus at different parts, exhibiting severe spatial-variant defocusing and making it impossible to determine the true shape of the ship. After being processed using the PGA and FMEPC methods, some parts of the defocused ships are refocused, while the rest are still defocused. When using Pelich's method, the range-variant defocusing is alleviated, but there are still many defocused scattering points. After applying the algorithm proposed in this paper, it can be observed that all seven ships show clear structural characteristics, making it easier to recognize them.

Table 5. The system parameters of Gaofen-3, SL mode.

Parameter	Value
Carrier frequency (GHZ)	5.4
Platform velocity (m/s)	7567
Band width(MHZ)	240
Pulse Width (μs)	45
Pulse repetition frequency (Hz)	3738

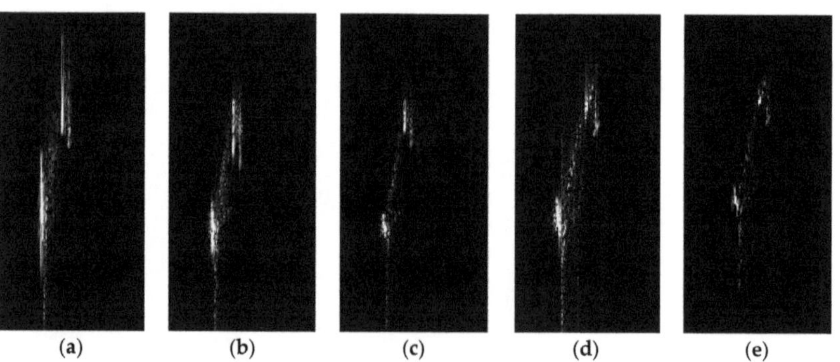

Figure 15. The performance of different methods on swing ship 1: (**a**) SAR subimage. (**b**) PGA. (**c**) FMEPC. (**d**) Pelich's method. (**e**) The proposed method.

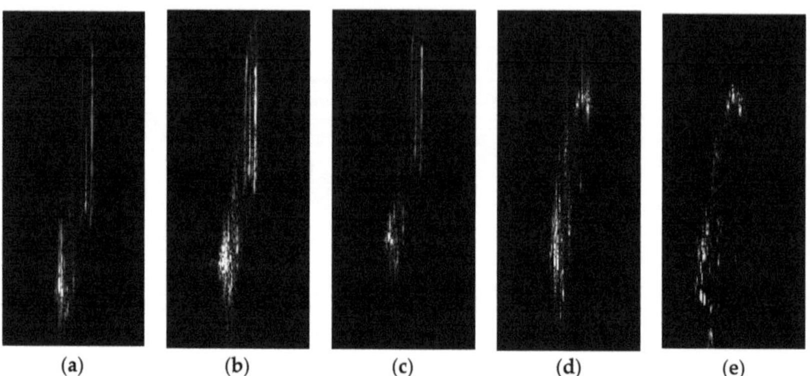

Figure 16. The performance of different methods on swing ship 2: (**a**) SAR subimage. (**b**) PGA. (**c**) FMEPC. (**d**) Pelich's method. (**e**) The proposed method.

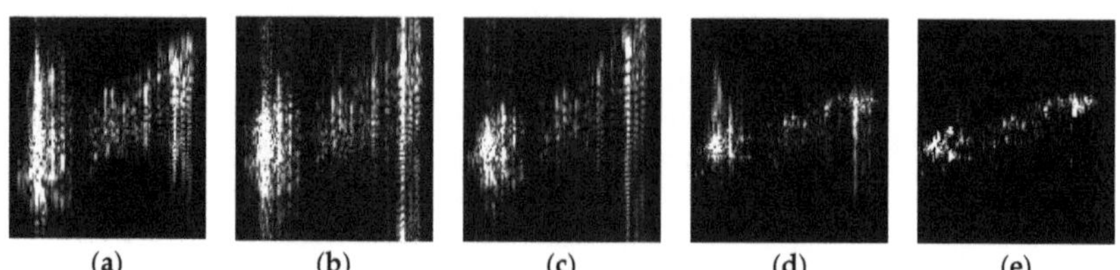

Figure 17. The performance of different methods on swing ship 3: (**a**) SAR subimage. (**b**) PGA. (**c**) FMEPC. (**d**) Pelich's method. (**e**) The proposed method.

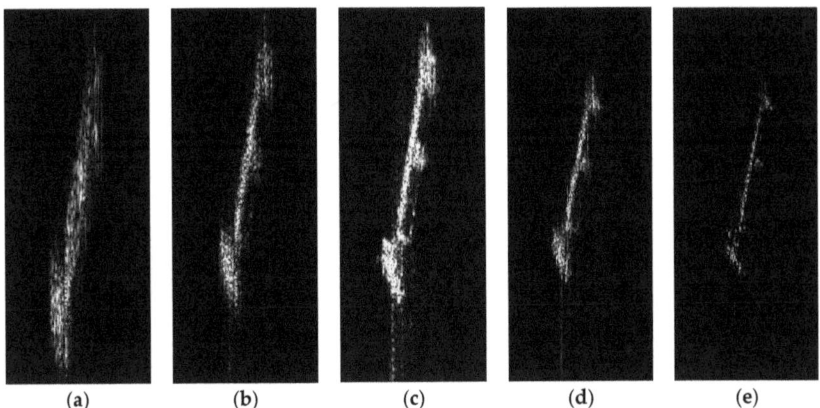

Figure 18. The performance of different methods on swing ship 4: (**a**) SAR subimage. (**b**) PGA. (**c**) FMEPC. (**d**) Pelich's method. (**e**) The proposed method.

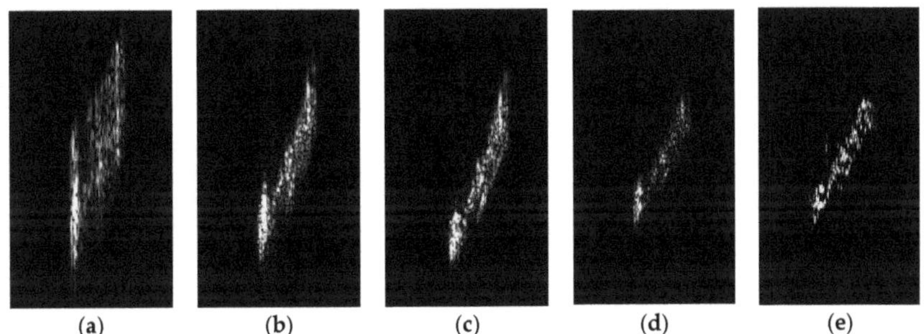

Figure 19. The performance of different methods on swing ship 5: (**a**) SAR subimage. (**b**) PGA. (**c**) FMEPC. (**d**) Pelich's method. (**e**) The proposed method.

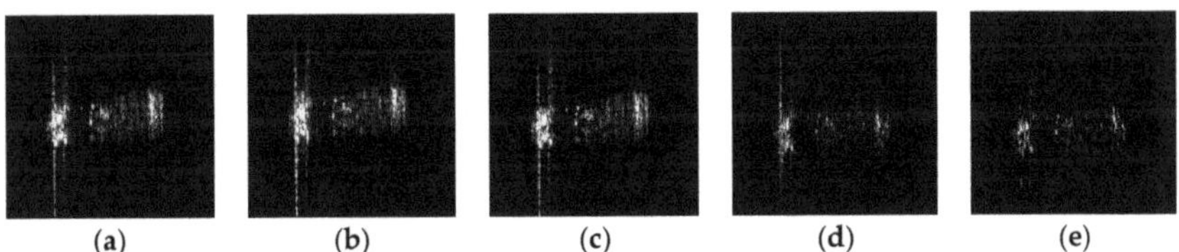

Figure 20. The performance of different methods on swing ship 6: (**a**) SAR subimage. (**b**) PGA. (**c**) FMEPC. (**d**) Pelich's method. (**e**) The proposed method.

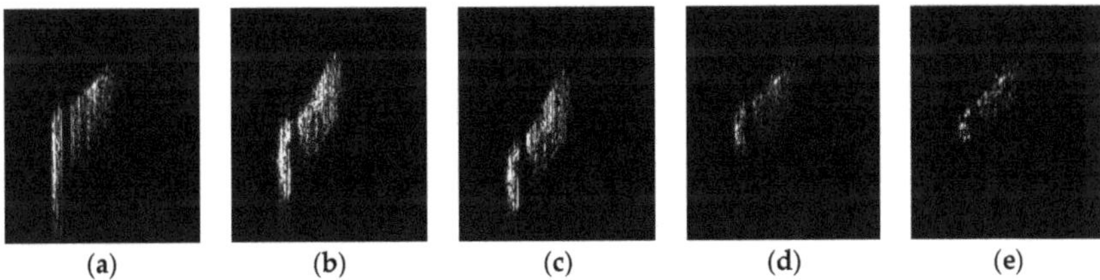

Figure 21. The performance of different methods on swing ship 7: (**a**) SAR subimage. (**b**) PGA. (**c**) FMEPC. (**d**) Pelich's method. (**e**) The proposed method.

In order to quantitatively evaluate focusing quality with different algorithms, the authors of this paper use image entropy as a metric. Image entropy can be used to characterize the aggregation of strong scattering points in an image. The greater the image entropy, the more severe the defocusing of ships, while a lower image entropy indicates a better focusing quality of ships. The definition of image entropy is as follows

$$E(I) = \sum_{m=0}^{M-1} \sum_{n=0}^{N-1} \frac{|I(m,n)|^2}{S} \ln \frac{S}{|I(m,n)|^2} \tag{35}$$

where $I(m,n)$ represents the complex scattering intensity of a SAR complex image, m represents the azimuth position, n represents the range position, and S represents the total energy of the image.

$$S = \sum_{m=0}^{M-1} \sum_{n=0}^{N-1} |I(m,n)|^2 \tag{36}$$

Table 6 shows the original image entropy of the seven SAR ship images and their image entropies after being processed using different algorithms. The average image entropy of the proposed algorithm is 5.94, which is an order of magnitude lower than the average image entropies obtained using other algorithms, indicating its superior focusing quality over other methods.

Table 6. The image entropy of different algorithms.

Real SAR Image	Ship1	Ship2	Ship3	Ship4	Ship5	Ship6	Ship7	Average
Original Image	7.94	6.94	9.22	7.19	8.03	7.27	7.98	7.80
PGA	7.24	6.35	8.75	6.97	7.40	7.24	7.36	7.33
FMEPC	6.51	5.82	8.30	6.52	6.95	7.22	7.23	6.94
Pelich's Method	6.53	6.22	8.00	5.99	6.80	7.14	6.81	6.78
Proposed Method	5.71	5.42	7.12	5.16	5.80	6.40	6.0	5.94

5. Conclusions

This research proposed a new method for refocusing swing ships in SAR imagery. We first derived the spatial-variant defocusing property of SAR swing ship images from SAR imaging principles. Based on the spatial-variant defocusing property, each azimuth line of a SAR swing ship image was modeled as an MC-LFM signal. The optimal rotation order for each MC-LFM signal was quickly calculated using FrAC, and then FrFT was performed on each LFM component individually in order to obtain the refocused azimuth lines. Finally, the refocused signal was used to replace the original azimuth lines of the SAR image. Experiments on both simulated and Gaofen-3 SAR images demonstrated that the proposed algorithm can effectively refocus the SAR images of 3D swing ships. It should be noted that, as the synthetic aperture time becomes longer, scattering points on the ship will become a nonlinear frequency modulation signal in the SAR image. Therefore, refocusing SAR images of swing ships with long synthetic aperture times needs to be studied in the future. In addition, the algorithm proposed in this paper can further improve the processing speed with parallel processing.

Author Contributions: Conceptualization, J.W.; methodology, J.W.; validation, J.W. and X.L.; formal analysis, J.W. and X.L.; investigation, J.W. and X.L.; resources, X.L.; data curation, X.L., Z.S. and K.J.; writing—original draft preparation, J.W. and X.L.; writing—review and editing, Z.S., X.L., K.J. and X.Z.; visualization, X.L.; supervision, K.J.; project administration, X.Z.; funding acquisition, X.L. All authors have read and agreed to the published version of the manuscript.

Funding: This work was jointly supported by the National Natural Science Foundation of China (62001480, 61971455), the Hunan Provincial Natural Science Foundation of China (2021JJ40684), and the Independent Research Fund of the Key Laboratory of Satellite Information Intelligent Processing and Application Technology (2022-ZZKY-JJ-10-02).

Data Availability Statement: Not applicable.

Acknowledgments: The authors would like to thank the pioneer researchers in SAR moving ship refocusing and other related fields.

Conflicts of Interest: The authors declare no conflict of interest.

References

1. Curlander, J.C.; McDonough, R.N. *Synthetic Aperture Radar*; Wiley: New York, NY, USA, 1991; Volume 11.
2. Cumming, I.G.; Wong, F.H. *Digital Processing of Synthetic Aperture Radar Data*; Artech House: Boston, MA, USA, 2005; pp. 108–110.
3. Kuang, G.; Gao, G.; Jiang, Y.; Lu, J.; Jia, C. *Theory, Algorithm and Application for Target Detection in Synthetic Aperture Radar*; Press of National University of Defense Technology: Changsha, China, 2007.
4. Zhan, R.; Cui, Z. Ship Recognition for SAR Scene Images under Imbalance Data. *Remote Sens.* **2022**, *14*, 6294. [CrossRef]
5. Zhang, L.; Leng, X.; Feng, S.; Ma, X.; Ji, K.; Kuang, G.; Liu, L. Azimuth-Aware Discriminative Representation Learning for Semi-Supervised Few-Shot SAR Vehicle Recognition. *Remote Sens.* **2023**, *15*, 331. [CrossRef]
6. Shao, Z.; Zhang, T.; Ke, X. A Dual-Polarization Information-Guided Network for SAR Ship Classification. *Remote Sens.* **2023**, *15*, 2138. [CrossRef]
7. Leng, X.; Ji, K.; Kuang, G. Ship Detection from Raw SAR Echo Data. *IEEE Trans. Geosci. Remote Sens.* **2023**, *61*, 5207811. [CrossRef]
8. Lin, Z.; Ji, K.; Leng, X.; Kuang, G. Squeeze and excitation rank faster R-CNN for ship detection in SAR images. *IEEE Geosci. Remote Sens. Lett.* **2018**, *16*, 751–755. [CrossRef]
9. Sun, Z.; Dai, M.; Leng, X.; Lei, Y.; Xiong, B.; Ji, K.; Kuang, G. An Anchor-Free Detection Method for Ship Targets in High-Resolution SAR Images. *IEEE J. Sel. Top. Appl. Earth Obs. Remote Sens.* **2021**, *14*, 7788–7816. [CrossRef]
10. Sun, Z.; Leng, X.; Lei, Y.; Xiong, B.; Ji, K.; Kuang, G. BiFA-YOLO: A novel YOLO-based method for arbitrary-oriented ship detection in high-resolution SAR images. *Remote Sens.* **2021**, *13*, 4209. [CrossRef]
11. Leng, X.; Ji, K.; Xiong, B.; Kuang, G. Complex Signal Kurtosis—Indicator of Ship Target Signature in SAR Images. *IEEE Trans. Geosci. Remote Sens.* **2022**, *60*, 1–12. [CrossRef]
12. Xiong, B.; Sun, Z.; Wang, J.; Leng, X.; Ji, K. A Lightweight Model for Ship Detection and Recognition in Complex-Scene SAR Images. *Remote Sens.* **2022**, *14*, 6053. [CrossRef]
13. Chen, Z.; Liu, C.; Filaretov, V.F.; Yukhimets, D.A. Multi-Scale Ship Detection Algorithm Based on YOLOv7 for Complex Scene SAR Images. *Remote Sens.* **2023**, *15*, 2071. [CrossRef]
14. Zhang, Y.; Lu, D.; Qiu, X.; Li, F. Scattering-Point-Guided RPN for Oriented Ship Detection in SAR Images. *Remote Sens.* **2023**, *15*, 1411. [CrossRef]
15. Lang, H.; Li, C.; Xu, J. Multisource heterogeneous transfer learning via feature augmentation for ship classification in SAR imagery. *IEEE Trans. Geosci. Remote Sens.* **2022**, *60*, 5228814. [CrossRef]
16. Lu, C.; Li, W. Ship classification in high-resolution SAR images via transfer learning with small training dataset. *Sensors* **2018**, *19*, 63. [CrossRef]
17. Wang, Y.; Wang, C.; Zhang, H. Combining a single shot multibox detector with transfer learning for ship detection using sentinel-1 SAR images. *Remote Sens. Lett.* **2018**, *9*, 780–788. [CrossRef]
18. Rostami, M.; Kolouri, S.; Eaton, E.; Kim, K. Deep transfer learning for few-shot SAR image classification. *Remote Sens.* **2019**, *11*, 1374. [CrossRef]
19. Lang, H.; Wu, S.; Xu, Y. Ship classification in SAR images improved by AIS knowledge transfer. *IEEE Geosci. Remote Sens. Lett.* **2018**, *15*, 439–443. [CrossRef]
20. Raney, R.K. Synthetic Aperture Imaging Radar and Moving Targets. *IEEE Trans. Aerosp. Electron. Syst.* **1971**, *7*, 499–505. [CrossRef]
21. Yang, Q.; Li, Z.; Li, J.; An, H.; Wu, J.; Pi, Y.; Yang, J. A Novel Bistatic SAR Maritime Ship Target Imaging Algorithm Based on Cubic Phase Time-Scaled Transformation. *Remote Sens.* **2023**, *15*, 1330. [CrossRef]
22. Sharma, J.J.; Gierull, C.H.; Collins, M.J. Compensating the effects of target acceleration in dual-channel SAR–GMTI. *IEE Proc. Radar Sonar Navig.* **2006**, *153*, 53–62. [CrossRef]
23. Ruegg, M.; Meier, E.; Nuesch, D. Capabilities of dual-frequency millimeter wave SAR with monopulse processing for ground moving target indication. *IEEE Trans. Geosci. Remote Sens.* **2007**, *45*, 539–553. [CrossRef]
24. Bethke, K.H.; Baumgartner, S.; Gabele, M.; Hounam, D.; Kemptner, E.; Klement, D.; Erxleben, R. Air-and spaceborne monitoring of road traffic using SAR moving target indication—Project TRAMRAD. *ISPRS J. Photogramm. Remote Sens.* **2006**, *61*, 243–259. [CrossRef]
25. Baumgartner, S.; Gabele, M.; Krieger, G.; Bethke, K.H.; Zuev, S. Traffic monitoring with SAR: Implications of target acceleration. In Proceedings of the European Conference on Synthetic Aperture Radar (EUSAR), Dresden, Germany, 16–18 May 2006; VDE Verlag GmbH: Berlin, Germany, 2006; pp. 1–4.
26. Chen, V.C.; Li, F.; Ho, S.S.; Wechsler, H. Micro-Doppler effect in radar: Phenomenon, model, and simulation study. *IEEE Trans. Aerosp. Electron. Syst.* **2006**, *42*, 2–21. [CrossRef]
27. Li, X.; Deng, B.; Qin, Y.; Wang, H.; Li, Y. The influence of target micromotion on SAR and GMTI. *IEEE Trans. Geosci. Remote Sens.* **2011**, *49*, 2738–2751. [CrossRef]
28. Liu, P.; Jin, Y.Q. A study of ship rotation effects on SAR image. *IEEE Trans. Geosci. Remote Sens.* **2017**, *55*, 3132–3144. [CrossRef]

29. Zhou, B.; Qi, X.; Zhang, J.; Zhang, H. Effect of 6-DOF Oscillation of Ship Target on SAR Imaging. *Remote Sens.* **2021**, *13*, 1821. [CrossRef]

30. Chen, J.; Xing, M.; Yu, H.; Liang, B.; Peng, J.; Sun, G.C. Motion compensation/autofocus in airborne synthetic aperture radar: A review. *IEEE Geosci. Remote Sens. Mag.* **2021**, *10*, 185–206. [CrossRef]

31. Moreira, A. Real-time synthetic aperture radar(SAR) processing with a new subaperture approach. *IEEE Trans. Geosci. Remote Sens.* **1992**, *30*, 714–722. [CrossRef]

32. Calloway, T.M.; Donohoe, G.W. Subaperture autofocus for synthetic aperture radar. *IEEE Trans. Aerosp. Electron. Syst.* **1994**, *30*, 617–621. [CrossRef]

33. Wahl, D.E.; Eichel, P.H.; Ghiglia, D.C.; Jakowatz, C.V. Phase gradient autofocus-a robust tool for high resolution SAR phase correction. *IEEE Trans. Aerosp. Electron. Syst.* **1994**, *30*, 827–835. [CrossRef]

34. Ye, W.; Yeo, T.S.; Bao, Z. Weighted least-squares estimation of phase errors for SAR/ISAR autofocus. *IEEE Trans. Geosci. Remote Sens.* **1999**, *37*, 2487–2494. [CrossRef]

35. Morrison, R.L. *Entropy-Based Autofocus for Synthetic Aperture Radar*; University of Illinois at Urbana-Champaign: Urbana, IL, USA, 2002.

36. Zeng, T.; Wang, R.; Li, F. SAR image autofocus utilizing minimum-entropy criterion. *IEEE Geosci. Remote Sens. Lett.* **2013**, *10*, 1552–1556. [CrossRef]

37. Zhang, S.; Liu, Y.; Li, X. Fast entropy minimization based autofocusing technique for ISAR imaging. *IEEE Trans. Signal Process* **2015**, *63*, 3425–3434. [CrossRef]

38. Huang, X.; Ji, K.; Leng, X.; Dong, G.; Xing, X. Refocusing moving ship targets in SAR images based on fast minimum entropy phase compensation. *Sensors* **2019**, *19*, 1154. [CrossRef] [PubMed]

39. Martorella, M.; Berizzi, F.; Haywood, B. Contrast maximisation based technique for 2-D ISAR autofocusing. *IEE Proc. Radar Sonar Navig.* **2005**, *152*, 253–262. [CrossRef]

40. Gao, Y.; Yu, W.; Liu, Y.; Wang, R. Autofocus algorithm for SAR imagery based on sharpness optimisation. *Electron. Lett.* **2014**, *50*, 830–832. [CrossRef]

41. Martorella, M.; Berizzi, F.; Giusti, E.; Bacci, A. Refocussing of moving targets in SAR images based on inversion mapping and ISAR processing. In Proceedings of the 2011 IEEE RadarCon (RADAR), Kansas City, MO, USA, 23–27 May 2011; pp. 068–072.

42. Huang, L.M.; Duan, W.Y.; Han, Y.; Chen, Y.S. A review of short-term prediction techniques for ship motions in seaway. *J. Ship Mech.* **2014**, *18*, 1534–1542.

43. Doerry, A.W. *Ship Dynamics for Maritime ISAR Imaging*; Sandia National Laboratories (SNL): Albuquerque, NM, USA; Livermore, CA, USA, 2008.

44. Chen, X.; Dong, Y.; Li, X.; Guan, J. Modeling of Micromotion and Analysis of Properties of Rigid Marine Targets. *J. Radars* **2015**, *4*, 630–638. [CrossRef]

45. Sommer, A.; Ostermann, J. Backprojection subimage autofocus of moving ships for synthetic aperture radar. *IEEE Trans. Geosci. Remote Sens.* **2019**, *57*, 8383–8393. [CrossRef]

46. Li, G.; Zhang, G.; Qin, H.; Liang, Y. A Non-Linearly Moving Ship Autofocus Method Under Hybrid Coordinate System. In Proceedings of the IGARSS 2020–2020 IEEE International Geoscience and Remote Sensing Symposium, Waikoloa, HI, USA, 26 September–2 October 2020; pp. 2471–2474.

47. Yu, J.; Liu, C.; Yu, Z. SAR Imaging of Swing Ships Based on Two-Dimensional Spatial Variance Correction. *IEEE J. Sel. Top. Appl. Earth Obs. Remote Sens.* **2022**, *15*, 9953–9962. [CrossRef]

48. Rigling, B.D. Image-quality focusing of rotating SAR targets. *IEEE Geosci. Remote Sens. Lett.* **2008**, *5*, 750–754. [CrossRef]

49. Darong, H.; Cunqian, F.; Ningning, T.; Yiduo, G. 2D spatial-variant phase errors compensation for ISAR imagery based on contrast maximisation. *Electron. Lett.* **2016**, *52*, 1480–1482. [CrossRef]

50. Jia, X.; Song, H.; He, W. A novel method for refocusing moving ships in SAR images via ISAR technique. *Remote Sens.* **2021**, *13*, 2738. [CrossRef]

51. Cao, R.; Wang, Y.; Zhao, B.; Lu, X. Ship target imaging in airborne SAR system based on automatic image segmentation and ISAR technique. *IEEE J. Sel. Top. Appl. Earth Obs. Remote Sens.* **2021**, *14*, 1985–2000. [CrossRef]

52. Lu, Z.J.; Qin, Q.; Shi, H.Y.; Huang, H. SAR moving target imaging based on convolutional neural network. *Digit. Signal Process* **2020**, *106*, 102832. [CrossRef]

53. Liu, Z.; Yang, S.; Gao, Q.; Feng, Z.; Wang, M.; Jiao, L. AFnet and PAFnet: Fast and Accurate SAR Autofocus Based on Deep Learning. *IEEE Trans. Geosci. Remote Sens.* **2022**, *60*, 5238113. [CrossRef]

54. Hua, Q.; Yun, Z.; Li, H.; Jiang, Y.; Xu, D. Refocusing on SAR ship targets with three-dimensional rotating based on complex-valued convolutional gated recurrent unit. *IEEE Geosci. Remote Sens. Lett.* **2022**, *19*, 4512405. [CrossRef]

55. Ozaktas, H.M.; Arikan, O.; Kutay, M.A.; Bozdagt, G. Digital computation of the fractional Fourier transform. *IEEE Trans. Signal Process* **1996**, *44*, 2141–2150. [CrossRef]

56. Sun, H.B.; Liu, G.S.; Gu, H.; Su, W.M. Application of the fractional Fourier transform to moving target detection in airborne SAR. *IEEE Trans. Aerosp. Electron. Syst.* **2002**, *38*, 1416–1424.

57. Pepin, M. A three-dimensional fractional Fourier transformation methodology for volumetric linear, circular, and orbital synthetic aperture radar formation. In *Algorithms for Synthetic Aperture Radar Imagery XXI*; SPIE: Baltimore, MD, USA, 2014; Volume 9093, pp. 26–38.

58. Clemente, C.; Soraghan, J.J. Range Doppler and chirp scaling processing of synthetic aperture radar data using the fractional Fourier transform. *IET Signal Process* **2012**, *6*, 503–510. [CrossRef]
59. Wang, W.Q. Moving target indication via three-antenna SAR with simplified fractional Fourier transform. *EURASIP J. Adv. Signal Process* **2011**, *2011*, 117. [CrossRef]
60. Baumgartner, S.V.; Krieger, G. SAR traffic monitoring using time-frequency analysis for detection and parameter estimation. In Proceedings of the IGARSS 2008–2008 IEEE International Geoscience and Remote Sensing Symposium, Boston, MA, USA, 7–11 July 2008; Volume 2, pp. II-25–II-28.
61. Wu, J.; Jiang, Y.; Kuang, G.; Lu, J.; Li, Z. Parameter estimation for SAR moving target detection using fractional Fourier transform. In Proceedings of the 2014 IEEE Geoscience and Remote Sensing Symposium, Quebec City, QC, Canada, 13–18 July 2014; pp. 596–659.
62. Yu, L.; Zhang, Y. Application of the fractional fourier transform to moving train imaging. *Prog. Electromagn. Res. M* **2011**, *19*, 13–23. [CrossRef]
63. Singh, J.; Datcu, M. SAR image categorization with log cumulants of the fractional Fourier transform coefficients. *IEEE Trans. Geosci. Remote Sens.* **2013**, *51*, 5273–5282. [CrossRef]
64. Pelich, R.; Longépé, N.; Mercier, G.; Hajduch, G.; Garello, R. Vessel refocusing and velocity estimation on SAR imagery using the fractional Fourier transform. *IEEE Trans. Geosci. Remote Sens.* **2015**, *54*, 1670–1684. [CrossRef]
65. Moghadasian, S.S. A Fast and Accurate Method for Parameter Estimation of Multi-Component LFM Signals. *IEEE Signal Process Lett.* **2022**, *29*, 1719–1723. [CrossRef]

Article

Anomaly-Based Ship Detection Using SP Feature-Space Learning with False-Alarm Control in Sea-Surface SAR Images

Xueli Pan [1,2,3], Nana Li [1,2], Lixia Yang [1,2,*], Zhixiang Huang [1,2], Jie Chen [1,2], Zhenhua Wu [1,2] and Guoqing Zheng [3]

[1] Key Laboratory of Intelligent Computing and Signal Processing, Ministry of Education, Anhui University, Hefei 230601, China; xlpan@ahu.edu.cn (X.P.); p20201070@stu.ahu.edu.cn (N.L.); zxhuang@ahu.edu.cn (Z.H.); jiechen@ahu.edu.cn (J.C.); zhwu@ahu.edu.cn (Z.W.)

[2] Information Materials and Intelligent Sensing Laboratory of Anhui Province, Anhui University, Hefei 230601, China

[3] East China Institute of Photo-Electron ICs, Suzhou 215163, China; zgq1980@126.com

*   Correspondence: 19002@ahu.edu.cn

Abstract: Synthetic aperture radar (SAR) can provide high-resolution and large-scale maritime monitoring, which is beneficial to ship detection. However, ship-detection performance is significantly affected by the complexity of environments, such as uneven scattering of ship targets, the existence of speckle noise, ship side lobes, etc. In this paper, we present a novel anomaly-based detection method for ships using feature learning for superpixel (SP) processing cells. First, the multi-feature extraction of the SP cell is carried out, and to improve the discriminating ability for ship targets and clutter, we use the boundary feature described by the Haar-like descriptor, the saliency texture feature described by the non-uniform local binary pattern (LBP), and the intensity attention contrast feature to construct a three-dimensional (3D) feature space. Besides the feature extraction, the target classifier or determination is another key step in ship-detection processing, and therefore, the improved clutter-only feature-learning (COFL) strategy with false-alarm control is designed. In detection performance analyses, the public datasets HRSID and LS-SSDD-v1.0 are used to verify the method's effectiveness. Many experimental results show that the proposed method can significantly improve the detection performance of ship targets, and has a high detection rate and low false-alarm rate in complex background and multi-target marine environments.

Keywords: superpixel (SP) processing cell; boundary feature; saliency texture feature; intensity attention contrast feature; clutter-only feature learning (COFL)

Citation: Pan, X.; Li, N.; Yang, L.; Huang, Z.; Chen, J.; Wu, Z.; Zheng, G. Anomaly-Based Ship Detection Using SP Feature-Space Learning with False-Alarm Control in Sea-Surface SAR Images. *Remote Sens.* **2023**, *15*, 3258. https://doi.org/10.3390/rs15133258

Academic Editor: Piotr Samczynski

Received: 29 May 2023
Revised: 18 June 2023
Accepted: 22 June 2023
Published: 24 June 2023

1. Introduction

With the development of high-resolution synthetic aperture radar (SAR) technology, we have widely applied SAR to various fields, both military and civilian, because of the conducting ability in any weather and condition. SAR has obvious advantages over optical and infrared sensors [1–3]. Therefore, the detection of ship targets in SAR images is an important application and has attracted a lot of attention and research in recent years.

The traditional pixel-level constant false-alarm rate (CFAR) algorithm is widely used. Scholars have successively put forward a series of CFAR methods, such as the cell-averaging CFAR (CA-CFAR) method, the smallest-of CFAR (SO-CFAR) method, the greatest-of CFAR (GO-CFAR) method, and the ordered statistic CFAR (OS-CFAR) method, etc. [4,5]. CFAR detection methods mainly depend on clutter statistical modeling, i.e., exploiting the grayscale characteristics between ships and the clutter background to acquire the target-detection decision. Typical sea-clutter statistical models mainly include Rayleigh, Weibull, Log-normal, K, Gamma, and generalized Gamma distribution (GΓD) models, etc. [6–12]. Later, the advent of the kernel density estimation (KDE) approach improved the goodness-of-fit (GoF) to the clutter background significantly [13]. However, due to the complexity of sea-clutter scattering, statistical modeling is intricate to build precisely, which will cause the performance of CFAR detectors using statistical characteristics to deteriorate and the false-alarm rate to rise. To adapt to different SAR scenes, researchers selected the best statistical model of clutter through the GoF analyses of diverse clutter statistical models so that the target-detection rate is excellent for various scenarios [14,15]. To improve the target-detection performance and reduce the false-alarm rate, some scholars have made a lot of improvements to the traditional pixel-level CFAR method and have achieved good results in target-detection performance [16–18].

When utilizing the pixel-level CFAR method for ship target detection, the speckle noise is likely to cause false alarms. Through superpixel (SP) segmentation, the ship target can be regarded as one or more connected regions, which can not only better preserve the ship target contour but also reduce the effect of speckle noise. A new SP-level CFAR detector was designed by Pappas et al., which could not only reduce the false-alarm probability of ship target detection but also better maintain the ship shapes [19]. Moreover, to improve ship-detection accuracy, SP-level CFAR detectors are widely used in SAR ship-detection tasks [20–22]. Due to the influence of adjacent targets and side lobes of ships in multi-target environments, clutter modeling is inaccurate and the target-detection performance is poor. Thus, to advance the goodness-of-fit of sea clutter and enhance ship-detection capability in multi-target scenarios, the SP-level CFAR detection method was presented based on the truncated Gamma statistic [23]. An improved SP-level CFAR detector was designed by Li et al., who considered weighted information entropy (WIE) as an SP statistical feature and adopted the coarse-to-fine detection idea to achieve two-stage CFAR detection of ship targets [24]. At present, attention contrast enhancement theory has been widely applied to target-detection tasks, and the saliency detection of ship targets has a great potential [25–29]. The SP-based local contrast measure (SLCM) method has been presented to detect ships hidden in the strong noise background efficiently [30]. Lin et al. employed the SP-based fisher vector for feature extraction, which can describe the deep difference between target and background and improve target-detection ability in SAR images with a low signal-to-noise ratio (SNR) [31]. However, due to the complexity of sea-surface environments, there is a significant impact on the detection performance of small ship targets with weak scattering while detecting targets based on traditional statistical characteristics. Therefore, it is necessary to study further saliency feature extraction in SAR images for achieving the efficient detection of ship targets in complex environments.

With the rapid development of machine learning, researchers have applied machine-learning algorithms in the radar field to achieve radar signal processing. Ship-detection methods based on machine learning mainly include two types, namely the detection algorithms based on traditional machine learning, and deep learning. As a typical traditional machine-learning model for a binary classification task, the support vector machine (SVM) is widely utilized in ship target detection on the sea surface. From the perspective of the SVM training model, the problem of ship target detection can be regarded as a binary classification problem between ship targets and sea-clutter backgrounds. Therefore, ship-detection algorithms based on the feature extraction of SAR slice images and SVM binary classification have been proposed successively, which have wide applications and excellent detection effects. He et al. designed the SVM-based detection method using the constructed gray-level co-occurrence matrix (GLCM) texture feature samples [32]. However, because the detection performance for the SVM classifier depends on the extracted feature property, the detection rate is limited to only using GLCM texture features to distinguish ships and clutter. To improve ship-detection performance in non-uniform sea conditions, the idea of coarse-to-fine detection was adopted by Xiong et al., who combined the SVM classifier and the maximum entropy thresholding method to achieve ship detection [33]. Later, Li et al. utilized the Relief method to choose the optimal feature combination from many polarimetric rotation domain features which can distinguish candidate pixels between targets and clutter and then combined it with a common SVM classifier to achieve accurate binary classification [34]. Furthermore, since interference (such as noise and ship side lobes, etc.) in complex environments is easy to cause false alarms, it is very significant to suppress false alarms for ship target detection in SAR images. Yang et al. presented a false-alarm removal method using one-class SVM (OC-SVM) for ship target detection [35]. The parameter adjustment of the traditional SVM training model needs to be finished manually and largely depends on experience, which easily leads to the unstable classification performance of SVM. Therefore, researchers have made a great many improvements based on conventional SVM. To reduce the false-alarm rate during docked ship detection in complex near-shore port environments, Zou et al. utilized genetic operators to improve the traditional particle swarm optimization algorithm, and adjusted SVM parameters based on the optimized genetic operator particle swarm optimization (GOPSO) algorithm. Although the optimized GOPSO-SVM algorithm has higher classification accuracy and fewer false alarms, there is still missed detection [36]. In addition, the algorithm for ship detection based on simulated annealing by fuzzy matching-SVM (SAFM-SVM) was provided, which could adaptively optimize the detector parameters and complete the self-adaptive feature screening [37].

With the development of deep learning in recent years, there are many ship-detection algorithms based on deep-learning networks. These algorithms are mainly divided into single-stage and two-stage detection algorithms. In terms of detection accuracy, two-stage methods are usually superior to single-stage. However, from the perspective of detection speed, the single-stage detection methods are faster. For the single-stage network structure, the typical You Only Look Once (YOLO) model is used for the ship target detection [38], and with the continuous improvement of network structure, scholars have presented a series of ship target detection methods based on improved YOLO algorithms, such as YOLOv3 [39], YOLOv5 [40], BiFA-YOLO [41] as well as some other improved YOLO models [42–44]. To improve the real-time performance of target detection in SAR images, based on the YOLOv2 model architecture, ref. [45] developed a new network architecture with fewer layers, namely YOLOv2-reduced. Its detection performance is similar to YOLOv2, but there is a significant improvement in detection speed. Region-based convolutional neural network (R-CNN) is the most basic two-stage network structure. Ship-detection algorithms based on faster R-CNN and many improved faster R-CNN have been proposed in recent years [46–48]. To further enhance the feature-extraction ability, Xia et al. concentrated mainly on the optimization of the backbone and neck parts of the ship-detection framework and designed the visual transformer ship-detection framework in SAR images based on contextual joint representation learning (CRTransSar) [49]. Zhou et al. provided the Doppler feature matrix fused with a multi-layer feature pyramid network (D-MFPN) which can effectively enhance the detecting capability of ship targets, especially for the moving ships [50]. Aiming at the problem of SAR image classification when only a few labeled data are available, the new framework to train a deep neural network for classifying SAR images by eliminating the need for a huge was designed by Rostami et al. [51]. Considering the importance of SAR ship-detection speed in practical applications, Zhang et al. [52] designed a new grid convolutional neural network (G-CNN) architecture, which mainly includes the backbone convolutional neural network (B-CNN) and the detection convolutional neural network (D-CNN). This method achieves ship target detection with high speed for SAR images while maintaining the accuracy of ship detection in actual applications. To solve the problem of ship detection in complex inshore and offshore scenarios, the ship-detection method using the high-resolution ship-detection network (HR-SDNet) was provided by Wei et al., who fully utilize the feature maps of high-resolution and low-resolution convolutions to design a novel high-resolution feature pyramid network (HRFPN), which has better detection accuracy and robustness [53]. In addition, Liu et al. [54] proposed the framework for exploring multi-scale ship proposals, mainly consisting of two stages: hierarchical grouping and proposal scoring, which can effectively solve the problem of significant ship-scale differences in SAR images, therefore improving ship-detection performance. Meanwhile, the physical scattering mechanism of the ship target is fused into the network model to improve the detection ability [55–57].

The motility of the sea surface results in complicated scattering characteristics which are related to radar parameters, sea state, etc. Ship detection is a great challenge in high-resolution sea clutter, especially in high-sea states. For detection methods of traditional machine learning and deep learning, two implicit conditions used effectively are that the training samples of clutter and ships are balanced, and samples need diversity. Ship detection encounters the two extreme unbalances on the sea surface, and the training samples of sea clutter can be ergodic in the feature space in a short time while ship targets are non-ergodic and difficult to obtain. The deep-learning method requires a large amount of data and high computational cost in the learning stage, while the detection stage is of low computational cost, which is unfavorable for fast implementation. Due to the sparsity of ship targets on the sea surface relative to sea clutter, the ship target samples can be regarded as outliers in the clutter sample space, and extracting ship targets from clutter background can be regarded as a problem of anomaly-based ship detection in the sea surface environment. The theory of anomaly detection has been widely applied in ship target detection tasks [58,59]. The feature space is critical for target-detection performance, and saliency feature space based on SP cells should be further explored for single-channel SAR intensity images. Therefore, we study the anomaly-based ship-detection method of SPs for clutter-only feature learning. The main work of this paper is as follows:

1. To enhance the feature representation capability in the SP cell, we construct a three-dimensional (3D) feature space that contains the boundary feature described by Haar-like, texture feature described by non-uniform LBP, and intensity attention contrast feature.
2. The clutter-only feature-learning (COFL) model with false-alarm control is developed in the anomaly-based detection decision based on the established feature space.
3. We execute extensive experiments on SAR datasets collected from different satellites, and it is obvious that our proposed method has a state-of-the-art feature discriminative ability and good detection accuracy.

The remainder of this paper is organized as follows: Section 2 introduces the main contents of our ship target detection method in detail. Section 3 compares and analyzes our method with other benchmark methods, and proves the effectiveness of the proposed method. In Section 4, we summarize the conclusions.

2. Detection Methodology

Because strong noise points generally occupy a small number of bright pixels in complex clutter backgrounds, it is easy to mistake strong noise points for target pixels when using conventional pixel-level methods to detect targets, which will cause a great many false alarms. Moreover, the weak scattering characteristics of small ships are likely to reduce the ship-detection rate. Therefore, this paper proposes an anomaly-based ship target-detection algorithm based on SP cells using the clutter-only feature space. First, preprocessing is carried out and consists of sea–land and SP segmentation. Second, we extract multi-feature from three aspects, namely the boundary feature, texture feature, and intensity attention contrast feature of the SP cell to enhance the information representation ability. At last, the ship target detection is realized using the detection model obtained through 3D feature-space training. Figure 1 presents the specific flow of anomaly detection in this paper.

2.1. Preprocessing Operation

2.1.1. Sea–Land Segmentation

In sea–land scenes, the land will affect ship-detection performance, as shown in Figure 2a–c, and the sea–land segmentation is a crucial step for ship detection. The maximum inter-class variance (OTSU) method is a non-parametric and unsupervised method proposed by Japanese Otsu in 1979, which can automatically select segmentation thresholds during image segmentation. Generally, the image is mainly divided into foreground and background according to the gray feature of the image. OTSU method obtains the best global threshold based on the maximum variance between classes of foreground and background in the image, which is simple to calculate and maximizes the separability of classes on grayscale [60].

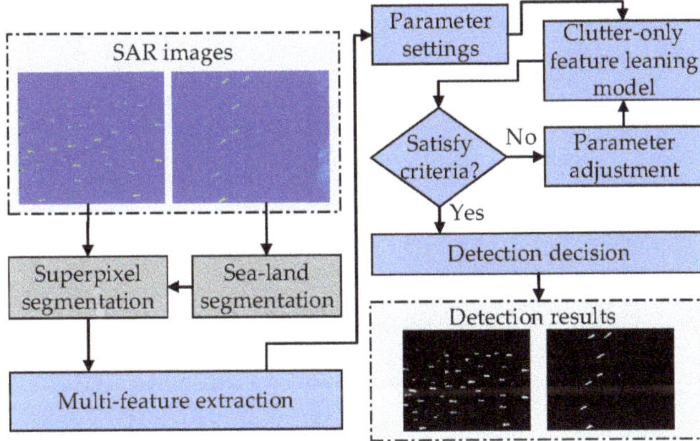

Figure 1. Flow chart of the proposed method.

Therefore, to accurately detect ship targets in the marine environment of the SAR image and eliminate the interference caused by the land, we utilize the OTSU method to complete the effective segmentation between the sea surface and the non-sea surface. Figure 2d–f show OTSU segmentation results of SAR images shown in Figure 2a–c. It can be seen that the OTSU method has an excellent segmentation effect and can accurately extract the sea area from the original SAR scenes, which is convenient for the subsequent SAR sea-surface ship target detection.

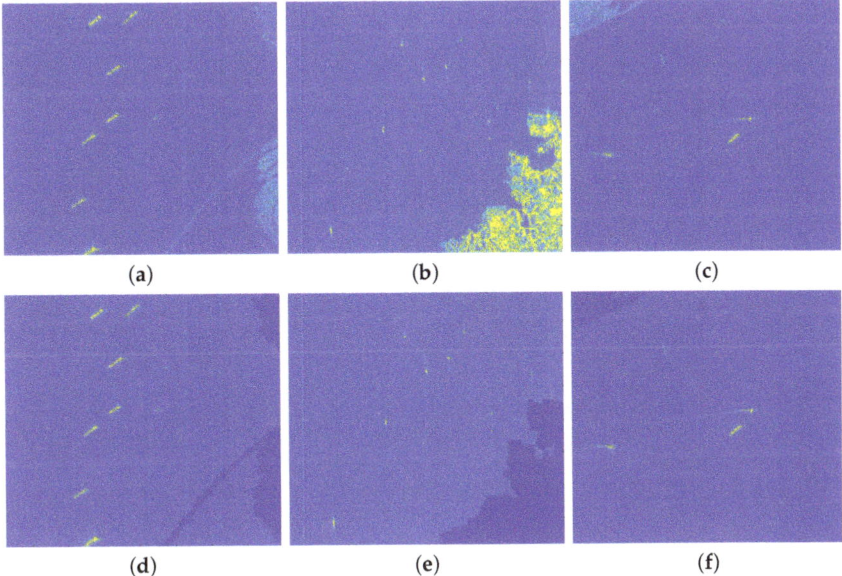

Figure 2. Sea–land segmentation results for different scenes. (**a**) scene 1; (**b**) scene 2; (**c**) scene 3; (**d**) result of scene 1; (**e**) result of scene 2; (**f**) result of scene 3.

2.1.2. SP Segmentation

SP segmentation agglomerates pixels with similar properties and locations into a cell, which can effectively suppress the influence of isolated clutter and speckle noise, and has become an important tool for computer vision. Simple linear iterative clustering (SLIC) is a common SP segmentation algorithm with excellent performance and fast segmentation speed by controlling the size and compactness of SP cells [24]. For single-channel SAR images, we adapt SLIC to 3D space $[g, x, y]^{T}$, where g represents the pixel intensity, x and y represent the spatial position information. The spatial proximity and intensity similarity between any two pixels are calculated as:

$$d_s = \sqrt{(x_u - x_v)^2 + (y_u - y_v)^2} \tag{1}$$

$$d_c = \sqrt{(g_u - g_v)^2} \tag{2}$$

where (x_u, y_u) and (x_v, y_v) represent the corresponding position information of pixel u and pixel v in the SAR image, respectively, g_u and g_v are the intensity values of pixel points u and v, respectively. The distance metric between any two pixels in a single-channel SAR image is expressed as [31]:

$$D = \sqrt{d_c^2 + (\frac{d_s}{S})^2 M^2} \tag{3}$$

where S denotes the SP size, M is an adjustment parameter. S is selected according to the actual detected SAR scenarios, and the optimal value range of M is 0.1~0.3 by extensive experimental analysis.

Figure 3b shows SP segmentation results with $S = 15$ and $M = 0.1$, which verifies that SP segmentation using the SLIC method can maintain the boundary contour of the ship target. When performing SP segmentation, a ship will likely be divided into multiple SP cells due to the discontinuity of ship targets, or a target SP cell contains clutter in the edge. Therefore, the feature-extraction stage needs to consider segmentation inaccuracy for the effect of ship-detection performance.

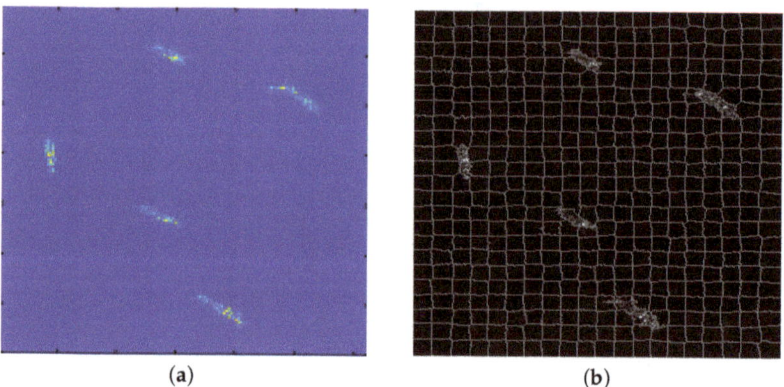

(a) (b)

Figure 3. SP segmentation result. (**a**) original SAR image; (**b**) segmentation result for (**a**).

2.2. Feature Extraction Based on SP

2.2.1. Boundary Feature Extraction Based on Haar-like

Considering the ambiguity of ship contour driving from noise and other factors, conventional edge-detected operators cannot work well in the edge extraction of ships. Haar-like is a feature description operator with excellent edge feature-extraction performance, which can be roughly divided into four types, namely edge feature, linear feature, center-surrounding feature, and diagonal feature [61]. Consequently, the proposed method exploits edge feature templates of Haar-like to achieve the boundary feature extraction as shown in Figure 4.

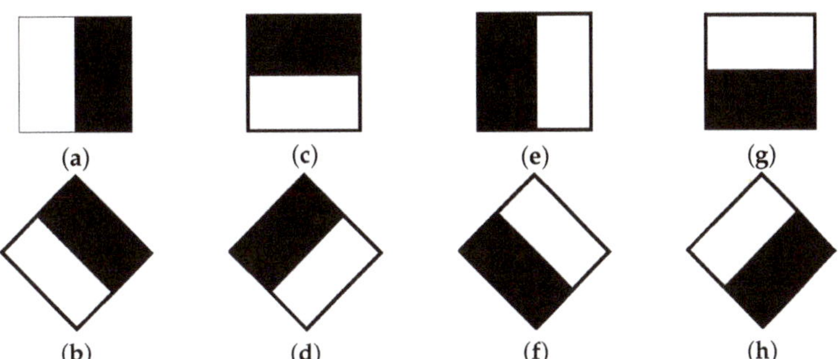

Figure 4. Haar-like edge features in different directions. (**a**) $\theta = 0°$; (**b**) $\theta = 45°$; (**c**) $\theta = 90°$; (**d**) $\theta = 135°$; (**e**) $\theta = 180°$; (**f**) $\theta = 225°$; (**g**) $\theta = 270°$; (**h**) $\theta = 315°$.

According to the description of Haar-like edge features in different directions, we utilize the normal direction of the closed curve for the SP cell corresponding to the Haar-like edge feature template as shown in Figure 5, in which a solid red line represents the closed boundary curve of the SP cell with counterclockwise direction, and the blue arrow points to the normal direction of the boundary point.

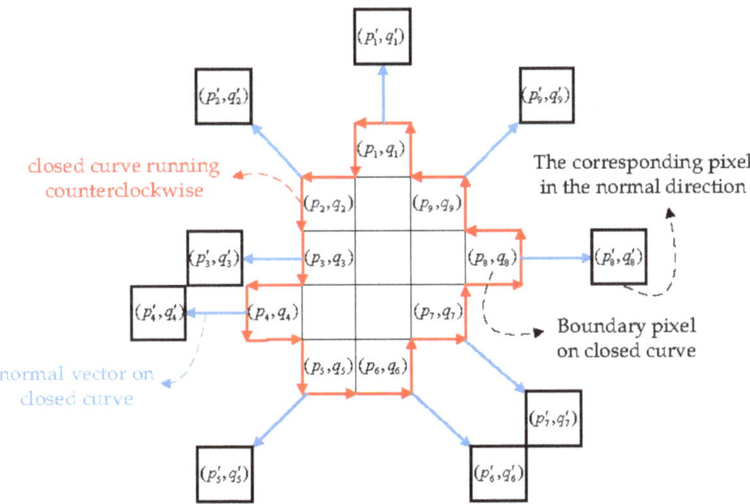

Figure 5. Schematic diagram about the normal direction of the boundary pixel point.

According to the normal direction of the boundary pixel point, the convolution operator F_θ based on the Haar-like theory to calculate any boundary pixel feature measure value of SP cells is used. The different normal directions θ are as follows:

- When $\theta = 0°$, $F_\theta = F_{\theta_1} = \begin{pmatrix} E_1 & -E_1 \end{pmatrix}_{1 \times 2L}$;

- When $\theta = 45°$, $F_\theta = F_{\theta_2} = \begin{pmatrix} 0 & -E_2 \\ E_2 & 0 \end{pmatrix}_{2L \times 2L}$;

- When $\theta = 90°$, $F_\theta = \left(-F_{\theta_1} \right)^{\mathrm{T}}$;

- When $\theta = 135°$, $F_\theta = F_{\theta_3} = \begin{pmatrix} -E_3 & 0 \\ 0 & E_3 \end{pmatrix}_{2L \times 2L}$;

- When $\theta = 180°$, $F_\theta = -F_{\theta_1}$;
- When $\theta = 225°$, $F_\theta = -F_{\theta_2}$;
- When $\theta = 270°$, $F_\theta = F_{\theta_1}^{\mathrm{T}}$;
- When $\theta = 315°$, $F_\theta = -F_{\theta_3}$.

where E_1 represents a $1 \times L$ matrix with all elements of 1, E_2 represents a $L \times L$ matrix in which the elements on a sub diagonal are all 1 and the other elements are all zero, E_3 stands for $L \times L$ identity matrix.

Therefore, the feature measure value at any boundary pixel point (p,q) for the SP cell can be defined as:

$$
\begin{aligned}
f_B(p,q) &= F_\theta \otimes I \\
&= \sum_m \sum_n F_\theta(m,n) I(p+m, q+n)
\end{aligned}
\tag{4}
$$

$$
\text{s.t.} \quad \theta = 0°, 45°, 90°, 135°, 180°, 225°, 270°, 315°
$$

where $\otimes$ denotes the convolution operation, $I(p,q)$ indicates the pixel intensity value at the boundary pixel point (p,q). The value ranges of m and n are different for different normal directions at boundary points.

Since a ship may be divided into several targets (as shown in Figure 3b), the first K maximum boundary pixel feature metric values are selected to extract the boundary feature value of the ship target SP, and the calculation criterion is as follows:

$$
f_1 = \frac{1}{K} \sum_{k=1}^{K} f_B^k
\tag{5}
$$

where f_B^k represents the boundary feature metric value at the k-th boundary pixel point (p_k, q_k) of the SP cell. Figure 6 shows the boundary feature values of different SP cells for Figure 3, and displays that values of ship SP cells are higher than clutter SP cells.

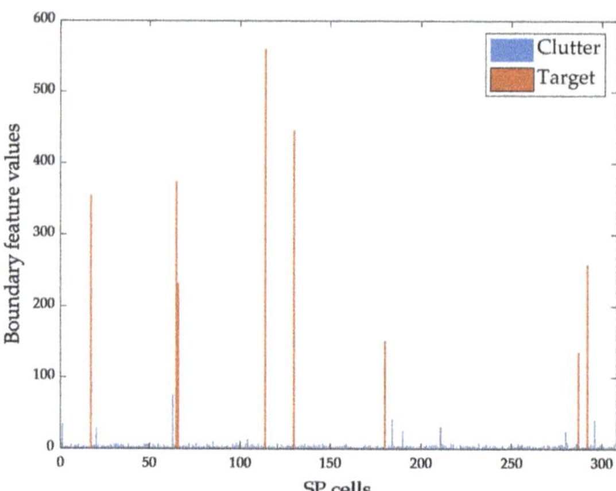

Figure 6. Boundary feature values of target and clutter SPs for Figure 3.

2.2.2. Saliency Texture Feature Extraction Based on Non-Uniform LBP

Due to the complexity of ship structure and differences in ship materials, the ship target scattering at different positions is differing, resulting in fluctuations of texture and intensity. This paper mainly aims at the texture difference between the ship target and clutter to extract texture features based on the SP cells, to realize the effective distinction between the ship and clutter.

The LBP feature operator is generally applicable to texture feature extraction of local images, and the LBP texture feature can be reflected through the differences between the center pixel and its surrounding pixels in the rectangular window. To represent the texture information of the region, the LBP value is calculated as [62]:

$$L_{bp} = \sum_{i=0}^{P-1} s(I_i - I_c) \cdot 2^i \qquad \text{s.t.} \quad s(I_i - I_c) = \begin{cases} 1, & I_i \geq I_c \\ 0, & I_i < I_c \end{cases} \qquad (6)$$

where I_c is the intensity value of center pixel, $I_i, i = 0, 1, \ldots, P-1$ represents the intensity value of i-th pixel in the local neighborhood, and there are P neighbor pixels in total.

Affected by the structure and material of ships, the scattering uniformity of the ship target is poor, and there is an obvious texture difference between the ship target and clutter. Therefore, to better represent the LBP texture feature of the ship SP, the local saliency texture feature measure value for ship SP is used as follows:

$$f_2 = \max \left\{ \log_2 \left(\frac{\overline{T}_{ex\_T}}{\overline{T}_{ex\_B_k}} + 1 \right) \right\} \times \overline{T}_{ex\_T} \qquad \text{s.t.} \quad k = 1, 2, \ldots\ldots, K_B \qquad (7)$$

where $T_{ex\_T}(p,q) = \sqrt{\sum_{j=0}^{L_1^2-2} |I_j - I(p,q)|^2}$ represents the texture value at pixel (p,q) of the target SP cell, $I(p,q)$ is the original intensity value at pixel (p,q) of the target SP cell, and $I_j, j = 0, 1, \ldots, L_1^2 - 2$ is the intensity value of j-th pixel in the local $L_1 \times L_1$ neighborhood. $\overline{T}_{ex\_T}$ is the average texture value of the center target SP cell and $\overline{T}_{ex\_B_k}, k = 1, 2, \ldots, K_B$ is the average texture value of the k-th neighboring clutter SP cell, K_B indicates the total number of clutter SP cells in the local neighborhood window. Moreover, with the increase of L_1 value, the non-uniformity of LBP texture becomes higher, which is conducive to the detection [27,63]. Figure 7 shows the saliency texture feature values of different SP cells for Figure 3 and presents a better texture distinction of ships and clutter.

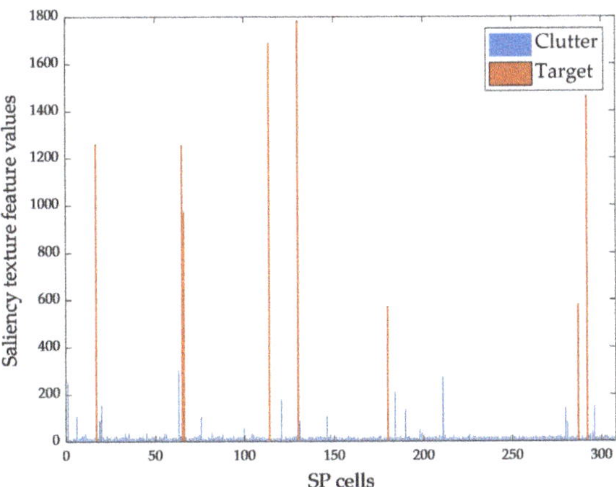

Figure 7. Saliency texture feature values of target and clutter SPs for Figure 3.

2.2.3. Attention Contrast Feature Extraction Based on Intensity Information

The scattering intensity of ships and clutter to electromagnetic waves is different. The energy of the ship target in SAR images is relatively concentrated, while the energy distribution of the clutter background is uniform. Therefore, inspired by the idea of local contrast measurement (LCM) [30], to achieve the saliency representation of the ship intensity feature, we use the metric value of intensity attention contrast enhancement for the ship target as:

$$f_3 = \max \left\{ \left(\frac{1}{K_I} \sum_{i=1}^{K_I} I_T^i \right) \times \frac{\mu_T}{\mu_{B_k}} \right\} \qquad \text{s.t.} \quad k = 1, 2, \ldots\ldots, K_B \qquad (8)$$

where I_T^i represents the *i*-th largest intensity value in the target SP cell, K_I is the number of maximal intensity values considered in the ship SP, μ_T is the average intensity value of the central target SP cell in the neighborhood window, and μ_{B_k}, $k = 1, 2, \ldots, K_B$ represents the average intensity value of the *k*-th surrounding clutter background SP cell in the neighborhood window. Figure 8 shows the intensity attention contrast feature values of different SP cells for Figure 3, and we can see that there are obvious intensity differences between ships and sea clutter.

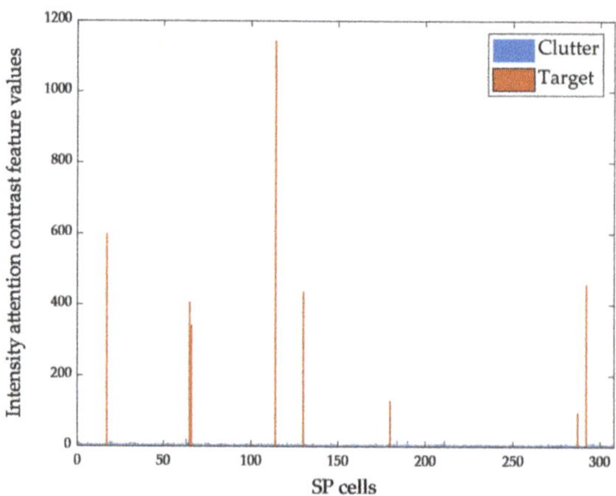

Figure 8. Intensity attention contrast feature values of target and clutter SPs for Figure 3.

As shown from Figures 6–8, there are significant differences for ship and clutter SP cells. Consequently, we construct a 3D feature vector based on the above three features. The *h*-th clutter SP feature sample is recorded as $f_h = [f_{1,h}, f_{2,h}, f_{3,h}]$, where $h = 1, 2, \ldots, H$, and H is the number of samples. The clutter-only feature training set F_H is expressed as:

$$F_H = \begin{bmatrix} f_1 \\ f_2 \\ \cdots \\ f_H \end{bmatrix} = \begin{bmatrix} f_{1,1} & f_{2,1} & f_{3,1} \\ f_{1,2} & f_{2,2} & f_{3,2} \\ \cdots & \cdots & \cdots \\ f_{1,H} & f_{2,H} & f_{3,H} \end{bmatrix} \tag{9}$$

Moreover, to limit the different feature values to the same scale, the 3D feature set F_H needs to be normalized. The normalized feature set F_H' is obtained by the following equation:

$$F_H' = \begin{bmatrix} f_1' \\ f_2' \\ \cdots \\ f_H' \end{bmatrix} = \begin{bmatrix} \frac{f_{1,1}-M_{in1}}{M_{ax1}-M_{in1}} & \frac{f_{2,1}-M_{in2}}{M_{ax2}-M_{in2}} & \frac{f_{3,1}-M_{in3}}{M_{ax3}-M_{in3}} \\ \frac{f_{1,2}-M_{in1}}{M_{ax1}-M_{in1}} & \frac{f_{2,2}-M_{in2}}{M_{ax2}-M_{in2}} & \frac{f_{3,2}-M_{in3}}{M_{ax3}-M_{in3}} \\ \cdots & \cdots & \cdots \\ \frac{f_{1,H}-M_{in1}}{M_{ax1}-M_{in1}} & \frac{f_{2,H}-M_{in2}}{M_{ax2}-M_{in2}} & \frac{f_{3,H}-M_{in3}}{M_{ax3}-M_{in3}} \end{bmatrix} \tag{10}$$

where M_{in1}, M_{ax1}, M_{in2}, M_{ax2}, M_{in3} and M_{ax3} represent the minimum and maximum values for different feature sets $\{f_{1,1}, f_{1,2}, \ldots, f_{1,H}\}$, $\{f_{2,1}, f_{2,2}, \ldots, f_{2,H}\}$, $\{f_{3,1}, f_{3,2}, \ldots, f_{3,H}\}$.

2.3. Detection Decision by COFL with False-Alarm Control

Extracting a ship from sea clutter can be regarded as a two-classification question, while SVM-based detection needs training samples of two classes to be balanced to obtain high detection accuracy, and the ship is the outlier on the sea surface. Based on this fact, we use the anomaly-detection theory to design the feature-learning metric. For the normalized 3D feature set $F'_H = \left\{ f'_i \in \mathbf{R}^3 : i = 1, 2, \ldots, H \right\}$, the decision hypersphere Ω with false-alarm control is determined as follows:

$$\min_{R,\xi} R^2 + C \sum_{i=1}^{H} \xi_i$$

$$\text{s.t.} \quad \left\| F(f'_i) - c \right\|^2 \leq R^2 + \xi_i, i = 1, \ldots, H \qquad (11)$$

$$\xi_i \geq 0, \forall i$$

$$\sharp \left\{ f'_j : f'_j \in \Omega \right\} = \left\lfloor H \times (1 - P_{fa}) \right\rfloor$$

where R is the radius of the decision hypersphere Ω, c is the center of Ω, C is the penalty factor, ξ_i represents the slack variable. $F(f'_i)$ is the mapping of f'_i in high-dimensional space, $\left\{ f'_j : f'_j \in \Omega \right\}$ represents the set of clutter samples contained in Ω, $\sharp \left\{ f'_j : f'_j \in \Omega \right\}$ represents the number of sample elements in set $\left\{ f'_j : f'_i \in \Omega \right\}$, P_{fa} stands for the probability of false alarm, and $\lfloor . \rfloor$ denotes a rounding downward symbol.

According to the classical single-class classification model support vector domain description (SVDD) [64], the original optimization problem can be simplified to Equation (12) using the Lagrange multiplier method and KKT condition.

$$\min_{\alpha} \sum_{i=1}^{H} \sum_{j=1}^{H} \alpha_i \alpha_j k_f(f'_i, f'_j) - \sum_{i=1}^{H} \alpha_i k_f(f'_i, f'_i)$$

$$\text{s.t.} \quad \sum_{i=1}^{H} \alpha_i = 1 \qquad (12)$$

$$0 \leq \alpha_i \leq C, \forall i$$

$$\sharp \left\{ f'_j : f'_j \in \Omega \right\} = \left\lfloor H \times (1 - P_{fa}) \right\rfloor$$

where $k_f(f'_i, f'_j)$ is the Gauss kernel function. By solving Equation (12), a hypersphere model Ω can be obtained and used as the detection decision. If a sample is contained in the hypersphere Ω, it is considered to be clutter, otherwise, it is determined as the target.

Algorithm 1 summarizes the implementation of our proposed SAR ship detector based on the COFL model with false-alarm control.

Algorithm 1 The proposed ship detector based on COFL with false-alarm control.

Input: The original SAR images.
Step-1: Preprocessing.
Step-1.1: Sea–land segmentation of input scene SAR images using the OTSU method.
Step-1.2: SP segmentation by SLIC algorithm.
Step-2: Multi-feature extraction based on SP cell, and three methods of SP feature extraction are as follows:
Step-2.1: Boundary feature value is calculated by Equation (5).
Step-2.2: Saliency texture feature value is calculated by Equation (7).
Step-2.3: Intensity attention contrast value is calculated by Equation (8).
Step-3: Construct normalized clutter-only feature set F'_H according to Equations (9) and (10), and then train the classification decision model based on Equation (12).
Step-4: Detection based on the trained decision hypersphere.
Output: Detection results.

---

3. Experimental Results and Analysis

The effectiveness of our proposed feature space and ship-detection performance is verified in this section using high-resolution SAR images from the public datasets HRSID [65] and LS-SSDD-v1.0 [66], in which all real targets have been labeled. Table 1 lists the different imaging parameters of SAR images from HRSID and LS-SSDD-v1.0 datasets. We select multi-target sea-surface scenes and sea–land scenes for experimental analysis.

Table 1. Imaging parameters of SAR images from HRSID and LS-SSDD-v1.0 datasets.

Dataset	Satellite	Imaging Mode	Incident Angle (°)	Resolution (m)	Polarization
HRSID	Sentinel-1, TerraSAR-X	SM, ST, HS	27.6~34.8, 20~45	0.5, 1, 3	HH, HV, VV
LS-SSDD-v1.0	Sentinel-1	IW	27.6~34.8	5 × 20	VV, VH

Note: SM: strip-map mode; ST: staring spotlight; HS: high-resolution spotLight; IW: interferometric wide-swath.

3.1. Effectiveness Analyses of Multi-Feature Extraction

To further verify that the proposed three-dimensional features (P3DF) in this paper can significantly distinguish ship target from the clutter background, we respectively present 3D feature spatial distributions constructed with P3DF and optimal three-dimensional features (O3DF) in [33] (O3DF-SVM-based method) about two SAR images shown in Figure 9a,b.

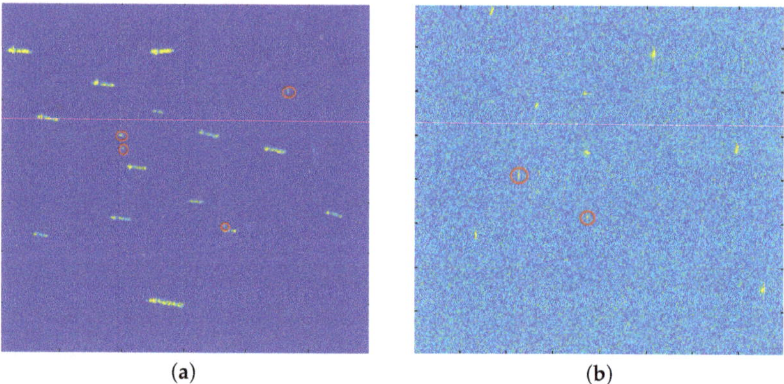

Figure 9. Different SAR scenes for feature-space analysis. (**a**) scene A; (**b**) scene B. Small targets are marked by red circles.

Figure 10a,b show 3D feature spaces constructed with the O3DF-SVM-based method for the training clutter samples and target samples of Figure 9a,b. Figure 10c,d present 3D feature spaces constructed with our method for clutter samples and target samples of Figure 9a,b. From all 3D feature spatial distributions, compared with multiple features extracted in the O3DF-SVM-based method, our proposed multi-features can better discriminate ships from clutter, especially for small ships, and it can be seen from Figure 10c,d that the feature distributions of ship targets are independent of the feature space of training clutter. Therefore, the 3D feature space proposed in this paper is more conducive to ship target detection.

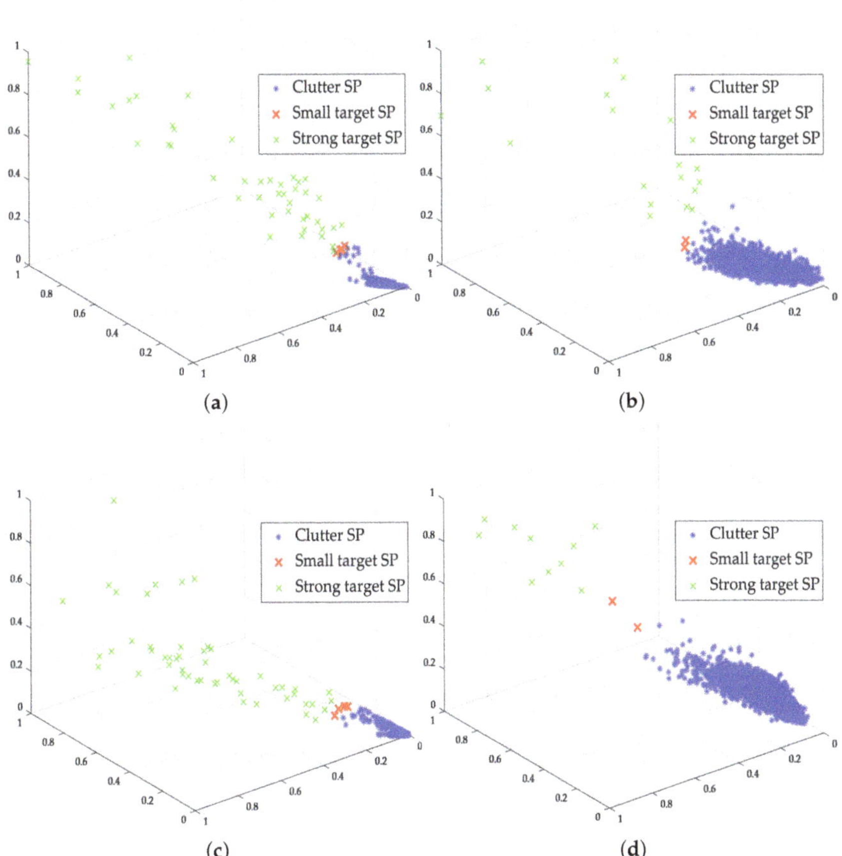

Figure 10. 3D feature distributions of SP cells for different scenes shown in Figure 9. (**a**) 3D feature space constructed with O3DF-SVM-based method for scene A; (**b**) 3D feature space constructed with O3DF-SVM-based method for scene B; (**c**) 3D feature space constructed with our method for scene A; (**d**) 3D feature space constructed with our method for scene B.

3.2. Performance Analyses of Target Detection

In this subsection, using many real SAR images from the public datasets HRSID and LS-SSDD-v1.0, target-detection experiments are carried out to validate ship-detection performance of the algorithm proposed in this paper for different scenes. This paper shows the detection effect of four different multi-target SAR images and compares our proposed method with CA-CFAR method [5], GΓD-CFAR method [15], improved SP-CFAR method [24], SLCM method [30], adaptive SP-CFAR [20], O3DF-SVM-based method [33] and P3DF-SVM-based method. Among all comparison methods involving false-alarm rates, such as CA-CFAR, GΓD-CFAR, improved SP-CFAR, and adaptive SP-CFAR methods, the false-alarm probability is 10^{-5}. Different multi-target scenes, such as the sea–land (Figure 11a–c), small targets (Figure 11b–d), and ship wake (Figure 11c), are presented in this paper. Real ship targets are marked with white rectangles.

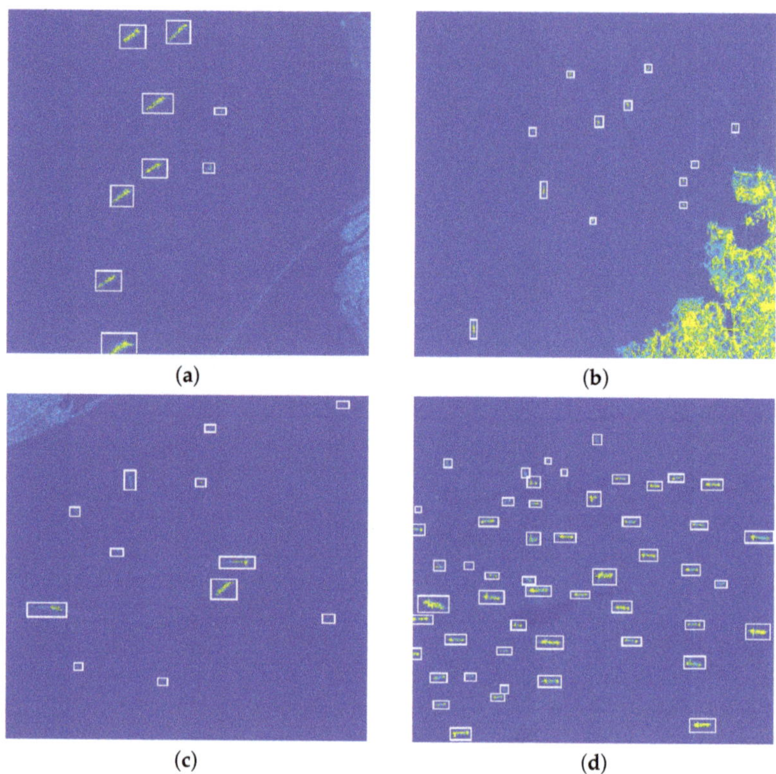

Figure 11. Original multi-target scenes. (**a**) scene 1; (**b**) scene 2; (**c**) scene 3; (**d**) scene 4.

Figures 12–15 present detection results of ship targets for seven comparison algorithms and our proposed algorithm in this paper, respectively. After the ship target clustering processing, we utilize red, green, and yellow rectangles to mark correctly detected targets, false alarms, and missed targets, respectively. From all detection results for CA-CFAR and GΓD-CFAR methods, it is easy to generate many false alarms because of the existence of interferences (such as speckle noise and ship wakes, etc.) which result in the inaccurate clutter statistical model. Compared with the pixel-wise detection method, the improved SP-CFAR method can reduce the probability of false alarms, but there are still a few false alarms due to the insufficient effect of strong interference suppression. From all detection results of SLCM and adaptive SP-CFAR methods, there is a phenomenon of missed detection for small targets with weak scattering characteristics (shown in Figures 12d,e–15d,e). However, compared with the SLCM method, the adaptive SP-CFAR method has the merit of fewer missed ships because the method adopts the non-local SP topology structure to avoid the overestimation of the threshold. The O3DF-SVM-based method can efficiently and accurately realize the binary classification of ship targets and clutter. However, in the actual SAR scenes, the number of ship target samples is very limited, which can produce the detection deterioration from the imbalance of training samples, and may produce a small number of false alarms (see in Figures 13f–15f). When using the P3DF-SVM-based method to detect target samples, ship-detection ability can be improved, but due to the limitation of SVM training decision, there are a few false targets as shown in Figures 13g–15g. However, from the detection results of our proposed method for Figures 13h–15h, the classification effect is better than SVM detection methods. Furthermore, in the uniform sea surface for Figure 11a, comparing the results of SVM methods with our proposed method in Figure 12f–h, we can find that ship-detection ability is the same. In addition, bright ship wakes usually appear with the moving ships as shown in Figure 11c, it is easy to be misjudged as ship targets. Comparing all detection results of Figure 14, it is indicated that our proposed algorithm can effectively reduce false alarms caused by ship wakes.

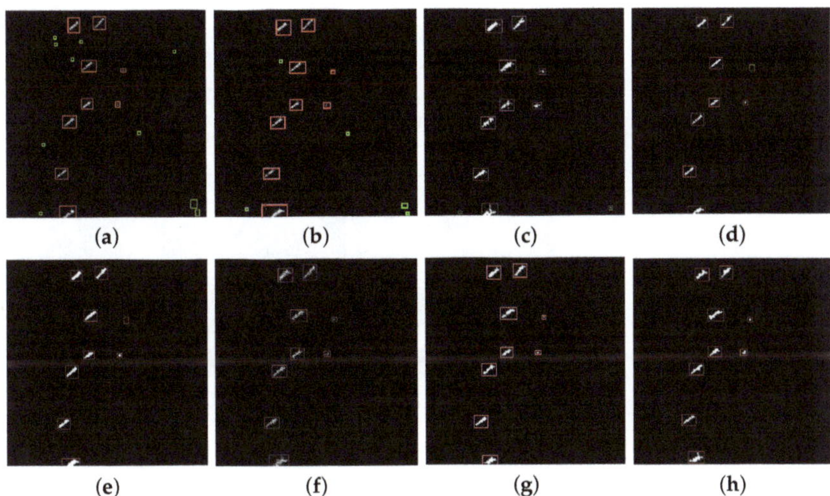

Figure 12. Different detection results of scene 1. (**a**) CA-CFAR method; (**b**) GΓD-CFAR method; (**c**) improved SP-CFAR method; (**d**) SLCM method; (**e**) adaptive SP-CFAR method; (**f**) O3DF-SVM-based method; (**g**) P3DF-SVM-based method; (**h**) proposed method.

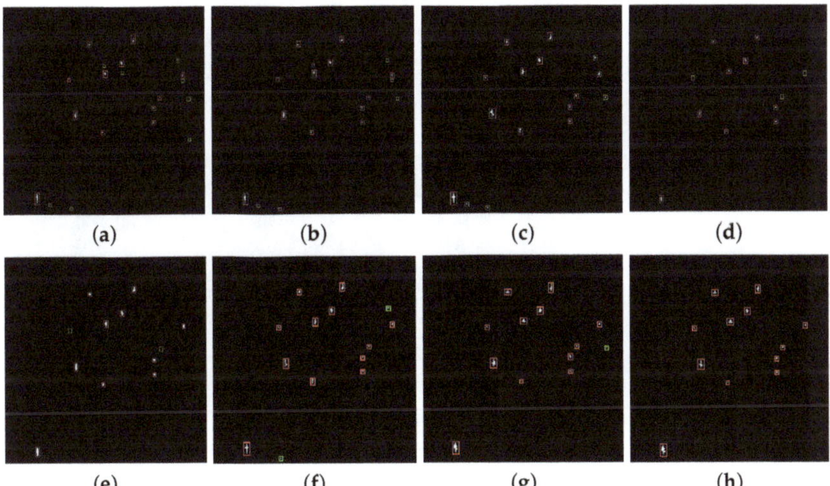

Figure 13. Different detection results of scene 2. (**a**) CA-CFAR method; (**b**) GΓD-CFAR method; (**c**) improved SP-CFAR method; (**d**) SLCM method; (**e**) adaptive SP-CFAR method; (**f**) O3DF-SVM-based method; (**g**) P3DF-SVM-based method; (**h**) proposed method.

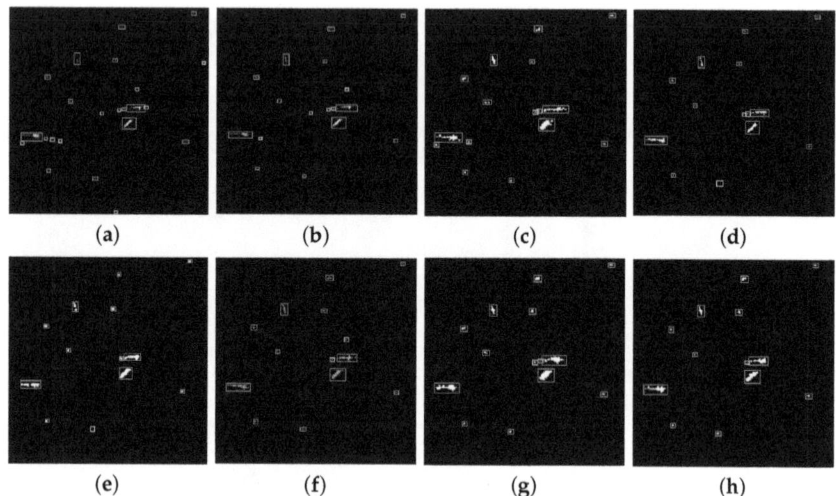

Figure 14. Different detection results of scene 3. (**a**) CA-CFAR method; (**b**) GΓD-CFAR method; (**c**) improved SP-CFAR method; (**d**) SLCM method; (**e**) adaptive SP-CFAR method; (**f**) O3DF-SVM-based method; (**g**) P3DF-SVM-based method; (**h**) proposed method.

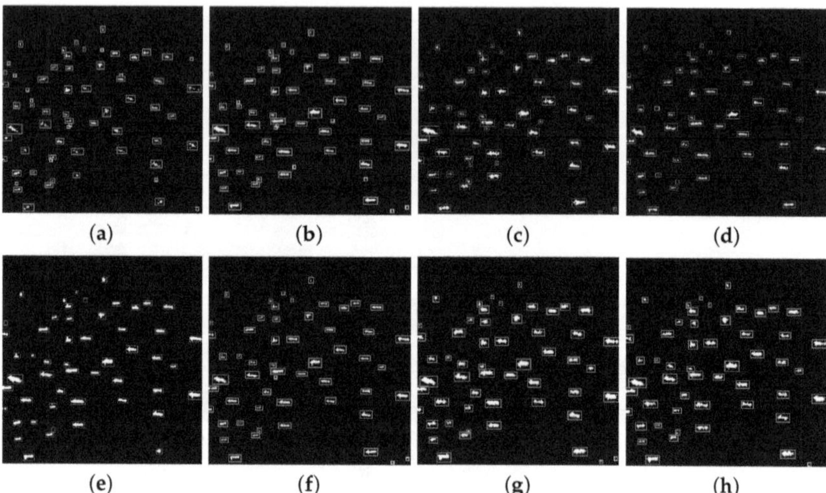

Figure 15. Different detection results of scene 4. (**a**) CA-CFAR method; (**b**) GΓD-CFAR method; (**c**) improved SP-CFAR method; (**d**) SLCM method; (**e**) adaptive SP-CFAR method; (**f**) O3DF-SVM-based method; (**g**) P3DF-SVM-based method; (**h**) proposed method.

Furthermore, the figure of merit (FoM) is adopted to evaluate the detection performance quantitatively and its calculation equation is:

$$FoM = \frac{N_d}{N_s + N_f} \tag{13}$$

where N_d, N_f and N_s are the number of correctly detected ship targets, false alarms, and real ships, respectively. The higher the FoM value, the better the target-detection performance.

FoMs of different methods are listed in Table 2. From Table 2, CA-CFAR and GГD-CFAR methods will bring a high false-alarm rate in complex multi-target SAR scenarios. The improved SP-CFAR method reduces the probability of false alarms. The SLCM method will cause some missed small ships with weaker scattering, and the adaptive SP-CFAR method can improve the ship target detection rate and reduce the probability of missed detection. Due to the limitation of ship training samples, the discrimination accuracy of the SVM model for strong interferences is limited, resulting in a small number of false alarms for the detection using the O3DF-SVM-based method and the P3DF-SVM-based method. However, the FoM values of the proposed method in Table 2 have more outstanding detection performance.

Table 2. FoMs of Different Methods in Different Scenes.

Scenes	Methods	N_d	N_f	*FoM*
Scene 1	CA-CFAR method	9	10	0.4737
	GГD-CFAR method	9	5	0.6429
	Improved SP-CFAR method	9	2	0.8182
	SLCM method	8	0	0.8889
	Adaptive SP-CFAR method	8	0	0.8889
	O3DF-SVM-based method	9	0	1
	P3DF-SVM-based method	9	0	1
	Proposed method	**9**	**0**	**1**
Scene 2	CA-CFAR method	12	8	0.6000
	GГD-CFAR method	12	6	0.6667
	Improved SP-CFAR method	12	4	0.7500
	SLCM method	9	0	0.7500
	Adaptive SP-CFAR method	10	0	0.8333
	O3DF-SVM-based method	12	2	0.8571
	P3DF-SVM-based method	12	1	0.9231
	Proposed method	**12**	**0**	**1**
Scene 3	CA-CFAR method	12	11	0.5217
	GГD-CFAR method	12	5	0.7059
	Improved SP-CFAR method	12	4	0.7500
	SLCM method	11	2	0.7857
	Adaptive SP-CFAR method	11	1	0.8462
	O3DF-SVM-based method	12	2	0.8571
	P3DF-SVM-based method	12	2	0.8571
	Proposed method	**12**	**1**	**0.9231**
Scen 4	CA-CFAR method	52	13	0.8000
	GГD-CFAR method	52	8	0.8667
	Improved SP-CFAR method	52	6	0.8966
	SLCM method	47	0	0.9038
	Adaptive SP-CFAR method	49	0	0.9423
	O3DF-SVM-based method	52	3	0.9455
	P3DF-SVM-based method	52	2	0.9630
	Proposed method	**52**	**1**	**0.9811**

In addition, by conducting target-detection experiments on many scene SAR images, the average performance and the standard deviation about FoM is shown in Table 3. From Table 3, the proposed algorithm has better detection performance and robustness compared to other comparative algorithms.

Table 3. The average values and standard deviation values of FoM for different algorithms.

	CA-CFAR Mrthod	GΓD-CFAR Method	Improved SP-CFAR Method	SLCM Method	Adaptive SP-CFAR Method	O3DF-SVM-Based Method	P3DF-SVM-Based Method	Proposed Method
Average values	0.6157	0.7127	0.8081	0.8312	0.8570	0.9090	0.9450	0.9724
Standard deviation values	0.1166	0.0949	0.0624	0.0548	0.0412	0.0387	0.0283	0.0200

To evaluate the detection performance of different methods more comprehensively, we use the receiver-operating-characteristic (ROC) curve to describe the relationship between the detection probability p_d and the observed false-alarm probability p_f. p_d and p_f are denoted, respectively, as:

$$p_d = \frac{N_{dtp}}{N_{tp}} \tag{14}$$

$$p_f = \frac{N_{dcp}}{N_{cp}} \tag{15}$$

where N_{tp} and N_{cp} represent the total number of target pixels and the total number of clutter pixels, respectively, N_{dtp} represents the number of correctly detected target pixels, and N_{dcp} is the number of pixels that falsely detect clutter pixels as target pixels.

Figure 16 presents the ROC curves in Figure 11a–d for different algorithms based on the real measured SAR data, which state the detection performance of all methods, and the observed false-alarm probability p_f in our method is the smallest under the same detection probability p_d, i.e., the proposed detection method in this paper outperforms other comparison methods in terms of ship-detection performance.

To verify ship-detection performance and computational complexity of the proposed method compared to the deep-learning-based methods, the YOLOv5-based detection method [40] is adopted for comparative analysis. Figure 17 presents the detection results for scene 2 and scene 4 (shown in Figure 11b,d), in which correctly detected targets and missed targets are still, respectively, marked by red and yellow rectangles. As shown in Figure 17, the YOLOv5-based detection method cannot effectively detect small targets, and our proposed method can improve the small ship target detection performance. In addition, the deep-learning detection method is data-driven, and the high-precision training model needs vast amounts of data, while our proposed anomaly-based detection method requires a small number of samples to train the detection decision model. Therefore, it is obvious that our proposed method has a lower computational cost in the learning stage. The average computational cost of the YOLOv5-based method and our proposed method in the testing stage are 57 ms and 42 ms, respectively, and it is clear that our method has an efficient execution capability.

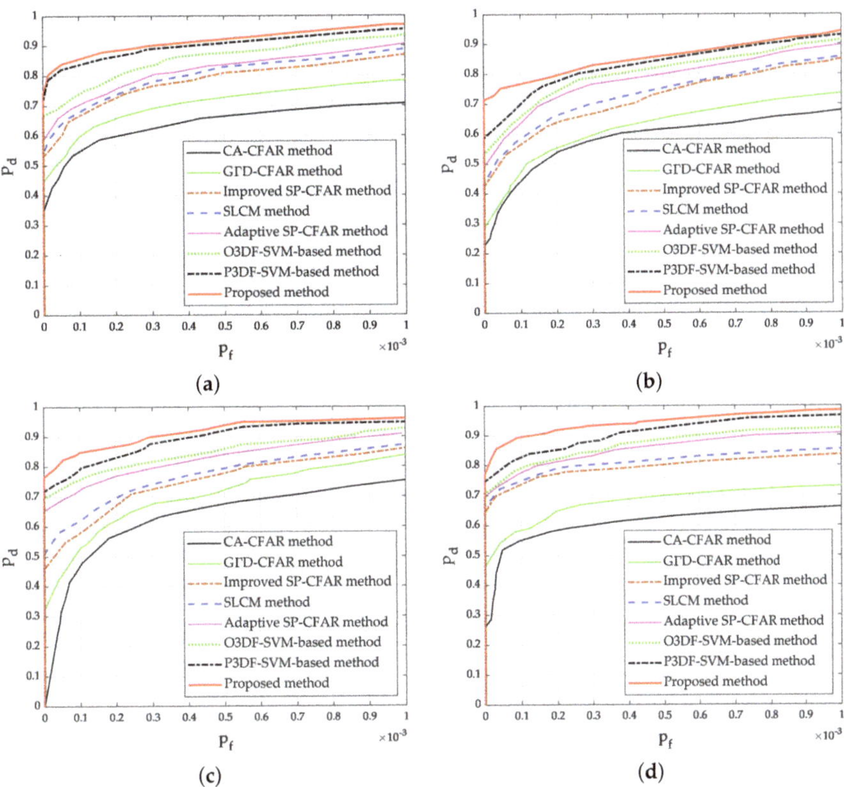

Figure 16. ROC curves for different methods. (**a**) ROC for Figure 11a; (**b**) ROC for Figure 11b; (**c**) ROC for Figure 11c; (**d**) ROC for Figure 11d.

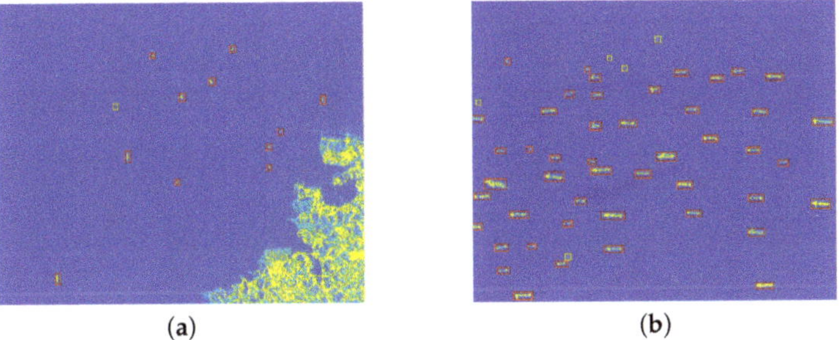

Figure 17. Detection results of different scenes based on the YOLOv5 model. (**a**) Detection result of scene 2 shown in Figure 11b. (**b**) Detection result of scene 4 shown in Figure 11d.

4. Conclusions

The paper investigates the anomaly-based ship-detection method of SP on the sea surface, and the effectiveness of the proposed algorithm is verified through huge amounts of multi-target SAR images. Due to the effect of island and speckle, preprocessing is indispensable and mainly includes sea–land and SP segmentation. To enhance the detection ability for different-scale ships, 3D feature spaces based on the boundary feature, saliency texture feature, and intensity attention contrast feature of the SP cell are designed. The anomaly-detection metric is built with flexible adjustment for the false-alarm rate. The COFL model with false-alarm control can efficiently and accurately classify ships and clutter SPs. Many experiments and comparative analyses of different methods prove that our algorithm can reduce the false-alarm rate in a complex multi-target environment, and at the same time significantly improves the detection performance of ship targets. However, there is often segmentation inaccuracy for ships of small sizes and weak scattering in complex scenes, which will affect the detection performance of small and weak ship targets. In the future, more effective SP feature-extraction methods and machine-learning models with more accurate classification accuracy will be further explored to improve the detection performance of ships of small sizes and weak scattering in complex scenes.

Author Contributions: Conceptualization, X.P.; methodology, X.P. and N.L.; software, N.L.; validation, N.L. and G.Z.; formal analysis, G.Z. and Z.W.; investigation, N.L. and X.P.; resources, L.Y.; data curation, X.P.; writing—original draft preparation, N.L.; writing—review and editing, N.L., X.P., L.Y., Z.H., J.C., Z.W. and G.Z.; supervision, L.Y., J.C. and Z.W. All authors have read and agreed to the published version of the manuscript.

Funding: This work was supported in part by the National Natural Science Foundation of China under Grant No. 62201004, in part by the Natural Science Foundation of Education Department of Anhui Province under Grant KJ2020A0030, in part by the Postdoctoral Fund of Anhui Province under Grant 2021B497 and in part by the Opening Foundation of the Key Laboratory of Intelligent Computing and Signal Processing under Grant 2020A009.

Institutional Review Board Statement: Not applicable.

Informed Consent Statement: Not applicable.

Data Availability Statement: Not applicable.

Acknowledgments: The authors would like to thank the reviewers for their valuable comments.

Conflicts of Interest: The authors declare no conflict of interest.

References

1. Allard, Y.; Germain, M.; Bonneau, O. Ship Detection and Characterization Using Polarime SAR Data. In *Harbour Protection Through Data Fusion Technologies*; Springer: Berlin/Heidelberg, Germany, 2009; pp. 243–250.
2. Xu, G.; Zhang, B.; Chen, J.; Xing, M.; Hong, W. Sparse synthetic aperture radar imaging from compressed sensing and machine learning: Theories, applications and trends. *IEEE Geosci. Remote Sens. Mag.* **2022**, *10*, 32–69. [CrossRef]
3. Zhang, B.; Xu, G.; Zhou, R.; Zhang, H.; Hong, W. Multi-channel back-projection algorithm for mmWave automotive MIMO SAR imaging with doppler-division multiplexing. *IEEE J. Sel. Top. Signal Process.* **2023**, *17*, 445–457. [CrossRef]
4. Sor, R.; Sathone, J.S.; Deoghare, S.U.; Sutaone, M.S. OS-CFAR based on thresholding approaches for target detection. In Proceedings of the 2018 Fourth International Conference on Computing Communication Control and Automation (ICCUBEA), Pune, India, 16–18 August 2018; pp. 1–6.
5. Tao, D.; Anfinsen, S.N.; Brekke, C. Robust CFAR detector based on truncated statistics in multiple-target situations. *IEEE Trans. Geosci. Remote Sens.* **2016**, *54*, 117–134. [CrossRef]
6. Ai, J.; Luo, Q.; Yang, X.; Yin, Z.; Xu, H. Outliers-robust CFAR detector of gaussian cutter based on the truncated-maximum-likelihood-estimator in SAR imagery. *IEEE Trans. Intell. Transp. Syst.* **2020**, *21*, 2039–2049. [CrossRef]
7. Almeida García, F.D.; Flores Rodriguez, A.C.; Fraidenraich, G.; Santos Filho, J.C.S. CA-CFAR detection performance in homogeneous Weibull clutter. *IEEE Trans. Geosci. Remote Sens. Lett.* **2019**, *16*, 887–891. [CrossRef]
8. Liu, N.; Sun, Y.; Ding, H.; Song, J. Comparative analysis of classical statistical models based on real sea clutter. *Comput. Simul.* **2017**, *34*, 448–452.
9. Li, W.; Zhang, Y.; Zhang, G. Sea clutter simulation research based on lognormal distribution. In Proceedings of the 2016 4th International Conference on Advanced Materials and Information Technology Processing (AMITP 2016), Guilin, China, 24–25 September 2016; pp. 545–548.
10. Li, B.; Liu, X.; Du, S.; Li, W. CFAR detector based on the identification of sea clutter distribution characteristics. *J. Phys. Conf. Ser.* **2022**, *2221*, 1–10. [CrossRef]
11. Saldanha, M.F.S.; Freitas, C.C.; Sant'Anna, S.J.S. Single channel SAR image segmentation using gamma distribution hipothesis test. In Proceedings of the 2012 IEEE International Geoscience and Remote Sensing Symposium (IGARSS), Munich, Germany, 22–27 July 2012; pp. 4323–4326.
12. Gao, G.; Ouyang, K.; Luo, Y.; Liang, S.; Zhou, S. Scheme of parameter estimation for generalized gamma distribution and its application to ship detection in SAR images. *IEEE Trans. Geosci. Remote Sens.* **2017**, *55*, 1812–1832. [CrossRef]

13. Zhou, H.; Li, Y.; Jiang, T. Sea clutter distribution modeling: A kernel density estimation approach. In Proceedings of the 2018 10th International Conference on Wireless Communications and Signal Processing (WCSP), Hangzhou, China, 18–20 October 2018; pp. 1–6.

14. Xin, Z.; Liao, G.; Yang, Z.; Zhang, Y.; Dang, H. Analysis of distribution using graphical goodness of fit for airborne SAR sea-clutter data. *IEEE Trans. Geosci. Remote Sens.* **2017**, *55*, 5719–5728. [CrossRef]

15. Qin, X.; Zhou, S.; Zou, H. A CFAR detection algorithm for generalized gamma distributed background in high-resolution SAR images. *IEEE Geosci. Remote Sens. Lett.* **2013**, *10*, 806–810.

16. Dai, H.; Du, L.; Wang, Y.; Wang, Z. A modified CFAR algorithm based on object proposals for ship target detection in SAR images. *IEEE Geosci. Remote Sens. Lett.* **2016**, *13*, 1925–1929. [CrossRef]

17. Ai, J.; Yang, X.; Yan, H. A local CFAR detector based on gray intensity correlation in SAR imagery. In Proceedings of the 2018 IEEE International Geoscience and Remote Sensing Symposium (IGARSS), Valencia, Spain, 22–27 July 2018; pp. 697–700.

18. Wang, C.; Wang, J.; Liu, X. A novel algorithm for ship detection in SAR images. In Proceedings of the 2019 IEEE International Conference on Signal Processing, Communications and Computing (ICSPCC), Dalian, China, 20–22 September 2019; pp. 1–5.

19. Pappas, O.; Achim, A.; Bull, D. Superpixel-level CFAR detectors for ship detection in SAR imagery. *IEEE Geosci. Remote Sens. Lett.* **2018**, *15*, 1397–1401. [CrossRef]

20. Li, M.; Cui, X.; Chen, S. Adaptive superpixel-level CFAR detector for SAR inshore dense ship detection. *IEEE Geosci. Remote Sens. Lett.* **2022**, *19*, 1–5. [CrossRef]

21. Li, T.; Peng, D.; Shi, S. Outlier-robust superpixel-level CFAR detector with truncated clutter for single look complex SAR images. *IEEE J. Sel. Top. Appl. Earth Obs. Remote Sens.* **2022**, *15*, 5261–5274. [CrossRef]

22. Zhang, L.; Zhang, Z.; Lu, S.; Xiang, D.; Su, Y. Fast Superpixel-Based Non-Window CFAR Ship Detector for SAR Imagery. *Remote Sens.* **2022**, *14*, 2092. [CrossRef]

23. Li, T.; Peng, D.; Chen, Z.; Guo, B. Superpixel-level CFAR detector based on truncated gamma distribution for SAR images. *IEEE Geosci. Remote Sens. Lett.* **2021**, *18*, 1421–1425. [CrossRef]

24. Li, T.; Liu, Z.; Xie, R.; Ran, L. An improved superpixel-level CFAR detection method for ship targets in high-resolution SAR images. *IEEE J. Sel. Top. Appl. Earth Obs. Remote Sens.* **2018**, *11*, 184–194. [CrossRef]

25. Yang, M.; Guo, C.; Zhong, H.; Yin, H. A curvature-Based saliency method for ship detection in SAR images. *IEEE Geosci. Remote Sens. Lett.* **2021**, *18*, 1590–1594. [CrossRef]

26. Wang, Z.; Wang, R.; Fu, X.; Xia, K. Unsupervised ship detection for single-channel SAR images based on multiscale saliency and complex signal kurtosis. *IEEE Geosci. Remote Sens. Lett.* **2022**, *19*, 1–5. [CrossRef]

27. Li, N.; Pan, X.; Yang, L.; Huang, Z.; Wu, Z.; Zheng, G. Adaptive CFAR method for SAR ship detection using intensity and texture feature fusion attention contrast mechanism. *Sensors* **2022**, *22*, 8116. [CrossRef]

28. Cheng, J.; Xiang, D.; Tang, J.; Zheng, Y.; Guan, D.; Du, B. Inshore ship detection in large-scale SAR images based on saliency enhancement and bhattacharyya-like distance. *Remote Sens.* **2022**, *14*, 2832. [CrossRef]

29. Liang, Y.; Sun, K.; Zeng, Y.; Li, G.; Xing, M. An adaptive hierarchical detection method for ship targets in high-resolution SAR images. *Remote Sens.* **2020**, *12*, 303. [CrossRef]

30. Wang, X.; Chen, C.; Pan, Z. Superpixel-based LCM detector for faint ships hidden in strong noise background SAR imagery. *IEEE Geosci. Remote Sens. Lett.* **2019**, *16*, 417–421. [CrossRef]

31. Lin, H.; Chen, H.; Jin, K.; Zeng, L.; Yang, J. Ship detection with superpixel-level fisher vector in high-resolution SAR images. *IEEE Geosci. Remote Sens. Lett.* **2020**, *17*, 247–251. [CrossRef]

32. He, G.; Xia, Z.; Chen, H.; Li, K.; Zhao, Z.; Guo, Y.; Feng, P. An adaptive ship detection algorithm for HRWS SAR images under complex background: Application to sentinel1a data. *Int. Arch. Photogramm. Remote. Sens. Spat. Inf. Sci.* **2018**, *3*, 497–503. [CrossRef]

33. Xiong, W.; Xu, Y.; Yao, L. A new ship target detection algorithm based on SVM in high resolution SAR image. *Remote. Sens. Technol. Appl.* **2018**, *33*, 119–127.

34. Li, H.; Cui, X.; Chen, S. PolSAR ship detection with optimal polarimetric rotation domain features and SVM. *Remote Sens.* **2021**, *13*, 3932. [CrossRef]

35. Yang, X.; Bi, F.; Yu, Y.; Chen, L. An effective false-alarm removal method based on OC-SVM for SAR ship detection. In Proceedings of the IET International Radar Conference 2015, Hangzhou, 14–16 October 2015; pp. 1–4.

36. Zou, B.; Qiu, Y.; Zhang, L. Docked ships detection using PolSAR image based on GOPSO-SVM. In Proceedings of the 2019 IEEE Radar Conference (RadarConf), Boston, MA, USA, 22–26 April 2019; pp. 1–6.

37. Zou, B.; Qiu, Y.; Zhang, L. Ship detection using PolSAR images based on simulated annealing by fuzzy matching. *IEEE Geosci. Remote Sens. Lett.* **2022**, *19*, 1–5. [CrossRef]

38. Jiang, S.; Zhu, M.; He, Y.; Zheng, Z.; Zhou, F.; Zhou, G. Ship detection with SAR based on Yolo. In Proceedings of the 2020 IEEE International Geoscience and Remote Sensing Symposium (IGARSS), Waikoloa, HI, USA, 26 September–2 October 2020; pp. 1647–1650.

39. Yu, H.; Li, Y.; Zhang, D. An improved YOLO v3 small-scale ship target detection algorithm. In Proceedings of the 2021 6th International Conference on Smart Grid and Electrical Automation (ICSGEA), Kunming, China, 29–30 May 2021; pp. 560–563.

40. Xu, X.; Zhang, X.; Zhang, T. SAR ship detection using YOLOv5 algorithm with anchor boxes cluster. In Proceedings of the 2022 IEEE International Geoscience and Remote Sensing Symposium (IGARSS), Kuala Lumpur, Malaysia, 17–22 July 2022; pp. 2139–2142.
41. Sun, Z.; Leng, X.; Lei, Y.; Xiong, B.; Ji, K.; Kuang, G. BiFA-YOLO: A novel YOLO-based method for arbitrary-oriented ship detection in high-resolution SAR images. *Remote Sens.* **2021**, *13*, 4209. [CrossRef]
42. Guo, Y.; Chen, S.; Zhan, R.; Wang, W.; Zhang, J. LMSD-YOLO: A lightweight YOLO algorithm for multi-scale SAR ship detection. *Remote Sens.* **2022**, *14*, 4801. [CrossRef]
43. Xu, X.; Zhang, X.; Zhang, T. Lite-YOLOv5: A lightweight deep learning detector for on-board ship detection in large-scene sentinel-1 SAR images. *Remote Sens.* **2022**, *14*, 1018. [CrossRef]
44. Tang, G.; Zhuge, Y.; Claramunt, C.; Men, S. N-YOLO: A SAR ship detection using noise-classifying and complete-target extraction. *Remote Sens.* **2021**, *13*, 871. [CrossRef]
45. Chang, Y.L.; Anagaw, A.; Chang, L.; Wang, Y.; Hsiao, C.Y.; Lee, W.H. Ship detection based on YOLOv2 for SAR imagery. *Remote Sens.* **2019**, *11*, 786. [CrossRef]
46. Ke, X.; Zhang, X.; Zhang, T.; Shi, J.; Wei, S. SAR ship detection based on an improved faster R-CNN using deformable convolution. In Proceedings of the 2021 IEEE International Geoscience and Remote Sensing Symposium (IGARSS), Brussels, Belgium, 11–16 July 2021; pp. 3565–3568.
47. Chai, B.; Chen, L.; Shi, H.; He, C. Marine ship detection method for SAR image based on improved faster RCNN. In Proceedings of the 2021 SAR in Big Data Era (BIGSARDATA), Nanjing, China, 22–24 September 2021; pp. 1–4.
48. Liu, D.; Gao, S. Ship targets detection in remote sensing images based on improved faster-RCNN. *J. Phys. Conf. Ser.* **2021**, *2132*, 1–6. [CrossRef]
49. Xia, R.; Chen, J.; Huang, Z.; Wan, H.; Wu, B.; Sun, L.; Yao, B.; Xiang, H.; Xing, M. CRTransSar: A visual transformer based on contextual joint representation learning for SAR ship detection. *Remote Sens.* **2022**, *14*, 1488. [CrossRef]
50. Zhou, Y.; Fu, K.; Han, B.; Yang, J.; Pan, Z.; Hu, Y.; Yin, D. D-MFPN: A doppler feature matrix fused with a multilayer feature pyramid network for SAR ship detection. *Remote Sens.* **2023**, *15*, 626. [CrossRef]
51. Rostami, M.; Kolouri, S.; Eaton, E.; Kim, K. Deep transfer learning for few-shot SAR image classification. *Remote Sens.* **2019**, *11*, 1374. [CrossRef]
52. Zhang, T.; Zhang, X. High-speed ship detection in SAR images based on a grid convolutional neural network. *Remote Sens.* **2019**, *11*, 1206. [CrossRef]
53. Wei, S.; Su, H.; Ming, J.; Wang, C.; Yan, M.; Kumar, D.; Shi, J.; Zhang, X. Precise and robust ship detection for high-resolution SAR imagery based on HR-SDNet. *Remote Sens.* **2020**, *12*, 167. [CrossRef]
54. Liu, N.; Cao, Z.; Cui, Z.; Pi, Y.; Dang, S. Multi-scale proposal generation for ship detection in SAR images. *Remote Sens.* **2019**, *11*, 526. [CrossRef]
55. Sun, Y.; Wang, Z.; Sun, X.; Fu, K. SPAN: Strong scattering point aware network for ship detection and classification in large-scale SAR imagery. *IEEE J. Sel. Top. Appl. Earth Obs. Remote Sens.* **2022**, *15*, 1188–1204. [CrossRef]
56. Gao, S.; Liu, H. Polarimetric SAR ship detection based on scattering characteristics. *IEEE J. Miniat. Air Space Syst.* **2022**, *3*, 197–203. [CrossRef]
57. Zhang, T.; Yang, Z.; Xing, C.; Zeng, L.; Yin, J.; Yang, J. Ship detection from polsar imagery based on the scattering difference parameter. In Proceedings of the 2020 IEEE International Geoscience and Remote Sensing Symposium (IGARSS), Waikoloa, HI, USA, 26 September–2 October 2020; pp. 1255–1258.
58. Wang, N.; Li, B.; Xu, Q.; Wang, Y. Automatic ship detection in optical remote sensing images based on anomaly detection and SPP-PCANet. *Remote Sens.* **2019**, *11*, 47. [CrossRef]
59. Zhai, L.; Li, Y.; Su, Y. A novel ship detection algorithm based on anomaly detection theory for SAR images. In Proceedings of the 2016 Progress in Electromagnetic Research Symposium (PIERS), Shanghai, China, 8–11 August 2016; pp. 2868–2872.
60. Chen, X.; Sun, J.; Yin, K.; Yu, J. Sea-land segmentation algorithm of SAR image based on Otsu method and statistical characteristic of sea area. *J. Data Acquis. Process.* **2014**, *29*, 603–608.
61. Abdullah, A.; Albashish, D. Empirical comparision on boosted cascade of Haar-like features to histogram of oriented gradients for person detection. In Proceedings of the 2021 International Conference on Electrical Engineering and Informatics (ICEEI), Kuala Terengganu, Malaysia, 12–13 October 2021; pp. 1–6.
62. Karanwal, S. Improved LBP and discriminative LBP: Two novel local descriptors for face recognition. In Proceedings of the 2022 IEEE International Conference on Data Science and Information System (ICDSIS), Hassan, India, 29–30 July 2022; pp. 1–6.
63. Ravi Kumar, Y.B.; Ravi Kumar, C.N. Local binary pattern: An improved LBP to extract nonuniform LBP patterns with Gabor filter to increase the rate of face similarity. In Proceedings of the International Conference on Cognitive Computing and Information Processing (CCIP), Mysuru, India, 12–13 August 2016; pp. 1–5.
64. Huang, W.; Lu, S.; Tang, X. A method using clustering and SVDD for quality detection. In Proceedings of the 2021 33rd Chinese Control and Decision Conference (CCDC), Kunming, China, 22–24 May 2021; pp. 4069–4072.

65. Wei, S.; Zeng, X.; Qu, Q.; Wang, M.; Su, H.; Shi, J. HRSID: A high-resolution SAR images dataset for ship detection and instance segmentation. *IEEE Access* **2020**, *8*, 120234–120254. [CrossRef]
66. Zhang, T.; Zhang, X.; Ke, X.; Zhan, X.; Shi, J.; Wei, S.; Pan, D.; Li, J.; Su, H.; Zhou, Y.; et al. LS-SSDD-v1.0: A deep learning dataset dedicated to small ship detection from large-scale Sentinel-1 SAR Images. *Remote Sens.* **2020**, *12*, 2997. [CrossRef]

Article

Locality Preserving Property Constrained Contrastive Learning for Object Classification in SAR Imagery

Jing Wang [1], Sirui Tian [1,*], Xiaolin Feng [1], Bo Zhang [2,3], Fan Wu [2,3], Hong Zhang [2,3] and Chao Wang [2,3]

[1] School of Electronic and Optical Engineering, Nanjing University of Science and Technology, Nanjing 210094, China; 121104010582@njust.edu.cn (J.W.); fengxiaolin@njust.edu.cn (X.F.)

[2] International Research Center of Big Data for Sustainable Development Goals, Beijing 100094, China; zhangbo@radi.ac.cn (B.Z.); wufan@aircas.ac.cn (F.W.); zhanghong@radi.ac.cn (H.Z.); wangchao@radi.ac.cn (C.W.)

[3] Key Laboratory of Digital Earth Science, Aerospace Information Research Institute, Chinese Academy of Sciences, Beijing 100094, China

* Correspondence: tiansirui@njust.edu.cn; Tel.: +86-138-0518-9826

Citation: Wang, J.; Tian, S.; Feng, X.; Zhang, B.; Wu, F.; Zhang, H.; Wang, C. Locality Preserving Property Constrained Contrastive Learning for Object Classification in SAR Imagery. *Remote Sens.* **2023**, *7*, 3697. https://doi.org/10.3390/rs15143697

Academic Editor: Dusan Gleich

Received: 26 April 2023
Revised: 21 June 2023
Accepted: 22 July 2023
Published: 24 July 2023

Abstract: Robust unsupervised feature learning is a critical yet tough task for synthetic aperture radar (SAR) automatic target recognition (ATR) with limited labeled data. The developing contrastive self-supervised learning (CSL) method, which learns informative representations by solving an instance discrimination task, provides a novel method for learning discriminative features from unlabeled SAR images. However, the instance-level contrastive loss can magnify the differences between samples belonging to the same class in the latent feature space. Therefore, CSL can dispel these targets from the same class and affect the downstream classification tasks. In order to address this problem, this paper proposes a novel framework called locality preserving property constrained contrastive learning (LPPCL), which not only learns informative representations of data but also preserves the local similarity property in the latent feature space. In LPPCL, the traditional InfoNCE loss of the CSL models is reformulated in a cross-entropy form where the local similarity of the original data is embedded as pseudo labels. Furthermore, the traditional two-branch CSL architecture is extended to a multi-branch structure, improving the robustness of models trained with limited batch sizes and samples. Finally, the self-attentive pooling module is used to replace the global average pooling layer that is commonly used in most of the standard encoders, which provides an adaptive method for retaining information that benefits downstream tasks during the pooling procedure and significantly improves the performance of the model. Validation and ablation experiments using MSTAR datasets found that the proposed framework outperformed the classic CSL method and achieved state-of-the-art (SOTA) results.

Keywords: synthetic aperture radar (SAR); automatic target recognition (ATR); contrastive self-supervised learning (CSL); instance-level contrastive loss; noise-induced estimation of mutual information (InfoNCE); locality preserving projections (LPP)

1. Introduction

Synthetic aperture radar (SAR) is a high-resolution coherent imaging radar. It is widely employed in commercial and military surveillance and reconnaissance as an active microwave remote-sensing technology [1]. As the number of operating SAR sensors and images to be interpreted continues to grow, automatic target recognition using SAR imagery (SAR ATR) has piqued the interest of many researchers [2].

Due to the specific electromagnetic imaging mechanism of SAR sensors, it is a considerable challenge to interpret SAR images without domain knowledge and obtain sufficient annotated data for target classification tasks. Recently, with the rapid development of unsupervised or self-supervised deep learning theory, robust representation learning with unlabeled data has received increasing attention from the SAR remote sensing community [3].

Numerous unsupervised models have been developed to mine discriminative features from unlabeled SAR images, including Auto-Encoders (AE) [4], Variational Auto-Encoders (VAE) [5], Generative Adversarial Networks (GANs) [6] and contrastive self-supervised learning (CSL) [7]. Although these methods can adaptively learn robust, informative representations without labels, they still have some issues that greatly deteriorate the performance of their learned features in downstream tasks. AEs and their variants, which learn representation by reconstructing the inputs from the encoded features, focus on the reconstruction performance rather than the discriminative capability of their learned features. Accordingly, there is no guarantee that these models' learned representations will contribute to downstream classification tasks [8]. As a specific AE, VAEs have these same drawbacks. Moreover, they usually assume that the variables follow Gaussian distributions with diagonal covariance matrices, which are quite different from the distributions of SAR images, leading to performance degradation in downstream classification tasks [9]. GANs provide an alternative method of learning the distributions of the inputs through the adversarial procedure with a generator and a discriminator. However, GANs encounter many problems in representation learning tasks related to, for instance, the complexity of their model architecture, the mode collapse problem, the non-convergence and instability of the training procedure, and their requirement for large amounts of data [10].

Recently, the CSL framework, which is a popular form of self-supervised learning, has been adapted to solve various vision tasks, including classification, object detection, and instance segmentation. Using unlabeled data, the CSL models are trained in instance discrimination tasks that discriminate pairs of positive (similar) inputs from a selection of negative (dissimilar) pairs minimizing the InfoNCE loss [11]. CSL models usually obtain a positive sample of the anchor by complex data augmentations and treat the other data in the dataset as negative samples. Due to its excellent performance in unsupervised representation learning, it has been utilized to extract discriminative features from unlabeled SAR images. However, in the absence of true labels, negative samples are often randomly sampled; these samples may have the same label. Furthermore, because SAR images belonging to the same category are highly similar, the CSL framework can dispel samples from the same class using the traditional InfoNCE loss. Both aspects mentioned above lead to a biased representation, greatly affecting these models' performance in downstream classification tasks. To alleviate this problem, a Debiased Contrastive Learning (DCL) [12] model based on the SimCLR [13] framework has been proposed, which exploits a decomposition of the true negative distribution to correct for the sampling of same-label data points. Although the DCL model significantly outperforms the traditional CSL methods, it still struggles to estimate the negative distribution.

To tackle this problem, in this paper, an alternative debiased approach based on the LPP constraint is proposed, called locality preserving property constrained contrastive learning (LPPCL). Similar to the DCL model, the proposed model adopts the SimCLR framework as its prototype due to its tractability and satisfactory results. The proposed method organically combines contrastive loss and the LPP constraint and is able to learn an informative representation of data while simultaneously maintaining the local similarity property in the feature space. Specifically, the LPP constraint is added to the cross-entropy form of the InfoNCE loss as a soft pseudo label, which can introduce additional data structure in order to alleviate the problem of biased sampling and promote performance in representation learning. Moreover, we used a multi-branch structure to produce more training samples through numerous stochastic data augmentations. Finally, we replaced the traditional average pooling approach with self-attention pooling to extract features more effectively.

The main contributions can be summarized as follows:

1. Prior knowledge of local similarity was embedded into the InfoNCE loss, which was reformulated in the cross-entropy form, reproducing the debiased contrastive loss that the intra-class relationship of nearby samples maintains in the feature space.

2. A multi-branch structure was devised to replace the traditional double-branch structure of CSL, significantly improving the sample diversity, the robustness of representations, and the stability of mutual information estimation.
3. The novel self-attention pooling was introduced to replace the global average pooling in the standard ResNet encoders, providing an adaptive informative feature extraction capability according to the characteristics of inputs and thus avoiding information loss caused by traditional hand-crafted pooling methods.

The rest of this manuscript is organized as follows: a literature review on the CSL method and its application in the SAR ATR is presented in Section 2. A brief summary of backgrounds, including those on LPP and the traditional SimCLR framework, is provided in Section 3, while the proposed new framework is presented in Section 4. Section 5 reports the results of validation and ablation experiments on the MSTAR dataset, with this being followed by the study's conclusion in Section 6.

2. Related Work

The CSL approach, which provides a means of enhancing the performance of unsupervised feature learning, is beginning to capture the interest of many researchers. The approach has two essential components: notions of positive and negative pairs of data points. Its training objective, typically the noise-contrastive estimation of mutual information [14], guides the learned representation to map positive pairs to nearby locations in the feature space, as well as negative pairs that are farther apart. Most recent CSL models fall into one of two categories: context-instance contrast (CIC) and instance-instance contrast (IIC) models [15]. CIC models [16,17] aim to represent the affiliation between a sample's local characteristics and its global context representation. However, IIC models analyze the relationships between various samples' instance-level local representations directly. For a variety of classification tasks, instance-level representation is more important than context-level representation, which was first studied in a paper on DeepCluster [18]. The authors suggested using clustering to generate the pseudo-label for every image, and the labels of these images were predicted by a discriminator. Instance discrimination-based approaches, on the other hand, eliminate the sluggish clustering stage and add useful image augmentation mechanisms. InstDisc [19] was a prototype model that used instance discrimination as its pretext task. Based on InstDisc, the MoCo [20] model further investigated the concept of utilizing instance discrimination and significantly increased the number of negative samples by a sample queue. With an online encoder and an offline encoder, it created momentum contrast learning by updating the weights of the offline encoder with the exponential moving average according to the parameters of the online encoder. The momentum contrast learning scheme played a critical role in preventing model collapse. To address the issue of requiring massive negative samples in MoCo, SimCLR [13] used an end-to-end training architecture with a large batch size. SimCLR possessed a two-branch structure that had encoders with shared weights. It also replaced the original linear layer with a projection head to change the dimensions of the representation. Based on SimCLR, Debiased Contrastive Learning (DCL) [12] proposed a debiased contrastive objective that corrected for the sampling of same-label examples without labels. It achieved this by taking the viewpoint of Positive-unlabeled Learning and exploiting a decomposition of the true negative distribution. InfoMin [21] incorporated additional research on enhancing positive samples, it reduced MI between augmented views in order to learn more effective representations. Furthermore, BYOL [22] was devised in a Siamese network framework to reject the need for negative samples in MoCo and SimCLR models. Additional research on SimSiam [23] revealed the significance of the stop-gradient strategy to avoid collapsing in CSL. To conclude, the CSL approach has better generalization performance and fault tolerance in classification tasks.

The developing CSL approach provides a novel method for learning useful features from unlabeled SAR images. However, the traditional contrastive learning method does not perform well on SAR ATR tasks; many efforts have been made to improve the CSL framework, and several new models have been devised. Wang et al. conducted SAR imagery classification on the basis of pseudo labels and contrastive learning to address the issue of the lack of labeled SAR images [24]. Zhou et al. proposed a limited data loss function (LDLF) that naturally integrated contrastive loss function and cross-entropy loss function. This aimed to solve the problem of overfitting in SAR ATR by combining strong supervision and weak supervision [25]. Based on the instance-level contrastive loss, Zhai et al. proposed batch instance discrimination and feature clustering (BIDFC), which can adjust the embedding distance in the feature space [26]. A contrastive domain adaption methodology was used by Bi et al. to reduce the disparity in distribution between a simulated and actual sparse SAR dataset [27]. The efficiency of multi-view contrastive loss was confirmed by Chen et al. [28]. They suggested the use of a self-supervised method to extract pixel-level feature representations from unlabeled SAR images. To increase the classification accuracy of SAR ATR for ships, Xu et al. modified the SimSiam framework, which is a classic CSL framework, and developed a new positive pair sampling method that considered polarization information [29]. Wang et al. proposed a mixture loss method consisting of contrastive loss and label propagation to investigate the global and local representations in SAR images [30]. Ren et al. proposed a Siamese feature embedding network and leveraged the CSL approach to train a low-dimensional feature space for feature extraction in SAR ATR [31]. The efficiency of CSL-based pre-training models for SAR and optical imagery classification was examined by Liu et al. They used registered examples that had structural consistency as contrastive pairs in the Siamese framework to acquire shared representations of SAR images [32]. For the task of SAR image scene classification, Xiao et al. forwarded a lightweight CSL framework to learn features by maximizing the similarity between augmented views [33]. Liu et al. proposed a clustering-based CSL model in order to map SAR images from pixel space to feature space, promoting node representation and information propagation in the network [34]. Additionally, Yang et al. forwarded a coarse-to-fine CSL framework to extract valuable representations for SAR image classification tasks at the pixel level [35]. Despite these advances, the CSL model in the SAR image processing domain does not perform as well as in the CV domain due to the lack of prior knowledge and biased sampling of negative samples.

3. Background

3.1. Locality Preserving Projections

The LPP algorithm [36] is a manifold learning strategy that maps high-dimensional data into a low-dimensional manifold where the local relationship of the data is preserved, utilizing the nearest-neighbor graph of the input data. The algorithm's objective function is shown below:

$$\min \sum_{i,j} \left(e_i - e_j \right)^2 S_{ij} \tag{1}$$

where $e_i = G^T x_i$, x_i denotes the original samples, G is the optimal projection matrix. The affinity matrix S represents the similarity between x_i and x_j, which is established by calculating the cosine distance between two points:

$$S_{ij} = \begin{cases} \frac{x_i^T x_j}{\|x_i\| \cdot \|x_j\|}, & \text{if } x_j \text{ is } k - \text{nearest neighbor of } x_i \\ 0, & \text{otherwise} \end{cases} \tag{2}$$

It is not difficult to determine from (1) and (2) that when the neighboring points x_i and x_j are mapped far away from each other, the loss function with the symmetric weights $S_{ij} = S_{ji}$ will incur a heavy penalty. Therefore, the purpose of minimizing this function is to ensure that the corresponding points will still be neighbors in the low-dimensional projection space if two points are considered to be close by in the high-dimensional space. To some extent, LPP optimally preserves the local neighborhood information of data.

3.2. Contrastive Learning and InfoNCE Loss

As a very popular self-supervised learning technique, CSL usually obtains effective image features from the multi-view perspective. In the CSL framework, two views of the input are generated independently by various image transformations. Based on the assumption that each view contains all the required information for the downstream tasks, the CSL network utilizes a dual-branch architecture to learn discriminative representations via an instance discrimination criterion named InfoNCE [11]. Accordingly, the features extracted from different views of the same instance are attracted together, while features extracted from the views of different instances are dispelled.

The SimCLR framework, a crucial baseline for current CSL methods, has a two-branch structure with shared weights in both the encoders and the projection heads of each branch. Instead of using a momentum encoder to update the memory bank, the model requires the batch size of the input data to be large enough to eliminate the requirement for a memory bank. The framework also uses more complicated data augmentation strategies and replaces the original linear layer with a projection head to change the dimensions of a given representation to enhance the effectiveness of contrastive learning. SimCLR is nearly the simplest model, and yet one of the models that achieved the SOTA results in various downstream tasks, including classification, semantic segmentation, and object detection. As shown in Figure 1, the core idea of SimCLR is to learn informative representations by solving an instance discrimination task that pulls together the positive samples from the same instance and pushes away the other $2(N-1)$ negative samples from different instances in the feature space. More specifically, positive samples are generally derived from various augmented images of the same instance, while negative samples are augmented data from different instances. The objective function of SimCLR is derived from the estimation of mutual information (MI) between the inputs and the learned representations. Intuitively speaking, a useful feature should contain as much input information as possible. To this end, an effective strategy is to maximize the MI between embedding features and the original image. However, it is infeasible to compute MI directly in most conditions. An alternative approach is to calculate the upper bound or low bound of MI as an approximal estimation. Recent work has introduced variational bounds with deep learning by using a variational distribution, $p_v(y|x)$, instead of the conditional distribution, $p_c(y|x)$, when calculating the lower bound of the MI:

$$I(X;Y) \geq E_{p_c(x,y)}[\log p_v(y|x)] + g(X) \tag{3}$$

where, X and Y are random variables in the data space and the learned embedding space, respectively; x represents the samples of X; y is the corresponding representation that has maximal MI with x subject to constraints on the mapping. Poole et al. combined the above unnormalized bound with multiple examples [11]. Then, the low bound of the MI can be maximized by minimizing the InfoNCE loss [37]:

$$\mathcal{L}_{InfoNCE} = \mathbb{E}\left[-log\frac{exp(sim(x_i, y_i)/\tau)}{\sum_{k=1}^{N} exp(sim(x_i, y_k)/\tau)}\right] \tag{4}$$

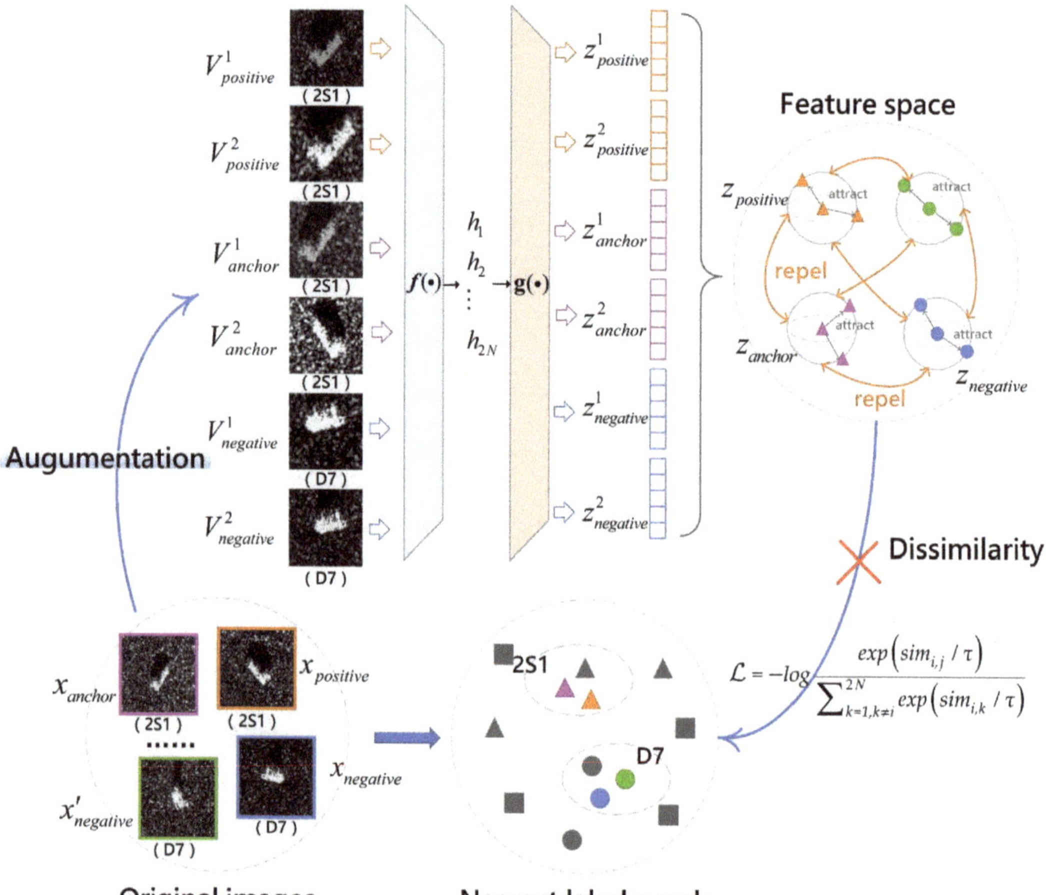

Figure 1. Basic concept of the SimCLR framework. V^1_{anchor} and V^2_{anchor} are two augmented views generated by the same anchor instance x_{anchor}, which are regarded as a positive pair by SimCLR, while all other views from different instances are regarded as negative samples, whether from the same class like $V^1_{positive}$ and $V^2_{positive}$ and from different class like $V^1_{negative}$ and $V^2_{negative}$, which is also the reason for biased sampling.

The loss function of SimCLR can be obtained by substituting the parameters into Equation (4):

$$\mathcal{L} = -\log \frac{exp\left(sim_{i,j}/\tau\right)}{\sum_{k=1,k\neq i}^{2N} exp\left(sim_{i,k}/\tau\right)} \tag{5}$$

with

$$sim_{i,j} = \frac{z_i^{\mathrm{T}} z_j}{\left(\parallel z_i \parallel \cdot \parallel z_j \parallel\right)} \tag{6}$$

where $z = g_\varphi(f_\theta(V))$, and the positive pair (z_i, z_j) represents the corresponding representation projection of the augmented views V_i and V_j. Specifically, this involves the data augmentation $t \sim \mathcal{T}$, $t' \sim \mathcal{T}$ that transforms the given example x_{anchor}, resulting in two corresponding views of the same example, denoted V^1_{anchor} and V^2_{anchor} (see Figure 1). Further, $f_\theta(\cdot)$ is the encoder network and $g_\varphi(\cdot)$ represents the projection head. The temperature coefficient τ is used to modulate the intensity of contrastive learning, which is dependent on confidence. Larger τ values result in a smoother probability distribution. By optimizing the objective function in (5), the SimCLR model can be easily trained.

In fact, by combining the positive sample z_j and negative samples $\left\{z_1, z_2 \ldots z_{2(N-1)}\right\}$ together as $\left\{z'_1, z'_2 \ldots z'_{2N-1}\right\} \triangleq \left\{z_j, z_1, z_2 \ldots z_{2(N-1)}\right\}$, the objective function of the SimCLR in (5) can be reformulated as a cross-entropy loss function with one-hot labels:

$$\mathcal{L} = -\sum_{q=1}^{2N-1} y_q log p_q \tag{7}$$

with

$$p_q = \frac{exp\left(S'_{i,q}/\tau\right)}{\sum_{k=1}^{2N-1} exp\left(S'_{i,k}/\tau\right)}$$
$$y_q = \begin{cases} 1, q = 1 \\ 0, otherwise \end{cases} \tag{8}$$
$$S'_{i,q} = \frac{z_i^T z'_q}{\left(\|z_i\| \cdot \|z'_q\|\right)}$$

where $\left[p_1, p_2 \ldots p_q \ldots p_{2N-1}\right]$ represents the prediction probability vector and $\left[y_1, y_2 \ldots y_q \ldots y_{2N-1}\right]$ is the corresponding one-hot pseudo label.

In a supervised case, p_q represents the probability that the sample belongs to each class and y_q denotes the true label. In unsupervised or self-supervised cases, however, the label y_q is generated at the instance level, i.e., each instance belongs to a separate semantic class. As a result, samples are not clustered according to their categories in the latent space. This issue is more serious when the samples of different categories are very similar, and the distance between individual samples is smaller than the distance between categories. As a result, samples of the same categories and different categories will all be dispelled, thus affecting the framework's classification accuracy in downstream tasks.

4. Methodology

4.1. Locality Preserving Property Constrained InfoNCE Loss

To solve the problems associated with the instance-level CSL mentioned above, we aim to introduce additional data structure information into our proposed model to assist its representation learning. The problem with instance-based contrastive learning is that it does not take into account the actual distribution of negative examples when sampling the negative samples, and all other samples except the anchors are regarded as negative samples. Such a sampling scheme leads to bias in the learned representations. Some features extracted from samples of the same class are regarded as negative samples, which seriously weakens the classification capability of the learned features. Given this problem, additional categorical information is required to alleviate the biases in the sampling of negative samples. Obviously, in the case of supervised learning, it is easy to introduce the labels as the categorical information to the cross-entropy form of the SimCLR loss function in (7), and thus biases can be eliminated.

However, in unsupervised representation learning, the prior knowledge we can obtain only comes from the distribution of the data itself. DCL [12] alleviates the biased sampling process by assuming the distribution of negative samples, which could be interfered with the improper prior distribution of the negative samples. Recalling the LPP algorithm, the local similarity relationship in the original data space is preserved in feature space since it accurately reflects the inherent clustering structure of the samples. Motivated by the LPP algorithm, in this study, we considered the local similarity of samples in the data space as the prior knowledge to improve the sampling of negative samples to improve the SimCLR InfoNCE loss.

To accomplish this, the adjacency matrix used for the local similarity measure in the LPP algorithm is introduced into the CSL framework, and its loss function in the cross-entropy form is modified. The aim of this is to preserve the local neighborhood relationship of the original data in the learned features. We first construct the LPP affinity matrix

S_{ij} of the original feature of the target image $[x_1, x_2 \ldots x_N]$, which will maintain the local similarity property in the feature space. For a given sample, the data in the original dataset, which has a local similarity relationship, is regarded as the framework's positive sample. Further, the modified contrastive loss in the cross-entropy form is as follows:

$$\mathcal{L} = -\sum_{i=1}^{2N}\sum_{j=1,j\neq i}^{2N} S_{ij} log \frac{exp(z_i \cdot z_j^T / \tau)}{\sum_{k=1,k\neq i}^{2N} exp(z_i \cdot z_k^T / \tau)} \tag{9}$$

$$S_{ij} = \begin{cases} \frac{x_i^T x_j}{\|x_i\| \cdot \|x_j\|}, & \text{if } x_j \text{ is } k - \text{nearest neighbor of } x_i \\ 0, & \text{otherwise} \end{cases} \tag{10}$$

where S_{ij} can be regarded as the soft pseudo label of the cross-entropy loss, and k represents the quantity of defined adjacent similar samples.

The core idea of the proposed locality preserving property-constrained contrastive learning is illustrated in Figure 2. The proposed model has more than one positive sample. This change comes from the introduction of the LPP algorithm's affinity matrix into the SimCLR framework. The positive samples consist of both the augmented pair V_{anchor}^i and V_{anchor}^j of the anchor instance x_{anchor} and the augmented views $V_{positive}^i$ and $V_{positive}^j$ of x_{anchor}'s adjacent sample $x_{positive}$. All these adjacent samples not only have a neighborhood relationship with x_{anchor} in the sample space but also quite possibly come from the same class as x_{anchor}. Augmented views of the anchor instance and its adjacent samples constitute a larger positive sample set to assist the representation learning. The other views from different classes are considered negative samples, like $V_{negative}^i$ and $V_{negative}^j$. Specifically, the model constructs the affinity matrix of the original images in the whole dataset to find the nearest first k samples that have higher local similarity properties to x_{anchor}. The soft pseudo label S_{ij} is introduced when calculating the contrastive loss value between x_{anchor} and its corresponding positive samples. The modified pseudo label alleviates the performance deterioration caused by the biased pseudo labels without true labels and induces the model to learn better feature representation. This leads to the embedding features of similar image instances being closer to each other and the embedding features of dissimilar image instances being dispelled in the feature space.

Furthermore, (9) can be transformed into the following form:

$$\begin{aligned} \mathcal{L} &= -\sum_{i=1}^{2N} S_{ip} log \frac{exp(z_i \cdot z_p^T / \tau)}{\sum_{k=1,k\neq i}^{2N} exp(z_i \cdot z_k^T / \tau)} - \lambda \sum_{i=1}^{2N}\sum_{j=1,j\neq i,p}^{2N} S_{ij} log \frac{exp(z_i \cdot z_j^T / \tau)}{\sum_{k=1,k\neq i}^{2N} exp(z_i \cdot z_k^T / \tau)} \\ &= \underbrace{-\sum_{i=1}^{2N} log \frac{exp(z_i \cdot z_p^T / \tau)}{\sum_{k=1,k\neq i}^{2N} exp(z_i \cdot z_k^T / \tau)}}_{\mathcal{L}_C} \underbrace{-\lambda \sum_{i=1}^{2N}\sum_{j=1,j\neq i,p}^{2N} sim(z_i, z_j) \cdot S_{ij}}_{\mathcal{L}_{LPP}} \end{aligned} \tag{11}$$

where $\mathcal{L}_C$ is standard contrastive loss. z_p is the positive example. λ is the weight coefficient; when it is set to 1, the loss function degenerates into (9). $\mathcal{L}_{LPP}$ can be regarded as the objective function of the LPP between different views in the feature space, where the Euclidean distance similarity is replaced by the normalization exponential cosine similarity between different views. Consequently, it demonstrates that the proposed model's objective function is a contrastive loss function constrained by an LPP term.

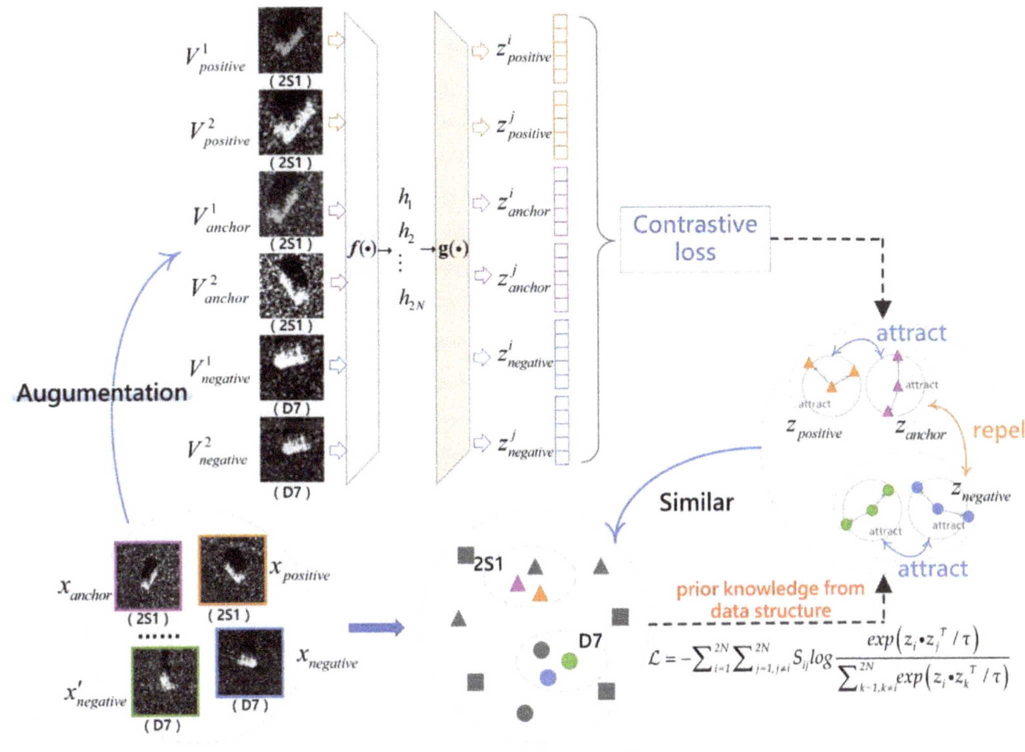

Figure 2. Basic architecture of the proposed model. In LPPCL, for augmented view V^i_{anchor}, the positive samples include not only augmented view V^j_{anchor} from the anchor instance x_{anchor}, but also views $V^i_{positive}$ and $V^j_{positive}$ of x_{anchor}'s adjacent sample $x_{positive}$, which quite possibly comes from the same class as x_{anchor}.

4.2. Multi-Branch Contrastive Learning

In some recent research [12,38], it is found that the robustness of representation learning and the performance of classification tasks could be improved to some extent by increasing the number of positive samples. A similar strategy is therefore included in the proposed model. In order to obtain a robust estimation of mutual information, we adopted a multi-branch structure instead of SimCLR's two-branch structure. As shown in Figure 3, in order to improve the model's representation of augmented views, we augment the original images $[x_1, x_2 \ldots x_N]$ for M(M > 2) times to obtain more views $[V_1^{1,\ldots,m,\ldots M}, V_2^{1,\ldots,m,\ldots M} \ldots V_N^{1,\ldots,m,\ldots M}]$ with $m \in [1, M]$. Obtaining the corresponding representation projection $[z_1^m, z_2^m \ldots z_N^m]$, we then calculate the loss of multiple groups using permutation and combination in pairs. Encoders and projection heads of all branches share their weights. The final loss function value is obtained after the average:

$$\mathcal{L} = \sum_{m,t=1;m\neq t}^{M}\left\{\sum_{i=1}^{N}\left[-\log\frac{exp\left(z_i^m\cdot\left(z_i^t\right)^T/\tau\right)}{E_k} - \lambda\sum_{j=1,j\neq i}^{N} S_{ij}\left(\log\frac{exp\left(z_i^m\cdot\left(z_j^m\right)^T/\tau\right)}{E_k} + \log\frac{exp\left(z_i^m\cdot\left(z_j^t\right)^T/\tau\right)}{E_k}\right)\right]\right\} \tag{12}$$

$$E_k = \sum_{k=1,k\neq i}^{N} exp\left(z_i^m\cdot\left(z_k^m\right)^T/\tau\right) + \sum_{k=1}^{N} exp\left(z_i^m\cdot\left(z_k^t\right)^T/\tau\right)$$

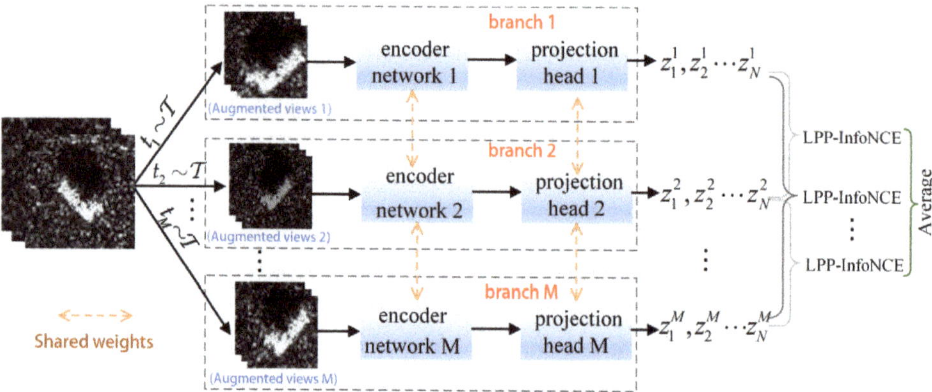

Figure 3. The multi-branch structure of the proposed model.

Our experimental results and some recent research [12,38] both demonstrated that the representation learning capability of the CSL models would be boosted by embedding more augmented views into the framework. The multi-branch structure not only enriched the number of positive and negative samples but also increased the diversity of samples, therefore significantly improving the robustness and discriminative capability of the learned features. As shown in Figure 2, in the embedding space, features extracted from different views of samples in the same class are attracted, while features learned from samples of different classes are dispelled so that the local data structure will be preserved. As illustrated in Algorithm 1, the overall loss to optimize the encoder can be formulated as:

Algorithm 1 The proposed LPPCL algorithm

---

input: batch size **N**, matrix S_{ij}, constant τ, m, the structure of f, g, $\mathcal{T}$.

for $\{x_k\}_{k=1}^N$ **do**

    obtain the corresponding affinity matrix S_{ij}

    draw M augmentation functions $t_m \sim \mathcal{T}$, $m \in [1, M]$

    for a in range (M):

        for b in range $(a + 1, M)$:

            for all $k \in \{1, \ldots, N\}$ **do**

                $V_k = t_a(x_k)$, $h_k = f(V_k)$, $z_k = g(h_k)$

                $V_{k+N} = t_b(x_k)$, $h_{k+N} = f(V_{k+N})$, $z_{k+N} = g(h_{k+N})$

            end for

$$\mathcal{L}+ = -\sum_{i=1}^{2N} log \frac{exp(z_i \cdot z_p^T / \tau)}{\sum_{k=1, k \neq i}^{2N} exp(z_i \cdot z_k^T / \tau)} - \lambda \sum_{i=1}^{2N} \sum_{j=1, j \neq i, p}^{2N} S_{ij} log \frac{exp(z_i \cdot z_j^T / \tau)}{\sum_{k=1, k \neq i}^{2N} exp(z_i \cdot z_k^T / \tau)}$$

        end for

    end for

    $\mathcal{L} = \frac{\mathcal{L}_{total}}{M(M-1)/2}$

    update f and g to minimize $\mathcal{L}$

end for

return f and g

4.3. Model Architecture

As shown in Figure 4, the model uses an elaborate network architecture inspired by self-attentive pooling [39]. In the first stage, we make use of a stochastic data augmentation module to transform the selected samples randomly and produce M correlated views $[V_1, V_2 \ldots V_M]$ of the same example. In this study, following the approaches in [12,13,25,26], we use the data augmentation methods comprising random cropping involving resizing, color jitter, random horizontal flip, random grayscale, and Gaussian blur to strengthen the richness of the augmentation combinations and the complexity of the pretext. Secondly, we adopt the ResNet50 architecture as the encoder network $f_\theta(\cdot)$ to map data into an embedding space, which can extract representation vectors $h = f_\theta(\cdot)$ from augmented samples. A convolutional layer with a 3×3 kernel is used before the ResNet module to reduce the model's computational complexity. ResNet50 utilizes an average pooling layer at the end of the convolutional layer to generate the feature vectors, which only correspond to one local area on the input feature maps. This pooling technique results in the loss of feature information to some extent. Therefore, we replace the global average pooling in the original ResNet50 network with self-attentive pooling.

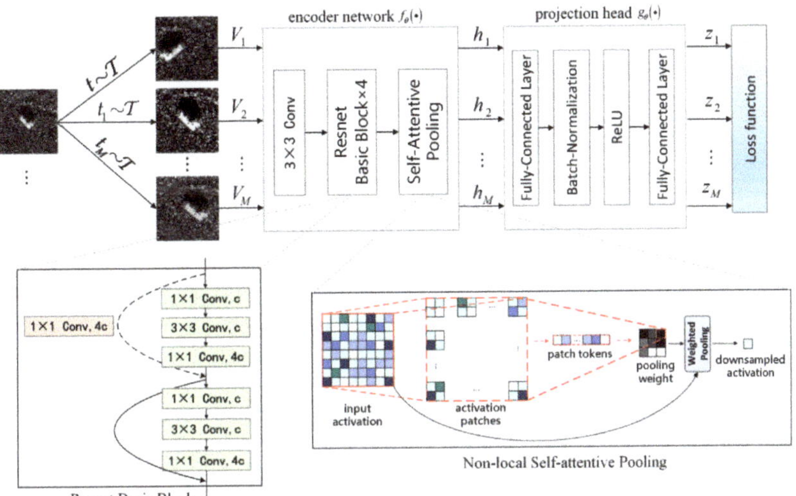

Figure 4. The overall framework of the model architecture. This improves the ResNet50 network and uses self-attentive pooling layers after the convolutional layers. The basic block of the Resnet network is shown in the lower left corner, where c denotes the number of channels. The module of the non-local self-attentive pooling is shown in the lower right corner, with the red box shown being the sensitive fields of pooling weights.

Figure 5 shows the unique pooling approach, which uses the multi-head self-attention module and weights the pooling layer using non-local information. At first, the patch embedding module is used to compact and encode local information of the input $x \in \mathbb{R}^{h \times w \times c_x}$, and the output $x_p \in \mathbb{R}^{\left(\frac{h \times w}{\varepsilon_p^2}\right) \times (\varepsilon_r c_x)}$ is a token series, where ε_p is the patch size and $\varepsilon_r c_x$ denotes the number of output channels. A learnable positional encoding is then added to x_p. Secondly, the multi-head self-attention module is employed to simulate the long-term interdependence between patch tokens, yielding a self-attentive token sequence x_{attn}. Q, K, and V represent three weight matrices of this module. x_{attn} maintains the size of x_p, which is $\left(b \times \frac{h \times w}{\varepsilon_p^2} \times \varepsilon_r c_x\right)$, where b represents the batch size. Thirdly, the spatial and channel information from x_{attn} is decoded in the spatial-channel restoration module, and the x_{attn} is restored to the same size as the input x. To be specific, the token sequence is first reshaped to $\left(b \times \varepsilon_r c_x \times \frac{h}{\varepsilon_p} \times \frac{w}{\varepsilon_p}\right)$, and then expanded to the original spatial resolution $(b \times \varepsilon_r c_x \times h \times w)$ by bilinear interpolation up-sampling. The number of channels is afterwards altered back to $(b \times c_x \times h \times w)$ via a convolution process. Finally, the down-sampled output feature map $\pi(x)$ from x_r is obtained in the weighted pooling module, which is utilized to produce the whole output activation map. The model's well-designed global view increases its capacity to capture long-term dependencies and aggregation characteristics in order to extract data features more effectively.

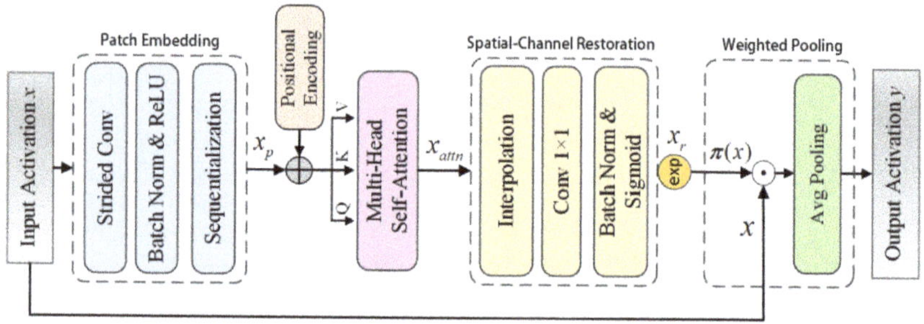

Figure 5. Non-local self-attentive pooling.

The representations are then further mapped to the space where contrastive loss is performed using a tiny neural network projection head $g_\varphi(\cdot)$. In this study, we adopt an MLP with two fully connected layers to obtain $z = g_\varphi(f_\theta(x))$. After the training is completed, we utilize the encoder $f_\theta(\cdot)$ and representation h for classification tasks. Although many unsupervised models often utilize a pre-training and fine-tuning scheme, which will fine-tune the representation networks when it is applied to the downstream classification tasks, the proposed model does not require such a scheme. As the proposed model can learn robust and discriminative representations for classification, it is not necessary to apply a fine-tuning process to the trained model to achieve acceptable results in downstream classification tasks.

5. Experiments and Results

5.1. Experimental Settings

5.1.1. Dataset Description

The MSTAR program provided the experimental dataset required to assess our proposed model. This dataset provides a large amount of SAR images from several kinds of military vehicles as a standard for assessing SAR ATR performance. The photographs and corresponding SAR images are illustrated in Figure 6. Table 1 lists specific details of the experimental dataset.

To thoroughly evaluate the performance of the proposed model, it was tested under two conditions: SOC and EOC. Under SOC, the training set images and the testing set images had the same serial numbers. However, under EOC, images of all the serial numbers were used to test the performance of the proposed method [40]. Following the setup of the EOC experimental method in [41], we used MSTAR datasets with different configurations for evaluation. Due to insufficient training samples, we did not evaluate the performance of the proposed model at the variance of the depression angle. In our experiments, the validations of the other CSL approaches were all carried out using the MSTAR dataset. The training set featured patches acquired at a 17° depression angle, and the testing set was made up of patches taken at a 15° depression angle. Among these, we used the images from BMP2-9563 and T72-132 as training samples for BMP2 and T72.

Figure 6. Photographs and corresponding SAR images of the MSTAR dataset.

Table 1. Detailed information on the MSTAR dataset.

Type	Serial No.	Number of Samples	
		17°	15°
2S1	b01	299	274
	9563	233	194
BMP2	9566	232	196
	c21	233	196
BRDM2	E-71	298	274
BTR60	K10yt7532	256	195
BTR70	c71	233	196
D7	92v13015	299	274
T62	A51	299	273
	132	232	196
T72	812	231	195
	s7	233	191
ZIL131	E12	299	274
ZSU234	d08	299	274

5.1.2. Experiment Setup

To reduce the data volume calculated while ensuring the input data size remained consistent, we used image crop processing based on the center of the target region to resize the input patches to the same shape, which was 64×64. In addition, many of the target patches had significant variations in target intensity, which obscured the target distinctions. Therefore, intensity normalization was used to map the pixel intensities into the range $[-1, 1]$.

A particular network was used for the proposed method, which is presented in Figure 4 in detail. In this experiment, we set the temperature coefficient τ to 1 and the number of the adjacent samples k to 5. The balance weight coefficient λ of the LPP constraint was set to 1. The branch number M was set to 3 in this study due to the limited memory of the GPU. The preprocessed training dataset was used to train the model, and the Adam optimizer, which had a starting learning rate of 0.0001, was used to optimize it. The exponential decay rates β_1 and β_2 were 0.9 and 0.999. The maximum number of iterations was 8000, and the size of the mini-batch was 64. After extracting features from the images, the classification performance was evaluated using a SoftMax classifier. The experiments were implemented on a PC with an Intel(R) Xeon(R) Sliver 4215R CPU @ 3.20 GHz, 256.0 G DDR4, and an NVIDIA GeForce RTX 3090 GPU. The proposed network was built using Pytorch 1.10.2.

The probability of correct classification (P_{cc}) serves as a measure to evaluate the performance of the model, which is equal to the number of samples recognized correctly divided by the number of total samples.

5.2. Recognition Results under SOC

The experiment conducted under SOC was a standard vehicle target classification problem. Table 2 shows the confusion matrix of the recognition results under SOC. The curve of training loss of the proposed model is shown in Figure 7. The matrix's diagonal lines display the proportion of successfully recognized images for each category. It can be seen that the recognition ratios of BTR60 and T72 were above 94%, the recognition ratios of the other categories were above 97%, and BTR70 attained a 100% recognition rate. The aggregate rate of recognition was 97.94%, which was obviously excellent. Through the structure of the neural network, some robust features were extracted successfully, proving that the proposed model performs excellently under SOC.

Table 2. Recognition results under SOC.

Class	2S1	BMP2	BRDM2	BTR60	BTR70	D7	T72	T62	ZIL131	ZSU234	P_{cc} (%)
2S1	266	0	0	1	0	0	0	7	0	0	97.08
BMP2	0	189	0	0	1	0	4	0	0	0	97.42
BRDM2	2	0	269	0	0	2	0	0	1	0	98.18
BTR60	0	0	3	184	0	2	0	5	0	1	94.36
BTR70	0	0	0	0	196	0	0	0	0	0	1
D7	0	0	0	0	0	272	0	1	0	1	99.27
T72	0	8	0	0	0	0	188	0	0	0	95.92
T62	1	0	0	1	0	0	0	269	1	1	98.53
ZIL131	1	0	0	0	0	0	0	0	270	3	98.54
ZSU234	0	0	0	0	0	0	0	1	2	271	98.91
Total											97.94

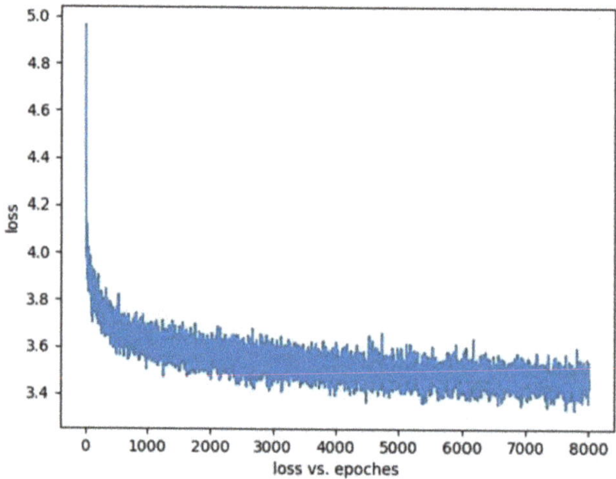

Figure 7. The training loss curve of the proposed model.

5.3. Recognition Results under EOC

Target recognition was more complicated under EOC due to its variable operational conditions. It was necessary to evaluate the robustness of the proposed model when faced with varying target configurations. Table 3 shows the classification accuracy of the proposed architecture under EOC. The recognition ratios of BMP2, BTR60, and T72 were above 94%, the recognition ratios of other categories were above 97%, and BTR70's recognition rate was 100%. The proposed method obtained a superior overall recognition performance of 97.31%. The recognition results prove that the proposed architecture can recognize targets with different configurations and show resilience to configuration variance.

Table 3. Recognition results under EOC.

Class	2S1	BMP2	BRDM2	BTR60	BTR70	D7	T72	T62	ZIL131	ZSU234	P_{cc} (%)
2S1	266	0	0	1	0	0	0	7	0	0	97.08
BMP2	0	561	0	0	3	0	22	0	0	0	95.73
BRDM2	2	0	269	0	0	2	0	0	1	0	98.18
BTR60	0	0	3	184	0	2	0	5	0	1	94.36
BTR70	0	0	0	0	196	0	0	0	0	0	1
D7	0	0	0	0	0	272	0	1	0	1	99.27
T72	0	24	0	0	0	0	558	0	0	0	95.88
T62	1	0	0	1	0	0	0	269	1	1	98.53
ZIL131	1	0	0	0	0	0	0	0	270	3	98.54
ZSU234	0	0	0	0	0	0	0	1	2	271	98.91
Total											97.31

5.4. Comparison with Reference Methods

In this section, the proposed method was compared with SimCLR and DCL. In addition, locality preserving property constrained DCL was also tested in a comparative experiment. When incorporating additional data cluster structure to contrastive loss, we also tried the second different form:

$$\mathcal{L} = \mathcal{L}_C + \lambda' \mathcal{L}'_{LPP} \tag{13}$$

with

$$\mathcal{L}'_{LPP} = \|S_{ij}' - S_{ij}\|$$

$$S_{ij}' = \begin{cases} \frac{z_i^T z_j}{\|z_i\| \cdot \|z_j\|}, & \text{if } z_j \text{ is } k - \text{nearest neighbor of } z_i \\ 0, & \text{otherwise} \end{cases}$$

$$S_{ij} = \begin{cases} \frac{x_i^T x_j}{\|x_i\| \cdot \|x_j\|}, & \text{if } x_j \text{ is } k - \text{nearest neighbor of } x_i \\ 0, & \text{otherwise} \end{cases} \tag{14}$$

where $\mathcal{L}_C$ represents the contrastive loss in DCL or SimCLR, $\mathcal{L}'_{LPP}$ denotes the distance between the affinity matrix of the corresponding network outputs and the affinity matrix of the original data, and λ' is used for balancing $\mathcal{L}_C$ and $\mathcal{L}'_{LPP}$, which we set to 0.1 for the experiment according to the ten-fold cross-validation.

Accordingly, the proposed model was compared with the following models: LPP-constrained SimCLR model in the form of cross entropy (LPPS-I), SimCLR model with a conventional LPP regularization (LPPS-II), LPP-constrained DCL in the form of cross entropy (LPPD-I) and DCL model with a conventional LPP regularization (LPPD-II). We obtained the results for the above methods by implementing these methods with concrete code. In addition, several recently proposed self-supervised learning approaches that achieved state-of-the-art results in SAR ATR were selected to be used in comparison experiments. These included three contrastive self-supervised learning models, i.e., the PL method [24], the CDA method [27], and the ConvT method [30]. In addition, four self-supervised learning methods were used, i.e., the TSDF-N method [9], the ICSGF method [10], the SFAS method [42], and the DKTS-N method [43]. The experimental results for the above methods were obtained from the corresponding references. The recognition performance of various approaches under SOC was then evaluated, as illustrated in Figure 8. Compared with the other approaches, our method recorded the highest P_{cc}, i.e., 97.94%; that is, it performed excellently in SAR image recognition. Further, the proposed method yielded features with superior classification performance, even compared with the DCL method, which achieved SOTA results for feature extraction, i.e., 96.74%. The main reason for this is that the proposed structure ensures that the model can learn an informative representation of data but also preserve the local similarity property in the latent feature space, which is overlooked by most CSL methods. The proposed objective function can diminish the influence of instance-level contrastive loss. In addition, applying the LPP affinity matrix constraint yielded better results for SimCLR than for DCL. For the loss function of SimCLR, the form of cross entropy was found to be more conducive to representation learning than the conventional LPP regularization method. However, for the loss function of DCL, the conventional LPP regularization method achieved better performance.

5.5. Further Analyses of Ablation Study

For an additional in-depth assessment of whether the proposed model achieves enhanced recognition performance using self-attentive pooling and a multi-branch structure, further studies are discussed in this subsection. Our ablation experiment analyzed the original SimCLR model (SimCLR), the multi-branch SimCLR model (m_SimCLR), and the SimCLR model with self-attentive pooling (SimCLR_p). It also analyzed four cases in which the locality-preserving property constraint was introduced, namely the locality-preserving property constrained original SimCLR learning (LPPCL), the multi-branch LPPS (m_LPPCL), the LPPS with self-attentive pooling (LPPCL_p), and the multi-branch LPPS with self-attentive pooling (m_LPPCL_p) models. Further, the recognition performance of these models under SOC was compared. Figure 9 displays the recognition rates of each model.

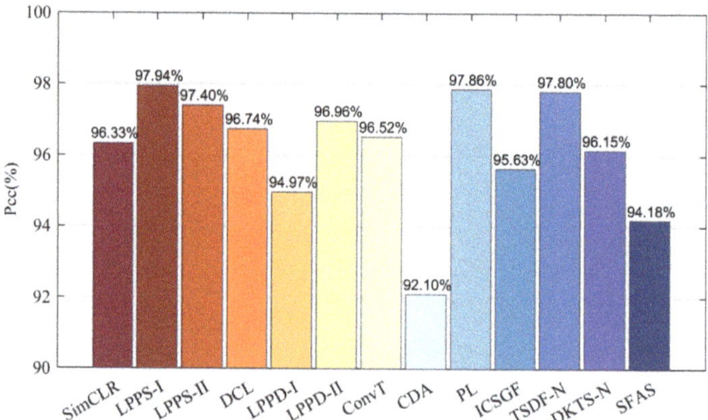

Figure 8. Recognition rates of different models.

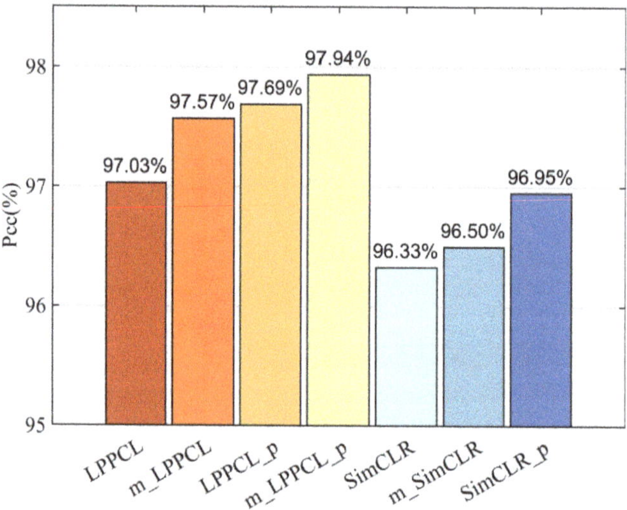

Figure 9. Recognition rates of different models.

It can clearly be seen that after introducing the local preserving property constraint to the SimCLR model, its classification accuracy was greatly improved, i.e., its P_{cc} increased from 96.33% to 97.03%. This proves that the introduction of additional local similarity relationships can effectively assist representation learning and alleviate the problem of biased sampling in instance-level contrastive learning. In the feature space, the embedding features of similar image instances were closer to each other, and the embedding features of dissimilar image instances were dispelled effectively. Further, the performance of the multi-branch structure was also better than that of the double-branch structure. The main reason for this is that the multi-branch structure can enrich the diversity of training samples through stochastic data augmentations, thus ensuring the robustness of the model, especially when there are limited samples. It is also beneficial to the calculation of contrastive loss and the estimation of mutual information. By incorporating the multi-branch structure in the traditional SimCLR model, the performance was improved by 0.17%, while in the LPP-constrained two-branch SimCLR model, the performance was improved by 0.54%. In addition, replacing average pooling with self-attention pooling can assist feature extraction. Unlike average pooling, self-attention pooling can extract features adaptively according to the structure of the data itself and thereby reduce the loss of feature information. Thus, more effective information is retained, which is beneficial to downstream tasks. By incorporating the self-attention pooling in the traditional SimCLR model, the performance was improved by 0.62%, while in the LPP-constrained two-branch SimCLR model, the performance was improved by 0.66%, and in the LPP-constrained multi-branch SimCLR model, the performance was improved by 0.37%. Finally, when the multi-branch structure and self-attention pooling were added together to the SimCLR model with LPP constraint, it recorded the highest P_{cc}, i.e., 97.94%, which proves that the proposed strategy can successfully enhance the model's performance.

5.6. Experiment with Noise Corruption

In reality, inevitable noise often corrupts measured SAR imagery. In this study, SAR images corrupted by different SNR ratios were used to assess the proposed model's robustness to noise. The original SAR images were considered to be noise-free, which was first transformed into the frequency domain with 2D-IDFT, and additive noises with SNRs ranging from 20 dB to −20 dB were applied to the transformed images to produce noise-contaminated images with SNRs specified in (15). The same imaging process was then used to convert the noisy data into the image domain.

$$NR(dB) = 10log_{10} \frac{\sum_{u=0}^{U-1} \sum_{v=0}^{V-1} |f(u,v)|^2}{HW\sigma^2} \tag{15}$$

where $f(u,v)$ is the RCS; σ^2 is the variance in the additive noise. H and W represent the height and width of the image. Figure 10 shows various noise-contaminated images with varying SNRs. The average experimental results for the proposed model, the SimCLR model, and the DCL model at different SNR levels are shown in Figure 11. As noise interference worsens, the recognition rates of all the models gradually decrease, with the proposed model achieving the highest rate at every SNR level. Even when the image is contaminated by the noise with SNR being 0dB, the geometric and scattering characteristics of the image are not seriously affected by the noise; the P_{cc} of the model is still above 89%, demonstrating the proposed model's resilience under considerable noise interference.

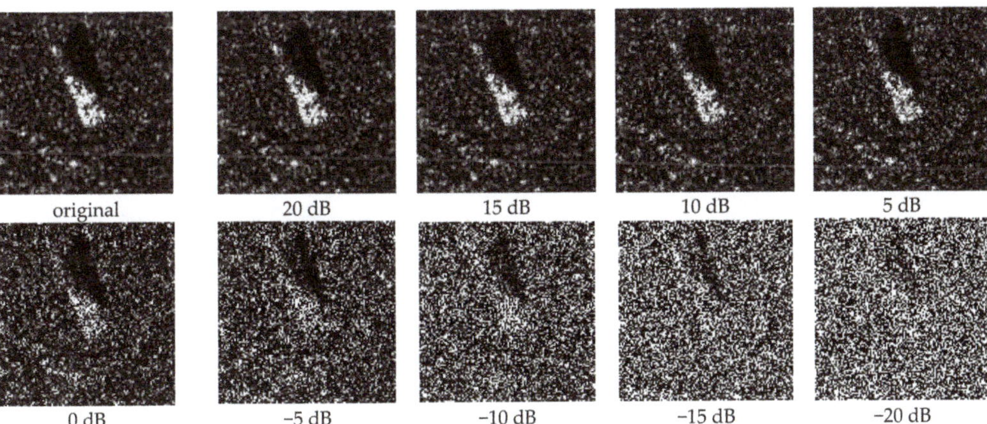

Figure 10. MSTAR data with different SNRs.

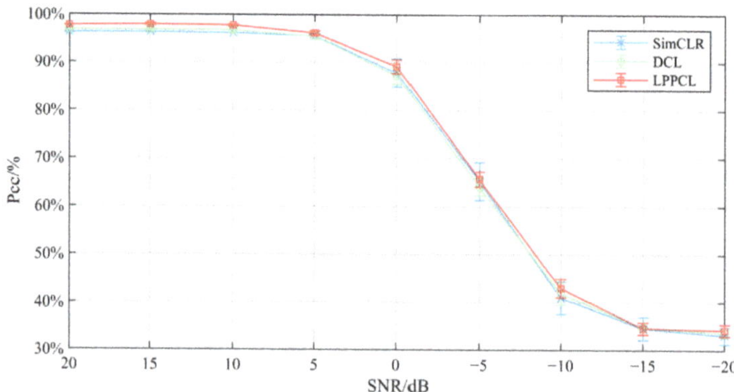

Figure 11. Recognition results at different SNR levels.

5.7. Experiment with Resolution Variance

Another important consideration when assessing the performance of a model is its robustness to resolution fluctuation. Therefore, we varied the resolution of the SAR images from 0.3 m × 0.3 m to 0.8 m × 0.8 m. Specifically, the sub-band was first retrieved when the spatial SAR images were transformed by 2D-IDFT into the frequency domain. The sub-band data were then resampled in the frequency domain using zero padding before being switched back to the spatial domain. Figure 12 displays several images at various resolutions. In our experiment, the model was trained using the original single-resolution training set, and then the performance of the model was tested on different resolution testing sets. The average experimental results for the proposed model, the SimCLR model, and the DCL model at different resolution levels are shown in Figure 13. It is demonstrated that the proposed model is still effective even with some deterioration in resolution. At all resolutions, the model achieves the highest classification rate when compared with other models. The proposed model's P_{cc} is still higher than 85% even when the resolution is 0.7 m × 0.7 m, proving the model's stability in the face of resolution variation.

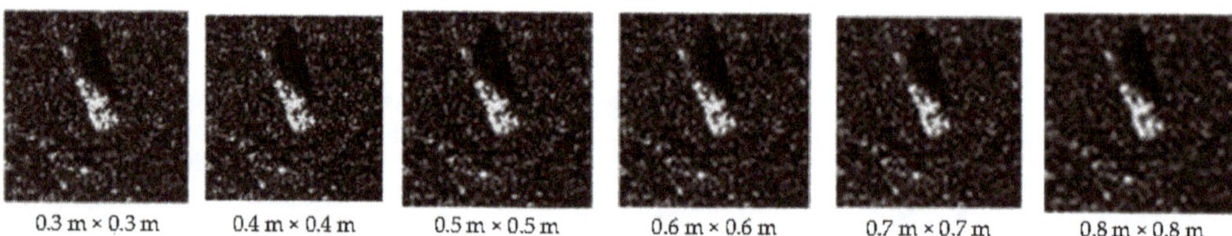

Figure 12. MSTAR data at different resolutions.

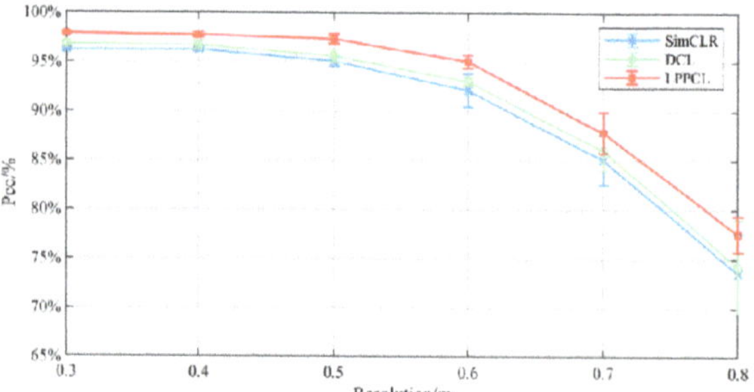

Figure 13. Recognition results at different resolutions.

6. Conclusions and Future Work

In this paper, a contrastive self-supervised representation learning model is proposed. This model provides an effective method for learning useful features from unlabeled SAR images. Its use of the local preserving property constraint method prevents performance deterioration caused by biased pseudo labels without true labels; rather, the model not only learns informative representations of data but also preserves the local similarity property in the latent feature space. Further, the proposed model's multi-branch structure increases the diversity of training samples, enhancing the model's capacity for representation. In addition, its use of self-attention pooling assists the model with learning informative features, which is conducive to downstream classification tasks. Experiments demonstrated that the proposed model outperformed most of the present CSL algorithms and obtained the SOTA performance without supervised knowledge.

Author Contributions: Conceptualization and Validation, J.W. and S.T.; methodology, S.T. and J.W.; software, J.W., F.W. and B.Z.; writing—original draft, J.W. and X.F.; writing—review and editing, C.W. and H.Z.; visualization, J.W. and X.F.; supervision, S.T. and C.W.; project administration, S.T.; funding acquisition, S.T. All authors have read and agreed to the published version of the manuscript.

Funding: This research was funded in part by the Key Program of National Natural Science Foundations of China under Grant No. 41930110.

Data Availability Statement: Not applicable.

Acknowledgments: The authors thank the US Air Force Research Lab for providing the public MSTAR data at: https://www.sdms.afrl.af.mil/index.php?collection=mstar (accessed on 1 September 2022). In addition, we are grateful to the anonymous referees for their instructive comments.

Conflicts of Interest: The authors declare no conflict of interest.

References

1. Wang, P.; Liu, W.; Chen, J.; Niu, M.; Yang, W. A High-Order Imaging Algorithm for High-Resolution Spaceborne SAR Based on a Modified Equivalent Squint Range Model. *IEEE Trans. Geosci. Remote Sens.* **2015**, *53*, 1225–1235. [CrossRef]
2. Kechagias-Stamatis, O.; Aouf, N. Fusing Deep Learning and Sparse Coding for SAR ATR. *IEEE Trans. Aerosp. Electron. Syst.* **2019**, *55*, 785–797. [CrossRef]
3. Qi, G.J.; Luo, J. Small Data Challenges in Big Data Era: A Survey of Recent Progress on Unsupervised and Semi-Supervised Methods. *IEEE Trans. Pattern Anal. Mach. Intell.* **2022**, *44*, 2168–2187. [CrossRef] [PubMed]
4. He, K.; Chen, X.; Xie, S.; Li, Y.; Dollár, P.; Girshick, R. Masked Autoencoders Are Scalable Vision Learners. In Proceedings of the 2022 IEEE/CVF Conference on Computer Vision and Pattern Recognition (CVPR), New Orleans, LA, USA, 18–24 June 2022; pp. 15979–15988. [CrossRef]
5. Kingma, D.P.; Welling, M. Auto-Encoding Variational Bayes. *arXiv* **2013**, arXiv:1312.6114.
6. Creswell, A.; White, T.; Dumoulin, V.; Arulkumaran, K.; Sengupta, B.; Bharath, A.A. Generative Adversarial Networks: An Overview. *IEEE Signal Process. Mag.* **2018**, *35*, 53–65. [CrossRef]
7. Le-Khac, P.H.; Healy, G.; Smeaton, A.F. Contrastive Representation Learning: A Framework and Review. *IEEE Access* **2020**, *8*, 193907–193934. [CrossRef]
8. Gromada, K. Unsupervised SAR Imagery Feature Learning with Median Filter-Based Loss Value. *Sensors* **2022**, *22*, 6519. [CrossRef] [PubMed]
9. Du, L.; Li, L.; Guo, Y.; Wang, Y.; Ren, K.; Chen, J. Two-Stream Deep Fusion Network Based on VAE and CNN for Synthetic Aperture Radar Target Recognition. *Remote Sens.* **2021**, *13*, 4021. [CrossRef]
10. Cao, C.; Cui, Z.; Cao, Z.; Wang, L.; Yang, J. An Integrated Counterfactual Sample Generation and Filtering Approach for SAR Automatic Target Recognition with a Small Sample Set. *Remote Sens.* **2021**, *13*, 3864. [CrossRef]
11. Poole, B.; Ozair, S.; van den Oord, A.; Alemi, A.A.; Tucker, G. On Variational Bounds of Mutual Information. *arXiv* **2019**, arXiv:1905.06922.
12. Chuang, C.Y.; Robinson, J.; Yen-Chen, L.; Torralba, A.; Jegelka, S. Debiased Contrastive Learning. *arXiv* **2020**, arXiv:2007.00224.
13. Chen, T.; Kornblith, S.; Norouzi, M.; Hinton, G. A Simple Framework for Contrastive Learning of Visual Representations. *arXiv* **2020**, arXiv:2002.05709.
14. Gutmann, M.; Hyvrinen, A. Noise-contrastive estimation: A new estimation principle for unnormalized statistical models. In Proceedings of the International Conference on Artificial Intelligence and Statistics, Sardinia, Italy, 13–15 May 2010.
15. Liu, X.; Zhang, F.; Hou, Z.; Mian, L.; Wang, Z.; Zhang, J.; Tang, J. Self-Supervised Learning: Generative or Contrastive. *IEEE Trans. Knowl. Data Eng.* **2023**, *35*, 857–876. [CrossRef]

16. Jing, L.; Tian, Y. Self-Supervised Visual Feature Learning With Deep Neural Networks: A Survey. *IEEE Trans. Pattern Anal. Mach. Intell.* **2021**, *43*, 4037–4058. [CrossRef]

17. Hjelm, R.D.; Fedorov, A.; Lavoie-Marchildon, S.; Grewal, K.; Bachman, P.; Trischler, A.; Bengio, Y. Learning deep representations by mutual information estimation and maximization. *arXiv* **2019**, arXiv:1808.06670.

18. Caron, M.; Bojanowski, P.; Joulin, A.; Douze, M. Deep clustering for unsupervised learning of visual features. In Proceedings of the European Conference on Computer Vision (ECCV), Munich, Germany, 8–14 September 2018; pp. 132–149.

19. Wu, Z.; Xiong, Y.; Yu, S.X.; Lin, D. Unsupervised Feature Learning via Non-parametric Instance Discrimination. In Proceedings of the 2018 IEEE/CVF Conference on Computer Vision and Pattern Recognition, Salt Lake City, UT, USA, 18–23 June 2018; pp. 3733–3742. [CrossRef]

20. He, K.; Fan, H.; Wu, Y.; Xie, S.; Girshick, R. Momentum Contrast for Unsupervised Visual Representation Learning. In Proceedings of the 2020 IEEE/CVF Conference on Computer Vision and Pattern Recognition (CVPR), Seattle, WA, USA, 13–19 June 2020; pp. 9726–9735. [CrossRef]

21. Tian, Y.; Sun, C.; Poole, B.; Krishnan, D.; Schmid, C.; Isola, P. What Makes for Good Views for Contrastive Learning? *arXiv* **2020**, arXiv:2005.10243.

22. Grill, J.B.; Strub, F.; Altche, F.; Tallec, C.; Richemond, P.H.; Buchatskaya, E.; Doersch, C.; Pires, B.A.; Guo, Z.D.; Azar, M.G.; et al. Bootstrap Your Own Latent: A New Approach to Self-Supervised Learning. *arXiv* **2020**, arXiv:2006.07733.

23. Chen, X.; He, K. Exploring Simple Siamese Representation Learning. In Proceedings of the Computer Vision and Pattern Recognition, Nashville, TN, USA, 20–25 June 2021.

24. Wang, C.; Gu, H.; Su, W. SAR Image Classification Using Contrastive Learning and Pseudo-Labels With Limited Data. *IEEE Geosci. Remote Sens. Lett.* **2022**, *19*, 4012505. [CrossRef]

25. Zhou, X.; Tang, T.; Cui, Y.; Zhang, L.; Kuang, G. Novel Loss Function in CNN for Small Sample Target Recognition in SAR Images. *IEEE Geosci. Remote Sens. Lett.* **2022**, *19*, 4018305. [CrossRef]

26. Zhai, Y.; Zhou, W.; Sun, B.; Li, J.; Ke, Q.; Ying, Z.; Gan, J.; Mai, C.; Labati, R.D.; Piuri, V.; et al. Weakly Contrastive Learning via Batch Instance Discrimination and Feature Clustering for Small Sample SAR ATR. *IEEE Trans. Geosci. Remote Sens.* **2022**, *60*, 5204317. [CrossRef]

27. Bi, H.; Liu, Z.; Deng, J.; Ji, Z.; Zhang, J. Contrastive Domain Adaptation-Based Sparse SAR Target Classification under Few-Shot Cases. *Remote Sens.* **2023**, *15*, 469. [CrossRef]

28. Chen, Y.; Bruzzone, L. Self-Supervised SAR-Optical Data Fusion of Sentinel-1/-2 Images. *IEEE Trans. Geosci. Remote Sens.* **2022**, *60*, 5406011. [CrossRef]

29. Xu, Y.; Cheng, C.; Guo, W.; Zhang, Z.; Yu, W. Exploring Similarity in Polarization: Contrastive Learning with Siamese Networks for Ship Classification in Sentinel-1 SAR Images. In Proceedings of the IGARSS 2022-2022 IEEE International Geoscience and Remote Sensing Symposium, Kuala Lumpur, Malaysia, 17–22 July 2022; pp. 835–838. [CrossRef]

30. Wang, C.; Huang, Y.; Liu, X.; Pei, J.; Zhang, Y.; Yang, J. Global in Local: A Convolutional Transformer for SAR ATR FSL. *IEEE Geosci. Remote Sens. Lett.* **2022**, *19*, 4509605. [CrossRef]

31. Ren, H.; Yu, X.; Wang, X.; Liu, S.; Zou, L.; Wang, X. Siamese Subspace Classification Network for Few-Shot SAR Automatic Target Recognition. In Proceedings of the IGARSS 2022-2022 IEEE International Geoscience and Remote Sensing Symposium, Kuala Lumpur, Malaysia, 17–22 July 2022; pp. 2634–2637. [CrossRef]

32. Liu, C.; Sun, H.; Xu, Y.; Kuang, G. Multi-Source Remote Sensing Pretraining Based on Contrastive Self-Supervised Learning. *Remote Sens.* **2022**, *14*, 4632. [CrossRef]

33. Xiao, X.; Li, C.; Lei, Y. A Lightweight Self-Supervised Representation Learning Algorithm for Scene Classification in Spaceborne SAR and Optical Images. *Remote Sens.* **2022**, *14*, 2956. [CrossRef]

34. Liu, F.; Qian, X.; Jiao, L.; Zhang, X.; Li, L.; Cui, Y. Contrastive Learning-Based Dual Dynamic GCN for SAR Image Scene Classification. *IEEE Trans. Neural Netw. Learn. Syst.* **2022**, 1–15. [CrossRef] [PubMed]

35. Yang, M.; Jiao, L.; Liu, F.; Hou, B.; Yang, S.; Zhang, Y.; Wang, J. Coarse-to-Fine Contrastive Self-Supervised Feature Learning for Land-Cover Classification in SAR Images With Limited Labeled Data. *IEEE Trans. Image Process.* **2022**, *31*, 6502–6516. [CrossRef] [PubMed]

36. Liu, M.; Wu, Y.; Zhao, Q.; Gan, L. SAR target configuration recognition using Locality Preserving Projections. In Proceedings of the 2011 IEEE CIE International Conference on Radar, Chengdu, China, 24–27 October 2011; Volume 1, pp. 740–743. [CrossRef]

37. Oord, A.; Li, Y.; Vinyals, O. Representation Learning with Contrastive Predictive Coding. *arXiv* **2018**, arXiv:1807.03748.

38. Cai, Q.; Wang, Y.; Pan, Y.; Yao, T.; Mei, T. Joint Contrastive Learning with Infinite Possibilities. *arXiv* **2020**, arXiv:2009.14776.

39. Chen, F.; Datta, G.; Kundu, S.; Beerel, P. Self-Attentive Pooling for Efficient Deep Learning. *arXiv* **2022**, arXiv:2209.07659.

40. Ross, T.D.; Worrell, S.W.; Velten, V.J.; Mossing, J.C.; Bryant, M.L. Standard SAR ATR evaluation experiments using the MSTAR public release data set. In *Algorithms for Synthetic Aperture Radar Imagery*; International Society for Optics and Photonics: Bellingham, WA, USA, 1998; pp. 566–573.

41. Tian, S.; Lin, Y.; Gao, W.; Zhang, H.; Wang, C. A Multi-Scale U-Shaped Convolution Auto-Encoder Based on Pyramid Pooling Module for Object Recognition in Synthetic Aperture Radar Images. *Sensors* **2020**, *20*, 1533. [CrossRef] [PubMed]

42. Wang, C.; Liu, X.; Huang, Y.; Luo, S.; Pei, J.; Yang, J.; Mao, D. Semi-Supervised SAR ATR Framework with Transductive Auxiliary Segmentation. *Remote Sens.* **2022**, *14*, 4547. [CrossRef]
43. Zhang, L.; Leng, X.; Feng, S.; Ma, X.; Ji, K.; Kuang, G.; Liu, L. Domain Knowledge Powered Two-Stream Deep Network for Few-Shot SAR Vehicle Recognition. *IEEE Trans. Geosci. Remote Sens.* **2022**, *60*, 5215315. [CrossRef]

MDPI

St. Alban-Anlage 66

4052 Basel

Switzerland

www.mdpi.com

Remote Sensing Editorial Office

E-mail: remotesensing@mdpi.com

www.mdpi.com/journal/remotesensing